THE EVOLUTION OF
X-RAY BINARIES

Bruno Rossi

1905 - 1993

Photograph courtesy of the MIT Museum

AIP CONFERENCE PROCEEDINGS 308

THE EVOLUTION OF X-RAY BINARIES

COLLEGE PARK, MD OCTOBER 1993

EDITORS:
STEPHEN S. HOLT
CHARLES S. DAY
NASA/GODDARD SPACE
FLIGHT CENTER

American Institute of Physics   New York

Authorization to photocopy items for internal or personal use, beyond the free copying permitted under the 1978 U.S. Copyright Law (see statement below), is granted by the American Institute of Physics for users registered with the Copyright Clearance Center (CCC) Transactional Reporting Service, provided that the base fee of $2.00 per copy is paid directly to CCC, 27 Congress St., Salem, MA 01970. For those organizations that have been granted a photocopy license by CCC, a separate system of payment has been arranged. The fee code for users of the Transactional Reporting Service is: 0094-243X/87 $2.00.

© 1994 American Institute of Physics.

Individual readers of this volume and nonprofit libraries, acting for them, are permitted to make fair use of the material in it, such as copying an article for use in teaching or research. Permission is granted to quote from this volume in scientific work with the customary acknowledgment of the source. To reprint a figure, table, or other excerpt requires the consent of one of the original authors and notification to AIP. Republication or systematic or multiple reproduction of any material in this volume is permitted only under license from AIP. Address inquiries to Series Editor, AIP Conference Proceedings, AIP Press, American Institute of Physics, 500 Sunnyside Boulevard, Woodbury, NY 11797-2999.

L.C. Catalog Card No. 94-76853
ISBN 1-56396-329-9
DOE CONF-9310265

Printed in the United States of America.

Preface

This is the fourth in a series of annual October Astrophysics Conferences in Maryland. These conferences are organized by astrophysicists at the Goddard Space Flight Center and the University of Maryland. The topic for each conference is selected by a permanent committee of senior scientific staff with the help of an International Advisory Committee, the current membership of which is:

Marek Abramowicz, Göteborg
Roger Blandford, Pasadena
Arnon Dar, Haifa
Alan Dressler, Pasadena
Riccardo Giacconi, München
Guenther Hasinger, Potsdam
Steve Holt, Greenbelt
Frank Kerr, College Park

Dick McCray, Boulder
Jim Peebles, Princeton
Sir Martin Rees, Cambridge
Vera Rubin, Washington
Joseph Silk, Berkeley
Rashid Sunyaev, Moscow
Alex Szalay, Budapest
Yasuo Tanaka, Tokyo
Scott Tremaine, Toronto

"The Evolution of X-Ray Binaries" was the subject chosen for this conference. In previous years we have been somewhat informal in the choice of a conference title, and in spite of the obvious suitability of titles like "Fatal Attraction" and "Liaisons Dangereuses," a few members of the Scientific Organizing Committee insisted that we utilize the straightforward title. The reader is left to guess which members of the committee exercised this veto:

Joe Dolan
Guenther Hasinger
Steve Holt
Tim Kallman
Walter Lewin
Fumiaki Nagase

Tom Prince
Jean Swank
Virginia Trimble
Frank Verbunt
Mike Watson
Nick White

The first half of the thirty-year history of X-ray astronomy was dominated by the study of X-ray binaries. The search for the nature of the bright X-ray sources in the plane of the galaxy was one of the great enterprises of astronomical research in the late sixties, and their detailed characterization was at the heart of the discipline into the mid-seventies. The onset of X-ray imaging at the end of that decade provided the means to move X-ray astronomy out of the galaxy, and much of the interest in X-ray measurements moved to extragalactic astronomy. The discovery of quasi-periodic oscillations rekindled some interest in galactic X-ray binaries in the eighties, but it was the connection of low mass X-ray binaries to millisecond pulsars that has been responsible for the rejuvenation of widespread interest in the subject. It was particularly gratifying, therefore, to receive the news (during the conference) that the Nobel Prize for Physics was being awarded to Joe Taylor and Russell Hulse for their work in the study of the first binary pulsar.

As usual, the conference began with two invited summaries reviewing the general subject (this year's were delivered by *Ed van den Heuvel* and *Walter Lewin*). It then proceeded through the next two days with a series of non-parallelled sessions, each devoted to a specific topic and led by a distinguished session chair. Each of these sessions featured two or three invited talks and an extensive discussion period, and may have also included one or two short contributions "promoted" from the poster papers by the session chair. All the chairs should be commended for keeping the activities lively and the speakers to their allotted times.

Marek Abramowicz	*Fred Lamb*	*Jean Swank*
George Clark	*Mario Livio*	*Ron Taam*
Joe Dolan	*Luigi Stella*	*Mike Watson*

The concluding session after lunch on Wednesday consisted of a pair of talks given by individuals who are not specialists in the discipline, but who were more than capable of performing the tasks put before them. *Virginia Trimble* delivered what has become now something of a tradition at these conferences: a seemingly irreverent (but always respectful) history of the subject. The conference was concluded with a thoughtful rapporteur presentation by *Julian Krolik*.

Thanks are due *Charles Day* and *Saeqa Vrtilek* for organizing the poster sessions, to *Susan Lehr* for all kinds of administrative help, and especially to *John Trasco* for making sure that everything that needed doing got done. Thanks also to all the participants who made it a memorable three days.

For this editor, who began his post-graduate research career studying X-ray binaries during the late sixties (but who has been less active in this area recently), this conference provided a marvelous combination of nostalgia and contemporary work on the cutting edge of astrophysical investigation.

Steve Holt
January 1994

Note added: The large fraction of the early work in X-ray astronomy was performed by the students and associates of Bruno Rossi, who established the group that discovered the first extra-solar X-ray source (Sco X-1) from a 1962 sounding rocket flight. Soon after its discovery, Sco X-1 was correctly described by Shklovskii as the prototype for the X-ray binary systems discussed at this conference. Bruno Rossi passed away shortly after the conference, and we dedicate these Proceedings to his memory.

TABLE OF CONTENTS

PREFACE   v
    Holt, S. S.

INTRODUCTIONS

Three Decades of X-Ray Astronomy from the Point of View of a Biased Observer   3
    Lewin, W. H. G.

Three Decades of X-Ray Binaries from the Point of View of a Theoretician   18
    van den Heuvel, E. P. J.

Some Events in X-ray Astronomy that Aren't in the Journals   39
    Gursky, H.

BLACK HOLES

Dynamical Evidence for Black Holes in Soft X-ray Transients and Other X-ray Binaries   45
    Cowley, A. P.

The Galactic Distribution of Black Holes in X-ray Binaries   53
    White, N. E.

Observational Constraints on the Models of Disk Accretion onto a Black Hole   61
    Grebenev, S. A., Sunyaev, R. A., and Pavlinsky, M. N.

A Systematic and Statistical Study of the X-Ray and Optical Light Curves of X-Ray Novae   67
    Chen, W., Shrader, C., and Livio, M.

Supermassive Black Holes with Stellar Companions   71
    Podsiadlowski, P., and Rees, M. J.

X-Ray Transitions of Black Hole Binaries and Variable α-Parameter Disks   75
    Luo, C., Meirelles, C., and Liang, E.

Application of Magnetic Braking Theories to Black Hole Binaries   79
    Mukai, K.

A Lower Limit on the Pair Density Ratio (z_+) in an Electron-Positron Pair Wind   83
    Moscoso, M. D., and Wheeler, J. C.

Analysis of the Assumption of Frequency Independence of
                              Light Velocity in a Gravitational Field   87
    Gall, C. A.

ING Photometry and Spectroscopy of the Optical Counterpart of
                              the X-Ray Transient GRO J0422+32   91
    Harlaftis, E., Jones, D., Charles, P., and Martin, A.

Multiwavelength Study of the X-Ray Nova GRO J0422+32   95
    *Shrader, C. R., Wagner, R. M., Hjellming, R. M., Han, X. H.,
and Starrfield, S. G.*

Optical Observations of the X-Ray Nova GRO J0422+32   99
    *Zhao, P., Callanan, P., Garcia, M. R., McClintock, J. E.,
Remillard, R. A., and Silber, A.*

Observations of GRO J0422+32 on High and Low Optical States   103
    *Bartolini, C., Guarnieri, A., Piccioni, A., Zampieri, G., Beskin, G.,
Neizvestny, S., Panferova, I., Plokhotnichenko, V., and Zhuravkov, A.*

The Mass of the Black-Hole in GS2023+338/V404 Cygni   107
    Casares, J., and Charles, P. A.

X-Ray Observations of the Black Hole Binary V404 Cygni in Quiescence   111
    *Wagner, R. M., Starrfield, S. G., Kreidl, T. J., Howell, S. B.,
and Hjellming, R. M.*

BATSE Observations of Nova Muscae 1991   115
    *Paciesas, W. S., Briggs, M. S., Pendleton, G. N., Harmon, B.A.,
Wilson, C.A., Zhang, S. N., and Fishman, G. J.*

A Broad Band X-ray Telescope Observation of the
                              Black Hole Candidate LMC X-1   119
    Schlegel, E. M.

The Mass of the Black Hole in A0620-00   123
    Shahbaz, T., Naylor, T., and Charles, P. A.

The Periodicities of Aql X-1 in Quiescence   127
    Martin, A. C., Charles, P. A., Callanan, P. J., and Harlaftis, E.

Black Holes vs Neutron Stars Among Sigma Sources   131
    *Ballet, J., Laurent, P., Lebrun, F., Paul, J., Roques, J. P.,
Mandrou, P., Malet, I., Schmitz-Fraysse, M. C., Churazov, E.,
Gilfanov, M., Sunyaev, R., Vikhlinin, A., Finoguenov, A.,
Dyachkov, A., Khavenson, N., and Sheikhet, A.*

The August 1993 Outburst of GRO J0422+32   135
    Castro-Tirado, A. J., Ortiz, J. L., and Gallego, J.

ACCRETION DISKS

X-Ray Energy Spectra From Accretion Disks in Black Hole Candidates   143
    Ebisawa, K.

Spectral Diagnostics and Accretion Disk Corona Reprocessing in X-Ray Binaries   155
    Kallman, T. R.

The Interpretation of Soft X-Ray Emission Lines in the Spectra of
    Accretion-Powered Sources: the Effects of Thermal Stability   166
    Hess, C. J., Paerels, F., Kahn, S. M., Liedahl, D. A., and Rogers, R. D.

A Close-up View of Interacting Binaries   171
    Rutten, R. G. M.

Spectral Study of GX339-4 in the Low Intensity State Observed with Ginga   181
    Ueda, Y., Ebisawa, K., and Done, C.

The Dipping X-Ray Binary X1916-05 Observed with Ginga   185
    Yoshida, K.

ROSAT Observations of Cataclysmic Variables   189
    van Teeseling, A., and Verbunt, F.

The Einstein Solid State Spectrometer and Monitor Proportional Counter
    Survey of Low Mass X-ray Binaries   193
    Christian, D. J.

The Cooling of the White Dwarf in OY CAR After 1992 Superoutburst   197
    Cheng, F. H., Marsh, T. R., Horne, K., and Hubeny, I.

The I Band Light Curve of the Low Mass X-ray Binary GX 9+9   201
    Haswell, C. A., and Abbott, T. M. C.

Emission Line Spectrum of an X-Ray Heated Accretion Disk in LMXB   205
    Ko, Y.-K., and Kallman, T. R.

Models of Accretion Disk Coronae in High Luminosity Systems   209
    Murray, S. D., Klein, R. I., Castor, J. I., and McKee, C. F.

The Effect of Irradiation on Outbursts in X-ray Novae   213
    Kim, S.-W., Wheeler, J. C., and Mineshige, S.

Low Frequency Oscillations from Accretion Disks in X-Ray Binaries   217
    Chen, X., and Taam, R. E.

On Turbulent Viscosity in Thin Accretion Disks   221
    Lee, G.

The Vertical Structure and Stability of Accretion Disks Surrounding
    Black Holes and Neutron Stars   225
    Milsom, J. A., Chen, X., and Taam, R. E.

The Ultraviolet Spectrum of 2S 0921-630   229
Schmidtke, P. C.

BINARY INTERACTIONS

Observations of Accreting Pulsars   235
Prince, T. A., Bildsten, L., Chakrabarty, D., Wilson, R. B., and Finger, M. H.

Coalescence of Neutron Star Binaries   245
Blaes, O.

Recent Observations of EXO 2030+375 with BATSE   255
Stollberg, M. T., Paciesas, W. S., Finger, M. H., Fishman, G. J., Wilson, R. B., Harmon, B. A., and Wilson, C. A.

BATSE Observations of GS 0834-430   259
Wilson, C. A., Harmon, B. A., Wilson, R. B., Fishman, G. J., and Finger, M. H.

Hard X-Ray Variability of Cygnus X-3   263
Matz, S. M., Grabelsky, D. A., Purcell, W. R., Ulmer, M. P., Johnson, W. N., Kinzer, R. L., Kurfess, J. D., and Strickman, M. S.

The Precessional Phase Dependent X-Ray Emission of SS433   267
Yuan, W., Kawai, N., and Matsuoka, M.

IUE Low and High Dispersion Spectra of the Be-X Star
BD+53° 2790 = 4U2206+54   271
Teodorani, M., Guarnieri, A., Bartolini, C., and Piccioni, A.

On the Nature of LSI+65° 010, the Optical Counterpart of the
X-ray Binary 2S0114+650   275
Minarini, R., Teodorani, M., Bartolini, C., Guarnieri, A., and Piccioni, A.

Optical Behaviour of the Be-X Binary V635 CAS = 4U 0115+63
Before and During a Transient Phase   279
Guarnieri, A., Bartolini, C., Teodorani, M., Silingardi, R., and Piccioni, A.

Six-year Photometry of BQ Cameloparadalis --
The Optical Counterpart of V0332+53   283
Mazeh, T., and Mendelson, H.

Orbital Period Changes in the Eclipsing Pulsar Binary PSR B1957+20   287
Arzoumanian, Z., Fruchter, A. S., and Taylor, J. H.

The Evolutionary Status of PSR 1259-63   291
Cominsky, L.

A Radio Pulsar-B Star Binary in the Small Magellanic Cloud 295
 Kaspi, V. M., Johnston, S., Manchester, R. N., Bailes, M.,
 Bell, J. F., Bessell, M., Lyne, A. G., and D'Amico, N.

A Search for Pulsar Companions to Runaway OB Stars 299
 Philp, C., Frail, D. A., Evans, C. R., and Leonard, P. J. T.

Hydrodynamical Instability and Orbital Evolution of Close Binary Systems 303
 Lai, D., Rasio, F. A., and Shapiro, S. L.

Evolution of LMBs and Asteroseismology 307
 Sarna, M. J., Lee, U., and Muslimov, A. G.

Excitation of Neutron Star Oscillation Modes During Binary Inspiral 311
 Reisenegger, A., and Goldreich, P.

Binary-Binary Collisions Involving Main-Sequence Stars, White Dwarfs and
 Neutron Stars in Globular Clusters 315
 Leonard, P. J. T., and Davies, M. B.

EVOLUTION

On the Origin of Low-Mass X-Ray Binaries 321
 Webbink, R. F., and Kalogera, V.

Recycled Radio Pulsars 331
 Lyne, A. G.

Dim X-ray Sources in Globular Clusters 339
 Grindlay, J. E.

Do Magnetic Fields of Neutron Stars Decay? 351
 Verbunt, F.

ROSAT Measurement of the Evolving Orbital Period in
 the LMXB EXO 0748-676 363
 Hertz, P., Ly, Y., Wood, K. S., and Cominsky, L. R.

UV Polarimetry of the X-Ray Binary Systems 4U1700-37, Vela XR-1
 & Cygnus XR-1 367
 Wolinski, K. G., Dolan, J. F., Boyd, P. T., Elliot, J. L.,
 Nelson, M. J., Percival, J. W., Townsley, L. C., and van Citters, G. W.

Lithium in Quiescent X-Ray Novae 371
 Charles, P. A., Casares, J., Martín, E. L., and Rebolo, R.

The Orbital Lightcurve of PSR 1957+20 375
 Callanan, P. J., van Paradijs, J., and Rengelink, R.

The Orbital Period Derivative of PSR 1957+20 379
  McCormick, P., Frank, J., King, A. R., and Rajasekhar, A.

A Study of Angular Momentum Loss in Binaries Using
                           the Free Lagrange Method 383
  Rajasekhar, A., Frank, J., and Whitehurst, R.

Do Quiescent Soft X-Ray Transients Contain Millisecond Radio Pulsars? 387
  Stella, L., Campana, S., Colpi, M., Mereghetti, S., and Tavani, M.

Evolutionary Scheme for Low-Mass Binary with Millisecond Pulsar 391
  Muslimov, A. G., and Sarna, M. J.

The Evolution of Evaporating Binary Pulsars 395
  McCormick, P., Frank, J., and King, A. R.

Evolution Versus Variability in Neutron Star Binaries 399
  Wijers, R. A. M. J.

The Fate of Thorne-Zytkow Objects 403
  Podsiadlowski, P., Cannon, R. C., and Rees, M. J.

Irradiation-Driven Evolution of Low-Mass X-Ray Binaries 407
  Tavani, M.

X-RAY PULSARS

The Properties of Accreting X-Ray Pulsars 415
  Parmar, A.

Emission Processes in X-ray Pulsars 429
  Harding, A. K.

Spin Evolution of Neutron Stars in Accretion Powered Pulsars 439
  Ghosh, P.

Discovery of the Hard X-Ray Pulsar GRO J1008-57 by BATSE 451
  Wilson, R. B., Harmon, B. A., Fishman, G. J., Finger, M. H.,
  Stollberg, M. T., Pendleton, G. N., Briggs, M., Prince, T. A.,
  Bildsten, L., Chakrabarty, D., Rubin, B. C., and Zhang, N. S.

BATSE Observations of 4U1538-52: a 530 Second Pulsar 455
  Rubin, B. C., Finger, M. H., Wilson, R. B., Fishman, G. J.,
  Meegan, C. A., Paciesas, W. S., Prince, T., Chiu, J.,
  and Chakrabarty, D.

Hard X-Ray Observations of A 0535+262 459
  Finger, M. H., Cominsky, L. R., Wilson, R. B., Harmon, B. A.,
  and Fishman, G. J.

Cyclotron Absorption in the High Energy Spectrum of Her X-1   463
    Kunz, M., Staubert, R., Gruber, D. E., Pietsch, W.,
    Trümper, J., Kaniovsky, S., and Sunyaev, R.

Flux and Spectrum Variability in Her X-1   467
    Leahy, D. A.

EXOSAT Studies of Dips in Hercules X-1   471
    Reynolds, A. P., and Parmar, A. N.

Observation of a Correlation Between Main-On Intensity and
Spin Behavior in Her X-1   475
    Wilson, R. B., Finger, M. H., Pendleton, G. N., Briggs, M.,
    and Bildsten, L.

HST/FOS Observations of Hot Gas During the Total Eclipse of
the Neutron Star in HZ Her/Her X-1   479
    Wachter, S., Anderson, S. F., Margon, B., and Downes, R. A.

Sub-synchronous Rotation and Tidal Lag in HD 77581/Vela X-1   483
    Wilson, R. E., and Terrell, D.

Long Term Multiwavelength Monitoring of High Mass X-ray Binaries   487
    Roche, P., Coe, M., Everall, C., Fabregat, J., Reglero, V., Prince, T.,
    Chakrabarty, D., Bildsten, L., Norton, A., Unger, S., and Buckley, D.

Dynamics of Accretion Shocks in AM Herculis Systems: Models
for High Mass White Dwarfs   491
    Wolff, M. T., Wood, K. S., and Imamura, J. N.

Effects of Scattering Atmospheres on the Pulse Profiles of
Accreting X-Ray Pulsars   495
    Sturner, S. J., and Dermer, C. D.

X-Ray Pulsar Hydrodynamics: Collisionless Shock Waves and
Steady State Infall Hydrodynamics   499
    Rose, W. K.

DYNAMICS

Rapid Variability in Neutron Stars and Black Holes --
Comparison and Attempt at Unification   505
    van der Klis, M.

Tidal Instabilities in Accretion Discs   515
    King, A. R.

Fractal Analysis of X-Ray Emission from Centaurus X-3 Observed with Ginga   522
    Kanetake, R., and Takeuti, M.

Precession and Long-Term Cyclic Behavior 525
    Smale, A. P.

An Ephemeris Update for X 1822-371 535
    Hellier, C., and Smale, A. P.

GX 5-1 and GX 17+2 with EXOSAT: New Insights in Two Z-Sources 539
    Kuulkers, E., van der Klis, M., Oosterbroek, T., van Paradijs, J.,
    and Lewin, W.

A Search for Chaos in the Rapid Burster 543
    Bockrath, M., DiStefano, R., and Rappaport, S.

Modelling Black Hole X-Ray Power Spectra 547
    Nowak, M. A.

Effect of New Opacities on the Disk Instability 551
    Takeuti, M., and Kanetake, R.

Searches for Millisecond Pulsations in Low-Mass X-Ray Binaries
                                                        with Ginga Data 553
    Vaughan, B., van der Klis, M., Wood, K. S., Norris, J. P., Hertz, P.,
    Michelson, P. F., van Paradijs, J., Lewin, W. H. G., and Mitsuda, K.

Microsecond Temporal Structure from X-Ray Binary Pulsars:
                                              Observability with XTE 557
    Orlandini, M., and Boldt, E.

The USA Experiment on the ARGOS Satellite:
            A Low Cost Instrument for Timing X-Ray Binaries 561
    Wood, K. S., Fritz, G., Hertz, P., Johnson, W. N., Lovelette, M. N.,
    Wolff, M. T., Bloom, E., Godfrey, G., Hanson, J., Michelson, P.,
    Taylor, R., and Wen, H.

WIND INTERACTIONS

X-ray Scattering in X-ray Binary Pulsars 567
    Nagase, F.

Hydrodynamics of Winds in High Mass X-Ray Binaries 578
    Blondin, J. M.

Properties of Interstellar Grains Derived from X-Ray Eclipse Observation ... 588
    Woo, J., Clark, G. W., and Nagase, F.

A Potential Cyclotron Line Signature in Low Luminosity X-Ray Pulsars ... 592
    Nelson, R. W., Wang, J. C. L., Salpeter, E. E., and Wasserman, I.

Near-Eddington Winds from Neutron Stars 597
    Nobili, L., Turolla, R., and Lapidus, I.

The Outflowing Regime of Quasi-Spherical Accretion on to X-Ray Objects
 and the Spin-Down Mechanism for Wind-Fed X-ray Pulsars 601
Illarionov, A. F., Igumenshchev, I. V., and Kompaneets, D. A.

Hot and Cold Atmospheres around Neutron Stars 605
Turolla, R., Zampieri, L., Colpi, M., and Treves, A.

X-RAY SOURCES IN EXTERNAL GALAXIES

Supersoft X-Ray Sources 611
Hasinger, G.

ROSAT PSPC Observation of M31 631
Supper, R.

Optical Identifications of M31 Sources 640
Magnier, E.

Low-Mass X-Ray Binary Models for the Supersoft X-Ray Sources
 in the Large Magellanic Cloud 650
Kylafis, N. D., and Xilouris, E. M.

ROSAT Observations of Globular Clusters 654
Johnston, H. M., Verbunt, F., and Hasinger, G.

The Galactic High Mass X-Ray Binary Population 658
Dalton, W. W., and Sarazin, C. L.

SUMMARIES

Creator's Records: The Pre-History of Single and Binary Neutron Stars 665
Trimble, V.

X-Ray Binaries: What Progress Have We Made? 673
Krolik, J. H.

Appendix A - Conference Programme 687

Appendix B - List of Attendees 691

Appendix C - Physical Constants 701

Author Index 705

Subject Index 709

Introductions

Three Decades of X-Ray Astronomy from the Point of View of a Biased Observer

Walter H. G. Lewin
MIT 37-627, Cambridge, MA 02139
lewin@space.mit.edu

ABSTRACT

In this talk I will review some of the highlights of three decades of X-ray astronomy. That leaves me with an average of 90 sec per year which is highly insufficient to do justice to the many exciting milestones. The choice I have made obviously reflects my own perceptions and biases. The following text is a somewhat edited version of my 45-min talk. To meet the page limit in these proceedings, I had to delete 15 of my 20 figures. However, all are referenced in the text so that the reader can look them up if (s)he so desires.

TEXT

We are here because at the end of the fifties Bruno Rossi suggested to Martin Annis and Riccardo Giacconi at AS&E to explore the X-ray window in search for sources outside our solar system. This led to the historic rocket flight in June 1962 (Riccardo Giacconi, Herb Gursky, Frank Paolini, & Bruno Rossi 1962); I am very pleased that Herb is in the audience.

During the first few years of X-ray astronomy the three most active groups were AS&E, NRL, and Lockheed. The first optical identification came from NRL (Friedman's group) in 1964. Using a lunar occultation of the Crab, they showed unambiguously that the Crab is an X-ray source (Bowyer et al. 1964).

Phil Fisher and co-workers of Lockheed submitted a paper in 1965 in which they compare their results with those of NRL. This is a vintage slide (Fig. 1) from my collection of historical highlights; it is not well preserved and hard to read. I made this slide in 1966. Table 4 (from Fisher et al. 1966) indicates that Cyg XR-1 is highly variable by about a factor of 5. Notice that no errors in the measurements were given (those were the good old days ...). Fisher writes:

"Because time variation in X-ray flux may some day be observed, it may be worth noting that the fluxes used for Table 4 were derived from three different rocket flights made over only a 5-month time interval."

Notice the XR. At the time the groups in Cambridge used exclusively an X, but other groups preferred XR. In the long run, the X prevailed.

1966 brought the optical identification of Sco X-1. Based on a huge error box of several square degrees, and in spite of the rainy season in Japan, Osawa Jugaku & co-workers of Tokyo Astronomical Observatory found the star. It turned out *later* that it was located in one of the two 1 x 2 arcmin error boxes which resulted from a measurement with the modulation collimator invented by Minoru Oda. I quote here from the historical paper by Sandage et al. (1966):

"Although frequently interrupted by clouds, which are prevalent in Japan during the rainy season, the observations gave $V = 12.6 \pm 0.2$ ..." ...."These results were communicated by cable to Giacconi, who relayed them by telephone to Palomar on June 23, P.S.T."

Observations with the 200 inch that same night confirmed the colors measured in Japan and revealed the *"fast flicker of the order of 2 percent (0.02 mag) in several minutes"*.

Based on an extrapolation of the X-ray spectrum down to optical wavelengths, Sco X-1 was expected to be a blue star of ~ 13 mag, and that is precisely what it turned out to be. Thus, for the wrong reasons, a correct prediction had been made; what a remarkable coincidence!

COMPARISON OF MEASURED FLUXES OF THE REPORTED X-RAY SOURCES

This Work (Uncorrected)			1964 Survey of Naval Research Laboratory*		Ratio of Counts $cm^{-2} sec^{-1}$ (This work corrected ÷ NRL)
Time from Launch (sec)	4–8-keV Counts $cm^{-2} sec^{-1}$	Assumed θ	Source	Counts $cm^{-2} sec^{-1}$	
178.5........	0.73	2°	Cyg XR-1	3.6	0.2
205.1........	0.86 ⎫	1	Sgr XR-2	1.5	1.1
208.2........	0:80 ⎭				
211.0........	1.39	8	Oph XR-1	1.3	1.5
212.6........	0.86	1	Sgr XR-1	1.6	0.6
218.3........	0.66	7	Sco XR-2	1.4	0.7
220.8........	1.09	0	Sco XR-3	1.1	0.9
....	11.0	0	Sco XR-1	18.7	0.6†

* Bowyer *et al.* (1965).
† A similar comparison to the data of Giacconi *et al.* (1964) yields a ratio of 0.9 for Sco XR-1; beryllium-window counters were used for both of the measurements.

Fig. 1. Table 4 of Fisher et al. 1965

1967 witnessed the first X-ray Nova. Australian scientists flew a rocket from Woomera on April 4 and discovered a source (almost as bright as Sco X-1) that was not detected during a previous flight. The source was pinpointed in the Southern Cross which made it an even more important (and emotional) discovery for the Aussies, as they carry this constellation in their flag. They named the source CRUX (Harries et al. 1967). It is sad for our Australian friends that more precise observations moved the source just outside the constellation Crux into Centaurus. The source is now known as Cen X-2. I remember how Ken McCracken was spreading the news jubilantly in Washington during the APS meeting; he even had preprints with him ... only a few weeks after the flight! The next slide (Fig. 2) shows how the X-ray flux decayed over a time span of several months.

X-ray novae are now generally called X-ray transients. To date we know of 70 transients (see Chapter 14 in *X-Ray Binaries* due for publication by Cambridge University Press in 1994). The mechanisms that cause transients are not fully understood yet; surely more than one mechanism is at work. The details are still unclear and controversial.

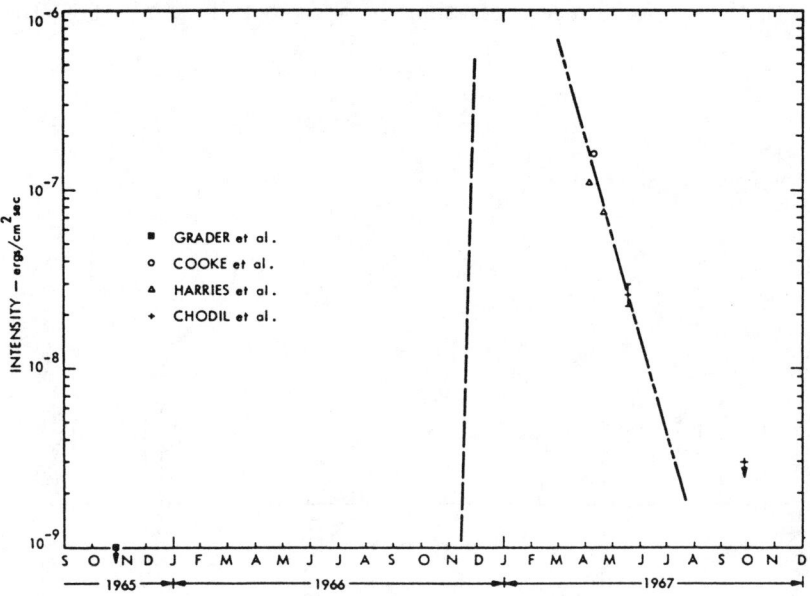

Fig. 2. Decay of Cen XR-2. The energy range is 2 – 5 keV. Observations on October, 28, 1965, and September 28, 1967, only set upper limits on the intensity. The dashed line in November 1966 is an estimate based on assumption that Cen XR-2 is a nova whose maximum occurred on December 1, 1966, ±3 months. This figure is from Chodil et al. 1968.

Variability was in the air. On October 15, 1967 Sco X-1 was caught with a smoking gun during a 7-hour balloon observation from Australia (next slide = Fig. 3). George Clark and I were fortunate to catch this flare. The source brightened in ~10 minutes by a factor of about four (20-30 keV). This could not have been discovered during rocket flights as they lasted typically only 5-10 min during which several sources were observed.

By the end of 1967 three kinds of variability were known: (i) the long-term variability as observed from Cyg X-1, (ii) the nova-like appearance and subsequent disappearance (Cen X-2), and (iii) the flare-like variability (Sco X-1) on a time scale of ten minutes.

Radio pulsars were discovered in 1967 (Hewish et al. 1968). I still remember after the Crab pulsar had been found that I had luncheon with Geoffrey Burbidge at MIT in February of 1969. I had a balloon flight in preparation; our time resolution was no better than a few seconds, and I asked Jeff whether it would perhaps be useful to convert our electronics to millisec timing to take a peek at the 33 msec Crab pulsar. Jeff started to laugh in his usual roaring way, and said that I would not see any X rays from any radio pulsar. He added: *"You don't understand anything about pulsars"*. He was right; I didn't, that's why I asked.

The NRL group, and independently Hale Bradt's group at MIT, detected two months later the 33 msec pulsations in X rays (Fritz et al. 1969; Bradt et al. 1969).

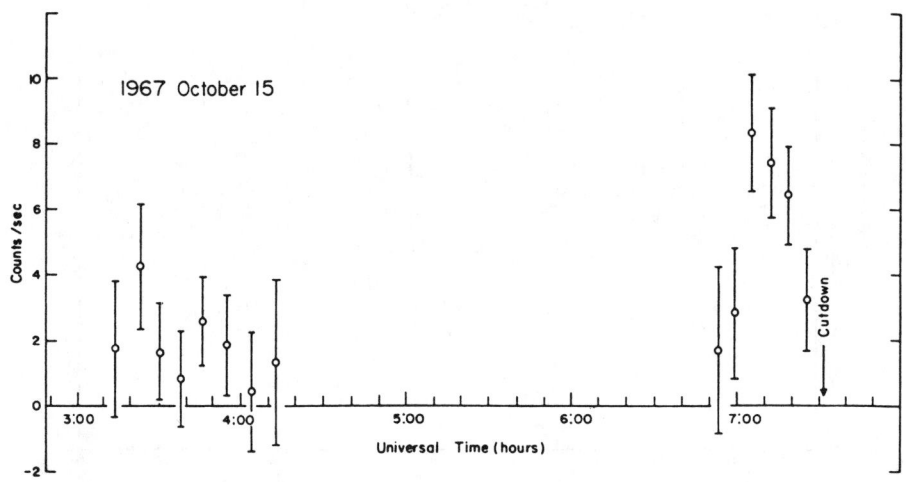

Fig. 3. Intensity of Sco X-1 (20 – 30 keV). The flare-up occurs shortly after 7:00 hours UT. This figure is from Lewin, Clark, & Smith (1967).

Larry Mertz (1967) suggested to use the modulation collimator in a rotating mode. This idea was pushed by Herb Schnopper (1968) which resulted in very accurate positions and in the discovery of new sources (obtained by Hale Bradt's group) in the general region of the galactic center. The sources were labeled with GX (e.g., GX 5-1, GX 3+1 ...). Later discoveries of new sources by the MIT balloon group (e.g GX 1+4, GX 301-2) followed this notation.

The nomenclature of sources is interesting. First we had the XR versus X notation, then the GX was introduced, and later sources were simply named by their RA and Dec. We say *Sco X-1*, not *1617-15*; if someone did, I wouldn't even know what (s)he was talking about. We say *GX 9+9, GX 17+2*, but *1820-30* and *1636-53*. Frankly, I like this. I do not like uniformity. This adds to the flavor that comes with evolution; it reflects the beauty of the pioneering efforts from the past.

Uhuru was launched on December 12, 1970. The pulsar fever was in the air! Riccardo Giacconi had earlier requested to NASA to update the timing of Uhuru to milliseconds, but unfortunately this request was turned down. The Uhuru group came with a daring splash in March 1971 when they reported that Cyg X-1 was a 73 msec pulsar (Oda et al. 1971). This was courageous as this was above Uhuru's Nyquist frequency. We now know that this result was not correct;

however, as Riccardo told me later, this accelerated the search for pulsars in the Uhuru data. Shortly afterwards in May 1971 they submitted their historic paper with the 4.8 sec pulsations in Cen X-3 (Giacconi et al. 1971). However, there was no observational evidence yet for the binary origin of X-ray sources which is the topic of this meeting.

X-ray astronomy had by now taken over almost all of astronomy. The next slide shows the cover page of ApJ Letters of August 15, 1971.

X-ray Observations of Virgo XR-1
I see the names of Mike Lampton, Stew Bowyer, and Bruce Margon.

X-rays from the Magellanic Clouds
I recognize the name of Fred Seward.

Measurement of the Location of the X-ray Source Cygnus X-1
I see the names of Miyamoto, Matsuoka, Nishimura, Minuro Oda.

X-ray Source Positions for Cyg X-1, Cyg X-2, and Cyg X-3
again I see Fred's name.

On the Location of Cyg X-1
Saul Rappaport and his student Roger Doxsey.

Radio Emission from X-ray Sources
heroic work by Bob Hjellming and Wade.

X-ray Observations of GX 17+2 from Uhuru
Harvey Tananbaum, Herb Gursky, Ed Kellog, and Riccardo.

A Search for Deuterium on Venus
this must have been an oversight. Chandrasekhar, who was the editor at the time, must have had a weak moment.

The all-important step to X-ray binaries was around the corner. Refined positions from Uhuru, and an MIT rocket flight (Rappaport, Zaumen & Doxsey 1971) led to the discovery of a variable radio source independently by Hjellming & Wade (1971), and Braes & Miley (1971); both groups associated the radio source with Cyg X-1. The very small error box of the radio position (arc seconds) contained a very bright B0 supergiant. The next slide (from Kristian et al. 1971) shows the error boxes and the bright star. It is interesting to note that Kristian et al. dismiss this bright star as the counterpart of Cyg X-1 for several reasons, among them that the star was only "*a normal B0 supergiant with no peculiarities*".

The observational breakthrough to X-ray binaries came from ground based optical observations made independently by Webster & Murdin (1972), and by Bolton (1971, 1972). They discovered the 5.6 day orbital period of the B0 supergiant, and they measured the radial velocity curve of the star. This led to the mass function, and with an assumed range of masses of the B0 supergiant, they both found high values for the mass of the companion.

I quote from Webster & Murdin: "*The mass of the companion being probably larger than $2M_\odot$, it is inevitable that we should also speculate that it might be a black hole*". Webster & Murdin's paper was submitted to Nature on November

17, 1971, and it appeared in print in the January 7, 1972 issue.

I quote from Tom Bolton's paper: *"The high energies of the X-rays imply that large accelerations are involved - accelerations such as might be produced by the intense gravitational field of a collapsed object. The lower limits placed on the secondary mass are too high for a white dwarf and probably rule out a neutron star. ... This raises the distinct possibility that the secondary is a black hole"*.

The next slide (Fig. 4) shows the radial velocity curve from Webster & Murdin.

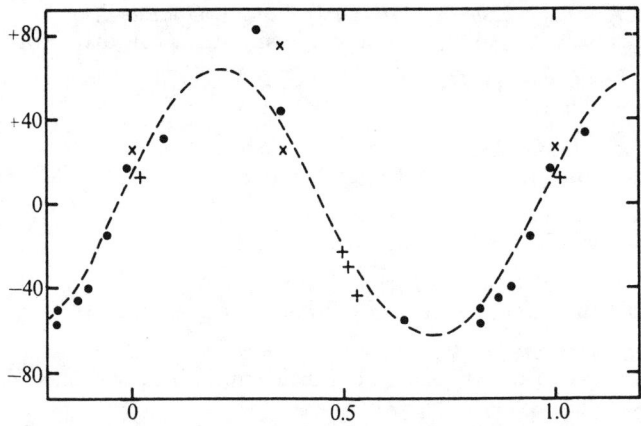

Fig. 4. The radial velocity of HD 226868 against phase in a 5.60 day period. Zero phase is defined to occur at Julian Day 2441163.597. ●, Measurements on photographic spectra at 62 Å/mm; ×, photographic spectra at 184 Å/mm; +, image tube spectra at 75 Å/mm. This figure is from Webster & Murdin, 1972.

It is unfortunate that the popular press and even the professional literature almost always credits Uhuru exclusively with the discovery of the binary nature of X-ray sources (see e.g. Scientific American, November 1993). Many authors even credit Uhuru with the black-hole discovery in Cyg X-1; that too is wrong. This is perhaps understandable in the light of what followed very shortly afterwards. Schreier et al. (1972) demonstrated that Cen X-3 was a binary. The source showed X-ray eclipses every 2.1 days, and the doppler shift of the 4.8 sec pulsations beautifully confirmed the binary nature. Their paper was submitted on January 21, 1972. This had such a splashing impact, that it probably overshadowed the milestone discovery of the binary nature of Cyg X-1. Some skeptics may have argued that the Cyg X-1 identification with the star was not iron clad. Be this as it may, this is "astro-sociology". Almost everyone, probably including those of you in the audience who were actively working in the field at the time have forgotten what happened, when it happened, and who did what*.

Note: After I gave my talk, several people came to me and questioned the accuracy of my dates. One of them even challenged the dates that I quoted during his talk. Of course, they agreed when I showed them copies of the key papers which I had with me. History is something very delicate, and people's memories are weak. Virginia Trimble gave a very nice talk on the last day of the conference. She talked about the history of neutron stars. There too people's perceptions have evolved far beyond "the truth", and inaccurate stories and legends begin to live a life of their own. They even make it into the professional literature, and they are quoted, again, and again, and again; so it goes ....

Uhuru carried the flag of the X-ray binaries even further with a second discovery of the binary nature of Her X-1; the paper was submitted April 26, 1972). Her X-1 (like Cen X-3) showed X-ray pulsations, X-ray eclipses, and the doppler shift of the pulsations. The next slide (Fig. 5) shows the pulsations of Her X-1 (from Tananbaum et al. 1972); the next slide (from Giacconi et al. 1973) shows the eclipses as well as the "on-state" in the 35-day period which is generally interpreted as being due to precession of the accretion disk around the neutron star (Katz 1973; Boynton, Crosa & Deeter 1980).

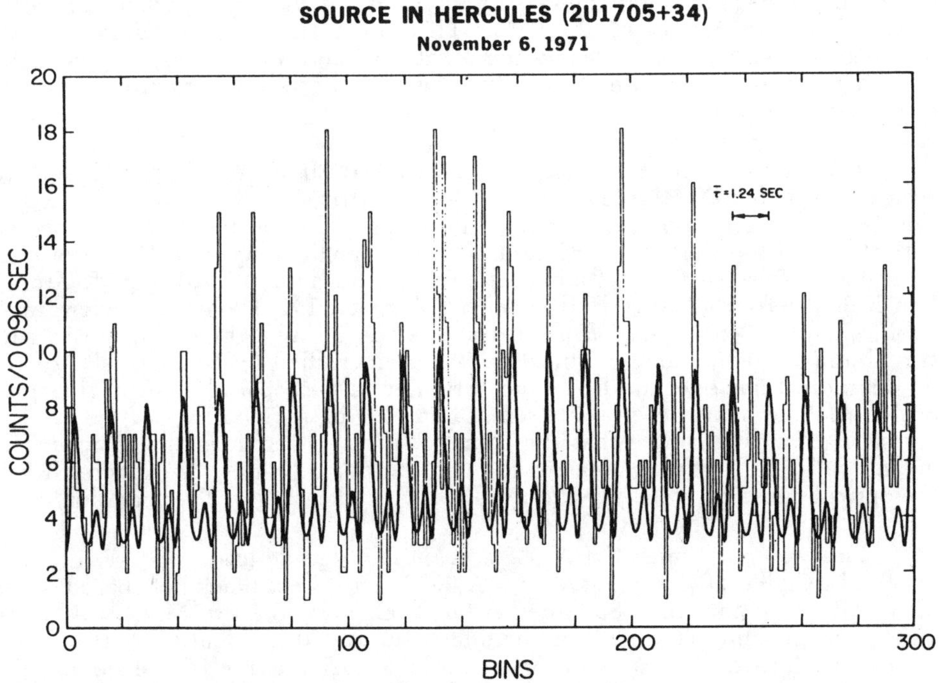

SOURCE IN HERCULES (2U1705+34)
November 6, 1971

Fig. 5. Uhuru data during a 100-sec pass over Her X-1 on 1971 November 6. The heavier curve is a minimum χ^2 fit to the pulsations of a sine function, its first and second harmonics plus a constant, modulated by the triangular response of the collimator. The functional fit is systematically below the peak counting rate partly due to the sharpness of the pulsing and partly due to the minimum χ^2 technique. This figure is from Tananbaum et al. 1972.

After 1972, X-ray pulsars became a big industry. Lots of work was done to study the pulse profiles, the changes in the profiles, the pulse periods, and the changes in these periods (spin-up and spin-down). In combination with optical studies, orbital parameters were measured as well as their changes, and, perhaps most intriguing of all, masses of neutron stars were measured. All this was a great stimulance to theoreticians who began to wonder about the formation and evolution of these systems as you will hear from the next speaker. The next viewgraph shows the latest compilation of the masses of neutron stars and black holes. This figure is from Chapter 2 in *X-ray Binaries*, Cambridge University Press (1994). The data points with exceedingly small errors (smaller than the size of the dots) are from timing measurements of two well known binary radio pulsars; the most famous of them is the Hulse-Taylor pulsar 1913+16 already discovered in 1974 which provided the first proof of the existence of gravitational radiation (Hulse & Taylor 1975).

Two days after I gave my talk it was announced that Joe Taylor and Russel Hulse had been awarded the Nobel Prize. How deserved! Though the prize, in my opinion, was long overdue.

The first believable X-ray line was an iron line discovered in 1974 by the Goddard group (Pete Serlemitsos, Elihu Boldt, Steve Holt, Rick Rothschild) in Cyg X-3 (Serlemitsos et al. 1975). This paved the road for measurements of chemical compositions, plasma temperatures, and, of course, doppler shifts of which Yasuo Tanaka will probably show you a stunning example obtained recently with ASCA.

A *bomb* went off in 1975; the transient A 0620-00 rose to prominence to even outshine Sco X-1. A nice present for our British colleagues, just prior to the X-ray conference in Leicester in August 1975. The outburst was discovered with the Ariel V observatory (Elvis et al. 1975); the light curve was monitored in several ways (also with SAS-3), in particular by the "famous" All-Sky Monitor of Lou Kaluzienski and Steve Holt (Kaluzienski et al. 1977). The next viewgraph (from Chapter 3 in *X-Ray Binaries*) shows the X-ray light curve; notice the two "bumps". The star was quickly optically identified by Boley et al. (1976); it was more than one hundred times brighter in the optical than during pre-outburst. The next viewgraph (from Chapter 2 in *X-ray Binaries*) shows the optical light curve. Notice that the "bumps" are also visible in the optical. The overall decay in X rays was faster than in the optical. That is expected when the optical emission is largely the result of X-ray heating of the accretion disk.

This was not the end of A 0620. Almost a decade later, Nick White and Frank Marshall (1984) suggested that A 0620 might be a black hole based on its ultra-soft spectrum. In 1986, when A 0620 (also called Nova Mon) was X-ray quiet, Jeff McClintock and Ron Remillard succeeded in identifying the donor with a K5 dwarf; they measured the radial velocity curve of the donor (the orbital period is 7.75 h) and concluded that the compact object is very likely a black hole (McClintock & Remillard 1986). The mass function is 3.18 ± 0.16 $M_\odot$, which makes the mass of the accretor probably in excess of 7 $M_\odot$ (for reviews see Chapters 2 and 3 of *X-Ray Binaries*, Cambridge University Press, 1994).

1975 held more surprises: X-ray bursts! They were discovered independently by Josh Grindlay & John Heise (ANS data; Grindlay et al. 1976), who are both in the audience, and by Belian, Conner & Evans (1976; Vela data). We were fortunate that SAS-3 was ideally suited to find burst sources (it was never designed with that in mind), and we discovered many burst sources in 1976 and 1977. Jean Swank and co-workers (OSO-7 data) were the first to derive the size of the emitting region from the blackbody spectra observed during a burst (Swank et al. 1977). The bursts from a given source recurred typically on time scales of several hours; the distinct spectral softening during burst decay (which could take anywhere from tens of seconds to several minutes) was clearly indicative of the cooling of the surface of a neutron star. During the decay the blackbody temperature decreases substantially, but the radius of the emitting region remains approximately constant (radii are typically 10 - 15 km; Hoffman, Lewin & Doty 1977a,b).

One bizarre burster was discovered with SAS-3 which, for good reasons, was called the Rapid Burster. In early March, 1976, it produced about thousand bursts per day (Lewin et al. 1976). This discovery, was very intriguing, but it spoiled the waters! Only one month earlier (in February), when Laura Maraschi was visiting MIT, she had suggested that the bursts resulted from thermonuclear flashes on the surface of a neutron star (Maraschi & Cavaliere, 1977). This was independently proposed by Stan Woosley and Ron Taam (1976). The problem with the Rapid Burster was that the large number of bursts could not possibly be due to thermonuclear energy as so much fuel could not be accreted in the available time; such accretion would have made itself known in the form of a very strong flux of X rays (gravitational potential energy), but that was not observed. The next viewgraph (from Chapter 4 of *X-Ray Binaries*) shows the repetitive bursts as they were discovered with SAS-3. Notice that the time between two bursts depends strongly on the fluence of the first. The larger the fluence, the longer is the time to the next burst; this is the mechanism of a relaxation oscillator. The next viewgraph shows data from Ginga (Dotani, private communication) where the bursts are almost of equal size and therefore the intervals between bursts are very regular. At first sight you would think that this is a periodic signal, but at closer inspection one can see that that is not the case.

In the fall of 1977 the Rapid Burster became the Rosetta stone when it was noticed that in addition to the rapidly repetitive bursts (which show no spectral softening during their decay), the Rapid Burster also emitted bursts with intervals of several hours, and those bursts did show the spectral softening (Hoffman, Marshall & Lewin 1978). The picture became immediately clear. The repetitive bursts (called type-II bursts) were due to spasmodic accretion as originally proposed, and the bursts with spectral softening and time intervals of hours (type-I bursts) were due to thermonuclear flashes on the surface of a neutron star. Due to this discovery and the theoretical work by Paul Joss (1978) on thermonuclear flashes, this picture was generally accepted in 1978.

Since several burst sources are located in globular clusters (including the Rapid Burster which is located in Liller I), this showed convincingly that globular clusters contain neutron stars. This was by no means obvious at the time, as it was believed that neutron stars would escape the globular clusters due to a kick velocity (∼100 km/sec) given at birth. The escape velocity from globular clusters is only a few tens of km/sec.

The bizarre behavior of the Rapid Burster is not well understood. The type II bursts are not the only peculiarity of the Rapid Burster. In fact there is so much to be told that I could give a whole one hour lecture on the Rapid Burster; it will probably bore most of you, but I would love it! It is an embarrassment that we understand so little about the Rapid Burster, not for the observers, but for theoreticians, of course. Thank goodness at least we know that the Rapid Burster is an accreting neutron star (type I bursts!). In the absence of the type-I bursts, no doubt some would declare it a "black hole" in a desperate attempt to hide our ignorance.

Type I bursts contain information on the mass and radius of neutron stars (Van Paradijs 1979; Goldman 1979). The bursts can be so strong that they reach the Eddington limit causing the photosphere of the neutron star to expand to hundreds of kilometers. The mass and radius of the neutron star can be measured separately from radius expansion data (Paczynski & Anderson 1986). This led to an industry to obtain independently the mass and radius of neutron stars, and thereby information on the equation of state of neutron-star matter. This, however, has not been too successful to date (for a review see Chapter 4 of *X-Ray Binaries*).

The so familiar distinction between Low-mass X-ray Binaries (LMXB) and High-mass X-ray binaries (HMXB which fall into two groups: the O & B supergiants, and the Be systems), grew slowly but steadily in the seventies (Tananbaum 1973; Canizares 1975; Jones 1977; Ostriker 1977; Milgrom 1978; Joss & Rappaport 1979). X-ray pulsars have hard spectra and massive companions which are optically bright, LMXB are optically faint (or invisible in the optical), their X-ray spectra are soft, and with few exceptions no pulsations are observed. Ostriker (1977) eluded to the very soft spectra of Cyg X-1 & Cir X-1 and created a "Black Hole" group. As already mentioned, in 1984 Nick White & Frank Marshall expanded on this.

The dividing line between LMXB and HMXB is quite sharp. In Chapter 14 of *X-Ray Binaries* you will find 125 LMXB listed and 69 HMXB. About 40 LMXB produce bursts, none of the known HMXB do.

In 1978, SS433 rose to prominence. It was a known radio source near the center of the supernova remnant W50, and a known X-ray source of modest strength. Bruce Margon and co-workers discovered the moving optical lines (Margon et al. 1979). In 1979 Andy Fabian and Martin Rees suggested a jet interpretation for these lines. Moti Milgrom (1979) suggested that the line movements would be periodic. A 164-day periodicity was discovered which is believed to be due to a precessing disk about the compact object which has an orbital period of about 13 days (it is still unclear whether the accretor is a neutron star or a black hole). The jet was resolved in the radio and in X rays. The next view graph shows the *equivalent* line doppler shift of the red-shifted and the blue-shifted lines during the first 15 months (taken from Margon 1983). Notice that at the times that the red-shifted and the blue-shifted lines have the same energy, the *equivalent* doppler shift is near 10,800 km/sec. This is due to the so-called transverse doppler shift which is simply *time dilation*; it is not due to doppler shift. This was first realized by Fabian and Rees (1979). The corresponding value for the Lorentz factor γ is ~ 1.036, which translates into a jet speed of 0.26c.

In the period 1983-1984, Bill Priedhorsky, Jim Terrell & Steve Holt discovered periods on time scales of a few hundred days using data from the Vela 5B satellite. Proper explanations for these periods are still missing. Best known are the periods of Cyg X-1 (~ 294 days), 1820-30 which is a burst source located in a globular cluster (~ 176 days), and 1916-05 (also a burster) with a period of about 199 days (Priedhorsky, Terrell & Holt 1983; Priedhorsky & Terrell 1984). The next viewgraph shows the long-period light curves of 1820-30 and 1916-05; 1820-30 is a binary with the shortest known orbital period of 11 min, and 1916-05 has an orbital period of ~ 50 min.

Lynn Cominsky and Kent Wood (1984) were the first to measure an orbital period using X-ray *dips* (~ 7.1 h period in 1659-29). Dips are the result of obscuration by the disk. *Dippers* became big business with EXOSAT (its orbital period of ~ 90 h was ideal to find dippers). The next figure is taken from Chapter 1 in *X-Ray Binaries*. It shows the periodic dips of 1755-33 (4.4 h), 1254-69 (3.9 h), and 1916-05 (0.83 h); the latter two are burst sources.

EXOSAT will certainly also be remembered for quasi-periodic oscillations (QPO). In 1984, in collaboration with MPE & Amsterdam, we proposed to search for millisecond pulsations in LMXB. The connection with the msec radio pulsars and the LMXB (you will hear several talks on this subject) requires that the neutron stars in most LMXB rotate at a spin rate in the range of several to tens of msec (Bisnovatyi-Kogan & Komberg 1974; Smarr & Blandford, 1976; Radhakrishnan & Srinivasan 1982; Alpar et al. 1982). Our 1984 observations showed no sign of such coherent pulsations! But ... in stead Michiel van der Klis found a broad peak (quasi-periodic oscillations) in the power density spectra of GX 5-1 near 30 Hz. The centroid frequency was strongly correlated with the source strength. The strength of the QPO was also associated with that of a low-frequency noise component (LFN). The next slide is a colored version of the one published in Nature (Van der Klis et al. 1985). You see how the source strength nicely tracks the centroid frequency of the QPO, and how the LFN strength goes hand in hand with that of the QPO.

This talk is about observations. I am lucky that I do not have to discuss the various theories for QPO. Ed van den Heuvel will undoubtedly do so. The most promising explanation for the intensity dependent QPO as described above (other forms of QPO were discovered later) was proposed by Ali Alpar & Jakob Shaham (1985) who first published their ideas in an IAU Circular.

The QPO research led to a sub-classification of LMXB into *Z* and *Atoll* sources by Günther Hasinger & Michiel van der Klis (1989). To date there are six Z sources and a dozen Atoll sources. Michiel will tell you more about this.

Research on black-hole candidates (BHC) received a large boost from the observations with Ginga (e.g., Tanaka 1992; Miyamoto et al. 1993) and Granat (e.g., Sunyaev et al. 1992). Of course, the optical observations, notably by Jeff McClintock, Ron Remillard, Anne Cowley, Casares & Phil Charles played a key role. During the life of Ginga three very intriguing transients were observed (2000+25 = QZ Vul, 2023+338 = V 404 Cyg, and 1124-68 = Nova Muscae 1991). I already showed the X-ray light curves on the same viewgraph that contained the lightcurve of A 0620-00. There will be two talks on BHC during this meeting, and you will see how the spectral hardness varies with source intensity, and no doubt you will also learn how erratic and bizarre the spectra can change (2023+338) on time scales of minutes, hours, and days.

The spectra of BHC are often decomposed in two components; one so-called multi-temperature disk blackbody component and a power-law component. The fits to the disk blackbody spectrum yield a value which in the model represents the radius of the inner disk. It is an observational fact that this radius in the case of the three BHC 1124-68, 2000+25 & LMC X-3 remains remarkably constant even though there are large changes in the flux (4 orders of magnitude in 1124-68, and at least 3 orders of magnitude in 2000+25). While the spectra soften during decay of the transients (1124-68 and 2000+25), and the blackbody temperature decreases from kT of $\sim$1 keV to $\sim$0.5 keV, this radius remains approximately constant. The next viewgraph (from Tanaka 1992) shows the flux levels of the three sources on time scales of 100 days (1124-68), 200 days (2000+25), and 2 years (LMC X-3). Notice the decay in flux of the two transients, and the constant values of $R_{in}(\cos\theta)^{1/2}/(D/kpc)$ km.

Our Japanese colleagues interpret the constancy of the radius as a measure (apart from an unknown constant which is larger than 1, and which will be different for different sources) of 3 times the Schwarzschild radius ($3R_g$). Once R_g is known, the mass of the compact object follows immediately (R_g = 3 km per $M_\odot$). The interpretation that the constant value of the blackbody radius is related to $3R_g$ is not universally accepted.

Nova Muscae (1124-68) was independently discovered with Ginga (Makino 1991) and Granat/Watch (Lund & Brandt 1991). During an $\sim$13 h observation on January 20, 1992, Goldwurm et al. (1992) observed a line near 500 keV (next view graph). They interpret this line as due to annihilation of positrons.

Characteristics, once believed to be unique to BHC, no longer are. These are: (i) the fast flickering in X rays (which goes back to the famous rocket flight by Rick Rothschildt and co-workers in the early seventies), (ii) an ultra-soft spectrum below $\sim$5 keV, and (iii) high-energy (>20 keV) tails. For reviews see Chapters 3 and 6 in *X-Ray Binaries*.

BHC are a growing industry, and we will learn much more about them in the years to come from their spectral and temporal variability. There will be two talks on BHC during this meeting. I want to specifically mention the nice work done during the past few years by the group in Osaka under the direction of Sigenori Miyamoto (see e.g. Miyamoto et al. 1993, and references therein).

What have I forgotten or left out? (i) cyclotron lines discovered in 1976 by Joachim Trümper and his co-workers during a balloon observation of Her X-1 (Trümper et al. 1978). This led to measurements of the magnetic field strength of accreting neutron stars. Fumiaki Nagase will talk about this in detail. (ii) Multiple faint X-ray sources in globular clusters. There will be two talks on that subject. (iii) Supersoft Sources. They were discovered in the LMC by Knox Long, David Helfand, & Grabelsky (1981) with the Einstein Observatory. Many have been added by recent work with ROSAT, and 15 have been found in M31 (there will be a presentation on this by Rodrigo Supper during this meeting). I made a dirty back on the envelope calculation and concluded that if 15 are seen in M31, there must be many hundreds hiding in M31 as the majority of these very soft sources are undetectable due to the extinction in M31, just as all these sources are completely undetectable in our own galaxy!

I have no X-tal ball vision, and it would be impossible for me to foresee what will happen during the next three decades. But I will make a modest attempt to list the issues that I consider important and which I am sure will get a lot of attention in the next five years.

- Spin periods of neutron stars in LMXB
- Equation of state (spectral lines in X-ray bursts)
- Interpretation of X-ray spectra (meaning of the various *components*)
- Character of BHC – Interpretation of spectra, Timing, Hi-energy tails
- Super Soft Sources
- Nature of the weak multiple X-ray sources in globular clusters and their possible connection to the msec radio pulsars, so abundant in these clusters
- Formation of LMXB outside globular clusters
- Long-term periods of order of a few hundred days
- B-fields of neutron stars, and their decay and dependence on spin period and amount of accreted matter – B-field decay – Cyclotron Lines – Z versus Atoll sources
- Transient mechanisms
- Spin-up and spin-down of neutron stars <=> Interior structure
- Orbital parameters and their changes (orbital evolution)
- Precessing disks (why do they precess?)
- Some odd balls: Cyg X-3, the Rapid Burster, SS433, and Cir X-1

I have addressed several of these issues in seven ASCA proposals. All seven proposals were rejected. If only the organizers of this conference had known this, they would probably not have invited me for this opening talk, as they would have considered me unfit for the job.

We are now going to listen to Ed van den Heuvel. I found an interesting statement in his 1968 PhD thesis (in Dutch) which I will translate for you. He argues that there are objections against the binary model for galactic X-ray sources as suggested by Skhlovski for reasons that the explosive mass loss when the neutron star was formed would "most probably" *(hoogstwaarschijnlijk)* disrupt the binary system.

If only the organizing committee had known this ....

REFERENCES

Alpar, M.A., Cheng, A.F., Ruderman, M.A., & Shaham, J., 1982, Nature, 300, 728
Alpar, M.A., & Shaham, J., 1985, Nature, 316, 239
Belian, R.D., Conner, J.P., & Evans, W.D., 1976, ApJ, 206, L135
Bisnovatyi-Kogan, G.S., & Komberg, B.V., 1974, Sov. Astr., 18, 217
Boley, F., Wolfson, R., Bradt, H., Doxsey, R., Jernigan, G., & Hiltner, W.A., 1976, ApJ, 203, L13
Bolton, C.T., 1971, Bull. Am. Astron. Soc., 3, 458, & 1972, Nature, 235, 271 (received January 3, 1972)
Bowyer, C.S., Byram, E.T., Chubb, T.A., & Friedman, H., 1964, Science, 146, 912
Boynton, P.E., Crosa, L.M., & Deeter, J.E., 1980, ApJ, 237, 169
Bradt, H., Rappaport, S., Mayer, W., Nather, R.E., Warner, B., MacFarlane, M., & Kristian, J., 1969, Nature, 222, 728
Braes, L.L.E., & Miley, G.K., 1971, Nature, 232, 246
Canizares, C., 1975, ApJ, 201, 589
Chodil, G., Mark, H., Rodrigues, R., & Swift, C.D., 1968, ApJ, 152, L45
Cominsky, L.R., & Wood, K.S., 1984, ApJ, 283, 765
Elvis, M., Page, C.G., Pounds, K.A., Ricketts, M.J., & Turner, M.J.L., 1975, Nature, 257, 656
Fabian, A.C., & Rees, M.J., 1979, MNRAS, 187, 13P
Fisher, P.C., Johnson, H.M., Jordan, W.C., Meyerott, A.J., & Acton, L.W., 1966, ApJ 143, 203
Fritz, G., Henry, R.C., Meekins, J.F., Chubb, T.A., & Friedman, H., 1969, Science, 164, 709
Giacconi, R., Gursky, H., Paolini, F.R., & Rossi, B.B., 1962, Phys. Rev. Lett., 9, 439
Giacconi, R., Gursky, H., Kellogg, E., Schreier, E., & Tananbaum, H., 1971, ApJ, 167, L67 (received May 17, 1971)
Giacconi, R., Gursky, H., Kellogg, E., Levinson, R., Schreier, E., & Tananbaum, H., 1973, ApJ, 184, 227
Goldman, I., 1979, A&A, 78, L15
Goldwurm, A., Ballet, J., Cordier, B., Paul, J., Bouchet, L., Roques, J.P., Barret, D., Mandrou, P., Sunyaev, R., Churazov, E., Gilfanov, M., Dyachkov, A., Khavenson, V., Kovtunenko, V., Kremnev, R., & Sukhanov, K., 1992, ApJ, 389, L79
Grindlay, J., Gursky, H., Schnopper, H., Parsignault, D., Heise, J., Brinkman, A.L., & Schrijver, J., 1976, ApJ, 205, L127
Harries, J.R., McCracken, K.G., Francey, R.J., & Fenton, A.G., 1967, Nature, 215, 40
Hasinger, G., & Van der Klis, M., 1989, A&A, 225, 79
Hewish, A., Bell, S.J., Pilkington, J.D.H., Scott, P.F., & Collins, R.A., 1968, Nature, 217, 709
Hjellming, R.M., & Wade, C.M., 1971, ApJ, 168, L21
Hoffman, J.A., Lewin, W.H.G., & Doty, J., 1977a, MNRAS, 179, 57P, and 1977b, ApJ, 240, L27
Hoffman, J.A., Marshall, H., & Lewin, W.H.G., 1978, Nature, 271, 630
Hulse, R.A., & Taylor, J.H., 1975, ApJ, 195, L51
Jones, C., 1977, ApJ, 214, 856
Joss, P.C., 1978, ApJ, 225, L123
Joss, P.C., & Rappaport, S.A., 1979, A&A, 71, 217
Kaluzienski, L.J., Holt, S.S., Boldt, E.A., & Serlemitsos, P.J., 1977, ApJ, 212, 203
Katz, J.I., 1973, Nature Phys. Sci., 246, 87
Kristian, J., Brucato, R., Visvanathan, N., Lanning, H., & Sandage, A., 1971, ApJ, 168, L91
Lewin, W.H.G., Clark, G.W., & Smith. W.B., 1968, ApJ, 152, L55
Lewin, W.H.G., Doty, J., Clark, G.W., Rappaport, S.A., Bradt, H.V., Doxsey, R., Hearn, D.R., Hoffman, J.A., Jernigan, J.G., Li, F.K., Mayer, W., McClintock, J.E., Primini, F., & Richardson, J., 1976, ApJ, 207, L95
Long, K.S., Helfand, D.J. & Grabelsky, D.A., 1981, ApJ, 248, 925
Lund, N., & Brandt, S., 1991, IAU Circ. No. 5161
Makino, F., 1991, IAU Circ. No. 5161

Maraschi, L., & Cavaliere, A., 1977, in: "Highlights in Astronomy", ed. E.A. Müller, Reidel, Dordrecht, Vol. 4, Part I, p. 127
Margon, B., 1983, in: "Accretion Driven Stellar X-Ray Sources", eds. W.H.G. Lewin & E.P.J. van den Heuvel, Cambridge University Press, p. 287 Margon, B., Ford, H.C., Katz, J.I., Kwitter, K.B., Ulrich, R.K., Stone, R.P.S., & Klemola, A., 1979, ApJ, 230, L41
McClintock, J.E., & Remillard, R.A., 1986, ApJ, 308, 110
Mertz, L., 1967, Symp. on Modern Optics Polytechnic Inst. Brooklyn, March 22-24, p. 787
Milgrom, M., 1978, A&A 67, L25
Milgrom, M., 1979, A&A 76, L3
Miyamoto, S., Iga, S., Kitamoto, S., & Kamado, Y., 1993, ApJ, 403, L39
Oda, M., Gorenstein, P., Gursky, H., Kellogg, E., Schreier, E., Tananbaum, H., & Giacconi, R., 1971, ApJ, 166, L1 (received March 22, 1971)
Ostriker, J.P., 1977, Ann. New York Ac. Sci., 302, 229
Paczynski, B., & Anderson, N., 1986, ApJ, 302,1
Priedhorsky, W.C., Terrell, J., & Holt, S.S., 1983, ApJ, 270, 233
Priedhorsky, W.C., Terrell, J., 1984, ApJ, 284, L17
Radhakrishnan, V., & Srinivasan, G., 1982, Curr. Sci., 51, 1096
Rappaport, S., Zaumen, W., & Doxsey, R., 1971, ApJ, 168, L17
Sandage, A.R., Osmer, P., Giacconi, R., Gorenstein, P., Gursky, H., Waters, J., Bradt, H., Garmire, G., Sreekantan, B.V., Oda, M., Osawa, K., & Jugaku, J., 1966, ApJ, 146, 316
Schnopper, H., 1968, ApJ, 161, L161
Schreier, E., Levinson, R., Gursky, H., Kellogg, E., Tananbaum, H., & Giacconi, R., 1972, ApJ, 172, L79 (received January 21, 1972)
Serlemitsos, P.J., Boldt, E.A., Holt, S.S., Rothschild, R.E., & Saba, J.L.R., 1975, ApJ, 201, L9
Smarr, L.L., & Blandford R., 1976, ApJ, 207, 575 Sunyaev, R., Aref'ev, V., Borozdin, K., Churazov, E., Efremov, V., Gilfanov, M., Kaniovsky, A., Kendziorra, E., Mony, B., Maisack, M., Staubert, R., Doeberainer, S., Englhauser, J., Pietsch, W., Reppin, C., Trümper, J., Skinner, G.K., Nottingham, M.R., Pan, H., Willmore, A.P., Brinkman, A.C., Heise, J., In't Zand, J.J.M., & Jager, R., 1991, in: "Frontiers of X-Ray Astronomy", eds. Y. Tanaka, & K. Koyama, Universal Acad. Press, Inc. Tokyo, p. 697
Swank, J.H., Becker, R.H., Boldt, E.A., Holt, S.S., Pravdo, S.H., & Serlemitsos, P.J., 1977, ApJ, 212, L73
Tanaka, Y., 1992, in: "Ginga Memorial Symp.", eds. F. Makino, & F. Nagase, p. 19
Tananbaum, H.D., 1973, IAU Symp. 55, eds. H. Bradt, & R. Giacconi, p. 9
Tananbaum, H., Gursky, H., Kellogg, E.M., Levinson, R., Schreier, E., & Giacconi, R., 1972, ApJ, 174, L143 (received April 26, 1972)
Trümper, J.E., Pietsch, W., Reppin, C., Voges, W., Staubert, R., & Kendziorra, E., 1978, ApJ, 219, L105
Van der Klis, M., Jansen, F., Van Paradijs, J., Lewin, W.H.G., Van den Heuvel, E.P.J., Trümper, J.E., & Sztajno, M., 1985, Nature, 316, 225
Van Paradijs, J., 1979, ApJ, 234, 609
Webster, B.L., & Murdin, P., 1972, Nature, 235, 37 (January 7 issue; received November 11, 1971)
White, N.E., & Marshall, F.E., 1984, ApJ, 281, 354
Woosley, S.E., & Taam, R.E., 1976, Nature, 263, 101

THREE DECADES OF X-RAY BINARIES, FROM THE POINT OF VIEW OF A THEORETICIAN

Edward P.J. van den Heuvel
Astronomical Institute and Center for High Energy Astrophysics,
University of Amsterdam, Kruislaan 403, 1098 SJ Amsterdam,
The Netherlands

ABSTRACT

The gathering of a wealth of precise quantitative observational data on X-ray binaries and recycled radio pulsars during the last decades has allowed the testing of many theoretical ideas on the physics, formation and evolution of neutron stars and black holes, which could never have been tested on single compact objects. Also, new observations often have led to new theories and models which again can be tested observationally. In this talk I give a personal view of this development over the past thirty years.

1. INTRODUCTION

My personal experience with X-ray astronomy covers some 21 years, *i.e.* about two-third of the period of extra-solar X-ray astronomy. In my perception the last thirty years of theoretical research in this field can be roughly divided into four epochs:

a. 1964-1971. The epoch of basic ideas and concepts

This epoch started with the suggestions by Salpeter (1964) & Zeldovitch (1964) that accretion of matter onto a compact object (in their papers: a supermassive black hole, accreting interstellar matter) provides an extremely powerful energy source and might produce the huge energy output of QSOs and AGNs. In section 2 I briefly describe some highlights of this epoch.

b. 1971-1978. The epoch of great discoveries, confirmation
and many new ideas and concepts

This epoch started with the discovery by UHURU of the pulsating binary X-ray sources (Schreier *et al.* 1972), which made a huge impact also in the community of theoreticians and was followed by an avalanche of new discoveries, by UHURU and subsequent X-ray satellites such as Copernicus, ANS, Ariel V, SAS-C, OAO-1. It was in this period that the concepts of high- and low-mass X-ray binaries arose, spin-up was discovered, globular cluster sources, bursters, Be/X-ray binaries and transients were discovered, the neutron star nature of bursters was established, etcetera.

At every Texas Symposium in these years (1972, 1974, 1976) many new and exciting discoveries were reported. What was so nice for the theoreticians in this time was that the many new discoveries triggered a lot of new theoretical work which itself led to new predicions, several of which were subsequently confirmed by the observations. There are not many fields of astrophysics where this is possible and it is very satisfying to see this happen. I will describe a number of such instances in section 3.

c. 1978-1982. The epoch of consolidation

Around 1978 the time of new great discoveries seemed over. Most things about X-ray binaries seemed to be known. Several important new satellites (HEAO-I, Einstein) were launched in this period and produced further confirmations of the general picture that had arisen in the preceding seven years and showed that also in other galaxies (M31, M33) populations of X-ray sources are present resembling those in our own Galaxy.

Peculiarly enough, in the field of radio pulsars - that at the time seemed to have hardly any connection with that of X-ray binaries - this was also a period in which little news seemed to be happening. Almost everything that was to be known about pulsars and neutron stars seemed to be known. How wrong this later proved to be!

d. 1982-present. The epoch of 'connection' between X-ray binaries and binary and millisecond radio pulsars

The discovery in 1982 of the 1.55 msec pulsar PSR 1937+21 (Backer et al. 1982), with its 10^4 times weaker magnetic field than usual and its 20 times faster spin than that of the then fastest known pulsar (the Crab pulsar) came as a great shock and suddenly made us realize that we understood much less about neutron stars and their structure and evolution, than had previously been thought.

The idea that X-ray binaries may later on produce spun-up rapidly rotating radio pulsars had already been put forward in the middle seventies (Bisnovatyi-Kogan & Komberg 1975; Smarr & Blandford 1976) but had up till then received little attention, except in connection with the origin of the Hulse-Taylor pulsar and of 0655+64 (Smarr & Blandford 1976; Srinivasan & Van den Heuvel 1978, 1982; Radhakrishnan & Srinivasan 1984).

The discovery of the millisecond pulsar suddenly put the 'recycling' of neutron stars by accretion in binary systems in the center of attention (Alpar et al. 1982; Radhakrishnan & Srinivasan 1982). The discovery, since 1985, of many millisecond and binary pulsars, together with the discovery of QPO (Van der Klis et al. 1985 (a,b)) have firmly established the connections between X-ray binaries and binary and millisecond pulsars.

Since the middle eighties, furthermore, many new and exciting discoveries in the field of X-ray astronomy (by EXOSAT, TENMA, Ginga and the Russian experiments) have put the study of compact stars in binaries again in a very prominent position in present-day astrophysics, making this one of the most active fields of our science.

In section 5. I summarize the, in my opinion, most important developments in this fourth epoch.

2. THE EPOCH OF BASIC IDEAS AND CONCEPTS: 1964-1971

Especially the Russian theoretical school made many impressive contributions in these years. Zeldovitch & Guseinov (1965) were the first to search for black holes in binary systems, by looking for spectroscopive binaries with massive unseen secondary stars. They did, however, not mention the possibility of X-rays in these first papers, but Zeldovitch & Novikov did so slightly later on, see Trimble (this volume).

The culmination of this early work was in Shklovskii's (1967) neutron-star binary model for Scorpius X-1. Shklovskii showed that the optical light in the system could not arise from the same source as the X-rays and that the X-ray energy distribution is consistent with thermal brems-strahlung from an optically thin plasma accreting onto a neutron star. Since the optical spectrum of the source resembles that of a cataclysmic variable (CV) and since no stellar spectrum is seen (implying that the star is faint) he postulated that the neutron star is in a binary and accretes matter from a low-mass companion. The proposition from my 1968 PhD thesis which Lewin (this volume) quoted, expresses the difficulty which I had, from the *evolutionary* point of view, with this model: I could not understand how a low-mass (solar type) companion could have remained bound after the supernova explosion in which the neutron star was formed. We knew then already that the progenitors of neutron stars are quite massive stars $\gtrsim 8 M_\odot$. If this star is in a binary, with a solar-type companion, one would (naively) expect that in the supernova explosion more than half of the system mass is ejected, such that, according to the virial theorem, the system will become unbound (Blaauw 1961).

Shklovskii apparently did not worry about the evolutionary history of the system, and history has proven him right not to have done so. I, however, did worry about this and indeed it has later become clear that it is very difficult for nature to make low-mass X-ray binaries (see Webbink, this volume). Their birthrate in the Galaxy is very low: there are only of order 10^2 in the Galaxy and their lifetime is expected to be on average $> 10^8 yrs$, which implies a galactic birthrate $< 10^{-6} yr^{-1}$ (Van den Heuvel 1983). This implies that not more than one out of every 10^4 supernovae succeeds in producing a neutron star binary that later evolves into a LMXB.

Since over 50 per cent of all stars are members of binaries, apparently an extremely rare set of initial system conditions has to be fulfilled to produce an LMXB (see Webbink, this volume) and in 1968 I (wrongly) thought this was plainly impossible.

Following Shklovskii's work, other important ideas and developments in this period were:
a. The construction of the first model of an accretion disk in a CV-binary, by Prendergast & Burbidge (1968), who at the time thought that strong galactic X-ray sources might be CVs.
b. The proposition by Lynden-Bell (1968) that quasars are supermassive black holes powered by disk-accretion.
c. The first detailed model for accretion onto a magnetized neutron star by Zeldovitch & Shakura (1969) showing that X-ray emission with a spectrum peaking in the $1 - 50 keV$ range is expected to arise from such a process.
d. The models of accreting magnetized neutron stars and accreting black holes by Shvartsman (1970 a,b, 1971).
e. The recognition by Setti & Woltjer (1970) that there is a class of intrinsically bright X-ray sources in the galactic bulge: they showed that the sources seen in the direction of the galactic center must indeed roughly be at the distance of this center, because if they were just a group of foreground sources (*i.e.* intrinsically

less bright) one would also expect there to be a large background population of fainter sources in the Galaxy, which is not observed.
f. The first detailed modelling of accretion disks around black holes by Shakura (1973, submitted in 1971).

3. 1971-1978: THE EPOCH OF GREAT DISCOVERIES, CONFIRMATION AND MANY NEW IDEAS AND CONCEPTS

This epoch started with the discovery of the binary character of Centaurus X-3. The 4.84 sec pulse period of this source had been discovered with UHURU in May 1971 and in the first paper on the source (Giacconi et al. 1971) its puzzling nature was described. It behaved very differently from radio pulsars, as it emitted only X-rays and sometimes its pulse period was somewhat shorter, at other times somewhat longer, in an apparently irregular way, and also the source at some times was completely absent for half a day. I remember Herb Gursky in the summer of 1971 giving a talk in Utrecht about these baffling characteristics: it behaved in a totally incomprehensible way.

But then early in January 1972, John Heise told me he had received a preprint from the UHURU-group (Schreier et al. 1972), reporting that Cen X-3 is an eclipsing binary with an orbital period of 2,087 days and an eclipse duration of about half a day. The strange pulse period variations appeared to be simply due to the Doppler effect produced by the motion of the pulsar around a star with a mass $\gtrsim 15 M_\odot$.

This was extremely exciting news and since I was interested in the evolution of binaries and John was interested in compact stars, we decided we should try to think of a model for this system and its evolutionary history. We thought at first that the eclipse duration together with the other observed parameters and the constraint that the companion fills its Roche lobe would allow us to set a limit to the orbital inclination: the long eclipse duration suggested that the Earth is almost in the orbital plane. However, John found that one can set no limit to the orbital inclination. In fact, every inclination between 0° and 90° is possible: each inclination corresponds to another mass ratio. The only thing he found is that a lower limit to the mass ratio M_{opt}/M_x can be set, which yielded an upper limit to the mass of the neutron star of about $0,5 M_\odot$ (with a rather large uncertainty; later, using more accurate representations of the Roche lobes a value of $\sim 1 M_\odot$ resulted).

I from my side realized that the neutron star must originally have been the more massive star of the two, since it had already finished its evolution and its companion had not. From what we knew of close binary evolution at the time, from the work of Paczynski (1967), Kippenhahn & Weigert (1967) and Plavec (1968) it was clear to us that a system with such a short orbital period, large-scale mass transfer must have taken place such that at the time of the supernova explosion, the more evolved star had become the less massive component of the system. This is a crucial point, as it explained why the system had not been disrupted by the explosive mass ejection: according to Blaauw's (1961) criterion, if less than half of the system mass is ejected, the system is not disrupted in the explosion (we roughly estimated the effects of the impact of the supernova shell onto the companion and these turned out to be small). It then was straightforward to construct a plausible model for the origin of the Cen X-3 system (as depicted in figure 1): the $16 M_\odot$ initial primary star overflows its Roche lobe after it has left the main sequence and transfers its $12 M_\odot$ hydrogen-rich envelope to the companion, leaving behind a $4M_\odot$ helium star. Due to the mass transfer, the companion increases in mass to $15M_\odot$ and becomes a practically newborn star, which needs 8 million years to terminate its hydrogen burning.

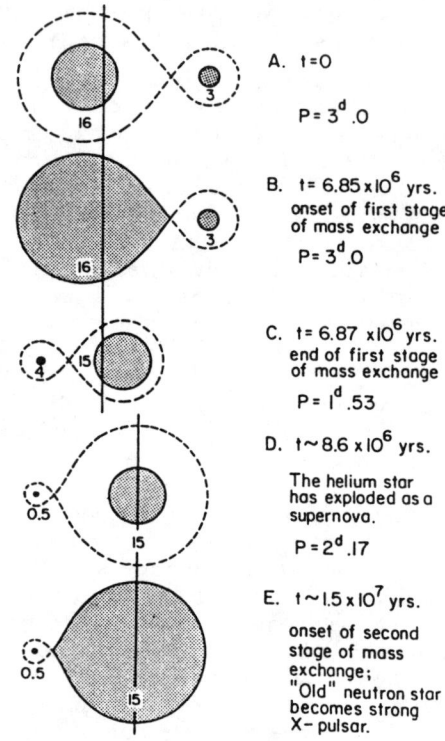

Fig. 1. Scheme for the evolutionary history of a high-mass X-ray binary, in this case: Cen X-3, by Van den Heuvel & Heise (1972).

The helium star is too heavy to leave behind a white dwarf. It never overflows its Roche lobe again and some 1.7 million years after the mass transfer explodes as a supernova. We assumed it to leave behind a $0.5M_\odot$ neutron star (as that was the most plausible mass estimate for Cen X-3 at the time) and $3.5M_\odot$ to be explosively ejected. This ejection does not disrupt the system but makes its orbit eccentric and imparts a runaway velocity of 90 km/sec to its center of gravity, as indicated in the figure. Some 6.5 million years later the $15M_\odot$ star leaves the main-sequence and overflows its Roche lobe, turning the system briefly into a bright X-ray source like Cen X-3. Although some details of this particular model need correction (Flannery 1977), its overall outline, in which large-scale mass transfer before the supernova is the key element, seems well established as the basic reason for the existence of the High Mass X-Ray Binaries (HMXB), as a normal phase in the evolution of every high mass close binary. That this picture is indeed straightforward is demonstrated by the fact that within one year it was also put forward independently by two other groups

(Börner et al. 1972, Tutukov & Yungelson, 1973).

Although the discovery of the binary character of the Cygnus X-1 system by Webster & Murdin (1972) slightly preceded that of the Cen X-3 system (cf. Lewin, this volume), the immediate impact of the latter on theoretical work in this field was very much larger, probably because the pulsar-character and the eclipses of Cen X-3 allowed a much more detailed quantitative understanding of the system, and of the physical processes going on in it.

Davidson & Ostriker (1973) introduced the concept of wind accretion. They, as well as Lamb et al. (1973) introduced the concept of Alfvén-radius and the latter authors also introduced the concepts of co-rotation radius and 'centrifugal barrier'. Further major contributions concerning wind accretion were made by McCray & co-workers. They showed (e.g.Hatchett & McCray 1977) that the wind around the X-ray source may become highly ionized, which might prevent its further radiative acceleration. Their calculations were for the 'static' case (the X-ray source not moving through the wind). Blondin & co-workers (see Blondin, this volume) in recent years carried out more-dimensional calculations for a moving X-ray source and showed that a photo-ionization wake can be formed behind the source, which may cover over 90 degrees of the orbit. Zuiderwijk et al. (1974) found spectral variations in the Halpha line of Vela X-1, which they interpreted as due to an accretion wake behind the source. However, recently Kaper & Hammerschlag-Hensberge (1993) showed by combining optical and UV spectra, that an interpretation in terms of a photo-ionization wake in the Hatchett-McCray/Blondin sense is the more plausible explanation here.

In these early days we ourselves wondered about the further evolution and fate of systems like Cen X-3. Fully developed thermal-timescale Roche-lobe overflow in such a system leads to a mass-transfer rate of $10^{-3} M_\odot/yr$. As their present transfer rates are below $10^{-8} M_\odot/yr$, this thermal timescale mass transfer obviously has not yet started here (Savonije 1983a). When it starts, conservation of mass and orbital angular momentum will inevitably lead to a dramatic shrinking of the orbit, by orders of magnitude. We considered two cases (Van den Heuvel & De Loore 1973): (i) the neutron star is able to accept all the matter transferred to it (e.g.due to very efficient neutrino cooling) and evolves into a black hole and: (ii) the neutron star cannot accept more than the Eddington-limit rate ($10^{-8} M_\odot/yr$) and ejects the remaining transferred matter from the system. Using plausible assumptions about the angular momentum carried away by the expelled matter we found that also in the latter case the orbital separation will decrease dramatically. Thus, the outcome in both cases is a very close system consisting of a compact object and a helium star (the helium core of the massive star). Since massive helium stars are known to be Wolf-Rayet stars, which have very dense high-velocity winds, we suggested that the 4.8-hour orbital period X-ray binary Cyg X-3 is such a post-spiral-in system (Van den Heuvel & De Loore 1973). Cyg X-3 has an X-ray light curve which shows that it is enveloped in a dense wind; also otherwise it is a very peculiar object: it exhibits gigantic synchrotron radio outbursts (Hjellming 1973), and shows 4.8-h infrared variations in phase with the X-ray variations (Becklin et al. 1973). These are characteristics never found in a LMXB, with which it has often been confused because of its short orbital period. It is right in the galactic plane, behind three hydrogen spiral arms, at a distance of about 10 kpc and completely obscured in the optical. The suggestion that it is a helium-star binary was recently confirmed by high-resolution infrared spectroscopy by Van Kerkwijk et al. (1992), which shows that it is a Nitrogen-type (WN7) Wolf-Rayet star, i.e. a massive helium star.

In 1974 Hulse & Taylor discovered the first binary pulsar, consisting of two neutron stars in a close and highly eccentric orbit (P=7h 45m, e=0.615). It was immediately clear that such a system is the natural later product of the close helium star plus neutron star binary (like Cyg X-3) resulting from the later evolution of a high-mass X-ray binary (Flannery & Van den Heuvel 1975; De Loore et al. 1975; Webbink 1975). Our 1973 calculations of the later evolution of HMXB did not yet involve the concept of common-envelope (CE) evolution, suggested by Ostriker (cf. Paczynski 1976), which was subsequently worked out by Paczynski(1976), Ostriker & Dupree (1975), Taam et al. (1978), Bodenheimer & Taam (1984), Taam & Bodenheimer (1992) and by Livio & co-workers (see Iben & Livio 1993). An important early CE-model was already that of Sparks & Stecher (1974), who considered the spiral-in of a white dwarf in the envelope of a red giant.

If the concept of CE-evolution is adopted, only HMXB with orbital periods longer than about 100d are expected to survive as binaries after spiral in, whereas in systems with shorter periods, like Cen X-3, the neutron star is expected to spiral completely into the core of the massive star, leading to the formation of a Thorne-Zytkov star (Thorne & Zytkov, 1975; 1977): a red supergiant with a neutron star in its center (cf. Podsiadlowski, this volume).

Although the work on CE-evolution is important, it is not clear as yet how it precisely links up with the observed evolution of HMXBs in which at present highly super-Eddington mass transfer is taking place, such as SS 433. Here one observes that the transferred matter is first stored in an enormous and highly luminous ($2.10^4 L_\odot$ in the optical) accretion disk, from where it is being ejected in the form of relativistic jets. As this matter carries relatively high specific angular momentum, the outcome of such evolution might well be similar to what we conjectured in our above-mentioned 1973 paper.

a. Disks and Jets

The SS433 system and its jets brings us to the pioneering works on accretion disks in the early seventies by Pringle & Rees (1972) and Shakura & Sunyaev (1973), followed by the discovery in 1974 and 1975 of instabilities that may occur in the very inner parts of disks around black holes (Lightman & Eardley, 1974, Thorne & Price, 1975). As shown by Thorne & Price, these instabilities might well be linked to the two states (soft vs. hard) of Cyg X-1.

An important prediction following from the calculations by Shakura & Sunyaev (1973) was that in the case of supercritical disk accretion onto a black hole the excess matter will be ejected in the form of jets perpendicular to the disk - just as observed five years later in the case of SS 433. Also Cyg X-3 has such (radio) jets (Geldzahler et al. 1983) and some CV-binaries as well.

b. Spin Evolution, Slow Pulsars, Be/Transients, Bursts

Already in 1973 it had become clear that the spin periods of Cen X-3 and SMC X-1 decrease on a timescale of only a few thousand years and that of Her X-1 on a timescale of $10^5 yrs$. These 'spin-ups' were readily explained in terms of disk accretion, which feeds angular momentum to the neutron star. Rappaport & Joss (1977) and Mason (1977) showed, by plotting P vs. $L^{6/7}$ that this spin-up can be understood in terms of accretion torques and that the observed relation fits the one theoretically expected for neutron stars, but not for white dwarfs, with their about one million times larger moments of inertia.

Ariel V, ANS and SAS-C were all launched in 1974. Ariel V discovered many transient sources, first of all the pulsating transients A1118-61 (P=405s) and A0535+26 (P=104s). These were briefly speculated to be close white dwarf plus neutron star binaries (like we now know to be the case in the 687-s orbital period X-ray binary 4U1820-30). However, SAS-C found the high- mass X-ray binary Vela X-1 (0900-40) to be a slow pulsar with a period of 283s (McClintock et al. 1976), which made it clear that the slow pulsars are most probably very slowly rotating magnetized neutron stars in high-mass binaries. With Ariel V and the X-ray package on the Copernicus satellite (which had been launched already in 1972), Jocelyn Bell & co-workers at Mullard discovered long pulse periods in an additional number of sources and around mid-1975 several slowly pulsing transients had been found to coincide closely with rather bright (9 to 11m) B-emission stars. Thus a new class of - highly variable - high-mass X-ray binaries was recognized and the causes of their variable nature were explained in terms of the variable nature of the mass loss of their Be companions (Maraschi et al. 1976).

The reasons why the rotation of these neutron stars had slowed down so much were, however, far from clear. It was realized that this probably had to be related to the fact that these sources are wind accretors for most of the time and several promising models for the slowdown were put forward (Kundt 1975; Illarionov & Sunyaev 1975; Davies & Pringle 1981), but only Kundt's model did not have difficulties in explaining the short (less than $10^5 yr$) timescale of the slowdown. However, in this magnetic-torque model it was not so clear where the lost angular momentum went. Wang (1979, 1981) was, however, able to put it in a physically self-consistent picture and obtained spindown timescales of order 10^4 to $10^5 yrs$ even in weak winds.

Models for the interaction of the neutron-star magnetosphere with an accretion disk, constructed by Ghosh & Lamb (1979), did show that a disk may produce spin-up as well as spin-down, depending on the accretion rate. At low accretion rates, the interaction with the disk can very rapidly spin down the rotation of the neutron star (see also Henrichs 1983). Thus, since during off stages - which occupy most of the time - the accretion rates in Be systems are low and since the slow Be-winds may carry sufficient angular momentum for the formation of a small disk, the slow rotation of these pulsars might, in principle, also be understood in terms of disk accretion. For an overview of spin evolution and the wide range of possible magnetospheric interactions, see Lipunov (1992).

c. Bursts

In 1975 Grindlay & Heise (Grindlay et al. 1976) working with the ANS satellite, discovered the first X-ray burst, from the source 4U1820-30 in the globular cluster NGC 6624. This discovery was quickly followed by the discovery of many burst sources by a SAS-C team led by Lewin, which intensively studied their temporal and spectral behaviour.

Maraschi & Cavaliere (1977) and Woosley & Taam (1976) independently suggested the bursts to be thermo-nuclear explosions of the matter accreted on the surface of a neutron star. (Outbursts with the right timescale had already been found in calculations by Hansen & Van Horn (1975), who had not realized that these might be observable as X-ray bursts). The subsequent calculations by Joss (1978) and others confirmed this model for the Type I bursts (the dominant type of bursts, observed in all burst sources).

That we must indeed be dealing here with neutron stars (and not with black holes, as the Harvard-group believed up till about 1980) was confirmed by the first radius measurement derived from the spectral softening during the decay of an abnormally long burst, carried out by Swank et al. (1977). Swank found that during the decay the blackbody radius of the source is that of a neutron star. This was confirmed by Hoffman et al. (1977ab) for 2 more sources.

Application of the same method to a collection of burst sources around the galactic center (*i.e.* with known distances and therefore known absolute burst luminosities) by Van Paradijs (1978) yielded average radii around 10 km, which made it clear that we are indeed dealing here with neutron stars.

d. Globular Cluster Sources

During the July 1973 Cambridge conference on the Physics and Astrophysics of Compact Objects (a two-week school of which unfortunately no proceedings were published), Herb Gursky one evening invited me over to his college room to discuss the possible optical identifications of UHURU-sources, on which he was to lecture one of the following days. He showed me that two UHURU sources coincided with globular clusters and that a third one also had a globular cluster in its (fairly large) error box. He pointed out to me how peculiar this was, since all globular clusters together have only a few times $10^7 M_\odot$, while the Galaxy as a whole has $10^{11} M_\odot$. So, the fact that two or three out of the about one hundred strong UHURU sources in the Galaxy coincide with globular clusters implies, as Herb pointed out to me, that per unit mass globular clusters contain over 10^2 times more strong X-ray sources than the Galaxy as a whole. I thought to remember that he later mentioned this also in one of his lectures (John Heise told me that he recalls that the peculiar overabundance of sources in globular clusters was vividly discussed by participants between the lectures).

The work of Clark (1975) with the SAS-C satellite, which produced error boxes much smaller than the size of the clusters, established with certainty the existence of the globular cluster sources as a class and showed that there are of order of a dozen of them in the Galaxy, indicating that the overabundance of strong sources in clusters is of order 10^3 (Katz 1975). To explain this abnormally large incidence Fabian et al. (1975) suggested X-ray binary formation by tidal capture of an an old neutron star by a cluster main-sequence star and Sutantyo (1975a) by direct collisions between old neutron stars and (sub-)giant stars in the cluster. These works led to many important new studies on the dynamics of dense stellar systems. In recent years it was found that the most efficient formation mechanism may be: exchange collisions between primordial binaries and old neutron stars (Sigurdsson, 1992; *cf.* Bhattacharya & Van den Heuvel 1991; Hut 1992).

e. Low-Mass X-Ray Binaries And Their Evolution

It was not until the detection of a 0.78d optical photometric period in Sco

X-1 by Gottlieb, Wright & Liller (1975) that we knew with certainty that Low Mass X-Ray Binaries do exist. However, since the detection of a faint F-type spectrum in Cyg X-2 by Kraft & Dumoulin (1967) and the discovery of the X-ray sources belonging to old stellar populations (*i.e.* in the Galactic Bulge and in globular clusters), most workers had already been convinced that the X-ray sources without bright optical counterparts are neutron stars (or black holes) with low-mass companions. The fact that the X-ray spectra of these non-pulsing old-population sources are somewhat softer (*i.e.* blackbody temperature about 4-8 keV) than those of the pulsating sources (kT over 10 keV) (*cf.* Ostriker 1977, Maraschi *et al.* 1977) suggested that here the accreting objects are non- (or weakly) magnetized neutron stars, an idea in line with the (then believed) indications from radio pulsars that the magnetic fields of neutron stars would decay on a timescale of a few million years (Gunn & Ostriker 1970). An alternative suggestion by Salpeter (1973) was that the bulge sources are white dwarfs with steady surface nuclear burning, which now is the going model for the supersoft (kT=20-50 eV) ROSAT sources (Van den Heuvel *et al.* 1992; Rappaport, this volume).

Concerning the origins of the LMXBs, Gursky at the above- mentioned 1973 conference suggested that the HMXB are connected with Type II supernovae and the LMXB with Type I supernovae. For the latter type of supernovae, Whelan & Iben (1973) had suggested as a model: accretion-induced collapse (AIC) of a massive white dwarf with a low-mass red-giant companion, *i.e.* a system similar to a symbiotic star. Although at various occasions objections have been raised against the formation of neutron stars in LMXB by AIC, I still think that the Whelan-Iben model is a very viable one, particularly since a low-mass giant companion can provide just the right accretion rate $(1 - 4 \cdot 10^{-7} M_\odot/yr)$ required to induce steady nuclear burning on the white dwarf surface, such that this star will grow in mass and collapse when it passes the Chandrasekhar limit (*cf.* Van den Heuvel *et al.* 1992). Renzini (1993) has recently revived this symbiotic-star model as a most viable one.

For the origin of the Her X-1 system AIC probably does not work, and a model in which the system originated from a massive binary system with a very small mass ratio (*e.g.* $15M_\odot + 2M_\odot$) seems most plausible (Sutantyo 1975b, 1992; Verbunt *et al.* 1990). Such a model also is the only one that works well for the origin of the LMXB in which the compact star is a black hole (Van den Heuvel & Habets 1984; De Kool *et al.* 1987), since the black holes in these systems have masses too large $(3 - 10 M_\odot)$ to have grown out of accreting neutron stars. A third model for the formation of LMXBs is the Eggleton-Verbunt (1985) triple star model (see Webbink, this volume). All these LMXB-formation models are exotic and therefore will occur only very rarely, as they should (see section 1).

Concerning the mechanisms driving the mass transfer in LMXB: for close systems the obvious mechanism seemed, by analogy to the CV- binaries: orbital angular momentum losses by gravitational radiation, for which the 'standard' theory was established by Faulkner (1971). However, in the early eighties it was realized that many of the LMXB show mass transfer rates one to two orders of magnitude higher than can be produced by this mechanism. The introduction by Verbunt & Zwaan (1981) of orbital angular momentum losses by magnetic braking (a mechanism already suggested by Eggleton (1976) for W UMa systems) provided a solution for this discrepancy (*cf.* Verbunt 1990).

For wider systems such as Cyg X-2 (Porb=9.8d) Webbink *et al.* (1983) and Taam (1983) showed that the internal nuclear evolution of the low-mass companion (a low-mass subgiant or giant with a small degenerate helium core) may drive a large and steady rate of mass transfer, which might explain the large X-ray luminosities of the bright Galactic Bulge sources. The final product of this evolution is a wide binary consisting of a neutron star a 0.2 to $0.4M_\odot$ helium white dwarf in a wide and circular orbit (see figure 2).

When the papers by Webbink *et al.* and Taam were in press, the 6ms pulsar PSR 1953+29 was discovered by Boriakoff *et al.* (1973), which is in a wide and circular binary (P=117 d) with an $0.26M_\odot$ white dwarf companion, *i.e.* exactly resembling the predicted final product of the evolution of a wide LMXB, as was immediately realized by several authors (Savonije 1983b; Paczynski 1983; Joss & Rappaport 1983). At present already eight of these wide binary pulsar systems are known (see section 5).

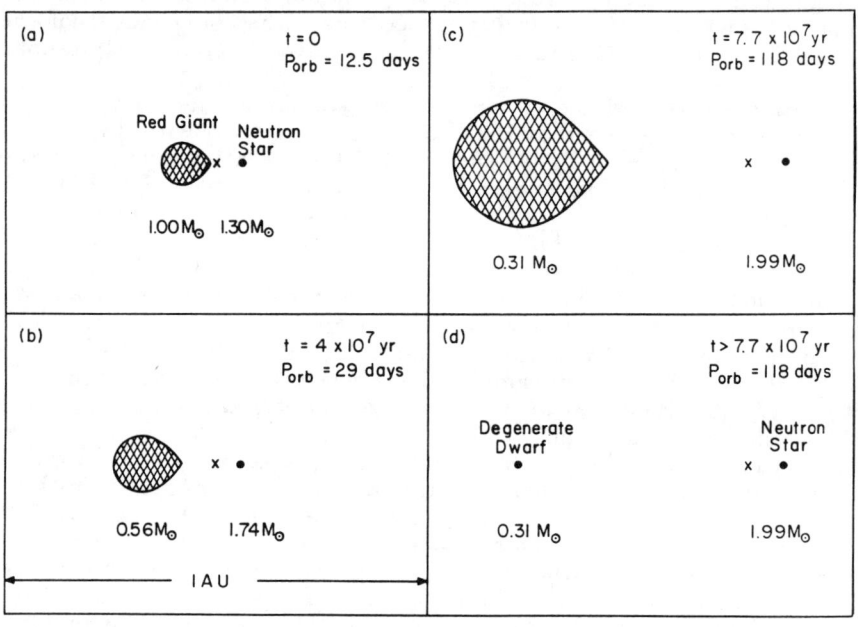

Fig. 2. Evolution of a wide LMXB-system (initial orbital period 12.5d) into a wide radio-pulsar binary (orbital period 118d) composed of a $0.31M_\odot$ white dwarf and a 'recycled' pulsar. The initial system consists of a low-mass red giant star which transfers matter to a neutron star on a nuclear timescale. Age since the onset of the mass transfer, orbital period and the masses of the componets are indicated in each frame. After Joss & Rappaport (1983).

4. 1978-1982: EPOCH OF CONSOLIDATION

In this period relatively few new fundamental discoveries were made, but the general picture established in the period 1971-1978 was further confirmed and important missing details filled in. The statistical mass determinations of the globular cluster sources derived from the accurate Einstein-measurements of the Ωdistributions of their distances to the centers of their globular clusters confirmed that they must be neutron-star binaries (*cf.* Grindlay *et al.* 1984), as was already expected from their burster characteristics. Also, largely due to the Einstein Observatory, the number of orbital periods detected for LMXBs increased from two to eight in this period, firmly establishing the existence of LMXBs as a class.

A further highlight was the detection by Einstein of many point X-ray sources in M31, with population characteristics roughly similar to those in our Galaxy: several dozen bright Bulge sources, two dozen globular cluster sources and a spiral-arm population of sources. This showed that, although in the details there may be some differences, the populations of sources found in our Galaxy are, apparently, quite characteristic for normal large spirals (*cf.* Long & Van Speybroeck 1983). Also many new sources were found in the Magellanic Clouds and other galaxies, like M33.

5. 1982 - PRESENT: EPOCH OF CONNECTION AND OF MANY NEW DISCOVERIES

a. Recycling And The Millisecond Pulsars

As mentioned in the Introduction, the idea of recycling was already suggested in the mid-seventies. Smarr & Blandford (1976) had suggested that the combination of abnormally weak B- field ($2.10^{10}G$) and abnormally rapid spin (P=0.059s) of PSR 1913+16 can be explained if this pulsar is the oldest neutron star in the system, which was spun up by accretion in an X-ray binary phase, after its magnetic field had partly decayed. Srinivasan & Van den Heuvel (1978, 1982) demonstrated that indeed its combination of B-field and spin could never have been reached by normal spindown evolution of a newborn neutron star, even if (as was then generally believed) its magnetic field had decayed on a timescale of order a few million years. They, as well as Blandford & DeCampli (1981), therefore argued that this pulsar must be recycled. They also showed that, given the magnetic dipole field strength, there exists a shortest possible rotation period to which a neutron star can be spun up and that PSR 1913+16 rotates at about twice that period, so the spin-up model is consistent with its field strength. As during the X-ray and spiral-in phases the orbit must have been completely circularized, the high orbital eccentricity of the system then implies (they argued) that a second supernova explosion must have taken place in the system, and therefore, the companion of the pulsar must also be a neutron star: the younger one of the two.

The reasons why this star is not observed as a pulsar are also obvious: newborn neutron stars are expected to have strong magnetic fields (10^{12} to $10^{13}G$) and will spindown very rapidly and become unobservable after only a few million years. On the other hand, in view of its weak magnetic field, the old recycled pulsar in the system spins down very slowly and will therefore remain observable for over one hundred million years. The old recycled one of the two is therefore the most likely one to be observed in a double neutron star system (Van den Heuvel & Taam 1984) and indeed also in the other two double neutron star systems known in the galactic disk this seems to be the case.

Up till the end of 1982 only three binary pulsar systems were known (1913+16, 0655+64 and 0820+02) and PSR 1913+16 was the only one for which the recycling model seemed solidly based. Nevertheless, already in 1981 Radhakrishnan & Srinivasan (1984) suggested that there is an entire class of 'recycled' radio pulsars (the term was introduced by them and their paper was the first in which the 'spin-up line' was drawn, see figure 3). They pointed out that there are at least two single pulsars (1541-52 and 1804-08) whose combination of (weak) B-field and spin period, like that of PSR 1913+16, can never have been reached by 'normal' spindown evolution of a newborn pulsar. By tracing back Ωthe evolution of these pulsars along the 'normal' evolutionary tracks (involving spindown and standard magnetic field decay) one finds a 'progenitor' region in the B vs. P diagram (or $\dot{P}$ vs. P diagram) that is empty, i.e. the B - P combinations of the progenitors do not exist among young pulsars. (This is even more so if fields do not decay spontaneously). They therefore concluded that these pulsars can have formed only through recycling in binary systems.

In the following year, 1982, the 1.55ms pulsar PSR 1937+21 was discovered (Backer et al. 1982), which is spinning twenty times faster than the Crab pulsar and has a magnetic field strength some 10^4 times weaker than that of young pulsars. Alpar et al. (1982) and Radhakrishnan & Srinivasan (1982) independently suggested that this is a recycled pulsar, originating through spin-up in a binary, in which subsequently the the donor star was lost.

This seemed an extremely bold suggestion at the time for at least two reasons: (i) it is spinning some 40 times faster than PSR 1913+16, the fastest recycled pulsar then known and (ii) the pulsar should have gotten rid of its companion. Nevertheless, within only half a year this model for making a millisecond pulsar was beautifully confirmed by the discovery of the 6 ms binary pulsar PSR 1953+29 by Boriakoff et al. (1983), already mentioned above, which is the perfect final product of the evolution of a wide LMXB.

At present some twenty binary radio pulsars with circular orbits and low-mass white dwarf companions are known, many of them millisecond pulsars and all of them presumably products of recycling in LMXBs (cf. Van den Heuvel & Bitzaraki 1994). They form the 'other' class of binary radio pulsars (cf. Van den Heuvel & Taam 1984).

b. Quasi-periodic Oscillations (QPO) in Bright LMXB

In 1984 Van der Klis et al. (1985ab), working with EXOSAT, discovered QPO with frequencies of 20 to 50 Hz in the X-ray emission of the bright Galactic Bulge source GX5-1. This discovery was important since the bright bulge sources and their relatives, the Sco X-1-type sources, had up till then never shown any systematic behaviour in their X-ray emission, even though they had been known for over twenty years.

Van der Klis found that the QPO-frequency of GX5-1 showed a clear correlation with the measured intensity of the X-ray emission, i.e. with the intrinsic luminosity of the source. Before publishing this discovery, we attempted in our group for several months to find an explanation for this clear correlation. Among the many possibilities we discussed, Jan van Paradijs at some point made the suggestion that this might be a beat-frequency phenomenon (which he knew to occur in some CV systems), but we did not follow-up on this suggestion. So, when the discovery of QPO was made public in an IAU circular (Van der Klis et al. 1985b), we had no satisfactory explanation for the frequency vs. intensity correlation, which was mentioned in the circular.

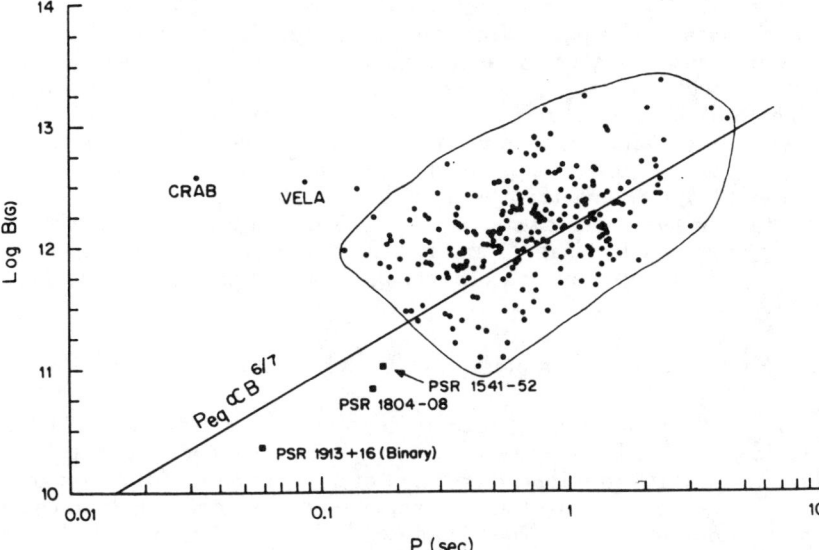

Fig. 3. Diagram of magnetic field strength B against pulse period P for radio pulsars, in which the spin-up line was introduced for the first time, by Radhakrishnan & Srinivasan (1984; paper presented and submitted in 1981, at the 2nd Asian-Pacific IAU Regional Meeting, Bandung). The authors pointed out that besides PSR 1913+16 two other pulsars, marked as squares, stand out from the crowd of normal pulsars by having weak fields and relatively short periods. They lie below the spin-up line and are argued by the authors to be recycled (single) pulsars.

As soon as our circular was out, two very clever theoreticians (Alpar & Shaham 1985) came out with the - to my knowledge - first theory ever published in an IAU circular. They showed that the observed frequency vs. intensity relation of GX 5-1 arises naturally as a 'beat-frequency' phenomenon between the spin frequency of a magnetized neutron star and the Keplerian frequency at the Alfven surface, if one assumes that the neutron star in GX 5-1 is weakly magnetized ($B \approx 10^9 G$) and has a spin period of about 10 ms, *i.e.* values very similar to those of the binary millisecond radio pulsars that are expected to have descended from wide (and therefore bright) LMXB systems.

Although it has since been found that the frequency vs. intensity relation of GX5-1 is characteristic for only one kind of QPO, there can be little doubt that the 'beat-frequency' explanation for this type of QPO must be the correct one. It, as well as models for the other types of QPO, has been worked out in detail by Lamb & co-workers (*cf.* Ghosh & Lamb 1991,1992).

c. Globular Cluster Pulsars

In 1985 it was realized that the best places to look for recycled pulsars are the globular clusters, because of their abnormally high incidence of X-ray binaries (Hamilton *et al.* 1985; Fabian & Verbunt priv.comm.). Hamilton *et* al. (1985) detected a radio source in the cluster M28 and Lyne et al. (1987) discovered that this source is a single 3ms-period pulsar. This discovery triggered an intensive search for pulsars in globular clusters by the Jodrell Bank/Parkes group and groups at Arecibo (Kulkarni, Prince, Backer, Wolszczan, Taylor & co-workers), which so far has resulted in the discovery of over 30 cluster pulsars.

The bulk of them shares the characteristics of the binary and millisecond pulsars in the galactic disk: abnormally weak magnetic fields and abnormally rapid spin (for a review: see Lyne, this volume). The relatively 'low' incidence of binaries (40%) among them is probably due to close encounters with other cluster stars in the dense cluster core, which ionize the wide binaries in which many of the recycled pulsars originated. The discovery of the cluster pulsars is one more nice example of a theoretical prediction which was confirmed by the observations.

d. Evaporation of Companions in LMXBs, Effects of X-Ray Heating

The puzzling absence of a companion in the first-discovered millisecond pulsar, PSR 1937+21, gave rise to a variety of models for getting rid of a companion. These vary from tidal disruption of a companion of very low mass (Ruderman & Shaham 1983; Hut & Paczynski 1984; Kluzniak 1992) to coalescence with a close massive white dwarf (Van den Heuvel & Bonsema 1984).

Ruderman, however, since 1986 suggested at several occasions that the heating of the companion in an LMXB, both by X-rays and by the relativistic wind from the pulsar that later appeared in the system, may cause the low-mass companion to be continuously evaporated, by driving a strong wind from its surface. (The existence of such a self-excited wind in the Her X-1 system had already been suggested by Davidson & Ostriker 1973; Arons 1973; and Basko & Sunyaev 1973, but later investigations showed that it is unlikely to work in that system). Ruderman's ideas were worked out in a number of papers (Kluzniak et al. 1988; Ruderman et al. 1989), the first of which were in print when Fruchter et al. (1988) discovered the eclipsing-binary millisecond radio pulsar PSR 1957+21, which has a companion of very low mass (less than $0.02 M_\odot$) which is in a state of continuous evaporation. A second system of this type, PSR 1744-24A, was discovered in the globular cluster Ter 5 by Lyne et al. (1990). The debate concerning whether or not the evaporation may indeed already start in the in the LMXB-phase, or starts only after the radio pulsar has appeared, is still going on (cf. Harpaz & Rappaport 1991; Podsiadlowski 1991; see especially Hameury et al. 1993).

e. Black Hole Systems with Low-Mass Donors; Supersoft ROSAT Sources

We already discussed the origin of the black hole sources with low-mass donors in section 3. An important discovery by ROSAT was that of the supersoft sources in the LMC with X-ray luminosities close to the Eddington limit and blak-body temperatures around 30 to 50 eV. At present already several dozen of these sources are known in the LMC, SMC, M31 and our Galaxy (cf. Hasinger, this volume). Their combination of luminosity and temperature allows one immediately to estimate - with Stefan-Boltzmann's law - their radii, which turn out to be about $10^4 km$, i.e. typical of a white dwarf, as was quickly realized by several people.

However, models in which the energy is purely generated by accretion onto a white dwarf (as suggested for example by Grindlay, 1991) suffer from the problem that the accretion rates required to produce their luminosity are so high ($10^{-5} M_\odot/yr$) that immediately a red- giant-like envelope will form around the the white dwarf, opaque to X-rays, such that the X-ray source will be immediately quenched. This problem was, however, overcome when it was realized that nuclear burning of hydrogen on the surface of a white dwarf generates per unit mass some 50 times more energy than accretion, and that at accretion rates in the range $(1 - 4) \cdot 10^{-7} M_\odot/yr$ the accreted matter goes into stable nuclear burning on the white dwarf surface, without appreciable radius expansion (Van den Heuvel *et al.* 1992).

With such accretion rates white dwarfs will produce supersoft sources with just the right luminosity and temperature. Models of white dwarfs with accretion rates in the above-mentioned range and steady surface burning turned out to already have been available in the CV-literature for over a decade (*e.g.*Iben 1982). We further refer to Hasinger (this volume) and Rappaport (this volume and references therein).

6. WHAT COULD WE HAVE PREDICTED AND WHAT NOT?

From the above it is clear that the field of X-ray binaries is a happy one for a theoretician since quite a number of phenomena and objects were predicted, and subsequently found by observers. Looking back, one might wonder which other phenomena or objects might have been predicted and which phenomena or objects are most unlikely to ever have been predicted.

Phenomena and objects in the first category are of the type of which one might say, to paraphrase M. Ruderman: 'With a little bit of hindsight I might have predicted this'. Such phenomena/objects relate in a logical way to already existing knowledge and, with a little bit of imagination, could have been predicted.

Characteristic for these phenomena/objects is that, as soon as they were discovered, the explanation for their existence was quite clear and was given by someone shortly afterwards.

Phenomena/objects in this category are:
- the existence of the High Mass X-Ray Binaries
- accretion from a wind in a HMXB
- the existence of the Hulse-Taylor-type close double neutron stars (once it was realized that HMXB must spiral-in)
- thermonuclear (Type I) X-ray bursts
- globular cluster X-ray sources
- the existence of the Be/X-ray binaries (once HMXB had been discovered)
- the existence of the binary radio pulsars with wide and circular orbits
- QPO with intensity-dependent frequency (once it was understood that millisecond pulsars descend from LMXB)
- the existence of supersoft ROSAT sources

Phenomena and objects in the second category are so unlikely or strange that even with the greatest imagination it would have been most unlikely that anyone would have predicted them. Characteristic for this category is that in a number of instances even many years after their discovery there still is no good (or completely accepted) explanation for their existence.

In this category belong the existence of:
- the Low Mass X-ray Binaries
- the slow X-ray pulsars
- single millisecond pulsars in the galactic disk
- the high velocities of radio pulsars
- the rapid burster
- gamma-ray bursts
- type I supernovae (and their possible relation to LMXB)

An intermediate case is formed by the millisecond pulsars. Although their existence is understandable in hindsight in terms of recycling, it would have taken a lot of courage to *predict* their existence by extrapolating by a factor 40 in pulse period from PSR 1913+16.

What would we dare to predict for the future?

An obvious case is that of binary pulsars consisting of a neutron star and a black hole. Possibly LMC X-3 and almost certainly Cyg X-1 will evolve into such a system. Pulsar-black hole systems are, however, expected to be much harder to find than double neutron stars, since in almost all cases the neutron stars in such systems are expected to have formed after the black hole. Hence, they are not recycled and therefore they will be strong- field short-lived pulsars (spindown lifetime a few million years), in contrast to the PSR 1913+16-pulsars, which are recycled weak-field pulsars with spindown-lifetimes of hundreds of millions of years. Since also the formation rate of black holes is expected to be an order of magnitude lower than that of neutron stars (*cf.* Van den Heuvel 1992b), the incidence of *observable* pulsar-black-hole systems may well be two orders of magnitude lower than that of systems like PSR 1913+16. (However, there is some hope that black-hole neutron star systems may have an order of magnitude smaller probability of disruption during formation, which may again increase their relative incidence by the same factor).

A further, in my view obvious, prediction is the existence of close double neutron stars with orbital periods of less than an hour. Such systems are expected to be formed at a higher rate than PSR 1913+16-like systems (Van den Heuvel 1992a; Tutukov & Yungelson 1993) since many Be/X-ray systems are expected to go through two spiral-in phases before the second supernova explosion. As, however, the orbits of these close systems decay much faster by GR-losses than those of the wider systems, the incidence of these very close systems is is probably lower than that of the latter systems, but not by a large factor. In view of the expected limits on their formation rates (*cf.* Van den Heuvel 1994) we would not be surprised if these systems would show up in surveys in the coming half decade. Also, we would expect more systems similar to PSR 0655+64 to show up, though with shorter as well as longer orbital periods. The nicest system of all to find would, obviously, be one consisting of two observable pulsars. This would allow one to double-check all relativistic effects which have been so beautifully observed in the Hulse-Taylor and Wolszczan pulsars.

REFERENCES

Alpar, M.A., Cheng, A.F., Ruderman, M., Shaham, J. 1982, Nature 300, 728
Alpar, M.A. & Shaham, J. 1985, IAU Circ.Nr. 4046 (21 March)
Arons, J. 1973, ApJ 184, 539
Backer, D.C., Kulkarni, S.R., Heiles, C., Davis, M.M. & Goss, W.M. 1982, Nature 300, 615
Basko, M.M. & Sunyaev, R.A. 1973, Ap.Sp.Sci. 23, 117
Becklin, E.E., Neugebauer, G., Hawkins, F.J., Mason, K.O., Sanford, P.W., Matthews, K. & Wynn-Williams, C.G. 1973, Nature Phys.Sci. 245, 302
Bhattacharya, D. & Van den Heuvel, E.P.J. 1991, Phys.Repts. 203, 1
Blaauw, A., 1961, Bull.Astron.Inst.Netherlands 15, 265
Bisnovatyi-Kogan, G.S. & Komberg, B.V. 1975, Sov.Astron. 18, 217
Blandford, R.D. & DeCampli, W.M.1981, in 'Pulsars', eds. Sieber, W. & Wielebinski, R., Reidel, Dordrecht, 371
Bodenheimer, P. & Taam, R.E. 1984, ApJ 280, 771
Boriakoff, V., Buccheri, R. & Fauci, F. 1983, Nature 304, 417
Borner, G., Meyer, F., Schmidt, H.U. & Thomas, H.-C. 1972, Paper presented at the meeting of the Astron.Gesellsch., Vienna
Clark, G.W. 1975, ApJ 199, L143
Davidson, K. & Ostriker, J.P. 1973, ApJ 179, 585
Davies, R.E. & Pringle, J.E. 1981, MNRAS 196, 209
De Kool, M., Van den Heuvel, E.P.J., Pylyser, E. 1987, Astr.Ap. 183, 47
De Loore, C., De Greve, J.P. & De Cuyper, J.P. 1975, Ap.Sp.Sci. 36, 219
Eggleton, P.P.1976, in 'Structure & Evolution of Close Binary Systems' (Eggleton, P.P. et al. eds.), Reidel, Dordrecht...
Eggleton, P.P. & Verbunt, F. 1985, MNRAS 220, 13P
Fabian, A., Pringle, J.E. & Rees, M.J. 1975, Mon.Not. RAS 172, 15P
Faulkner, J. 1971, ApJ 170, L99
Flannery, B.P. 1977, Ann.N.Y.Acad.Sci. 302, 36
Flannery, B.P. & Van den Heuvel, E.P.J. 1975, Astr.Ap. 39, 61
Fruchter, A.S., Stinebring, D.R. & Taylor, J.H. 1988, Nature 333, 237
Geldzahler, B.J. et al. 1983, ApJ 273, L65
Ghosh, P. & Lamb, F.K. 1979, ApJ 234, 296
Ghosh, P. & Lamb, F.K. 1991, in 'Neutron Stars, Theory and Observation' (eds. Ventura, J. & Pines, D.) Kluwer, Dordrecht, p.363
Ghosh, P. & Lamb, F.K. 1992, in 'X-ray Binaries and Recycled Pulsars' (eds. Van den Heuvel, E.P.J. & Rappaport, S.) Kluwer, Dordrecht, p.487
Giacconi, R., Gursky, H., Kellogg, E., Schreier, E. & Tananbaum, H. 1971, ApJ 167, L67
Gottlieb, E.W., Wright, E.L. & Liller, W. 1975, ApJ 195, L33
Grindlay, J.E. 1991, Pers.comm. ITP workshop 'Neutron Stars in Binary Systems', Santa Barbara, January
Grindlay, J.E., Gursky, H., Schnopper, H., Parsignault, D., Heise, J., Brinkman, A.L. & Schrijver, J. 1976, ApJ 205, L125
Grindlay, J.E. et al. 1984, ApJ 282, L13
Gunn, J.E. & Ostriker, J.P. 1970, ApJ 160, 979
Hameury, J.-M., King, A.R., Lasota, J.-P. & Raison, F. 1993, Astr.Ap. (in press)
Hamilton,T.T., Helfand, D.J. & Becker, R.H. 1985, Astr.J. 90, 606

Hansen, C.J. & Van Horn, H.M. 1975, ApJ 195, 735
Hatchett, S. & McCray, R. 1977, ApJ 211, 552
Harpaz, A. & Rappaport, S. 1991, ApJ 383, 739
Henrichs, H.F. 1983, in 'Accretion Driven Stellar X-ray Sources' (eds. Lewin, W.H.G. & Van den Heuvel, E.P.J.) Cambridge Univ. Press, p.303
Hjellming, R.M. 1973, Science 182, 1089
Hoffman, J.A., Lewin, W.H.G. & Doty, J. 1977a, ApJ 217, L23 and 1977b, MNRAS 179, 57p
Hut, P. 1992, in 'X-Ray Binaries and Recycled Pulsars' (eds. Van den Heuvel, E.P.J. & Rappaport, S.) Kluwer, Dordrecht, p.317
Hut, P. & Paczynski, B. 1974, ApJ 284, 675
Iben, I. 1982, ApJ 259, 244
Iben, I. & Livio, M. 1993, PASP (in press)
Illarionov, A.F. & Sunyaev, R.A. 1975, Astr.Ap. 39, 18
Joss, P.C. 1978, ApJ 225, L123
Joss, P.C. & Rappaport, S. 1983, Nature 304, 419
Kaper, L. & Hammerschlag-Hensberge, G. 1993, Astr.Ap. (in press)
Katz, J.I. 1975, Nature 253, 698
Kippenhahn, R. & Weigert, A. 1967, Zeitschr.Ap. 65, 251
Kluzniak, W. 1992 in 'X-Ray Binaries and Recycled Pulsars' (eds. Van den Heuvel E.P.J. & Rappaport, S.) Kluwer, Dordrecht, p.413
Kluzniak, W., Ruderman, M. & Tavani, M. 1988, Nature 334, 225
Kraft, R.P. & Dumoulin, M.H. 1967, ApJ 150, L183
Kundt, W. 1975, Phys.Lett. 57A, 195
Lamb, F.K., Pethick, C.J. & Pines, D. 1973, ApJ 184, 271
Lightman, A.P. & Eardley, D.M. 1974, ApJ 187, L1
Lipunov, V.M.1992, 'Astrophysics of Neutron Stars', Springer, Heidelberg, 322pp
Long, K.S. & Van Speybroeck, L.P. 1983 in 'Accretion Driven Stellar X-Ray Sources' (eds. Lewin, W.H.G. & Van den Heuvel, E.P.J.) Cambridge Univ. Press, p.117
Lynden-Bell, D. 1968, Nature 223, 690
Lyne, A.G., Brinklow, A., Middleditch, J., Kulkarni, S.R., Backer, D.C., Clifton, T.R. 1987, Nature 328, 399
Lyne, A.G., Manchester, A.N., D'Amico, N., Stavely-Smith, L., Johnston, S., Lim, J., Fruchter, A.S., Goss, W.M., Frail, D. 1990, Nature 347, 650
McClintock, J.E., Rappaport, S., Joss, P.C., Bradt, H., Buff, G., Clark, G.W., Hearn, D., Lewin, W.H.G., Matilsky, T., Mayer, W. & Primi, F. 1976, ApJ 206, L99
Maraschi, L. & Cavaliere, A. 1977, in 'Highlights in Astronomy' (ed. Muller, E.A.) Reidel, Dordrecht, vol.4, Part I, p.127
Maraschi, L., Treves, A. & Van den Heuvel, E.P.J. 1986, Nature 259, 292
Maraschi, L., Treves, A. & Van den Heuvel, E.P.J. 1987, ApJ 216, 819
Mason, K.O. 1977, MNRAS 178, 81P
Ostriker, J.P. & Dupree, R. 1975 (preprint)
Ostriker, J.P. 1977, Ann. N.Y. Acad. Sci. 302, 229
Paczynski, B. 1967, Acta Astr. 17, 355
Paczynski, B. 1976, in 'Structure and Evolution of Close Binary Systems' (P.P. Eggleton et al. eds.) Reidel, Dordrecht, p.76
Plavec, M. 1968, Advances Astr.Ap. 6, 201
Podsiadlowski, P. 1991, Nature 350, 136
Prendergast, K.H. & Burbidge, G.R. 1968, ApJ 151, L83
Pringle, J.E. & Rees, M.J. 1972, Astr.Ap. 21, 1
Radhakrishnan, V. & Srinivasan, G. 1982, Current.Sci. 51, 1096
Radhakrishnan, V. & Srinivasan, G. 1984, in 'Proc. 2nd Asian Pacific IAU

Regional Meeting' held at Bandung, Indonesia, 1981, eds. Feast, M.W. & Hidayat B., Tira Pustaka, Jakarta, p.423
Rappaport, S. & Joss, P.C. 1977, Nature 266, 683
Renzini, A. 1993 (preprint)
Ruderman, M. & Shaham, J. 1983, Nature 304, 425
Ruderman, M., Shaham, J. & Tavani, M. 1989, ApJ 336, 507
Salpeter, E.E. 1964, ApJ 140, 796
Salpeter, E.E. 1973, in 'X- and Gamma-ray Astronomy' IAU Symp.Nr.65 (eds. Bradt, H. & Giacconi, R.), Reidel Dordrecht, p.135
Savonije, G.J. 1983a, in 'Accretion Driven Stellar X-ray Sources' (eds. Lewin, W.J.G. & Van den Heuvel, E.P.J.) Cambridge Univ.Press, p.343
Savonije, G.J. 1983b, Nature 304, 422
Schreier, E., Levinson, R., Gursky, H., Kellogg, E., Tananbaum, H. & Giacconi, R. 1972, ApJ 172, L79
Setti, G. & Woltjer, L. 1970, Ap.Sp.Sci. 9, 185
Shakura, N.I. 1973, Sov.Astr. 16, 756 (orig.submitted 1971)
Shakura, N.I. & Sunyaev, R.A. 1973, Astr.Ap. 24, 337
Shklovskii, I. 1967, ApJ 148, L1
Shvartsman, V.F. 1970a, Astr.Zh. 47, 824 (Sov.Astr. 14, 662, 1971)
Shvartsman, V.F. 1970b, Radiofyzika 13, 1852
Shvartsman, V.F. 1971, Sov.Astr. 15, 377
Sigurdsson, S. 1992, PhD Thesis Caltech
Smarr, L.L. & Blandford, R. 1976, ApJ 207, 574
Sparks, W.M. & Stecher, T.P. 1974, ApJ 188, 149
Srinivasan, G. & Van den Heuvel, E.P.J. 1978, paper presented at the 9th Texas Symp. Rel.Ap., München (unpublished)
Srinivasan, G. & Van den Heuvel, E.P.J. 1982, Astr.Ap. 108, 143 (paper submitted 1980)
Sutantyo, W. 1975a, Astr.Ap. 44, 227
Sutantyo, W. 1975b, Astr.Ap. 41, 47
Sutantyo, W. 1992, in 'X-ray Binaries and Recycled Pulsars' (eds. Van den Heuvel, E.P.J. & Rappaport, S.) Kluwer, Dordrecht, p.293
Swank, J.H., Becker, R.H., Boldt, E.H., Holt, S.S., Pravdo, S.H., Serlemitsos, P.J. 1977, ApJ 212, L73
Taam, R.E. 1983, ApJ 270, 694
Taam, R.E. & Bodenheimer, P. 1992, in 'X-ray Binaries and Recycled Pulsars' (eds. Van den Heuvel, E.P.J. & Rappaport, S.) Kluwer, Dordrecht, p.281
Taam, R.E., Bodenheimer, P. & Ostriker, J.P. 1978, ApJ 222, 269
Thorne, K.S. & Price, R.H. 1975, ApJ 195, L101
Thorne, K.S. & Zytkov, A.N. 1975, ApJ 199, L19
Thorne, K.S. & Zytkov, A.N. 1977, ApJ 212, 832
Tutukov, A.V. & Yungelson, L.R. 1973, Nautsnie Inform. 27, 58
Tutukov, A.V. & Yungelson, L.R. 1993, MNRAS 260, 675
Van den Heuvel, E.P.J. 1983, in 'Accretion Driven Stellar X-ray Sources' (eds. Lewin, W.H.G. & Van den Heuvel, E.P.J.) Cambridge Univ.Press, p.303
Van den Heuvel, E.P.J. 1992a, in 'Xray Binaries and Recycled Pulsars' (eds. Van den Heuvel, E.P.J. & Rappaport, S.) Kluwer, Dordrecht, p.233
Van den Heuvel, E.P.J. 1992b, Proc.Sat.Symp.Nr. 3, Int.Space Yr.Conf. München (ESA ISY-3, July 1992) ESTEC, Noordwijk, p.29
Van den Heuvel, E.P.J. 1994, in 'Interacting Binaries' (Proc. 22nd Saas Fee School) Springer, Heidelberg, p.263
Van den Heuvel, E.P.J., Bhattacharya, D., Nomoto, K. & Rappaport, S. 1992,

Astr. Ap. 262, 97
Van den Heuvel, E.P.J. & Bitzaraki, O. 1994, Nature (submitted)
Van den Heuvel, E.P.J. & Bonsema, P.F.J. 1984, Astr.Ap. 139, L16
Van den Heuvel, E.P.J. & De Loore, C. 1973, Astr.Ap. 25, 387
Van den Heuvel, E.P.J. & Heise, J. 1972, Nature Phys.Sci. 239, 67
Van den Heuvel, E.P.J. & Habets, G.M.H.J. 1984, Nature 309, 698
Van den Heuvel, E.P.J. & Taam, R.E. 1984, Nature 309, 598
Van der Klis, M., Jansen, F., Van Paradijs, J.A., Lewin, W.H.G., Van den Heuvel, E.P.J., Truemper, J. & Stajno, M. 1985a, Nature 316, 225
Van der Klis, M., et al. (same authors), 1985b, IAU Circ.Nr. 4140
Van Kerkwijk, M.H., Charles, P.A., Geballe, T.R., King, D.L., Miley, G.K., Molnar, L.A. & Van den Heuvel, E.P.J. 1992, Nature 355, 703
Van Paradijs, J.A. 1978, Nature 274, 650
Verbunt, F. 1990, in 'Neutron Stars and their Birth Events' (ed. Kundt, W.) Kluwer, Dordrecht, p.179
Verbunt, F., Wijers, R.A.M.J. & Burm, H. 1990, Astr.Ap. 234, 195
Verbunt, F. & Zwaan, C. 1981, Astr.Ap. 100, L7
Wang, Y.M. 1979, Astr.Ap. 74, 253
Wang, Y.M. 1981, Astr.Ap. 102, 36
Webbink, R.F. 1975, Astr.Ap. 41, 1
Webbink, R.F. 1992, in 'X-ray Binaries and Recycled Pulsars' (eds. Van den Heuvel, E.P.J. & Rappaport, S.) Kluwer, Dordrecht, p.269
Webbink, R.F., Rappaport, S. & Savonije, G.J. 1983, ApJ 270, 678
Whelan, J. & Iben, I. 1973, ApJ 186, 1007
Woosley, S.E. & Taam, R.E. 1976, Nature 263, 101
Zeldovich, Ya.B. 1964, Sov.Phys.Dokl. 9, 246
Zeldovich, Ya.B. & Guseinov, O.Kh. 1965, Dokl.Acad.Nauk USSR 162, 791
Zeldovich, Ya.B. & Shakura, N.I. 1969, Sov.Astr. 13, 175
Zuiderwijk, E.J., Van den Heuvel, E.P.J. & Hensberge, G. 1974, Astr.Ap. 35, 353

SOME EVENTS IN X-RAY ASTRONOMY THAT AREN'T IN THE JOURNALS

Herbert Gursky
Naval Research Laboratory, Code 7600, Washington, DC 20375-5352

ABSTRACT

I was fortunate to have been actively involved in x-ray astronomy from its inception. As is frequently the case, the discoveries did not always procede as neatly as the journal articles would lead one to believe.

1. INTRODUCTION

Drs. Van den Heuvel and Lewin have provided excellent summaries of the first 10 or 15 years of x-ray astronomy, so it frees me from the need to do so. In the delivered version of his paper, Dr. Van den Heuvel finished with a curious remark that fortuitously provides a theme for my own paper. He said, "There are people who are smart and people who are not so smart." He said this in a reference to an idea on binary evolution that had been advanced by Jacob Shaham and that he had overlooked, implying that Shaham was in the smart category and that he was in the other category. Of course nothing is further from the truth. What Dr. Van den Heuvel probably meant to say was, "Sometimes the best of us do things that are not so smart." I will illustrate this by discussing the background of some of the scientific advances that were cited in the introductory articles.

2. THE DISCOVERY FLIGHT

The group of us at American Science and Engineering (AS&E) who were responsible for developing the flight experiment were trained as physicists. Both Riccardo Giacconi and I had worked in cosmic rays and Frank Paolini in nuclear physics. We had good experience with scintillation counters, but little with the geiger counters that we were using as our x-ray detectors. And it showed. We could not build counters that lasted more than a few weeks. This was a serious, but not a fatal flaw, since we did have access to the rocket payload up until shortly before the launch. We simply took a portable vacuum system into the field with us and refilled and replaced counters as we needed to.

I discovered a far more serious flaw just a few days before the discovery flight, that the counters were sensitive to fluorescent lights. We were using thin mica as the window material, and the material was quite transparent. There was no question about what was happening. When the room lights were on, the counting rate was high. When the lights were off or the counters were shaded, the counting rate was low. I was personally very sensitive to the influence of fluorescent lights on electronic equipment. As a graduate student I had spent several months struggling with a problem that I eventually traced to the lab's fluorescent lights. In that case the equipment wouldn't work properly unless the lights were on. Incidentally we also were unaware that a conducting layer should be applied to the inside of insulating material such as mica when it's used as part of gas counter wall in order to avoid field distortions. But apparently our windows were small enough so that the electric field was not seriously perturbed.

I found a simple solution involving a classical technique. I coated the windows with carbon black produced with a candle that was burning in an enclosed space, in our case an inverted coffee can. Such burning is relatively inefficient and produced thick smoke. Allowing the smoke to escape from a hole at the edge of the can and flow over the counters produced a nice black layer on the windows. Such a coating is quite strong since a small amount of residual unburned wax from the candle makes the carbon particles adhere together.

The flight was of course successful and, as one says, the rest was history. But not quite. I need to explain that our principal target was the moon and we were hoping to see fluoresence radiation caused by solar x-rays striking the surface. The rocket was spinning and traced out a circle on sky. We flew the rocket so that the moon, which was full, was in that circle. An optical sensor that was coaligned with one of the counters recorded the position of the moon.

As soon as the rocket reached altitude and the doors ejected, it was apparent that a strong source of radiation was present. At one point in the spin, the counting rate obviously increased. It was even obvious to one of the technicians who said, "There it is, there's the moon". But instead of elation, my emotion was more distress. The moon had no business being bright in x-rays. The best we had hoped for was a small effect revealing itself when all the data were summed, but we never should have seen the moon show up in each spin. What I was seeing was at least two orders of magnitude higher than we expected. Of course, I immediately knew the answer--the carbon had flaked off the mica windows and the counters were happily recording, not x-rays from the moon, but reflected solar ultraviolet radiation. Fortunately the counting rate was so high that I could actually tell that its peak was not quite aligned with the moon, a fact that I quickly confirmed after the flight. But it was a still a week or so before I was able to confirm that nothing funny was happening to produce an offset between where we thought the instruments were pointing and where the signals were appearing in the record. Only then was I able to eliminate the moon as a source of the excess radiation.

Now we had to understand the results. Our knowledge of traditional astronomy was minimal. I could find my way around the Nautical Almanac since I had to set the local time of the rocket flight in relation to the moon's position and to sidereal time. Frank Paolini was responsible for projecting our data onto the celestial sphere and used Norton's Star Atlas as his principal reference. I don't believe that we yet knew of Allen's Astrophysical Quantities.

We were fortunate in having a close relation with Bruno Rossi and George Clark of MIT. Rossi was the Chairman of the Board of AS&E and Clark was one of the board members. At that time their work at MIT was focussed on measuring high energy γ-rays as an extension of their traditional cosmic ray investigations. It was at Rossi's urging that AS&E had begun its program in x-ray astronomy.

One of MIT's principal efforts was the Bolivian Air Shower Joint Experiment (BASJE) which was a collaborative effort between MIT and Japanese institutions to look for high energy gamma rays that should result from the decay of neutral pions produced by cosmic rays in the galaxy. The Japanese scientists included Minoru Oda and Satio Hayakawa who happened to be in the US at the time of the flight. They and Rossi and Clark clustered around the x-ray data. George Clark found an answer for what we were seeing. BASJE had been in operation for some months and the initial data had revealed a peak of gamma radiation at the galactic center of about 4 sigma significance. Their data also showed a less intense peak in Cygnus. These peaks coincided more or less with what we saw in x-rays. But the crucial point was that the flux in energy units in gamma rays was about the same as the flux we observed in x-rays. Obviously BASJE was seeing directly the gamma rays produced by neutral pion decay and we were seeing the x-ray synchrotron radiation produced by electrons produced by charged pion decay. Clark wrote up the result for publication at about the same time as our paper announcing the x-ray observations. But the paper never did get published, which was probably a good thing since the peaks in the BASJE data eventually disappeared.

3. GLOBULAR CLUSTERS

Clark was victimized by a common problem with statistical fluctuations. I had a more bizarre problem with statistics that relates directly to my having found x-ray sources in globular clusters as reported by Dr. Van den Heuvel. As part of the data analysis effort following the launch of UHURU, I had done a study of the association of x-ray sources with stars listed in the SAO catalog. Cyg X-1 had been revealed to be coincident with a 9th magnitude star, whereas the earlier identifications of Sco X-1 and Cyg X-2 had been with 12-13th magnitude stars. I reasoned simply that if there was one Cyg X-1 there should be others and that their optical counterparts should be present in the SAO catalog which listed stars to at least 10th magnitude. In fact I got a positive result; there was a significant excess of bright stars at the x-ray locations. I published those results and recall being especially pleased with the paper since it was one of the few research efforts that that I had done entirely on my own.

The next year, at about the time of the Cambridge NATO Conference referenced by Van den Heuvel, I decided to repeat the study since we had more X-ray sources and better positions. This time I was even more successful. Not only did I see excess numbers of associations, I found several cases where two SAO stars were associated with single X-ray sources. I even had one case where there were three stars present! This was a physical impossibility and the only way I could imagine such an occurence was that the X-ray source was associated with a member of a cluster of stars and the associated bright stars were simply other members of the cluster. I went to a catalog of star clusters--I don't remember its name, but it's one published by a Czech observatory and consists of file cards in a wooden box. Naturally none of my sources showed up among the open clusters, but I mindlessly kept going through all the other categories of clusters. When I came to the globular clusters I found the objects that I then discussed with Dr. Van den Heuvel and reported at the conference. But I never did publish the result since I couldn't understand why I got that funny result with the bright stars, and I no longer trusted statistical arguments as they related to stellar objects. Some time later, George Clark independently found, and published, the association of x-ray sources and globular clusters.

4. PULSATING SOURCES

Dr. Lewin told about the high time variability that was discovered in Cygnus X-1 from UHURU data. We had observed Cygnus X-1 as part of our general strategy of observing the highest priority targets as early as possible in the program. Cyg X-1 was one such target. It was bright in x-rays and well suited for optical and radio observations. We had tried, unsuccessfully, a few years earlier to obtain an optical identification. Minoru Oda arranged to visit the U. S. and participate in the analysis of data during the early phase of the mission. He chose to work on Cyg X-1 and discovered that its x-ray emission was highly variable on very short time scales. Interestingly I remember that work differently from how it was described by Dr. Lewin. He had reread what was actually published at the time. Relying on memory, I recalled the short time variability but had forgotten how hard we had tried to make the object 'pulse'.

Shortly after finishing the Cyg X-1 work, I was in Giacconi's office and he was musing on the success of the mission up to that point. He was especially pleased with Cyg X-1 since he personally was heavily involved in the analysis of the data and its subsequent interpretation. He said that we had probably found all the important new results that Uhuru was capable of uncovering. I don't remember how I responded, but I do recall muttering something to the contrary.

The following week Ethan Schreier came around to our offices with the stripchart record of a traversal of Centaurus X-3 that showed an unmistakable sign of periodic x-ray emission. Cen X-3 was the first of the binary x-ray pulsars to be seen; their discovery proved to be the single most outstanding result from Uhuru, certainly this group would say so.

There is no deep message that I was trying to develop in presenting this material, beyond the one in the introduction that all of us do or say some things that in hindsight are not very smart.

Black Holes

DYNAMICAL EVIDENCE FOR BLACK HOLES IN SOFT X-RAY TRANSIENTS AND OTHER X-RAY BINARIES

Anne P. Cowley
Physics & Astronomy Dept., Arizona State Univ., Tempe, AZ 85287
cowley@vela.la.asu.edu

ABSTRACT

Although various observational properties have been suggested as indicators of a black hole in a binary system (*e.g.* rapid X-ray fluctuations, ultra-soft X-ray spectra, etc.), the most certain way to determine the mass is to measure it dynamically, using observed radial velocities of the companion star. This paper discusses the X-ray binaries for which such observational evidence for a stellar black hole is presently available.

1. INTRODUCTION

In this paper I will review the dynamical evidence for the existence of stellar black holes. Although many writers cautiously refer to them merely as 'black-hole candidates', I will call the well-established cases simply 'black holes', and leave the 'candidate' status to the questionable systems. In the first section I will make some general remarks about determination of stellar masses from radial velocity data, and in the second section I will discuss the properties of individual systems which appear to contain a black hole. There are still so few well-established black-hole systems that global properties are not clearly identified, but there is mounting evidence that the soft-X-ray transients may be a class of low-mass black-hole binaries.

2. MEASURING STELLAR MASSES DYNAMICALLY

It is well known that in a spectroscopic binary system one can measure the individual stellar masses in the special case where both the orbital inclination to the observer's line-of-sight and the mass ratio of the stellar components are known. One could determine the orbital inclination from the light curve, especially if the system eclipses, and the mass ratio can be derived from the velocity amplitudes when spectra of both stars are measured. However, in a more typical case, only one spectrum is detected and the inclination is unknown. Then, one can only determine the *mass function*, $f(M)$, using the orbital period, P, and semi-amplitude, K_{opt}, of the optical star's radial velocity curve. Here we will designate the mass of the unseen star as M_X, since in this paper we are discussing only X-ray binaries.

$$f(M) = (M_X \sin i)^3 / (M_{opt} + M_X)^2 \propto K^3 P \qquad (1)$$

Note that the value of the mass function is the minimum mass of the unobserved star, M_X, if the optical star has no mass and the orbit is viewed edge-on ($i = 90°$). In general, the mass function is much smaller than the actual mass of the unseen star. Therefore, if a large mass function is measured, one can be certain that the unseen star has at least that mass.

It is generally agreed that neutron star masses cannot exceed $\sim 3M_\odot$. Thus, the strongest case for a black hole is when the measured mass function is greater than three solar masses. In addition, generally one is able to make some intelligent guess at the value of M_{opt} from the star's spectral type and luminosity, so that M_X can be estimated. In cases where M_X turns out to be much larger than M_{opt}, one can be quite confident that a black hole must be present, since the inferred high-mass object is not detected spectroscopically. Thus, the most well-established black holes are in systems where:

$$f(M) \geq 3 \quad and/or \quad M_X > M_{opt}$$

2.1 Possible Mine Fields in Determining the Mass Function

As with most measurements, there are many places one can go astray. The most obvious sources of error are in determining the value of the period, P, or the radial-velocity semi-amplitude, K. When trying to identify the orbital period one has to be aware of possible alias periods, sometimes due to the spacing of the observations themselves, and beat periods if more than one periodicity is present in the system. Errors in K can arise from a multitude of effects including measuring errors, hour-angle and guiding effects, blending of absorption and emission lines, non-uniform brightness of the star causing the mass-center and photo-center not to be coincident, tidal distortions of the star, rotation effects, and many others. Even small errors in K are important since it enters the mass function as K^3. While none of these possible difficulties precludes an accurate determination of the mass function, they do require the observer be extremely careful.

Assuming that both P and K are well determined, to measure the unseen star's mass one still must make some assumptions about both M_{opt} and i. If the system does not eclipse (none of the 'best' black-hole binaries does), one generally models the optical light curve to obtain information about the inclination, i, and mass ratio $q = M_X/M_{opt}$. Much work has been done on modeling ellipsoidal light variations from tidally distorted stars, and there are many papers in the literature applying this technique to X-ray binaries (*e.g.* Kuiper *et al.* (1988) for LMC X-3, to mention one). A general discussion of the method is given by Shore (1994). Suffice it to say that there are many complications in this type of analysis too. Finally, it has been found empirically in X-ray binaries that often the optical star is undermassive for its temperature and luminosity (*e.g.* Hutchings 1982). This is because the evolutionary history of stars in X-ray binaries is far different from that of single stars. Thus, assuming the optical star's mass from its spectral appearance may not be appropriate. The importance of this complication should not be overlooked.

Thus, while the method of dynamical mass determination is straightforward, in practice it is full of difficulties. In spite of this, there are some well-studied X-ray binaries which apparently contain black holes. A detailed review of the properties of these systems has been given by Cowley (1992).

3. DISCUSSION OF INDIVIDUAL SYSTEMS

Black holes are found in both the massive X-ray binaries (MXRB) and the low-mass systems (LMXB). The LMXB black-hole systems for which companion-star radial velocities have been measured are all soft X-ray transients. However, the converse has not been demonstrated, namely that all soft X-ray transients contain black holes. Table 1 lists the X-ray binaries for which dynamical evidence of a black hole is available. The table is divided into the MXRB and LMXB. In each group, the systems listed below the dotted lines have some difficulties and should not be considered as established black-hole binaries, but only as 'candidates'.

TABLE 1. Black Holes with Dynamical Studies

System	Opt. Star	P_{orb}	Orbital Lt. Curve	$f(M)$ ($M_\odot$)	M_X ($M_\odot$)	Problems / References
MXRB:						
Cyg X-1 V1357Cyg	O9.7 Iab	5.6^d	ellipsoidal	0.25	10-5	$M_{opt} > M_X$ 2, 3, 9, 15, 24
LMC X-3 0538-64	B3 V	1.7^d	ellipsoidal	2.3	≥ 6	1, 5, 14
weaker cases:						
LMC X-1 0540-70	O7 pec.	4.2^d	—	0.14	4-10	i.d. uncertain; $M_{opt} > M_X$; 12, 20
SS433 V1343 Aql	pec. em.	13.1^d	eclipse??	—	~10	em. lines meas. 7, 8, 16, 22
LMXB:						
V616 Mon 0620-00	K3/5 V	7.8^h	ellipsoidal	2.9	~4	10, 11, 13, 17, 18, 25
Nova Mus '91 1124-68	K3 V	10.4^h	ellipsoidal	3.1	4-6	21, 25
V404 Cyg 2023+34	K0 IV	6.5^d	ellipsoidal	6.3	8-15	star vs. orbit size in conflict; 4, 23, 25
weaker cases:						
CAL 87 0547-71	disk em. lines only	10.7^h	eclipse	—	≥ 4	em. lines meas. 6, 19

References: (1) Bochkarev et al. 1988, (2) Bolton 1972, (3) Bolton 1975, (4) Casares et al. 1992, (5) Cowley et al. 1983, (6) Cowley et al. 1990, (7) Crampton & Hutchings 1981, (8) D'Odorico et al. 1991, (9) Gies & Bolton 1986, (10) Haswell & Shafter 1990, (11) Haswell et al. 1993, (12) Hutchings et al. 1987, (13) Johnston et al. 1989, (14) Kuiper et al. 1988, (15) Lloyd & Walker 1989, (16) Margon 1984, (17) Marsh et al. 1993, (18) McClintock & Remillard 1986, (19) Naylor et al. 1989, (20) Pakull & Angebault 1986, (21) Remillard et al. 1992, (22) Stewart et al. 1987, (23) Wagner et al. 1992, (24) Webster & Murdin 1972, (25) papers in this conference volume

3.1 Massive X-ray Binaries (MXRB)

<u>Cyg X-1</u>: This was the first X-ray binary to be recognized as probably containing a black hole (Webster & Murdin 1971, Bolton 1972). The velocity curve of its O9.7 Iab primary is well observed (*e.g.* Gies and Bolton 1982) but gives a small value of the mass function ($f(M) = 0.25 M_\odot$). The secondary star has a lower mass than the optical primary, so that if Cyg X-1 were not an X-ray source we would assume the secondary was a fainter, normal star. However, the presence of strong X-ray emission indicates the secondary is either a neutron star or black hole, and its mass is clearly too large to be neutron star. The determination of the secondary's mass has a number of associated uncertainties. One must assume a value for the O star's mass, without any knowledge of how much mass it may have already lost. The ellipsoidal variations are small, indicating the system is viewed at a low angle of inclination ($\sim 30°$), introducing further uncertainty into the mass determination. Finally the distance of Cyg X-1 is not well known. All of these factors reduce the accuracy with which the mass of the black hole can be determined. If one adopts a primary-star mass of ~ 30 $M_\odot$ (no mass lost during its evolution), then the X-ray source mass lies between 10 – 15 $M_\odot$.

<u>LMC X-3</u>: Of the MXRB, this system is the most convincing case for a black-hole binary. Because the unseen star is more massive than the B3 V optical star, there is no possibility that the X-ray source is a normal star 'hidden' by the light of the B star. Thus, even without knowledge of the X-ray emission, one would conclude that this system must contain a black hole. The best mass estimates are $M_{opt} \sim 6 M_\odot$ and $M_X \sim 10 M_\odot$. The system does not eclipse in either optical light or X-rays (Cowley *et al.* 1983), but ellipsoidal variations show LMC X-3 is viewed at an inclination of $\sim 70°$ and indicate that $M_X \sim 4 - 11 M_\odot$ (Bochkarev *et al.* 1988, Kuiper *et al.* 1988.)

<u>LMC X-1</u>: This highly luminous X-ray source has long been suspected of containing a black hole, partially because of its ultra-soft X-ray spectrum (White & Marshall 1984). The main difficulty with LMC X-1 is its optical identification. The *Einstein* HRI position (Long *et al.* 1981) lies between two possible counterparts which are separated by $6''$: namely, R148, a B5 supergiant, and a peculiar O7 star known as '#32' (Cowley *et al.* 1978). Either star is a possible candidate, but #32 is more likely as it shows a spatially resolved He III nebula surrounding the star which may be ionized by the X-ray source (Pakull & Angebault 1986). Velocity measurements of #32 indicate an orbital period of 4.2 days and a mass function, $f(M) = 0.14 M_\odot$ (Hutchings *et al.* 1987). Adopting $M_{opt} \sim 20 M_\odot$ leads to a black-hole mass in the range $4 \leq M_X \leq 10 M_\odot$. Recent *ROSAT*-HRI observations have not resolved the positional uncertainty. In fact, the *ROSAT* position is still approximately equidistant from the two stars, but further from both. To clearly establish the identity of LMC X-1, correlated X-ray/optical variations need to be found. These should be searched for before the identification of this source and its recognition as black-hole binary can be confirmed.

<u>SS433</u>: This enormously peculiar system with relativistics jets has been written about extensively (see review by Margon 1984). It is included in Table 1 mostly for completeness. The velocity curve in the 13-day orbit is measured from emission lines which are assumed to be formed in an accretion disk near

the degenerate star (Crampton & Hutchings 1981). Since the spectrum of the primary star itself is not measured, this mass determination is not definitive. Furthermore, higher quality spectra have recently been used to show the masses are consistent with a neutron-star binary system (D'Odorico *et al.* 1991).

3.2 Low-Mass X-ray Binaries (LMXB) and Soft X-ray Transients

<u>0620−00 = V616 Mon:</u> The optical star identified with this soft X-ray transient is a recurrent nova which had outbursts in 1917 and 1975. The 1975 X-ray outburst was discovered by *Ariel V* (Elvis *et al.* 1975), and the simultaneous optical brightening was found by Boley *et al.* (1976). When the system faded sufficiently to detect the secondary star, McClintock & Remillard (1986) showed that the radial velocity curve of the K3 dwarf star gave a mass function of $f(M) = 3.18 M_\odot$, already larger than a neutron-star mass. A recent study of the system by Marsh *et al.* (1993) concludes that the mass of the black hole is in the range $3.3 \leq M_X \leq 4.2 M_\odot$. Analysis of the ellipsoidal light variations supports this range of masses.

<u>Nova Mus '91 = GS1124−68:</u> This is another soft X-ray transient with properties very similar to 0620−00. Its X-ray outburst was discovered independently by *Ginga* (Makino *et al.* 1991) and WATCH/*Granat* (Lund & Brandt 1991). Velocity measurements of the K3 V secondary star show a radial velocity curve with a semi-amplitude of K = 409 km s^{-1}. This results in a mass function $f(M) = 3.1 \pm 0.4 M_\odot$ and a black-hole mass $\sim 4 - 6 M_\odot$. Also, as in 0620−00, the light curve from the tidally distorted secondary star is consistent with this interpretation.

<u>V404 Cyg = GS2023+338:</u> This is yet another soft X-ray transient (Makino *et al.* 1989) which has now been found to be a binary system probably containing a black hole. Casares *et al.* (1992) measured the radial velocity of the late-type secondary star and found the orbital period to be ∼6.5 days. This relatively long period (at least for an LMXB), combined with the measured velocities, gives a mass function $f(M) = 6.3 M_\odot$ and a black-hole mass in the range $8 - 15 M_\odot$. Although photometric observations show ellipsoidal variations with this same period (Wagner *et al.* 1992), there is a substantial brightness variation (∼0.2 mag.) during a single night whose range is as large as the orbital modulation. One wonders if the orbital period could be shorter, with the 6.5-day periodicity resulting from an alias or beat period. With this long period there is a conflict between the size of the secondary star and the size of the orbit. In order to drive the mass transfer, one assumes that the secondary is near to filling its Roche lobe. However, with a 6.5-day period, this star must be K0 giant. From the observed magnitude and interstellar absorption in the direction of V404 Cyg one can estimate the distance of the system, which turns out to be so large that the peak L_X is excessively high, well above the Eddington limit. Casares *et al.* (1993) argue that consistency can be obtained if the secondary is a K0 subgiant. A nice confirmation of this suggestion would be to show spectroscopically that a luminosity class IV star is present.

<u>CAL 87:</u> CAL 87 is one of the recently identified 'supersoft' X-ray sources in the Large Magellanic Cloud. It shows a pronounced optical eclipse every 10.6 hours (Naylor *et al.* 1989, Cowley *et al.* 1990). Recent *ROSAT* observations show a shallow X-ray eclipse (Schmidtke *et al.* 1993). This interesting binary

is listed below the dotted line in Table 1 (indicating 'weaker cases') because the velocities are based on emission lines which are presumed to come from the accretion disk near the compact star. This disk dominates the optical light so that the spectrum of the secondary star is not detected. Cowley et al. (1990) reported the presence of a cool star spectrum in observations made at minimum light, but later this was found to be contamination by a field star $< 1''$ away (Cowley et al. 1991). To obtain masses from the emission lines, one must make reasonable estimates of the mass of the unseen, donor star. Such an analysis leads to $M_X \geq 4M_\odot$, but one worries whether non-orbital motions, such as a flow between the stars, might effect the emission-line profiles and hence the measured velocities. Furthermore, recent theoretical arguments have been made that the 'supersoft' sources contain white dwarfs, not neutron stars or black holes (e.g. van den Heuvel et al. 1992).

Other Systems: There are several other systems which are not included in Table 1, but which potentially may contain black holes. The most frequently mentioned are other soft X-ray transients. GS2000+25 was discovered as a soft X-ray transient by *Ginga* (Tsunemi et al. 1989). It is identified with the variable star QZ Vul. Chevalier & Ilovaisky (1993) have shown its color and orbital light curve are consistent with a tidally distorted G/K dwarf in an 8.2-hour binary system with a more massive companion. The similarity of the light curve to that of 0620−00 and Nova Mus '91 suggests this system should be studied further. However, obtaining radial velocities will be challenging since the system is $\sim 20^{th}$ magnitude.

In this conference Remillard (1993) presented a poster paper showing that the X-ray transient H1705−25 (Griffiths et al. 1978) may be a black-hole candidate. The optical counterpart, Nova Oph '77, appears to show an orbital period of P$\sim$13 hours. Preliminary spectral analysis of this very faint (m $\sim$21) star indicates that it could have a large velocity amplitude, like the other transient systems discussed above. We await further analysis of these data.

4. SUMMARY

The total number of binary systems presently known or suspected to contain black holes is very small. However, this small number may not fairly represent the population of such objects in the Galaxy and the Magellanic Clouds. If X-ray binaries are only detectable for $\sim 10^4$ years (van den Heuvel 1983), their rarity partially reflects the short lifetime when they are continuously radiating at their Eddington luminosity. We note in this regard, only $\sim$150 neutron-star systems are presently detected in the Galaxy.

The soft X-ray transients already represent half of the known black holes, and there are indications that others may soon be confirmed. Inoue (1991) has estimated that the Milky Way may contain a few thousand such systems, based on the number found in outburst per year and the distance to which they are detected. If a substantial percentage of these (perhaps even all!) contain black holes, the total galactic population of black-hole binaries could be substantial.

It is of interest to note the relatively high percentage of black holes and 'candidates' in the Large Magellanic Cloud (see Table 1). In total, there are only seven luminous X-ray binaries in the LMC. The evidence that three of them (approximately half) may be black-hole systems is remarkable. All are 'steady' sources (LMC X-3, LMC X-1, and CAL 87). It is perhaps not surprising that no soft X-ray transients have yet been found in the LMC. If factors like environment and metallicity do not affect the rate of outburst, then one could estimate the expected number of such LMC systems from the mass ratio of the two galaxies. Since the LMC/Galaxy mass ratio is ~ 0.1, at most one might expect only a few soft transients per decade in the LMC. Unfortunately, even if such systems were found they would be too faint at minimum light to be studied spectroscopically with present-day instrumentation.

ACKNOWLEDGEMENTS

This work has been supported by NSF and NASA. I thank wish to thank my colleagues Paul Schmidtke, John Hutchings, & David Crampton.

REFERENCES

Bochkarev, N.G., Syunyaev, R.A., Khruzina, T.S., Cherepashchuk, A.M., & Shakura, N.I. 1988, Sov. Astron. AJ, 32, 405
Boley, F, Wolfson, R., Bradt, H., Doxsey, R., Jernigan, G., & Hiltner, W.A. 1976, ApJ Lett, 203, L13
Bolton, C.T. 1972, Nature Phys. Sci., 240, 124
Bolton, C.T. 1975, ApJ, 200, 269
Casares, J., Charles, P.A., & Naylor, T. 1992, Nature, 355, 614
Casares, J., Charles, P.A., Naylor, T., & Pavlenko, E.P. 1993 MNRAS, in press
Chevalier, C. & Ilovaisky, S.A. 1993, A&A, 269, 301
Cowley, A.P. 1992, ARA&A, 30, 287
Cowley, A.P., Crampton, D., & Hutchings, J.B. 1978, AJ, 92, 195
Cowley, A.P., Crampton, D., Hutchings, J.B., Remillard, R., & Penfold, J.E. 1983, ApJ, 272, 118
Cowley, A.P., Schmidtke, P.C., Crampton, D., & Hutchings, J.B. 1990, ApJ, 350, 288
Cowley, A.P., Schmidtke, P.C., Crampton, D., Hutchings, J.B., & Bolte, M. 1991, ApJ, 373, 228
Crampton, D. & Hutchings, J.B. 1981, ApJ, 251, 604
D'Odorico, S., Oosterloo, T., Zwitter, T., & Calvani, M. 1991, Nature, 353, 329
Elvis, M., Page, C.G., Pounds, K.A., Ricketts, M.J., & Turner, M.J.L. 1975, Nature, 257, 656
Gies, D.R. & Bolton, C.T. 1986, ApJ, 304, 371
Griffiths, R.E., Bradt, H., Doxsey, R., et al. 1978, ApJ Lett, 221, L63
Haswell, C.A., Robinson, E.L., Horne, K., Stiening, & Abbott, T.M.C. 1993, ApJ, 411, 802
Haswell, C.A. & Shafter, A.W. 1990, ApJ Lett, 359, L47
Hutchings, J.B. 1982, in *Galactic X-ray Sources*, ed. P. Sanford, P. Laskarides, J. Salton, [Chichester: Wiley & Sons Ltd.], 3
Hutchings, J.B., Crampton, D., Cowley, A.P., Bianchi, L., & Thompson, I.B. 1987, AJ, 94, 340
Inoue, H. 1991, in *Frontiers of X-ray Astronomy*, ed. T. Tanaka & K. Koyama, [Tokyo: Universal Academy Press], 291
Johnston, H.M., Kulkarni, S.R., & Oke, J.B. 1989, ApJ, 345, 492
Kuiper, L., van Paradijs, J., & van der Klis, M. 1988, A&A, 203, 79
Lloyd, C. & Walker, E.N. 1989, in *Proc. 23rd ESLAB Symp.; ESA SP-296*, ed. J. Hunt & B. Battrick, [Paris: ESA Publ.], 511

Long, K.S., Helfand, D.J., & Grabelsky, D.A. 1981, ApJ, 248, 925
Lund, N. & Brandt, S. 1991, IAU Circ., No. 5161
Makino, F. et al. 1989, IAU Circ., No. 4782
Makino, F. et al. 1991, IAU Circ., No. 5161
Margon, B. 1984, ARA&A, 22, 507
Marsh, T.R., Robinson, E.L., & Wood, J.H. 1993, MNRAS, in press
McClintock, J.E., & Remillard, R.A. 1986, ApJ, 308, 110
Naylor, T., Callanan, P., Machin, G., & Charles, P.A. 1989, IAU Circ., 4747
Pakull, M.W. & Angebault, L.P. 1986, Nature, 322, 511
Remillard, R.A. 1993, this conference
Remillard, R.A., McClintock, J.E., & Bailyn, C.D. 1992, ApJ Lett, 399, L145
Schmidtke, P.C., McGrath, T.K., Cowley, A.P., & Frattare, L.M. 1993, PASP, 105, 863
Shore, S.N. 1994, in *Interacting Binary Stars: 22nd Saas Fee Adv. Astrophys. Course*, eds. H. Nussbaumer & A. Orr, [Berlin: Springer-Verlag], 1
Stewart, G.C., Watson, M.G., Matsuoka, M., Brinkmann, W., Jugaku, J., et al. 1987, MNRAS, 228, 293
Tsunemi, H., Kitamoto, S., Okamura, S., & Roussel-Dupré, D. 1989, ApJ Lett, 337, L81
van den Heuvel, E.P.J. 1983, in *Accretion-Driven Stellar X-ray Sources*, ed. W.H.G. Lewin & E.P.J. van den Heuvel, [Cambridge: Cambridge Univ. Press], 303
van den Heuvel, E.P.J., Bhattacharya, D., Nomoto, K., & Rappaport, S.A. 1992, A&A, 262, 97
Wagner, R.M., Kreidl, T.J., Howell, S.B., & Starrfield, S.G. 1992, ApJ, 401, L97
Webster, B.L. & Murdin, P. 1972, Nature, 235, 37
White, N.E. & Marshall, F.E. 1984, ApJ, 281, 354

The Galactic Distribution of Black Holes in X-ray Binaries

N.E. White
Code 668, Laboratory for High Energy Astrophysics,
NASA Goddard Space Flight Center, Greenbelt, MD 20771.
white@lheavx.gsfc.nasa.gov

ABSTRACT

The X-ray spectra of accreting black holes in X-ray binaries show an *ultra-soft/ultra-hard* spectral combination that is distinct amongst the galactic population of X-ray binaries. This spectral signature is used to identify new black hole candidates, BHC. A total of 23 BHC are listed (including six that have been well established from optical radial velocity measurements). In 20 BHC systems the binary companion star is late type, and of these 15 are transient systems. Only 3 BHC are associated with early type companions. The galactic distribution of these black hole candidates, BHC, is compared with that of the entire population of low mass and high mass X-ray binaries. The transient BHC are not strongly clustered at the galactic center, nor have any been found in a globular cluster. This suggests the transient BHC systems are population I.

1. INTRODUCTION

The existence of accreting black holes in X-ray binaries has now been well established. In six systems optical radial velocity measurements have demonstrated that the mass of the compact object is greater than the maximum mass of a neutron star (Cowley 1994 and references therein). Table 1 summarizes the parameters of these six systems. The black hole X-ray binary, BHXRB, systems can be classified based on the mass donor companion star. The early type BHXRB Cyg X-1, LMC X-3 and LMC X-1 have an OB star companion. They are steady sources that are a regular feature of the X-ray sky. The late type BHXRB A0620-00 (Nova Mon), GS2023+338 (V404 Cyg) and GS1121-68 (Mova Mus) contain a G or K companion. They are transient X-ray sources with outbursts lasting typically a few months. During outburst the optical counterparts show a nova-like brightening, in quiescence the companion is seen and the binary parameters can be determined (*e.g.*McClintock and Remillard 1986).
The measurement of the mass function of the binary system from optical radial velocity spectroscopy provides the only sure method to determine whether the compact object is a black hole. But optical measurements are difficult and in many cases impossible if the system lies in a region of high optical reddening, or is simply too far away. If a unique combination of X-ray properties can be established for the known black hole systems, then it should be possible to identify other BHC amongst the general population of X-ray binaries. Several attempts to do this have been made (Ostriker 1977; White and Marshall 1984; Tanaka and Lewin 1994). If an X-ray binary source exhibits X-ray bursts or X-ray pulsations, then it almost certainly contains a neutron star (or perhaps a white dwarf in some cases). For other sources the X-ray spectrum is the key factor in determining if the compact object is a BHC or a neutron star. In this review the unique spectral properties are discussed and used to generate a list of

Table 1: BHXRB established from Optical Radial Velocity Measurements.

Source	Cyg X-1	LMC X-3	LMC X-1	Nova Mon A0620-00	V404 Cyg GS2023+338	Nova Mus GS1121-68
Max Flux (mJy)	1.0	0.04	0.03	50	20	6
Transient	No	No	No	Yes	Yes	Yes
D (Kpc)	2.5	55	55	1	3	3
Log(L_x) (erg/s)	37.3	38.5	38.4	38.0	38.5	38.0
Period (d)	5.6	1.7	4.2(?)	0.32	6.47(+0.24?)	0.43
Companion	O9.7Iab	B3V	O7-9III(?)	K5V	G9V-K0III	K0V-K4V
f(M) $M_\odot$	0.25	2.3	0.12(?)	2.91	6.26	3.1
BH Mass ($M_\odot$)	16	9	4(?)	9	12	6
Discovery	1972	1982	1983	1986	1992	1992
Ref	1,2	3	4	5	6	7

References: 1) Webster and Murdin 1972; 2) Bolton 1972; 3) Cowley et al. 1983; 4) Hutchings et al. 1987; 5) McClintock & Remillard (1986); 6) Casares et. al. (1992); 7) McClintock, Bailyn and Remillard (1992).

BHC. The overall galactic distribution of BHC is investigated for the first time.

2. Spectral Signatures

The X-ray spectrum of Cyg X-1, the first BHXRB to be discovered (Webster and Murdin 1972; Bolton 1972), shows distinctive high-low state behavior (e.g. Coe, Engel and Quenby 1976). In its high state, the 1-10 keV spectrum is very soft (kT~1 keV) with a hard high energy tail. In the low state the luminosity of the very soft component is much reduced and the hard power law (photon index~ 1.4) spectrum dominates. This power law component extends up to several hundred keV, above which it gradually turns over (Sunyaev and Trümper 1979). The high-low state behavior of Cyg X-1 was suggested as a signature of accreting black holes that could be used to identify other systems (e.g. Ostriker 1977). An X-ray color-color diagram derived from the HEAO-1 A2 all sky survey demonstrated that a few X-ray binaries have *ultra-soft* spectra with a characteristic temperature of 1 keV, similar to the high state Cyg X-1 spectrum, and it was suggested such a soft spectrum might be a signature of a BHC (White and Marshall 1983). Among these *ultra-soft* sources was LMC X-3, a then newly discovered BHXRB (Cowley et al. 1983) and LMC X-1 which was latter shown to be a likely BHXRB (Hutchings et al. 1987).

The list of BHC gathered using the *ultra-soft* spectral signatures contained some transient X-ray sources, including A0620-00/Nova Mon (White, Kaluzienski and Swank 1983). Subsequently McClintock and Remillard (1986) determined the radial velocity variation of the companion star in A0620-00 and established the system to very probably contain a black hole. This gave more confidence in using the spectral signature as a way to establish BHC. Another spectral feature of the *ultra-soft* sources is that they also show a hard tail, that is comparable to the power law component seen from Cyg X-1 in its low state (Coe,

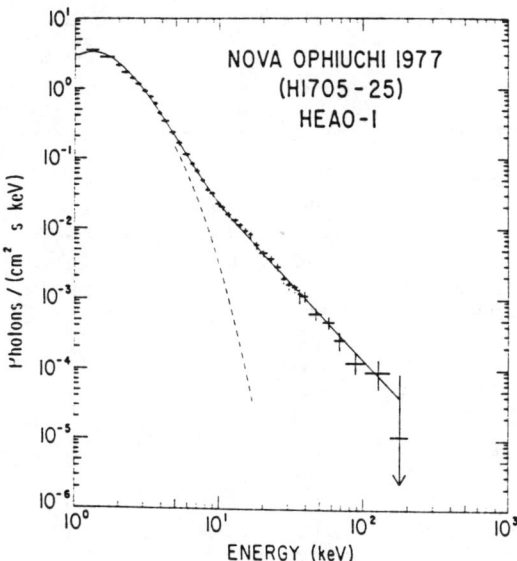

Figure 1: The inferred incident spectrum of the X-ray transient system Nova Oph (H1705-25), approximately 1 month after maximum measured by HEAO 1 A2 and A4 in 1977. This figure is taken from Wilson and Rothschild (1983). The dashed line illustrates the contribution from the *ultra-soft* component.

Engel and Quenby 1976; Wilson and Rothschild 1983). The spectrum of Nova Oph 1977 measured by HEAO 1 is shown in Figure 1, and well illustrates the *ultra-soft/ultra-hard* spectral signature. At energies > 10 keV this high energy tail becomes dominant and makes the sources appear to be *ultra-hard* compared to the typical galactic population. Sunyaev et al. (1991b) suggest this hard *Cyg X-1-like* power law tail, extending out to several hundred keV is a unique signature of accreting black holes. Over the past decade 1-2 of these *ultra-soft/ultra-hard* X-ray transients have been discovered each year (see e.g. Tanaka 1989; Tanaka and Lewin 1994). Subsequent optical observations of two of them (V404 Cyg and Nova Mus) have established that they too are BHXRB, as predicted by the *ultra-soft/ultra-hard* signature (Casares, Charles, and Naylor 1992; McClintock, Bailyn and Remillard 1992).

In steady sources (Cyg X-1 and GX339-4) the *ultra-soft* and power law components seem to be anti-correlated. In the transient systems the *ultra-soft* component usually dominates early on, during the 1-10 keV peak flux, and only later does the hard power law tail take over. Rapid variability is commonly seen, usually associated with hard component. The *ultra-hard* component can be modelled by a Comptonized spectrum, where soft photons are up-scattered on hot electrons (Sunyaev and Trümper 1979). The *ultra-soft* component can be well fit by an optically thick accretion disk spectrum (see Ebisawa 1994 and refs therein). In some a 511 keV annihilation line is occasionally seen (e.g. Bouchet et al. 1991; Sunyaev et al. 1991a).

Table 2 lists all transient X-ray binaries that have exhibited either an

Table 2: LMXRB Transients with ultra-soft/ultra-hard spectra

Name	Alt Name	P_{orb}(d)	Outburst Year	Interval (yr)	ref
A 0620-00	Nova Mon	0.32	1917 1975	58	1,2
GS 2023+33	V404 Cyg	6.47	1938 1989	51	3,4
4U 1543-47			1971 1983 1992	9-12	5,6,7
4U 1630-47			1971 1972 1974 1976 1984	1.6-8	8,9,10
A 1524-62	Tra X-1		1974 1990	16	11
H 1743-32	XT 1741-322		1977	-	12,13,14
H 1705-25	Nova Oph		1977	-	15,16
EXO 1846-031			1985	-	17
GS 1354-645			1987	-	18
GS 2000+25	Nova Vul	0.34	1988	-	19
GS 1826-24			1988	-	20
GRS 1124-68	Nova Mus	0.43	1991	-	21,22
GRO J0422+32		0.21	1992	-	23
GRS 1009-45			1993	-	24
GRO J1719-24	GRS 1716-249		1993	-	25,26

References: 1) Elvis et al. (1975); 2) Eachus et al. (1976); 3) Makino (1989); 4) Casares et al. (1992); 5) Matilsky et al. (1972); 6) Kitamoto et al. (1984); 7) Harmon et al. (1992); 8) Jones et al. (1976); 9) Priedhorsky & Holt (1987); 10) Parmar et al. (1986); 11) Kaluzienski et al. (1975); 12) Kaluzienski & Holt (1977); 13) Doxsey et al. (1977); 14) White et al. (1984); 15) Kaluzienski & Holt (1977); 16) Griffiths et al. (1977); 17) Parmar et al. (1993); 18) Makino (1987); 19) Makino (1988a); 20) Makino (1988b); 21) Makino (1991); 22) Lund & Brandt (1991); 23) Paciesas et al. (1992); 24) Lapshov et al. (1993); 25) Ballet et al. (1993); 26) Harmon et al. (1993).

ultra-soft/ultra-hard or an *ultra-hard* spectral signature. There are now 15 such systems. Some have been seen to recur, at intervals between 600 day and 58 yr. The orbital periods are typically centered around 0.3d, with the exception of V404 Cyg where it is 6.5d. There are persistent (although in some cases highly variable) sources which also show *ultra-soft* and/or *ultra-hard* signatures They are listed in Table 3. Three of these are established BHXRB from optical radial velocity measurements. Two of the persistent systems (4U 1755-338 and 4U1957+11) show an *ultra-soft* spectrum, but no *ultra-hard* high energy tail, and another two show the opposite (1E 1740.7-2942 and Gr 1758-258). These latter four systems have unusual spectra compared to the galactic population, but its less clear if they are to be considered BHC as opposed to simply odd-ball systems.

The *ultra-soft/ultra-hard* criterion seems to be a proven indicator of BHC. In contrast X-ray pulsars exhibit flat power law spectra up to 10-20 keV, which then decay rapidly (*cf.* White, Nagase and Parmar 1994). A few pulsars do show a soft excess (*e.g.* Her X-1; McCray and Lamb 1976), but these can be distinguished by the fact they are pulsars, and the high energy spectrum is quite different. Between giving the presentation at the meeting, and writing this article for the proceedings, one persistent *ultra-soft* system, 4U0142+614, has been removed from the BHC list. Israel, Mereghetti, and Stella (1993) report

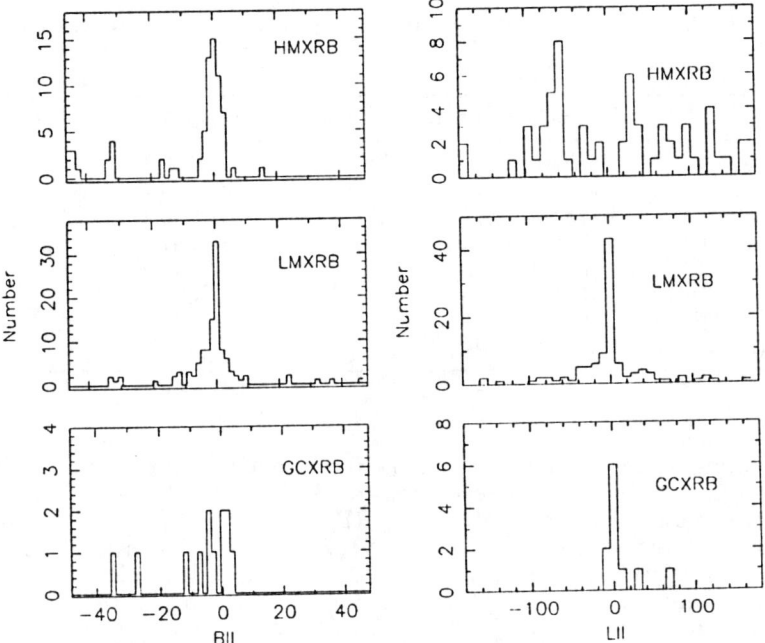

Figure 2: The Galactic distribution of X-ray binaries. The upper plot shows High Mass X-ray Binaries, the middle, Low Mass X-ray Binaries and the lower plot Low Mass X-ray Binaries in Globular Clusters. The longitude distribution only includes systems at $bii\pm30°$

the discovery of low amplitude 8.69s pulsations, the signature of a neutron star or white dwarf. The *ultra-soft* excess in this system maybe similar to that seen from Her X-1, and indicates that the *ultra-hard/ultra-soft* combination is a more definitive BHC signature.

Low magnetic field neutron star systems exhibit either exponential spectra with a characteristic temperature of 5–10 keV, or power law spectra with a photon index of 2–3 (*cf.* White, Nagase and Parmar 1994). In their low luminosity state some X-ray burst sources show a power law spectrum that seems to extend to high energies (*e.g.* Mitsuda *et al.* 1989), however, the photon index in these cases is steeper than seen from the BHC. Another strange case is Cir X-1, which was on early lists of BHC (*e.g.* Ostriker 1977). X-ray bursts were discovered from Cir X-1 by Tennant, Fabian and Shafer (1986). Cir X-1 is a pathological X-ray binary source that exhibits many different and spectacular properties, so the fact that it shows a 1–10 keV spectral signature similar to Cyg X-1 must be taken in the context of the overall properties.

3. The Galactic Distribution of BHC

Figure 2 and 3 show the galactic distribution of the BHC and the X-ray binary population as a whole (compiled from a catalogue of X-ray binaries by van Paradijs 1994). Figure 2 shows the distribution of all X-ray binaries as a function

Table 3: Persistant Emission Sources with *ultra-soft/ultra-hard* spectra

Name	Type	Signature	P_{orb}(d)	d(kpc)	High State	Low State
Cyg X-1	HMXRB	US/UH	5.6	2.5	37.3	36.6
LMC X-3	HMXRB	US/UH	1.7	55	37.9-38.5	no
LMC X-1	HMXRB	US/UH	4.2(?)	55	37.5-38.4	no
GX 339-4	LMXRB	US/UH	0.62	4	37.2-37.9	<36.2
1E 1740.7-2942	LMXRB	UH	?	7	no	36.6
Gr 1758-258	LMXRB	UH	?	?	no	?
4U 1755-338	LMXRB	US	0.19	?	?	?
4U 1957+11	LMXRB	US	0.39	?	?	?

of galactic longitude (lii) and latitude (bii). To exclude the effects of the LMC the longitude distribution was restricted to within $bii\pm30°$. The X-ray binaries are shown separately for the late (LMXRB), early type (HMXRB) and those in globular clusters (GCXRB). The LMXRB are clustered around the galactic center, with a very sharp spike at the galactic center itself. This distribution has been interpreted as indicating the LMXRB are part of an old stellar population, *i.e.* population II (van Paradijs 1983; Cowley *et al.* 1987). The HMXRB are, in contrast confined along the galactic plane, with no clustering at the galactic center, consistent with a population I distribution.

In Figure 3, the galactic distribution of the BHC are shown for the transient systems (Table 2), and all BHC systems (Tables 2 and 3). The systems considered are effectively all late type, since two of the three early systems lie in the LMC. The transient BHXRB (15 candidates), the largest and most robust class of BHC, show no clustering towards the galactic center. In particular there is no evidence for a sharp spike centered within 5° of the galactic plane that is so prominent for the LMXRB. The BHC population as a whole shows a peak towards the galactic center, caused by the inclusion of GX339-4, 1E 1740.7-2942, Gr 1758-258 and 4U 1755-338 (Table 3). The black hole candidacy of these sources is not so secure because each system is unique and does not form a coherent class like the transient BHC. A total of 10% of the LMXRB population are in globular clusters, so at least 1-2 transient BHC should be similarly located if they are a related population; to date no BHC has been identified in a globular cluster.

4. Discussion

The *ultra-soft/ultra-hard* X-ray spectral signature of accreting black holes has been used to identify 17 BHC in X-ray binaries. These, combined with the six established from optical radial velocity measurements, brings the total number of BHC to 23, ~10% of the X-ray binary population. A notable feature is that most BHC have late type companions (~20 of 23 systems) *i.e.* about 20% of the LMXRB population. A large fraction of the LMXRB BHC (15 of 20 systems) are transient systems. Compared to the overall population of LMXRB, there is a deficiency of the transient BHC at the galactic center, in the galactic

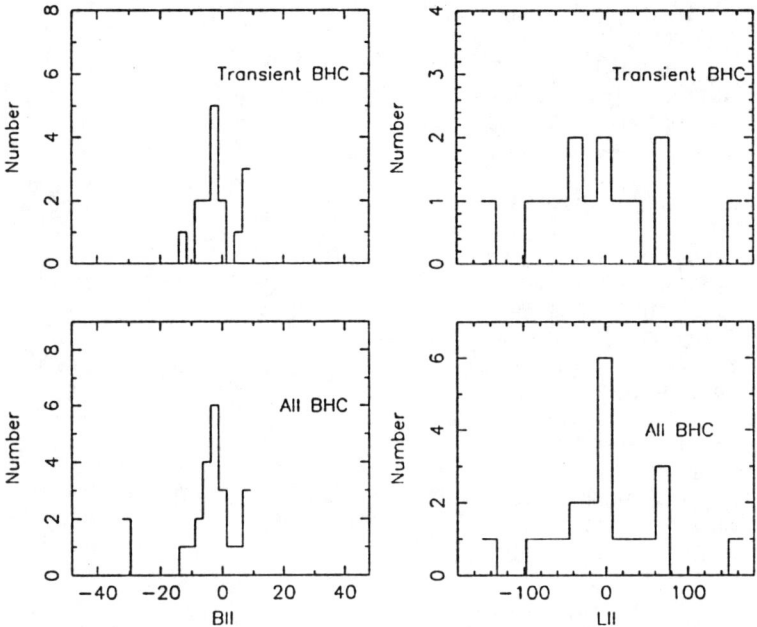

Figure 3: The Galactic longitude and latitude distribution of black hole candidates. The upper plot shows the transient Low Mass X-ray Binaries and the lower plot the distribution for all black hole candidates. The longitude distribution only include systems at $bii\pm30°$.

plane and in globular clusters. This conclusion is based on a relatively small number of objects, and as more transient systems are found, different trends may develop. There may be a bias in the distribution since there is a tendancy to detect the closest, brightest transient BHC systems. But at the current time there does appear to be a difference in the galactic distribution of the LMXRB containing BHC vs. those containing neutron stars. This suggests the LMXRB BHC are more closely aligned with a population I distribution and that the black hole progenitors were population I massive OB stars, with a late type companion. The current close orbital configuration probably arose because of a spiral-in when the OB star atmosphere expanded and enveloped the low mass companion (Thorne and Zytkow 1977).

REFERENCES

Ballet et al. (1993), IAUC 5874
Bolton, C.T., 1972, Nature, 235, 271
Bouchet, L. et al. 1991, ApJ, 383, L45
Casares, J., Charles, P.A., & Naylor, T., 1992, Nature, 355, 614
Coe, M.J., Engel, A.R., & Quenby, J.J., 1976, Nature, 259, 544
Cowley, A.P., et al. 1983, ApJ, 272, 118
Cowley, A.P., et al. 1987, ApJ, 296, 299

Cowley, A.P., 1994, these procedings
Doxsey et al. (1977), IAUC 3113
Eachus et al. (1976), ApJ, 203, L17
Ebisawa, K. 1994, these procedings
Elvis et al. (1975), Nature, 257, 656
Griffiths et al. (1977), IAUC 3110
Harmon et al. (1992), IAUC 5504
Harmon et al. (1993), IAUC 5874.
Hutchings, J.B. et al. 1987, AJ, 94, 340
Israel, G.L., Mereghetti, S., & Stella, L., 1993, IAUC no 5889
Jones et al. (1976), ApJ, 364, 664
Kaluzienski et al. (1975), ApJ, 201, L121
Kaluzienski & Holt (1977), IAUC 3099 & 3106
Kaluzienski & Holt (1977), IAUC 3104
Kitamoto et al. (1984), PASJ, 36, 799
Lapshov et al. (1993), IAUC 5864
Lund & Brandt (1991), IAUC 5161
Makino (1987), IAUC 4342
Makino (1988a), IAUC 4587 & 4600
Makino (1988b), IAUC 4653
Makino (1989) IAUC 4782 & 4786
Makino (1991), IAUC 5161
Matilsky et al. (1972), ApJ, 174, L53
McClintock, J.E., & Remillard, R. 1986, ApJ, 308, 110
McClintock, J.E., Bailyn, C., & Remillard, R. 1992, ApJ, 399, L145
McCray, R. & Lamb, F.K., 1976, ApJ, 335, 755
Mitsuda, K., et al. 1989, PASJ, 41, 97
Ostriker, J.P. 1977, Ann. New York Ac. Sci., 302, 229
Paciesas et al. (1992), IAUC 5580
Parmar et al. (1986), ApJ, 304, 664
Parmar et al. (1993), A&A, 279, 179
Priedhorsky & Holt (1987), Space Sc. Rev., 45, 291
Sunyaev, R.A., & Trümper, J. 1979, Nature, 279, 506
Sunyaev, R.A., et al. 1991a, ApJ, 383, L49
Sunyaev, R.A., et al. 1991b, Sov. Astron. Lett, 17(2), 123
Tanaka, Y. 1989, in: Two Topics in X-ray Astronomy, 23rd ESLAB Symposium, Bologna, Italy, (ESA, SP296), p. 3
Tanaka, Y. & Lewin, W.H.G., 1994, in: X-ray Binaries, ed W.H.G. Lewin, J., van Paradijs & E.P.J. van den Heuvel (Cambridge:CUP)
Tennant, A.F., Fabian, A.C., and Shafer, R.A. 1986, MNRAS, 219, 871
Thorne, K.S. and Zytkow, A.N., 1977, AJ, 212, 832
Van Paradijs, J., 1983, in: Accretion Driven X-ray Sources, ed. W.H.G. Lewin and E.P.J. van den Heuvel (Cambridge: CUP) p. 189
Van Paradijs, J., 1994, in: X-ray Binaries, ed W.H.G. Lewin, J., van Paradijs & E.P.J. van den Heuvel (Cambridge: CUP)
Webster, B.L. & Murdin, P., 1972, Nature, 235, 37
White, N.E., & Marshall, F.E., 1984, ApJ, 281, 354
White, N.E., Kaluzienski, J.L., & Swank, J.H., 1984, in: High Energy Transients in Astrophysics, ed S.E. Woosley (AIP, New York) p. 31
White, N.E., Nagase, F., & Parmar, A., in: X-ray Binaries, ed W.H.G. Lewin, J., van Paradijs & E.P.J. van den Heuvel (Cambridge: CUP)
Wilson, C.K., and Rothschild, R.E. 1983, ApJ, 274, 717

OBSERVATIONAL CONSTRAINTS ON THE MODELS OF DISK ACCRETION ONTO A BLACK HOLE

Sergei A. Grebenev, Rashid A. Sunyaev & Mikhail N. Pavlinsky
Space Research Institute, Profsoyuznaya 84/32, 117810 Moscow, Russia

ABSTRACT

Observations of Galactic black hole candidates performed with the ART-P telescope aboard *Granat* in 1990-1992 focused our attention on studying of the sources spectral variability vs. the accretion rate. We show that in contrast with the current scenario of disk accretion onto a black hole the hard spectral state of the sources always occurs at smaller accretion rate than the soft state. The arguments are presented that hard X-rays detected during the "very high" and "low" states of black hole candidates have a distinct origin.

1. INTRODUCTION

Two emission components (soft and hard) observed in X-ray/γ-ray spectra of many stellar mass black hole candidates are usually explained in terms of two physically distinct regimes of disk accretion onto a black hole (Shakura & Sunyaev, 1973). The soft component bright at the energies $h\nu \leq 8$ keV is assumed to be produced in the outer optically opaque disk region. It gives a dominant contribution to the source overall luminosity only at small accretion rates ($\dot{m} = \dot{M}/\dot{M}_{ed} \ll 1$ where $\dot{M}_{ed}$ is the Eddington critical accretion rate). On the contrary, the hard X-ray emission originates in the hot ($kT_e \sim 50$ keV) optically thin plasma of the innermost region. The observed spectrum extended up to $h\nu \sim 200$ keV is formed due to the inverse Compton scattering of low energy photons by high temperature electrons (Shapiro *et al.*, 1976; Sunyaev & Trümper, 1979). In the case of accretion onto a Schwarzschild black hole, the hard component appears in the spectrum if the accretion rate $\dot{m}$ exceeds $1/70 \, (\alpha m_*)^{-1/8}$ (Shakura & Sunyaev, 1973). Here $m_* = M/10 \, M_\odot$ is the black hole mass, α - the turbulence parameter.

Long-term observations of several Galactic black hole candidates performed with ART-P/*Granat* in 1990-1992 provided a unique opportunity to study spectral variability of the sources vs. the accretion rate. In particular, X-ray luminosity of GX339-4 was measured during its different states. The drop by a factor of 10 in X-ray brightness of 1E1740.7-2942 was discovered. However the most spectacular was the behaviour of Nova Muscae 1991 (GRS1124-68). The source exhibited step-by-step three distinct spectral states (two-component, soft and hard) while its overall luminosity declined gradually. It was a great surprise for us that in contrast with the existing accretion disk models the hard state of these sources always occurred at smaller accretion rates than the soft one.

2. X-RAY OBSERVATIONS OF BLACK HOLE CANDIDATES

Bimodal X-ray behaviour was first discovered in Cyg X-1, a most reliable and best studied black hole candidate. More than 90% of the time the source spends in a "low" state of intensity exhibiting a hard power-law spectrum with the photon index $n \sim 1.6$. During a transient "high" state the flux in the 2-10 keV band increases by a factor of 5 and the spectrum steepens to $n \sim 4$ (Tananbaum et al., 1972). It is difficult to say in which of the two states the source overall luminosity is higher. The low-state spectrum integration shows that the hard X-ray band contains the major part of the luminosity, $\sim 3 \times 10^{37}$ $(d/2.5 \text{ kpc})^2$ ergs s^{-1} where d is the distance to the source (see e.g. Grebenev et al., 1993 for the results of Granat measurements in the 3-200 keV energy band). This is roughly 50 times lower than the critical Eddington luminosity for a 10 $M_\odot$ black hole ($L_{ed} \simeq 1.3 \times 10^{39}$ m_* ergs s^{-1}), i.e. $\dot{m} \leq 0.02/m_*$. In the high state the contribution of the soft $h\nu \leq 2$ keV X-ray and EUV spectral bands becomes significant, so the determination of the bolometric luminosity is complicated. As a rule high-state observations of Cyg X-1 were incomplete and performed in the narrow 2-10 keV band.

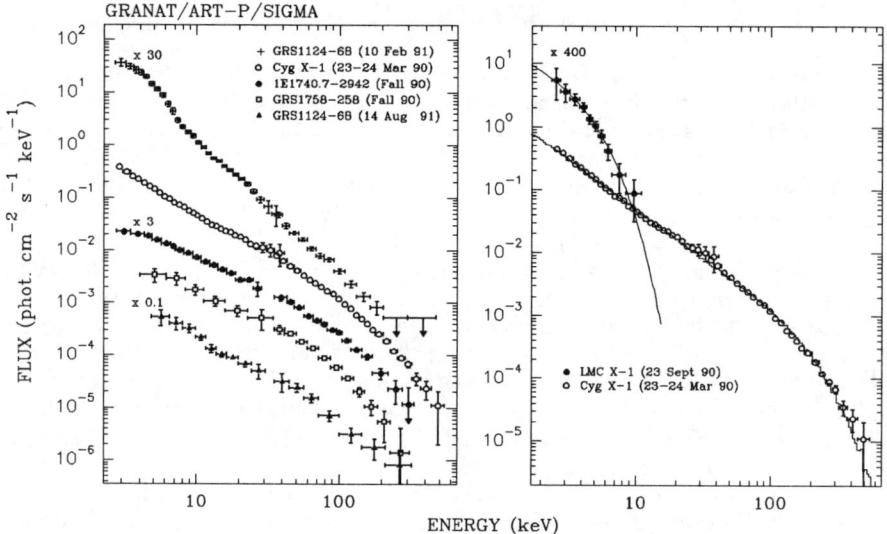

Fig. 1. The hard-state spectra of several black hole candidates (left panel) and the soft-state spectrum of LMC X-1 (right panel) obtained with Granat and given in comparison with the spectrum of Cyg X-1, the most reliable black hole candidate.

Recently two other X-ray sources were suspected to be good potential black hole candidates (Sunyaev et al., 1991). Those are 1E1740.7-2942, the hardest source in the Galactic center region, and GRS1758-258, a source discovered with Granat in the vicinity of GX 5-1. Both sources have a Comptonized Cyg X-1 type spectrum (see left panel in Fig.1) slightly flatter in the standard X-ray band ($n \simeq 1.4$). Their 3-200 keV luminosity is normally also very close to that of Cyg X-1, 2.7 and 1.9×10^{37} $(d/8.5 \text{ kpc})^2$ ergs s^{-1} for 1E1740.7-2942 and GRS1758-258 respectively (Grebenev et al., 1993). Note that no essential spectral change was detected during the drop of 1E1740.7-2942 intensity occurred in the beginning of 1991. The source remained in the hard state in spite of the total luminosity decrease by a factor of 10 (Pavlinsky et al., 1993).

Unlike these Galactic sources, LMC X-1 and LMC X-3, two black hole candidates located in the Large Magellanic Cloud, spend most of the time in the soft state. In Fig.1 (right panel) the typical photon spectrum of LMC X-1 obtained with ART-P in the fall of 1990 is presented in comparison with that of Cyg X-1. It was scaled for clarity to the 2.5 kpc distance of Cyg X-1 (by 400 times !). According to White & Marshall (1984) the luminosity of both sources is equal to $\sim 2 \times 10^{38}$ $(d/50 \text{ kpc})^2$ ergs s^{-1} in the 0.5-60 keV band. Thus it corresponds to the accretion rate $(\dot{m} \sim 0.2/m_*)$ one order of magnitude higher than that typical for the hard-state sources.

Due to the dispersion in $\dot{M}_{ed}$ values of individual sources caused by the obvious variety in black hole masses the observation of spectral transitions from the same source should be most indicative (from the point of view of the spectral hardness-accretion rate analysis). GX339-4 provides us with such a possibility. In 1990-1992 all the three known spectral states of the source (soft, hard and off) were investigated with ART-P (Grebenev et al., 1991, 1993). The spectra measured during these states are given in the left panel of Fig.2. The luminosity observed above 3 keV was equal to 0.5, 1.9 and 0.07×10^{37} $(d/4 \text{ kpc})^2$ ergs s^{-1} in the soft, hard and off (3-σ upper limit) states respectively. It is important to emphasize that during the soft state only a small part of the total disk luminosity was detectable in the rather hard 3-35 keV energy band of ART-P. Approximation of the GX339-4 spectrum with the disk black-body model allows its overall luminosity to be evaluated using the best-fit parameters, the maximum black-body temperature of the disk, kT_{max}, and the radius where it is reached, r_{max}. Such an estimate, $L_B \simeq 40\, r_{max}^2 \sigma T_{max}^4 \simeq 5.5 \times 10^{37} \cos(i/60°)^{-1}$ ergs s^{-1} (Grebenev et al., 1991) where i is the disk inclination and σ is the Stefan-Boltzmann constant, shows that more than 90% of the disk soft-state luminosity is emitted in the soft $h\nu \leq 2$ keV X-ray and EUV bands*. The measurements performed with *Ginga* in 1988 during the very high state of GX339-4 demonstrated that its 1-37 keV luminosity could reach 6×10^{37} ergs s^{-1} (Miyamoto et al., 1991). In this very high state the source had a two-component spectrum.

2. BRIGHT X-RAY NOVAE

The observations carried out with X-ray instruments aboard *Mir-Kvant*, *Ginga*, *Granat* and *Compton Observatory* in 1987-1993 indicated that bright X-ray novae flared up in the few kpc vicinity of the Earth approximately once a year. These sources are likely to form the most numerous class of the Galactic binaries containing black holes (Grebenev et al., 1992). Studying of X-ray novae in quiescent when the accretion disk contributes nothing to the optical light allowed the mass functions for the most of them to be determined. The mass lower limits of the compact objects in A0620-00, GS2000+25, GRS1124-68 and GS2023+338 were estimated to be greater than $3 M_\odot$, the maximum mass of a neutron star.

Nova Muscae 1991 (GRS1124-68) was intensively studied with ART-P during 9 months after the outburst (Grebenev et al., 1992). The observed spectra

---

* The contribution of the soft X-ray and EUV bands to the source hard-state luminosity can be evaluated in the same manner. Let us assume that low-energy photons with the black-body spectrum are responsible for 1/4 (3-σ upper limit) of the GX339-4 photon flux, I_s $[phot \cdot cm^{-2} s^{-1} keV^{-1}]$, measured in the softest ART-P channel at $E_s \simeq 3$ keV. Maximum luminosity of the soft source occurs if the black-body temperature $kT_{bb} = E_s/4$. Thus $L_{bb} \leq 2\pi/15(\pi e/4)^4\, 4\pi d^2 E_s^2 (I_s/4) \cos(i)^{-1} \simeq 9 \times 10^{36}$ $(d/4 \text{ kpc})^2 \cos(i/60°)^{-1}$ ergs s^{-1}.

can be divided up into 3 groups distinctive by both the spectral type and observing epoch (the right panel of Fig.2). The spectrum was either two-component with a bright disk black-body component and a steep ($n \simeq 2.5$) hard power-law tail (days 8-33 after the outburst), or soft, similar to the high-state spectrum of GX339-4 (day 94 - the only *Granat* observation occurred that time), or hard, characterized by $n \sim 1.6$ and resembling the low-state spectra of Cyg X-1 or GX339-4 (days 143-219). The described spectral variability reproduced mainly irregular changes in the hard emission component. The soft 3-8 keV flux decreased exponentially with the e-folding time of about 35-40 days during the whole time of observations with the exception of the day ~ 70 when a sudden flux increase by 2.3 times ("kick" in the light curve) was detected with *Ginga* (Ebisawa et al., 1994). It is natural to assume that the monotonous decay of the soft X-ray light curve (and the observed succession of spectral transitions) reflects a gradual decrease of $\dot{m}$. The 3-200 keV luminosity of GRS1124-68 was equal to 3.7, 0.23 and 0.26×10^{37} $(d/3 \text{ kpc})^2$ ergs s^{-1} in the very high, high and low states respectively (Grebenev et al., 1992; Sunyaev et al., 1992). For the luminosity in the off-state (observed after the source switch-off) the 3-σ upper limit, 8×10^{35} $(d/3 \text{ kpc})^2$ ergs s^{-1}, was obtained. According to *Ginga* performing measurements in the softer $h\nu \geq 1$ keV band the source luminosity was at the level of 30 and 6×10^{37} $(d/3 \text{ kpc})^2$ ergs s^{-1} nearly about the dates when the very high and high states were observed with ART-P (Ebisawa et al., 1994).

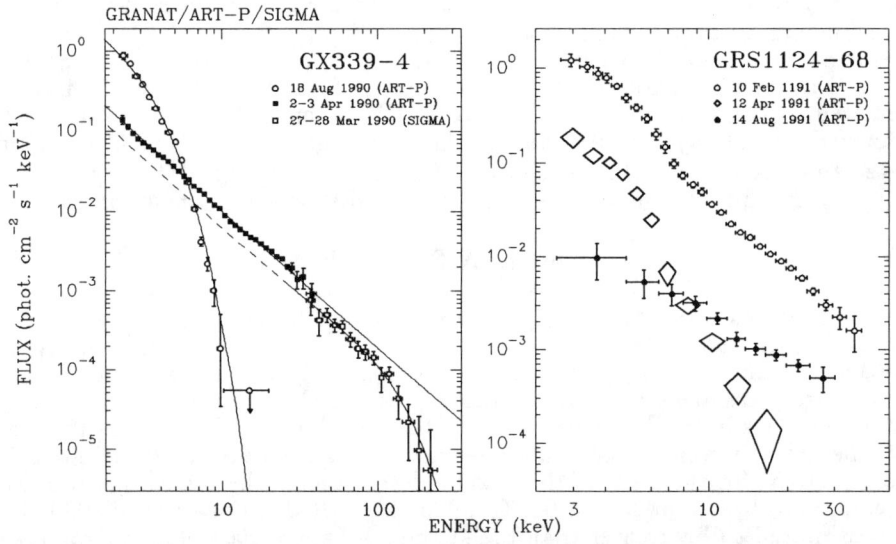

Fig.2. Photon spectra of GX339-4 and Nova Muscae (GRS1124-68) measured with *Granat* during the various states of the sources

3. ACCRETION DISK MODELS VS. OBSERVATIONS

The results of observations discussed above can be summarized as follows:

Depending on $\dot{m}$ stellar mass black hole candidates spend most of the time in one of the following 4 spectral states: very high state occuring at $0.2 \leq \dot{m} \leq 1$ and characterized by the two-component spectrum, high state ($0.02 \leq \dot{m} \leq 0.2$, ultrasoft spectrum), low state ($0.001 \leq \dot{m} \leq 0.02$, hard Comptonized spectrum) or off-state ($\dot{m} \leq 0.001$, unknown spectrum). Thus, in contrast with the current scenario of disk accretion onto a black hole, the low (hard) state of the sources always occurs at smaller accretion rates than the soft one.

It is typically assumed that the hard X-ray spectrum of black hole candidates is produced in the hot optically thin region of the disk, in which ions have temperatures significantly exceeding electron temperatures and give the dominant contribution to the pressure. Shapiro et al. (1976) pointed out that such "two-temperature" region could be formed in the inner part of the standard cool disk at high accretion rates when the radiation pressure becomes equal to the gas pressure, $\dot{m} \geq 1/70\,(\alpha m_*)^{-1/8}$ (hereafter it is adopted that $r \simeq 6 r_g$ corresponding to the maximum energy release, r_g is the gravitational radius). The secular (Lightman & Eardley, 1974) and thermal (Shakura & Sunyaev, 1976) instabilities present in this radiation-pressure dominated part should swell it to the hot gas-pressure dominated optically thin region.

The data of X-ray observations considered above can not be explained in the framework of this model. The observations unambiguously show that the disk becomes really hot (the hard state) only at small accretion rates and that this hot region should occupy the extended part of the disk (hard X-rays take away at least 70% of the total disk luminosity). The two-temperature solution formally valid outside the inner disk region as well as within it can still be the best solution to explain the observed spectrum and the structure of the hard state. However the reason why the disk turns out to be in the hot state rather than in the cool one in this case is uncertain. Moreover under the conditions when the whole disk is hot the production of soft photons which are necessary for effective disk cooling by Comptonization becomes rather a complicated problem.

The hard spectral component observed from black hole candidates in the very high state differs drastically from that in the low state. It is steeper ($n \sim 2.5$), contains only a small ($\leq 10\%$) portion of the total disk luminosity and is likely formed within the disk region restricted in size and being under physically distinct conditions. Note that at high accretion rates typical for the state ($\dot{m} \geq 0.2$) the two-temperature solution is not longer valid and self-consistent because it formally leads to the geometrically thick and radiation-pressure dominated disk. We assume that in this case hard X-rays are produced in the inner compact ($r \leq 90 r_g\,\dot{m}^{64/93} m_*^{2/93} \alpha^{34/93}$) optically thin region of the disk. Such a region arises within the radiation-pressure dominated part of the standard cool disk at high accretion rates, $\dot{m} \geq 1/14\, m_*^{-1/32} \alpha^{-17/32}$ (Shakura & Sunyaev, 1973). The observed hard X-ray spectrum could be formed in this high temperature region due to Comptonization of low-energy photons.

Finally we propose the following receipt to reconcile accretion disk models with X-ray observations:

1). *very high/two-component spectral state* ($\dot{m} \geq 1/14\, m_*^{-1/32} \alpha^{-17/32}$) - The disk consists of the cool optically thick extended outer region producing soft X-rays with the modified black-body spectrum (Thomson scattering dominates in opacity) and the hot optically thin radiation-pressure supported inner region producing hard X-rays.

2). *high/soft state* - The whole disk is cool and optically thick. In its central ($r \leq 130 r_g\,\dot{m}^{16/21} (\alpha m_*)^{2/21}$) part the radiation pressure exceeds the gas one whenever $\dot{m} \geq 1/70\,(\alpha m_*)^{-1/8}$ (Shakura & Sunyaev, 1973). The change in the disk color temperature occurred near the outer boundary of this radiation-pressure dominated region can be responsible for the "kick" observed in the light curves of X-ray novae.

3). *low/hard state* - The whole disk (or its significant part) is hot and optically thin supported by the ion pressure and keeping the electron temperature nearly independent of r, $\dot{m}$ and α (the "two-temperature" solution).

4). *off-state* - Currently there is no available information. The "off-state" could mean the source transition to another spectral state, *e.g.* again to the cool standard disk emitting in the unobservable soft X-ray and EUV bands.

We are still far from understanding of all details of disk accretion onto a black hole. Reporting here very briefly the results of the existing accretion disk models comparison with X-ray observations and giving in outline the possible scenario for disk accretion, we are planning to present the details of our analysis in a subsequent paper (Grebenev & Sunyaev, 1994).

REFERENCES

Ebisawa, K., *et al.* 1994, ApJ, submitted
Grebenev, S., Sunyaev, R., Pavlinsky, M., & Dekhanov, I. 1991, Sov.Astr.Lett., 17, 985
Grebenev, S., Sunyaev, R., & Pavlinsky, M. 1992, Sov.Astr.Lett., 18, 11; in Workshop on Nova Muscae 1991, ed. S.Brandt (Lyngby: DRI), p.19
Grebenev, S., *et al.* 1993, A&AS, 97, 281
Grebenev, S., & Sunyaev, R. 1994, ApJ, in preparation
Lightman, A.P., & Eardley, D.M. 1974, ApJ, 187, L1
Miyamoto, S., *et al.* 1991, ApJ, 383, 784
Pavlinsky, M., Grebenev, S., & Sunyaev, R., 1993, ApJ, in press.
Shakura, N.I., & Sunyaev, R.A. 1973, A&A, 24, 337
Shakura, N.I., & Sunyaev, R.A. 1976, MNRAS, 175, 613
Shapiro, S.L., Lightman, A.P., & Eardley, D.M. 1976, ApJ, 204, 187
Sunyaev, R.A., & Trümper, J. 1979, Nature, 279, 506
Sunyaev R., *et al.* 1991, A&A, 247, L29
Sunyaev R., *et al.* 1992, ApJ, 389, L75
Tananbaum, H., Gursky, H.H., Kellog, E., Giacconi, R., & Jones, C. 1972, ApJ, 177, L5
White, N.E., & Marshall, F.E. 1984, ApJ, 281, 354

A SYSTEMATIC AND STATISTICAL STUDY OF THE X-RAY AND OPTICAL LIGHT CURVES OF X-RAY NOVAE

Wan Chen[1] Chris Shrader[2] and Mario Livio[3,4]

[1] USRA/Goddard Space Flight Center, Code 661, Greenbelt, MD 20771
[2] CSC/Goddard Space Flight Center, Code 668, Greenbelt, MD 20771
[3] Space Telescope Science Institute, Baltimore, MD 21218
[4] Dept. of Physics, Technion, Haifa 32000, Israel

ABSTRACT

We have collected data from the literature of the past two decades, and are carrying out for the first time a systematic, statistical study of all the documented long-term x-ray and optical light curves of x-ray nova outbursts. This study can yield information which is critical to our understanding of these systems, as well as provide us with the necessary empirical ingredients for detailed theoretical modeling efforts. In this paper we discuss the basic properties of the light curves. Some distinguished features not well-known before are also identified and briefly discussed.

1. INTRODUCTION

As a subclass of low-mass x-ray binaries (LMXBs), the x-ray nova (also called soft x-ray transient) systems have generated a great interest in recent years, because dynamical studies indicate that a large fraction of them contain a black hole (BH). In fact, all the dynamical BH candidates found in LMXBs are x-ray novae (XNs) [1].

An XN system usually remains in a low state (quiescence) with little detectable x-ray radiation. However, episodic large accretion events may force the system into a major outburst to become temporarily (sometimes) the brightest source in the x-ray sky. The soft x-ray flux we observe during a XN outburst traces the accretion rate through the inner edge of the accretion disk about the compact object, while the optical flux reflects the physical conditions at the outer edge of the disk in its high state. Thus, the x-ray and optical light curves of XN outbursts contain valuable information on the accretion process and the time dependent behavior of the accretion disk during outburst.

The historical XNs, discovered mostly by the Vela and Ariel 5 satellites before 1984, have been studied by White et al. [2] and van Paradijs & Verbunt [3]. Since 1988, the all-sky x-ray monitors on board Ginga, GRANAT, and recently on GRO, have been detecting bright x-ray novae at a rate of approximately one per year. Now it is high time for a systematic investigation of the XN phenomenon. We have started to collect all the existing data of XNs [2]–[5], including both the historical and the newly discovered ones, for a statistical study and an attempt to confront it with the current theoretical thinking about XNs.

2. BASIC X-RAY NOVA PROPERTIES

During quiescence which occupies most of their life time, majority XNs are not detected in x-rays. Their faint secondaries dominate in the optical. Occasionally, large accretion events onto the compact object take place and lead to a sudden increase of x-ray flux by more than 2–3 orders of magnitude within a few days, followed by an exponential decay with an e-folding time scale of ∼ 30–60 days [4]. In the optical, XNs may also brighten by more than 7 magnitudes (now disk emission dominates) and then fade away more slowly than the x-rays.

The x-ray light curves of XNs can thus be characterized by a fast increase and a much slower decay. After reaching the peak flux, some systems stayed at the peak for an extended period of time (∼ a few to 20 days), thus showing a plateau in their light curves. In other cases, the x-ray flux almost immediately starts to decline after the peak.

In several cases, the decay of the x-ray flux was interrupted by secondary flares which have been called in the literature "secondary maxima" [6], or "glitches" [7].

In Table 1 we list some preliminary results of the basic properties of some well-known XN light curves.

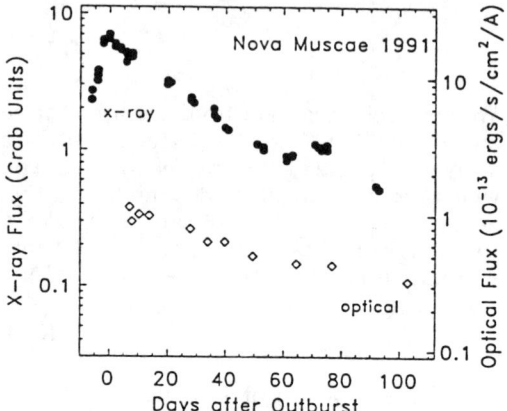

Fig. 1. X-ray and optical light curves of X-ray Nova Muscae 1991.

3. SOME DISTINGUISHED FEATURES

Recurrence Time Scales. Many XNs have been detected over the last two decades to have more than one outburst. One system, A0620-00, has been found to have a similar outburst in the optical in 1916. The reported recurrence time scale in different systems varies by a factor of 100, from 200 days to 60 years. A closer look at the recurrent XNs (Table 1), however, reveals that there are two types of outbursts, "major" ones and "minor" ones, differing by about an order of magnitude in their peak x-ray fluxes. The recurrence time scale between major outbursts is long (≫ 10 years), while that between minor outbursts is much shorter.

Major and Minor Outbursts. Although much brighter, a major outburst

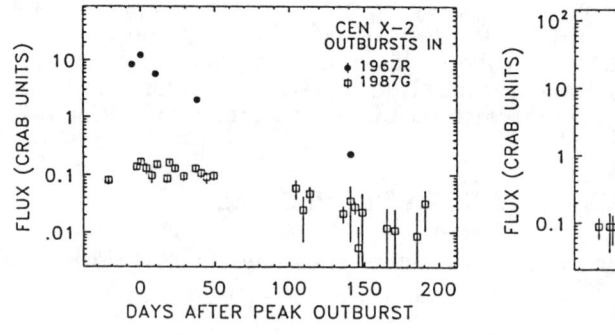

Fig. 2.

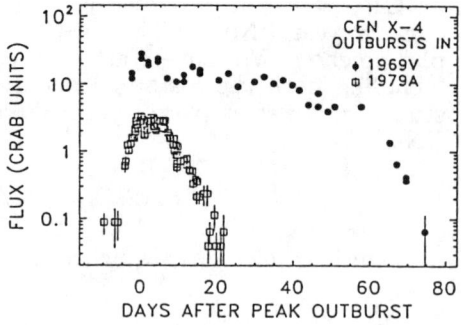

Fig. 3.

does not always decay more slowly (or faster) than a minor one, as shown in Fig. 2 and Fig 3. On the other hand, the minor outbursts from the same system are almost identical (Fig. 4). We point out that the most convincing BH candidate XNs (A0620-00, X-ray Nova Muscae 1991, GS2000+25, and V404 Cyg) have not yet been observed to display minor outbursts. Are their minor outbursts always too weak to be detected? Or they do not have minor outbursts at all?

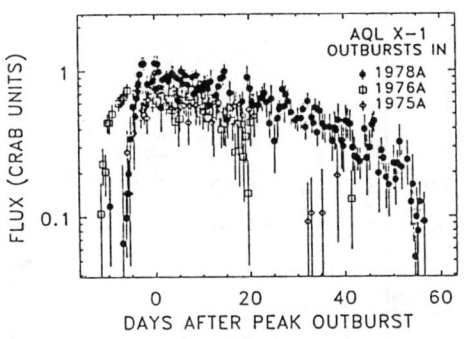

Fig. 4.

4. DISCUSSIONS

As an initial step in a major effort to study systematically the XN outbursts, we have shown in this work that the x-ray/optical light curves of XN outbursts are characterized by a variety of distinguishing features. Some of them (major andminor outbursts, plateaus, and secondary maxima) have not been well recognized before. Much future work will be required in terms of the identifcation of all the light curve characteristics, the establishment of their relation to system parameters and in theoretical modeling.

The reason for the occurrence of major and minor outbursts in a single source is unknown, but it is perhaps worth noting that in the disk instability models for XNs, calculations do suggest that there could be many minor outbursts occurring between two major outbursts.

It is still not clear why there are plateaus in the early stages of the outbursts and secondary maxima in the later decay phases. We suspect that the x-ray heating of the accretion disk early on in the outburst may contribute to a prolonged peak, and the secondary maxima are due to mass transfer events caused by x-ray heating of the secondary star during the main outburst [6]-[7].

REFERENCES

[1] Cowley, A. 1992, ARA&A 30, 287
[2] White, N.E., Kaluzienski, J.L., & Swank, J.H. 1984, in High Energy Transients in Astrophysics, ed. S. E. Woosley (New York: AIP), 31
[3] van Paradijs, J., & Verbunt, F. 1984, in High Eenergy Transients in Astrophysics, ed. S.E. Woosley (New York: AIP), 49
[4] Bradt, H.V.D., & McClintock, J.E. 1983, ARA&A 21, 13
[5] van Paradijs, J., & McClintock, J.E. 1993, in X-Ray Binaries, eds. W. Lewin, J. van Paradijs, & E. van den Heuvel (Cambridge: Cambridge Univ Press), in press
[6] Chen, W., Livio, M., & Gehrels, N. 1993, ApJ 408, L5
[7] Augusteijn, T., Kuulkers, E., & Shaham, D. 1993, A&A, in press

Table 1. Basic Properties of X-ray Novae

Source	Type	Year	Energy Band	T_{rise} (days)	T_{decay} (days)	$T_{plateau}$ (days)	Peak F_x (Crab)	Peak m_v (mag)	Note
GRO J 0422+32	BH?	1992	BATSE	5 ± 1	70 ± 10	15 ± 5	3		XN Per '92
A 0620-00	BH	1916	V-band					12.1	
		1975	Ariel 5	2.4 ± 0.2	37 ± 1	< 2	50	11.2	XN Mon '75
			B-band		96 ± 17				
GS 1124-68	BH	1991	Ginga	3.0 ± 0.3	39 ± 9	< 2	3	13.6	XN Mus '91
1353-64	BH?	1967	Rockets		30 ± 5	< 10	20		Cen X-2, BW Cir
		1987	Ginga	5 ± 2	41 ± 5	?	0.2	16.9	
1456-32	NS	1969	Vela 5	2.8 ± 4.2	12 ± 17	48 ± 5	25		Cen X-4, V822
		1979	Ariel 5	2.1 ± 0.4	5 ± 0.5	7 ± 3	4		
			V-band	?	8.2 ± 2			12.8	
1524-62	BH?	1974	B-band	6 ± 2	5 ± 2	10 ± 3	0.9	17.5	TrA X-1
4U 1543-47	BH?	1971	UHURU	?	65 ± 30	20 ± 5	15		
		1975	Ariel 5	18 ± 5	77 ± 22	?	15	14.9	
		1992	BATSE	0.8 ± 0.3	1.4 ± 0.4	< 1	?		
3U 1630-47	BH?	1970	UHURU	10	106 ± 233		?		
1742-289	NS	1975	Ariel 5	5 ± 2	55 ± 10	< 3	5		N Oph '77
1908+005	NS	1975	Ariel 5	4.5 ± 2.9	21 ± 7	20 ± 3	1.2	14.8	Aql X-1, V1333 Aql
		1976	Ariel 5	5.2 ± 3	27 ± 13	16 ± 2	1.1		
		1978	Ariel 5	2.3 ± 0.7	31 ± 5	15 ± 3	1.3		
GS 2000+25	BH?	1988	Ginga	2.5 ± 0.3	37 ± 4	< 3	10	16.2	XN Vul '88, QZ Vul
GS 2023+338	BH	1989	Ginga	5 ± 2	35 ± 10			12.7	V404 Cyg

SUPERMASSIVE BLACK HOLES WITH STELLAR COMPANIONS

Philipp Podsiadlowski and Martin J. Rees
Institute of Astronomy, Cambridge CB3 0HA, U.K.

ABSTRACT

Low-mass stars in close orbits around supermassive black holes in active galactic nuclei (AGN) may provide an important signature for testing the existence of supermassive black holes. Here we discuss how such systems can form by being "ground down" by their interaction with an AGN accretion disk and discuss the peculiarities of such binaries with extreme mass ratios. In particular, we show that, in the case of a low-mass main-sequence companion, the orbit has to be circularized to avoid tidal disruption. If the companion is filling its Roche lobe, mass loss will occur through the outer Lagrangian point rather than the inner because of irradiation effects (this may also be important for some binary pulsars like PSR 1957+20). Finally, we discuss possible signatures of such systems.

1. INTRODUCTION

While it is widely believed that accretion onto supermassive black holes (with mass $M \sim 10^6 - 10^8 \, M_\odot$) is responsible for the observed activity of active galactic nuclei (AGN), so far this has not been conclusively proven. Here, we investigate the possibility that a star in a close orbit around a supermassive black hole may provide an observable signature of such systems (also see Syer, Clarke & Rees 1991; Sikora & Begelman 1992; King & Done 1993; Tsuruta et al. 1993).

We consider a supermassive black hole of mass M, surrounded by a standard accretion disk and orbited by a star of mass m and radius R. In order for the star to affect the accretion flow onto the black hole, the star has to be sufficiently close to the black hole that the amount of disk material the star interacts with per orbit is comparable to the central accretion rate, $\dot{M}$, i.e., $\Sigma(r) \pi R^2 / P(r) \sim \dot{M}$, where $\Sigma(r)$ and $P(r)$ are the surface density and Keplerian period at radius r, respectively[†]. For a standard, Newtonian α-disk, dominated by gas pressure and electron scattering, this defines a characteristic separation, $r_{\rm ch}$, at which the star can modulate the accretion flow:

$$r_{\rm ch} \simeq 2 \, {\rm AU} \, R_1^{20/21} \, M_{10^7}^{1/3} \, \dot{M}_{10^{-3}}^{-4/21} \, \alpha^{-8/21}. \qquad (1)$$

Here, $R_1 \equiv R/R_\odot$, $M_{10^7} \equiv M/10^7 \, M_\odot$, $\dot{M}_{10^{-3}} \equiv \dot{M}/10^{-3} \, M_\odot \, {\rm yr}^{-1}$, and α is the viscosity parameter. This separation must lie outside the separation,

$$r_{\rm RL} \simeq 2 \, {\rm AU} \, R_1 \, M_{10^7}^{1/3} \, m_1^{-1/3}, \qquad (2)$$

[†] This criterion for observable effects is probably sufficient, but may not be necessary; indirect effects, e.g., the screening of the central black hole by material thrown out of the disk may already be important at lower surface densities.

at which the star would fill its Roche lobe ($m_1 \equiv m/M_\odot$). The orbital period at this characteristic separation is $\sim 8\,\mathrm{hr}$ for our canonical parameters, and the corresponding Keplerian velocity is $v_\mathrm{ch} \sim 0.2c$, where c is the speed of light.

2. THE FORMATION OF SUPERMASSIVE BLACK-HOLE BINARIES

2.1 The "grinding-down" process

Syer et al. (1991) have shown that stars can be "ground down" into short-period orbits by their repeated interactions with a black-hole accretion disk. A characteristic grinding-down time scale, t_grind, can be defined as $t_\mathrm{grind}/P \equiv m/\Delta m\, x$, where Δm is the amount of disk material the star interacts with per orbit, $x \equiv q/a$ the ratio of the separation at which the star passes through the disk to the semimajor axis of the orbit. For a standard, outer accretion disk,

$$t_\mathrm{grind} \simeq 6 \times 10^7\,\mathrm{yr}\, m_1\, R_1^{-2}\, M_{10^7}^{-3/4}\, \dot{M}_{10^{-3}}^{-7/10}\, a_{10^3\mathrm{AU}}^{9/4}\, \alpha^{4/5}\, x_{0.1}^{7/4}. \qquad (3)$$

This time scale is generally shorter than the gravitational radiation time scale, $t_\mathrm{grav} \simeq 10^5\,\mathrm{yr}\, m_1^{-1}\, M_{10^7}^{-2}\, a_{2\mathrm{AU}}^4$, even for $a = 2\,\mathrm{AU}$ where $t_\mathrm{grind} \sim 3 \times 10^3\,\mathrm{yr}$. Equation 3 defines, within a factor of a few, the characteristic time scale on which the semimajor axis, the eccentricity and the inclination of the orbital plane with respect to the disk plane decrease. Syer et al. (1991) find that the ordering of these three time scales is $a/\dot{a} < e/\dot{e} < \cos i / d\cos i/dt$. This implies that the orbit may shrink and be circularized without being dragged completely into the plane of the disk. The maximum distance, a_max, from which a star can be dragged in is given by the condition that the evolutionary time scale of the star, T, is less than t_grind. This yields

$$a_\mathrm{max} \simeq 10^4\,\mathrm{AU}\, m_1^{-4/9}\, R_1^{8/9}\, M_{10^7}^{1/3}\, \dot{M}_{10^{-3}}^{14/45}\, T_{10^{10}}^{4/9}\, x_{0.1}^{-7/9}\, \alpha^{-16/45}, \qquad (4)$$

where $T_{10^{10}} \equiv T/10^{10}\,\mathrm{yr}$. This is to be compared with the semi-major axis of orbits accessible by two-body encounters within the central stellar cluster, $a_\mathrm{cl} \sim 9\times 10^3\,\mathrm{AU}\, M_{10^7}\, v_\mathrm{esc,1000}^{-2}$, where $v_\mathrm{esc,1000} \equiv v_\mathrm{esc}/10^3\,\mathrm{km\,s^{-1}}$. The minimum separation, a_min, is determined by the condition that the star fills its Roche lobe (eq. 2). Applying these relations to different types of stars, we find:

- If the star is a low-mass main-sequence star ($m = 1\,M_\odot$, $R = 1\,R_\odot$, $T = 10^{10}\,\mathrm{yr}$), its orbit can be ground down from a typical cluster orbit to r_ch for any value of α.
- If the star is a single massive star ($m = 10\,M_\odot$), it cannot be dragged in during its main-sequence phase ($R = 10\,R_\odot$, $T = 10^7\,\mathrm{yr}$), but during its subsequent red-supergiant phase ($R = 500\,R_\odot$, $T = 10^6\,\mathrm{yr}$). After it has lost its supergiant envelope (either due to tidally induced dissipation [see below] or Roche-lobe overflow) and has been transformed into a helium star, it can be dragged in further only if $\alpha \lesssim 10^{-4}$.
- However, if the massive star is a member of a close binary and becomes a helium star due to binary interaction, it can be dragged into a very close orbit in a multi-step process for any value of α, using the gas drag caused by the primary in its supergiant phase, the secondary in its main-sequence phase and the primary in its helium-star phase (in this order).

2.2 Tidal dissipation

If the orbit of the companion is eccentric, orbital energy is dissipated in the outer parts of the companion. The tidal energy deposition rate, $\dot{E}_{\rm tid}$, averaged over an orbital period, for a non-corotating companion can be estimated as

$$\dot{E}_{\rm tid} \simeq 10^{42}\,{\rm erg\,s^{-1}}\,f_{0.1}\,g(e)\,M_{10^7}^{5/2}\,R_1^5\,a_{2{\rm AU}}^{-15/2}, \qquad (5)$$

where $f_{0.1} \equiv f/0.1$ and $g(e) \gtrsim 1$; f depends on the ratio of the duration of the tidal kick (at pericenter) to the dynamical time scale of the star and is ~ 0.1 if the two time scales are comparable (Lee and Ostriker 1986). The rate at which the deposited energy is dissipated cannot exceed the star's Eddington rate at the photosphere, $L_{\rm edd} \simeq 0.5 \times 10^{38}\,{\rm erg\,s^{-1}}\,m_1\,\kappa^{-1}$, where κ is the photospheric opacity, without inducing an evaporative wind. However, for a low-mass star at $r_{\rm ch}$, the tidal deposition rate would exceed the star's Eddington luminosity by several orders of magnitude. Thus, such a star can only exist in a close orbit if the orbit is nearly circular; otherwise it would have been tidally disrupted in its past. For a helium star the outcome is less clear, since its radius is smaller and its dynamical time scale shorter (which decreases f substantially). These effects will decrease $\dot{E}_{\rm tid}$ by several orders of magnitude making it comparable to or less than the larger Eddington luminosity of a helium star. Thus, the orbit of a helium star could still be eccentric at $r_{\rm ch}$.

The tidal circularization time scale, evaluated when the star is close to filling its Roche lobe, can be estimated as

$$\frac{t_{\rm circ}}{P} \sim 10^7 \left(\frac{M_{10^7}}{m_1}\right)^{2/3} f_{0.1}^{-1}\,[g(e)]^{-1}. \qquad (6)$$

For small eccentricities, this time scale is comparable to $t_{\rm grind}$, while for large eccentricities ($e \sim 0.5$), tidal circularization will dominate. Unlike the case of an ordinary binary where the mass ratio is of order unity, tidal circularization is relatively inefficient even when the star is close to filling its Roche lobe.

3. IRRADIATION EFFECTS

The Roche geometry of the irradiated star in a binary with an extreme mass ratio is affected by two effects. One, the potential difference between the inner and the outer Lagrangian points, L_1 and L_2, which scales with m/M, will be greatly reduced and be of order a few eV, comparable to the thermal energy of protons at the photosphere. Two, irradiation pressure on the irradiated side will become important and modify the shape of the star's equipotential surface. The potential difference between the two Lagrangian points can be estimated as

$$\Delta U \equiv U_2 - U_1 \simeq \frac{Gm}{a}\left[\frac{2}{3} - \gamma\left(\frac{m}{3M}\right)^{1/3}\right], \qquad (7)$$

where $\gamma = L\kappa/4\pi cGm = L/L_{\rm edd}^{\rm star}$, is the ratio of the black-hole luminosity, L, over the Eddington luminosity of the companion star. (Equation 7 is a good approximation if $m/M \ll 1$ and $\gamma m/M \ll 1$.) In the case of a luminous supermassive black hole (where $\gamma \sim 10^4 - 10^6$), ΔU is negative; hence, we expect that, in this case, the critical surface for the companion is the equipotential through the outer Lagrangian point and that a star filling this equipotential will lose mass through L_2 rather than L_1, leading to the formation of an excretion disk rather than an accretion disk.

Equation 7 is also applicable in the case of some ordinary binaries. For example in the case of the binary millisecond pulsar PSR 1957+20, the "black-widow" pulsar (Fruchter et al. 1988), $L_{\text{edd}}^{\text{star}} \simeq 10^{36}\,\text{erg}\,\text{s}^{-1}\,m_{0.02}\,\kappa^{-1}$, and $L \sim 2 \times 10^{35}\,\text{erg}\,\text{s}^{-1}$. Thus, γ will be larger than 2/3 if the photospheric opacity on the irradiated side is larger than ~ 10. If this is the case, mass loss may also occur through the outer Lagrangian point if the companion is filling its Roche lobe or is emitting a "slow" evaporative wind.

Irradiation effects can also affect the internal structure of a low-mass star if the irradiation temperature exceeds the ionization temperature for hydrogen ($\sim 10^4$ K; Podsiadlowski 1991). This will be the case if the orbit of the companion is inside a characteristic radius, $r_{\text{irr}} \sim 200\,\text{AU}\,L_{10^{44}}^{1/2}$, where $L_{10^{44}}$ is the X-ray luminosity of the black hole in units of $10^{44}\,\text{erg}\,\text{s}^{-1}$. At this radius, the grinding-down time scale is typically longer than the Kelvin-Helmholtz time scale for the outer layer of a solar-type star. Then, the star will have time to expand (typically by a factor of 2–4) to its new (irradiated) equilibrium state (note that this will decrease t_{grind} by roughly an order of magnitude and increase r_{ch}).

4. OBSERVABLE SIGNATURES

The most likely signature for a close companion to the black hole is a periodic modulation of the accretion flow due to the interaction of the star with the accretion disk. At the radius $r_{\text{ch}} \sim 2\,\text{AU}$, the companion will draw out of the disk a similar amount of material during each orbit as is accreted by the black hole. We expect that part of this material will be accreted by the black hole in at most a few orbits, possibly after forming a short-lived, oblique torus in a transient phase. Thus, the luminosity from the central object should be modulated with a period determined by the injection period of material into tight orbits, i.e. half the orbital period of the companion. In addition, this material is likely to be optically thick to X-rays and therefore provide a complicated screen of the central X-ray source (Sikora & Begelman 1992). The geometry of this screen will vary with the orbital period and depend, unlike the mass-feeding rate, on the Lense-Thirring precession phase (Lense & Thirring 1918). Both of these effects are likely to be of similar importance at r_{ch} and could provide a detectable, observable signature of such systems, i.e. a strictly periodic modulation of the X-ray light curve with a period of several hours to (at most) a few days.

REFERENCES

Fruchter, A. S., Stinebring, D. R. & Taylor, J. H. 1988, Nature, 333, 237
King, A. R. & Done, C. 1993, MNRAS, 264, 388.
Lee, H. M. & Ostriker, J. P. 1986, ApJ, 310, 176
Lense, J. & Thirring, H. 1918, Physik Z., 19, 156
Podsiadlowski, Ph. 1991, Nat, **350**, 654
Sikora, M. & Begelman, M. C. 1992, Nat, 356, 224
Syer, D., Clarke, C. J. & Rees, M. J. 1991, MNRAS, 250, 505
Tsuruta, S., Sivron, R., Caditz, D. & Leighly, K. 1993, preprint

X-RAY TRANSITIONS OF BLACK HOLE BINARIES AND VARIABLE α-PARAMETER DISKS

Chuan Luo, Cesar Meirelles & Edison Liang
Department of Space Physics and Astronomy
Rice University, Houston, TX 77251-1892

INTRODUCTION

The theory of limit cycle has been quite successful in applying to dwarf novae and there have also been efforts to explain X-Ray Novae in the context of instability caused by the S-shaped curve in the surface density-accretion rate plane of a Shakura-Sunyaev disk (Minishige & Wheeler[1]). However, for this kind of standard disk model it is not possible to produce a hard spectrum above 100 keV since its typical temperature is below 10^7 K. Here we try to explore the possibility to use a variable viscosity parameter α to introduce a limit cycle similar to that of dwarf novae in the hybrid disk model based on the work by Liang and Wandel[2]. The idea is that we should have an S-shaped curve leading a transition from a cold, super low state to a hot, optically thin high state using the equations proposed by Liang and Wandel[2] for disk analysis.

A variable α is not something artificial though the actual value as a function of other physical quantities is still an unsettled issue. It is reasonable to expect α to depend on the accretion rate $\dot{M}$ since physically we just cannot expect that the value of α should be constant[3,4] when the accretion rate ranges from very low to that of a maximum rate which, in some cases, could correspond to a super-Eddington luminosity. Here we discuss a probable way to address the limit cycle issue by varying α as a function of accretion rate to study the surface density-accretion rate plot for a steady disk solution.

DISK STRUCTURE

The solutions for the steady disk remains unchanged in mathematical forms even though we adopt a variable α in our equations but it introduces a possible way to work out the S-shaped curve in the Σ-$\dot{M}$ or surface density-accretion rate plane. As we can see, the criteria for stability also remain the same as the constant α case because for the disk to be stable we should have the same requirement for the viscous stress $\frac{dW_{r\varphi}}{d\Sigma} > 0$, which leads to stability conditions for the disk.

Based on the Σ-$\dot{M}$ plot, we can also perform stability analysis for both thermal and secular modes and get similar results for the topology of the curve as in the constant α case.

One plausible assumption for α is that when the accretion rate is very low, α should not be quite different from that corresponding to molecular viscosity, which is round 10^{-6}. When turbulence set in, the viscosity increases (though always less than unity, of course) and at the other extreme when the material begins to pile up around the black hole, α diminishes again. Therefore it is quite helpful to understand how the Σ-$\dot{M}$ plot changes with α.

The plots of Σ-$\dot{M}$ at constant α are a bunch of tilted inverted U-shaped curves[5], each corresponding to a different α value. Therefore the plot of Σ-$\dot{M}$ with changing α is a curve passing from curve to curve in the whole congruence, the positive slope being secularly stable. As we trace the curve, we notice that there is an S-shaped curve very similar to that discussed by people working on dwarf novae. We suggest that the disk can experience similar behaviour for x-ray novae, except that when the disk evolves to the turning point of the Σ-$\dot{M}$ curve, where the instability set in, the disk will have a accretion rate jump to several order of magnitudes higher, at which point the disk is still unstable and again may shift to the hot, optically thin branch. Afterwards α will adjust to the bigger value, when the disk becomes deplete with matter after this burst. Then the disk can switch back to the Shakura-Sunyaev branch after the accretion rate drops to the turning point of the cool branch and conclude the cycle.

DISCUSSION

Here we make some estimates of the time scale for all the transitions. The time scale for the jump from low state to the superhot state consists of two steps, i. e. from the cool, stable state to an unstable, radiation pressure dominated high state, which change quickly to the hot, optically thin state. The first step takes place within a time scale comparable to the drift time scale, which is much longer than the second process when the disk evolves on an instability time scale corresponding to a mixed mode of thermal and secular instabilities, both of which are a function of α. For a thin disk, we have the

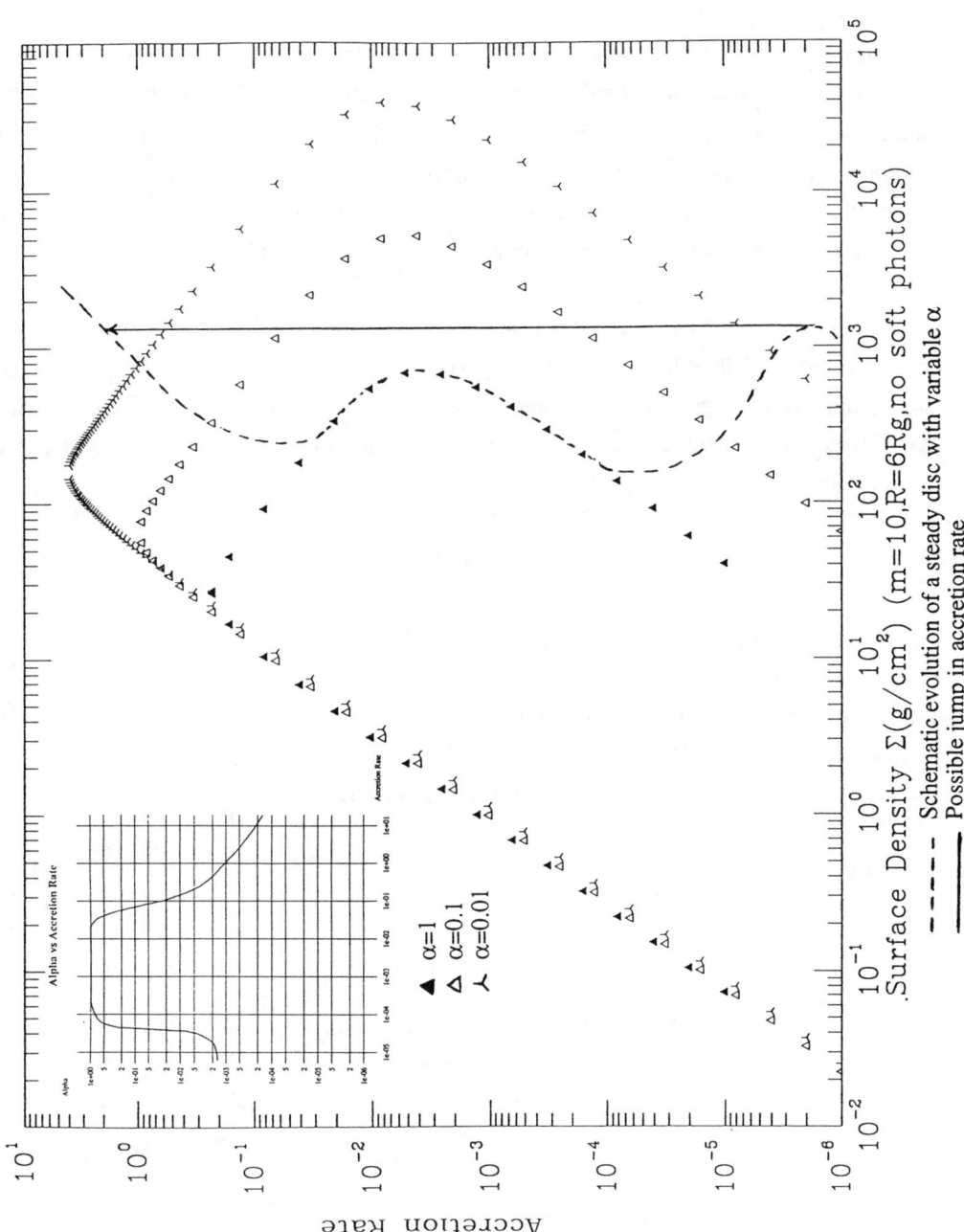

drift time scale $\approx (R/H)^2 t_H/\alpha$ while the thermal time scale is t_H/α. If the transition takes place at the time when the α value deviates significantly from 10^{-6} or that induced by molecular viscosity, the time scale is about days which is right at the range of observed rising time of x-ray novae. The resultant spectrum looks like a very hot plasma producing unsaturated comptonization spectrum with soft photons. And the hard spectral tail can also be accounted for by using the model proposed by Liang and Wandel[2].

All of the above calculations are done with local analysis and thus are very preliminary especially when we are dealing with a global instability-induced transition of state in the disk configuration. We could make our argument more relevant only after we have developed a global scheme for disk evolution. This work was partially supported by NASA grant No. NAG 5-1547.

REFERENCE

1 S. Mineshige and J. Wheeler, 1989, Ap. J. **343**, 241.
2 Liang, E. P. & Wandel, A., 1991, Ap. J., **376**, 746.
3 Abramowicz, M. A. et al, 1988, Ap. J., **332**, 646.
4 Blaes, O. M., 1987, MNRAS, **227**, 975.
5 Luo, C. & Liang, E. P., 1993, MNRAS (to be published).

APPLICATION OF MAGNETIC BRAKING THEORIES TO BLACK HOLE BINARIES

Koji Mukai
Code 668, NASA Goddard Space Flight Center, Greenbelt, MD 20771.
mukai@heagip.gsfc.nasa.gov

ABSTRACT

Many of the recently-discovered soft X-ray transients are also black hole candidates; I explore the possibility that magnetic braking, a plausible driving mechanism of mass transfer in these systems, may naturally lead black hole systems to become transients. I show that, regardless of the details, magnetic braking leads to a lower accretion rate for black hole primary with masses substantially greater than 1.4 $M_\odot$ than for neutron star binaries.

1. BLACK HOLE CANDIDATES AMONG SOFT X-RAY TRANSIENTS

V616 Mon (A0620−00), one of the best known Soft X-ray Transients (SXTs), is now regarded as a very strong Black Hole Candidate (BHCs). In recent years, increased monitoring of the X-ray sky (by, e.g., Ginga and GRANAT) has led to discoveries of more SXTs: QZ Vul (GS 2000+25), V404 Cyg (GS 2023+38), GU Mus (GS 1124−68) and GRO J0422+32 among others These are all BHCs based on X-ray properties and/or dynamical evidence (Cowley 1994). The incidence of BHCs among recent SXTs is high enough to lead one to ask why black hole binaries might produce a higher proportion of transient sources than neutron star systems.

Source	P_{orb} (days)	M_2 (ZAMS)	M_2 (TAMS)	Spectral Type	Mass Function
V404 Cyg	6.473				6.26±0.31
GU Mus	0.4333	1.22	0.46	K0–K4	3.07±0.40
QZ Vul	0.34	0.91	0.34		
V616 Mon	0.3230	0.85	0.32	K5–K7	3.18±0.16
GRO J0422+32	0.2212	0.51	0.19		

Table 1: Relevant properties of SXT/BHC systems.

Given P_{orb}, one can derive the properties of the secondary using Roche geometry, Kepler's 3rd law and a mass-radius relationship appropriate for the mass losing star. $R_2 = M_2^{0.88}$ for Zero Age Main Sequence (ZAMS) implies $M_2 = 0.070\, P_{orb}^{1.22}$ (where M_2 and R_2 are the mass and the radius of the secondary in Solar units, respectively). Terminal Age Main Sequence (TAMS) stars may have a 70% greater radius than a ZAMS thus $M_2 = 0.027\, P_{orb}^{1.22}$. Thus the two

M_2 columns in Table 1 represents the possible range of secondary mass if it is on the main sequence. The estimated spectral types of the secondaries in GU Mus and V616 Mon are consistent with this assumption, somewhat evolved from zero-age but not yet at the rapidly evolving sub-giant stage. The orbital period of V404 Cyg, however, is too long to permit a main sequence secondary, probably indicating a radically different evolutionary scenario; this system is excluded from further discussion in this paper.

2. MAGNETIC BRAKING

To sustain accretion in close binaries, the secondary must expand or the Roche-lobe must contract. Gravitational radiation can explain $\dot{M} \sim 10^{-10}$ $M_\odot$ yr^{-1} or less, which is not sufficient to power the luminous low mass X-ray binaries (LMXBs). However, there is another mechanism for angular momentum loss.

A secondary with convective envelope loses angular momentum through stellar wind coupled to the magnetic field lines; the secondary is tidally locked so orbital angular momentum is lost. This mechanism (magnetic braking) is widely accepted as the driving mechanism for cataclysmic variables (CVs) with periods in the 3–10 hr range. Details, however, remain uncertain (Shafter 1992) and various forms of parameterization are are used by different groups, thus this may be regarded as a family of theories rather than a single theory.

Should the same mechanism be operating in LMXBs as well as CVs? There does not seem to be a compelling reason why this should not be the case. The Roche lobes of many LMXBs, including many of the SXT/BHCs, are probably too small to contain sub-giant or giant secondaries, those stars whose rapid nuclear evolution could drive a high rate of mass-transfer. Moreover, magnetic braking is known to predict accretion rates for LMXBs that are correct to an order of magnitude (Patterson 1984).

In the rest of this paper, I will explore the implications of a scenario in which evolution of SXT/BHCs is driven by the combination of magnetic braking and gravitational radiation.

3. PRIMARY MASS DEPENDENCE

Magnetic braking models with conservative mass transfer gives:

$$\dot{M} \propto \frac{P^\gamma}{\alpha[(5/3 + \xi)M_1 - 2M_2]}$$

where α and ξ are parameters of the magnetic braking law, while γ depends on the above and on the mass-radius relationship, and M_1 is the mass of the primary (Shafter 1992). From this numerical expression, it follows that magnetic braking models predict a smaller $\dot{M}$ for a greater M_1.

This is a general prediction of the magnetic braking models, regardless of the details of the braking law, and can be understood from the following argument. Properties of the secondary (M_2, rotation rate etc.) for a given orbital period are similar, regardless of M_1. Therefore the secondary loses the same amount of angular momentum in absolute terms, regardless of M_1. The relative importance of this, is less for a higher value of M_1 (i.e., smaller fraction of orbital angular momentum lost), thus the lower accretion rate.

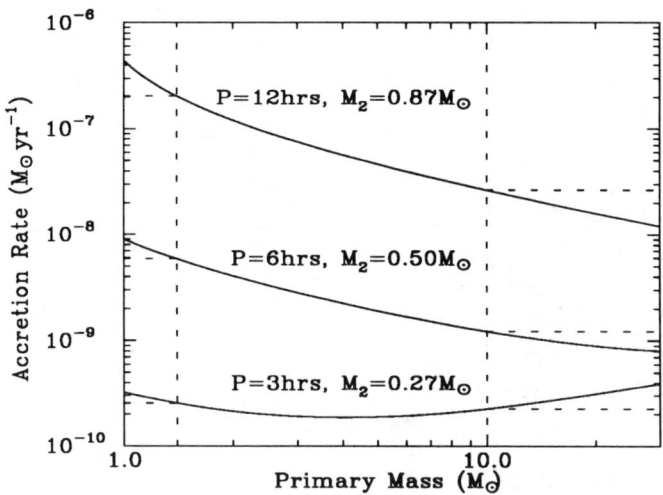

Fig. 1. Mass transfer rate predicted by Patterson's model.

In Fig. 1., I plot $\dot{M}$ as a function of M_1 for several values of P_{orb} and associated M_2 that are reasonable for that period, according to Patterson's magnetic braking model plus gravitational radiation. The latter becomes important for periods near 3 hrs, and the dependence on M_1 then is not simple; at longer periods, however, the predicted $\dot{M}$ is much smaller for a system containing a 10 $M_\odot$ black hole primary than for a system with a 1.4 $M_\odot$ neutron star.

4. DISK INSTABILITY MODEL

Disk instability is a possible mechanism for the outburst of SXTs (Huang & Wheeler 1989; Mineshige & Wheeler 1989). Disk instability models have been studied in detail in the context of CVs. Above a critical accretion rate $\dot{M}_{crit}$ which depends on the orbital period, the disk becomes hot and stable; irradiation of the disk can have the same effect. I plot, in Fig. 2, $\dot{M}$ as a function of P_{orb} for three values of M_1 (using Patterson's formula for combined magnetic braking plus gravitational radiation), as well as $\dot{M}_{crit}$ according to Shafter (1992). The weak dependence of the latter on M_1 is represented by the three lines which are almost indistinguishable on this plot.

It can be seen that, for a $10 M_\odot$ primary, the predicted $\dot{M}$ is always lower than $\dot{M}_{crit}$ over the period range studied here. Thus disk instability can in principle operate for such a system. Even though this in itself does not prove much (for example, neutron star systems with $P_{orb} < 7 hrs$ also satisfy this condition), it is certainly encouraging that a potential driving mechanism for accretion, without any extra added assumptions, predicts a lower $\dot{M}$ for black hole binaries than for neutron systems.

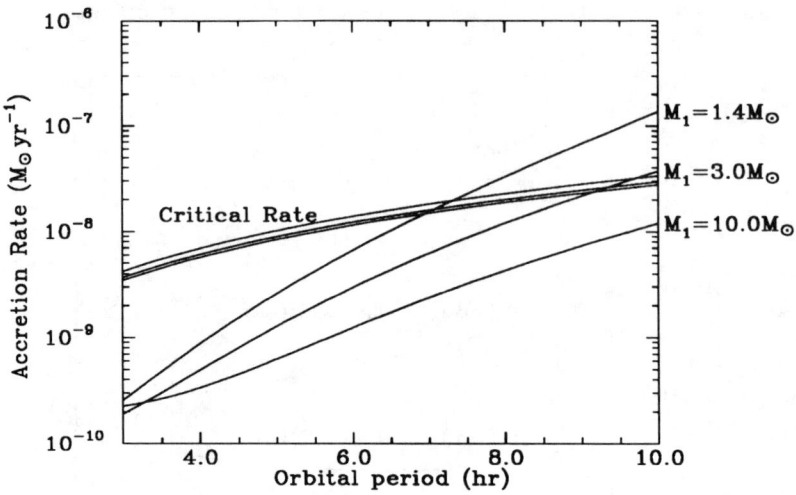

Fig. 2. $\dot{M}$ and $\dot{M}_{crit}$ as functions of P_{orb}

5. CONCLUSIONS

Magnetic braking is likely to be operating in LMXBs, including SXTs, although nothing presented in this paper excludes the presence of other mechanisms. I have shown that a consequence of any magnetic braking theories is a lower $\dot{M}$ for black hole binaries than neutron star systems. This may perhaps explain the high incidence of BHCs among SXTs. It would seem worthwhile to investigate the consequences of magnetic braking further for LMXBs in general and SXT/BHCs in particular.

REFERENCES

Cowley, A. P. 1994, these proceedings.
Huang, M. & Wheeler, J. C. 1989, ApJ **343,** 229
Mineshige, S., & Wheeler, J. C. 1989, ApJ 343, 241
Patterson, J. 1994, ApJSuppl 54, 443
Shafter, A. W. 1992, ApJ **394,** 268
Verbunt, F., & Zwaan, C. 1981, A&A 100, L7

A LOWER LIMIT ON THE PAIR DENSITY RATIO (z_+) IN AN ELECTRON-POSITRON PAIR WIND

Michael D. Moscoso and J. Craig Wheeler
Department of Astronomy, University of Texas at Austin, Austin, TX 78712
mdm@astro.as.utexas.edu; wheel@astro.as.utexas.edu

ABSTRACT

We derive a constraint on the pair density ratio, $z_+ = n_+/n_p$ in an electron-positron pair wind flowing away from the central region of an accretion disk around a compact object under the assumption of a coupling between electrons, positrons and protons. The observed annihilation flux per unit volume is used to determine a minimum mass loss rate per unit area, $\dot{M}_*$, for a given pair density ratio at the base of the streamline. The requirement that $\dot{M}_* < \dot{M}_{*,Edd}$ (the mean Eddington mass loss rate per unit area) then places a lower limit on the pair density ratio at the base of the wind, $z_{+,min}$.

The GRANAT/SIGMA experiment observed a positron annihilation line in Nova Muscae 1991. The narrow width and redshift of the line suggest that the pair production and annihilation regions are physically distinct. We hypothesize that an electron-positron pair wind transports the pairs from the production to the annihilation region and calculate $z_{+,min}$.

1. INTRODUCTION

The study of electron-positron pair winds from stellar black hole candidate systems has several observational motivations. The presence of electron-positron pairs in some of the black hole candidate systems is evidenced by observations of a narrow, redshifted positron annihilation line in Nova Muscae 1991 in the "very high" state (Sunyaev et al. 1992) and of a broad, redshifted positron annihilation line in the Galactic center source 1E1740.7-2942 (Bouchet et al. 1991). In addition, positron annihilation radiation may have been observed in Cyg X-1 (Ling & Wheaton 1989). For a more detailed discussion of the observational evidence for electron-positron pairs in the black hole candidate systems see Moscoso & Wheeler (1994).

There is observational evidence for outflows from some of the black hole candidate systems. Radio observations of 1E1740.7-2942 have found the source to be at the center of an aligned double radio jet (Mirabel et al. 1992). Several black hole candidates have undergone radio outbursts (Hjellming & Penninx, 1991). A more detailed discussion of the evidence for outflows is given in Moscoso & Wheeler (1994). With these motivations, we consider electron-positron pair winds from black hole candidate systems.

2. CONSTRAINT ON z_+ FOR PAIR ANNIHILATION IN A WIND

Under the assumption that there is an electron-positron pair wind flowing away from the inner disk region and that all of the annihilation is occurring in the wind, a lower limit can be placed on the pair density ratio in the wind.

A Lower Limit on the Pair Density Ratio

We assume that for a steady state flow the positrons, electrons and protons move with a common velocity v through a fiducial area A, then $\dot{n} = A n v$ and $z_+ = \dot{n}_+/\dot{n}_p$. Here $\dot{n}_+$ and $\dot{n}_p$ are the number of positrons and protons injected at the base of the streamline per unit volume per sec respectively and $z_+ \equiv n_+/n_p$ is the pair density ratio. This assumes that there is some coupling between the electrons, positrons and protons by either magnetic fields or plasma effects. If the pairs can stream freely through the gas independent of the ions, then this assumption is invalid. The number of positrons that flow into the volume where the annihilation occurs must exceed the number that annihilate in that volume. The minimum value of $\dot{n}_+$ in the flow is thus given by the observed annihilation flux per unit volume, $\dot{n}_{+,min} = \frac{2\pi D^2 F_{line}}{V_{ann}}$, where F_{line} is the observed 511 keV line flux, D is the distance to the source and V_{ann} is the annihilation volume which is taken to be a sphere of radius R. For a fully ionized hydrogen plasma, the minimum mass loss rate per unit area for a given pair density ratio at the base of the streamline is given by:

$$\dot{M}_{*,min} = \dot{n}_{+,min} h \left(\frac{1}{z_{+,0}} (m_p + m_e) + 2 m_e \right), \quad (1)$$

where h is a characteristic length scale which we take to be the height of the disk and $z_{+,0}$ is the pair density ratio at the base of the streamline. The assumed coupling between the electrons, positrons and protons modifies the Eddington luminosity, L_{Edd} for the plasma since the opacity is changed by the inclusion of the pairs. Using the modified Eddington luminosity, we can obtain a mean Eddington accretion rate per unit area:

$$\dot{M}_{*,Edd} = \frac{L_{Edd,o}}{\eta c^2 A} \times \frac{\left(1 + \frac{m_e}{m_p}(1 + 2z_+)\right)}{(1 + 2z_+)}, \quad (2)$$

where η is the efficiency, A is the surface area of the wind base and $L_{Edd,o}$ is the Eddington luminosity for an ionized hydrogen plasma. If there is no non-local deposition of energy at the base of the streamlines, then it is required that $\dot{M}_{*,min}$ not exceed $\dot{M}_{*,Edd}$ and this places a lower limit on z_+. Using eqs. (1) and (2) this is given by:

$$z_{+,min} = \frac{2\dot{N}_{+,min} - \frac{L_{Edd,o}}{\eta c^2} + \left[\left(\frac{L_{Edd,o}}{\eta c^2} - 2\dot{N}_{+,min} \right)^2 + 8\dot{N}_{+,min} \left(\frac{L_{Edd,o} m_e}{\eta c^2 m_p} - 2\dot{N}_{+,min} \right) \right]^{1/2}}{4 \left(\frac{L_{Edd,o} m_e}{\eta c^2 m_p} - 2\dot{N}_{+,min} \right)} \quad (3)$$

where $\dot{N}_{+,min} = m_p A h \dot{n}_{+,min}$, is the minimum total volume-integrated annihilation rate.

The expression for the annihilation rate as a function of positron number density can be inverted to give the minimum positron number density in the annihilation region, $n_{+,min}$ as:

$$n_{+,min} = \sqrt{\frac{8\dot{n}_{+,min}}{3\sigma_T c \left(1 + \frac{1}{z_{+,min}}\right)}}. \quad (4)$$

The minimum Thomson optical depth can be estimated by $\tau_{min} \sim 2n_{+,min}\sigma_T R$.

3. APPLICATION TO NOVA MUSCAE 1991

As noted in § 1, a transient redshifted electron-positron annihilation line was observed in Nova Muscae 1991. If the redshift is of gravitational origin, then the annihilation region is located at $r_{ann} = 8.2 \pm 4.6 r_s$ (Chen et al. 1993). If the annihilation region is close to the black hole, then the narrow line width becomes very interesting because there are many physical processes which may contribute to the line width such as thermal, turbulent and Doppler broadening. For pure thermal broadening, the temperature of the annihilation region must be $T < 3 \times 10^7$ K or 2.5 keV. This low temperature implies that for thermal pair production, the pair production and annihilation regions must be physically distinct from one another.

There are several interesting issues which are raised at this point. Assuming thermal pair production, it is reasonable to identify the pair production region with the location of the hard X-ray component. There is evidence that the optically thick accretion disk may extend down to the last stable circular orbit in Nova Muscae 1991 in the "very high" state (see Moscoso & Wheeler 1994 and references therein). Thus thermal pair production can only occur above the disk midplane or in the disk midplane near the inner edge of the disk. If the pair production region in Nova Muscae 1991 is near the inner edge of the disk, then a mechanism is required to transport the pairs from the production to the annihilation region. Given the observational evidence for outflows and the necessity of a mechanism for transporting electron-positron pairs from a hot production region to a relatively cool annihilation region, we hypothesize that the transport mechanism for the pairs is an optically thick (to $\gamma - \gamma$ pair production) wind. We note that for Nova Muscae 1991 the upper limit to the optical depth to photon-photon pair production is of order unity (Moscoso & Wheeler, 1994). If the wind is optically thick to pair production, then at some distance from the production region the wind will turn optically thin and allow the annihilation radiation to escape. In this picture, the observed redshift of the annihilation radiation would correspond to the radius from the central object where the wind turned optically thin. The observed narrow width of the annihilation line may be accounted for by the cooling of the annihilating pairs via adiabatic expansion and Compton cooling. The observed annihilation line in such a wind would still be subject to broadening from the plasma motions in the wind. The details of this broadening depend on the configuration of the streamlines in the wind which are not yet known. A continued flow of the wind may provide a possible particle source for the expanding synchrotron bubbles, though the particle acceleration mechanism remains unclear.

We now apply the results of § 2 to Nova Muscae 1991 where the observed annihilation line flux was $F_{line} = (6.3 \pm 1.5) \times 10^{-3}$ $photons\ cm^{-2}\ s^{-1}$ (Sunyaev et al. 1992). We take $M = 3.3$ $M_\odot$, $r_{out} = 8r_s$, $r_{in} = 3r_s$, $R = 8r_s$ and $D = 8$ kpc. Assuming $h/r_{in} = 10^{-2}$ and $\eta = 0.06$ for a Schwarzschild black hole, eqs. (1)-(4) yield: $z_{+,min} = 0.013$, $n_{+,min} = 1.7 \times 10^{17} cm^{-3}$, and $\tau_{min} \sim 0.78$. For a Kerr black hole, $\eta = 0.4$ and we then have: $z_{+,min} = 0.094$, $n_{+,min} = 4.6 \times 10^{17} cm^{-3}$ and $\tau_{min} \sim 2.1$. Note that these results are sensitive to the geometry of the inner disk region, in particular the characteristic length, h. From these results we note that for this particular choice of parameters, the winds are not necessarily pair-dominated since z_+ could be appreciably less than 1. In both the Schwarzschild and Kerr cases the Thomson scattering optical depth is also of order unity. Therefore the pairs in the wind will be subject to Compton cooling. Thus it is plausible that a narrow, redshifted positron annihilation line could be produced in an electron-positron pair wind. The source of the pairs could be thermal pair production associated with the region where the hard power law is produced, perhaps at $r \lesssim 3r_s$. If the hard power law itself arises from a substantially non-thermal process then the production and annihilation of the pairs must be reconsidered, but these general properties of a wind which transports the pairs may still hold.

In future work, the limits obtained above will be compared with pair equilibrium cloud studies to determine what parts of parameter space are excluded and to obtain constraints on the physical conditions within the pair production region.

4. SUMMARY

In this paper we have noted the observational evidence for the presence of electron-positron pairs and for outflows in the black hole candidate systems. We have derived a lower limit on the pair number density ratio at the base of an electron-positron pair wind. This lower limit can be compared with previous pair equilibrium cloud studies to constrain the physical parameters within the pair production region. We have reviewed previous constraints placed on the annihilation region in Nova Muscae 1991. The pair production and annihilation regions in Nova Muscae 1991 must be physically distinct if the pair production is occurring thermally. Motivated by this, we hypothesized that the pair annihilation is occurring in an optically thick electron-positron pair wind and calculated the minimum values of the pair density ratio, the positron number density and the scattering optical depth.

ACKNOWLEDGEMENTS

This research is supported in part by NASA Grants NGT-50853 and NAGW-2975, NSF Grant AST-9115143 and a grant from the Texas Advanced Research Program.

REFERENCES

Bouchet, L., et al. 1991, ApJ, 383, L45
Chen, W., Gehrels, N. & Cheng, F.H. 1993, ApJ, 403, L71
Hjellming, R.M. & Penninx, W. 1991, in Particle Acceleration near Accreting Compact Objects, eds. J. van Paradijs, M. van der Klis and A. Achterberg (Amsterdam: North Holland)
Ling, J.C. & Wheaton, W.A. 1989, ApJ, 343, L57
Mirabel, I.F. et al. 1992, IAU 5477
Moscoso, M.D. & Wheeler, J.C. 1994, in Interacting Binary Stars, ed. A.W. Shafter, (PASP: Provo)
Ramaty, R., Leventhal, M., Chan, K.W. & Lingenfelter, R.E. 1992, ApJ, 392, L63
Sunyaev, R., et al. 1992, ApJ, 389, L75
Svensson, R. 1990 in Physical Processes in Hot Cosmic Plasmas, eds. Brinkman, Fabian & Giovannelli, (Amsterdam: Kluwer)

ANALYSIS OF THE ASSUMPTION OF FREQUENCY INDEPENDENCE OF LIGHT VELOCITY IN A GRAVITATIONAL FIELD

Clarence A. Gall
Universidad del Zulia, INPELUZ-POSTGRADO, Apartado # 98
Maracaibo, Venezuela.

ABSTRACT

The concept of the black hole was derived from the Schwarzschild metric which was then the only exact and reliable solution of Einstein's field equations($R_{\mu\sigma} = 0$). That metric gives a coordinate velocity for electromagnetic radiation (EMR) that goes to zero at the Schwarzschild radius, thereby resulting in the black hole. This analysis shows that inherent in the Schwarzschild solution is the assumption that the velocity of EMR in a gravitational field is independent of wavelength and frequency. An alternate exact solution is derived making the opposite assumption. The resulting metric's properties include: a) the coordinate velocity of light is frequency dependent and does not predict a black hole and b) the mass parameter has dimensions of volume unlike Schwarzschild's geometric mass parameter. The implications of these results are discussed.

1. INTRODUCTION

In his general theory of relativity, Einstein set up physical laws that would be valid in noninertial as well as in inertial frames of reference [1-3]. For a gravitational field in a matter free region of space, the solution of the tensor field equations ($R_{\mu\sigma} = 0$), was required to define a line element whose metric coefficients would then describe the space. The Schwarzschild line element [4] derived over seventy five years ago is an exact solution of these field equations for the gravitational field exterior to a spherically symmetric body. This derivation involves symmetry arguments based on the idea that the gravitational field decreases with the increase in distance (r) from the center of the spherically symmetric gravitating body. However, for any spherical surface defined by a given value of r the field is constant. There are different approaches to deriving the Schwarzschild line element. One of the simplest [2] is to set up the line element in spherical polar coordinates :

Assumption of Frequency Independence

$$ds^2 = e^{p(r)}c^2dt^2 - e^{h(r)}dr^2 - r^2d\theta^2 - (r^2\sin^2\theta)d\phi^2 \qquad (1)$$

The use of the exponential functions $e^{p(r)}$ and $e^{h(r)}$ is meant to ensure that the signs of the metric coefficients are not affected in the derivation by using intrinsically positive functions. This form satisfies the symmetry requirements and yields the familiar metric:

$$ds^2 = \left(1 - \frac{2m}{r}\right)c^2dt^2 - \frac{dr^2}{1 - \frac{2m}{r}} - r^2d\theta^2 - (r^2\sin^2\theta)d\phi^2 \qquad (2)$$

where m is the geometric mass. The concept of the black hole follows when the coordinate velocity of EMR in the radial direction is defined along a null geodesic for which $ds^2 = 0$ so that:

$$\frac{dr}{dt} = \left(1 - \frac{2m}{r}\right)c \qquad (3)$$

Hence the velocity of EMR goes to zero for $r = 2m$, the so called Schwarzschild radius.

There is however an inherent assumption in the Schwarzschild solution, which appears to have been overlooked ever since the original derivation of this line element. It is simply that the line element begs the question as to whether the velocity of EMR in a gravitational field is a function of wavelength and frequency. The form of equation (1) does not address this question. Indeed by omission, it assumes that there is no such dependence! The resulting line element of equation (2) therefore shows no dependence.

2. AN ALTERNATIVE DERIVATION

Is it valid then to assume that the velocity of EMR in a gravitational field is independent of wavelength and frequency? It is then very unlike the velocity of EMR in a material medium. For the sake of argument, it will be assumed that the velocity of EMR in a gravitational field is a function of wavelength and frequency. Then a possible form in which to seek a solution [3] which would express this dependence as well as satisfy Schwarzschild's symmetry arguments is:

$$ds^2 = \frac{\lambda_g^2 \nu_r^2}{c^2}c^2dt^2 - \frac{c^2}{\lambda_g^2 \nu_r^2}dr^2 - r^2d\theta^2 - (r^2\sin^2\theta)d\phi^2 \qquad (4)$$

where λ_g [$= f(r)$] is the wavelength of the EMR and ν_r is the corresponding value of the frequency at a distance r from the center of the gravitating body. Since the values of λ_g and ν_r are taken in squared form, then the signs of the metric coefficients are also not affected as before. A variational problem can then be set up in the usual way [2-3] :

$$\delta \int \left\{ \frac{\lambda_g^2 \nu_r^2}{c^2} (\dot{x}^\circ)^2 - \frac{c^2}{\lambda_g^2 \nu_r^2} \dot{r}^2 - r^2 \dot{\theta}^2 - (r^2 \sin^2\theta) \dot{\phi}^2 \right\} ds = 0 \quad (5)$$

where $x^\circ = ct$ and $\dot{}$ means d/ds. The Christoffel symbols of the second kind are obtained from the Euler Lagrange equations of motion and then substituted into the field equations. These yield two differential equations whose solution gives the relationship between λ_g and ν_r :

$$\lambda_g = \frac{c}{\nu_r} \left(1 - \frac{\alpha \nu_r^2}{c^2 r} \right)^{1/2} \quad (6)$$

where α is a constant of integration. Substituting in equation (4) gives the following line element :

$$ds^2 = \left(1 - \frac{\alpha \nu_r^2}{c^2 r} \right) c^2 dt^2 - \left(\frac{dr^2}{1 - \frac{\alpha \nu_r^2}{c^2 r}} \right) - r^2 d\theta^2 - (r^2 \sin^2\theta) d\phi^2 \quad (7)$$

3. DISCUSSION

This metric has the same type of r dependence as the Schwarzschild metric. However, the coordinate velocity of EMR in the radial direction is :

$$\frac{dr}{dt} = \left(1 - \frac{\alpha \nu_r^2}{c^2 r} \right) c \quad (8)$$

This is an interesting result because the velocity of EMR in a gravitational field is now a function of its frequency (and implicitly its wavelength)! For a given value of r called r_c, a cutoff frequency can be defined for which all higher frequencies are trapped. However the lower frequencies can still escape as they have a real positive coordinate velocity :

$$v_r \geq \left(\frac{c^2 r_c}{\alpha}\right)^{1/2} \quad \text{Frequencies that are trapped.}$$

$$v_r < \left(\frac{c^2 r_c}{\alpha}\right)^{1/2} \quad \text{Frequencies that can escape.}$$

This result negates the black hole concept since the mathematical singularity of the Schwarzschild metric does not appear here! The coordinate velocity is a real physical velocity and coincides with the vacuum velocity of light on the earth's surface (ignoring rotational effects). Its value should decrease as the frequency increases. Unlike the one dimensional geometric mass (m) this integrating constant (α) has dimensions of volume! A gravitating mass (M) is related to its volume (V) by its density (d) , so that α may be written as :

$$\alpha = \beta V = \beta \frac{M}{d} \tag{9}$$

where β is a dimensionless constant. This makes more physical sense! Equation (6) couples the electromagnetic and the gravitational fields. The constant β is then a dimensionless coupling constant. It has been estimated to be $\cong 3 \times 10^{-36}$.

This analysis implies that Eddington [5] was correct when he stated "I think there should be a law of Nature to prevent a star from behaving in this absurd way!" (i.e. ending in a black hole). The law is that matter has mass and occupies **volume** not length! Unfortunately, he confronted Chandrasekhar and not Schwarzschild!

REFERENCES

Einstein, A., Lorentz, H. A., Weyl, H., & Minkowski, H. 1952, The Principle of Relativity, (New York: Dover) p.113, 117, 142-145
Adler, R., Basin, M., & Schiffer, M. 1975, Introduction to General Relativity, (New York: Mc Graw Hill) p. 145-151, 151, 167-169, 185-196, 222-225
Gall, C. A. 1987, Matter, Gravity and Spin, (New York: Vantage Press) p.37-47
Schwarzschild, K. 1916, Sitzber. Preuss. Akad. Wiss. Berlin, 189 - 196
Wali, K. C. 1982, Physics Today, 10, 33

ING Photometry and Spectroscopy of the optical counterpart of the X-ray transient GRO J0422+32

Emilios Harlaftis, Derek Jones
Royal Greenwich Observatory, 38780-Santa Cruz de La Palma, Spain

and

Phillip Charles and Andrew Martin
Astrophysics, Oxford University, Keble Rd, Oxford, OX1 3RH

ABSTRACT

We present new photometric observations of the GRO transient J0422+32 which show a two-hour and a five-hour variation of only 0.1 magnitude in amplitude. Spectroscopic observations during the late decline of the August 1992 outburst and during the August 1993 outburst show variability in the double-peaked Balmer and HeII profiles on a timescale of hours and reveal red-shifted emission components at two epochs.

1. INTRODUCTION

The hard X-ray/soft γ-ray transient GRO J0422+32 was discovered by the BATSE instrument on board the Compton Gamma Ray Observatory on 5 August 1992 (IAUC 5580). Soon after, the optical counterpart was identified with a star of 13th magnitude, Nova Per 1992 (15 August; IAUC 5588). No X-ray or gamma-ray emission had ever been recorded before from this object and there is only a faint star on the Palomar Sky Survey plates at approximately V $\sim$ 20-21 mag. This establishes J0422+32 to be a low-mass system because of the large increase in its brightness from quiescence. The optical and ultraviolet spectra also showed J0422+32 to be typical of other LMXBs (Shrader et al. 1992). The X-ray properties and their comparison with other black-hole candidates (such as Cyg X-1, V404 Cyg and Nova Mus 1991) quickly established Nova Per 1992 as the latest galactic black-hole candidate (with X-ray variability on very short timescales and a hard X-ray spectrum).

2. PHOTOMETRIC OBSERVATIONS

Photometric observations with the 1.0m Jacobus Kapteyn telescope (JKT) on La Palma (part of the Isaac Newton Group of telescopes, thereafter ING) were undertaken in the V-band on 28 October 1992 with the aim of looking for short-term variability. A further 91 CCD R-band images were taken with the 2.5m Isaac Newton Telescope (INT) during the week 12-17 November with the aim of searching for any longer-term periodicities. Aperture photometry was performed on the debiased and flat-fielded images using two reference stars which were of a similar brightness to J0422+32 during our observations and had similar colours. The light curve was constructed with respect to the average of these two reference stars. Inspection of the two reference stars showed that they

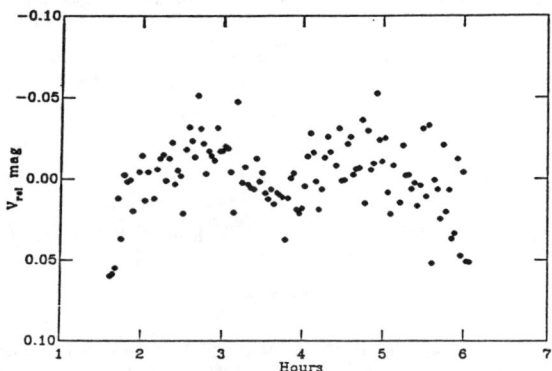

Fig. 1. V-band JKT observations on 28 October 1992 showing a two hour and a five hour variation.

were constant, but with random scatter of the order of two hundredths of a magnitude within the estimated errors due to Poissonian noise. In Fig. 1, the V-band data from 28 October 1992 show a variation close to five hours of amplitude 0.1 of a magnitude. A shorter variation at two hours is also seen. The four and a half hour duration of the observation is shorter than the photometric variation found by Kato et al. (1993). This probably explains the dominance of the 2.13 ± 0.1 hour variation over the 5.32 ± 0.2 hour variation in the periodogram. The window function of the data was also examined and showed no significant peaks near either two or five hours.

The R-band data of November has a much longer time base of 4 nights, and so has greater phase coverage than the October data. Short period variations of 0.1 magnitudes can be seen on each night as well as changes from night to night (Harlaftis et al. 1994, in preparation). The periodogram of this dataset was produced using the CLEAN algorithm in order to remove the effects of the window function. A single, strong peak at 5.27 ± 0.1 hours was found.

Kato et al. (1993) claim to have observed regular minima every 5.090 ± 0.001 hours which they presume to be the orbital period. Their data shows these small ∼ 2 hour modulations that we see, superimposed on top of an overall five hour sinusoidal variation. The nature of this modulation is important as Kato et al. interpret it as a superhump (Whitehurst 1988) at a slightly longer period than their presumed orbital period; from this using the formula of Mineshige et al. (1992), which gives the relation between the binary mass ratio and the superhump period excess, they obtain a lower limit to mass of the compact object of 2.9 $M_\odot$, which leads to the possibility of J0422+32 being a black hole candidate. However, we have to await for the quiescent state to deduce the orbital period with certainty. During outburst, photometric variabilities of unclear nature can occur as in the case of V404 Cyg (Casares et al. 1991).

3. SPECTROSCOPIC OBSERVATIONS

Optical spectra were obtained with the 2.5m INT (and IDS spectrograph) and the 4.2m WHT (and ISIS spectrograph) during February, March and August 1993 (described in Harlaftis et al. 1994, in preparation). Here, we report on the most interesting spectra from the WHT (Fig. 2). A spectrum obtained with ISIS on the WHT on 19 February 1993 (V ∼ 14.7 mag) shows the Balmer lines

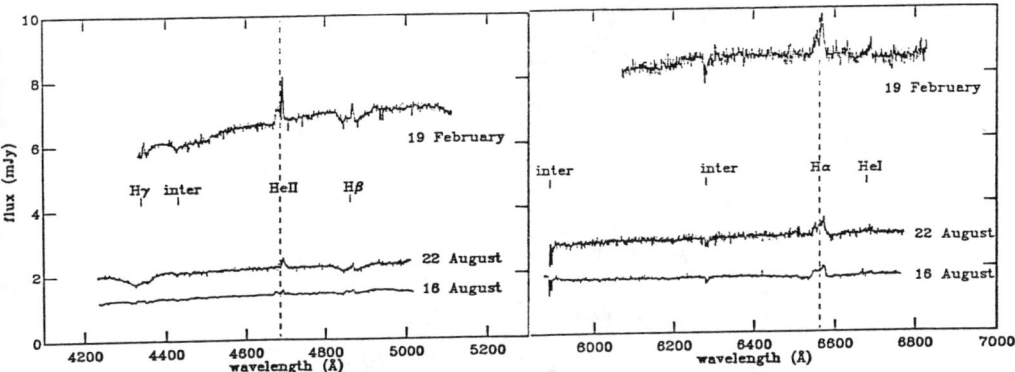

Fig. 2. Blue and red spectra obtained with the 4.2-m William Herschel telescope and the ISIS spectrograph at La Palma on 19 February 1993 and 17 and 23 August 1993 at a resolution of ~1.5Å. See text for details.

to have a narrow emission feature and other complex structure, with Hβ and Hγ both superposed on broad, shallow absorption features. The FWZI (~5000 km/s) is typical of that expected from accretion disks. The dominant line in the WHT spectrum is HeII 4686Å with an equivalent width of ~2Å. This emission line arises from X-ray reprocessing into optical wavelengths and indicates that the source is still X-ray active but decaying in strength because of the disappearance of the 4640-4650Å Bowen blend since November 1992 (Shrader et al. 1992).

The He II profile is complex and is dominated by a narrow emission component at +414±10 km/s relative to the line centre. Irradiation of the secondary star can in principle generate such emission. However, we did not detect this sharp emission with the INT on 2 March 1993, which is somewhat puzzling, nor was it detected by Shrader et al. (1992) in October 1992. Using the ephemeris of Kato et al. (1993; photometric minimum at UT 14:47 on 30 December) we find a phase of 0.94±0.24 for the spectrum obtained on 19 February 1993. We estimate the colour excess of the system to be E(B-V)=0.20 from the interstellar features at 4430Å and 6813Å using the empirical relation of Herbig (1975; the EW of 4430Å is 900mÅ and that of 6813Å is 400mÅ) which is consistent with that found by Shrader et al. (1992).

Most interestingly, the object was found in optical outburst again on 12 August 1993 (IAUC 5842). We obtained three spectra with ISIS on the WHT on 17 August when the object was at V~15.4 (IAUC 5851) and one spectrum on 23 August. The spectra are featureless except for the Balmer, He and some interstellar lines. The continuum is three times weaker than that of February, but slightly bluer. We observe a 57% increase in the blue continuum (50% in the lines) and a 71% increase in the red (67% in Hα) between 17 and 23 August. The maximum of the outburst was reached on 14 August (IAUC 5842 and 5844) and such flux increase as observed on 23 August could be accommodated

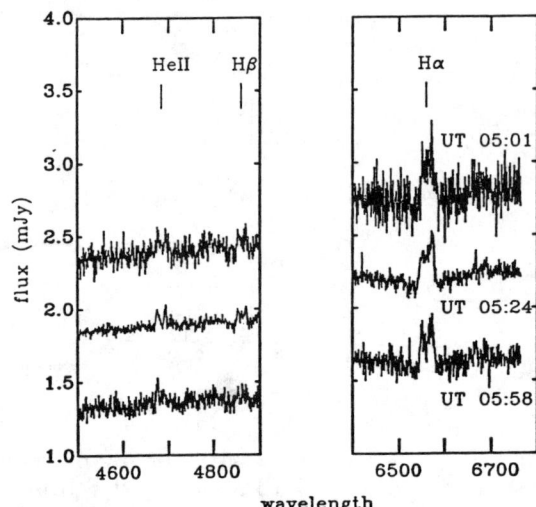

Fig. 3. Variation of the HeII, Hβ and Hα lines on a timescale of ∼1/2 hour on 17 August 1993. An offset has been applied to the spectra for clarity.

with a minor mass-transfer burst from the secondary close to that date. The flux delay between the blue and red wavelenghts is also supportive of such a suggestion. The emission components of the Balmer lines show a double-peaked structure, usually a signature of an accretion disk at high inclination. Short-term profile variation in the lines is evident particularly in Hα on 17 August, which is generally related to S-wave behaviour (Wade & Ward 1985). There is no HeII on 14 August which suggests that the photoionization of the disk had not started (Castro-Tirado et al., 1993), whereas its double-peaked profile is evident on 17 August. The HeII line changes from a double-peaked profile to a single-peaked component between 17 and 23 August.

The single-peaked emission on 23 August is red-shifted by 545±30 km/s. Such a large shift suggests the secondary star as the most likely location of the emission. During outburst, the radiation from the hot disk can irradiate the atmosphere of the secondary star (Marsh & Horne 1990). If this is the case, then a lower limit of the mass function of $3.7M_\odot$ is implied for the compact object for an orbital period of 5.1 hours. However, we will have to obtain phase-resolved observations in order to determine the mass function precisely.

REFERENCES

Castro-Tirado et al. 1993, *The 4th Annual Astrophysics Conference in Maryland*, The Evolution of X-ray Binaries, in press
Casares et al. 1991, MNRAS, 250, 712
Herbig, G. H. 1975, ApJ., 196, 129
Kato, T., Mineshige, S., & Hirata R. 1993, PASJ, in press
Marsh, T. R. & Horne, K. 1990, ApJ., 349, 593
Mineshige, S., Hirose, M., Osaki, Y. 1992, PASJ, 44, L15
Shrader, C. R. et al. 1993, *1st Compton Observatory Symposium*, Maryland, in press
Wade, R. A. & Ward, M. J. 1985, in Interacting Binaries, p 129
Whitehurst, R. 1988, MNRAS, 232, 35

Multiwavelength Study of the X-Ray Nova GRO J0422+32

C.R. Shrader

Laboratory for High Energy Astrophysics, NASA/GSFC/CSC

R. Mark Wagner

Department of Astronomy, Ohio State University

R.M. Hjellming and X.H. Han

National Radio Astronomy Observatory

S.G. Starrfield

Department of Physics and Astronomy, Arizona State University

INTRODUCTION

The *BATSE* experiment on-board the Compton Gamma Ray Observatory (*CGRO*) reported the detection of an new source of hard X-rays in the direction of the constellation Perseus on 5 August 1992 (Paciesas et al. 1992). The transient was subsequently classified an X-ray nova (XRN), and designated GRO J0422+32 (herein J0422+32). In this contribution, we describe some results of a multi-wavelength campaign designed to study J0422+32 at ultraviolet, optical, and radio wavelengths. Our coverage began approximately 10 days post-discovery and has continued through the decline phase. A definitive astrometric determination of the optical counterpart, the progenitor, long- and short- term photometric variability including possible periodicity, distance estimates, energetics and comparisons to previous XRN are discussed.

THE OPTICAL COUNTERPART AND PROGENITOR

On August 15.0, Castro-Tirado et al. 1992 identified an optical counterpart of J0422+32 with a new V = 13.25 mag star that appeared near the Granat/WATCH all-sky monitor X-ray error circle but was absent on the POSS. Subsequent optical spectra at the Perkins 1.8-m telescope and ultraviolet spectroscopy with the International Ultraviolet Explorer (*IUE*) of this candidate (Wagner et al. 1992, Shrader et al. 1992) confirmed its identification as the optical counterpart, but astrometry of the object based on both the *IUE* and Perkins telescope encoders were substantially different from the position reported by Castro-Tirado et al. 1992. Pavlenko, Shlyapnikov, and Castro-Tirado (1992) reported the presence of a weak star on the red (E) print with a magnitude of about 16.5 ± 0.7. If this star was the progenitor of the nova then the outburst amplitude would have been about 3.7 R mag. A radio counterpart was detected by Han and Hjellming 1992 with the VLA which was consistent with the *IUE* position. Shortly thereafter, McCrosky 1992 and Mueller 1992 independently reported optical positions of the nova that differed by 0".4 of time in right ascension and 2".7 in declination. Mueller noted the absence of the progenitor on the POSS.

We examined the Lowell Observatory copy of the POSS O and E prints as well as the digitized STScI/GSSS plates seeking confirmation of the progenitor. We confirm the presence of a stellar object near the location of the nova on the E print, however, it is near the E plate limit of about 20th mag and not 16.5 mag as suggested by Pavlenko et al.(1992). An object also appears to be present on the O print at the same location and near the plate limit of about 21st mag, but its appearance is marred by a defect. The position of this stellar object was measured with respect to 30 primary astrometric standards. Results indicated that the position of this object is consistent with that of the nova to within several seconds of arc, but the situation remained ambiguous.

We obtained a CCD image in the V band on 20 July 1993 after J0422+32 had subsided into a brief, near–quiescence at 19th mag. This frame easily shows the nova and a companion 5″ to the northeast (Figure 1). Astrometry of the nova and companion were performed based on this frame and the astrometric reference grid that was established from our previous analysis of the POSS. We find that our optical position for the nova, $\alpha\ \delta$ ($J2000$): $04^h21^m42.79^s$, $+32°54'27.2''$ is in excellent agreement with the IUE position and the radio source as observed with the VLA.

Re-examination of the POSS suggests that the quiescent magnitude of J0422+32 is probably below the limiting magnitude of the survey, i.e. fainter than $R \simeq 20$, thus the amplitude of the present optical outburst was ≥ 7 mag. If J0422+32 was similar to other XRN such as V616 Mon, or Nova Muscae 1991, then its quiescent magnitude might just be below the limit of the POSS. We note that on the STScI digitized version of the POSS E plate, there is a marginal excess at the location of J0422+32, however, other excesses of similar significance are not identified with any stellar objects on our CCD frames which are complete to well below the POSS limit.

LIGHT CURVES

The optical light curve of J0422+32 covering the period from 10-days to 13-months post outburst is shown in Figure 2. Our light curve has been assembled from visual and digital measurements reported on the IAUCs, digital photometry reported directly to us, and our own CCD photometry. The optical identification was made 10 days after the discovery by $BATSE$ so we have no information regarding the optical brightness during the X-ray rise to a maximum of about 3 Crab (20–300 keV by day 11. If the optical and X-ray maxima coincided then the peak optical brightness was probably about 13.2 mag. The decline up to day 210 can be represented by an exponential with an e–folding time of 170 days.

After day 210, the optical brightness dropped quickly; subsequent to solar conjunction it was visible at $\sim 19th$ mag. The form of the light curve up to day 360 is similar to that observed for A0620-00. The most spectacular feature of the optical light curve is a second major outburst beginning near day 370: J0422+32 brightened from $V \simeq 19$ mag to $V \simeq 15.4$ mag in less than 5 days. A second, major outburst with similar characteristics, following the initial decline to quiescence has not to our knowledge been observed before in any XRN. The amplitude of the second outburst is consistent with an extrapolation exponential decay law observed earlier in the outburst. After day 390, the brightness again declined with the same slope observed earlier indicating that the second outburst concluded. Indications are that the optical brightening was not accompanied by a dramatic increase in X-rays, as ASCA detected a flux of approximately 1 mCrab on the 0.5-10 keV band (Tanaka 1993) and $BATSE$ apparently did not see the source reappear in preliminary analysis of the data (Harmon 1993).

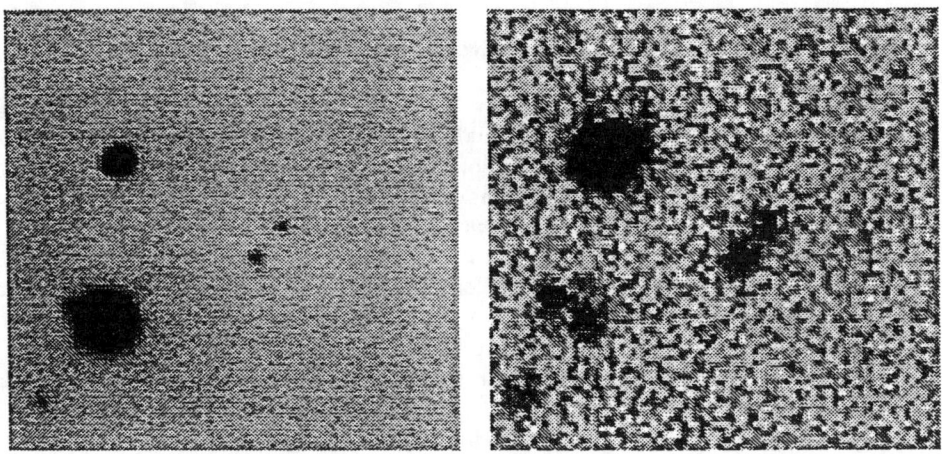

Figure 1 V-band CCD frames obtained on(a) on 9/15/92. (b) 07/20/93. Here J0422+32 is seen at about mag 19. The faint companion about 5" NE is seen on both frames, as well as on the POSS.

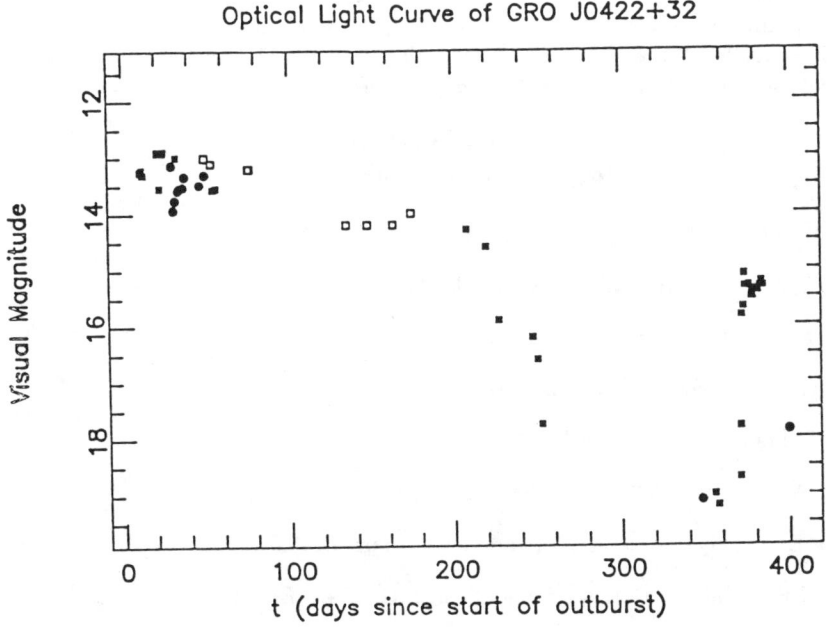

Figure 2. Long term optical light curve for J0422+32. Including our CCD photometry (circles), digital photometry reported to us (open squares), and measurements reported in IAUCs (filled squares).

DISTANCE AND ENERGETICS

Given a color excess of $E(B-V) = 0.4$ derived from of the 2200 Å interstellar feature (Shrader et al. 1992) and our estimate of the magnitude of the progenitor on the POSS, the intrinsic apparent magnitude of the progenitor is approximately 19. If we assume that the light in quiescence is dominated by a K V star, as in a number of other XRN, this would imply a distance of about 2.4 kpc. Alternatively, we can obtain an estimate of the distance based on the linear relationship between the equivalent width of the NaD lines and the distance. Based on 7 of our best optical spectra, we measure a mean NaD line equivalent width of 1.2 ± 0.2 Å, leading surprisingly to a distance of 2.4 ± 0.4 kpc. Thus, a distance of 2-3 kpc would not be unreasonable to assume.

Our distance estimate allows us to estimate the outburst luminosity and a make comparison with other X-ray novae. At maximum, the apparent visual magnitude of J0422+32 was $\simeq$ 13th mag. Given a color excess of 0.4 mag and a distance of 2.4 kpc, we find that the absolute visual magnitude at maximum was $\simeq -0.2$ mag. For comparison, the absolute visual magnitude of A0620-00 (1.1 kpc) and Nova Muscae 1991 ($\geq$ 3 kpc) were $\simeq -0.4$ and $\leq +0.1$ mag respectively, thus the these three XRNs are similar as far as optical luminosities are concerned.

At higher energies, Harmon et al. 1992 et al. reported that at maximum light, the 40-230 keV intensity of J0422+32 was about 3 Crab, implying an integrated flux of $\sim 5.1 \times 10^{-8}$ erg cm^{-2} s^{-1}. Assuming a distance of 2.4 kpc, we find an average peak 40-230 keV luminosity of $\sim 3 \times 10^{37}$ erg s^{-1}. An extrapolation onto the 2-10 keV band implies a luminosity $\sim 5 \times 10^{37}$ erg s^{-1} making its nominal $L_x/L_{opt} \simeq 10^2$, typical of persistent LMXBs. We note however, that this assumes the absence of a soft excess component, so this should be considered a lower limit.

REFERENCES

Castro-Tirado, A.J., et al. 1992, *IAU Circ.* no. **5588**.

Han, X.-H., and Hjellming, R.M., 1992, *IAU Circ.* no. **5593**.

Harmon, B.A. et al. 1992, *IAU Circ.* no. **5584**.

Harmon, B.A., 1993, private communication. and Hirata R., *IAU Circ.* no, **5704**.

McCrosky, R.E., 1992, *IAU Circ.* **5597**.

Mueller, J. 1992, *IAU circ.* no. **5597**.

Paciesas, W.S., et al. 1992, *IAU Circ.* no. **5580**.

Pavlenko, E.P., Shlyapnikov, A.A., and Castro-Tirado, A.J., 1992, *IAU circ.* no. **5594**.

Shrader, C.R., Wagner, R.M. and Starrfield, S.G. 1992, *IAU Circ.* no. **5591** .

Tanaka, Y., 1993, *IAU Circ.* no. **5851**.

Wagner, R.M., et al. 1992, IAU *Circ. No.* **5589**.

OPTICAL OBSERVATIONS OF THE X-RAY NOVA J0422+32

Ping Zhao, Paul Callanan, Michael R. Garcia, and Jeffrey E. McClintock
Harvard-Smithsonian Center for Astrophysics, Cambridge, MA 02138
Ronald A. Remillard, Center for Space Research, MIT, Cambridge, MA 02138
Andy Silber, Astronomy Dept., University of Washington, Seattle, WA 98195

ABSTRACT

We present the results of a photometric and spectroscopic study of the optical counterpart of the X-ray nova J0422+32, a recently discovered black hole candidate. Its light curve shows a similar decay pattern to that of other novae black hole candidates, but with a slower decay rate. Its rapid decline started 240 days after the outburst. Its current magnitude is R $\sim$ 19.4, which is still brighter than its quiescent value. Intensive photometry in January 1993 shows evidence of the 5.1 hour modulation reported by other groups. However, this modulation is not present in data taken in October 1992 and October 1993. Outburst spectra of J0422+32 are remarkably similar to those of Nova Muscae 1991 (a strong black-hole candidate) including double peaked emission lines of Hα, Hβ, Hγ, HeII 4686Å, HeI 5876Å and 6678Å.

1. INTRODUCTION

X-ray novae, also known as Soft X-ray Transients (SXTs), are a subset of Low Mass X-ray Binaries (LMXBs) that are characterized by a dramatic increase in X-ray intensity ($\sim 10^5 - 10^7$) during outburst, with a subsequent decay period that lasts several months. This extreme behavior makes them ideal for studying the accretion processes occurring in compact binary systems. These systems have provided us with our most secure evidence for the existence of black holes in the Galaxy.[1] A remarkable and exciting result from a recent study of the 13 known X-ray novae is that 9 (69%) of them are (strong or probable) black-hole candidates.[2]

J0422+32 is an X-ray nova discovered in outburst by GRO/BATSE on 5 August 1992.[3] Its primary is a probable black hole because its X-ray spectrum extends to about 600 keV and is similar to the spectrum of Cyg X-1.[4,5] Since September 1992 we have made extensive photometric and spectroscopic observations of the optical counterpart of J0422+32, which brightened by more than 7 magnitudes during outburst. We used the MMT, 1.2-m and 1.5-m telescopes on Mt. Hopkins and the MDM 1.3-m telescope on Kitt Peak. We present here an up-to-date optical light curve, and photometric and spectroscopic data obtained during outburst.

2. OBSERVATIONS

X-ray: On 5 August 1992, J0422+32 was detected by GRO/BATSE at a level of 0.2 Crab in the 20–300 keV band.[3] It increased to 3 Crabs within a few days. It stayed at that level for a few more days and then started to decay. Its 1/e decay time (time for its brightness to fall to 1/e of the maximum) in the above X-ray band is 44 days.[6] For comparison, A0620-00 (V616 Mon, a strong black-hole candidate) had an X-ray (2–20 keV) 1/e decay time of 37d.[7] The X-ray flux of J0422+32 is currently less than 0.001 Crab.

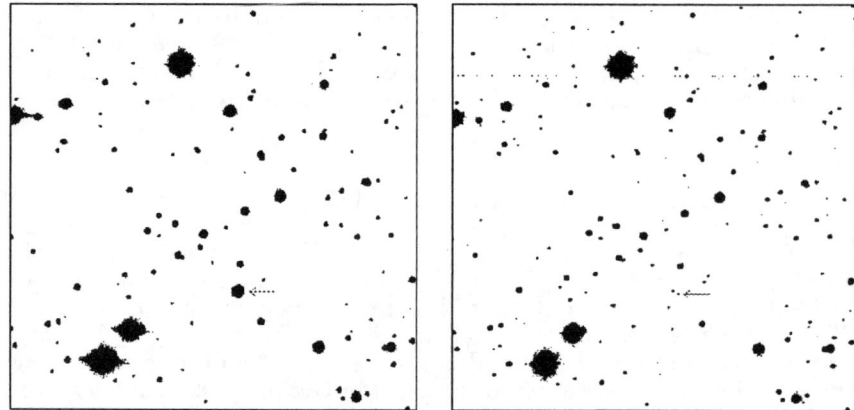

Figure 1. J0422+32 R images (5.3×5.3 arcmin^2) taken with the 1.2-m telescope on 11/18/92 (left) and 11/23/93 (right), 104 and 475 days after the outburst, respectively. Arrows point to J0422+32.

Optical: The optical counterpart of J0422+32 was discovered on 15 August 1992.[8] From a quiescent V magnitude >19.5, as determined from the STScI "Quick V" survey, it reached a maximum brightness of V ∼ 12.6.[9] Archival plate searches show no evidence of its outburst since 1928.[8,10] J0422+32 has the optical and ultraviolet characteristics expected for an X-ray nova in outburst.[11,12]

Photometric Study: We have been monitoring the decay of J0422+32 in both V and R bands with the 1.2-m telescope since 18 November 1992, 100 days after its outburst. We also made continuous observations on a number of nights in both V and R bands with the 1.2-m and the MDM 1.3-m telescopes to search for its orbital period. The results include:

1) Decay Light Curve: Figure 1 shows 2 R images of J0422+32 taken during the decay period. Figure 2 shows the R light curve. For the first 220 days (8/5/92 – 3/13/93) after outburst, the light curve is characterized by a very gradual drop in brightness (dV/dt = 0.0054 mag/day). This is the slowest optical decay rate ever reported for an X-ray nova. Its V band 1/e decay time is 190d. (For comparison, A0620-00 had a V band decay rate of 0.017 mag/day for the first 165 days and a 1/e decay time of 63d.) A rapid decline started 240 days after outburst (3/27 – 4/13/92) with a decay rate of dV/dt = 0.032 mag/day). There is a reported brief optical outburst during August 10 - 16, 370 days after its first outburst.[13] A second period of rapid decline started 20 September 1993. Currently (11/23/93) the magnitude is R ∼ 19.4 and quite variable (ΔR ∼ 0.5). Pre-outburst POSS data suggest R(quiescence) > 20, so that J0422+32 has not yet returned to quiescence 475 days after outburst.

2) Orbital Period: For the short term photometry, a total of 1779 V and 298 R images were obtained during the period October 1992 – October 1993. The data taken October 1992 show no clear evidence for a coherent periodic modulation. The data from January 1993 show 5.13 hours periodic modulation, which agrees with other reports.[14,15] However, the data from October 1993 do not show clear evidence for this 5.13 hour period. A secure determination of the orbital period will require that observations of J0422+32 be made in quiescence.

Spectroscopic Study: The outburst spectra of J0422+32 were taken at the MMT, the 1.5-m telescope and the MDM 1.3-m telescope during the period September – November 1992. Figure 3 shows typical spectra with 12Å and 5Å

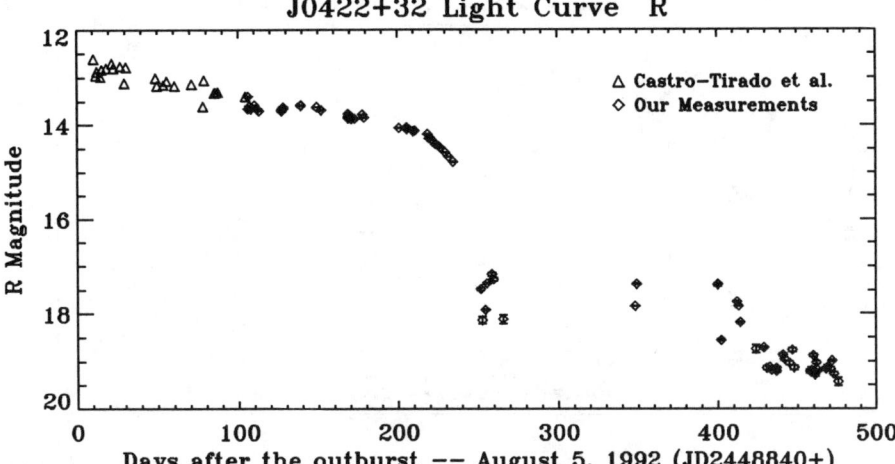

Figure 2. J0422+32 R band light curve, combining our data with the first 100 days of data by Castro-Tirado et al.[8] Our last data point was taken 11/23/93, 475 days after outburst. We believe that J0422+32 has still not returned to its pre-outburst state (R > 20).

resolution (FWHM). The following features were observed:

1) **12Å Resolution Spectra:** Interstellar lines due to the Na-D doublet at ~ 5890Å (EW=1.2 ± 0.2Å), the blend at 5780Å, 5797Å, and 5788Å (EW=0.5 ± 0.1Å), and the blend at 6269Å and 6282Å (EW=0.6 ± 0.1Å) are visible and indicate a reddening of E(B-V)= 0.4±0.1. Visible emission lines include Hα, Hβ, HeII 4686Å, HeI 5876Å and 6678Å. Equivalent widths are $\lesssim$ few Å, as is typical of LMXB in outburst. The Balmer decrement Hα/Hβ = 2.5. These spectra, which were taken three months after outburst appear remarkably similar to Nova Muscae 1991 a few days after outburst.[16] One notable difference is the weakness here of the NIII Bowen fluorescence lines at 4640Å, which were much stronger in earlier spectra of J0422+32.[17] This weakness was also noted in spectra taken by others on Feb 19.[18] The very blue continuum reddens slightly from 11/19 to 11/21, as the flux decreases.

2) **5Å Resolution Spectra:** Visible emission lines include Hβ, Hγ, and HeII 4686Å. All emission lines are double peaked with separations of 700−1000 km/sec and FWZI of 2000−2500 km/sec, similar to that seen in the Balmer lines of Nova Muscae 1991 in quiescence.[19] In 11 out of 12 cases, which clearly show double-peaked lines, the blue peak is always stronger. The velocity of the center of the Hβ and HeII 4686Å lines varies between 50 and 310 km/sec; and between −100 to 160 km/sec, respectively. Note the weakness of the N III Bowen blend at 4640Å (see above). All of the visible H-Balmer lines (Hβ through Hϵ) show wide (3000 km/sec FWZI) and shallow absorption. The emission reversals are on the blue edge of these absorption lines, therefore creating the appearance of an "inverse P-Cygni" profile. We note that as we go towards the blue, the absorption lines remain strong relative to the emission lines. The continuum level decreases between Oct 4 and Oct 9, but then increases by a factor of ~ 2 by Nov 24. The absorption lines of the later Balmer series are particularly prominent at this higher flux level.

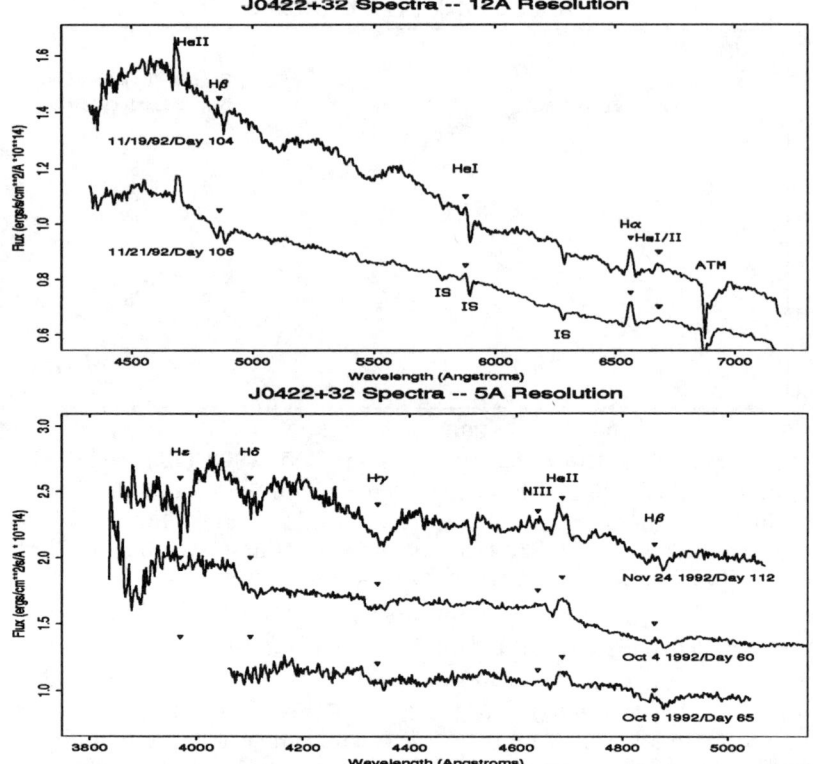

Figure 3. J0422+32 Spectra with 12Å (top) and 5Å (bottom) resolution taken with MMT and MDM in October and November 1992.

REFERENCES

1. McClintock, J.E. & Remillard, R.A. 1986, ApJ. 308, 110
2. van Paradijs, J. & McClintock, J.E. 1993, to appear in "X-ray Binaries", editors W.H.G. Lewin et al., Cambridge: Cambridge University Press
3. Paciesas, W.S. & Briggs, M.S. 1992, IAU Circ. 5580
4. Cameron, R.A. et al. 1992, IAU Circ. 5587
5. Sunyaev, R. et al. 1992, IAU Circ. 5593
6. Compton Observatory Newsletter Vol.3 No.1, 4, 1993
7. Ciatti, F. et al. 1977, A&A 56, 311
8. Castro-Tirado, A.J. et al. 1993, A&A 276, L37
9. Castro-Tirado, A.J. et al. 1992, IAU Circ. 5588
10. Shao, C.Y. 1992, IAU Circ. 5606
11. Wagner, R.M. et al. 1992, IAU Circ. 5589
12. Shrader, C.R. et al. 1992, IAU Circ. 5591
13. Filippenko & Matheson 1993, IAU Circ. 5842
14. Kato, T. et al. 1992, IAU Circ. 5676
15. Chevalier, C. & Ilovaisky, S.A. 1993, IAU Circ. 5692
16. Della Valle, M., Jarvis, B.J., & West, R.M. 1991, A&A 247, L33
17. Shrader et al. 1993, Proceedings of the Compton Symposium, St. Louis
18. Harlaftis, E.T. & Chartes, P.A. 1993, IAU Circ. 5728
19. Remillard, R.A., McClintock, J.E. & Bailyn, C.D. 1992, ApJ. 399, L145

OBSERVATIONS OF GRO J0422+32
ON HIGH AND LOW OPTICAL STATES

C. Bartolini, A. Guarnieri, A. Piccioni, G. Zampieri
Dipartimento di Astronomia, Università di Bologna, Italy
bartolini@alma02.bo.astro.it

G. Beskin, S. Neizvestny, I. Panferova, V. Plokhotnichenko, A. Zhuravkov
Special Astrophysical Observatory, Russia
nws@sao.sovam.com

ABSTRACT

We hereby present spectroscopic and photometric observations of the Nova Persei 1992 = GRO J0422+32 obtained at the Special Astrophysical Observatory and at the Bologna Observatory. Our observations show that different outbursts are driven by different mechanisms of accretion. We have interpreted the optical activity we observed on an ultrashort time scale as indicative of non-thermal processes in accretion structures around the compact object.

1. INTRODUCTION

The high energy source GRO J0422+32 was discovered in the X-ray band on August 5, 1992 by Paciesas et al. (1992), and optically identified by Castro-Tirado et al.(1992) as a 12th mag transient star called Nova Persei 1992. The X-ray spectrum presented a soft component and a hard tail, as is characteristic of X-ray novae, a subclass of low mass X-ray binaries which probably contain black hole candidates. The X-ray flux showed quasi-periodic oscillations (Kouveliotou et al. 1992, Vikhlinin et al. 1992, Pietsch et al. 1992). The optical light curve displayed second and third secondary maxima approximately 32 and 167 days after the outburst. The average decline rate between the maxima was about 0.01 mag/day. The characteristics of the X-ray and optical light curve make the Nova Per 1992 similar to the V616 Mon and the Nova Mus, supporting the hypothesis that it is a black hole candidate.

In August 1993 the Nova Per 1992 underwent a new major optical burst attaining a luminosity of about 15 mag (Castro-Tirado & Ortiz 1993, Filippenko & Matheson 1993). The star remained at a high luminosity level for about 15 days, then underwent a very steep decline. In early September, 1993, the star's V magnitude was about 19 (Beskin et al. 1993).

2. OBSERVATIONS

Nova Per 1992 was observed in 1992 and 1993 at the Special Astrophysical Observatory (SAO) with a TV spectral scanner, a UBVR photometer, and the hard/software MANIA (Multichannel Analysis of Nanoseconds Intensity Alteration) complex mounted on the 6-m BTA telescope and with the 1-m telescope equipped with CCD, and at the Bologna Observatory with the fast double head photometer and a Boller & Chivens 26767 grating spectrograph with a CCD sensor, mounted on the 1.5-m telescope.

Considering the spectroscopic and photometric behaviour of the Nova Per 1992 during the fading phase from August 1992 to July 1993, we distinguish, somewhat arbitrarily, high and low optical states.

2.1 The first optical decline

a) The high optical state. We consider Nova Per 1992 in "high optical state" when its V magnitude is ≤ 15.5; this happened during the first 6 months after the outburst of August 1992. In this period the spectra of the star showed a blue continuum variable on time scales of days or hours (Fig. 1) and some weak emission features (H and He).

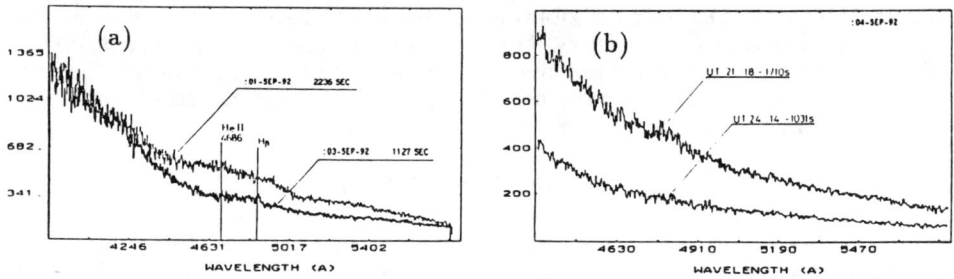

Fig. 1. (a) Variability of the continuum and of the emission lines of the Nova Per 1992 on time scale of days. On September 1, 1993, Hβ was probably in absorption, and He II 4686 was not visible; on September 3 both the lines had become clearly in emission. (b) Variability of the continuum and of the emission lines of the Nova Per 1992 on a time scale of hours on September 4, 1993.

The luminosity of Nova Per 1992 was irregularly variable in different color bands (B, V, R) on time scales ranging from 4 ms to 200 s (Fig. 2a). An example of a short flare is given in Fig. 2b.

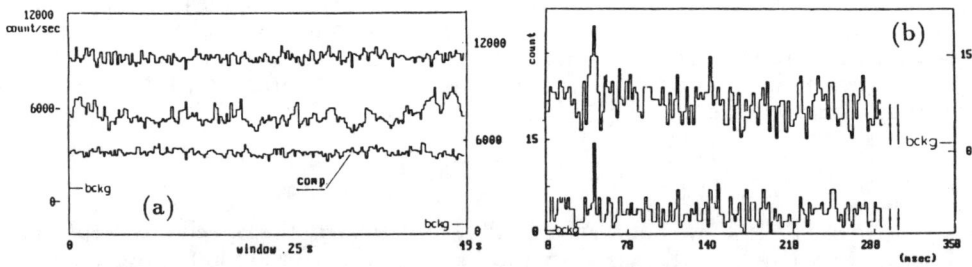

Fig. 2. (a) Absence of microvariability of Nova Per 1992 during the low optical state on July 21, 1993 (top). Microvariability of Nova Per 1992 during the high optical state on January 18, 1993 (middle). The behaviour of a constant star of approximately the same magnitude is shown for comparison (bottom). On the left and right scales the photon counts of the Nova and of the comparison star respectively are indicated. (b) Flash of the Nova Per 1992 showing a rise time of 4ms, observed on January 18, 1993 in V light (top) and in B light (bottom).

The shortest flares showed rising times of 4-40 ms, which allows us to establish an upper limit of $10^8 \div 10^9$ cm on the size of the flares origin regions. In the temporal range of 100 ns to 1 ms it is highly probable that variability was absent. A preliminary analysis of the observations taken with the double head photometer shows a flat power spectrum in the range between 5 s and 15 min.

Our double head photoelectric observations on December 27 and 29, 1992 (Bartolini et al., 1993), confirm the 5-hour periodic variations reported by Kato et al. (1992) and by Chevalier & Ilovaisky (1992). Nevertheless the modulation, whose amplitude was of about 0.35 mag, can be fairly well represented by a two-maxima sinusoidal curve with a doubled period. On February 13, 1993 the periodic variation was still present but the amplitude decreased to 0.08 mag.

b) The low optical state. During the "low optical state" (V magnitude >15.5) Nova Per 1992 does not show any significant variability (Fig. 2a), at least in the time interval from 100 ns to 200 s we have analyzed thus far.

A sudden passage from the high to low state occurred in 20 s on December 29, 1992 (Bartolini et al. 1993).

2.2 The outburst of August 1993

In the second week of August 1993 Nova Per 1992 underwent a second big optical outburst, the characteristics of which were quite different from those of August 1992: the 1993 maximum presented a flat shape that lasted at least 15 days, followed by a rapid drop at a rate of not less than 0.16 mag/day.

We observed Nova Per 1992 with the TV scanner of the SAO 6-m telescope on September 11 and 12, 1993 when the V magnitude of the star was fainter than 18. The spectra (Fig. 3a,b) showed some emission features including Hα, Hβ and probably He I 5876 (but not He II 4686) on a red continuum. There were indications that He I 5876 (and perhaps Hα) was variable on a time scale of hours.

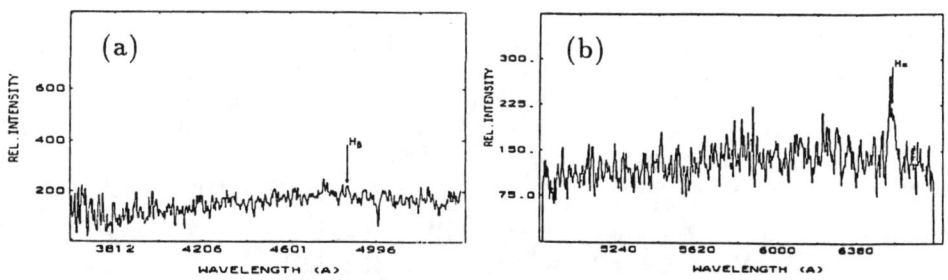

Fig. 3. (a) Spectrum of Nova Per 1992 in the blue region taken with the 6-m SAO telescope on September 11, 1993. Hβ in emission is probably present. (b) Spectrum of Nova Per 1992 in the red region taken with the 6-m SAO telescope on September 12, 1993. Hα in emission is evident. HeI 5876 is also in emission.

3. CONCLUSIONS

The ultrashort optical flares we detected in the high state can be linked to the dimensions of the origin regions (which we found to be the typical ones of accretion structures) and to the emission mechanism of the radiation. Assuming a distance of the object larger than 2 Kpc, the flares are indicative of a brightness temperature $T_b > 10^8 \div 10^9$ K (Bartolini et al. 1993), which proves the existence of a non-thermal mechanism for the generation of the photons. We believe that the microvariability present in the high state could be the result of a fragmentation in the accreting structure, where the non-thermal optical emission is generated. This fragmentation is absent in the low state. It can also be noted that a variability on a time interval 2 ms $\div$ 1 min was observed in the X-ray band near the X-ray maximum of Nova Cyg 1989 = GS 2023+338 (Kitamoto et al. 1989).

It has recently been proposed (Chen, Livio & Gehrels 1993) that disk instability and mass transfer instability could be responsible, respectively, for the main and the following bursts in the X-ray novae which are black holes candidates. This hypothesis is supported by the recent behaviour of Nova Per 1992. As we have already emphasized, the outburst of August 1993 is completely different from the one on August 1992.

Moreover, the spectra we acquired approximately one month after the onset of the optical outburst of August 1993 are very different from those we obtained one month after the optical outburst of August 1992. We conclude that there are strong indications of a different mechanism for the two big outbursts which occurred one year apart in Nova Per 1992. Knowledge of the behaviour of the source in other energy bands is crucial for checking the nature of these mechanisms.

ACKNOWLEDGEMENTS

This work was partially funded by a grant from the Italian Ministero per l'Universitá e la Ricerca Scientifica e Tecnologica.

REFERENCES

Bartolini, C., Guarnieri, A., Piccioni, A., Beskin, G., & Neizvestny, S. 1993, Proc. Les Diablerets Integral Workshop, ApJ S. S. in press
Beskin, G., Kajsin, S., Neizvestny, S., Panferova, I., Plokhotnichenko, V., Tikhonov, A., Ugrumov, A., Zhuravkov, A., Bartolini, C., Cosentino, G., Guarnieri, A., & Piccioni, A. 1993, IAUC 5863
Castro-Tirado, A. J., Pavlenko, P., Salyapikov, A., & Gershberg R. 1992, IAUC 5588
Castro-Tirado, A. J., & Ortiz, J. 1993, IAUC 5842
Chen, W., Livio, M., & Gehrels, N. 1993, ApJ, 408, L5
Chevalier, C., & Ilovaisky, S. A. 1992, IAUC 5644
Filippenko, A.V., & Matheson, T. 1993, IAUC 5842
Kato, T., Mineshige, S., & Hirata, R. 1992, IAUC 5676
Kitamoto, S., Tsunemi, H., Miyamoto, S., Yamashita, K., Mizobuchi, S., Nagakawa, M., Dotani, T., & Makino, F. 1989, Nature, 342, 518-520
Kouveliotou, C., Finger, M. H., Fishman, G. J., Meegan, C. A., Wilson, R. B., & Paciesas, W., S. 1992, IAUC 5592
Paciesas, W.S., & Briggs, M. S. 1992, IAUC 5580
Pietsch, W., Haberl, F., Gehrels, N., & Petre, R. 1993, A&A, 273, L11-L14
Vikhlinin, A., Finoguenov, A., Sitdikov, A., Sunyaev, R., Goldwurm, A., Paul, J., Mandrou, P., & Techine, P. 1992, IAUC 5608

THE MASS OF THE BLACK HOLE IN GS2023+338/V404 CYGNI

J. Casares[1,2] & P. A. Charles[1]
1.- Department of Astrophysics, Nuclear Physics Lab., Oxford OX1 3RH, UK.
2.- Instituto de Astrofísica de Canarias, E-38200 La Laguna (Tenerife), Spain.
19464::jcv; 19464::pac

ABSTRACT

New high resolution (0.8 Å) Hα spectroscopy provides further constraints on the orbital parameters of the black hole candidate V404 Cyg. Our updated mass function is now 6.08 ± 0.06 M$_\odot$ which confirms V404 Cyg as the most important black hole candidate yet found. We also resolve the rotational broadening of the secondary star which yields the lowest mass ratio ever measured ($q = M_c/M_x = 0.050$). The mass of the black hole is now restricted to the range 7-29 M$_\odot$.

1. INTRODUCTION

V404 Cyg is the optical counterpart of the X-ray transient GS2023+338, discovered by the Ginga satellite at the start of its outburst in May 1989 (Makino et al. 1989). Two years later, Casares et al. (1992a) discovered absorption features from the late-type companion and measured its radial velocity curve. The implied mass function (6.3 ± 0.3 M$_\odot$) is more than two times higher than that of earlier black hole candidates, thus providing the best evidence yet for a galactic black hole. Later Wagner et al. (1992) reported the discovery of ellipsoidal modulations in their I-band light curve giving further support to the spectroscopic ephemerides of Casares et al. (1992a). However, there is evidence for a 0.24 days modulation in the radial velocities of the Hα emission whose interpretation remains unclear (see Casares et al. 1992b, 1993). On the other hand, the detection by Martín et al. (1992) of the LiIλ6707.8 line in the spectrum of the optical companion suggests the action of spallation processes in the vicinity of the black hole. Since then ^{7}Li has also been detected in the companion stars of the soft X-ray transients A0620-00 and Cen X-4 (see Charles et al. 1994).

2. OBSERVATIONS

V404 Cyg was observed with the ISIS spectrograph attached to the 4.2m William Herschel telescope on the nights of 4-10 July 1992. A special dichroic (with a wavelength cut-off at 7500Å) was used, enabling the Hα region to be observed simultaneously with the CaII IR triplet (whose results will be presented elsewhere). The principal aim of the project was to measure the rotational broadening of the absorption lines so we chose a resolution of 0.8 Å (FWHM) at Hα. We decided to take individual exposures of 1800s (0.003 phase resolution), giving a total of $\sim$ 9 spectra per night. Unfortunately, a system failure occured on the night of 6 July and only a single spectrum could be obtained. The seeing varied from 0.8 to 1.3 arcsec, except for the first when it rose to 2 arcsec. Three more hours of data were obtained on the night of 12 August 1993 as part of the Service programme.

3. UPDATED EPHEMERIDES

In order to measure the radial velocities and the rotational broadening of the companion star a grid of 15 template stars was obtained with the same instrumental configuration as for V404 Cyg. These cover a range of spectral types from G8 to K2 in three different luminosity classes (V to III), seven of which are also radial velocity standards.

Individual radial velocities were extracted through cross-correlation with the K0IV template HR8857, which gave the lowest χ^2 in the least-square sine fit. In order to extend our time-base we have combined the new radial velocities with those reported in Casares et al. (1992a). Our 4 year data-base now allows the orbital period to be refined to 6.4714 ± 0.0001 days and the velocity semi-amplitude, $K_c = 208.5 \pm 0.7$ kms^{-1}. The mass function is therefore $\frac{(M_x \sin i)^3}{(M_x + M_c)^2} = 6.08 \pm 0.06$ M$_\odot$ which represents a factor of 5 improvement in accuracy compared to our previous results. The systemic velocity and zero epoch are -0.4 ± 2.2 km s^{-1} and 2448813.873 ± 0.004 respectively. Figure 1 presents the radial velocity curve folded on the new ephemerides.

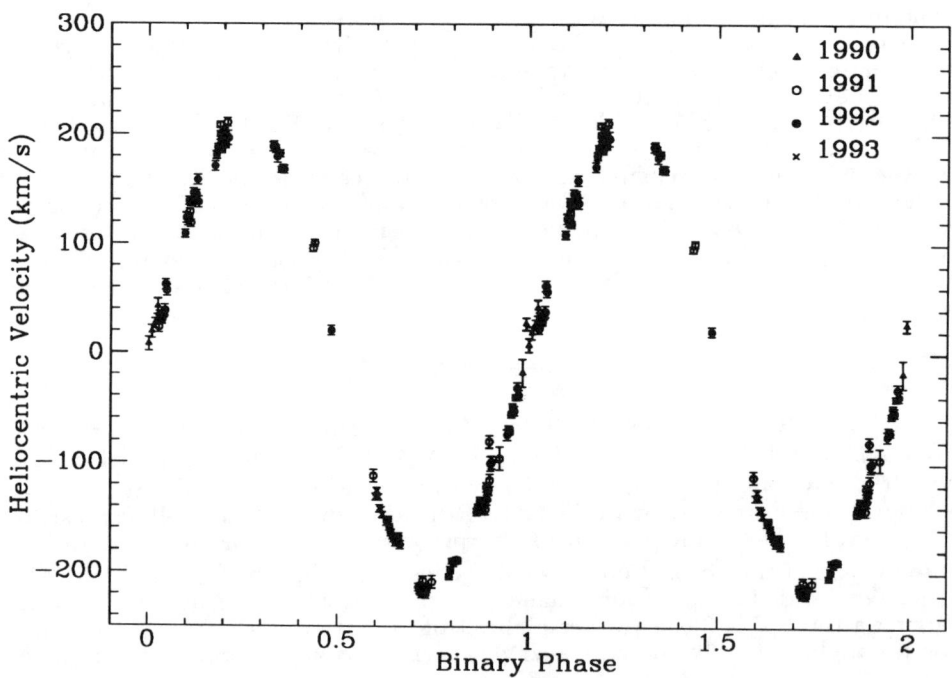

Fig. 1. Radial velocity curve of the secondary of V404 Cyg folded on the ephemerides given in the text. Each symbol indicates data sets obtained in different years.

4. ROTATIONAL BROADENING

We have subtracted different broadened versions of the K0IV template from the doppler corrected sum of our V404 Cyg high resolution spectra. The broadened templates were also multiplied by a variable factor to account for the continuum excess of the accretion disc, prior to subtraction. The minimum χ^2 was achieved for the spectrum broadened by 36 km/s. This could be improved by performing a parabolic fit to all the χ^2 values giving $V_{\rm rot}\sin i = 36.2 \pm 1.3$ km/s. The same analysis was applied to the remaining 14 templates yielding $V_{\rm rot}\sin i$ always in the range 35-38 km/s. This defines a gaussian distribution characterized by a mean of 36.5 ± 1.2 km/s, which is the value we have adopted for the rotational broadening hereafter. The reliability of this result is demonstrated in fig.2.

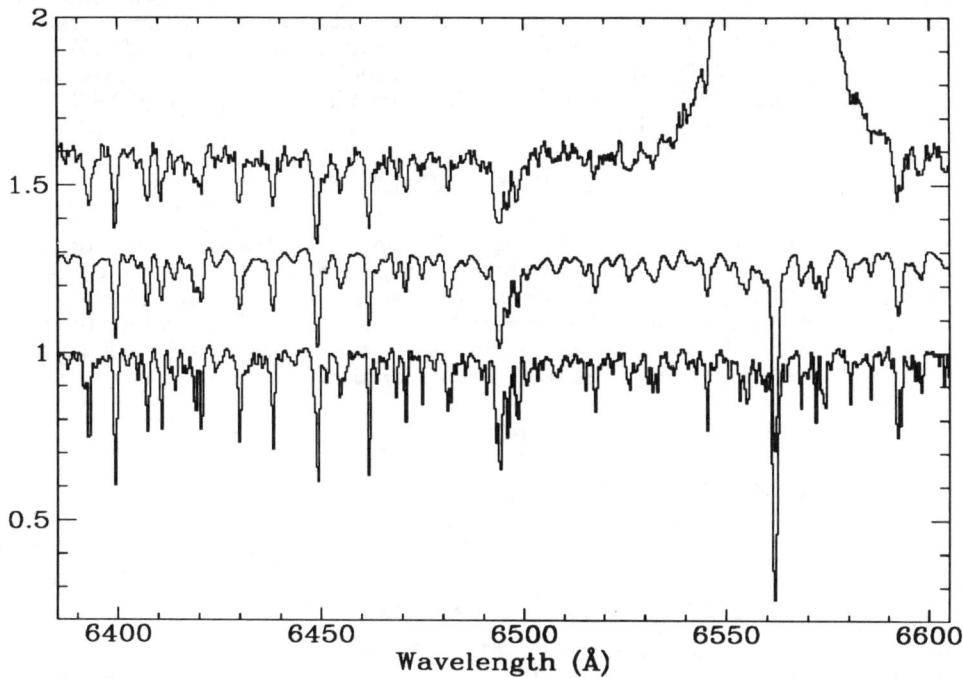

Fig. 2. Doppler corrected sum of V404 Cyg (top) compared to the raw spectrum of the K0IV template HR8857 (bottom) and a version broadened by 36.5 km/s (middle).

The binary mass ratio $q\ (= M_c/M_x)$ is known to be related to the rotational broadening through the formula (Wade & Horne 1988)

$$V_{\rm rot}\sin i = 0.462\ K_c\ q^{1/3}(1+q)^{2/3}. \qquad (1)$$

Substituting our values of $V_{\rm rot}\sin i$ and K_c in eq.1 we find $q = 0.050 \pm 0.005$. The compact object in V404 Cyg is thus 20 times more massive than its companion.

5. DISCUSSION

Our 4 year data base of the X-ray transient V404 Cyg in quiescence enables us to confirm and refine our previous ephemerides. We derive an updated mass function of 6.08 ± 0.06 M$_\odot$, the highest yet measured. This is so ar in excess of the maximum allowed mass for a rotating neutron star (Friedman et al. 1986) that V404 Cyg unequivocally provides the best dynamical evidence for a black hole.

Furthermore, the use of high resolution spectroscopy has been succesful in resolving the absorption lines of the optical companion and thereby measuring its rotational broadening. This has provided the first measurement of the mass ratio in V404 Cyg, and hence a very tight constraint on the system parameters. Combining this result with constraints on the inclination and secondary's mass gives us a well-defined range for M_x. A severe upper limit to the inclination comes from the absence of X-ray eclipses at the outburst peak (Tanaka 1989) which leads to i$\leq 80°$. On the other hand, we assume an upper limit of 1.32 M$_\odot$ to the mass of the K0IV companion as implied by its stripped-giant evolution (King 1993). The combination of these two constraints with $q = 0.050 \pm 0.005$ gives a black hole mass of 7-29 M$_\odot$.

ACKNOWLEDGEMENTS

We thank T.R. Marsh for the use of his optimal extraction routines, the MOLLY analysis package and his valuable support and suggestions. We also thank Renné Rutten for supporting the 1993 service observations. The Isaac Newton Group of telescopes are operated on the island of La Palma by the Royal Greenwich Observatory in the Spanish Observatorio del Roque de los Muchachos of the Instituto de Astrofísica de Canarias.

REFERENCES

Casares, J., Charles, P.A. & Naylor, T. 1992a, Nature, 355, 614
Casares, J. & Charles, P.A. 1992b, MNRAS, 255, 7
Casares, J., Charles, P.A., Naylor, T. & Pavlenko, E.P. 1993, MNRAS, in press
Charles, P.A., Casares, J., Martín, E.L. & Rebolo, R. 1994, these proceedings
Friedman, J.L., Ipser, J.R. & Parker, L. 1986, ApJ, 304, 115
King, A.R. 1993, MNRAS, 260, L5
Makino, F. & the Ginga Team 1989, IAUC 4782
Martín, E.L., Rebolo, R., Casares, J. & Charles, P.A. 1992, Nature, 358, 129
Tanaka, Y. 1989, in Proc. 23rd ESLAB Symp. on Two-Topics in X-Ray Astronomy, J.Hunt & B.Battrick eds. ESA Publ. Division, p.1
Wade, R.A. & Horne, K. 1988, ApJ, 324, 411
Wagner, R.M., Kreidl, T.J., Howell, S.B. & Starrfield, S.G. 1992, ApJ, 401, L97

X-ray Observations of the Black Hole Binary V404 Cygni in Quiescence

R. M. Wagner[1], S. G. Starrfield[2], T. J. Kreidl[3], S. B. Howell[4], and R. M. Hjellming[5]

[1] Department of Astronomy, Ohio State University, 174 West 18th Avenue, Columbus, OH 43210
[2] Department of Physics and Astronomy, Arizona State University, Tempe, AZ 85287-1504
[3] Lowell Observatory, 1400 West Mars Hill Road, Flagstaff, AZ 86001
[4] Planetary Science Institute, 620 N. 6th Avenue, Tucson, AZ 85705
[5] National Radio Astronomy Observatory, P.O. Box O, Socorro, NM 87801

ABSTRACT

The results of a ROSAT observation of V404 Cygni are described and discussed. The 0.2–2.4 keV light curve of V404 Cyg exhibits substantial variability on time scales of less than a day. The observed X-ray spectrum can be described equally well by either a blackbody, power law, or a thermal bremsstrahlung continuum and interstellar absorption column densities in excess of $log N_H (cm^{-2}) \simeq 22.1$. Both optical and radio measurements suggest that $log N_H \simeq 21.7$ so that some internal absorption of X-rays, presumably in the accretion disk surrounding the compact object, is required to fully explain the observations.

1. INTRODUCTION

X-ray novae are a subclass of low mass X-ray binaries (LMXBs) which undergo brief episodic outbursts during which their X-ray luminosity can increase by factors of $\sim 10^6$ relative to long intervals of quiescence. During their outbursts, the X-ray and optical properties of X-ray novae are similar to those of bright persistent LMXBs. The outburst of an X-ray nova is thought to be caused by a sudden increase in the mass accretion rate onto a neutron star or black hole. After the peak of the outburst, the X-ray flux declines exponentially to a nearly non detectable level. The light of the system between outbursts and in quiescence is dominated by an optically bright accretion disk and spectrum of the donor or secondary star. Both mass transfer (Hameury et al. 1986) and accretion disk instability (Huang and Wheeler 1989) models have been proposed to modulate the mass transfer rate. Recent optical observations of X-ray novae in quiescence indicates that these objects may represent the most promising source of new black hole candidates.

Major observational questions regarding these systems are the shape of the X-ray spectrum and the X-ray luminosity during their periods of quiescence. For example, mass transfer instability models require a substantial amount of hard X-rays during quiescence to efficiently heat the companion star in order to initiate a mass transfer instability and thus produce an outburst. The X-ray luminosity in quiescence for Cen X-4 has been measured to be $(1.5 - 4.2) \times 10^{33} erg\ s^{-1}$ (van Paradijs et al. 1987), but this system contains a neutron star and thus may not be characteristic of black hole X-ray novae. The quiescent soft X-ray luminosity of the black hole X-ray nova A0620-00 (McClintock and Remillard 1986) has been measured to be below $10^{32} erg\ s^{-1}$ (Long, Helfand,

and Grabelsky 1981) and this result suggests that the X- ray luminosities of X-ray novae could be very low in quiescence.

In this contribution, we report the results of X-ray observations of the X-ray nova and black hole candidate V404 Cyg (= GS 2023+338) in quiescence which may be of great importance in our efforts to understand the mechanisms which drive the outbursts of X-ray novae.

2. RESULTS

On 1992 November 6–7, we obtained X-ray observations of V404 Cyg (outburst days 1265–1266) in the 0.2–2.4 keV energy band with the ROSAT satellite and Position Sensitive Proportional Counter (PSPC). A total on–source integration time of 16.111 ksec was achieved with the PSPC in an unfiltered mode. Our X-ray observations cover spectroscopic phases (Casares, Charles, and Naylor 1992) 0.40 − 0.57, where phase zero corresponds to inferior conjunction of the secondary star. Thus, these observations were obtained near superior conjunction of the secondary star and inferior conjunction of the accretion disk. V404 Cyg was detected easily and was found to be coincident with the optical and radio positions (Wagner et al. 1991) to within 1.5" in right ascension and 3.5" in declination. A total of 389 ± 21 net counts or an average count rate of 0.024 ± 0.001 $counts$ s^{-1} were obtained for V404 Cyg.

The 0.2–2.4 keV light curve of V404 Cyg during the time span of our observations is shown in Figure 1. For this analysis, the data have been averaged into 600 sec bins in order to achieve a sufficient signal–to–noise ratio. Our observations show that at this time the source decreased in intensity by a factor of 10 over an interval of ≤ 0.5 day. The light curve appears to resemble the decay of a flaring event that began sometime just before the start of our observations or indicates an anisotropy of the disk X–ray surface brightness that is modulated by the orbital motion. In addition, inspection of the highest intensity bins shows variability of factors of ~ 2 on time scales of ~ 30 minutes. The sampling of the data and faintness of the source precludes further time series analysis.

The average 0.2–2.4 keV spectrum of V404 Cyg is shown in Figure 2. The spectrum at energies below about 1.3 keV is strongly attenuated by the interstellar medium. The neutral hydrogen absorption column density in this direction has been estimated to be $log N_H (cm^{-2}) \simeq 21.7$ as determined by continuum fitting of our optical spectra, the strength of several interstellar absorption lines (Casares et al. 1991; Wagner et al. 1991), and the results of a 21 cm neutral hydrogen absorption experiment at the VLA (Han and Hjellming 1992). The effects of interstellar absorption and the small energy range of the spectrum make detailed analysis of the intrinsic shape difficult. We have, however, modeled these data assuming blackbody, power law, and thermal bremsstrahlung continuua.

In our initial set of models, we assumed that $log N_H$ was given by the optical and radio measurements. With this assumption, fits of the data to blackbody and power law continuua were poor and resulted in relatively large values of the reduced χ^2. The largest discrepancy occurred between 0.9 and 1.5 keV in both cases. Thermal bremsstrahlung fits did not converge for any reasonable value of the temperature.

In our second set of models, we varied the absorbing column density as well as the continuum parameters during the fitting. We find that these three model continuua can fit the observed data equally well in a statistical sense, with a reduced χ^2 a factor of ~ 2 better than those models for which the absorbing column density was fixed to the optical/radio value. All of these models require an absorbing column density in excess of $log N_H \simeq 22.1$. The best-fit blackbody spectrum is shown in Figure 2 and it requires kT = 0.21 ± 0.05 keV and $log N_H = 22.17 \pm 0.15$. The best-fit thermal bremsstrahlung model requires a temperature of 0.33 keV and $log N_H = 22.24$. A fit to a power law requires an energy index of 6 and an absorbing column density of $log N_H = 22.36$. The total observed 0.2–2.4 keV flux was $2.95 \times 10^{-13} erg\ cm^{-2}\ s^{-1}$.

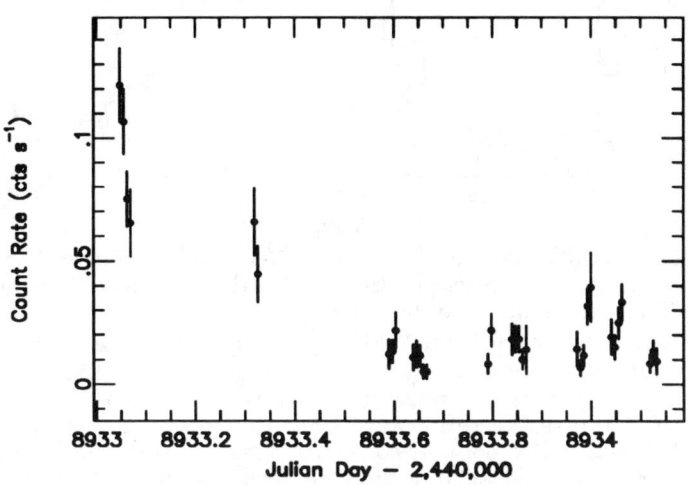

Fig. 1. X-ray light curve of V404 Cyg obtained 1992 November 6–7 with the ROSAT/PSPC in the 0.2–2.4 keV band. Substantial variability of V404 Cyg on ≤ 0.5 day time scales while in quiescence is evident.

Assuming a distance of 3.5 kpc (Wagner et al. 1992) and the blackbody spectrum described above, we derive a total 0.2–20 keV luminosity of $\simeq 8 \times 10^{33} erg\ s^{-1}$ for V404 Cyg in quiescence. If nearly all of the X-rays are produced near the inner edge of the accretion disk, then our X-ray luminosity corresponds to an accretion rate of $\simeq 10^{14} g\ s^{-1} = 1.4 \times 10^{-12}\ M_\odot yr^{-1}$ in quiescence, assuming an efficiency for converting mass to radiative energy of 10%.

3. DISCUSSION

Our X-ray observations of V404 Cyg represent the first unambiguous detection of a black hole X-ray nova in quiescence. Recent spectroscopic observations by Casares, Charles, and Naylor (1992) demonstrated that V404 Cyg consists of a compact object with a mass in excess of 6.3 $M_\odot$ and a late-type secondary star in a binary system with an orbital period of 6.473 days. The discovery of ellipsoidal light variations by Wagner et al. (1992) suggested that the mass of the compact object might be as high as 19 $M_\odot$ and that the secondary star must be evolved in order to avoid an excessively large X-ray luminosity during the 1989 outburst. The spectroscopic and photometric results imply that the compact object must be extremely massive and probably a black hole for any reasonable orbital inclination or mass of the secondary star.

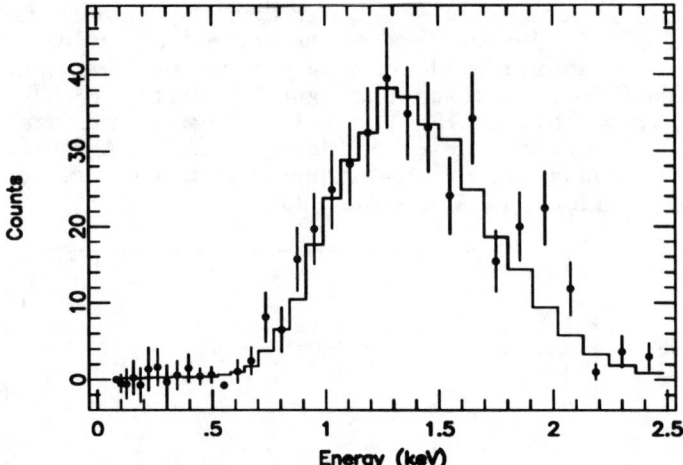

Fig. 2. Observed soft X-ray spectrum of V404 Cyg (filled circles) and the best-fit blackbody spectrum. Neutral hydrogen absorbing column densities in excess of $logN_H \simeq 22.1$ are required to describe the spectrum.

The observed X-rays from V404 Cyg in quiescence are most likely due to accretion onto the compact object at a rate of $\sim 10^{-12}$ $M_\odot yr^{-1}$. This rate is not necessarily equal to the mass flow rate through the disk at an arbitrary radius or even the mass accretion rate from the companion star. The present data preclude any detailed comparisons with theoretical model X-ray spectra of relativistic accretions disks surrounding stellar mass black holes (Fu and Tamm 1990 and references therein). In addition, our observations indicate that the emergent X-ray spectrum of V404 Cyg is relatively soft during quiescence. This suggests that the 1989 outburst was not initiated by a mass transfer instability in which a substantial hard X-ray ($kT \geq 10$ keV) luminosity is required to irradiate the companion star. The orbital separation and thus the solid angle subtended by the secondary, as viewed from the compact object, also preclude a strong X-ray heating effect in quiescence. Thus, based on our observations, the lack of a significant hard X-ray flux and the probable presence of a large, extended, and cold neutral region in the accretion disk suggests that an accretion disk instability is a likely model for the outburst of V404 Cyg and possibly other X-ray novae as well.

REFERENCES

Casares, J., et al. 1991, MNRAS, 250, 712
Casares, J., Charles, P.A., & Naylor, T. 1992, Nature, 355, 614
Fu, A. & Taam, R.E. 1990, ApJ, 349, 553
Hameury, J.M., et al. 1986, A&A, 162, 71
Han, X. & Hjellming, R.M. 1992, ApJ, 400, 304
Huang, M. & Wheeler, J.C. 1989, ApJ, 343, 229
Long, K.S., Helfand, D.J., & Grabelsky, D.A. 1981, ApJ, 248, 925
McClintock, J.E. & Remillard, R.A. 1986, ApJ, 308, 110
van Paradijs, J, Verbunt, F., Shafer, R.A., & Arnaud, K.A. 1987, A&A, 182, 47
Wagner, R.M., et al. 1991, ApJ, 378, 293
Wagner, R.M., et al. 1992, ApJ, 401, L97

BATSE OBSERVATIONS OF NOVA MUSCAE 1991

W. S. Paciesas, M. S. Briggs, G. N. Pendleton
University of Alabama in Huntsville

B.A. Harmon, C.A. Wilson, S.N. Zhang,[1] G. J. Fishman
NASA/Marshall Space Flight Center

ABSTRACT

The *Compton* Burst and Transient Source Experiment (BATSE) detected hard x-ray flux from Nova Muscae 1991 (GS/GRS 1124–68) during an interval of ~100 days beginning ~130 days after the January 1991 main outburst. The light curve during this secondary outburst is roughly symmetric, reaching a maximum around mid-July 1991 at an intensity of ~15% of the peak intensity during the main outburst. The hard x-ray spectrum displays a soft-to-hard evolution during the rise to maximum; the post-maximum spectral evolution is less well determined. We compare our observations with those of the *GRANAT*/SIGMA experiment, which covered the initial outburst well but missed most of the secondary outburst.

1. INTRODUCTION

Nova Muscae 1991 was discovered independently as an x-ray transient (GS/GRS 1124–68) in January 1991 by monitors on *Ginga* (Makino & the *Ginga* team 1991) and *GRANAT* (Lund & Brandt 1991). In low-energy x-rays ($E < 20$ keV), the source reached a maximum intensity of ~8 Crab on 15 January (Kitamoto et al. 1992). Subsequently, it decayed almost exponentially with a time constant of ~31 days for ~75 days, at which time a secondary outburst occurred, followed by a further exponential decay with a slightly longer time constant (~37 days). The power density spectrum showed flat low-frequency noise (below ~2 Hz) and variable-frequency quasi-periodic oscillations at higher frequencies (~3–10 Hz) (Tanaka, Makino & Dotani 1991; Grebenev et al. 1992). The spectrum was "ultrasoft" at the lowest energies with an apparently power-law tail extending to several hundred keV (Gil'fanov et al. 1991). A narrow emission line at 481 ± 22 keV was detected during a 13 hour SIGMA observation on 20–21 January; this was interpreted by Gil'fanov et al. as a red-shifted positron annihilation line. The power-law tail diminished in intensity faster than the ultrasoft component during the first few months. Nova Muscae was not detected in high-energy x-rays in April, but was present during SIGMA observations in May and August at ~10% of the January maximum.

An optical nova was detected within the x-ray error circle (Della Valle, Jarvis & West 1991a, 1991b). Subsequent monitoring showed it to decay at a rate similar to the low-energy x-rays. A binary orbital period of 10.42 hours was found, from which a mass function of $3.1 \pm 0.4\,M_\odot$ was determined (McClintock, Bailyn & Remillard 1992), making the source a leading black hole candidate.

The BATSE instrument on the Compton Gamma Ray Observatory (CGRO)

---

[1] Also Universities Space Research Association

functions as an all-sky monitor by detecting Earth occultations of low-energy gamma-ray sources (Harmon et al. 1992). CGRO was launched in April 1991, too late to observe the main outburst of Nova Muscae. Nevertheless, we used the occultation method to search for subsequent emission and found that Nova Muscae was detectable by BATSE during an interval of ~100 days, beginning just before the May 1991 detection by SIGMA.

2. OBSERVATIONS

The BATSE Earth occultation methodology has been described by Harmon et al. (1992). We obtained spectra in 16 energy channels spanning the range 20–2000 keV by measuring differences in detector counting rates as Nova Muscae rose above or set below the horizon. We fit the data for each occultation in each energy channel with a model consisting of a source term superimposed on a time-varying background. Potential interference from other sources was handled as follows: A catalog of known bright or potentially bright sources is maintained as part of BATSE mission operations. All such sources whose location was within 60° of the axis of any detector which also saw Nova Muscae were treated as possible interfering sources. If the occultation time of any of the possible interfering sources fell within 10 s of a Nova Muscae occultation step, that step was eliminated from further analysis. Otherwise, terms were included in the fit for all possible interfering sources with an occultation step within the fit region (±110 s around each Nova Muscae occultation step).

We searched for emission from Nova Muscae during the interval 28 April–31 October 1991 (TJD 8374–8560). However, during portions of this time, including the interval prior to 5 May, the source was too far out of the CGRO orbit plane to produce usable occultations. Nevertheless, it was clear from inspection of the initial results that a significant signal was detected from Nova Muscae during much of the time when usable occultations were available. In order to produce a light curve, we summed the individual occultation count rates over 3 day intervals and deconvolved the spectra using conventional forward-folding. For this purpose we assumed a power-law input spectrum with fixed number index $\alpha = -2$.

The upper panel of Figure 1 shows the light curve derived from the BATSE data. The temporal evolution is characterized by a slow, approximately linear rise beginning around 15 May, leading to a single broad, symmetric maximum around 13 July, and a decay below our detection limit around 1 September. For comparison, we show in the same figure the SIGMA observations (Gil'fanov et al. 1991). It can be seen from the figure that the maximum flux during our observations was ~15% of the primary maximum. We note that the secondary maximum fell approximately midway between the last two SIGMA observations.

We investigated the evolution of the spectrum during the secondary outburst by summing the data over longer intervals (but never longer than one CGRO viewing period, usually two weeks), and then fitting various models to the resulting count spectra. A single power-law model was found to be an adequate fit in all intervals, although alternative models such as optically thin thermal bremsstrahlung and Sunyaev-Titarchuk Comptonization could not in general be ruled out. In the lower panel of Figure 1 we show the evolution of the power-law index during our observations compared with the SIGMA results. During the secondary outburst, the hardness increases as the intensity rises to a maximum; the trend after the maximum is unclear because of limited statistics. Gil'fanov et al. (1991) noted that their data were consistent with a clustering

Fig. 1. Intensity and spectral evolution of the entire Nova Muscae outburst, combining both BATSE and SIGMA data, the latter from Gil'fanov et al. 1991. The flux is integrated over 35–100 keV.

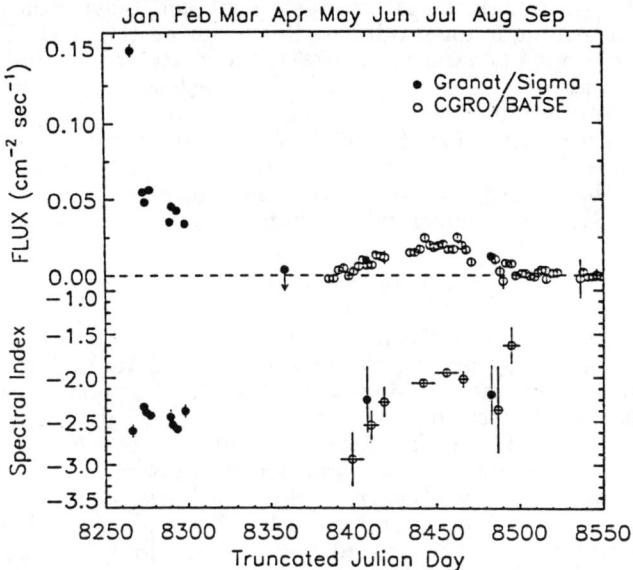

Fig. 2. Images of a region of the sky near Nova Muscae around the time of the secondary maximum (27 June–7 August). The images were produced using the radon transform and maximum-entropy inversion (Zhang et al. 1993). The energy intervals are 25–55 keV and 55–110 keV for the left and right panels, respectively.

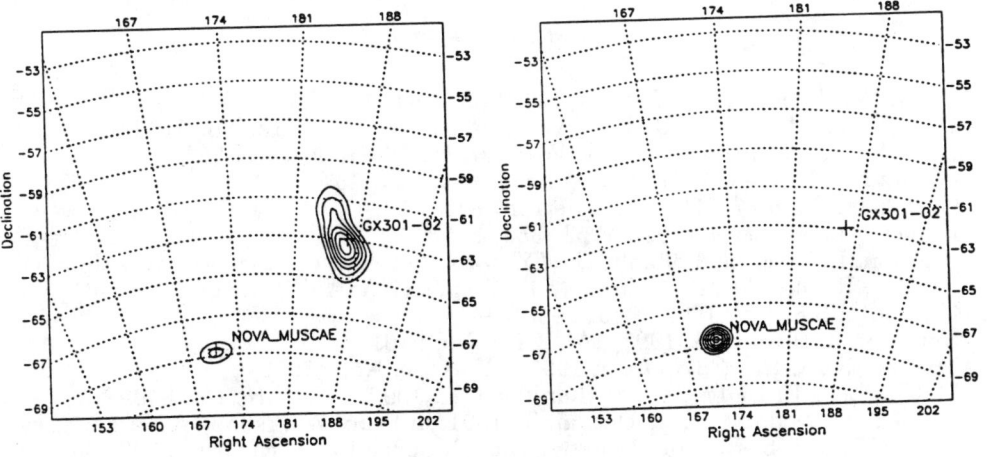

around an average index $\alpha \approx -2.5$ with no trend throughout the outburst. The BATSE data, however, show a significantly harder spectrum ($\alpha \approx -2$) around the secondary maximum and possibly an even harder spectrum just before the source disappears (although the latter may be subject to systematic errors).

Recently, a technique for producing images using the BATSE occultation data has been developed (Zhang et al. 1993). We generated a number of images of the region around Nova Muscae to verify that the latter was in fact the source of any emission detected in our occultation fitting, and that all significant nearby sources had been accounted for. Samples of these are shown in Figure 2, from which one can see that no other source besides Nova Muscae is visible above 55 keV. Below that energy the only other significant source is the recurrent transient pulsar GX 301-2, which was included in our analysis.

3. DISCUSSION

Two main theoretical mechanisms have been proposed to explain the outbursts of black-hole X-ray novae: the disk thermal instability model and the mass transfer instability model (Hameury et al. 1990 & refs. therein). Chen, Livio & Gehrels (1993) attempted to explain the entire light curve by a combination of these ideas: a main disk instability outburst is followed by a second outburst (in soft x-rays) due to mass transfer by evaporation of the outer layers of the secondary by x-ray heating. A third outburst also results from mass transfer, this time due to more prolonged heating and expansion of the secondary's convective layer. The second outburst in Nova Muscae was clearly seen in late March in soft x-rays but not in hard x-rays (Kitamoto et al. 1992). The hard x-ray secondary maximum which we observed would then correspond in the above scenario to the third outburst. However, if both of the secondary maxima simply represent episodes of increased mass accretion through the disk, then it is necessary to explain their completely different spectral characteristics. Future models must consider the distinctly different temporal behaviors of the hard power-law and ultrasoft spectral components in black-hole x-ray novae.

REFERENCES

Chen, W., Livio, M., & Gehrels, N., 1993, ApJ, 408, L5
Della Valle, M., Jarvis, B. J., & West, R. 1991a, IAU Circ. No. 5165
Della Valle, M., Jarvis, B. J., & West, R. M. 1991b, Nature, 353, 50
Gil'fanov, M., et al. 1991, Sov. Astron. Lett., 17, 437
Grebenev, S., et al. 1992, Sov. Astron. Lett., 18, 11
Hameury, J.-M., et al. 1990, ApJ, 353, 585
Harmon, B.A., et al. 1992, in The Compton Observatory Science Workshop, ed. C. R. Shrader, N. Gehrels & B. Dennis (NASA CP-3137), 69
Kitamoto, S., et al. 1992, ApJ, 394, 609
Lund, S., & Brandt, S. 1991, IAU Circ. No. 5161
Makino, F., & the *Ginga* team 1991, IAU Circ. No. 5161
McClintock, J., Bailyn, C., & Remillard, R. 1992, IAU Circ. No. 5499
Tanaka, Y., Makino, F., & Dotani, T. 1991, in Proc. Workshop on Nova Muscae 1991, ed. S. Brandt (Lyngby: Danish Space Research Inst.), 125
Zhang, S. N., Fishman, G. J., Harmon, B. A., & Paciesas, W. S. 1993, Nature, 366, 245

A *Broad Band X-Ray Telescope* Observation of the Black Hole Candidate LMC X-1

Eric M. Schlegel and the *BBXRT* Team
Code 668, Laboratory for High Energy Astrophysics
NASA-Goddard Space Flight Center, Greenbelt, MD 20771
and Universities Space Research Association
Internet: eric@heasfs.gsfc.nasa.gov

Abstract

We present the *BBXRT* spectrum of the black hole candidate LMC X-1. The spectrum can not be fit by a simple model, but requires a soft disk blackbody component and a power law tail, confirming earlier studies. The blackbody disk component is essentially unchanged since the *Ginga* measurement in 1987. We report a 95% confidence detection of weak emission features at ~5.1keV and ~7.3keV; if attributed to Fe I Kα at 6.39keV, then the redshift is ~0.19. No quasi-periodic behavior is found in the data at this epoch.

1. Introduction

LMC X-1, the first identified X-ray source in the LMC (Johnston *et al.* 1979), is a black hole candidate (BHC). The X-ray spectrum can not be fit by a single-component model, but requires at least two components: an ultrasoft blackbody with kT ~0.8keV, and a hard power law with a photon index of ~2.5 (Ebisawa *et al.* 1989). A quasi-periodic oscillation (QPO) of 0.075Hz was discovered in *Ginga* data (Ebisawa *et al.* 1989), associated with the hard spectral component. The ultrasoft component appears to be a characteristic unique to BHCs. Theoretically, this is expected, as a general relativistic accretion disk has a lower blackbody temperature than its Newtonian counterpart. The optical counterpart of LMC X-1 is still not firmly established (Cowley 1991). The probable counterpart is an O7 binary with an orbital period of ~4 days, yielding a lower mass limit for the compact object of ~2.6$M_\odot$ (Hutchings *et al.* 1983), larger than the theoretical upper mass limit for a neutron star of ~2.5$M_\odot$ (e.g., Shapiro & Teukolsky 1983). Details of the *BBXRT* observation are in Schlegel *et al.* (1994).

2. BBXRT Observations

The *Broad Band X-Ray Telescope* (Serlemitsos *et al.* 1992) flew on the shuttle *Columbia* as one component of the multi-telescope *Astro-1* mission. Attempts to point at LMC X-1 were made on five occasions during the nine day mission, but, unfortunately, none with on-axis pointings. With relatively large off-axis angles and a point spread function with ~1' FWHM, most photons were detected in an outer pixel. We use only the most populated pixels in our analysis (Table 1 in Schlegel *et al.* (1994)). This approach is justified by the rather high count rate from LMC X-1 (~15-18 counts s^{-1}), and the low total background (typically, 0.5-0.8 counts s^{-1}). With these considerations, we have six pixels from three pointings. Unfortunately, only 2 or 3 pixels have relatively good signal-to-noise, strongly constraining our conclusions. The target points were all made resolution of the *BBXRT* detectors. Assuming that the features are emission lines, we used Gaussians to fit the lines in the three pixels. The features are present on more than one day of the mission. The lines are significant at the 95% level for 2 parameters of interest (line energy and normalization). We remain confident these features are not instrumental because nothing similar is present in other *BBXRT* sources and no known X-ray emission lines from abundant elements exist in this energy region (Kortright 1986). The features do not match the broad, red-shifted iron lines found in *EXOSAT* data (see Schlegel et al. 1994 for references).

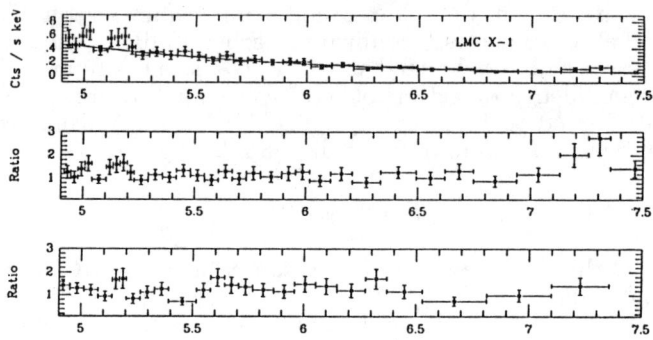

Fig. 2 - The data plotted as the ratio of the data/model.

If the features are real, two possibilities emerge. First, the lines could be unrelated, in which case the 5.1keV line could be red-shifted FeI Kα 6.39keV line. The calculated redshift is $\sim$0.19, with higher redshifts resulting from choices of more highly ionized species of iron. Second, we could attribute the features to an iron Kα line from an X-ray photoionized accretion disk as in Matt et al. (1993). In their figure 6, for an extended source geometry, a double-horned line profile results for an inclination angle of 70°. The red-shifted horn is present at about 5keV, and the blue-shifted horn at $\sim$7.3keV. The observed fluxes in the two lines are approximately equal, as opposed to the models which show the blue-shifted horn to be the stronger component. A crude simulation using the *BBXRT* response matrix shows that even though the effective area is decreasing above 6keV, the decline is probably not so steep to invert the red-blue asymmetry sufficiently. An *ASCA* observation should provide better quality data to confirm the lines' presence and to study the potential asymmetry.

Ebisawa et al. (1989) discovered a 0.075Hz QPO in a segment of their *GINGA* data. The QPO was present in 1987 April, but was not present in data obtained in three months later. The QPO was apparently associated with the hard spectral component, as that component changed its behavior between the two observation epochs, while the behavior of the soft blackbody disk component was nearly constant over the same interval. We computed the power spectral density for only one segment (number 4), a 1200-s pointing, of the *BBXRT* data to search for any QPO activity. The power spectrum was created from the data divided into 6 segments (M = 6) and every 2 bins were averaged (W = 2) (van der Klis 1989). The power spectrum was flat (Figure 3a); a fit gives χ^2 $\sim$2.7. The power spectrum was fitted with a power law, for the continuum, and a Gaussian outside of the South Atlantic Anomaly, but during shuttle daylight (Sun angles ~90°, and Earth angles ~$90\text{-}95^O$). A daytime background spectrum, scaled by the guard rate for that pointing, was used to subtract the background. Residual background contamination exists below ~0.6keV, but given the moderate column to LMC X-1 (~0.5×10^{22} cm^{-2}, Ebisawa 1991), there is little detected source flux below ~0.7keV. Off-axis angles were estimated by ray-tracing, as aspect solutions did not exist.

3. Results

Figure 1 presents the observed spectrum and fitted model of one pixel for clarity only. Spectral fitting was done using the *BBXRT* response matrix calibration (Weaver 1993) in which fits to spectra of the Crab show broad, ~3% amplitude residuals remaining, plus ~7% residuals at ~1.38keV and the gold M edge at ~2.2keV. A single component model does not describe the observed LMC X-1 spectrum, but two-component models do. The best fit model is a 'multi-color blackbody disk' (e.g., Mitsuda *et al.* 1984) plus a power law tail, shown in Figure 1. The disk blackbody appears to have been relatively constant from the *Ginga* measurements made in 1987 (kT ~0.91keV) (Ebisawa *et al.* 1989). One also expects fluorescent iron lines from an accretion disk surrounding a black hole (e.g., Matt *et al.* 1993). No iron line is present in the 6.5keV region. Schlegel *et al.* (1994) assigned an energy-dependent upper limit to the presence of an iron line. The limits are all below literature values (e.g., White & Marshall 1984).

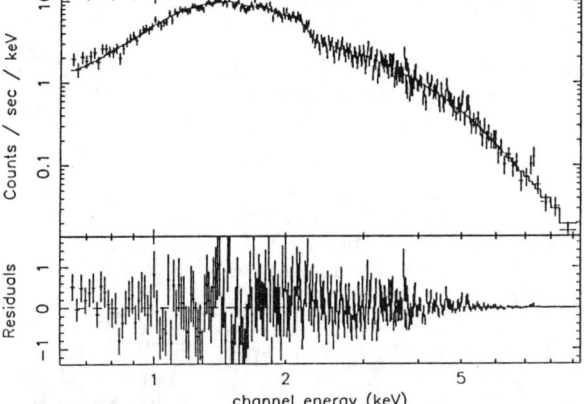

Fig. 1 - The *BBXRT* spectrum (log flux scale) and residuals (linear scale) of the fit to LMC X-1. The fitted column is 5.8×10^{21} cm^{-2}; the photon index is 2.18±0.37; the disk blackbody temperature is 0.83±0.07keV.

We turn now to the possible detection of two small emission-like features present at ~5.1keV and ~7.3keV. Figure 2 shows an expanded view of the data and the model in the 4.6 to 7.6keV region, showing one pixel as residuals from the continuum fit and as a ratio of data/model. The other pixel is shown only as data/model. These line-like features are present in the three pixels with the highest signal-to-noise ratio. The features are admittedly not strong and are narrow, being consistent with a zero width line broadened by the finite energy component to simulate the QPO found by Ebisawa et al. (Figure 3b). A fit with this model would not converge. The parameters were reset to the Ebisawa et al. values and the normalization of the line component was then increased until $\Delta\chi^2$ went up by 2.71 (90%) and 6.63 (99%). The resulting signal strengths were 1.2% and 1.6% which are $\sim 8.5\sigma$ and $\sim 6.4\sigma$ lower than the signal found by Ebisawa et al.. We interpret these values to mean that no QPO or periodic behavior was found at any frequency between 0.01 and 4Hz. The QPO is clearly a transient phenomenon, having been discovered in only one *Ginga* dataset. The long-term behavior needs additional observations to be understood.

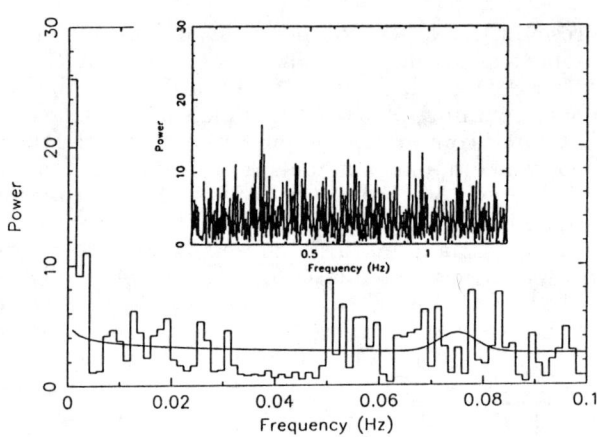

Fig. 3 - a) A search for a QPO in the *BBXRT* data. b) An expansion of Fig. 3a in the region surrounding the Ebisawa et al. QPO of 0.075Hz.

References

Cowley, A. P. 1991, *Frontiers of X-ray Astronomy*, ed. Y. Tanaka & K. Koyama, (Universal Academy Press Inc.), p344
Ebisawa, K. 1991, Ph.D. thesis, University of Tokyo
Ebisawa, K., Mitsuda, K., & Inoue, H. 1989, PASJ, 41, 519
Ebisawa, K., Mitsuda, K., & Hanawa, T. 1991, ApJ, 367, 213
Hutchings, J. B., Crampton, D., & Cowley, A. P. 1983, ApJ, 275, L43
Johnston, M. D., Bradt, H. V., & Doxsey, R. E. 1979, ApJ, 233, 514
Kortright, J. B. 1986, *X-ray Data Booklet*, ed. D. Vaughn, (Berkeley: Lawrence Berkeley Laboratory), pp2-12
Matt, G., Fabian, A. C., & Ross, R. R. 1993, MNRAS, 262, 179
Mitsuda, K. et al. 1984, PASJ, 36, 741
Schlegel, Eric M. et al. 1994, ApJ, xxx, xxx
Serlemitsos, P. J., et al. 1992, *Proc. of the 28th Yamada Conference on X-ray Astronomy*, ed. Y. Tanaka & K Koyama, (Tokyo: Univ. Acad.), p221
Shapiro, S. & Teukolsky, S. 1983, *Black Holes, White Dwarfs, and Neutron Stars*, (New York: Wiley Interscience)
van der Klis, M. 1989, *Timing Neutron Stars*, ed. Ögelman, H. & van den Heuvel, E. P. J., (Netherlands: Kluwer Acad. Pub.), p27
Weaver, K. A. 1993, Ph.D. thesis, Univ. of Maryland, in preparation

THE MASS OF THE BLACK HOLE IN A0620-00

T. Shahbaz[1], T. Naylor[1], P.A. Charles [1]
[1] Department of Physics, Keele University, Keele, Staffordshire, ST5 5BG.
[2] Department of Astrophysics, Nuclear Physics Building, Keble Road, Oxford OX1 3RH.

ABSTRACT

We determine mass-ratio/inclination solutions for A0620-00 by fitting the quiescent infrared light curve with an ellipsoidal model. Combining these solutions with the rotational broadening and the K velocity of the secondary star obtained by Marsh, Robinson & Wood (1993), allows us to determine the mass of the system components. We find the most likely values are $M_1 = 10\ M_\odot$ and $M_2 = 0.6\ M_\odot$. We also find that the infrared colour of the secondary star is consistent with it being a main sequence star. When used in conjunction with Bailey's relation and our estimates to the radius of the secondary star, we find the most probable value for the distance to A0620-00 is 1050 pc.

1. INTRODUCTION

The best black hole candidates are now in the class of systems called the low mass X-ray binary transients. These are systems where a low mass stars is losing mass to a neutron star or black hole. They undergo rare but very luminous outbursts during which it is impossible to observe the secondary star. In quiescence however, the accretion disc does not dominate the observed flux thus providing us with a perfect opportunity to observe the secondary star.

The optical light curves of A0620-00 show a doubled humped modulation, which is due to the changing aspect presented to the observer by a tidally and rotationally distorted secondary star, orbiting the compact object. Here we present the results of an infrared study of A0620-00 during quiescence.

2. OBSERVATIONS AND DATA REDUCTION

Infrared data for A0620-00 in quiescence were taken at the 3.8m United Kingdom Infrared Telescope during January 1990 using IRCAM, and reduced using the standard procedure (see McLean 1987). Using the standard convention where phase 0.0 corresponds to when the secondary star is in front of the compact object, the K band data were then folded using the ephemeris of McClintock & Remillard (1986, hereafter MR), and then averaged into bins, of approximately 0.05 in phase (Fig. 1).

2. THE INCLINATION OF THE SYSTEM

The optical light curves of the SXTs show a doubled humped modulation (see MR and Haswell et al. 1993), which are are contaminated by flux from the accretion disc. At V it is $\sim 40\ \%$ and deceases to $\sim 6\ \%$ by the I band (Oke 1977, MR and Marsh et al. 1993).

Infrared photometry allowed us to determine the uncontaminated double humped light curve (Fig. 1), which we then fitted with the ellipsoidal model to obtain solution in the (q,i) plane (Fig. 2) in a similar method to that used for Cen X-4 (Shahbaz, Naylor & Charles 1993.) We find that the system is confined to lie in the inclination range $31-54°$ (90 % confidence, Shahbaz,Naylor & Charles 1994).

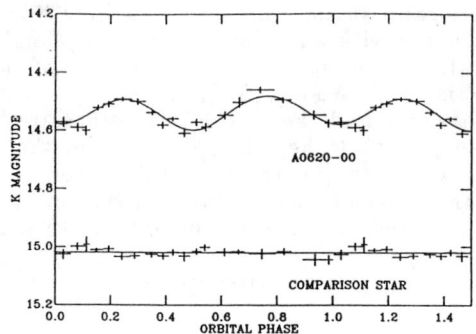

Fig. 1. K band light curve of A0620-00 with model fit.

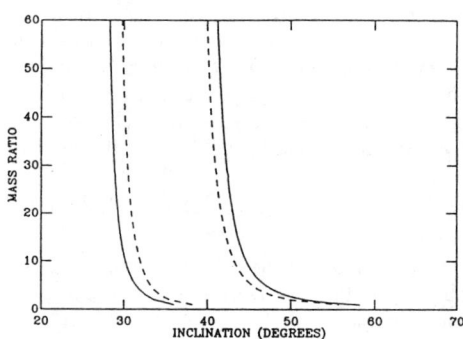

Fig. 2. The 68 (inner ellipse) and 90 % (outer ellipse) confidence regions in the (q,i) plane.

3. THE MASS OF THE BINARY COMPONENTS

Rotating neutron stars can have a mass up to 3.8 $M_\odot$, (Friedman & Ipser 1987) so the case for a black hole in A0620-00 based on the mass function alone is not conclusive. The only sure way to prove this is to determine its mass. Using $v \sin i$ and K from Marsh et al. (1993) we find $q \sim 16$, which when combined with our (q,i) solution gives $i = 37°$. Using a Monte Carlo error propagation technique we computed a probability density function for the masses of the system components (Fig. 3). We find the most likely value for the component masses to be $M_1 = 10.0\ M_\odot$, $M_2 = 0.6\ M_\odot$ and the 90 % confidence region to be $M_1 = 5.1 - 28.8\ M_\odot$ and $M_2 = 0.2 - 1.9\ M_\odot$.

We then calculate the allowed ranges for a and R_2 using Kepler's Law and Paczynski's formula (Paczynski 1971) respectively (see Table 1). We find that we can rule out the case where the compact object is a non-rotating neutron star at the 99 % confidence level.

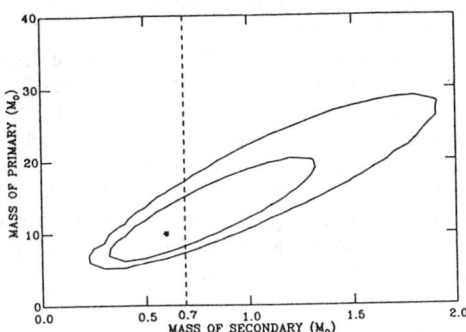

Fig. 3. This shows the 68 and 90 % confidence regions for solutions in the $M_1 - M_2$ plane.

Parameter	Most likely value	Lower limit	Upper limit	New Upper limit*
Primary mass M_1 ($M_\odot$)	10.0	5.1	28.8	17.1
Secondary mass M_2 ($M_\odot$)	0.6	0.2	1.9	0.7
Secondary radius R_2 ($R_\odot$)	0.8	0.5	1.1	0.8
Orbital separation a ($R_\odot$)	4.3	3.4	6.1	5.2

* Obtained from the observed spectral type of the secondary.

Table. 1. Derived system parameters at the 90 % confidence level.

4 THE EVOLUTIONARY STATUS OF THE SECONDARY STAR

The observed infrared colour supports the secondary being a main sequence star. We dereddened the infrared fluxes using a colour excess of $E_{B-V}=0.35$ (Wu et al. 1983) and assuming the accretion disc contamination in the infrared is negligible, we obtain $(J-K)_o=0.70$, which is is consistent with the colour of a K3/4 main sequence star (Koorneef 1983).

We can further constrain the mass of the secondary star obtained from the Monte Carlo simulation from its observed spectral type, a K3/4 star. The secondary must be less massive than a main sequence K3/4 star of 0.7 $M_\odot$. Combining this upper limit to the mass of the secondary with the allowed mass range for the compact object from Fig. 3, we obtain a new upper limit to the mass of the black hole to be $M_1 < 17.1$ $M_\odot$ at the 90 percent confidence level. If however, the accretion disc contamination is significant in the infrared then our value for the inclination is a lower limit and thus we obtain a upper limit to the mass of the compact object.

5. THE DISTANCE TO A0620-00

We estimate to distance to A0620-00 using Bailey's relation (Bailey 1981), which relates the star's surface brightness in the K band (S_K) and its effective temperature. Since this relation is relatively insensitive to the evolution of the secondary, it can be used to determine the distance. Using the observed mean V band magnitude of 18.2 (Haswell et al. 1993) and allowing for the accretion disc contribution (40 percent at V, see MR) and reddening we obtain V=17.5 for the secondary alone. Similarly for our observed K magnitude of 14.5, we obtain $K_o = 14.4$ (assuming no accretion disc contamination). Given that $(V - K)_o = 3.1$, $S_K = 3.8$ (Ramseyer 1993) and using the limits to the radius of the secondary star (see Table 1), we limit the distance to lie in the range 660< d <1450 pc, with the most probable value being 1050 pc.

5. CONCLUSION

From the infrared study of the ellipsoidal variations of the secondary star during quiescence, we have determined the mass of the system components in A0620-00. We find that the most likely masses for the compact object and secondary star are 10 and 0.6 $M_\odot$ respectively. The mass of the compact object is well in excess of the maximum mass for a neutron star and so places it amongst the strongest black hole candidates. The observed infrared colour is consistent with that of a K3/4 main sequence star. Using Bailey's relation in conjunction with our estimates to the radius of the secondary star, we limit the distance to lie in the range 660-1450 pc, with the most probable value being 1050 pc.

REFERENCES

Bailey, J., 1981, MNRAS, 219, 619
Friedman, J.L., Ipser, J.R., 1987, ApJ, 314, 594
Haswell C.A., Robinson E.L., Horne K., Stiening R.F., Abbott T.M.C., 1993, ApJ, 411, 801
Koorneef J., 1983, A&A, 128, 84
Marsh, T.R., Robinson, E.L., Wood, J.H., 1993, MNRAS, in press
McLean, I.S., 1987, in Infrared Astronomy With Arrays, eds Wynn-Williams C.G., Becklin E.E. University of Hawaii, p180
McClintock, J.E., Remillard, R.A., 1986, ApJ, 308, 110
Oke, J.B., 1977, ApJ, 217, 181
Paczynski, B. E., 1971, ARA&A, 9, 183
Ramseyer, T.F., 1993, ApJ, in press
Shahbaz, T., Naylor, T., & Charles, P.A., 1993, MNRAS, in press
Shahbaz, T., Naylor, T., & Charles, P.A., 1994, MNRAS, submitted
Wu, C.C., Panek, R.J., Holm, A.V., Schmitz, M., 1983, PASP, 95, 391

THE PERIODICITIES OF AQL X-1 IN QUIESCENCE.

A.C. Martin, P.A. Charles
Astrophysics Department, Keble Rd, Oxford, OX1 3RH.
acm@oxds02.astro.ox.ac.uk, pac@oxds02.astro.ox.ac.uk

P.J. Callanan
Center for Astrophysics, 60 Garden St, Cambridge, Massachussetts, 02138, USA.

AND

E. Harlaftis
RGO, Santa Cruz de La Palma, Tenerife, Canary Islands.

1. INTRODUCTION

Aql X-1 is a recurrent soft X-ray transient which was optically identified by Thorstensen et al., (1978) with a faint blue variable star. Its behaviour lies between the ordinary LMXBs which have steady X-ray and optical emission and the soft X-ray transients which are normally quiescent and only occasionally switch into the 'on' state (every $\sim$ 10-50 years). Aql X-1 has regular outbursts every $\sim$ 1 year, lasting about 1 month (Charles et al., 1980), and goes from a V magnitude of 19 to 15-16, in its high state resembling ordinary LMXB systems. But there had previously been no evidence for any short term variations, except for an uncomfirmed report by Watson (1976) of a 1.3 day 10% X-ray modulation.

Czerny et al. (1987) detected X-rays in quiescence with the Einstein MPC with a (1-10 keV) flux of $2.1 \pm 0.2 \times 10^{-11}$ erg cm^{-2} s^{-1} indicating that accretion continues at a low level in quiescence. However, van Paradijs et al. (1987) did not detect Aql X-1 with EXOSAT, with an upper limit to the (1-4.5 keV) flux of 1.4-3.0 x 10^{-12} erg cm^{-2} s^{-1}. If both of these results are correct then the quiescent X-ray flux varies by at least a factor three.

Optical spectroscopy revealed that the secondary is a late type (G7-K3) star (Thorstensen et al., 1978). X-ray bursts have been detected with the Einstein and Hakucho satellites (Czerny et al., 1987) indicating that the compact object in Aql X-1 is a neutron star. Attempts to find the orbital period revealed variations, but no firm period, until Ilovaisky et al. (1991) announced a 18.97 hour period.

However, if P $\sim$ 19 hours, we get $\bar{\rho} = 0.3$ g cm^{-3} for the secondary to fill its Roche lobe (King, 1988). If the secondary is a K0 main sequence star, then $\bar{\rho}$ = 1.8 g cm^{-3}. This implies that the radius of the secondary, R_2, must be about twice that of a main sequence star in order to fill its Roche lobe. It is therefore likely to be a sub-giant as the secondary must have a smaller average density than a typical main sequence star (see King, 1993).

2. PHOTOMETRY.

We obtained V and R band images with the SAAO 1m and the RCA CCD from June 26th - July 3rd. The observations were undertaken in order to search for an ellipsoidal modulation at half the (presumed orbital) 19 hour period of Chevalier et al., (1991). In quiescence the contribution of the disc should be greatly reduced and hence modulation due to the secondary should dominate the light curve. Instead we see a smooth variation ($\sim$ 0.4 mag), near 18.5 hours which may be related to the outburst modulation (possible residual X-ray heating?).

The program PHOT in IRAF was used to perform aperture photometry with an 8 pixel radius aperture and centroid centering, and the sky fitting used an inner annulus radius of 8 pixels with a width of 10 pixels and 3 sigma rejection.

3. RESULTS

The light curves show short term variations at the level of $\sim$ 0.4 magnitudes. Initially, both data sets (V band from June 26-30 and R band from 5 June 30 - July 3) were analysed separately for periodicities using a discrete fourier transform (DFT).

The power spectrum for the R band data (see figure 1) has several peaks at 99.909, 30.303, 18.2 and 13.5 hours, all of these except the 18.2 hour period being due to our data sampling. The light curve folded on 30.3 hours, showed clustering at four different phases which would increase the power at that frequency. The 13.5 hour peak is a 24 hour alias of the 30.3 hour peak, and the 90.9 hour peak also showed some clustering in its folded light curve. Thus only the 18.2 hour modulation seems to be real, but is swamped by aliasing in our data.

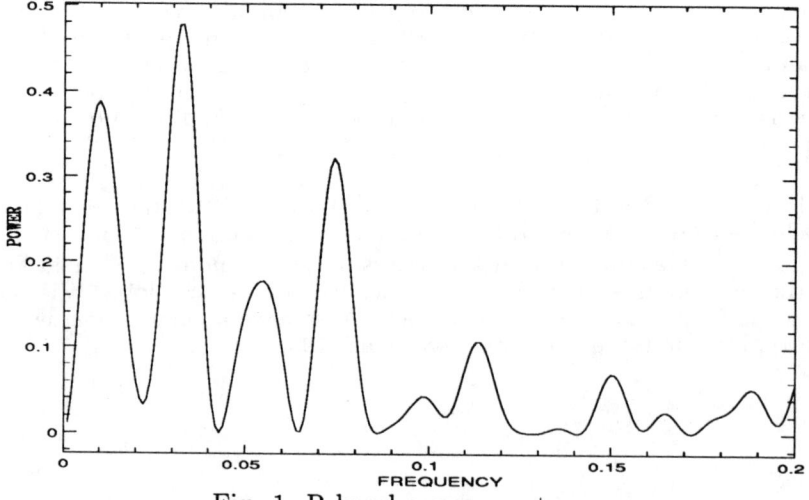

Fig. 1. R band power spectrum.

The V band data showed little power at these frequencies, and was dominated by power below 2 hours, indicating that the time series was too noisy to obtain a significant result.

The two sets of data were then combined by calculating the mean of each and using this to normalise the two time series. The resulting light curve is shown in figure 2, and its fourier transform in figure 3. It shows power at 90.909, 30.303, 18.519 and 12.987 hours. The 90 and 30 hour peaks show clustering and the 12.99 peak is a close 24 hour alias of 30 hours. The window function (figure 4) shows power at 1 day, with an associated harmonic at 12 hours.

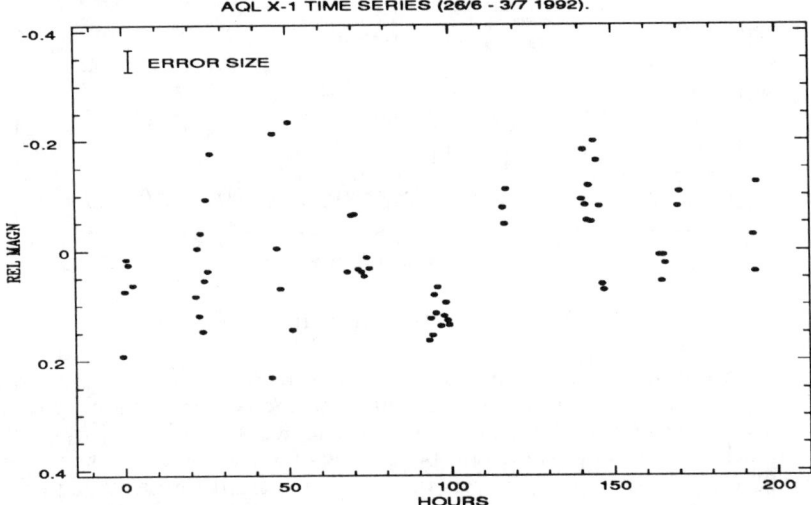

Fig. 2. Mean subtracted V and R light curves.

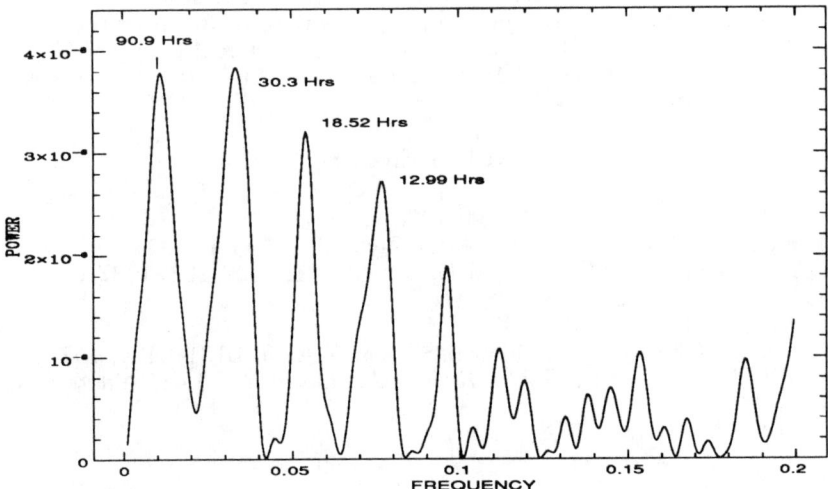

Fig. 3. Power spectrum of complete dataset.

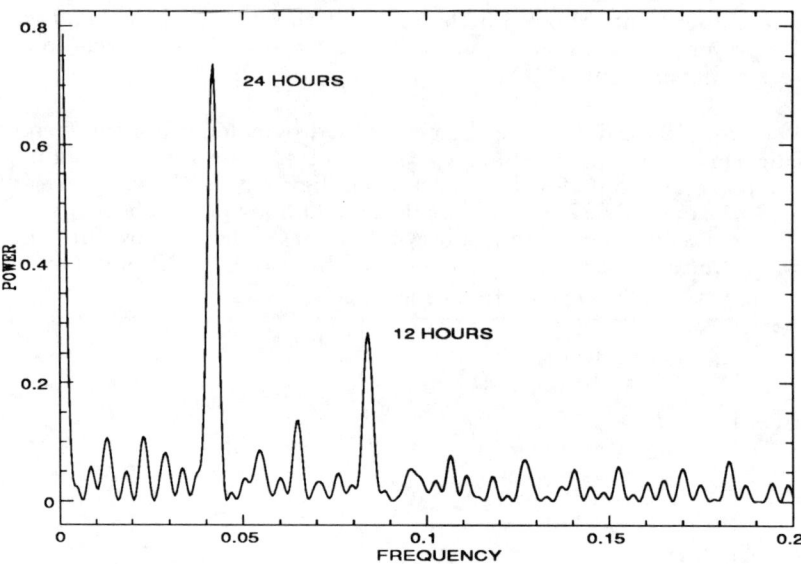

Fig. 4. Window function of complete dataset.

Our data shows clear variations of 0.4 magnitudes each night, and power at 18.5 hours. But there is no significant periodicity seen at half this period from the expected ellipsoidal modulation of the K0 secondary. Sinusoids were artificially added at this period, and became visible in the power spectrum at a level of 0.02 magnitudes, providing us with an upper limit for any real periodicity.

However, the periodicities in the data are swamped by aliases and power due to the window function, also noise in the data causes power to leak into the alias frequencies from the true periodicity. In order to improve the statistical quality of the data (given the faintness of the target and the crowded field) we plan to re-analyse using DAOPHOT, thereby giving us the best constraint on any periodicities in the data.

REFERENCES

Charles P.A. et al., Ap. J., 237, 154, 1980
Chevalier C., Ilovaisky S.A., Astr. Astrophys., 251, L11-L13, 1991
Czerny M., Czerny B. & Grindlay J.E., Ap. J., 312, 122-133, 1987
King, QJRAS, 29, 1, 1988
King, MNRAS, 260, L5-L6, 1993
Thorstensen J., Charles P.A., Bowyer S., Ap. J., 220, L131-L134, 1978
van Paradijs J., Verbunt F., Shafer R.A. & Arnaud K.A., Astr. Astrophys., 182, 47, 1987
Watson M.G., MNRAS, 176, 19p, 1976

BLACK HOLES VS NEUTRON STARS AMONG SIGMA SOURCES

J. Ballet, P. Laurent, F. Lebrun, J. Paul
DAPNIA/Sap, CE Saclay, 91191 Gif sur Yvette Cedex, FRANCE

J.P. Roques, P. Mandrou, I. Malet, M.C. Schmitz-Fraysse
CESR-CNRS/UPS, 9 av. du colonel Roche, 31029 Toulouse Cedex, FRANCE

E. Churazov, M. Gilfanov, R. Sunyaev, A. Vikhlinin
A. Finoguenov, A. Dyachkov, N. Khavenson, A. Sheikhet
IKI, Profsouznaya 84/32, 117810 Moscow, RUSSIA

ABSTRACT

X-ray binaries form the bulk of hard X-ray sources, in the 30 - 300 keV range. More than twenty of them have been detected by SIGMA. Suspected black hole systems seem to be particularly numerous among the sources emitting above 100 keV, and their spectra appear harder than those of neutron star systems. This amounts to a spectral distinction between the two classes of X-ray binaries. The nature of the companion (high mass or low mass) does not affect this conclusion.

1. INSTRUMENT

The French telescope SIGMA (Paul et al.1991, Mandrou et al.1991, Leray et al.1991) was launched on board the Soviet space observatory GRANAT on December 1st, 1989, and is still operating in October 1993. It is the first coded aperture telescope working in hard X-rays on a satellite. Its nominal range is 35 to 1300 keV, but most sources are detected only below 300 keV. Its angular resolution of 15-20', and its locating accuracy of 2' (for strong sources), ensure a clear distinction between nearby sources, and in most cases a straightforward identification of new sources. Those properties are maintained over the 11.5° x 10.6° field of view (at half sensitivity). Its spectral resolution improves from 20% at 60 keV to 10% at 300 keV, and its 3σ broad band sensitivity is $1.5~10^{-5}$ ph cm^{-2} s^{-1} keV^{-1} (or 30 mCrab) at 100 keV for a typical observation (1 day). More than two dozen sources, mainly X-ray binaries harbouring either a black hole or a neutron star, have been detected. This paper is an updated summary of a previous review on a similar subject (Ballet et al.1993).

2. BLACK HOLE SYSTEMS

The prototype of black hole X-ray binaries is Cyg X-1, whose spectrum (in the low state) is characterized by a power law of photon index close to -2.0, cut off above 200 keV. Such a spectrum is usually interpreted, following Shapiro et al.(1976) and Sunyaev & Titarchuk (1980) as the result of the Comptonization (upscattering) of soft (around 1 keV or less) photons in a thick ($\tau \simeq 3$) hot (around 50 keV) electron gas. Another possibility is Comptonization in a hotter (around 150 keV) optically thin electron gas (Haardt et al.1993). The soft photons are presumably emitted in optically thick parts of the disk, while the hot electrons would reside in a hot corona or inner torus.

Apart from Cyg X-1 itself (Salotti et al.1992), which SIGMA always caught in the same low state, GX 339-4, also suspected to harbour a black hole, was repeatedly detected by SIGMA (Bouchet et al.1993). It exhibited large brightness variations, correlated with a spectral hardening. Other sources, such as 1E 1740.7-2942 or GRS 1758-258 (Sunyaev et al.1991a), are thought to be accreting black holes because of the similarity of their spectra to Cyg X-1, although nothing is known of their mass or their companion. All those sources, except Cyg X-1, were seen to vary by more than a factor of five (down to undetectable), with a typical timescale of a few weeks. A transient broad high energy (400-500 keV) feature was detected in 1E 1740.7-2942 (Bouchet et al.1991, Sunyaev et al.1991b), probably related to an outburst of pair formation.

A very important class of X-ray binaries for black hole studies is X-ray novae (White et al.1984). In three members of this class, A0620-00 (McClintock & Remillard 1986), V404 Cyg (Casares et al.1992) and GRS 1124-684 (Remillard et al.1992), the mass of the compact object was shown to be larger than 3 $M_\odot$, more than the neutron star limit. Their strong hard X-ray emission, although the spectra differ from one source to another, also put them on a par with other black hole candidates such as Cyg X-1.

SIGMA has now observed three of them. The first (GRS 1124-684) appeared in Musca in January 1991, and was monitored from the earliest times of the flare (Sunyaev et al.1992, Goldwurm et al.1992). On one occasion SIGMA detected an emission line lasting more than 10 hours, centered around 480 keV, slightly below the energy of electron-positron annihilation. It was the first such line observed in the spectrum of an X-ray nova. Another such source (GRO J0422+32) appeared in Perseus in August 1992 (Paciesas et al.1992). It was at the time the brightest hard X-ray source in the sky ($\simeq$ 3 Crab). By analogy with the other sources of this class, it is expected to be a new black hole system, and this should be proved by optical studies as soon as the disk luminosity is low enough. The third such nova (GRS 1716-249) was discovered by SIGMA in September 1993 (Gilfanov et al.1993).

3. NEUTRON STAR SYSTEMS

Neutron stars form the bulk of X-ray binaries seen below 20 keV. However their spectra steepen earlier than those of black holes, with cutoff energies around 20 keV (White et al.1988). SIGMA has detected hard tails from several X-ray bursters (Barret & Vedrenne 1993). Those are relatively weak ($\simeq 10^{37}$ erg/s) X-ray binaries. They showed variability on timescales of a few days. On the other hand none of the bright ($\simeq 10^{38}$ erg/s, close to the Eddington limit) X-ray binaries has been detected, implying that their hard X-ray emission is less than 1% of the total.

The source in the globular cluster Terzan 2 is a case in point. If its identification with the X-ray burster X 1724-308, proposed by Barret et al.(1991), is correct then it was in 1990 by far the most luminous of the neutron star systems in hard X-rays (Table 1), exceeding even most black hole systems. Alternatively the hard X-ray emission may have arisen in another (black hole) system within the cluster. Note that the spectrum is actually measured up to 200 keV only, so that its extrapolation to 500 keV leads to a large uncertainty.

4. BLACK HOLES VS NEUTRON STARS

Table 1: High energy luminosity (in units of 10^{35} erg/s) and photon index of accreting binary systems in outburst (extended from Laurent et al.1993). A question mark means that the distance is very uncertain. The identification of the source within Terzan 2 is doubtful. The Crab nebula, powered by its rotating magnetic field, is included only for comparison.

Nature	Name	$L_{150-500\ keV}$	$\Gamma_{40-200\ keV}$
black hole	Cygnus X-1	$\simeq 20$	2.1
black hole ?	GX 339-4	$\simeq 30$	2.2 - 2.9
black hole ?	GRS 1915+105	> 3	2.5 ± 0.3
black hole ?	1E 1740.7-2942	$\simeq 40$	1.8 - 2.2
black hole ?	GRS 1758-258	$\simeq 20$	1.8 - 2.5
black hole	GRS 1124-684	$\simeq 50$	2.0 - 2.7
black hole ?	GRO J0422+32	$\simeq 115$?	2.2
black hole ?	TrA X-1	> 3	1.8 ± 0.7
neutron star	4U 1700-377	$\simeq 5$	3.0 ± 0.2
pulsar	OAO 1657-415	≤ 1	
pulsar	GX 1+4	≤ 2	3.0 ± 0.4
neutron star	KS 1731-260	≤ 2 ?	2.9 ± 0.8
neutron star	GX 354-0	≤ 3	3.0 ± 0.2
neutron star	A 1742-294	≤ 5	3.0 ± 0.5
globular cluster	Terzan 1	≤ 10	3.2 ± 0.6
globular cluster	Terzan 2 (average)	≤ 20	3.3 ± 0.4
globular cluster	Terzan 2 (outburst)	$\simeq 190$?	1.7 ± 0.5
fast pulsar	Crab Nebula	$\simeq 40$	2.1

Table 1, comparing spectra of known neutron stars and black holes, depicts a clear difference between the two sets of sources: the black hole spectra are clearly harder around 100 keV, and the variations within the two sets are secondary to this major difference. The comparison of luminosities (taking distances into account) above 150 keV also clearly tells black holes from neutron stars: the luminosity of neutron star systems above 150 keV was always less than a few 10^{35} erg/s (upper limits), compared with positive detections at a few 10^{36} erg/s for black hole systems. Note that the brightest neutron star system (4U 1700-377) maintained this luminosity for a few hours only, whereas the black holes' outbursts last a few weeks.

In classical X-rays, Tanaka (1989) proposed to characterize black holes as sources able to emit more than 10^{37} erg/s with no thermal excess around 1 keV (thought to arise at the heated surface of neutron stars). An additional criterion might then be that they should also be able to radiate more than 10^{36} erg/s above 150 keV. The reason for that (Sunyaev et al.1991a) could be that the same soft photons appearing in bright neutron star systems (and due to the hard X-ray illumination) act as a cooling agent or regulator (through Comptonization) for the hot electrons, and forbid them to reach temperatures larger than 20 keV.

The mass of the companion plays only a secondary role (probably through the way matter is transferred) in the hard X-ray emission, emitted close to the compact object; examples of high mass and low mass binaries were found within both black hole and neutron star systems. The hard X-ray spectra of high mass systems seem somewhat harder than those of low mass systems, and their variations are faster.

In summary (Ballet et al.1993), hard X-ray features which seem to be characteristic of stellar mass black hole systems are:
- A hard spectrum (photon index around -2.0) around 100 keV.
- A luminosity in outburst larger than 10^{36} erg/s above 150 keV.
- Relatively slow (timescale of weeks) large amplitude variations.
- Transient high energy ($\geq$ 400 keV) features.
- Low frequency ($\leq$ 1 Hz) quasi-periodic oscillations.

ACKNOWLEDGEMENTS

We acknowledge the paramount contribution of the SIGMA Project Group of the CNES Toulouse Space Center to the overall success of the SIGMA mission. We thank the staff of the Lavotchin Space Company, the Babakin Space Center, the Baikonur Space Center and the Evpatoria Ground Station for their unfailing support.

REFERENCES

Ballet J. et al.1993, Proc. of 27th ESLAB Symposium (Noordwijk), in "Frontiers of space and ground based astronomy", in press
Barret D. et al.1991, ApJ (Letters) 379, L21
Barret D., & Vedrenne G. 1993, in "The multi-wavelength approach to γ-ray astronomy" (Les Diablerets), to be published in ApJ Suppl.
Bouchet L. et al.1991, ApJ (Letters) 383, L45
Bouchet L. et al.1993, ApJ 407, 739
Casares J., Charles P.A.,& Naylor T. 1992, Nature 355, 614
Gilfanov M. et al.1993, this conference
Goldwurm A. et al.1992, ApJ (Letters) 389, L79
Haardt F., Done C., Matt F.,& Fabian A.C. 1993, ApJ (Letters) 411, L95
Laurent P. et al.1993, Adv.Sp.Res. 13 (12), 139
Leray J.P. et al.1991, 22nd ICRC (Dublin) 2, 495 (OG 10.1.11)
McClintock J.E.,& Remillard R.A. 1986, ApJ 308, 110
Mandrou P. et al.1991, in "Gamma-ray line astrophysics" (Saclay), eds Durouchoux P. and Prantzos N., AIP Conf. 232, 492
Paciesas W.S., Briggs M.S., Harmon B.A., Wilson R.B.,& Finger M.H. 1992, IAU Circ. 5580
Paul J. et al.1991, Adv.Sp.Res. 11 (8), 289
Remillard R.A., McClintock J.E.,& Bailyn C.D. 1992, ApJ (Letters) 399, L145
Salotti L. et al.1992, A&A 253, 145
Shapiro S.L., Lightman A.P.,& Eardley D.M. 1976, ApJ 204, 187
Sunyaev R.A.,& Titarchuk L.G. 1980, A&A 86, 121
Sunyaev R. et al.1991a, A&A 247, L29
Sunyaev R. et al.1991b, ApJ (Letters) 383, L49
Sunyaev R. et al.1992, ApJ (Letters) 389, L75
Tanaka Y. 1989, Proc. of 23rd ESLAB Symp. (Bologna), eds Hunt J. and Battrick B., ESA SP 296, 1
White N.E., Kaluziensky J.L.,& Swanck J.H. 1984, in "High energy transients in astrophysics" (Santa Cruz), ed Woosley S.E., AIP Conf. 151, 31
White N.E., Stella L.,& Parmar A.N. 1988, ApJ 324, 363

THE AUGUST 1993 OUTBURST OF GRO J0422+32

Alberto J. Castro-Tirado
Danish Space Research Institute, Gl. Ludtoftevej 7, DK-2800 Lyngby, Denmark

Laboratorio de Astrofisica Espacial y Fisica Fundamental
P.O. box 50727, 28080, Madrid, Spain

Jose Luis Ortiz
Consejo Superior de Investigaciones Cientificas.

Instituto de Astrofisica de Andalucia.
P.O. box 3004, 18080 Granada, Spain

Jesus Gallego
Departamento de Astrofisica. Universidad Complutense, Madrid, Spain

ABSTRACT

One year after the X-ray outburst, the optical counterpart (and possible black hole candidate) of the type II Soft X-ray Transient GRO J0422+32 underwent an unexpected one-month outburst beginning on August 10, 1993. We suggest that the Mass Transfer Instability model can explain the event on the basis of the observational facts presented here.

INTRODUCTION

Low Mass X-ray Binaries (LMXBs) are systems formed by a low-mass companion and a compact object. A subclass of LMXBs are the Soft X-ray Transients (SXTs). In these systems, sporadic outbursts are produced due to some mechanism for which some mass is transfer onto the compact primary via an accretion disk. There are two types of SXTs. In the Type I, the compact object is a weakly magnetized neutron star, like Cen X-4, Aql X-1, etc. Their lifetime is shorter and exhibit X-ray bursts. In the Type II (sometimes designed as X-ray Novae although they differ significantly from the classical optical novae in their nature), the compact object is likely to be a black hole. They brighten in the course of few days to become one of the brightest sources in the X-ray sky, then declining over the following months, with a smaller secondary X-ray/optical maximum 2-3 months after the outburst. They also present a very characteristic X-ray spectrum, often dominated by an ultrasoft X-ray component and a hard X-ray tail.

This Type II class comprises so far of the order of ten objects. Two newly discovered X-ray sources, GRS 1716-24[1,2] and GRS 1009-45[3] possibly belong to this class. But here we will focus on the previous type II SXT, GRO J0422+32 (sometimes referred as Nova Persei 1992). It was first detected in X-rays on 1992 August 5 by the BATSE experiment on GRO[4]. The source was observed with the all-sky monitor WATCH on GRANAT since August 11[5]. The optical counterpart was discovered ten days later at the Crimean Astrophysical Observatory[6]. Since then, the object has been observed at other wavelengths. The results of the optical observations we performed one year after the main outburst are presented here.

INSTRUMENTATION

Observations of the counterpart of GRO J0422+32 were carried out on the 1.5 m Spanish telescope at the German-Spanish Calar Alto Observatory. The CCD used was a THX CCD (1024x1024 pixels), that yielded a 5.46 x 5.46 square field. Typical exposures were 300 s for the V filter. More than 60 CCD images were analyzed. The data were reduced using the software package FIGARO. The images were processed to eliminate the eloctronic bias and flat field corrected to remove the pixel-to-pixel sensitivity variations. The optical light curve was done using the differences in flux between the object and 5 field stars. The net counts were integrated within a 9.6" aperture. The seeing was during most of the time of the order of 1.0" (FWHM 3 pixels). Two spectra covering the region 4,500-7000 A were taken on Aug 14 at the Cassegrain focus of the 2.2 m telescope. The integration times were 10 and 45 minutes.

OBSERVATIONS

The first set of CCD images were taken on August 10.105, and showed GRO J0422+22 as a faint star at V = 18.7. The object was not observed until August 13.17, when it was found to be at V = 15.1, indicating that GR0 J0422+32 underwent an unexpected increase in brightness[7,8] by 3.6 magnitudes. Two spectra were taken on August 14.11 (Figure 1), but no emission lines were detected on the blue continuum, like HeII (4686 A) and NIII (4640 A), that are characteristics of the optical counterparts of these objects.[9]

Figure 2 shows the GRO J0422+32 optical light curve in the V photometric band during the August 1993 outburst. Other V magnitudes reported in the IAU Circulars have been included. The optical maximum was probably reached approximately on August 14, (V = 15.1). After that, a constant decrease in brightness (with a decay rate of about 0.05 mag/day) is seen during the next four days, prior to a rather constant phase lasting several days. On August 31 and September 1, the source was still bright (V = 15.5-15.6, ref. 10). When our observations were resumed in September 10, GRO J0422+32 was found again to be in a fainter state (V = 18.2), as was already noticed in Sept. 8.37 (ref. 11), implying that sometime during the beginning of September the source had decreased in brightness very rapidly (at least, 0.3 mag/day). A value of V = 18.99 was found on September 11 [12,13], similar to the state prior to the outburst.

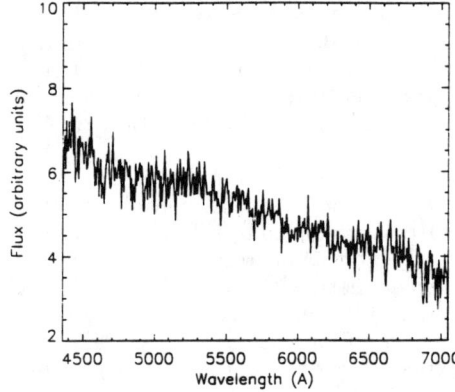

Figure 1. The blue featureless continuum of GRO J0422+32 on August 14, 1993. (See discussion on the text).

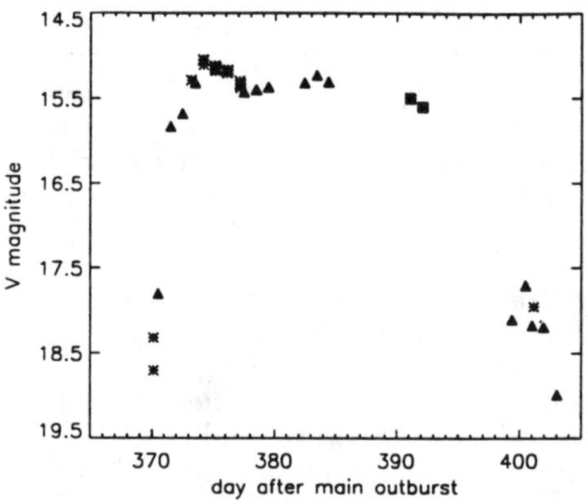

Figure 2. The light curve of GRO J0422+32 during the August 1993 outburst. Stars refer to observations taken with the 1.5-m telescope at Calar Alto. Other values (triangles) are taken from the IAU Circulars and squares from ref. 10.

DISCUSSION

Both the overall optical and X-ray light curves for GRO J0422+32 since August 1992 [14,15] resemble the light curves of other type II SXTs[16]. After the main outburst in August 1992, and the expected secondary maximum that took place in December 1992[17], GRO J0422+32 returned in July 1993 to a faint state (V=19.0, ref. 18) prior to the August 1993 outburst. This episode is the main difference to other type II SXTs.

There are two compiting models for explaining the outbursts in the type II SXTs. One of them is the Mass Transfer Instability model (MTI; ref. 19 and references therein), where the outburst is due to a sudden increase of the mass transfer rate of the companion star. The other one is the Disk Thermal Instability model (DTI; ref. 20, 21 and references therein), where the accretion disk in the quiescence state undergoes a thermal ionization instability of hydrogen and helium which leads to the so called limit-cycle (or monocycle) behaviour of the disk. It implies long recurrence times (>62 years for GRO J0422+32, ref. 22).

Let us first try to understand the onset of the main outburst in August 1992. We suggest that it could be explained by the DTI model. Due to the thermal instability, hard X-rays are produced in the inner regions of the accretion disk. And when they are reprocessed in the outer regions, soft X-rays and optical emission are originated. An observational effect is a delay of both soft

X-rays and optical emission, with respecting to hard X-rays, as was detected with WATCH[15]. On the other hand, the MTI model would explain[23] the secondary maximum in December 1992 as the result of the heating of the secondary by the X-rays produced during the previous three months time: a sudden increase of mass from the secondary towards the compact object via the accretion disk is originated. This caused the second outburst.

But what could trigger the 1993 August outburst? During the first months in 1993, the source followed a gradual decline. The above mentioned faint state in July 1993, was not the real quiescence state (V = 20-20.5, ref. 24), meaning that possibly during 100 days the compact object was accreting from the disk at a low accretion rate and the companion was constantly heated by the X-ray emission. This caused a new mass transfer instability episode. In this case and according to the MTI model, we should have seen an increase of the optical light, previous to the emission of the X-rays, as expected for the instability beginning in the outer parts of the disk. And this is what our observations revealed: an increase of 3.6 magnitudes in 4 days, without any high accretion rate (lack of He II and N III emission lines in the spectrum showed on Fig. 1). The detection of soft X-rays in August 26 by ASCA [25] could indicate that a denser accretion process begun after Aug 14 and lasted several days as the aparition of weak He lines in the optical spectrum revealed[26,27].

CONCLUSIONS

We have reported here an unusual outburst for GRO J0422+32 on August 1993. This is the first time that a rather bright optical outburst has been recorded in a type II SXT after the main X-ray outburst and secondary maximum. It can be explained by the MTI model as the result of the heating of the secondary by the X-rays produced during the three previous months. We suggest that a low accretion process may trigger an outburst not as bright as the main one that ocurrs every tens of years. Whether these "minioutbursts" are usual or not among type II SXTs during the long intervals between the main outbursts will be known only if deep photographic plates are continuosly taken for sky patrol archives* and continuous monitoring of the optical counterparts is carried out. In case of a positive detection, a quick follow-up response will be required in order to study its emission throughout all the spectrum. This will lead to a better understanding of the involved mechanisms in the low-rate accretion processes.

Acknowledgements. We are grateful to Hr. Frank, K. Birkley, T. Vives and M. Moles for the facilities given at Calar Alto Observatory for the observation of GRO J0422+32; to A. Vitores and M. Cordero for their helping in taking the spectrum, and to J. M. Gomez-Forrellad and J. Guarro for follow-up observations. Discussions with N. Lund, S. Brandt and J. Fabregat were very fruitful.

* the Sonneberg Observatory Plate Archive is the only one that is still in operation[28]. Most of the plates reach 14 magnitude, which allow important archival studies, like the one we carried out for GRO J0422+32. Unfortunately, the Observatory is threatening of closing in few years and the taking of plates would end up. International support by the scientific community would be desirable.

REFERENCES

1. Ballet, J. et al., IAU Circ. 5874 , (1993).
2. Harmon, B. A., et al., 1993, IAU Circ. 5874 , (1992).
3. Lapshov, I., Sazonov, S. and Sunyaev, R., IAU Circ. 5864 , (1993).
4. Paciesas, W. S., Briggs, M. S., Harmon, B. A. and Wilson, R. B., IAU Circ. 5580 , (1992).
5. Castro-Tirado, A. J., Brandt, S. and Lund, N, IAU Circ. 5587 , (1992).
6. Castro-Tirado, A. J., Pavlenko, E. P., Shlyapnikov, A. A., Gershberg, R., Hayrapetian, V., Brandt, S., IAU Circ. 5588 , (1992).
7. Castro-Tirado, A. J. and Ortiz, J. L., IAU Circ. 5842 , (1993).
8. Filippenko, A. V., and Matheson, T., IAU Circ. 5842 , (1993).
9. Bradt, H.V, and McClintock, J., ARA&A **21**, 30 (1983).
10. Guarro, J., private communication , (1993).
11. Zhao, P, Callanan, P., Garcia, M, and McClintock, IAU Circ. 5860 , (1993).
12. Beskin, G. et al., 1993, IAU Circ. 5863 , (1992).
13. Guarnieri, A. et al., This Proceedings , (1993).
14. Castro-Tirado, A. J., PhD Thesis, University of Copenhagen , (1993).
15. Castro-Tirado, A. J., Brandt, S., Lund, N. et al., to be submitted to A&A , (1994).
16. Tanaka, Y., Ginga Memorial Symposium (Makino, F. and Nagase, F. (ed.), 1992), p. 25.
17. Harmon, B. A., Fishman, G. J. and Paciesas, W. S., IAU Circ. 5685 , (1992).
18. Chevalier, S. and Ilovaisky, S. A., IAU Circ. 5844 , (1993).
19. Hameury, J.M., King, A.R. and Lasota, J.P., ApJ **353**, 585 (1990).
20. Mineshige, S., and Wheeler, J. C., ApJ **343**, 241 (1989).
21. Ichikawa, S. and Osaki, Y., PASJ **44**, 15 (1992).
22. Castro-Tirado, A. J., Pavlenko, E. P., Shlyapnikov, A. A., Brandt, S., Lund, N. and Ortiz, J.L., A&A **276**, L37 (1993).
23. Chen, W., Livio, M. and Gehrels, N., ApJ **408**, L5 (1993).
24. Shrader, C. et al., This Proceedings , (1993).
25. Tanaka, Y. and the ASCA team, IAU Circ. 5851 , (1993).
26. Filippenko, A. V., and Matheson, T., IAU Circ. 5860 , (1993).
27. Harlaftis, E. et al, This Proceedings , (1993).
28. Brauer, H.-J. and Fuhrmann, B., The Messenger **68**, 24 (1992).

Accretion Disks

X-RAY ENERGY SPECTRA FROM ACCRETION DISKS IN BLACK HOLE CANDIDATES

Ken Ebisawa
code 668, Laboratory for High Energy Astrophysics, NASA/Goddard Space
Flight Center, Greenbelt, MD 20771, U.S.A.

ABSTRACT

I will review major accretion disk spectral models which have been used to fit observed X-ray spectral of black hole candidates and low mass X-ray binaries. These spectral models are compared in the light of recent observations of luminosity/spectral variations of black hole candidates. Blackbody disk models can explain the observed luminosity/spectral variations in terms of variation of the mass accretion rate and constant mass of the central object, although the local emission of accretion disk will not be a blackbody. Because of the Comptonization, color temperature of the local emission will be significantly larger than the effective temperature, and success of the blackbody disk model suggests the color to the effective temperature ratio (T_{col}/T_{eff}) is constant along the disk radius and disk luminosity variations. Precise calculation of accretion disk spectral models by Titarchuk (1993) taking account of Comptonization and radiative transfer reproduced the observed results successfully.

1. INTRODUCTION

Galactic X-ray binary sources are powered by mass accretion onto compact objects. Gravitational energy is released and converted to X-ray photons in the accretion disks and on the neutron star surfaces. In the case of black hole binaries, all the X-rays are emitted from accretion disks, hence black hole candidates are most suitable laboratory to study X-ray emission from accretion disks. It is well-known that black hole candidates have two distinct X-ray spectral states; namely, the soft spectral state and the hard spectral state (for a review, e.g., Tanaka 1989, Inoue 1992). Soft state energy spectra are composed of the soft, thermal component and the hard, power-law like component which are respectively dominant below and above ~ 10 keV. On the other hand, hard state spectra are expressed with a single power-law function over 2 to 30 keV (or much higher energies) and its reflection component by the accretion disk. The soft component in the soft state is considered emission from optically thick, cool accretion disks around black holes, whereas the hard-tail component and the power-law spectra in the hard state are probably from optically thin, hot region. In this paper, I review accretion disk spectral models which have been used to fit observed soft components of black hole candidates and low mass X-ray binaries[1]. I compare those accretion disk models in the light of recent observational results of black hole candidates, and try to conjecture correct accretion disk spectral models.

---

[1]In this paper, the low mass X-ray binaries denote binaries of neutron stars, not black holes, and low mass companions.

2. REVIEW OF ACCRETION DISK SPECTRAL MODELS

In this section, I review, from practical point of view, several accretion disk spectral models which have been frequently used to fit observed X-ray energy spectra of black hole candidates and low mass X-ray binaries. Similar review of accretion disk models can be found in White, Stella and Parmar (1988).

Firstly I will show several fundamental formulae for accretion disks which hold regardless of minute difference of the spectral models.

As long as an accretion disk is optically thick to true absorption ($\tau_{eff} \equiv \sqrt{\tau_{abs}\tau_{sct}} \gg 1$ in the vertical direction), energy flux from unit surface of the disk at a given radius R is written as (e.g., Shakura and Sunyaev 1973),

$$Q(R) = \frac{3GM\dot{M}}{8\pi R^3}\left(1 - \sqrt{\frac{R_{in}}{R}}\right), \qquad (1)$$

where $M, \dot{M}$ are the mass of the central object and the mass accretion rate, and R_{in} is the innermost disk radius. The *effective temperature* of the disk is defined as $T_{eff} \equiv (Q/\sigma)^{1/4}$, so

$$T_{eff}(R) \equiv \left\{\frac{3GM\dot{M}}{8\pi\sigma R^3}\left(1 - \sqrt{\frac{R_{in}}{R}}\right)\right\}^{1/4}. \qquad (2)$$

Note that this is the definition of the *effective temperature*, and holds even if local emission is not a blackbody, as long as the disk is optically thick to true absorption. We should pay attention to the difference of the *effective temperature* and the *color temperature* of the emission. The temperature we may obtain through spectral fitting (e.g., through spectral shape) is the color temperature, and this will be different from the effective temperature unless local emission is a blackbody. This difference will be important when interpreting disk parameters obtained though model fitting (see below).

Accretion disk spectra may be obtained by integrating local emission along the disk radius from R_{in} to R_{out}. With the local emission being $I(M, \dot{M}, r, E)$ [photons/s/cm^2/keV/str], which will be in general a function of the mass, mass accretion rate and disk radius (and maybe the viscous parameter α), the observed disk spectra may be written as,

$$f(E)\, d\Omega[\text{photons/s/cm}^2/\text{keV}] = \frac{\cos\theta}{d^2}\int_{R_{in}}^{R_{out}} 2\pi r\, I(M, \dot{M}, r, E)\, dr, \qquad (3)$$

where d is the distance to the source and θ is the inclination angle (0 for the face-on disk).

2-1. Blackbody disk models

In the simplest case that the local emission is blackbody $B(T, E)$, the disk emission is written as,

$$f(E)\, d\Omega = \frac{\cos\theta}{d^2}\int_{R_{in}}^{R_{out}} 2\pi r\, B(T(R), E)\, dR, \qquad (4)$$

where $T(R) = T_{eff}(R)$ is given by (2). Putting $R_{in} = 3R_S = 6GM/c^2$ and $R_{out} = \infty$, the blackbody disk model will be also written as,

$$f(M, \dot{M}, E)\, d\Omega$$
$$= \frac{\cos\theta}{d^2} \left(\frac{6GM}{c^2}\right)^2 \int_{(R/R_{in})=1}^{\infty} 2\pi\, (R/R_{in})\, B\left(T_{eff}(R/R_{in}, M, \dot{M}), E\right)\, d(R/R_{in}), \tag{5}$$

where

$$T_{eff}(R/R_{in}, M, \dot{M}) \propto \dot{M}^{1/4} M^{-1/2} (R/R_{in})^{-3/4} \left(1 - \sqrt{\frac{R_{in}}{R}}\right)^{1/4}. \tag{6}$$

Note that the blackbody disk model is a function of solely M and $\dot{M}$, besides the normalization depending on d and θ^2.

If we neglect the $\sqrt{R_{in}/R}$ term in (2), which is appropriate when $R \gg R_{in}$, $T(R) \propto R^{-3/4}$, hence $dT/T = -\frac{3}{4}(dR/R)$. In such a case, the disk spectra may be written as[3],

$$f(E)\, d\Omega = \frac{8\pi R_{in}^{*2} \cos\theta}{3\, d^2} \int_{T_{out}}^{T_{in}} \left(\frac{T}{T_{in}}\right)^{-11/3} B(E, T)\, \frac{dT}{T_{in}}, \tag{7}$$

where R_{in}^* is the radius where the radial integration starts and T_{in} is the temperature there. This simplified formula, with $R_{in}^*\sqrt{\cos\theta}$ and T_{in} being free parameters (T_{out} can be set zero practically), was introduced by Mitsuda et al. (1984) (see also Mitsuda 1984), and has been also used for fitting observed energy spectra.

The disk temperature (2) takes the maximum

$$T_{eff}^{(max)} = 1.2\, \text{keV}\, (\dot{M}/10^{18}\text{g/s})^{1/4} (M_\odot/M)^{1/2} \text{ at } R_{max} = (7/6)^2 R_{in}, \tag{8}$$

and emission within R_{max} is not significant and $T(R) \propto R^{-3/4}$ is a good approximation for $R > R_{max}$. Hence, disk spectra of the forms (4) and (7) are nearly identical by regarding R_{in}^* and T_{in} in (7) as R_{max} and $T_{eff}^{(max)}$ in (8). Therefore, in the following, we do not explicitly distinguish the blackbody disk spectra of the forms (4) and (7).

Mitsuda et al. (1984) found energy spectra of bright low mass X-ray binaries are composed of a stable soft component (accretion disk origin) and a variable hard component (neutron star origin), and the soft component can be fitted with the blackbody disk model. The blackbody disk model has been also applied for the soft component of several black hole candidates. If we compare

---

[2]In the "XSPEC" spectral fitting package, which is widely used in the X-ray astronomy community, this model is available with the name "disk". Be careful that the unit of the inner disk radius is $3\, R_S$ instead of R_S as written in the manual, and the normalization of the model is incorrect and a half of the real one (at least up to the version 8.34).

[3]This is the "diskbb" model in XSPEC.

the temperatures T_{in} of the blackbody disk model, those for black hole candidates (typically $0.7 - 1.0$ keV) are systematically lower than those for low mass X-ray binaries (typically ~ 1.4 keV) (e.g., Inoue 1985; Makishima et al. 1986; Yaqoob, Ebisawa and Mitsuda 1993), corresponding to the "ultra-soft" spectral criterion for black hole candidates (White and Marshall 1984). The lower disk temperature may be explained with larger mass of the central objects; as a matter of fact, $T_{in} \approx T_{eff}^{(max)} \propto M^{-1/2}$ for a given mass accretion rate (see [8]).

It should be noticed that the blackbody disk model itself will not be correct for X-ray energy spectra; actually, local emission will not be a blackbody in the inner region of accretion disks because the temperature is so high and electron scattering is dominant (e.g. Shakura and Sunyaev 1973). Furthermore, the blackbody disk model would yield too small mass and/or too large mass accretion rate exceeding the Eddington limit (e.g., Czerny et al. 1986; White, Stella and Parmar 1988). These problems may be solved by introducing difference of the color temperature and the effective temperature of the emission (see the next section)[4].

2-2. "Diluted blackbody" disk models

In the inner region of accretion disks in X-ray binaries, electron scattering will be dominant and emergent X-ray energy spectra will be affected by Comptonization (e.g., Shakura and Sunyaev 1973). Local emission from such Comptonization dominant atmosphere will yield the Wien peak if $y \equiv (4kT_e/mc^2)\tau_e^2 \gtrsim 1$, and may be approximated, within X-ray energy range, with "diluted Planckian" in which color temperature of the emission is significantly higher than the effective temperature, and the normalization is smaller than the Planck function by the factor $(T_{eff}/T_{col})^4$ (e.g., Ebisuzaki, Hanawa and Sugimoto 1984; London, Taam and Howard 1986). Hence, a more realistic accretion disk spectrum, which we may call *diluted blackbody disk model*, may be obtained by putting $(T_{eff}/T_{col})^4 B(T_{col}, E)$ in place of the local emission in (3);

$$f'(M, \dot{M}, E)\, d\Omega \approx \frac{\cos\theta}{d^2} \int_{R_{in}}^{R_{out}} 2\pi R \left(\frac{T_{eff}}{T_{col}}\right)^4 B\left(T_{col}(R, M, \dot{M}), E\right) dR. \quad (9)$$

If the color to the effective temperature ratio T_{col}/T_{eff} is constant along the disk radius, this can be rewritten as (putting $R_{out} = \infty$ and measuring the radius in R_{in} as before),

$$f'(M, \dot{M}, E)\, d\Omega = \frac{\cos\theta}{d^2} \left(\frac{6G(T_{eff}/T_{col})^2 M}{c^2}\right)^2$$

$$\times \int_{(R/R_{in})=1}^{\infty} 2\pi (R/R_{in})\, B\left(T_{eff}(R/R_{in}, (T_{eff}/T_{col})^2 M, \dot{M}, E)\right) d(R/R_{in}), \quad (10)$$

where we used

$$T_{col}(R/R_{in}, M, \dot{M}, E) = \left(\frac{T_{col}}{T_{eff}}\right) T_{eff}(R/R_{in}, M, \dot{M}, E)$$

---

[4]It should be pointed out that Mitsuda [1984] has already discussed difference of the color temperature and the effective temperature extensively in the same work he introduced the blackbody disk model.

$$= T_{eff}(R/R_{in}, (T_{eff}/T_{col})^2 M, \dot{M}, E),$$

from (6). Comparing (5) and (10), we can see that, for a given energy spectrum, the diluted blackbody disk model yields $(T_{col}/T_{eff})^2$ times larger mass than is derived with the blackbody disk model. If the mass is fixed instead, the mass accretion rate will be $(T_{eff}/T_{col})^4$ times smaller for the diluted blackbody disk model to attain the observed color temperature (see [6]). Thus, difference between the color temperature and the effective temperature is very important to interpret accretion disk model parameters.

Hanawa (1989) calculated accretion disk spectra assuming the local emission is diluted blackbody and taking account of special/general relativistic effects. The model was applied to several black hole candidates and low mass X-ray binaries, and yielded reasonable mass and mass accretion rate with $(T_{col}/T_{eff}) \approx$ 1.5 (Ebisawa, Mitsuda and Hanawa 1991; Kitamoto et al. 1992a; Ebisawa et al. 1993).

2-3. Sunyaev and Titarchuk Comptonization models

Sunyaev and Titarchuk (1980) calculated Comptonized X-ray energy spectra emerging from completely ionized plasma in which photon sources are distributed. An analytical formula is obtained when the electron temperature is much higher than the energy of soft photons, and this formula has been used to fit observed X-ray spectra. The Comptonized energy spectrum is determined by the number of soft photons, electron temperature T_e and electron scattering optical depth of the plasma τ_e. Sunyaev and Titarchuk Comptonization model was applied to the *hard state* energy spectra of Cyg X-1, and $T_e \approx 27$ keV and $\tau_e \approx 5$ are derived (Sunyaev and Trümper 1979; Sunyaev and Titarchuk 1980).

White et al. (1985) and White et al. (1986) have shown energy spectra of bright low mass X-ray binaries are fitted with the sum of the Comptonized spectrum (probably the accretion disk component – not specific to a disk geometry though) and a blackbody emission (neutron star surface/boundary layer emission) [5]. Typical parameters for bright low mass X-ray binaries are $T_e \approx 3.5$ keV and $\tau_e \approx 10$ (White, Stella and Parmar 1988). Sunyaev and Titarchuk Comptonization model was also used for energy spectra of low luminosity low mas X-ray binaries and soft state black hole candidates (Treves et al. 1988; White, Stella and Parmar 1988). Typical parameters for the soft state black hole candidates are $T_e \approx 1$ keV and $\tau_e \approx 20$ (Treves et al. 1988; White, Stella and Parmar 1988).

It should be noticed that the Sunyaev and Titarchuk Comptonization model parameters for soft state black hole candidates imply the initial assumption of the model is broken. As a matter of fact, at $T_e \approx 1$ keV, heavy elements will not be fully ionized and the plasma should be optically thick to true absorption with the large column density of $N_H \approx 10^{25}$ cm^{-2}. Thereby Sunyaev and Titarchuk Comptonization model is not correct for soft state black hole candidate. Those low temperature and high column density suggest absorption of photons cannot be negligible as well as Comptonization, and more accurate

---

[5] For the discussion about spectral models for bright low mass X-ray binaries, especially about comparison of the models by White et al. 1985 and Mitsuda et al. 1984, see Makishima and Mitsuda 1985; White, Stella and Parmar 1988; Mitsuda et al. 1989; Makishima et al. 1989.

treatment of radiative transfer should be necessary to calculate X-ray emission from accretion disks (see section 2-5).

2-4. Modified blackbody disk models

In the disk region where electron scattering is dominant but the Comptonization is negligible ($y \lesssim 1$, that is, energy transfer between photons and electrons is negligible), the local emission will become the "modified blackbody", which may be expressed with the form,

$$I(E) \propto (\kappa_{ff}(E)/\kappa_{es})^{1/2} B(T, E), \qquad (11)$$

where $\kappa_{ff}(E)$ and κ_{es} are respectively the free-free and electron scattering opacity, and T is the *electron temperature* (Felten and Rees 1972; Illarionov and Sunyaev 1972). The free-free opacity for a given energy is the function of the electron temperature and density of the matter, and both of which are obtained by solving dynamical equations of accretion disks. Thus, $I(E)$ is calculated as a function of $M, \dot{M}, r$ and the viscous parameter α, and the accretion disk spectra are obtained from (3).

Modified blackbody disk spectra depend on assumptions of accretion disk structure through the formulae for the electron temperature and density of the matter. In Shakura and Sunyaev (1973), the viscous tensor is assumed proportional to the total pressure ($\equiv$ gas pressure + radiation pressure), and the modified blackbody disk model will be correct for "intermediate region" of the disk ($25 R_{in} \lesssim R \lesssim 800\, R_{in}$), by contrast Comptonization should be significant in the inner region of the disk. On the other hand, Nannurelli and Stella (1989) considered the case that the viscous tensor is proportional only to the gas pressure, and they showed that Comptonization is not significant even in the inner region of the disk, and the modified blackbody disk model will be appropriate for X-ray emission from such accretion disks [6].

White, Stella and Parmar (1988) applied the two modified blackbody disk models to low mass X-ray binaries and black hole candidates, and reported that the modified blackbody disk models yield too large, super-Eddington mass accretion rate for bright low mass X-ray binaries, but could explain energy spectra of soft state black hole candidates successfully with reasonable disk parameters.

2-5. Precise Comptonization disk model

In order to calculate X-ray emission from electron scattering dominant atmosphere precisely, the radiation transfer equation as well as dynamical equations of the accretion disk has to be solved. Such calculation was made for X-ray burst spectra by many authors (e.g., Ebisuzaki 1987; Titarchuk 1988; Pavlov, Shibanov and Zavlin 1991), and also recently carried out for accretion disk spectra (e.g., Ross, Fabian and Mineshige 1993; Titarchuk 1993). Here, I will briefly explain the Comptonized accretion disk model by Titarchuk (1993) (hereafter simply "the Comptonization disk model"); results of application of the model to the real data are shown in the next section.

---

[6] The modified blackbody disk models by Shakura and Sunyaev (1973) and Nannurelli and Stella (1989) are available in XSPEC with the names "disko" and "diskm", respectively.

Figure 1: Local emission of the Comptonization disk model by Titarchuk (1993) at three different radii ($R_{max} \equiv (7/6)^2 R_{in}$ is the radius where the effective temperature becomes the maximum). For comparison, the blackbody with T_{eff} and the diluted blackbody with T_{col} are shown together.

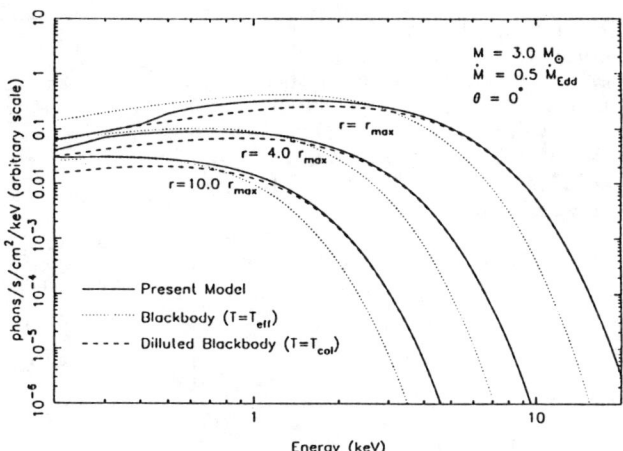

Figure 2: Comptonization disk spectra by Titarchuk (1993) for $M = 3M_\odot$ and three different mass accretion rates.

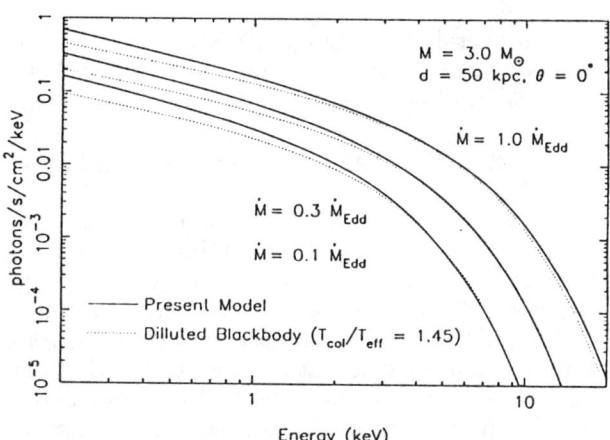

In the Comptonization disk model, vertical dynamical equilibrium and Kompaneets equation (vertical radiative transfer) are considered for each disk radius, and approximate solutions are obtained analytically with the Eddington approximation. Local emission is obtained for given M and $\dot{M}$ for each radius, and accretion disk spectra are obtained by summing up the local spectra according to (3). In figure 1, we show local emission of the Comptonization disk model, together with blackbody emission with T_{eff} and the diluted blackbody with T_{col}. As expected, the local emission of the disk is much harder than the blackbody with the same effective temperature, and approximated with the diluted blackbody.

Figure 2 shows the Comptonization disk spectra for three different mass accretion rate, together with diluted blackbody disk model with $(T_{col}/T_{eff}) = 1.45$ for comparison. It can be seen that the diluted blackbody model is a good approximation of the Comptonization disk model within 2 – 10 keV where standard proportional counters are most sensitive.

The ratio of the color temperature and the effective temperature is shown as a function of the disk radius for different mass accretion rates (figure 3).

Figure 3: The ratio of the color temperature and the effective temperature for the Comptonization disk spectra by Titarchuk (1993) for three different mass accretion rates as a function of the disk radius.

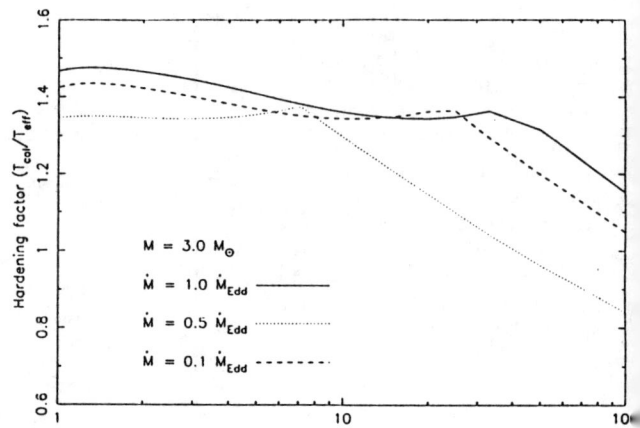

Within several R_{in}, where most X-ray photons originate, T_{col}/T_{eff} does not vary much with the radius. The mass accretion rate (disk luminosity) dependence is small too, and less than ~ 10 % for $\dot{M} = 0.1\ \dot{M}_{Edd}$ to $1.0\ \dot{M}_{Edd}$.

3. APPLICATION TO OBSERVATIONS AND DISCUSSION

Commonly, X-ray spectral study of accretion disks is carried out by fitting (theoretical) models to observed data of black hole candidates and low mass X-ray binaries. Because of the featureless energy spectra and rather poor energy resolution of the detectors, it often happens that the same data are fitted with several different accretion disk models. Therefore, in order to judge correct accretion disk models, rather than comparing details of spectral shapes, it will be more effective to study spectral variations. Black hole candidates often show large luminosity/spectral variations, and correct accretion disk models should be able to explain the spectral variations in a physically reasonable manner.

Figure 4 is an intensity-hardness diagram of LMC X-3 observed with GINGA (Ebisawa et al. 1993). LMC X-3 is always in the soft state and shows periodic luminosity variations by the factor of ~ 4 (Cowley et al. 1991). The dotted line in figure 4 is a trace of the blackbody disk model with a constant innermost radius ($\approx$ constant mass of the central object; section 2-1) and variable temperatures ($\approx$ variable mass accretion rate). It is obvious that the spectral variation is successfully explained with the blackbody disk model. Similar characteristic spectral variations, namely, constant innermost radius and variable temperatures of the blackbody disk model, are also found from black hole candidate novae GS2000+25 (Tanaka 1989), GS1124-68 (Kitamoto et al. 1992b) and EXO1846-031 (Parmar et al. 1993).

Sunyaev and Titarchuk Comptonization model could fit the energy spectra of LMC X-3, but did not yield such an invariant against the spectral variation as the innermost radius of the blackbody disk model (Ebisawa et al. 1993). Both the Shakura and Sunyaev (1973) and Nannurelli and Stella (1991) modified blackbody disk models could not reproduce the constant mass of the central

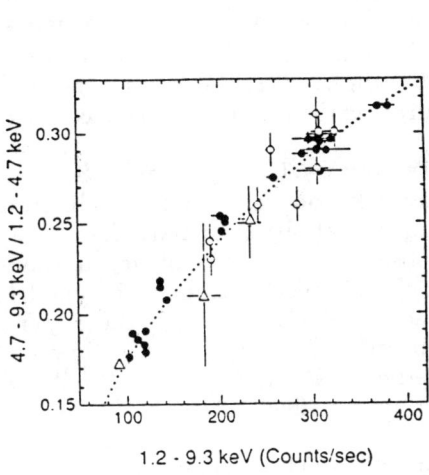

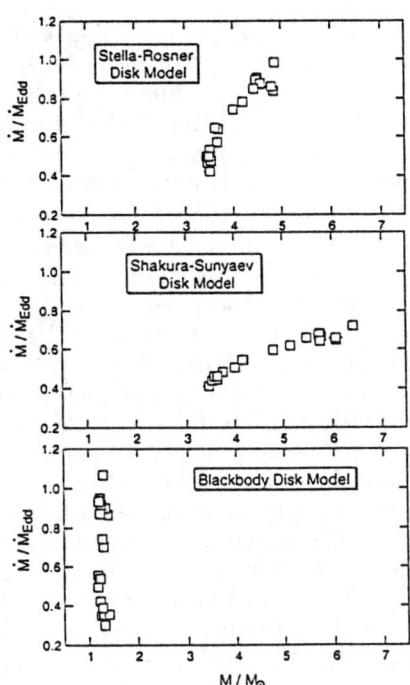

Figure 4: Intensity-hardness diagram of LMC X-3 observed with GINGA. The dotted line is a trace of the blackbody disk model with a constant innermost radius and variable temperatures.

Figure 5: Results of the spectral fitting for LMC X-3 spectra taken with GINGA with the blackbody disk model and Shakura and Sunyaev (1973) and Nannurelli and Stella (1991) modified blackbody disk models. Distance to the source and inclination angle are assumed 50 kpc and 0° respectively.

Figure 6: Results of application of the Comptonization disk model by Titarchuk (1993) to LMC X-3 data.

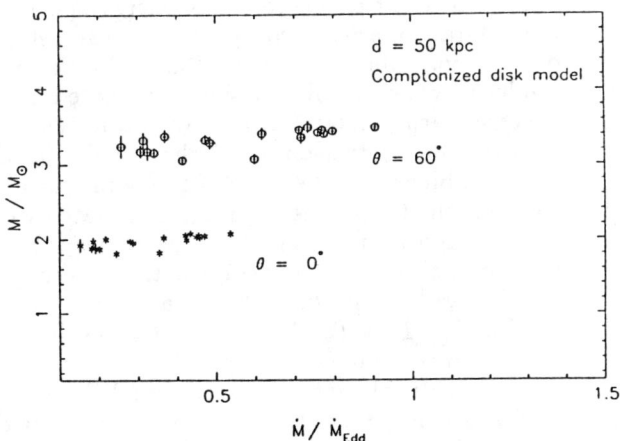

object either, and instead they yielded a systematic correlation between the mass and mass accretion rate (figure 5).

Considering the difference of the color temperature and the effective temperature, success of the blackbody disk model to explain the spectral variation with the constant mass of the central object means the color to the effective temperature ratio is invariable against the disk luminosity variations ($\propto$ mass accretion rate variations) (see section 2-3). The unrealistic $M - \dot{M}$ correlation with the modified blackbody disk model implies the T_{col}/T_{eff} becomes larger for higher disk luminosity (higher disk temperature), as the atmosphere is becoming transparent for absorption.

As is found in figure 5 (and also pointed out by Czerny et al. 1984 and White, Stella and Parmar 1988), the blackbody disk model gives super-Eddington mass accretion rate and too small mass of the central object of LMC X-3 compared with the lower limit obtained from observations of the binary motion ($\sim 7M_\odot$; Cowley et al. 1983). If we consider the diluted blackbody disk model instead, the mass becomes larger by the factor $(T_{col}/T_{eff})^2$, and the super-Eddington problem is solved. Additionally, the mass depends on the inclination angle, and relativistic effects will reduce the derived mass slightly. For $(T_{col}/T_{eff}) = 1.5$ and $\theta = 60°$, and taking account of relativistic effects, $M \sim 5.0 M_\odot$ is derived at $d = 50$ kpc (Ebisawa et al. 1993).

The Comptonization disk model by Titarchuk (1993) was applied to LMC X-3 (figure 6). This model has only two parameters M and $\dot{M}$ besides the distance and the inclination angle. The model yields a constant mass against the variation of the mass accretion rate, which is due to invariability of the (T_{col}/T_{eff}) value over the disk luminosity change (figure 3). The mass for $\theta = 0°$ is $\sim 2M_\odot$ and larger than $\sim 1.2 M_\odot$ by the blackbody disk model, because of $(T_{col}/T_{eff}) > 1$ due to Comptonization. For $\theta = 60°$, the mass becomes more reasonably larger (still smaller than the optical lower limit though), and the mass accretion rate does not exceed the Eddington limit.

4. SUMMARY

In this paper, I reviewed several accretion disk spectral models which have been used to fit observed X-ray energy spectra of black hole candidates and low mass X-ray binaries. I compared those accretion disk models in the light of recent observations of spectral variations of black hole candidates. The blackbody disk model in which local emission is assumed blackbody can explain the observed spectral variations in terms of variation of the mass accretion rate and constant mass of the central object, although the local emission of accretion disks will not be a blackbody because of the dominance of electron scattering. This result suggests the local emission is approximated with the diluted blackbody in which color temperature is significantly higher than the effective temperature due to Comptonization, and the color to the effective temperature ratio is invariant over the disk radius and luminosity variations. As a matter of fact, if we use diluted blackbody $(T_{eff}/T_{col})^4 B(E, T_{col})$ for the local emission with $(T_{col}/T_{eff}) \approx 1.5$, we obtain reasonable mass and mass accretion rate.

Precise calculation of accretion disk spectral models was made by Titarchuk (1993) solving vertical radiation transfer taking account of Comptonization.

This model successfully explain the observational results; the local emission of the model was very similar to the diluted blackbody in the X-ray energy range, and (T_{col}/T_{eff}) was virtually constant over the disk radius and the disk luminosity variation.

Modified blackbody disk models could not yield a constant mass of the central object for the spectral variation of black hole candidates, but instead unrealistic $M - \dot{M}$ correlation was found. This is mainly because, in these models, (T_{col}/T_{eff}) increases with the luminosity as disk is becoming transparent to absorption. Modified blackbody disk models will not be correct for X-ray energy spectra of black hole candidates and low mass X-ray binaries, since Comptonization is not taken into account.

Nannurelli and Stella (1991) showed that, if the viscous tensor is proportional only to the gas pressure, Comptonization is not significant in the inner region of accretion disks, and modified black body disk models are correct for X-ray emissions. This does not happen, and therefore it may be likely that the viscous tensor is proportional to the total pressure (gas pressure and radiation pressure), not only to the gas pressure. This result will impact our understanding of accretion disks in X-ray binaries.

ACKNOWLEDGMENT

The author acknowledge to Drs. K. Mitsuda, L. M. Ozernoy, L. G. Titarchuk and N. E. White for variable discussion.

REFERENCES

Cowley et al. 1983, ApJ, 272, 118
Cowley et al. 1991, ApJ, 381, 526
Czerny, B. et al. 1986, ApJ, 311, 241
Ebisawa, K., Mitsuda, K. and Hanawa, T., 1991 ApJ, 367, 213
Ebisawa et al. 1993, ApJ, 403, 684
Ebisuzaki, T. 1987, PASJ, 39, 287
Ebisuzaki, T., Hanawa, T. and Sugimoto, D. 1984, PASJ, 36, 551
Felten, J. E. and Rees, M. J. 1972, A&A, 17, 226
Hanawa, T. 1989, ApJ, 341, 948
Inoue, H. 1985, SpScRev 40, 317
Inoue, H. 1992, in *Frontiers of X-ray Astronomy*, p.291, Universal Academy Press, Inc., Tokyo, Japan
Illarionov, A. F. and Sunyaev, R. A. 1972, SovAstrAJ, 16, 45
Kitamoto, S., Tsunemi, H. and Roussel-Dupre, D. 1992a, 391, 220, ApJ
Kitamoto, S. et al. 1992b, ApJ, 394, 609
London, R. A., Taam, R. E. and Howard, W. M. 1986, ApJ, 306, 170
Makishima, K. and Mitsuda, K. 1985, in *Japan-U.S. Seminar on Galactic and Extragalactic Compact X-ray Sources*, p.127, ISAS, Japan
Makishima et al. 1986, ApJ, 308, 635
Makishima et al. 1989, PASJ, 41, 531

Mitsuda, K. 1984, Doctoral Thesis, Univ. of Tokyo (ISAS RN 251)
Mitsuda, K. et al. 1984, PASJ, 36, 741
Mitsuda, K. et al. 1989, PASJ, 41, 97
Nannurelli, M. and Stella, L. 1989, 226, 343, A&A
Parmar, A. N. et al. A&A, 1993, 279, 179
Pavlov, G. G., Shibanov, Y. A. and Zavlin, V. E. 1991, MNRAS, 253, 193
Ross, R. R., Fabian, A. C. and Mineshige, S. 1992, MNRAS, 258, 189
Shakura, N. I. and Sunyaev, R. A. 1973, A&A, 24, 337
Sunyaev, R. A. and Titarchuk, L. G. 1980, A&A, 86, 121
Sunyaev, R. A. and Trümper, 1979, Nature, 279, 506
Tanaka 1989, *Proc. 23rd ESLAB Symp. Two Topics in X-ray Astronomy*, vol.1, p.3, ESA.
Titarchuk, L. G. 1988, SovAstronLett 14(3), 229
Titarchuk, L. G. 1993, in preparation
Treves, et al. 1988, ApJ, 325, 119
White, N. E. and Marshall, F, 1984, ApJ, 281, 354
White, N. E., Peacock, A. and Taylor, B. G. 1985, ApJ, 296, 475
White, N. E. et al. 1986, MNRAS, 218, 129
White, N.E., Stella, L. and Parmar, A. N. 1988, ApJ, 324, 363
Yaqoob, T., Ebisawa, K. and Mitsuda, K. 1993, MNRAS, 264, 411

Spectral Diagnostics and Accretion Disk Corona Reprocessing in X-ray Binaries

T. R. Kallman

NASA/Goddard Space Flight Center, Code 665, Greenbelt, Maryland 20771, U.S.A.

ABSTRACT

Emission lines are valuable probes of the conditions in the accretion flows associated with binary X-ray sources. In this review I describe examples of emission lines in defferent spectral bands, review the observations, and discuss what they indicate about the conditions in binary X-ray sources. These lines are interpreted using an X-ray illuminated accretion disk model; The structure and dynamics of the heated disk, its spectral signatures, and the major unsolved theoretical issues surrounding them are discussed.

1. INTRODUCTION

Reprocessing of the continuum spectrum can provide valuable diagnostic information about X-ray binaries. This can lead to better understanding of these objects, since the reprocessor is likely to be associated with the gas flow which fuels the X-ray source, or the source of that gas. Reprocessed features are ubiquitous in X-ray binaries, and we cannot claim an understanding of these objects without accounting for the reprocessed spectra. Furthermore, the study of reprocessed spectra is in many ways easier, or at least the models are likely to have more sensitive tests, than is true for the study of other aspects such as the continuum shape or variability.

Reprocessing can occur in a variety of sites associated with X-ray binaries, icluding an accretion disk, the companion star, an accretion stream, an alfven shell, a wind from the companion or from the disk, a static corona near the disk, diffuse interstellar gas, or some other material in the vicinity.

This review will discuss the formation of discrete spectral features resulting from illumination of accreting gas by an X-ray source and will illustrate their importance through several examples. Emphasis will be given to the study of low mass X-ray binaries (LMXB's), although many of the same principles and techniques may be applied to massive X-ray binaries (MXRB) as well.

2. ON THE IMPORTANCE OF PHOTOIONIZATION

We begin by reviewing some of the important physical processes associated with reprocessing and line emission in LMXB's. We consider an idealized LMXB consisting of an accreting neutron star with luminosity $L \simeq 10^{38} \mathrm{ergs}^{-1}$, and an orbital period of ~hours, corresponding to an orbital separation of $\leq 10^{11}$ cm, so that an estimate of the X-ray flux in the system is $F \geq 10^{15}$ erg cm^{-2} s^{-1}. The gas density in the accretion flow can be crudely estimated by assuming a mass accretion rate necessary to fuel the X-ray source and free-fall velocity for the incoming material: $n \geq \dot{M}/(4\pi R^2 v_{freefall} m_H) = 2 \times 10^{11} L_{38} \eta_{0.1}^{-1} R_{11}^{-3/2}$ cm, where L_{38} is the luminosity scaled by 10^{38} erg s^{-1}, R_{11} is the radius scaled by

10^{11} cm, and $\eta_{0.1}$ is the accretion efficiency scaled to 0.1.

The fate of an ion in such an environment can be predicted as follows. First we write the rates (per volume) for various processes affecting ionization and recombination for a fictitious hydrogenic ion with nuclear charge Z: photoionization, photoionization heating, radiative recombination, radiative recombination cooling, and collisional ionization.

$$R_{PI} \sim x_0 < \frac{F_X}{\varepsilon} \sigma_{PI} > n \sim 4 \times 10^{16} x_0 F_{15} n_{11} \varepsilon_{10}^{-1} Z^{-2} \text{cm}^{-3} \text{s}^{-1}$$

$$R_{PIHT} \sim x_0 < F_X \sigma_{PI} > n \sim 6 \times 10^8 x_0 F_{15} n_{11} Z^{-2} \text{ergcm}^{-3} \text{s}^{-1}$$

$$R_{RR} \sim x_+ < v_{electrons} \sigma_{RR} > n^2 \sim 3 \times 10^9 x_+ n_{11}^2 T_4^{-1/2} Z^2 \text{cm}^{-3} \text{s}^{-1}$$

$$R_{RRCL} \sim < v_{electrons} \sigma_{RR} > n^2 kT \sim 3 \times 10^9 x_+ n_{11}^2 T_4^{1/2} Z^2 \text{ergcm}^{-3} \text{s}^{-1}$$

$$R_{CI} \sim x_0 < v_{electrons} \sigma_{CI} > n^2 \sim 10^{14} x_0 n_{11}^2 T_4^{1/2} Z^{-4} e^{-\Delta\varepsilon/kT} \text{cm}^{-3} \text{s}^{-1}$$

In these equations x_0 and x_+ are the fractional abundances of the neutral and ionized ion species, the angle brackets ($<>$) denote an average over the relevant ensemble: either the photon energy distribution for photoionization, or the electron velocity distribution for collisional processes, σ is the cross section for the relevant process, T is the electron temperature, ε is the photon energy, n is the gas density, $\Delta\varepsilon$ is the ionization threshold energy, and the subscrips refer to values scaled by the subscripted quantity (cgs units). Equating the photoionization and recombination rates gives:

$$\frac{x_+}{x_0} = \frac{<F\sigma_{PI}/\varepsilon>}{n<\sigma_{RR}v>}$$

which shows that the ion fractions depend on the ratio of cross sections (suitably averaged) and on the ratio F/n. This quantity, the ionization parameter, is widely used in the form $\xi = 4\pi F/n$ (Tarter, Tucker and Salpeter, 1968). The previous expression can be rewritten $x_+/x_0 = 10^2 T_4^{1/2} Z^{-4} \xi$, and setting the left hand side equal to 1 gives $\xi_{eq} = 10^{-2} Z^4 T_4^{-1/2}$. Equating the photoionization heating and recombination cooling rates then gives $T_{equ} \simeq \varepsilon/k$, where ε is the mean energy ejected per photoionization. Numerical models show that this quantity is approximately $\varepsilon_{photo} \simeq \Delta\varepsilon_{threshold}/5$, so that $T_{equ} \simeq 3 \times 10^4 Z^2$ K. For iron, this temperature turns out to be 2×10^7 K, and for oxygen 2×10^6 K. Substituting into the expression for collisional ionization shows that the ratio of photoionization to collisional ionization is $\simeq 7 \times 10^{-3} \xi^{-1} T_4^{1/2} Z^{-2}$; photoionization is the dominant ionization mechanism for $\xi \geq 10^{-3}$.

Since the ionization balance depends on ξ, the existence of a given ion species as implied by the observation of a line gives a limit on the gas density: using the expression above for ξ_{eq} we derive $10^{12}, 10^{14}, 10^{15}\ F_{15}$ cm^{-3} as lower limits to the density required for existence of Fe XXVI, O VIII, and He II, respectively. If these densities are multiplied by the size of the binary system to get a crude column density, it is clear that reprocessing gas in an X-ray binary will have a column density of at least 10^{24} cm^{-2} and so will be optically thick to Thomson scattering when viewed in transmission. Since it is likely that (at least part of) the continuum is viewed directly without being Comptonized, this suggests that the gas responsible for line emission is viewed in reflection rather than in transmission.

The most important physical processes affecting line emission are recombination, collisional excitation, and inner shell fluorescence. The rates of the first two are proportional to the rates shown above for total recombination and collisional ionization, and the rate for inner shell fluorescence is proportional to the photoionization rate. The first two depend on n^2. The innershell ionization rate depends on Fn, but when the ionization equilibrium condition is substituted it has the same scaling behavior as the recombination line emission rate. Then the luminosities of lines emitted by these processes may be expressed as an average emissivity and an emission measure for gas under the appropriate conditions, i.e. $L_{line} = <j/n^2> EM$, where j/n^2 is the emissivity per hydrogen atom and the emission measure is $EM = \int n^2 dV$. Plausible values for emissivities and line luminosities are $j/n^2 \simeq 7 \times 10^{-24} x_+ T_4^{1/2} Z^2$ ergcm^{3}s^{-1} and $L_{line} \sim 10^{35}$ ergs^{-1}, so that $EM \simeq 10^{58}$ cm^{-3}. Since we know an upper limit on the volume of the emitting gas provided by the binary size, $V \leq 10^{33}$ cm^{-3}, we can derive a limit on the density of the emitting gas of $n \geq 10^{12} EM_{58}^{1/2} R_{11}^{-3}$ cm^{-3}. This can be compared with the density limits from the ionization parameter in the previous paragraph: the two conditions can only be satisfied (for $Z <<26$) if the flux is much less than the fiducial value used so far, i.e. $F \sim 10^{13}$ ergcm^{-2}s^{-1} for O VIII and $F \sim 10^{12}$ ergcm^{-2}s^{-1} for He II. Thus, the reprocessing gas must be somehow prevented from seeing the full intensity of the X-rays from the neutron star. One way for this to occur is if the reprocessing gas is in a slab with the X-rays incident at a grazing angle.

A final constraint on the line emitting gas comes from the opacity that line photons encounter as they escape the emission region. The cross section for resonant scattering is $\sigma_{resonance} = 10^{-13} T_4^{-1/2} Z^{-2} A^{1/2}$ cm^2, for a Lα analog line with Doppler broadening from an ion with atomic weight A. Resonant scattering results in line destruction if an ion can be deexcited by electron collisions during a scattering event (other destruction mechanisms include continuum absorption, Compton scattering and Auger destruction; we ignore them in this discussion). The probability per scattering for this to occur is $\tau_{thermalization}^{-1} = 3 \times 10^{-5} T_4^{-1/2} Z^{-6} n_{11}$, so that the effective hydrogen column density that a photon can escape from is $\tau_{thermalization}/\sigma_{resonance} = 3 \times 10^{15} T_4^{-1} Z^8 y^{-1}$ cm^{-2}, where y is the fractional abundance of the line emitting ion relative to hydrogen. This limits the emission measure to $EM \leq 2\pi R^2 n y^{-1} \tau_{thermalization}/\sigma_{resonance} = 2 \times 10^{51} T_4^{-1} Z^8 y^{-1}$ cm^{-3}. This limit is clearly most stringent for lines from low

Z ions such as He II, and when taken together with the estimates of emissivities given in the previous paragraph suggests that the escape of such line photons may need to be somehow facilitated, for example by turbulence which broadens the line and thereby reduces the opacity to resonance scattering.

The results of these simple estimates can be summarized as follows: Under the hypothesis that the line emitting gas is contained within a region comparable in size to the binary separation, (i) Photoionization is the likely to determine the ionization distribution; (ii) The line emitting gas must be viewed in reflection rather than in transmission; (iii) The flux of continuum X-rays from the neutron star which are incident on the line emitting gas is less than the full amount which is available, so that the line emitting gas views the continuum source either at a grazing angle or through some (other) obscuring gas; (iv) Thermalization limits the efficiency of line emission for UV lines unless the line emitting gas is more extended than has been assumed or the lines are broadened by some mechanism other than thermal motions. All these arguments suggest a heated accretion disk as the origin of the line emission, and we now provide a brief discussion of the physics associated with such a disk.

3. ACCRETION DISK CORONA

The properties of accretion disk coronae have been studied by a number of authors, beginning with Begelman, McKee and Shields (1982), Begelman and McKee (1982), and White and Holt (1982). The corona arises naturally as a result of X-ray heating of a hydrostatic gas The temperature of X-ray heated gas depends only on the shape of the X-ray spectrum and on the ionization parameter. In situations where the gas pressure is specified, such as in a hydrostatic atmosphere above a star or disk, this quantity is defined as the ratio of the X-ray flux, F (in energy units over the band 1 - 1000 Ry), to the gas pressure P, according to $\Xi = F/Pc$ (c.f. Krolik, McKee, and Tarter, 1981). For ionization parameter values greater than a critical value, Ξ_c^*, the gas temperature is determined primarily by the effects of Compton scattering. For ionization parameters less than a critical value Ξ_h^*, the gas temperature is determined by the effects of atomic processes; the temperature in this case is $T_{atomic} \sim 10^4$ K. These critical ionization parameters have values $\Xi_h \simeq 0.3(T_{IC}/10^8 K)^{-3/2}$ and $\Xi_c^* \simeq 10$ (although various effects such as conduction, non-equilibrium effects, strong Compton cooling by the UV radiation from the disk and other sources of heating may modify these estimates). Near the interface between the corona and the disk photosphere must exist a chromosphere and transition region in which X-ray heating is important, but is not sufficiently strong to heat the gas to temperatures comparable to T_{IC}. These regions must dominate the optical and UV appearance of X-ray illuminated accretion disks. It is straightforward to estimate the density at the base of the corona and a crude upper limit on the vertical column density of the corona. Such estimates show (e.g. Kallman and White, 1988) that at radii smaller than a critical value, $R \leq R_{thick} = 10^9 \text{cm} L_{37}^2 f^2 T_{IC7}^{-1}$ the Thomson depth through the corona exceeds unity. Since the incident X-rays can't penetrate to Thomson depths much greater than unity, the corona in this region must have a mean ionization parameter which is greater than Ξ_c^*. Thus, in the inner disk regions the corona will be Compton thick with high ionization parameter, and in th outer disk regions it will be Compton thin with ionization parameter Ξ_c^*.

4. LINE OBSERVABLES

The utility of emission lines as diagnostic tools can be illustrated by considering the three most important observable quantities associated with a given line: the energy ε, width $\Delta\varepsilon$, and line luminosity or equivalent width EW. Line energy depends on the gas temperature or degree of ionization (and possibly also on relativistic effects associated with line broadening). For example, the K line of iron has an energy in the range 6.4 – 7.1 KeV depending on the ionization state of iron (c.f. Hirano, et al., 1987), and which ionization state dominates depends on the ionization parameter or gas temperature. The line width depends on any of a variety of broadening mechanisms, including blending of multiple line components with differing intrinsic energies ($\Delta\varepsilon/\varepsilon \leq 0.1$), virial motions of the emitting gas ($\Delta\varepsilon/\varepsilon \simeq 0.05(R/10^8 cm)^{-1/2}$, where R is the distance from the source of gravity), Compton scattering ($\Delta\varepsilon/\varepsilon \simeq 0.008\varepsilon_{KeV}\tau_{Th}$, where ε_{KeV} is the photon energy in KeV and τ_{Th} is the optical depth to Thomson scattering; Posdyakov, Sobol' and Sunyaev, 1979), or general relativistic effects ($\Delta\varepsilon/\varepsilon \simeq 3 \times 10^{-3}(R/10^8 cm)^{-1} f(\Omega, R_{disk})$ where $f(\Omega, R_{disk})$ is a function of the viewing angle and disk inner and outer radii; Fabian, Rees, Stella and White, 1989). The natural widths or thermal doppler widths of most emission lines are too small to be measured by the current detectors. The line luminosity or equivalent width depends on the amount of emitting material (emission measure) and the emissivity.

For LMXB's emission line energies, widths, and equivalent widths have been measured for emission lines in a variety of wavelength bands; we will discuss the 'hard' (2 – 10 KeV) X-rays, 'soft' (0.5 – 2 KeV) X-rays, and ultraviolet from the brightest of these objects, Cyg X-2 and Sco X-1.

5. IRON K EMISSION LINES

The iron K line at 6.4 – 7. KeV is the dominant emission feature in the hard X-ray spectra from objects ranging from active galaxies to cataclysmic variables. This line is unique in both its strength and in the fact that it is the strongest emission feature in the X-ray spectrum which can be formed in low temperature, neutral gas. The iron line from low mass X-ray binaries (LMXRB) has been observed by most X-ray astronomy experiments. The results from the EXOSAT observatory compiled by White, Peacock and Taylor (1985) show that for Sco X-1 the line centroid energy is $\simeq 6.7 KeV$, the width (Full Width Half Maximum) is $FWHM \simeq < \Delta\varepsilon > \simeq 1 KeV$, and the equivalent width is EW$\simeq$50 eV. Similar results are found for other LMXB's (White, et al., 1986); the best example is Cyg X-2, for with BBXRT found a centroid energy, width, and equivalent width of $6.71^{+}_{-}0.2$ KeV, 971^{+505}_{-376}eV, and $60^{+}_{-}27$ eV, respectively (Smale, et al., 1993). This measurement was obtained with much higher energy resolution instrument than the previous measurements and provides confirmation of the previous results.

These results imply that the dominant ionization state in the emitting gas is high (Fe XXIII or greater) and therefore that the ionization parameter in the emitting region is $\xi \geq 10^3$ erg cm s^{-1}. From the limit on the line width we can derive an emission radius of $\leq 10^7$ cm if the broadening is due to virial motions, a Thomson depth ≥ 3 if the broadening is due to Compton scattering, or a radius $\leq 10^{6.5}$ cm if the broadening is due to gravitational effects.

More detailed inferences come from comparison with models for the ionization and density structure in an accretion disk corona. Kallman and White (1988) integrated the hydrostatic equation in the vertical direction above an X-ray heated disk, and thereby calculated the temperature, density, ionization structure, and emitted spectrum as a function of height above the disk midplane and radius from the central object. This confirmed the predictions of the accretion disk corona theory outlined above: at small radius or when the illumination is strong the 'hot' corona (i.e. where $T \simeq T_{IC}$) can become optically thick to Compton scattering. If so, virtually all the illuminating X-rays will be stopped in the corona and the predominant iron line emission will be at energies greater than 6.4 KeV. If not, incoming X-rays will penetrate through the hot corona to the disk photosphere and below and make 6.4 KeV iron line photons. This effect can be suppressed if the incident photons have grazing incidence angles to the disk (Ko and Kallman, 1993). The observations thus set a sensitive upper limit on the flux of X-rays normally incident on the disk at large radii: $f \leq 0.01$ for $R \geq 10^8$ cm, assuming a 1 $M_\odot$ neutron star radiating near the Eddington limit. Fits to the models of Pozdnyakov, Sobol and Sunyaev (1988) show that the width of the line implies a Thomson optical depth of approximately 3 if the line width is due to Compton scattering. This appears difficult to reconcile with an accretion disk corona, since a corona cannot have a Thomson depth greater than unity. However, such estimates are based on fits to Comptonization through spherical clouds; a plane parallel scattering medium can produce a comparable broadening for $\tau_{Thomson} \simeq 1$. This is because many of the photon trajectories lie in the plane of the slab, thus allowing longer paths and more scatterings than would be expected based on the vertical optical depth. Figure 1 shows an example of a fit of a Comptonized profile to the iron line region of Cyg X-2 as observed by BBXRT, demonstrating that an acceptable fit to the line width can be found for $\tau_{Thomson} \simeq 1$ and $kT \simeq 1 KeV$ if the intrinsic energy of the line is 6.9 KeV.

Rotation can also account for the observed width of the iron line if $R \leq 10^7$cm; the true broadening mechanism can be distinguished by making further observations with resolution comparable to BBXRT or better. Additional tests for detailed models comes from searches for continuum features in the vicinity of the iron K line, such as the K edge (e.g. Vrtilek, Soker, and Raymond, 1993) or the Compton reflected continuum (e.g. Lightman and White, 1988). Both of these are likely be highly dependent on geometry: the inclination angle of the disk relative to our line of sight, and the angle of incidence of the illumination onto the disk. Thus it may not be surprising that we have no clear detections of these features so far from LMXB's viewed at moderate inclination, such as Cyg X-2 or Sco X-1. Detection of Compton reflected components from other classes of objects, such as black-hole candidates, may be an indication of relatively high incidence angle of X-rays onto the disk and is probably also aided by the presence of hard power law spectra in these objects.

6 SOFT X-RAY LINES

The wavelength range between 0.5 - 2 KeV contains a large number of emission lines which are useful diagnostics of the emitting gas in X-ray binaries. However, the high spectral density of lines necessitate spectrometers with higher sensitivity and spectral resolution than can be obtained with proportional counters, and so detections of such lines have so far been limited to the small number of sources which are bright enough to be observed by the gratings on the

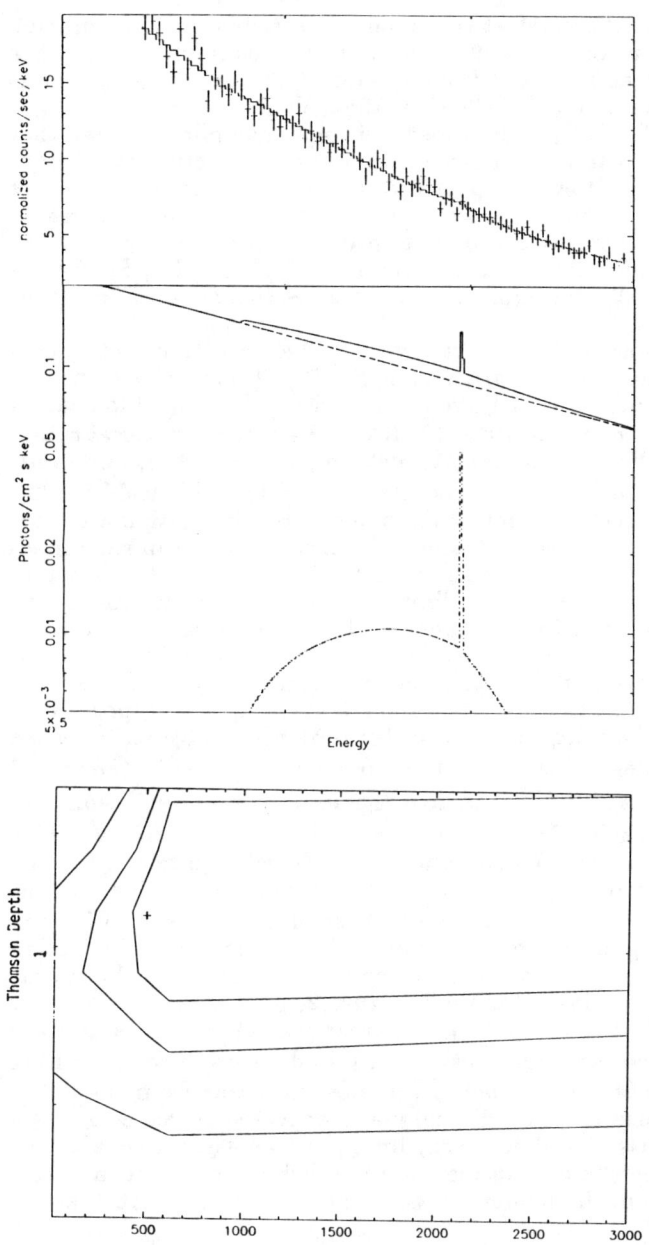

Figure 1. Fit of a Comptonized profile (solid curve) to the iron line region of Cyg X-2 as observed by BBXRT (crosses). The iron line is comptonized by a plane-parallel slab with Thomson depth of 1 from the midplane to the edge and with the photon source at the midplane, and the scattering medium is assumed to have electron temperature $kT_e = 1$ KeV. The intrinsic energy of the iron line is assumed to be 6.9 KeV. Upper panel: Count spectrum; Middle panel: Flux spectrum, showing contribution of continuum and Comptonized line; Lower panel: Confidence contours for fit of Comptonized iron line profile as a function of Thomson depth and electron temperature.

Einstein and EXOSAT satellites. Most commonly detected is the line emission due to the L shell transitions of iron and to the K shell transitions of medium-Z elements such as nitrogen and oxygen from several of the brightest low mass X-ray binaries (Kahn, Seward and Chlebowski, 1982; Vrtilek et al., 1986a, b).

The interpretation of soft X-ray line spectra is more complicated than that of the Iron K lines owing to the much larger number of competing transitions and excitation mechanisms. However, past grating observations of these lines can be described as follows: line energies in the range 0.8 – 1.6 KeV region (detections at shorter wavelengths are uncertain due to contamination of the data by unscattered photons) due to ions including N VII, O VIII, Fe XVII – XVIII, and Fe XXIII – XXIV; line equivalent widths 1 – 10 eV; and line widths $\Delta\varepsilon/\varepsilon \leq 0.1$.

Attempts to fit the observed soft X-ray grating spectra with optically thin collisional equilibrium models (e.g. Raymond and Smith, 1977) have been unsuccessful, although they do suggest that there is a broad distribution of ionization states and temperatures in the range $10^6 - 10^7$ K in the emitting material rather than a single set of conditions (Kallman, Vrtilek and Kahn, 1989), and emission measures EM$\sim 10^{58} - 10^{59}$ cm^{-3}. As pointed out by Liedahl et al., (1989), models which include the effects of photoionization and recombination are more likely to fit the data. If so, the lower emission efficiencies inherent in such models will increase the implied emission measures by a factor $\sim 3 - 10$ relative to the collision-dominated models. From the limit on the line width we can derive an emission radius of $\geq 10^7$ cm if the broadening is due to virial motions, or a Thomson depth ≤ 12 if the broadening is due to Compton scattering.

Figure 2 shows an attempt at fitting the spectrum of Cyg X-2 observed with the *Einstein* objective grating spectrograph (OGS) using a superposition of photoionization models. The solid curve shows the data after subtraction of the best-fitting continuum model. The strongest features are near 11–14Å (885–1130 eV), and 16–17 Å (730–775 eV). Also notable is the absence of line emission near 15Å (830 eV) and the presence of only a weak feature near 19Å (653 eV). This last is the energy of the Lα line of O VIII, which is produced with relatively high efficiency over a wide range of ionization parameter by photoionization models. The 16Å feature can be accounted for by the L shell lines of Fe XVII, and the shorter wavelength (higher energy) lines by L lines from higher ionization stages of iron. These exist within a range of ionization parameters $\log(\xi)$=1-3. However, as shown by the dashed curve in figure 2, photoionization models are not capable of reproducing the contrast between the strong lines and the 15Å gap. Furthermore, the emission measure required to produce these lines is EM$\sim 10^{58}$ cm^{-3}, which is comparable to the most available from the entire accretion disk corona. Since most of the corona is expected to have $\log\xi \geq 5$, the gas responsible for emission of low energy lines must be associated with the transition region separating the hot corona from the disk photosphere, and this gas is expected to have a much smaller emission measure (e.g. Paerels, et al., 1993).

7. UV LINES

Only a fraction of the low mass X-ray binaries are accessible to UV observations owing to the effects of interstellar reddening and the concentration of these sources toward the galactic center. On the other hand, UV spectra af-

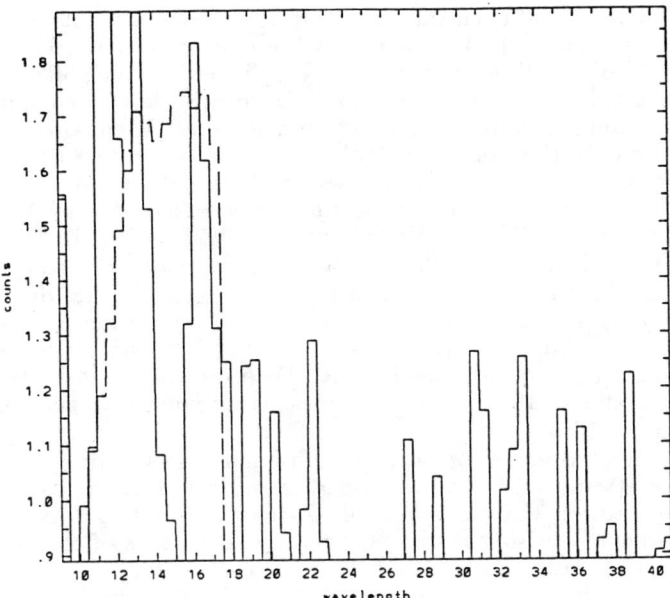

Figure 2: Fits to the observed line spectrum from the X-ray binary Cygnus X-2 observed by the grating spectrograph on the *Einstein* (HEAO-2) satellite (solid curve). The dashed curve represents an attempt at modelling this spectrum using a superposition of model photoionized spectra at various ionization parameters.

ford greater sensitivity and spectral resolution than do X-ray spectra, and also form a more homogeneous dataset owing to the longevity of the IUE satellite. The interpretation of UV line spectra is as complicated as the soft X-ray lines, although the theory for their interpretation has been better developed through the study of emission line nebulae and active galaxies. UV emission lines from LMXB have energies in the range 7 – 10 eV due to ions including He II, C IV, N IV, N V, O V, and Si IV; line equivalent widths 0.01 – 0.1 eV; and line widths $\Delta\varepsilon/\varepsilon \leq 0.01$. The properties of these lines have been discussed by Willis et al. (1980), Kallman et al. (1991), and Vrtilek, et al. (1992) for Sco X-1, by Blair et al. (1984) for Cen X-4, and by Chiapetti, et al. (1983) and Vrtilek et al. (1989) for Cyg X-2. The limit on the line width implies an emission radius of $\geq 10^{10}$ cm if the broadening is due to virial motions, or a Thomson depth ≤ 120 if the broadening is due to Compton scattering. Under the coronal assumption, the maximum available line emissivities for the UV lines are greater than that for the X-ray lines by a factor ~ 10; the emission measures required to account for the line intensities are $\sim 10^{57} - 10^{58}$ cm^{-3}.

Models for heated accretion disks by Raymond (1993) and Ko and Kallman (1993) have attempted to fit the strengths and ratios of the various strong UV lines. They showed that for plausible X-ray illuminating fluxes, $f \simeq 0.1$ near the outer edge of an LMXB disk, the relative strengths of the strong lines (C IV $\lambda1550$, N V $\lambda1240$, Lα, He II$\lambda1640$, and Si IV $\lambda1240$) can be accounted for relatively easily. However, the absolute strengths predicted by the models are smaller than those observed by a factor ≥ 50 (the uncertainty in this number is dominated by uncertainty in the interstellar reddening). One reason for this is the resonant trapping of the UV lines as discussed in section 2. Furthermore, the UV continuum is adequately fitted by models for X-ray heated disk photospheres in both its strength and shape using X-ray illuminating fluxes similar to those used in modelling the lines. This suggests either the presence of an additional emitting region or gas component to augment the disk line emission, or a mechanism to suppress the resonant trapping of the lines or augment the heating near the disk photosphere. It is interesting to note that line emission from gas outside the binary system is subject to the upper limits on the semi-forbidden line C III] $\lambda1909$, implying that the gas density is greater than $\simeq 10^{11}$ cm^{-3}, this then limits the radius to $\leq 10^{12.5}$ cm (Raymond, 1993).

8. SUMMARY

The study of emission lines and reprocessed emission is likely to receive increased attention in the future due to the launch of ASCA and future high spectral resolution X-ray spectroscopy experiments. Of particular interest is the variability behavior of the reprocessed emission, since some line emission properties have already been observed to vary in a way which is related to the X-ray spectral state and QPO properties (Vrtilek, et al., 1991). Heated accretion disk models have been shown to provide a convenient framework for discussing the observations, but the existing models are not yet capable of adequately reproducing the observed spectra. Emission associated with heating of the secondary star has not been studied in as much detail as that due to the disk, in part because of the lack of obvious orbital modulation of line emission. Limits on such emission are of interest, however, owing to suggestion that heating may play an important role in the evolution of low mass X-ray binaries.

REFERENCES

Blair, W., P., Raymond, J.C., Dupree., A.K., Wu., C.C., Holm, A.C., and Swank, J.H., 1984, *Ap. J.*, **278**, 270.
Chiapetti, L., Maraschi, L., Tanzi, E.G., and Treves, A., 1983, *Ap. J.*, **265**, 354.
Fabian, A.C., Rees, M.J., Stella, L., and White, N.E., 1989, *M.N.R.A.S.*,**238**, 739.
Field, G. B., 1965, *Ap. J.*, **142**, 531.
Hasinger G., van der Klis M., Ebisawa K., Dotani T., Mitsuda K., 1990, *A&A*, **235**, 131.
Hirano, T., Hayakawa, S., Nagase, F., Masai, K., and Mitsuda, K., 1987, *P.A.S.J.*, **39**, 619.
Kahn, S., Seward, F.D., and Chlebowski, T., 1984, *Ap. J.*, **283**, 286.
Kallman, T.R., and McCray, R. A., 1982, *Ap. J. Supp.*, **50**, 263.
Kallman, T., and White, N.E., 1989, *Astrophys. J.*,**341**,955.
Ko, Y, and Kallman, T., 1993, *Ap. J.*,,,submitted.
Krolik, J. H., McKee, C. F., and Tarter, C. B., 1981, *Ap. J.*, **249**, 422.
Liedahl, D., Kahn, S.M., Goldstein, W., and Osterheld, A., 1990, *Ap. J.*, **350**, 37.
Lightman A.P., White T.R., 1988, *ApJ*, **335**, 57.
Paerels, F., et al., 1993,,,this volume.
Pozdnyakov, L.A., Sobol, I.M., and Sunyaev, R.A., 1979, *Astron. and Astrophys.*, **75**, 214.
Raymond, J., 1993, *Ap. J.*,, in press.
Raymond J.C., and Smith, B. H., 1977, *Ap. J. Supp.*, **35**, 419.
Smale, A., et al., 1993, *Ap. J.*, **410**, 796.
Tarter, C. D., Tucker, W., and Salpeter, E. E., 1967, *Ap. J.*, **156**, 943.
Vrtilek, S.D., Kahn, S.M., Grindlay, J.E., Helfand, D.J., and Seward, F.E., 1986a, *Ap.J.*, **307**, 698.
Vrtilek, S.D., Helfand, D.J., Halpern, J.P., Kahn, S.M., and Seward, F.D., 1986b, *Ap. J.*, , .
Vrtilek, S.D., Raymond, J.C., Verbunt, F., and Penninx, W., 1989,,, .
Vrtilek S.D., Raymond J.C., Garcia M.R., Verbunt F., Hasinger G., Kürster M., 1990, *A&A*, **235**, 162.
Vrtilek S.D., Soker N., Raymond J.C., 1993, *ApJ*, **404**, 696.
White, N.E., and Holt, S.S., 1982, *Ap. J.*, **257**, 318.
White, N.E. et al., 1986, *M.N.R.A.S.*, **218**, 129.
White, N.E., Peacock, A., and Taylor, B.G. 1985, *Ap. J.*, **296**, 475.
Willis, A., et al., 1980, *Ap. J.*, **237**, 596.

THE INTERPRETATION OF SOFT X-RAY EMISSION LINES IN THE SPECTRA OF ACCRETION-POWERED SOURCES: THE EFFECTS OF THERMAL STABILITY

C.J. Hess, F. Paerels, S.M. Kahn
Department of Physics, UC Berkeley

D.A. Liedahl, R.D. Rogers
Lawrence Livermore National Laboratory

ABSTRACT

X-ray line emission in the 0.5-1.5 keV band has been detected in the spectra of a growing number of accretion-powered X-ray sources. The emission appears to be dominated by strong lines from the Fe L ions, although there are several peculiarities to the emission structure. In particular, the Fe L lines appear not to be accompanied by K-shell lines of He- and H-like oxygen, which should be several times stronger for most relevant emission mechanisms. If the lines are produced in a photoionized medium surrounding the central source, then the particular species observed may be strongly affected by the thermal stability of the irradiated gas. The coupling between the thermal balance and the ionization balance introduces a strongly non-linear relation between abundance and strength of the emission lines for the important elements. The absence of strong oxygen emission may simply be due to the fact that the H- and He-like oxygen line emission regions are located entirely in a thermally unstable phase of the medium, as would be predicted to occur for relatively small departures from standard abundances. We explore the fundamental thermal stability properties of an X-ray photoionized plasma, and their dependence on elemental abundance, and we discuss the obvious importance of these properties for any prospects of determining the chemical composition of accreting gas in accretion-powered sources from the X-ray emission spectrum.

1. INTRODUCTION

In the past 20 years, improvements in satellite-borne X-ray detectors have allowed progressively finer resolution of soft-X-ray (0.2-1.0 keV) spectra. While the identification of individual emission lines is rarely possible in archival data, much can be inferred from the analysis of coarser spectral features. Observations with the Einstein OGS and SSS and the EXOSAT TGS of X-ray binaries (XRB) and active galactic nuclei (AGN) can shed light on the thermal, ionization, and density structures of accretion flows, which might be quantifiable using higher-quality data from ASCA, AXAF and XMM. Figure 1 shows an example EXOSAT TGS spectrum of Sco X-1. Models for X-ray photoionized gas as well as for gas in collisional equilibrium predict strong emission line features due to O VIII Lyα and Fe L $3s \rightarrow 2p$ to appear simultaneously. In fact, a broad Fe L feature appears in the 700-900 eV band, while there appears to be no corresponding strong O VIII Lyα feature at 653 eV (Paerels et al. 1994). Similarly, the Einstein SSS spectrum of Markarian 509, an AGN, (Turner et al. 1991) shows a prominent Fe-L feature at about 0.8 keV but appears to lack strong O VIII Lyα. This pattern also appears in spectra of the X-ray binaries Cyg X-2 and 1820-30 and of Seyfert 2 galaxy NGC 1068 (Vrtilek et al. 1991; Marshall et al. 1993).

The simultaneous requirements of ionization and thermal equilibrium in X-ray photoionized gas may produce solutions in which some of the strongest expected line emitting ion species are located in a thermally unstable phase of the medium, so that their emission lines may be absent from the spectrum. In that case, one expects the precise location and extent of the unstable phase of the medium to be very sensitive to the parameters that control the intensity of the most important cooling lines. We have used a photoionization code to study the mechanisms which control the thermal stability properties, specifically their dependence upon elemental abundance and incident photoionizing spectrum, to explore the effects on the X-ray line emission spectrum.

2. CALCULATIONS

We used the photoionization code XSTAR (Kallman and McCray 1982) to model X-ray irradiated line-emitting gas. We chose a 10 keV bremsstrahlung incident ionizing spectrum, characteristic of LMXB continuum spectra. We calculated total radiative heating and cooling as well as the relative and absolute ion abundances, and the contribution to heating and cooling of each ion, as a function of temperature and ionization parameter $\xi = L/(nR^2)$. Assuming standard cosmic abundances, we calculated the locus of points where total radiative heating balances total cooling, shown in figure 2. To the left of the curve, cooling dominates; the opposite holds to the right. Along most of the curve, when a parcel of gas is heated (cooled) away from thermal balance, the increased cooling (heating) forces the gas back to its original state. This is, qualitatively, the condition for thermal stability. Likewise, gas is unstable where the temperature falls within the range $5.2 < \log T < 5.7$, where the curve doubles back in an "S" shape. Gas which is perturbed away from thermal balance in this range is driven away from its original state to a hotter or cooler stable temperature. The unstable zone is marked in figure 2 with dark lines. For nominal parameters (a 10 keV bremsstrahlung incident spectrum and cosmic abundances), Fe XVII, XVIII, and XIX peak in the unstable zone and therefore the $3s \rightarrow 2p$ line features emitted following recombination onto these ions cannot be produced in the source under these conditions. In addition, the O VIII abundance peaks in a stable zone and we should expect strong O VII 590 eV triplet emission following recombination onto O VIII. Neither of these two characteristics seems to agree with the observed spectra.

The onset of instability is obviously controlled by the relative abundances of the important coolants oxygen, iron, and helium (Hess etal. 1994). Since the ionization balance and the thermal balance are coupled in the case of X-ray photoionized gas, as opposed to collisionally ionized gas, changing the abundances may therefore have a strongly non-linear effect on the strength of important cooling lines from these elements. We demonstrate this sensitivity by arbitrarily changing the oxygen abundance. Figure 3 shows the "heating = cooling" curves for eight different values of oxygen abundance. Lowering the oxygen abundance raises the temperature where instability sets in, while raising the O abundance lowers that temperature. As an example of how this affects the observed emission spectrum, we show in Figure 4a the thermal balance curve, for a case where the oxygen abundance is increased by a factor of five while iron is decreased by a factor 2. The boundaries of stable and unstable temperature zones are marked by vertical lines. Plotted in figures 4b and 4c are the ion abundances of iron and oxygen as a function of $\log T$, on the same scale as figure 4a. Oxygen VIII peaks in an unstable zone, while Fe XVII and XVIII now occur in a region of stability. We would therefore expect strong recombination emission from Fe XVI,XVII (recombination onto Fe XVII and XVIII). At the same time, O VII emission should be suppressed. Likewise, O VIII Lyα is excited by recombination onto O IX, which is also partially located in an unstable phase, and could be substantially reduced as compared to the standard abundance case.

3. CONCLUSION

We have demonstrated that the abundances of oxygen and iron largely control the onset of thermal instability of X-ray photoionized gas, and that the dependence of emission line strength on abundance for these important coolants is strongly nonlinear. This suggests the possibility of explaining certain observed peculiarities in the soft X-ray line emission spectra of accretion powered sources, such as apparent low ratio's of equivalent widths in O VIII Lyα to Fe L $3s-2p$, as being due to relatively small abundance shifts. This underscores the importance of a quantitative understanding of this effect, which we intend to develop in the near future, for the measurement of abundances from the soft X-ray emission spectra.

REFERENCES

Hess, C.J., Kahn, S.M., Paerels, F.B., & Kallman, T.R., 1994, in preparation
Kallman, T.R. & McCray, R., 1982, ApJ Suppl. Series, 50, 263
Marshall, F.E. *et al.* 1993, ApJ, 405, 168
Paerels, F.B., *et al.* 1994, in preparation
Turner, T.J., *et al.* 1991, ApJ, 381, 85
Vrtilek, S.D. *et al.* 1991, ApJ Suppl. Series, 76, 1127

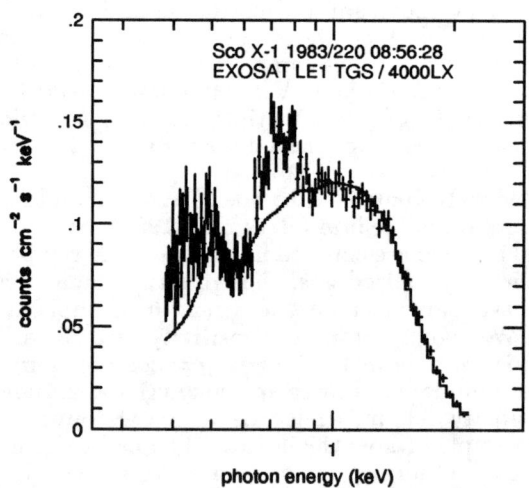

Fig. 1. An EXOSAT TGS spectrum of Scorpius X-1 shows a prominent Fe L line feature at 0.7-0.9 keV, but possibly only weak O VIII emission at 0.653 keV.

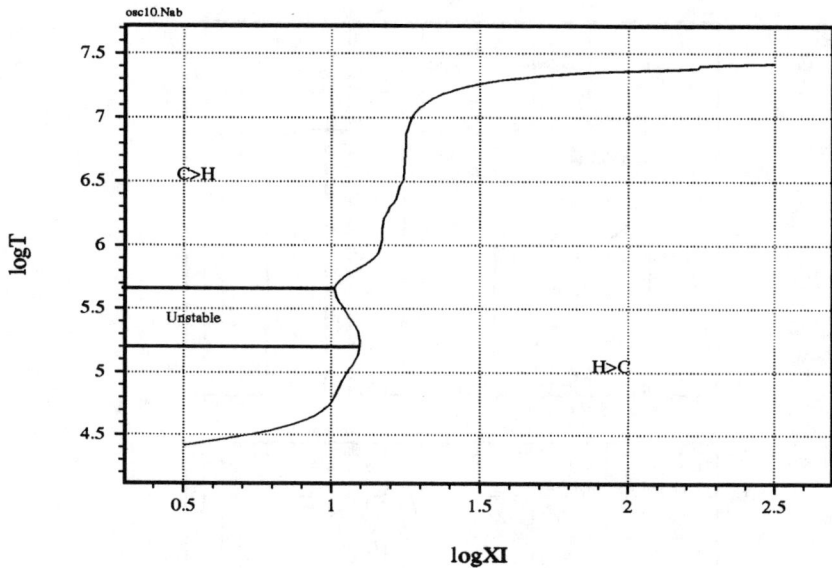

Fig. 2. The locus of points in T-Ξ space where radiative heating and cooling balance. The unstable phase is marked by heavy black lines.

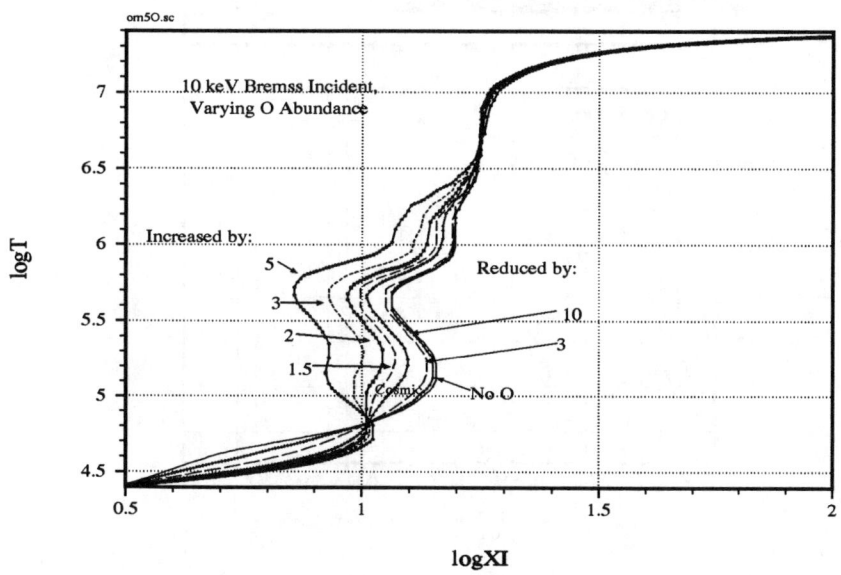

Fig. 3. The location of the thermal instability depends sensitively upon oxygen abundance. We show "*heating = cooling*" curves for 8 choices of oxygen abundance.

170   The Interpretation of Soft X-Ray Emission Lines

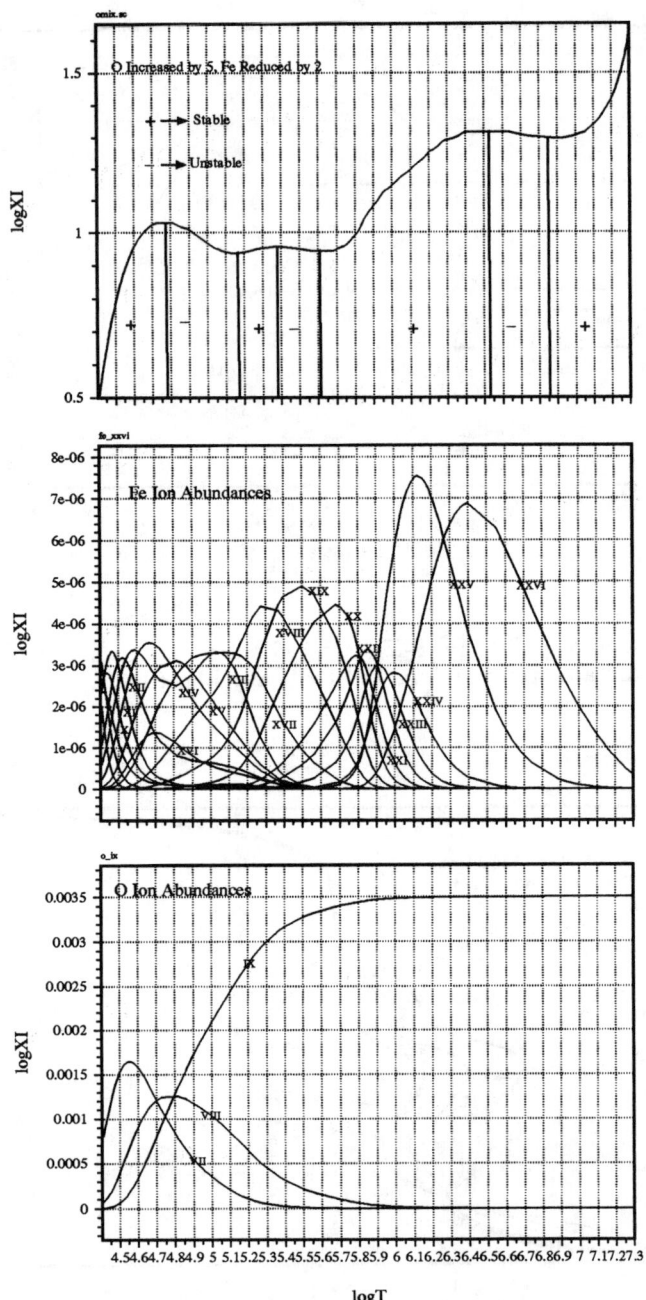

Fig. 4. (a) shows the "*heating* = *cooling*" curve where oxygen is increased by a factor of five, and iron is decreased by 2. The stable and unstable phases are marked. (b) Fe ion abundances. (c) O ion abundances.

A CLOSE-UP VIEW OF INTERACTING BINARIES

René G. M. Rutten

Netherlands Foundation for Research in Astronomy,
Dwingeloo, The Netherlands
and
Royal Greenwich Observatory, Apartado 321,
E-38780 Santa Cruz de La Palma, Canary Islands, Spain

rgmr@ing.iac.es

ABSTRACT

I review a number of ways how eclipses of accretion disks and white dwarfs in cataclysmic variable stars have been utilized to provide a detailed view of these interacting binary star systems. Furthermore, a number of recent developments are highlighted, such as the possibility to obtain spectral information across the face of the accretion disk, and a new technique to image the secondary star in the light of spectral lines is described.

1. INTRODUCTION

In the following pages I will highlight some recent developments in data analysis techniques which allow an unprecedented view of the accretion disks (with angular resolutions of ~ 0.00001 arcsec on the sky !) and secondary stars in cataclysmic variables (CVs). In particular, attention will be payed to the latest results obtained with the eclipse mapping technique for imaging accretion disks, and to the development of Roche tomography, a new tool which allows 3-D imaging of the secondary star in the light of spectral lines. These developments illustrate the steady improvement of tools we have available for the detailed study of close binary stars. Primarily used in the study of CVs, the same tools may also be applied to low-mass X-ray binaries.

The quality of our 'close-up' view of accretion disks and secondary stars begins to match the detail already offered by theoretical models. I will describe below how it is now possible to obtain spectra from different parts of the accretion disk from eclipse light curves, and how this facilitates detailed comparison with spectral model calculations. Another example concerns the secondary star for which it has become possible to (i) obtain its spectrum from eclipse light curve analysis, and (ii) construct a three-dimensional image of this Roche lobe filling star from phase-resolved spectra. These developments offer new possibilities to constrain temperature, luminosity and irradiation of the cool, mass losing star and may even help us to understand magnetic braking through the detection of magnetically active regions.

These recent developments are by no means finalized and their outcome and applicability are as yet unclear. In order to place them in an historic context I will give a short account of related earlier developments, giving credit to the pioneers on whose effort we continue to build. A number of excellent (review) articles have been published on these matters in which more detailed descriptions of the different techniques and their results are presented. In this paper I will limit myself only to basic principles, as far as they are needed to appreciate the new developments. Finally, I note that with the limited space available here I allow myself to give a biased view, and I have had to leave out a number of other interesting developments.

2. UTILIZING ECLIPSES

Eclipsing systems provide a valuable tool for the detailed study of accretion disks. In spite of many years of observational and theoretical investigations the structure of accretion disks is still not well understood. To a large extent this is caused by the lack of spatially-resolved information on disks. Since an accretion disk encompasses a large range of physical conditions, from the hot inner disk to the cool tenuous outer disk, integrated disk light often yields confusing results. Eclipsing systems, by their very nature, provide the opportunity to study accretion disks and even the compact objects in great detail. In this section I will give an overview on how eclipse information has been used to provide close-up views of CVs.

2.1 timing eclipses

When photometric observations with high time resolution of eclipsing CVs became available it was realized that some systems show complex eclipse light curves. In particular in quiescent dwarf novae the eclipses provided clear evidence for two distinct eclipse features, which were soon recognized as due to two confined light sources – the white dwarf and the disk's bright spot – which are regularly obscured by the cool secondary star (see e.g. Nather & Robinson 1974). This double eclipse thus provides evidence for the existence of structure in accretion disks such as the bright spot, where the material accreted from the secondary collides with the disk. Timing of the eclipse of the bright spot constrains its position in the orbital plane as well as its size.

Accurate timing of the eclipse of the bright spot in the eclipsing binary OY Car was recently employed by Hessman et al. (1992) to trace out the shape of the accretion disk. OY Car is a dwarf nova which exhibits super outbursts. These super outbursts last longer and are brighter than normal dwarf nova outbursts. During super outbursts so called superhumps appears in the light curve which have has a period slightly in excess of the orbital period. It was suggested by Vogt (1982) that superhumps could result from an asymmetric precessing disk.

The nature of the bright spot implies that its location coincides with that of the disk rim. By carefully timing eclipse ingress and egress events Hessman et al. (1992) was able to trace out the shape of the accretion disk during a super outburst. He found that indeed the disk was asymmetric as predicted by the superhump model.

2.2 imaging accretion disks

There is more to eclipses than just timing of conjunction events. The shape of an eclipse light curve contains information on the brightness structure across the face of the accretion disk. An early attempt to quantify this structure by modelling the light curve was published by Frank et al. (1981, see also Frank

& King 1981), whose model consisted of a symmetric disk component and a bright spot. Their model only comprised of a few parameters and, although not perfect, was able to produce a reasonable fit to the eclipse light curve. However, this approach assumed a particular brightness profile for the disk, based on the Shakura & Sunyaev (1973) steady-state disk model, whereas it would be more elegant to somehow deduce the brightness profile from the eclipse data.

In 1985 Horne devised a more flexible method to fit eclipse light curves which does not require any assumption regarding the brightness across the disk, and yields much better fits. This method, known as eclipse mapping, is based on the fact that for a given binary star geometry one can calculate which part of the disk is visible at each orbital phase, and hence the eclipse light curve can be calculated for any given disk brightness distribution. Vice versa, with the disk divided up into many elements and with each element assigned an intensity, the brightness of each element can be adjusted so that the predicted eclipse light curve fits the measured light curve. Such a set of brightness values may consist of, say, a few thousand elements, and represents an image of the accretion disk. The large number of elements, with each intensity acting as a free parameter, implies that the fitting problem is not fully constrained and hence many solutions exist which fit the light curve equally well. To make a choice a 'smoothness' parameter for the map is introduced as an additional constraint and the smoothest solution, the one which possesses least structure, is adopted as the final solution. The smoothness parameter is the so-called entropy, which is maximized to find the smoothest, maximum-entropy, solution. (The maximum-entropy method is now a well established and successful tool in solving inverse problems; see Skilling & Bryan 1984, and Skilling & Gull 1985).

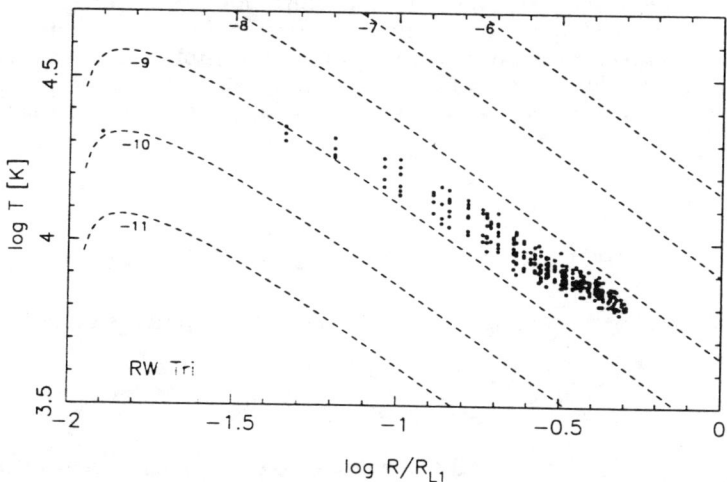

Figure 1. Effective temperature versus radial distance from the center of the disk in RW Tri. Each dot refers to an element on the disk, reconstructed with the eclipse mapping technique. The dashed lines indicate the theoretical steady-state disk $T_{\mathrm{eff}}(R)$ dependence for different values of $\dot{M}$ (numbers indicate $\log \dot{M}$ in $M_\odot/yr$).

Broad band light curves of several eclipsing binaries have been analyzed with the eclipse mapping technique: nova-like systems (Horne & Stiening 1985, Rutten et al. 1992a, Baptista & Steiner 1991), as well as dwarf novae in different stages of their outburst cycle (Horne & Cook 1985, Wood et al. 1986, Warner & O'Donoghue 1988, Wood et al. 1989, Rutten et al. 1992b). The main new result from these investigations has been the determination of the disk's radial dependence of the effective temperature, $T_{\text{eff}}(R)$, which offers a critical test for the theory of steady state accretion disks. Moreover, the theoretical dependence of $T_{\text{eff}}(R)$ on the mass accretion rate, $\dot{M}$, allows a reliable estimate of $\dot{M}$. An example is shown in figure 1 for the disk in the nova-like variable RW Tri; for a review of the results see Horne (1993).

Until recently eclipse mapping had only been employed to analyze broadband or white-light photometric data since such data is relatively easy to obtain with the required high time resolution and high signal-to-noise. However, there is no reason why the same technique should not be used on data of much higher wavelength resolution, as I will show in section 3.1 .

2.3 the compact object

If the orbital inclination is high the white dwarf in a CV will be eclipsed by the secondary star. Particularly in quiescent systems, where the accretion disk is not very luminous, the light curve clearly shows the eclipse of the white dwarf. In much the same manner as the accretion disk is analysed, the shape of the eclipse of the central object may be utilized to learn about the white dwarf's surface and its immediate surroundings. Since the central, mass accreting object is relatively small the eclipse occurs rapidly and hence very high time resolution is required to allow such analysis. Wood (1987) and Wood et al. (1989), for example, utilized the shape of the eclipse to constrain the size and temperature of the central object, consisting of a white dwarf and possibly a luminous boundary layer. From this analysis they could even distinguish between the presence and absence of an equatorial belt as the boundary layer between the disk and the white dwarf.

3. FURTHER USE OF ECLIPSE INFORMATION

In this section I will highlight some new developments in data analysis techniques.

3.1 eclipse mapping moving on

As described in Section 2.2 the eclipse mapping technique, developed by Horne (1985), has been used to reconstruct broad-band images of accretion disks. With the current generation of efficient spectrographs and CCD detectors it has become feasible to obtain spectroscopic data of photometric quality and with sufficient time resolution to resolve the eclipse. From such data accurate light curves in narrow wavelength bands, even for individual spectral lines, can be obtained. From these narrow-band light curves the brightness distribution across the face of the disk can be reconstructed in far better spectral resolution than before.

Such spectrophotometric data have been obtained on the nova-like cataclysmic variable UX UMa. By analysing the shape of the eclipse light curves in many wavelength intervals Rutten et al. (1993a, b) have obtained maps of the disk's spatial brightness distribution as a function of wavelength, from 3600 Å to 1 μm. With these disk images at many wavelengths it is possible to reconstruct the spectrum for any part of the disk ! Rutten et al. reconstructed the spectra

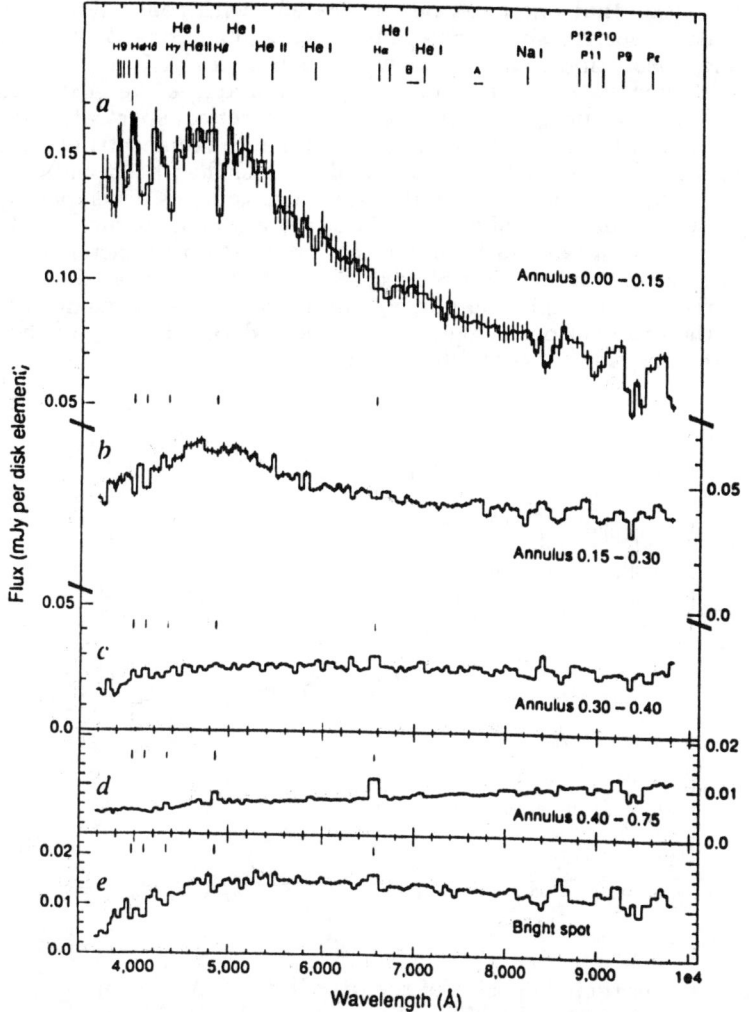

Figure 2. Reconstructed spectra for selected parts of the disk in UX UMa. From top to bottom spectra are plotted of annuli going out to progressively larger radii. The bottom spectrum is that of the bright spot region. Note that the lower two spectra are plotted on an expanded scale. The main spectral features are marked.

at different radii in the disk. Their reconstruction, shown in Figure 2, reveal a blue spectrum in the inner disk with Balmer absorption lines (with the exception of Hα), and a substantially redder outer disk spectrum with the Balmer lines in emission. These findings are in qualitative agreement with basic models for accretion disks which predict an optically thick inner disk and optically thin outer regions (Williams 1980, Tylenda 1981). Also, the bright spot exhibits an optically thick spectrum which appears hotter than other parts of the disk at the same radial distance; hence, the bright spot is indeed a hot spot. These spectra emphasize that one must be very careful with conclusions drawn from disk integrated spectra, since some line profiles may consist of a mixture of emission and absorption components as is the case in UX UMa.

The detailed reconstruction of the spectrum across the face of the accretion disk will help to bridge the gap between theoretical spectral models of the accretion disk and observations. Figure 3 compares the spectrum from the inner disk of UX UMa with a spectral model which has been computed using the radiative transfer code for accretion disk atmospheres developed by Kříž & Hubeny (1986; see also Hubeny 1990). The grey atmosphere model calculations shown here have been tailored to match the binary star parameters of UX UMa. The resulting spectrum has been binned to match the observed data in order to ease comparison. Although preliminary, the reasonable agreement between the model and the observations suggest that a more detailed study of the physical conditions in accretion disks will be possible.

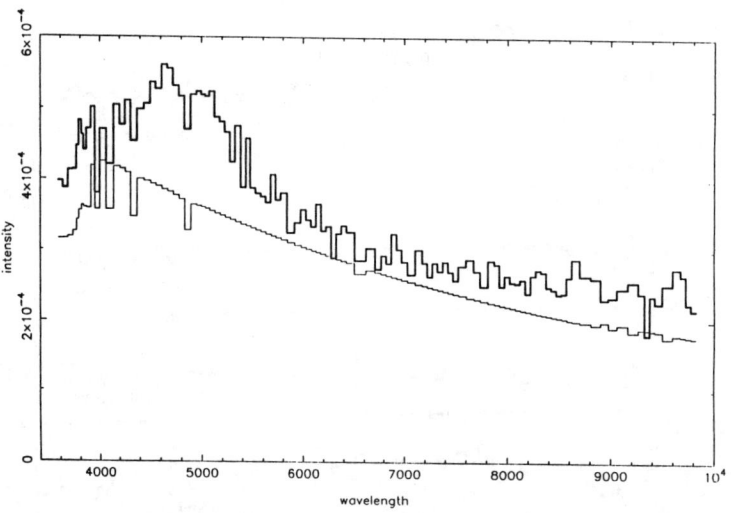

Figure 3. Reconstructed spectra of the disk in UX UMa covering the annulus in the inner disk from 0.1 to 0.2 R_{L1} (bold line), together with the theoretical grey model spectrum (thin line).

Whereas spectral eclipse mapping can now be used to map the line brightness distribution across the disk, another technique, Doppler tomography (Marsh

& Horne 1988) makes a map of the line in velocity coordinates in the corotating orbital plane. Doppler tomography interprets the observed variations of the line shape through the orbit as being due to Doppler shifts. The spatial emission-line map obtained from eclipse mapping can be compared with the corresponding Doppler maps in velocity space. In this way it will be possible to constrain the velocity field of the material in the disk for suitable emission line objects. Clearly, this will open up new ways to obtain information about the physics of accretion disks through fundamental tests which may show, for instance, whether or not the material in the disk follows a Keplerian velocity field.

3.2 the secondary star's spectrum

The original eclipse mapping method assumes that the secondary star is dark. If, however, the contribution from the secondary star is appreciable, then spurious structure will appear on the reconstructed disk image. Turning the argument around, Rutten et al. (1992b, 1993a) showed that the contribution of the secondary star (or of any other uneclipsed light source) can be deduced from the eclipse light curve. In this way the spectrum of the secondary star was deduced from their spectrally-resolved eclipse light curves of UX UMa. This red spectrum, plotted in Figure 4, is reminiscent of that of a late K- or early M-type star. The Balmer lines and the Balmer continuum jump appear in emission and originate most likely in an extended uneclipsed region above the disk. The spectral type of the secondary star estimated from this spectrum agrees remarkably well with what is found from other independent estimates, suggesting that eclipse mapping also offers a new tool for studying the spectrum of the secondary star.

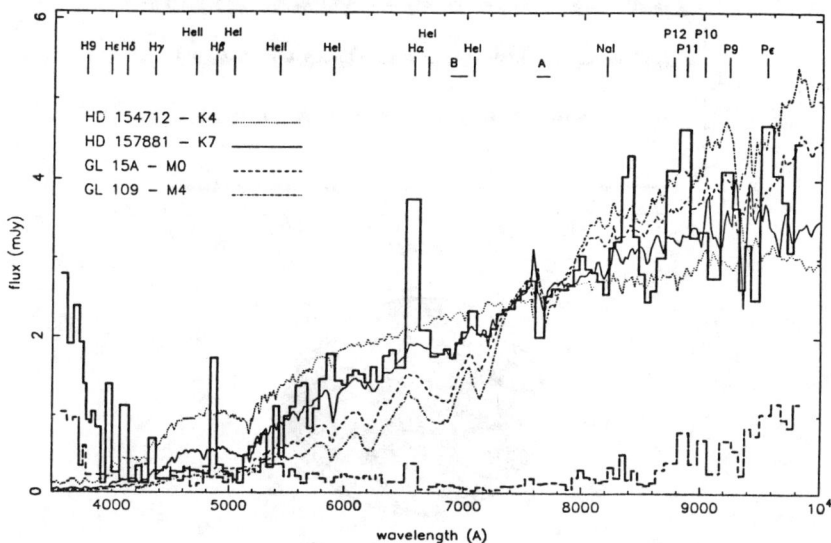

Figure 4. Spectrum of the uneclipsed light (bold line), and its uncertainty (bold dashed line). Spectra of four cool dwarf stars are also shown, after being scaled to give the same flux at 7400 Å.

4 MAPPING THE SECONDARY STAR

This section describes a new technique, Roche tomography, which allows mapping of the surface of the secondary star from phase-resolved line spectra (see Rutten & Dhillon 1994). The secondary stars in CVs reveal themselves through narrow emission lines or absorption lines. The observed line profile from the secondary star is the sum of line radiation emerging from different parts of its visible hemisphere. Hence, the observed line shape results from the combination of the intrinsic line profile from all visible parts of the stellar surface, Doppler shifted by different amounts as the star is revolving around the center of mass, and attenuated by different amounts due to projection effects and limb darkening.

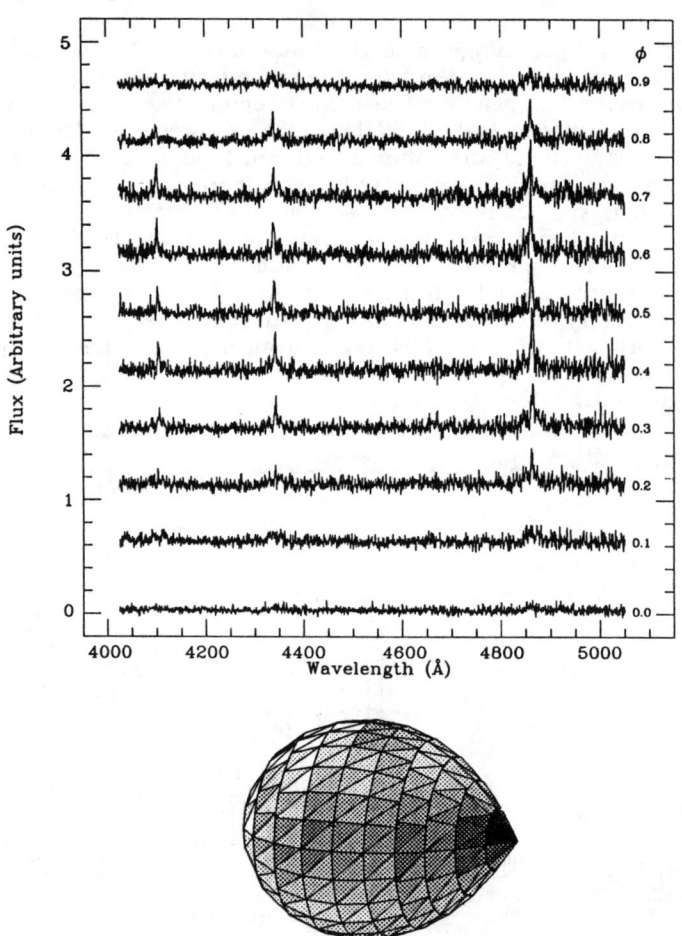

Figure 5a - top. Trailed spectrum of DW UMa in a low state. 5b - bottom. Reconstruction of the Roche surface of the secondary star in DW UMa resulting from the trailed spectrum in Fig. 5a.

In principle, the integrated line strength from different parts of the stellar surface can be recovered from phase-resolved spectra. To reconstruct the line brightness distribution over the surface of the star, the Roche surface is first approximated by a large number of 'tiles' of approximately equal size, acting as surface elements. Each element is assigned a specific integrated line intensity perpendicular to its surface. Then, for each surface element, the radial velocity and visibility throughout the orbital cycle are calculated from the binary star parameters. Fitting the observed phase-resolved spectra involves adjusting the line intensities of all surface elements. The maximum-entropy criterion is used to optimize the fit to the data, yielding the smoothest possible intensity distribution on the star.

Roche tomography is similar to techniques which image star spots by interpreting time dependent features in the line profile as resulting from surface inhomogeneities which are Doppler shifted by stellar rotation (Vogt & Penrod, 1983). The technique described here not only uses Doppler shifts, but also includes line intensity variations through the binary orbital cycle, and hence exploits all the available time-dependent spectral information. Both emission and absorption lines can be imaged in this way.

As an example, Figure 5a shows a trailed spectrum of DW UMa in a low state (taken from Dhillon et al. 1993) with a prominent narrow emission line from the secondary star. Note that at orbital phase zero, when the white dwarf is at the far side of the secondary star, the narrow emission line component has virtually disappeared. This data has been used to reconstruct the emission line intensity distribution on the secondary star. The result is shown in Figure 5b, where it is seen that the section of the star close to the inner Lagrange point is much brighter than other parts of the star, probably due to irradiation from a bright spot or the inner disk.

Although the example shown above exploits the fact that DW UMa was in somewhat unusual low state which revealed its secondary star, such a low state is by no means required to be able to see the secondary star. The cool star may be seen even in systems with a relatively bright disk provided suitable spectral features are chosen. Spectroscopy at (near) infra red wavelengths, where the cool star dominates, will prove to be very useful in this respect.

A detailed mapping of the surface of secondary stars in CVs is of interest for a number of reasons. Irradiation of the stellar surface by the (inner) disk or bright spot may be measured in this way, and provides information on the radiation field from the irradiating source. Furthermore, when irradiation is not accounted for the radial velocities of lines from the secondary star do not reflect the velocity of the center of mass of the star. Consequently, binary star parameters deduced from it will be affected. Imaging the intensity distribution over the surface of the star allows correction for these systematic errors, thus yielding a more reliable determination of the binary star parameters.

Another aspect which makes Roche tomography potentially interesting is that surface inhomogeneities may reveal magnetic activity. Magnetic fields on the secondary stars are thought to play an important role in the evolution of CVs as angular momentum is drained from the system through magnetic braking, which results from an ionized stellar wind streaming out along the magnetic field lines. This mechanism operates on the Sun, and the correlation between stellar rotation and age in single solar-type stars provides strong observational evidence that magnetic braking also operates on other cool stars (see Skumanich 1972). The same mechanism will drive CVs to ever shorter periods and is thought to be the main factor responsible for sustaining mass transfer between the stellar components (see Verbunt 1984). This theoretical conjecture, although very appealing, has actually never been corroborated by observational evidence, such as the presence of a stellar wind or magnetic activity on the star. A close-up view of the secondary stars in CVs through Roche tomography may provide crucial observational evidence for the magnetic braking model and thus for the evolution of close binary systems.

ACKNOWLEDGEMENTS

I thank Vik Dhillon for his comments on the manuscript.

REFERENCES

Baptista R., & Steiner J. E. 1991, A&A 249, 284
Dhillon V. S., Jones D. H. P., & Marsh T. R. 1993, MNRAS, in press
Frank J., & King A. R. 1981, MNRAS 195, 227
Frank J. et al. 1981, MNRAS 195, 505
Hessman F. V., Mantel K.-H., Barwig H., & Schoembs R. 1992, A&A 263, 147
Horne K. 1985, MNRAS 213, 129
Horne K. 1993, in Accretion Disks in Compact Stellar Systems, ed. J. C. Wheeler, World Scientific Publishing Company
Horne K., & Cook M. C. 1985, MNRAS 214, 307
Horne K., & Stiening R. F. 1985, MNRAS 216, 933
Hubeny I. 1990, Ap J. 351, 632
Kříž S., & Hubeny I., 1986, Bull. Astron. Inst. Czechosl. 37, 129
Marsh T. R., & Horne K. 1988, MNRAS 235, 269
Nather R. E., & Robinson E. L., 1974, ApJ 190, 637
Rutten R. G. M., & Dhillon V. S., 1994, in preparation
Rutten R. G. M., Dhillon V. S., Horne K., & Kuulkers E., 1993a, A&A, in press
Rutten R. G. M., Dhillon V. S., Horne K., Kuulkers E., & van Paradijs J. 1993b, Nature 362, 518
Rutten R. G. M., Kuulkers E., Vogt N., & van Paradijs J. 1992b, A&A 265, 159
Rutten R. G. M., van Paradijs J. & Tinbergen J. 1992a, A&A 260, 213
Shakura N. I., & Sunyaev R. A. 1973, A&A 24, 337
Skilling J., & Bryan R. K. 1984, MNRAS 211, 111
Skilling J., & Gull S. F. 1985, in Maximum-Entropy and Bayesian Methods in Inverse Problems, eds. C. R. Smith, W. T. Grandy, Reidel, p. 83
Skumanish A. 1972, ApJ 171, 565
Tylenda R. 1981, Acta Astron. 31, 127
Verbunt F. 1984, MNRAS 209, 227
Vogt N. 1982, ApJ 252, 653
Vogt S. S. & Penrod G. D. 1983, PASP 95, 565
Warner B., & O'Donoghue D. 1988, MNRAS 233, 705
Wood J. H. 1987, MNRAS 228, 797
Wood J. H., Horne K., Berriman G., & Wade R. A. 1989, Ap J. 341, 974
Wood J. H., Horne K., Berriman G., Wade R. A., O'Donoghue D., & Warner B. 1986, MNRAS 219, 629
Williams R. E. 1980, ApJ 235, 939

SPECTRAL STUDY OF GX339–4 IN THE LOW INTENSITY STATE OBSERVED WITH GINGA

Yoshihiro Ueda[1], Ken Ebisawa [1,2] and Chris Done[3]
1) Institute of Space and Astronautical Science, 3-1-1, Yoshinodai, Sagamihara, Kanagawa, 229 Japan
2) code 668, Laboratory for High Energy Astrophysics, NASA/Goddard Space Flight Center, Greenbelt, MD 20771, U.S.A.
3) Univ. of Leicester, University Road Leicester, LE1 7RH, U.K.

ABSTRACT

Energy spectra of the black hole candidate GX339–4 in the low intensity state were observed on four occasions through 1989 to 1991 with the Large Area Counters onboard the GINGA satellite. The spectra showed significant deviations from a power-law, with an iron K_α emission line at $\sim$ 6.4 keV and broad iron K-edge structure above $\sim$ 7 keV. The energy spectra above $\sim$ 4 keV were successfully explained with a reflection model, in which a part of the incident X-rays with a power-law spectrum is Compton reflected by optically thick matter. The line equivalent width with respect to the reflection component decreases as the source flux increases, consistent with an increase in the ionization state of the material so that resonant absorption followed by Auger ionization depletes the line.

1. INTRODUCTION

The Large Area Counters onboard the GINGA satellite observed GX339–4 on five occasions. In early September 1988, the intensity was $\sim$ 800 mCrab, which is the highest ever observed. On the other four occasions, GX339–4 was in the low (or off) state with the intensity from $\sim$ 3 mCrab to $\sim$ 160 mCrab. In this paper, we briefly report the energy spectra of GX339–4 in these four low state observations. Comprehensive results and discussion are presented in a separate paper (Ueda, Ebisawa and Done 1993, PASJ accepted).

2. OBSERVATIONS AND RESULTS

Log of the observations is shown in table 1. In order to study energy spectra, we used only the data taken with the 48 energy channel mode (MPC1 or MPC2 mode) in the energy range 1.2 – 37 keV. Six energy spectra were created corresponding to the six datasets in table 1, subtracting the background.

Considering the low signal-to-noise ratio and possible uncertainty in the background subtraction at high energies, data above 28.7 keV (22.1 keV for the August 1989 data) were excluded from the analysis. Firstly, a simple absorbed power-law, with the photon index and the column density of neutral matter being free parameters, was fit to the data. Besides the large deviation above $\sim$ 5 keV, there is a significant excess below $\sim$ 3 keV in almost all the data, together with derived N_H values being systematically lower than the interstellar column toward GX339-4 ($\log N_H = 21.82$; Vrtilek et al. 1991). This suggests the existence of a soft excess component, which may be identical to that observed by Ilovaisky et al. (1986). In order to minimize the influence of the soft-excess component, we truncate the data below ~ 4 keV and concentrate on the study of the energy spectra above this energy.

Table1: Observations of GX339-4 in the low state

No.	Start	End	Exposure (sec)	Intensity (mCrab)
1	1989/08/28 11:33	1989/08/28 16:33	6400	3
2	1989/09/28 15:23	1989/09/28 20:24	3008	18
3	1989/09/29 19:23	1989/09/30 23:20	10880	18
4	1990/04/07 16:48	1990/04/07 16:55	384	100
5	1991/09/11 03:35	1991/09/11 05:26	1952	160
6	1991/09/12 00:54	1991/09/12 01:11	384	160

Table2. Results of Spectral Fitting with the Reflection Model

No.	N_{power}†	Γ	f‡	ξ	E_{line}(keV)	χ^2/d.o.f	d.o.f.
1	1.70 ± 0.29	1.79 ± 0.12	$0.7^{+0.7}_{-0.5}$	0§	6.4§	1.86	24
2	7.41 ± 0.05	1.69 ± 0.04	$0.30^{+0.16}_{-0.14}$	< 537	6.69 ± 0.33	1.34	28
3	$8.54^{+0.38}_{-0.36}$	1.67 ± 0.03	0.22 ± 0.09	< 485	6.48 ± 0.28	1.37	28
4	$73.3^{+4.8}_{-4.6}$	1.91 ± 0.04	$0.74^{+0.22}_{-0.19}$	370^{+320}_{-180}	6.43 ± 0.40	0.85	27
5	$89.4^{+4.3}_{-4.5}$	1.78 ± 0.03	0.58 ± 0.11	340^{+250}_{-150}	$6.32^{+0.42}_{-0.47}$	1.00	27
6	88.6 ± 5.1	1.77 ± 0.03	0.41 ± 0.13	400^{+560}_{-230}	6.24 ± 0.53	0.82	27

All the errors are single-parameter 90 % confidence limits.
Broad intrinsic line width is assumed (1 $\sigma = 0.5$ keV).
†: Normalization of the direct component in the unit of 10^{-2} photons s^{-1} cm^{-2} keV^{-1}.
‡: Relative normalization of the reflection component to the direct component.
§: Fixed.

We thus restricted energy range of the spectral fitting to 4.1 – 28.7 keV (4.1 – 22.1 keV for the 1989 August data). In this energy range, spectral fitting was made again with a power-law function with the column density of the neutral matter fixed to the interstellar value. Figure 1a shows the results of fitting this model to one of the datasets (Sep. 1991; No. 5 in table 1, $\Gamma = 1.64$, reduced χ^2 = 9.5 for d.o.f. = 31). Clearly, the residuals indicate an emission line at ~ 6.4 keV, an absorption edge above ~ 7 keV, both seemingly associated with those of iron, and an excess at high energies > 20 keV. Similar spectral features are observed in all the other spectra, although they are not obvious in 1989 August data because of insufficient statistics.

We found the observed spectra can be fitted with the disk reflection model by Done et al. (1992) including emission lines. Free parameters of this reflection model are the relative normalization of the reflection component to the direct one f (where $f = 1.5$ is appropriate for a face on, solar abundance, flat disk),

Table 3: Iron Line Equivalent Widths

Data No.	EW to the Total (eV)	EW to the Reflection Component (eV)
2	82 ± 35	2600 ± 1100
3	77 ± 29	3400 ± 1300
4	83^{+49}_{-46}	580^{+340}_{-330}
5	60 ± 35	520 ± 300
6	58^{+50}_{-37}	630^{+540}_{-400}

Broad intrinsic line width is assumed ($1\sigma = 0.5$ keV).

the ionization parameter ξ, and temperature T of the material. The abundance is fixed to the cosmic value and the source is assumed to radiate isotropically. In all the fits, the temperature was fixed at $T = 10^5$ K.

We tried narrow ($1\sigma = 0.05$ keV) and broad ($1\sigma = 0.5$ keV) intrinsic line widths, but both yielded similar χ^2 values, so the line width is not constrained from the current data being limited by the counter energy resolution. Results of the fit with the reflection model and the broad lines are shown in table 2, and one of the spectra is shown in figure 1b. Equivalent widths of the iron emission lines to the total and to the reflection component are shown in table 3.

4. DISCUSSION

The line equivalent width with respect to the reflected component does decrease with increasing source intensity (table 3), suggesting that there are significant changes in the ionization structure of the disk. By contrast, the line equivalent width with respect to the total continuum (direct component + reflection component) shows no significant variability.

The mean line equivalent width relative to the reflection component for datasets 2–3 is 2.9 ± 1.2 keV, consistent with either neutral material at small inclinations or moderate ionization over the whole range of allowed inclination (Matt et al. 1992; Matt, Fabian and Ross 1993). However, the mean line equivalent width with respect to the reflected component for datasets 4–6 is significantly lower at 560 ± 350 eV. This can be explained if there is a change in ionization state of the disk, such that iron is predominantly below Fe XVIII in datasets 2–3 and above in 4–6. Then, in the higher ionization state where the iron L shell is not fully occupied, resonant absorption of the K_α line can lead to Auger ionization rather than re-emission of the K_α photon, strongly suppressing the line (Ross and Fabian 1993).

From the changing line equivalent width with respect to the reflected continuum, there is evidence that the predominant ionization state of iron is below Fe XVIII for a source intensity of 18 mCrab and below, and above FeXVIII for 100 mCrab and above. This is supported by the observed ionization parameter in the brightest spectra (4–6), $\xi \sim 370$, corresponding to FeXX. The fainter spectra (1–3) have intensity about a factor of ten or less, so the photo-ionizing flux and hence ξ decreases by the same factor, predicting $\xi \sim 37$, consistent with the observed values. Thus the changing equivalent width of the iron line is

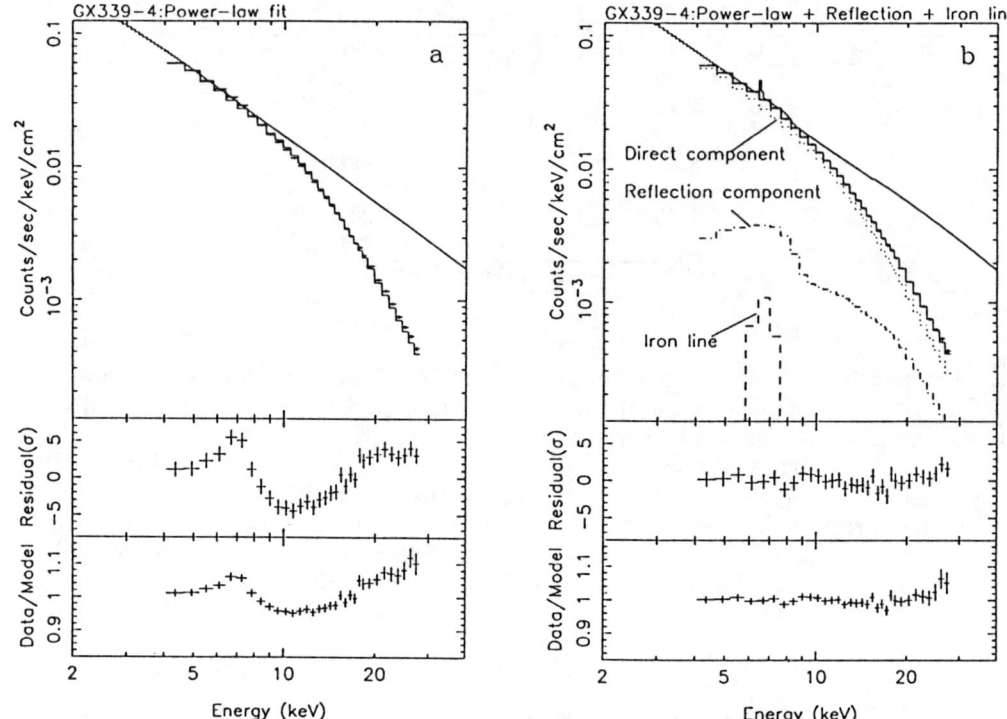

Figure 1: (a) Energy spectrum of GX339-4 in Sep. 1991 (data No. 5 in table 1) fitted with a power-law function in 4.1 – 28.7 keV. (b) Result of the model fitting for the same spectrum with a power-law function, a reflection component, and a narrow iron emission line.

consistent with the predictions of photo-ionization of the material by the central source, which is also found in the low state of X1608-522 (Yoshida et al. 1993).

ACKNOWLEDGMENTS

The authors thank Profs. F. Makino and Makishima, and Drs. K. Yoshida and F. Haardt for helpful discussions, and acknowledge Drs. I. Asaoka, T. Dotani and S. Takano for variable comments on the LAC collimator response calibrations. Y. U. and C.D. acknowledge financial support for the Japan Society of Promoting Science, and the Science and Engineering Research Council, respectively.

REFERENCES

Done, C., et al. 1992, ApJ, 395, 275
Ilovaisky, S. A., et al. A&A 1986, 164, 67
Matt, G., Fabian, A. C. and Ross, R. 1993, MNRAS, in press
Matt, G., Perola, G. C., Piro, L. and Stella, L. 1992, A&A, 257, 63
Ross, R. R. and Fabian, A. C. 1993 MNRAS, 261, 74
Vrtilek, S. D., et al. 1991, ApJ, 76, 1127
Yoshida, K. et al. 1993, PASJ, 45, 605

The Dipping X-Ray Binary X1916-05 Observed with Ginga

Kenji Yoshida
Faculty of Engineering, Kanagawa University,
3-27-1 Rokkakubashi, Kanagawa-ku, Yokohama 221, Japan
yoshida@astro.isas.ac.jp

ABSTRACT

The dipping low-mass X-ray binary X1916-05 was observed with Ginga on four occasions from 1987 to 1990. Although the averaged X-ray dip period was 50.00 min for 3 years, it was found that this period varied sinusoidally with a time scale of around 5 d. In addition to its period, the dip duration and depth also varied on the same time scale. Energy spectra in the X-ray dips consist of a direct component from the X-ray source and an absorbed component from intervening matter. From the spectral and timing analysis, it is suggested that the direct component in X-ray dips originates from the partial covering of the X-ray emission region by the absorbing matter. Given that the dips arise from clouds in Keplerian motion at the disk's outer edge partially eclipsing the X-ray source, the finite size of the X-ray emission region is derived to be $(1.2\pm0.9)\times10^8$ cm from the transition time between the quiescent state and X-ray dips.

1. INTRODUCTION

The X-ray source X1916-05 is known to be an X-ray burst source and an X-ray dipping source. The X-ray light curves of X1916-05 have periodic absorption dips (Walter et al. 1982; White and Swank 1982; Smale et al. 1989). The period ranges from 49.7 - 50.1 min. On the other hand, the optical counterpart of X1916-05 has been identified with a V=21st magnitude blue object and it shows an optical modulation with a period of 50.459 min (Grindlay et al. 1988).

X-ray dips are a powerful probe for investigating various physical phenomena. In the data, we searched for the beat period between the X-ray dip period and the optical period. Such a beat period would indicate another periodic motion in addition to the binary motion. Furthermore, if the velocity of the absorbing matter is known, the time scales of the variability in the dips could for the first time provide the size of the central X-ray source. Thus, X1916-05 is a good target to study the physical structure around a neutron star.

2. OBSERVATIONS AND RESULTS

X1916-05 was observed with the Large Area proportional Counter (LAC; Turner et al. 1989) onboard Ginga in May 1987, Sep. 1987, Sep. 1988, and Sep. 1990. We calculated the period of the dip by the folding method in the energy range of 1.2-9.2 keV for Sep. '87, Sep. '88. and Sep. '90 observations, separately. The reduced chi-square values have maxima at the period of 50.02 ± 0.05, 49.98 ± 0.07 and 50.00 ± 0.02 min, respectively. The time gaps between the three observations are so large that it is difficult to combine the three data sets coherently. The optical period of 50.459 min simultaneously observed in 1990 observation (Grindlay 1991) differs significantly from the period of 50.00 min.

In Fig. 1-(a) the phase of the dip in the case of a 50.00 min period is plotted as a function of time for the Sep. '90 observations and changes slightly at the sinusoidal period of 6.5±1.1 d. Dividing the data into three groups, (A) through (C), the periods of each group are 49.93 ± 0.07, 50.08 ± 0.06 and 49.90 ± 0.08 min with the folding method, respectively. Changes of the obtained periods are consistent with the sinusoidal modulation. To determine the time variability of the dip structure, we divided the Sep. '90 data into eleven sets in order of observation time and folded each data set with the period of 50.00 min. The durations and depths of the dips obtained from the folded light curves are plotted in Fig. 1-(b) as a function of time, where the dip depth is the minimum flux normalized by the quiescent flux level, and seem to have a periodic modulation with a period of about 5 d. These results are all consistent with the suggestion that the apparent properties of the X-ray dips vary with a time-scale of about 5 d. The time variation of ∼ 5 d is close to the beat period between the X-ray dip period (50.00 min) and the optical period (50.459 min).

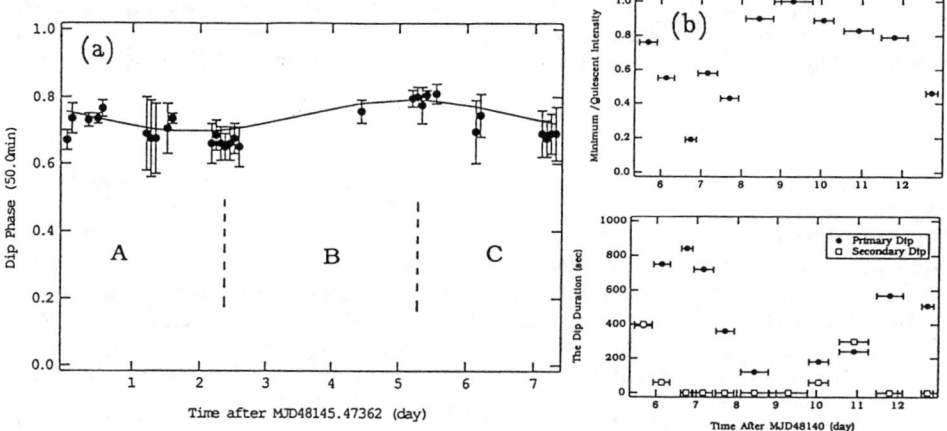

Fig. 1. (a) The time variation for the phase of the dip with a 50.00 min period. (b) The time variation of the duration and depth of the dips.

The observed dip spectrum can be represented by the sum of a direct and an absorbed component; $Af_0(E) + Bf_0(E)\exp(-\sigma N_H)$, where $f_0(E)$ is a normalized direct-spectrum from the X-ray source, A and B are normalization factors, σ is a photoelectric absorption cross-section and N_H is the hydrogen column density. To study dip spectra further in detail, we constructed intensity-selected spectra during the transition from the quiescent state to X-ray dips and dips to quiescence. Next we performed the one-dimensional simulations of X-ray dips and compared it with the intensity-selected spectra. The absorber gradually obscures the X-ray emitter from the sight (see Fig. 2-(a)). This process includes scattering of X-rays out of the line of sight. Fig. 2-(b) shows the fitting parameters of the simulated spectra with the best-fitting parameters of the observed spectra. The intensity-selected spectra by the simulation can quantitatively reproduce the fitting parameters of the observed spectra. The ratio of the X-ray emission region size (l) to the gradient size of the absorber column density (a) is derived to be 2±1 from the simulations.

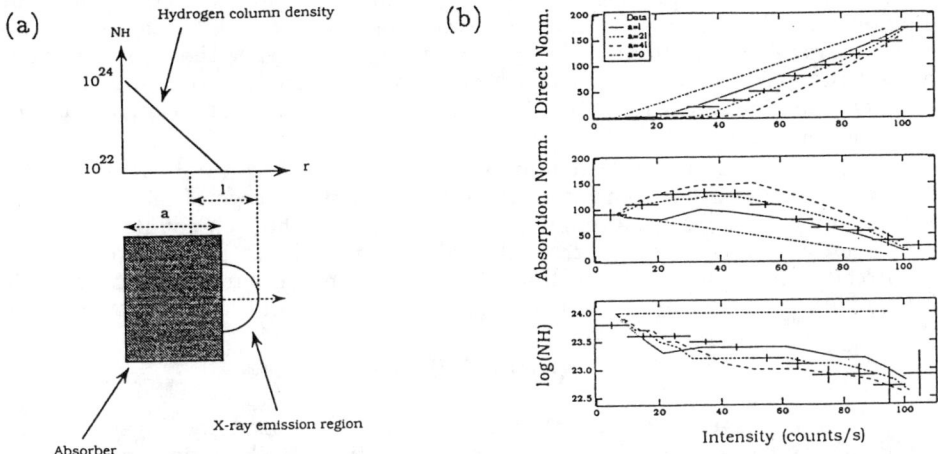

Fig. 2. (a) A schematic view of a partial covering model by homogeneous matter. (b) The best-fit parameters of the partial covering model to the intensity-selected spectra during the transition between the quiescent state and X-ray dips (crosses). The best-fit parameters of the simulated model spectra with the intensities are also shown for comparison.

There are two possible ways of producing a dip spectrum with a direct and an absorbed component. One is partial covering of the X-ray emission region. The other is a spectral mixing effect due to rapid changes in the column density. The latter results if the timescale for variability of the absorbing material in the line of sight is shorter than the integration time.

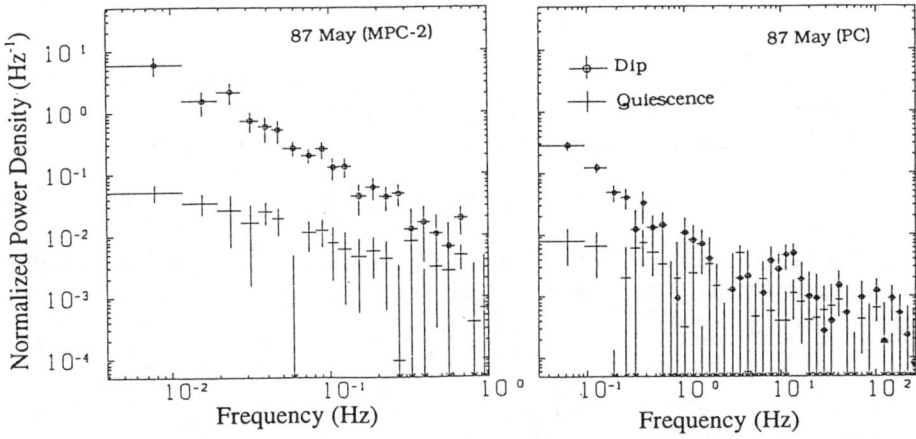

Fig. 3. Normalized dip and quiescent power-density spectra.

For the study of short-term variations in the X-ray dips, we derived the average power spectrum densities for X-ray dips from the May '87 observations (Fig. 3). The quiescent and dip power spectra were normalized by the square of the average intensities subtracted for background and then they were compared to one another. The quiescent power spectrum is well-fitted with a power-law function. On the other hand the dip power spectrum is well-fitted with a sum of two power-law functions: one power-law function with the same index as that for the quiescent spectra, which describes quiescent variation and the other with the index of 1.6, which describes dip variation. Assuming that the power-law function of $P(f) \propto f^{-1.6}$ continues to infinite frequency above 1.0 Hz, the r.m.s. amplitude of 1.6 ± 0.6 counts s^{-1} (1.2–5.7 keV) is obtained due to dip variation. On the other hand, for example in the flux range of 50–60 counts/s (1.2–37 keV), the direct component should be larger than 15 counts/s (1.2–5.7 keV) from fitting of the intensity-selected spectra. Hence it is suggested that the direct component in X-ray dips originates from the partial covering of the X-ray emission region by the absorbing matter, not by the rapid time variation.

4. DISCUSSION

Since short term variations below about 1 s caused by X-ray dips are negligible, we can obtain the typical transition time of the quiescent states to the dips, by investigating transition times with an integration time of 0.5 s. We examined the transition time of the quiescent states to the dips, using the Sep. '87 observations. The transition time is defined as the fall time from 90% to 10% or the rise time from 10% to 90% of the average quiescent flux. The distribution of the transition time has a most probable value of 2.0 ± 0.4 s and a mean value of 3.6 ± 2.5 s.

In the case of partial covering by homogeneous matter, the size of the X-ray emission region is derived. The averaged transition-time (t_{trans}) of 3.6 ± 2.5 s is represented in the following formula: $v t_{trans} = l + a$, where v is a mean velocity of clouds. Assuming the mean velocity of clouds is 1×10^8 cm s^{-1}, the size of the X-ray emission region is semi-empirically evaluated as

$$l = (1.2 \pm 0.9) \times 10^8 \text{ cm}, \tag{1}$$

using the ratio (2 ± 1) of the X-ray emission region size (l) to the gradient size of the absorber column density (a) from the simulation. This result suggests that the optically thin hot plasma near the neutron star extends out to $10^7 - 10^8$ cm.

ACKNOWLEDGEMENTS

The author is grateful to Drs. G. Zylstra and T. Yaqoob for carefully reading the manuscript and useful comments.

REFERENCES

Grindlay, J. E., et al. 1988, ApJ (Letters), 334, L25
Grindlay, J. E. 1991, private communication.
Turner, M.J.L., et al. 1989, PASJ, 41, 345
Smale, A.P., et al. 1989, PASJ, 41, 607
Walter, F. M., et al. 1982, ApJ (Letters) 253, L67
White, N. E., & Swank, J. H. 1982, ApJ (Letters) 253, L61

ROSAT OBSERVATIONS OF CATACLYSMIC VARIABLES

André van Teeseling and Frank Verbunt
Astronomical Institute, Utrecht University
P.O. Box 80 000, 3508 TA Utrecht, The Netherlands

1. INTRODUCTION

After 20 years of observing x-rays from non-magnetic cataclysmic variables the origin and nature of the x-rays is still not conclusively established (Córdova 1994). Most popular is the idea that the x-rays are emitted by the boundary layer between the accretion disk and the white dwarf (Pringle & Savonije 1979). Competing models are that (part of) the x-rays are emitted by an accretion disk wind or by an extended hot optically-thin corona. EINSTEIN observations of cataclysmic variables in quiescence showed that the (0.1–3.5 keV) x-ray luminosities are generally in the range 10^{30}–10^{32} erg s^{-1} and the temperatures in the range 1–5 keV (Córdova et al. 1981; Eracleous et al. 1991). Here we present some results of ROSAT observations of cataclysmic variables. A more detailed analysis of the observations will be published elsewhere (Van Teeseling & Verbunt 1994).

2. OBSERVATIONS

We analysed x-ray observations of 10 cataclysmic variables of which 6 are SU UMa dwarf novae, 3 are nova-like variables and 1 is a double-degenerate system (Table 1). The sources were observed with the Position Sensitive Proportional Counter on board of ROSAT. The observed cataclysmic variables were selected on their x-ray brightness in the ROSAT survey or in earlier observations.

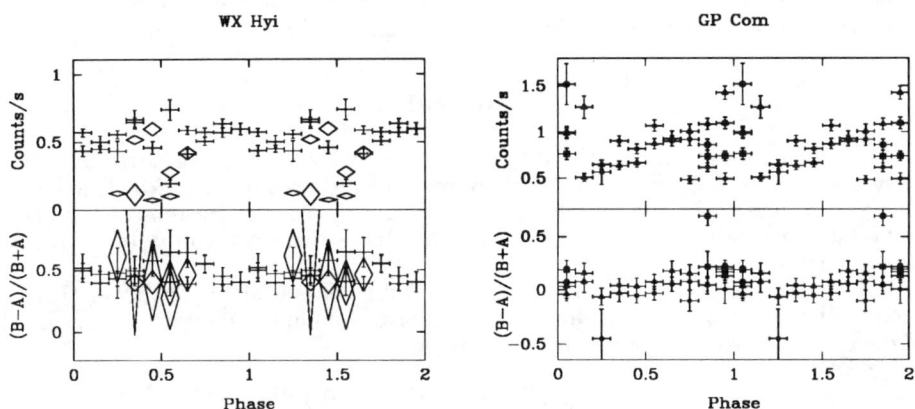

Figure 1: Folded light curves and hardness ratios of WX Hyi and GP Com. Same symbols refer to the same observation interval. The upper 4 diamonds in the orbital light curve of WX Hyi are in the rise to the first outburst and the lower 4 diamonds are the decline of the second outburst.

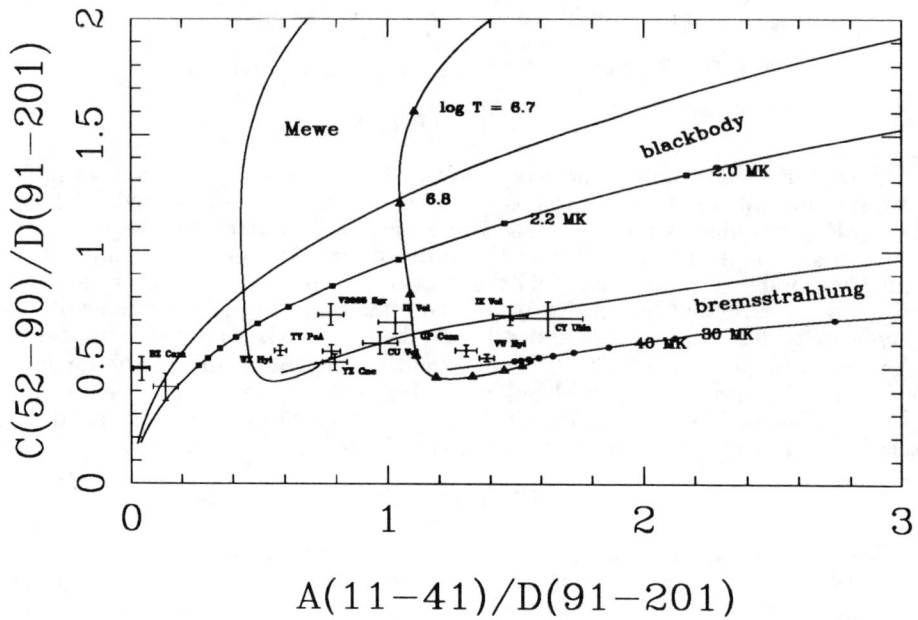

Figure 2: ROSAT PSPC count ratios of the analysed cataclysmic variables, with 1-σ error bars. Also plotted are the predicted count ratios for updated Mewe et al. (1985) spectra, blackbody spectra and thermal bremsstrahlung spectra. The theoretical spectra form bands with the right boundary corresponding to zero interstellar absorption and the left boundary corresponding to a hydrogen column density of $n_H = 10^{20}\,\text{cm}^{-2}$.

3. VARIABILITY

There is count rate variability in almost every source. In several of them the variability is not consistent with purely orbital variability. All dwarf novae have a hardness ratio consistent with a constant spectrum, despite of the strong count-rate variability. There is significant hardness ratio variability of IX Vel and GP Com (Fig. 1).

WX Hyi was detected during the rise and decline of an outburst (Fig. 3). The count rate during the decline was a factor of 6 below the quiescent level. The hardness ratio was not significantly different.

4. NATURE OF THE SPECTRA

The count ratios of the spectra are shown in Fig. 2. Some spectra cannot be fitted to a single-temperature spectrum. This indicates a range of temperatures in the x-ray emitting region or the presence of an unresolved emission feature at $\sim 1\,\text{keV}$ (e.g. Fe L fluorescence). The temperatures of the x-rays in quiescence are generally in the range 1–5 keV and the (0.1–2.4 keV) x-ray luminosities in the

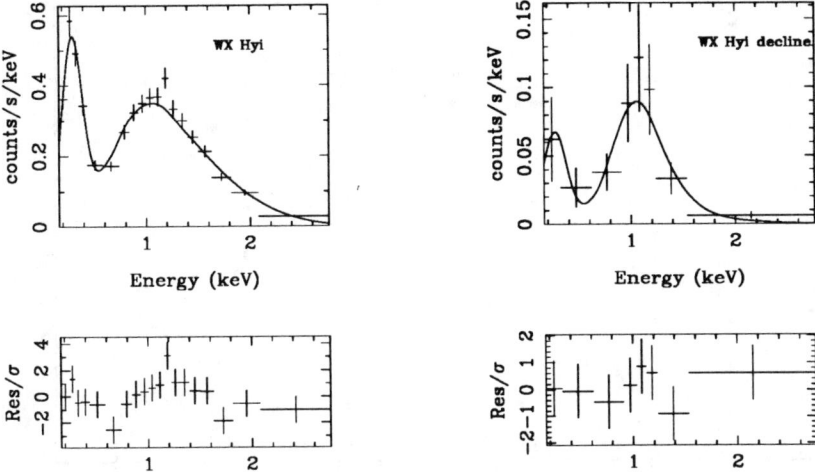

Figure 3: Fits of optically thin spectra to the observed spectrum of WX Hyi in quiescence and decline of an outburst.

ROSAT PSPC band are in the range 5×10^{30}–$10^{32}\,\mathrm{erg\,s^{-1}}$.

From the high emission measures we conclude that, although the continuum is optically thin, absorption will be important in strong resonance lines. Therefore, the nature of the spectrum is between a coronal model and a pure thermal bremsstrahlung model.

The ratio of X-ray flux to ultraviolet flux (taken from Verbunt 1987) is much less than unity and decreases with increasing period (Fig. 4). Adding the optical flux to the ultraviolet flux does not change this. This indicates that the boundary layer does not emit half of the total accretion energy as X-rays, in contradiction to the idea that half of the total accretion energy is emitted by the accretion disk in the ultraviolet and optical, and that the remaining part must be emitted by the boundary layer as X-rays.

REFERENCES

Belloni, T., Verbunt, F., Beuermann, K., etal. 1991, A&A, 246, L44
Córdova, F. A. 1994, in X-ray binaries, eds. W. H. G. Lewin, J. van Paradijs & E. P. J. van den Heuvel (Cambridge: Cambridge University Press)
Córdova, F. A., Mason, K. O., & Nelson, J. E. 1981, ApJ, 245, 609
Eracleous, M., Halpern, J., & Patterson, J. 1991, ApJ, 382, 290
Mewe, R., Gronenschild, E. H. B. M., and Van den Oord 1985, A&AS, 62, 197
Pringle, J. E. & Savonije, G. T. 1979, MNRAS, 187, 777
Van Teeseling, A. & Verbunt, F. 1994, A&A, in preparation
Verbunt, F. 1987, A&AS, 71, 339

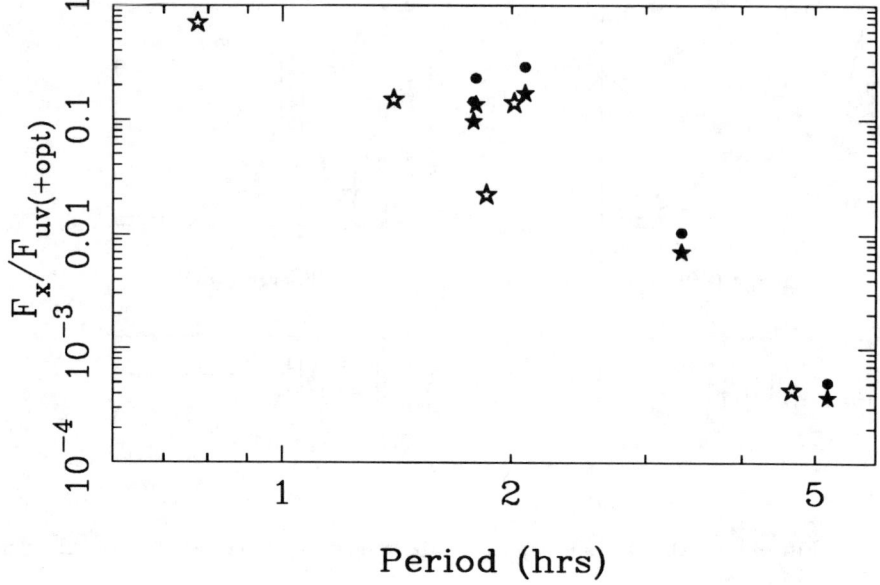

Figure 4: The ratio of x-ray flux to (912–3000Å) ultraviolet flux (•) and to ultraviolet+optical flux (⋆). For the open symbols the ratio is derived from only the optical magnitude.

Table 1: ROSAT PSPC observations of cataclysmic variables

source	subclass[a]	date	exp. (s)	counts/s[b]	HR^c
BZ Cam	aN	1992 Sep 29	6118	0.081 ± 0.004	0.95 ± 0.07
		1993 Sep 3	4594	0.068 ± 0.004	0.83 ± 0.08
YZ Cnc	SU	1991 Apr 3	2452	0.641 ± 0.016	0.31 ± 0.03
GP Com	dd	1991 Dec 19	6314	0.826 ± 0.012	0.09 ± 0.01
VW Hyi[d]	SU	1990 Jul 22-24	9614	1.275 ± 0.012	0.05 ± 0.01
WX Hyi	SU	1991 Nov 30-Dec 22	8292	0.515 ± 0.008	0.46 ± 0.02
		1991 Dec 16	881	0.080 ± 0.011	0.57 ± 0.16
TY PsA	SU	1992 May 7-8	10071	0.310 ± 0.006	0.34 ± 0.02
V3885 Sgr	UX	1992 Oct 16	10381	0.163 ± 0.004	0.38 ± 0.03
CY UMa	SU	1991 Dec 4	9395	0.096 ± 0.004	0.03 ± 0.04
CU Vel	SU	1992 May 31-Jun 1	8256	0.151 ± 0.005	0.25 ± 0.03
IX Vel	UX	1991 Apr 13-14	5164	0.523 ± 0.010	0.08 ± 0.02
		1993 Apr 13	2797	0.508 ± 0.014	0.24 ± 0.03

[a] aN = anti dwarf nova, UX = UX UMa nova-like variable, SU = SU UMa dwarf nova, dd = double degenerate
[b] photon channels 11–235 after background subtraction
[c] $HR = (B - A)/(B + A)$ with $A = 11$–41 and $B = 52$–201 after background subtraction
[d] see also Belloni et al. (1991)

THE *EINSTEIN* SOLID STATE SPECTROMETER AND MONITOR PROPORTIONAL COUNTER SURVEY OF LOW MASS X-RAY BINARIES

Damian J. Christian
Univ. of Maryland, Dept. of Astronomy, College Park, MD 20742
damian@astro.umd.edu

ABSTRACT

The HEAO-2 *Einstein* solid state spectrometer (SSS; 0.5-4.5 keV) and monitor proportional counter (MPC; 1.2-20.0 keV) carried out an extensive survey of 50 low mass X-ray binaries (LMXB). Simultaneous SSS plus MPC spectra, selected on the basis of their intensity, were fit with a set of simple and complex spectral models. For all the sources, including Eddington-limited bulge sources, bursters, dippers, the soft spectrum black hole candidates, and a few transients in decline, the spectra could be fit acceptably with combinations of thermal bremsstrahlung and blackbody spectra or a Comptonized spectrum and a blackbody. The results rule out optically thick disk models for the bright (Z) sources and for the bursters power law models are unacceptable. The SSS can confirm only the strongest of previously reported low energy emission lines due to OVIII or Fe L transitions. The data does not support a unique physical interpretation.

1. INTRODUCTION

Low mass X-ray binaries (LMXB) are some of the most luminous galactic X-ray sources, and generally have high signal-to-noise spectra that have promised an explanation of their underlying physical processes. Only recently have high resolution data, more frequent and longer observations allowed a more complete understanding to come to fruition. Many LMXB exhibit dips, flaring with an increase in intensity of a factor of two, quasi-periodic oscillations (QPO), and X-ray bursts. LMXB can be simply classified as consisting of a class with luminosities of $\sim 10^{38}$ ergs/s, and a less luminous class of a few 10^{36}-5×10^{37} ergs/s (Parsignault and Grindlay 1978; Ponman 1982). EXOSAT observation have shown these sources are better classified on the basis of their color-color diagrams, the bright sources showing a strong pattern resembling a 'Z' with correlations to QPO and spectral behavior, and a group with less distinct patterns, the Atoll sources.

The large X-ray luminosities of LMXB are understood as accretion onto neutron stars from late type (<1 $M_\odot$) stars that have filled their Roche lobes. However, over the limited bandpass of many X-ray instruments, continuum models with very different physical interpretation have fitted the data equally well. The ambiguity in the spectral form of LMXB has arisen in part because counts at low energies are diminished by absorption by the interstellar medium, and high energy counts are lost due to lack of instrument effective area. The SSS and MPC of the *Einstein* observatory (HEAO-2) obtained simultaneous observations of 50 LMXB between November 1978 and October 1979 (January 1, 1979=JD 244 3873.5). This combination offers a bandpass of 0.5-20.0 keV with the ability to test for soft components, measure the interstellar column density, and detect at least any strong low energy line emission or absorption features.

2. ANALYSIS

Spectral Models

Combined SSS and MPC spectra were fit with simple and complex continuum models. Fitting was performed with the 'Bspec' spectral fitting package (HEAO-2 analysis software), and Xspec. The photoelectric cross section for the absorbing column density along the line of sight was taken from work by Morrison and McCammon (1983). Spectral models included: a power-law (PL), blackbody (BB), optically thin thermal bremsstrahlung (TB), a physically thin and optically thick accretion disk model (Disk), an approximation to unsaturated Comptonization (USC; $\sim E^{-\Gamma}\exp(-E/kT)$), and Comptonization of the form of Sunyaev and Titarchuk (1980).

3. RESULTS

Model Fits

Single Component Fit to the SSS+MPC were generally poor for PL, BB, and Disk models. USC gave the best overall fits to the entire set.

Two component fits (addition of a BB) generally improved χ^2 for all classes. The following points are notable. Bright sources are least well fit with BB+Disk models. Bursters are least well fit with BB+PL. Almost all sources are reasonably well fit ($\chi^2 \leq 2$, corresponding to discrepancies of a few percent, which are not much more then remaining uncertainties in the response; Christian 1993) with BB+USC or BB+TB.

BB+Disk model parameters for fits to the bright LMXB often had disk temperatures of 0.7-1.5 keV (soft component) and blackbody temperatures near 2 keV (hard component). Mass accretion rates derived from the fits were not super-Eddington (0.2-5.5 x 10^{17} gm/s) as found by White, Stella, and Parmar (1988) for EXOSAT ME data. However, spectra show the disk model produces too few high energy counts to fit the data. This fact could only be determined for sources with good statistics (e.g Sco X-1 and GX 5-1), because of systematic errors of the MPC channels above 10 keV.

A two component model of BB+TB has been found to be an adequate description of the data and provides a phenomenological model with which to compare the sources to other experiments (Swank and Serlemitsos 1985; Shulz, Hasinger, and Trumper 1989). Ranges of fit parameter are shown in Table 1. Many blackbody radii derived from the fits are near 10 km for the bright sources. The fractional contribution of the BB is plotted as a function of luminosity in figure 1. It shows the bright sources dominating at higher luminosities and BB fractions. Bursters typically have a contribution of less then 20%.

Table 1. Typical BB+TB Fit Parameters

Class	BB radius (km)	kT_BB (keV)	EM_60 cm^{-3}	kT_TB (keV)	%BB
Bright	2 - 20	0.8 - 2.6	0.6 - 12.0	5 - 10 keV	16 - 60
Burster	~2	2.0-2.5	3.0 - 6.0	7 - 10	15 - 20
Dippers	1 - 14	0.2 - 2.6	0.1 - 1.0	~15	10 - 99
Soft	3 - 30	<1.3	0.05 - 21.0	<3	30 - 70

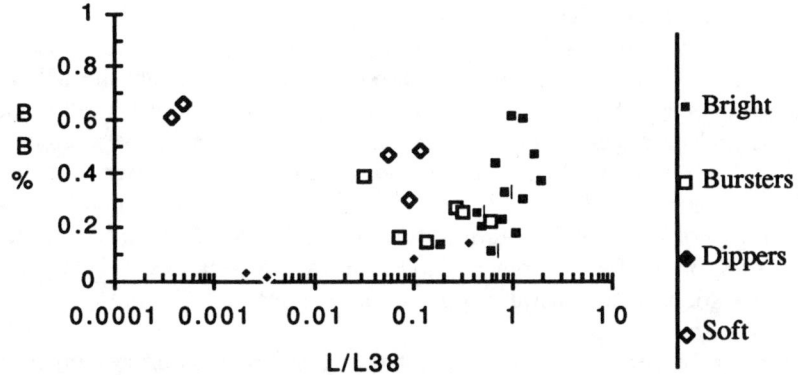

Figure 1. The percentage blackbody for the BB+TB model plotted as a function of luminosity in unit of 10^{38} ergs/s.

The USC model with $\Gamma = 1.2$ to 1.4 is similar to a TB spectrum (i.e. TB with a Gaunt factor $\sim E^{-0.4}$ in the 2-10 keV range) and with $\Gamma = -2$ is a Wien spectrum. Thus with variations of Γ, USC can describe both an optically thin plasma and optically thick emission. The bright sources had Γ ranging from -0.2 to 1.0 with cutoff temperature of 3-7 keV and luminosities ranging from 3×10^{37} erg/s (GX 3+1) to 1.3×10^{39} ergs/s (GX 5-1). The bursters have Γ ranging from 1 to 2 with cutoff temperature of 5-10 keV, and a wide range of luminosities (10^{36} ergs/s (0512-40) to 7.7×10^{37} ergs/s (Ser X-1)). There are 4 soft sources in this sample with a variety of Γ and cutoff energies less than 3 keV, and luminosities between $\sim 4 \times 10^{35}$ ergs/s (X0142+69) and 2×10^{37} ergs/s (GX 339-4). Luminosity of the dippers range from 10^{34} ergs/s (4U 2129+47) to 10^{38} ergs/s (1624-49). Γ is plotted as a function of luminosity in Figure 2.

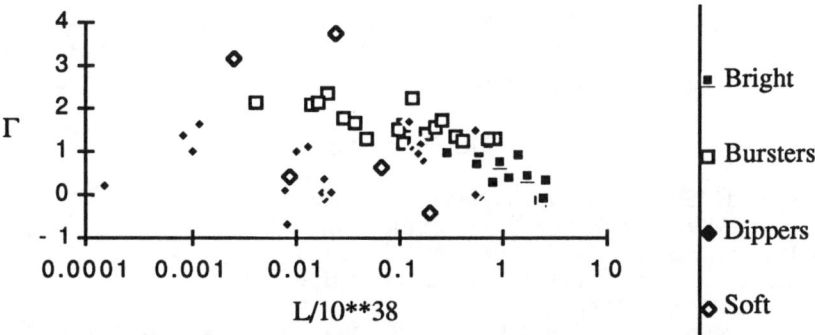

Figure 2. Γ plotted as a function of luminosity for the USC model.

Fitting the data with the formal solution for the scattering of cool photons off hot electrons (Sunyaev and Titarchuk 1980) offered no improvement over the USC form with similar or slightly worse reduced χ^2. Temperatures ranged from 1.1 keV (X1755-338) to 3.8 keV (Cyg X-2), and moderately large optical depths (8 - 24).

Comparison with the EXOSAT ME

BB+TB fits to time selected spectra of the bright LMXB were folded through the EXOSAT ME response matrix. This produced the counts that the ME would have observed in energy bands used by Shulz, Hasinger, and Trumper (1989) and the same softness and hardness ratios were generated. The states of the sources during the SSS+MPC observations then could be compared to the states observed with EXOSAT. Thus we determined GX 5-1 and GX 340+0 were on the horizontal branch, Cyg X-2 was on the normal branch, and Sco X-1, Sco X-2, GX 17+2, and an earlier Cyg X-2 observation on the flaring branch. Atoll sources, GX 9+9, GX 9+1, GX 3+1, and GX 13+1 were found with generally larger softness ratios.

Changes in Spectral Parameters

For the bright sources we attempted to resolves temporal changes in terms of spectral model components or parameters. For example, the BB+TB model changes in spectral parameters of Sco X-2 as it moved up the FB were complex. The fraction of blackbody contribution increased from 0.20 to 0.46 with changes in both kT_{BB} and BB radius. The TB temperature and the TB emission measure were observed to be correlated with bremsstrahlung flux. GX 17+2 showed similar FB behavior.

Low Energy Line Emission

The SSS only detected significant line emission in a few sources, namely X0614+091, Sco X-1, Cyg X-2, and Cyg X-3 (Christian 1993). Several bursters, e.g. X1636-536, X1735-44, and Ser X-1 had 3σ detections based on line energies previously reported by the higher resolution OGS (Vrtilek et al. 1991).

4. DISCUSSION

A unique physical model does not emerge from the data, however at least two physically distinct regions are suggested. An optically thick boundary layer, which is well described as a 1-2 keV blackbody, and an optically thin, possibly extended region, which is well described by TB or a form of Comptonization. However the data can not distinguish between the standard two component models (BB+TB or BB+USC) and models based on a spherically symmetric distribution of gas with varying optical depth (Lamb 1989; Ponman, Foster, and Ross 1990).

REFERENCES

Christian, D. J. 1993, Univ. of Maryland, Ph.D. Thesis
Lamb, F. K. 1989, Proc. 23rd ESLAB, Symposium 1, 215
Morris, R., and McCammon, D. 1983, ApJ, 270, 119
Parsignault, D. R., and Grindlay, J. E. 1978, ApJ, 225, 970
Ponman, T. J. 1982, MNRAS, 201, 769
Ponman, T. J., Foster, A. J., and Ross, R. R. 1990, MNRAS, 246, 287
Schulz, N. S., Hasinger, G., and Trumper, J. 1989, A&A, 225, 48
Sunyaev, R. A. and Titarchuk, L. G. 1980, A&A, 86, 121.
Swank, J. H., and Serlemitsos, P. J. 1985, in Galactic and Extragalactic Compact X-ray Sources, ed. Y. Tanaka and W.H.G. Lewin (Tokyo:ISAS), 175
Vrtilek, S. D., McClintock, J. E., Seward. F. D., Kahn, S. M., and Wargelin, B. J. 1991, ApJS, 76, 1127
White, N. E., Stella, L., and Parmar, A. 1988, ApJ, 324, 363

THE COOLING OF THE WHITE DWARF IN OY CAR AFTER 1992 SUPEROUTBURST

F.H. Cheng
STScI, 3700 San Martin Drive, Baltimore, MD 21218.
cheng@stsci.edu

T.R. Marsh
University of Oxford, Keble Road, Oxford, OX1 3RH, England.
19464::TRM

Keith Horne
Sterrekundig Instituut, University of Utrecht, Netherlands and
STScI, 3700 San Martin Drive, Baltimore, MD 21218.
horne@fys.ruu.nl

I. Hubeny
Code 681, NASA/GSFC, Greenbelt, MD 20771.
hubeny@stars.gsfc.nasa.gov

ABSTRACT

HST observations of the eclipsing dwarf nova OY Car after its 1992 April superoutburst are used to isolate ultraviolet spectra (1150–2500Å at 9.2Å FWHM resolution) of the white dwarf, the accretion disk, and the bright spot. The white dwarf spectra have a Stark-broadened photospheric Lα absorption, but are veiled by a forest of blended Fe II features that we attribute to absorption by intervening disk material. Spectral fits give white dwarf temperatures changing from $\sim$ 19500 K just after outburst to $\sim$ 17400 K around three months after outburst; The temperature of intervening disk material is $\sim$ 8600 K–9800 K; the velocity dispersion of the intervening disk material is $\sim$ 60–70 km/s. Fitting results also show that the decay time of white dwarf temperature is $\sim$ 27 days, that is much shorter than $\sim$ 687 days in dwarf nova WZ Sge.

1. INTRODUCTION

The Hubble Space Telescope (HST) has high ultraviolet sensitivity and fast photon-counting spectrographs that make it possible to study eclipsing cataclysmic variables in the ultraviolet after a superoutburst, providing new information about the spectral evolution of the white dwarf and the evolution of the physical parameters in the intervening disk material (or so called "Fe II curtain"). Based on time-resolved ultraviolet spectra of 7 eclipses of the dwarf nova OY Car obtained with HST's Faint Object Spectrograph (FOS), we decompose the data into white dwarf, bright spot, and accretion disk components and fit the observed white dwarf spectra using a model atmosphere in local thermodynamic equilibrium (LTE) for solar abundance and $\log g = 8$ observed through a veil of cooler solar abundance LTE gas that produces a forest of Fe II absorption features.

2. HST OBSERVATIONS

The *HST* observations were scheduled as a target of opportunity and started on 1992 April 11 following notification that a superoutburst had started on April 7 by the variable star observers of the Royal Astronomical Society of New Zealand (Bateson, private communication). The outburst lasted until April 24. We observed one eclipse during quiescence on 1991 December 4, four months prior to the superoutburst.

We used the Faint Object Spectrograph (*FOS*) on the *HST* to observe 16 epochs with the first 9 covering the superoutburst, and the last 7 extending 2.5 months into the quiescent interval after outburst. The *HST* exposure times were 4.74 seconds long with 1.2 seconds dead time between exposures. We used G160L grating which gave wavelength coverage from 1150 to 2500 Å at 9 Å FWHM resolution, comparable to the short wavelength cameras if IUE, for details, look for Marsh *et al.* (1993).

We decomposed the last 7 epochs' data, which were observed after the outburst, into white dwarf, bright spot, and accretion disk contributions each with a distinct spectrum and light curve. A simple numerical model was used to compute light curves for the eclipse of each component by the Roche lobe of the companion star (see details in Horne *et al.* 1993). The dotted curves in Fig. 1 are the decomposed white dwarf spectra at 7 epochs after the superoutburst.

3. MODELLING THE WHITE DWARF SPECTRUM

We used the same method by Horne *et al.* (1993), and tried models which pass the white dwarf spectrum through a slab of "Fe II Curtain". Since we view the white dwarf at an inclination of 83°, the line of sight passes through the upper atmosphere of the accretion disk, and hence we expect absorption by gas near the outer rim of the disk. The solid curves in Fig.1 represent the best fits to the white dwarf spectra at 7 epochs. The white dwarf tempratures and the physical conditions in the "Curtain" obtained from the best fit are shown in Fig. 2, where T_{wd} is the white dwarf temprature, T_{cur}, n_e, ΔV, N_H and ΔR are the temperature, electron density, velocity dispersion, column density of hydrogen nuclei and the length of the intervening material in the "Curtain" respectively. Our model WD spectra were normalized to 1 $R_\odot$ radius and 1 kpc distance, the 7th pannel in Fig.2 shows an averaged scale factor of 9.5×10^{-3} was used in the spectral fitting. The observations of 1991 December 4, taken 98 days after a *normal* outburst (1991 August 28) and 125 days before the *superoutburst* (1992 April 7), are plotted for convenience only 10 days before the outburst (see 'square' in Fig. 2). Fig. 2 shows that the white dwarf temperatures change from $\sim$ 19500 K just after outburst to $\sim$ 17400 K three months after outburst. The temperature of intervening disk material is $\sim$ 8600 K–9800 K; the velocity dispersion of the intervening disk material is $\sim$ 60–70 km/s.

4. TEMPERATURE DECAY TIME

Fig. 3a shows the temperature curve of the white dwarf in OY Car. If the temperature of the white dwarf in OY Car decays as follows:

$$T_{wd} = T_\infty + \Delta T e^{-(t-t_o)/\tau} , \qquad (1)$$

we find a decay time $\tau \sim$27 days, temperature excess ΔT $\sim$6000 K, and back-

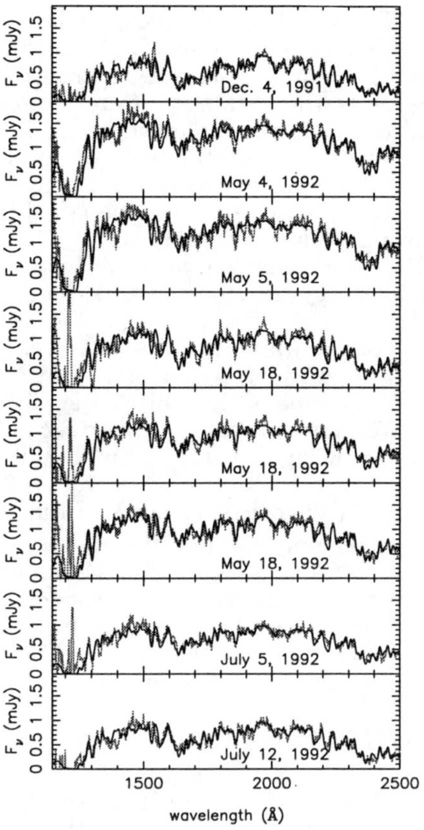

F. H. Cheng et al. 199

Fig. 1

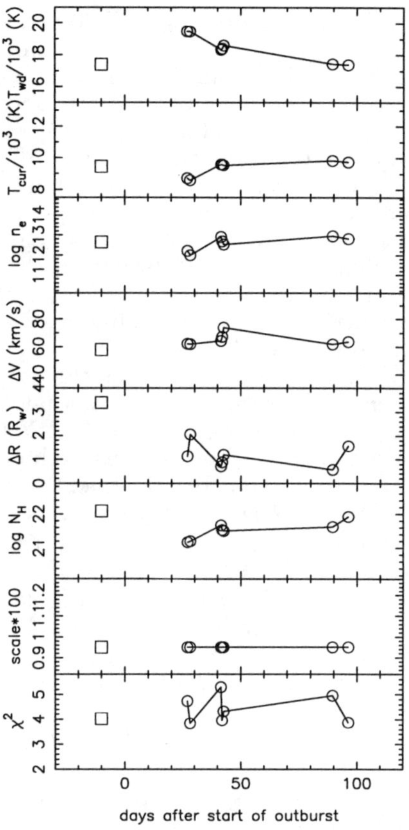

Fig.2

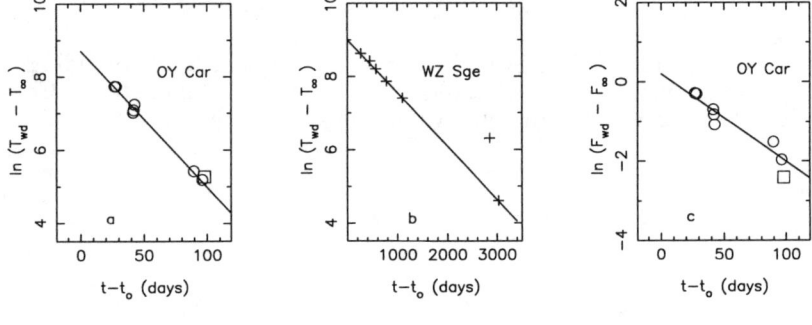

Fig. 3

ground temperature $T_\infty = 17200$ K, where T_∞ was chosen to make the last data point in Fig. 3a fall on the regression line. Therefore, we have $T_{wd} \sim 23200$ K, when $t=t_o$. For comparison between a normal outburst and a superoutburst, We put the data point of 1991 December 4 at the 98th day after the *superoutburst* in Fig. 3a (see 'square'), we find the data point almost falling in the straight line, it seems no obvious difference in the temperature decay between a normal outburst and a superoutburst if the temperature excess ΔT is the same in both cases, otherwise the temperature decay must be slower if the ΔT during a normal outburst is smaller than that during a superoutburst.

Sparks *et al.* (1993) gave the WD temperature curve of WZ Sge (see Fig. 3b), which was obtained from the UV energy distribution and the Stark-broadened Lyα absorption line profiles. We find $\tau \sim 687$ days, $\Delta T \sim 8100$K, $T_\infty = 12400$ K, and $T_{wd} \sim 20500$ K when $t=t_o$ for the white dwarf in WZ Sge. The white dwarf temperatures at the start of outburst in OY Car and in WZ Sge are nearly the same, but the temperature decay time of the white dwarf in WZ Sge is much longer than that in OY Car. We must mention that the second to last data point in Fig. 3b was excluded in measuring the decay time, because it is so far above the straight line.

Marsh *et al.* (1993) also gave a mean UV flux density curve (from 1300 to 1900 Å) of white dwarf in OY Car after the superoutburst, we find a decay time $\tau \sim 45$ days (see Fig.3c). As in Fig.3a, we also put the 1991 December 4 data point at the 98th day in Fig.3c, it is obvious that the flux density of 1991 December 4 is below the straight line. The reason is that the light from WD must pass throgh the curtain, the curtain after the superoutburst in OY Car has smaller column density N_H than that after the normal outburst.

ACKNOWLEDGEMENTS

This work has been funded by NASA grant GO-2380 from the STScI (which is operated by AURA under NASA contract NAS5-26555), by the NASA LTSARP grant NAGW-2678, and by the STScI Visitors Program.

REFERENCES

Horne, K., Marsh, T.R., Cheng, F.H., Hubeny, I. & Lanz, T., 1993, ApJ, in press

Marsh, T.R., Horne, K., Cheng, F.H., Bruch, A., O'Donoghue, D., & Thomas, G., in preparation.

Sparks, W.M., Sion, E.M., Starrfield, S.G. & Austin, S., Proceedings of Cataclysmic variables and Related Physics, 2nd Technion Haifa Conference, Eds. O.Regev & G.Shaviv, Ayalon Offset Ltd., Haifa, 1993. p.96

THE I BAND LIGHT CURVE OF THE LOW MASS X-RAY BINARY GX 9+9

Carole A. Haswell
STScI, 3700 San Martin Dr., Baltimore, MD 21218.
haswell@stsci.edu

Timothy M.C. Abbott
ESO, Cassilla 19001, Santiago 19, Chile.
tabbott@eso.org

ABSTRACT

The I-band light curve of GX 9+9 shows a roughly sinusiodal modulation. The best fitting sine-wave has a period and formal error of 4.1744 ± 0.0002 hours. When the data are folded on this period, however, clear variations in the shape of the light curve are apparent from night to night. The modulation in the mean light-curve has a semi-amplitude of 10.6%.

1. INTRODUCTION

A 4.19 ± 0.02 hour X-ray period for the bright LMXB GX 9+9 was reported by Hertz and Wood (1988). Schaefer (1990) found a roughly sinusiodal modulation of period 4.198 ± 0.028 hours in B band photometry. In both cases the period was interpreted as the orbital period. Schaefer (1990) suggests a model in which the optical modulation is due primarily to reprocessing on two locations in the system. The most important contribution is expected from the bulge on the disk rim associated with the impact of the accretion stream; a smaller contribution arises from the surface of the companion star. We present I band photometry of GX 9+9 taken with the 82" telescope at McDonald Observatory, and preliminary discussion of its interpretation.

2. OBSERVATIONS

The McDonald Observatory Time Series Photometer was used (Abbott 1993), yielding runs of high duty-cycle, equisampled, differential photometry. The target was observed in 1989 on the nights of 1, 2, and 4 May, and 8 June. GX 9+9 is one of a trio of close neighbours, and photometry for the field was extracted using DAOPHOT within IRAF. An example of the resulting photometry, that obtained on 4 May 1989, for GX 9+9 and its close neighbours is shown in Figure 1. GX 9+9 is clearly variable, and the scatter in the light curve of the fainter of the neighbours shows that the error bars are realistic.

3. DISCUSSION

We searched the data for periods by computing χ^2 for sine curve fits for trial periods satisfying 0.1 day $\leq$ P $\leq$ 0.5 day. Figure 2 shows results: the best fitting sine wave has a period and formal error of 4.1744 ± 0.0002 hours. The data are folded on this period in Figure 3(a), which clearly shows that the variation is not a simple sinusoid, and the shape of the light curve varies from night to night. Hence, the formal error in the period determination underestimates the true uncertainty. Since it is not certain that the photometric minimum will occur at the same phase for observations made in two quite different pass bands, we did not attempt to incorporate the epoch of minimum light from Schaefer's data in our refinement of the orbital period.

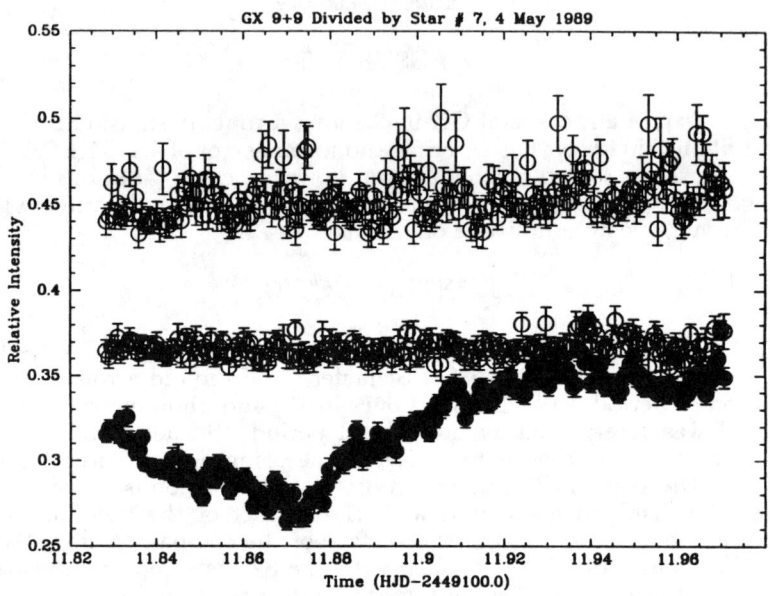

Fig. 1. The photometry of GX 9+9 (filled circles) and its neighbours (open circles). The fainter of the two neighbours is the closest to GX 9+9. The photometry for the brightest star is clearly poorer, resulting from a faint, fourth overlapping star, and possibly from the presence of a bad pixel on the CCD detector.

The mean semi-amplitude of the light curve is 10.6%; Schaefer (1990) found a semi-amplitude of 9% in the B-band. The slightly higher amplitude found here may indicate a change in the accretion geometry: if the bulge on the edge of the disk thickens to subtend a larger solid angle at the central X-ray source, then the percentage modulation will increase. Alternatively, the degree of modulation in the light curve may be higher in the I band than the B band. This would arise if the orbitally-modulated component(s) of the light curve are redder than the non-modulated component(s). Simultaneous B and I band photometry is required to test this hypothesis.

The system appears to have a fairly extreme mass ratio (Schaefer 1990): $M_1 \sim 1.4 M_\odot$, $M_2 \sim 0.4 M_\odot$; hence $q \sim 0.28$. This motivates an entirely different model for the optical modulation: it could be a persistent superhump, as observed in the old nova V603 Aquilae (Patterson and Richman 1991). The interpretation of the light curve will be greatly aided by good radial velocity data, which will allow the binary phases of the photometric minima to be determined.

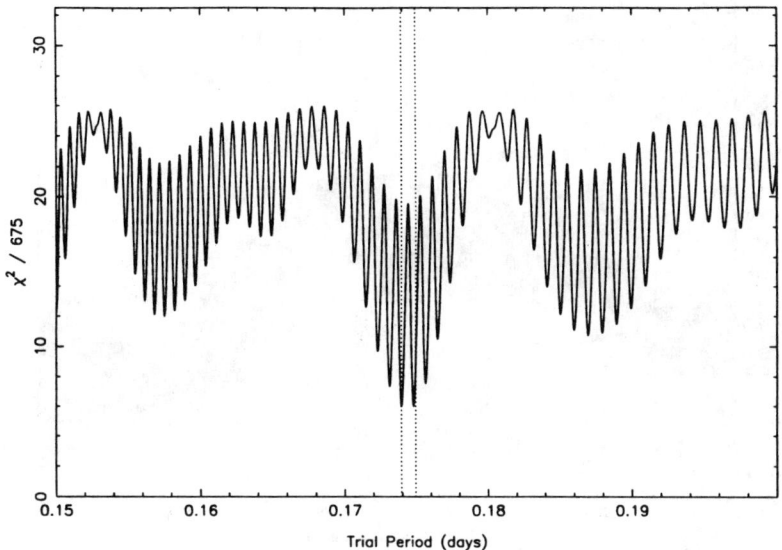

Fig. 2. The results of fitting sine curves to the I band photometry of GX 9+9. The best fit period (smallest reduced χ^2) is P=0.173932(8) days; this is indicated with a dashed line. The slightly longer period reported by Schaefer (1990) is also indicated. Schaefer's period is very close to an alias of the best period found here. This alias produces almost as good a fit as the best period.

ACKNOWLEDGEMENTS

This work was supported by NASA grant NAGW-2678. Helpful discussions with Joe Patterson and Rob Robinson are appreciated.

REFERENCES

Abbott, T.M.C. 1993, Ph.D. Thesis, University of Texas.
Hertz, P. and Wood, K.S. 1988, Ap. J., 331, 764.
Patterson, J. and Richman, H. 1991, PASP, 103, 735.
Schaefer, B.E. 1990, Ap. J., 354, 720.

204 The I Band Light Curve

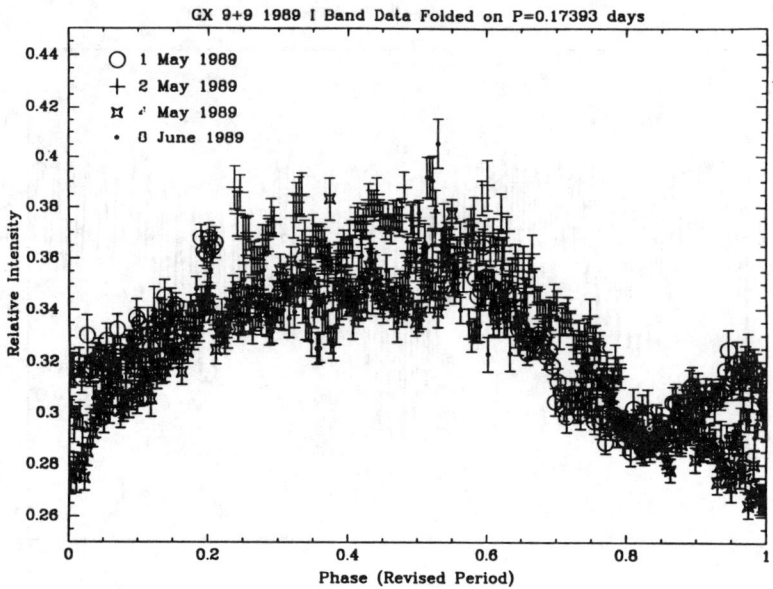

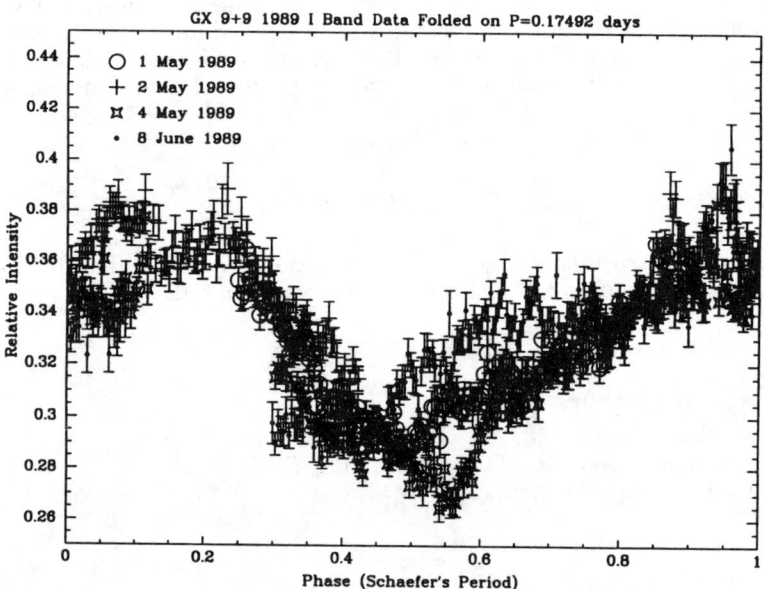

Fig. 3. (a) The data folded on the best-fitting period. The phase is reported with respect to the epoch HJD 2446973.7287071 from Schaefer (1990). (b) As above, except the data is folded on Schaefer's period.

EMISSION LINE SPECTRUM OF AN X-RAY HEATED ACCRETION DISK IN LMXB

Yuan-Kuen Ko
Center for EUV Astrophysics, University of California, Berkeley, CA 94720;
Code 665, NASA Goddard Space Flight Center, Greenbelt, MD 20771.

and

Timothy R. Kallman
Code 665, NASA Goddard Space Flight Center, Greenbelt, MD 20771.

ABSTRACT

We investigate the structure of the accretion disk corona (ADC) formed by X-ray heating in low mass X-ray binaries. Emission line spectra formed in the ADC ranging from optical to hard X-ray energies are calculated and compared with the observational data. These models are made by non-local thermodynamic equilibrium (non-LTE) calculations of ion and level populations and include a large number of atomic processes for ten cosmically abundant elements. Transfer of radiation is treated by using the escape probability formalism. Comparing the model results and the observational data can yield important information of the illumination factor, disk geometry and the X-ray source in LMXBs.

1. INTRODUCTION

Line emission is a common property of low mass X-ray binaries (LMXBs). It is plausible to suggest that these emission lines are formed in the accretion disk if X-rays from the central compact object can illuminate, photoionize and heat up the disk gas. One famous observational evidence for X-ray heating of the disk is the gradual and partial eclipse of X-rays at optical minimum when the companion star blocks the line of sight of the compact object in certain high inclination LMXBs (White & Holt 1982; McClintock et al. 1982). This indicates that there exists an accretion disk corona (ADC) caused by X-ray heating which serves as an extended X-ray source above the accretion disk. Due to temperature inversion in the ADC, emission lines can be formed in the corona and the underlying transition region and chromosphere above the disk photosphere (Kallman & White 1989). Studying these emission lines and comparing with the observations will then be a potential test for the effects of X-ray heating of the accretion disk in compact binary systems, hence help to understand the physical pictures of such systems.

2. CALCULATION METHOD

We calculate the vertical structures of the ADC by dividing the disk into rings of concentric annuli. At a certain annulus of radius R from the central compact object, the vertical structure of the ADC above the disk photosphere is calculated one-dimensionally in the vertical direction. The illuminating X-ray flux at R is

$$F_x(R) \equiv \frac{f(R)L_{x0}}{4\pi R^2}$$

where L_{x0} is the X-ray luminosity emitted from the center of the disk, and $f(R)$ is the illumination factor at R which we take as a free parameter.

For each vertical depth at one annulus, we make non-LTE calculations of the temperature, density, ionization, ion level populations and the emitted spectrum. The vertical structure is then calculated self-consistently under the condition of hydrostatic equilibrium. The boundary conditions at the bottom of the ADC are adopted from Ko & Kallman (1991).

3. RESULTS

Fig.1 shows the vertical temperature structure for two models as a function of height, z. The vertical structure of the ADC can be divided into three zones: (1) the high-T zone where Compton heating balances with Compton cooling and bremsstrahlung cooling, (2) the mid-T zone where Compton heating is balanced by bremsstrahlung cooling and the net atomic cooling from the highest ionized ions and (3) the low-T zone where the main heating and cooling sources are atomic.

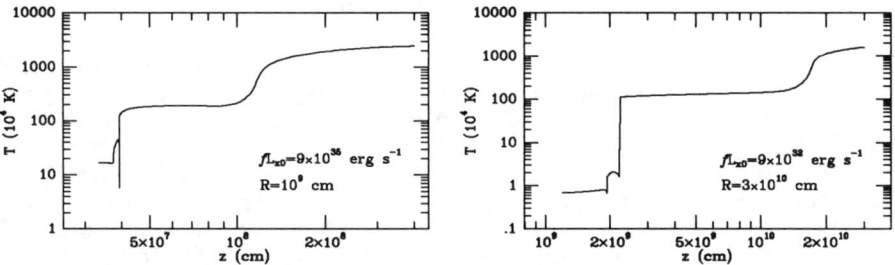

Fig.1. The vertical structure of the ADC for two models.

Fig.2 shows the emission spectrum for two models. The reprocessing of X-rays to the optical and UV continnum is clearly seen. We find that UV lines are formed in the outer disk regions ($R > 10^9$ cm) at the low-T zone just above the disk photosphere. Hard X-ray Fe lines are formed in the inner disk regions ($R < 10^8$ cm) in the mid-T zone except for the 6.4 KeV K-shell fluorescence lines which are formed in the disk photosphere in the outer disk regions.

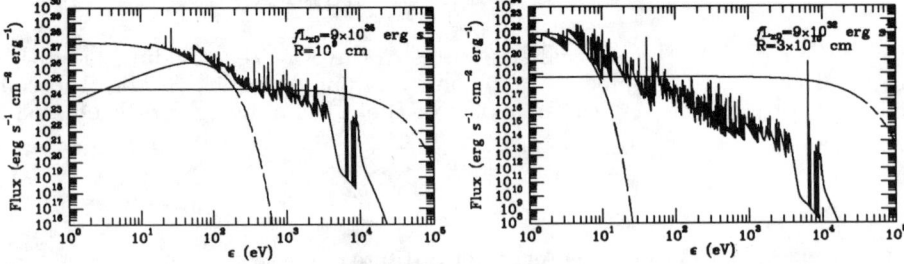

Fig.2. The emission spectrum for two models. A line resolution of $\varepsilon/\delta\varepsilon = 1000$

is assumed. Also plotted on these graphs are the input X-ray spectrum and the disk blackbody spectrum at these radii.

To compare with the observations, we need to assume the amount of illumination at different radii, i.e. $f(R)$. Table 1 lists our choice of 5 sets of $f(R)$. The predicted spectrum is thus $L_\epsilon = 2\pi \int_{R_{in}}^{R_{out}} F_\epsilon(R) R dR$ where $F_\epsilon(R)$ is the flux emerging from the ADC at R and we take $R_{in} = 10^7$ cm and $R_{out} = 3 \times 10^{10}$ cm in our calculation. Fig.3 shows the predicted spectrum for case 5 folded on the X-ray spectrum from the center of the disk for $L_{x0} = 10^{38}$ erg s^{-1} (short dashed curve) and the spectrum from the disk photosphere (long dashed curve).

R			$f(R)Lx0$		
	case 1	case2	case3	case4	case5
10^7	9×10^{35}	9×10^{36}	9×10^{35}	9×10^{34}	9×10^{36}
10^8	9×10^{35}	9×10^{36}	9×10^{35}	9×10^{34}	9×10^{36}
10^9	9×10^{35}	9×10^{35}	9×10^{35}	9×10^{34}	9×10^{35}
3×10^9	9×10^{35}	9×10^{34}	9×10^{34}	9×10^{34}	9×10^{35}
10^{10}	9×10^{35}	9×10^{33}	9×10^{34}	9×10^{34}	9×10^{35}
3×10^{10}	9×10^{35}	9×10^{32}	9×10^{35}	9×10^{34}	9×10^{35}

Table 1. The choice of $f(R)Lx0$ for the five cases. R is in cm and $f(R)Lx0$ is in erg s^{-1}.

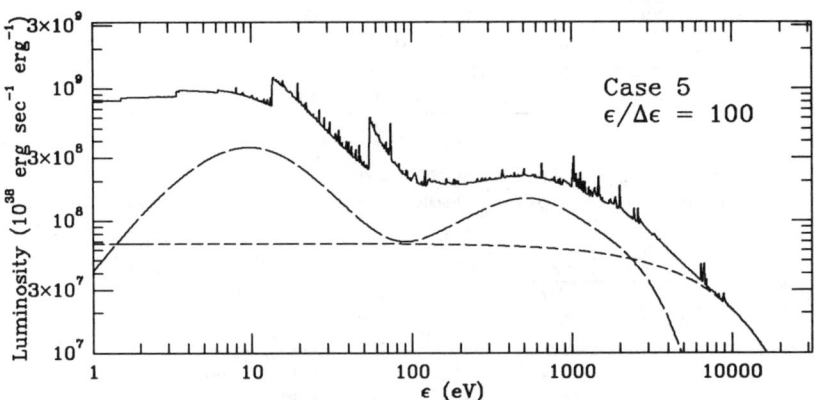

Fig.3. The predicted spectrum of case 5.

Fig.4 shows the predicted hard X-ray Fe line fluxes at earth compared with the observational data of Sco X-1. Table 2 lists the calculated equivalent widths of those lines. We find that cases 2 and 5 fit the data well which implies that the illumination at inner radii must be large ($f > 0.1$) for Sco X-1. The exceptionally large 6.4 KeV line flux may imply that X-rays are incident on the disk with an oblique angle in the outer disk regions where these lines are formed. It is because in this case less X-rays will be available to produce K-shell fluorescence lines.

Table 3 lists the predicted strong UV line fluxes at earth compared with the data of Sco X-1. We notice that our models predict UV line ratios well but the line fluxes are about 100 times smaller (cases 1 and 5) than the dereddened data of Sco X-1. There are several possibilities to get more UV line fluxes from our model, e.g., more illumination at outer radii, larger disk size, contribution from the 'hot spot' and additional soft components of the illuminating spectra. Therefore further investigation is needed to explore these possibilities to account for the observed UV line fluxes.

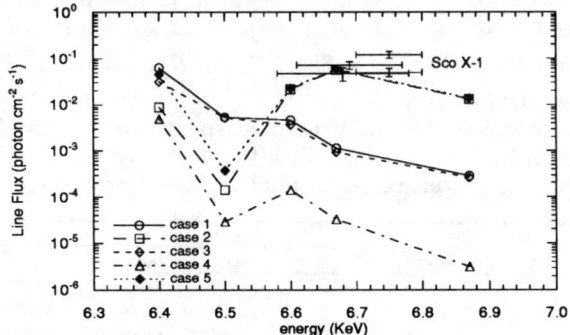

Fig.4. The predicted Fe line fluxes for the five cases against the observational data of Sco X-1. (White, Peacock & Taylor 1985, Hirano et al. 1987)

line	ε (KeV)	case 1	case 2	case 3	case 4	case 5
Fe XXVI Lyα	6.87	0.165	7.65	0.150	0.00182	7.88
Fe XXV K	6.67	0.617	31.3	0.508	0.0177	31.4
Fe XXIII-XXI K	6.6	2.48	11.34	1.98	0.0742	11.86
Fe XX-XVIII K	6.5	2.77	0.0735	2.66	0.015	0.194
sum	...	6.03	50.4	5.30	0.109	51.33
Fe XVII-III K	6.4	30.8	4.40	15.6	2.45	22.64

Table 2. Fe line equivalent widths (eV) for the five cases. The sum are for the first four groups of lines only. The observed values for Sco X-1 are 25 eV for high state and 50 eV for low state (White, Peacock & Taylor 1985).

line	case 1	case 2	case 3	case 4	case 5	Sco X-1
N V λ1240	0.277	0.0144	0.128	0.0523	0.272	16.8
O V λ1371	0.0421	0.00117	0.0270	0.00682	0.0418	3.4
Si IV λ1400	0.0740	0.00353	0.0354	0.0187	0.0740	10.
C IV λ1550	0.485	0.0224	0.174	0.0594	0.482	36.1
He II λ1640	0.0464	0.00678	0.0216	0.00992	0.0480	4.8
N IV λ1718	0.0152	4.37 X 10^{-4}	0.00556	0.00178	0.0152	3.5

Table 3. The predicted UV line fluxes at earth (in 10^{-11} erg s^{-1} cm^{-2}). Also shown are the observed values of Sco X-1 (dereddened) averaged over all intensity states assuming $\bar{D}$ = 1.5 kpc (Kallman, Raymond & Vrtilek 1991).

REFERENCES

Hirano, T., Hayakawa, S., Nagase, F., Masai, K. & Mitsuda, K. 1987, PASJ, 39, 619.
Kallman, T. R. & White, N. E. 1989, ApJ, 341, 955.
Kallman, T. R., Raymond, J. C. & Vrtilek, S. D. 1991, ApJ, 370, 717.
Ko, Y. & Kallman, T. R. 1991, ApJ, 374, 721.
McClintock, J. E., London, R. A., Bond, H. E. & Grauer, A. D. 1982, ApJ, 258, 245.
White, N. E. & Holt, S. S. 1982, ApJ, 257, 318.
White, N. E., Peacock, A. & Taylor, B. G. 1985, ApJ, 296, 475.

MODELS OF ACCRETION DISK CORONAE IN HIGH LUMINOSITY SYSTEMS

S. D. Murray, R. I. Klein
Dept. of Astronomy, 601 Campbell Hall, Univ. of California, Berkeley, CA 94720

J. I. Castor
Lawrence Livermore National Lab., L-58, P.O. Box 808, Livermore, CA 94550

C. F. McKee
Dept. of Physics, Univ. of California, Berkeley, CA 94720

ABSTRACT

We present results from a potentially powerful new method for modelling two-dimensional radiative transfer in anisotropic systems. The method is highly efficient, allowing inclusion of radiative attenuation in dynamical simulations. Comparisons are made with a detailed radiative transfer code. As a first application, we compute self-consistent models of Compton-heated coronae in systems with luminosities up to 0.6 L_{Edd}. Lower luminosity models are used to compare with the results of earlier work.

1. INTRODUCTION AND NUMERICAL METHOD

The work described here represents the second step towards the goal of modelling the structure and dynamics of Compton-heated accretion disk coronae and winds. The optically-thin case has already been modelled (Woods et al. in preparation). Here, we present static models of coronae which include the effects of attenuation. The final phase will be to couple the radiative transfer method described here with the hydrodynamic program.

The driving factor in developing the code is the need for great efficiency and vectorization, due to the long integration times already required by the hydrodynamic program. This eliminates the use of programs which solve the full transport equations, which, while accurate, are computationally expensive. We must, however, retain the essential physics of radiative transfer in systems where the opacity varies greatly, and is highly anisotropic.

The radiation field is split into the direct component received from the central source, and the scattered radiation. This split improves the accuracy with the which the optically thin and thick limits are treated. The direct radiation is computed by calculating τ along rays from the central source. In principle, the identities and path lengths of the zones through which each ray passes needs to be calculated only once, and can then be stored in a 3D array. The memory requirements of this approach, however, limits its usefulness to systems containing fewer than about 10^4 cells.

For larger systems, the ray data must be computed each time. We compute, for a ray to a given zone with indices (J, K), first the vertical position, Z, corresponding to each R(j), where $1 \leq j \leq J$, crossed by the ray. The value of Z is then analytically inverted to find k, the vertical index of the corresponding zone. The process is then repeated to find the values of R and j for each Z(k) with 1≤k≤K. The zone indices, j and k, and coordinate pairs are then stored in 1D arrays, indexed by i = j + k - 1, such that they are automatically ordered. Stepping through the array, using the opacity for each cell then gives τ. This method is highly vectorizable, and also allows each ray to be weighted based upon analytic estimates of the opacity in the first zone. This latter feature may be especially important in hydrodynamic simulations where the central region may not be completely resolved.

The scattered radiation is calculated using Flux-Limited Diffusion (FLD) (Levermore & Pomraning 1981). The entire grid is computed simultaneously, using the Incomplete Cholesky Conjugate Gradient method (Kershaw 1978). A simpler directional split would yield less vectorized code. The FLD approach should yield results for the energy density of radiation within a factor of two of those given by more accurate radiative transfer methods, with the worst results being for optical depths of order unity.

We have compared the method described above with those from the 2D version of ALTAIR, which solves the full transport equations (Klein et al. 1989). In the test problem, the opacity $\chi=0.2$ per unit distance. The grid runs from 0 to 20 in R, and from 0 to 1 in Z. R = 0 is the symmetry axis, and radiation is allowed to free stream from the other boundaries. The central source is placed at (R, Z) = (0, 0.5), such that the optical depth from the central source to the edge varies from 0.1 to 4, testing the full range of behaviour of the ray tracking/FLD method. We find that the results from the two codes agree extremely well, to better than 30% throughout the grid. This is especially impressive, given that the ALTAIR calculation required $\sim 10^4$ rays, and about 6 cpu hours on a Cray YMP, while the ray tracking/FLD method required only about 10 seconds.

2. ISOTHERMAL CORONAL MODELS

As a first application of the method described above, we have computed self-consistent models of static, isothermal accretion disk coronae. For comparison, the models shown here make the same assumptions as those of Ostriker, McKee, & Klein (1991 OMK). That study performed accurate calculations of radiative transfer, under the assumption that the scattered component consisted only of singly-scattered radiation. This assumption limited their results to luminosities $L/L_{Edd} \lesssim 0.1$. Here, we use their models as a further check upon our method, and also present results at higher luminosities.

The coronae are assumed to be isothermal in Z, but not R. The value of T(R) is determined by the Compton temperature, T_m, of the radiation field at the disk surface, which includes direct, scattered, and disk contributions, with the disk component given by a simple alpha model. Due to its lower characteristic temperature, the disk radiation acts as a coolant, quenching the formation of a corona at small R. The coronal temperature is assumed to be $T_m/2$, about the value expected at the base of the corona, where Compton and bremsstrahlung processes dominate the heating and cooling. Detailed simulations find that this holds to within a factor of two vertically in the corona.

The gas density at the base of the corona is determined by the radiation flux and the minimum ionization parameter at which the hot gas can exist. The flux and temperature distributions, coupled with the assumption of hydrostatic equilibrium, determine the opacity as a function of R and Z.

The need for hydrostatic equilibrium limits the models to $\xi \equiv R/R_{IC} \lesssim 0.1$, where R_{IC} is the radius at which gas at the Compton temperature of the central source has a sound speed equal to the escape velocity (Begelman, McKee, & Shields 1983).

Our procedure is then to use an initial guess at the coronal structure, using the analytic results of OMK. The resultant radiation field is then calculated, and used to find a new opacity distribution, and the process repeated to convergence. The converged opacities are then used as an initial guess for the next higher luminosity, and the process repeated.

Some of our results are presented in the figure below. Shown is the radial variation at the base of the corona of the dimensionless total radiative flux, $f_t = f_{dir} + f_{scat}$, where

$$f_i \equiv \frac{4\pi J_i}{L/4\pi R^2},$$

and L is the luminosity of the central source. A direct comparison between our results and those of OMK is given by the models with $L/L_{Edd}=0.064$. The agreement is good, lending confidence in the results at higher luminosities.

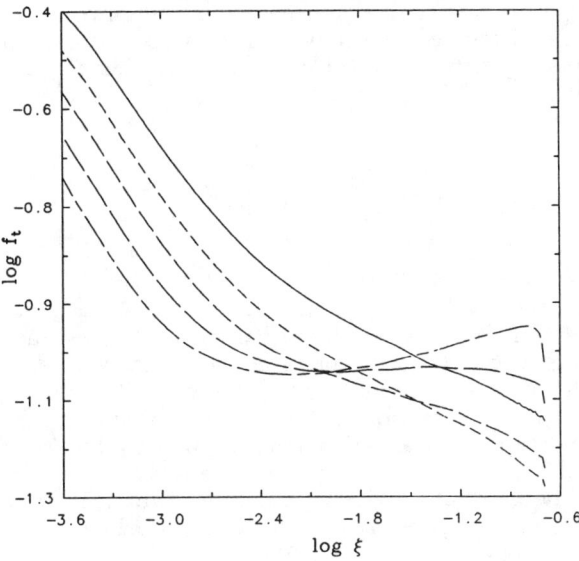

Fig 1. Radial dependence of f_t for $L/L_{Edd} = 0.064$ (solid), 0.112 (short dashes), 0.194 (medium dashes), 0.339 (long dashes), and 0.590 (long/short dashes).

One immediate conclusion drawn from the models at high luminosities is that substantial coronae do exist, and form a roughly continuous extension of the results at lower luminosity. Qualitatively, then, it appears that the multiply-scattered radiation acts to compensate in part for the loss of direct radiation. For the results shown here at $L \gtrsim 0.1 L_{Edd}$, we find that the density at the base of the corona varies as

$$\tilde{\chi} \equiv n_e \sigma_T R_{IC} \approx 0.1 \left(L/L_{Edd} \right) \xi^{-2},$$

where $\xi = R/R_{IC}$. At lower luminosities, the coefficient is ≈ 0.5.

This continuity with the results for lower luminosities is contrary to some expectations that the inclusion of multiple scattering might have significant effects. It also disagrees with earlier results (London 1985), which indicated that shadowing from the inner regions prevents the formation of coronae at high luminosities. The difference with the earlier work may result, at least in part, from the inclusion of disk radiation. The coronae are not, therefore, responsible for the shadowing inferred by fits to iron K lines (Kallman 1990).

There is a qualitative change in the behaviour of f_t with R at high L. At low L, it shows a central peak, flattening toward large R. At higher L, the flattening is replaced by a shallow minimum, and a slow rise outward. This can be understood qualitatively given that, at high L, the radiation reaching the disk is dominated by photons scattered near one coronal scale height, and then transported vertically downwards. The radiative transfer can then be modelled using the Eddington approximation, with a source term at $\tau=0$, and total optical depth determined by the scale height and the radiation reaching the disk. We find that this reproduces f_t to within 50%, and the minimum corresponds to the point where the angle between the surface of the corona, as defined by the scale height, and a ray from the central source is a minimum.

One possible observable effect of accretion disk coronae are in modifying the underlying disk spectra, due to Compton scattering of photons during passage through the corona. The energy of a photon is significantly altered if the Compton y parameter

$$y = \frac{4kT}{m_e c^2} \max(\tau_\perp, \tau_\perp^2) > 1.$$

From the results above, we see that y>1 only at $\xi > 7 \times 10^{-4}$ (L/L$_{Edd}$). Even in the highest luminosity model presented here, y<0.1 for $\xi > 3 \times 10^{-3}$. Coronal veiling would therefore become dominant only in the highest luminosity systems.

In upcoming work, we will combine the radiative transfer method presented here with hydrodynamic models. This will allow us to explore the dynamics and structure of accretion disk coronae and winds at high luminosities, where many important dynamical effects may occur. The results of the static models presented here can be used to compute the spectra of irradiated accretion disks, with the irradiation given by the mean intensities at the base of the coronae. This will also be presented in upcoming work.

REFERENCES

Kallman, T. R. 1990, in Accretion-Powered Compact Binaries, ed. C. W. Mauche, (Cambridge: Cambridge Univ. Press), p. 325
Kershaw, D. S. 1978, J. Comp. Phys., 26, 43
Klein, R. I., Castor, J. I., Greenbaum, A., Taylor, D., & Dykema, P. G. 1989, JQSRT, 41, 109
Levermore, C. D. & Pomraning, G. C. 1981, ApJ, 248, 321
London, R. A. 1985, in Cataclysmic Variables and Low-Mass X-Ray Binaries, ed. D. Q. Lamb & J. Patterson (Dordrecht: D. Reidel), p. 121
Ostriker, E. C., McKee, C. F. & Klein, R. I. 1991, ApJ, 377, 593

THE EFFECT OF IRRADIATION ON OUTBURSTS IN X-RAY NOVAE

Soon-Wook Kim and J. Craig Wheeler
Astronomy Department, University of Texas, Austin, TX 78712

Shin Mineshige
Astronomy Department, Kyoto University, Sakyo-ku, Kyoto 606-01, Japan

ABSTRACT

We study the disk instability and the effect of irradiation on accretion disks around black holes in X-ray novae outbursts. The disk instability theory is proposed as the mechanism for the cause of outbursts in X-ray novae. The mass flow rate into the black hole predicted by the disk instability theory is consistent with observations in both outburst and quiescence. Furthermore, a strong correlation between the optical and soft X-rays in outburst can be naturally explained by the disk instability theory. We adopt a simple model for the direct irradiation of the whole disk by X-rays from the innermost disk and examine its effects on the outburst in the context of the disk instability model. The effect of irradiation influences both optical and soft X-ray light curves and the disk evolution. In outburst, the disk become brighter and hotter in both optical and soft X-ray bands, and both the duration of the outburst and the recurrence time are lengthened.

1. INTRODUCTION: DISK INSTABILITY IN X-RAY NOVAE

The disk instability theory was first proposed to explain outbursts in dwarf novae (Osaki 1974). The essence of the disk instability theory is that the matter accreted from an external source (e.g., a companion star in a binary system) is accumulated during quiescence, and, as the density and temperature increase, an instability associated with opacity changes when hydrogen and helium are ionized results in a sudden heating and brightening of the disk. Heating waves propagating either inward or outward in the disk lead to outburst. Cooling fronts always beginning in the outer disk cause a return to quiescence. The disk instability theory provides a self-consistent account of the basic properties of dwarf nova outbursts (for reviews, see Osaki 1993 and references therein). It is widely believed that the disk instability plays an important role in the outburst evolution of accretion disks in dwarf novae. It is inevitable, furthermore, that the ionization of hydrogen and helium, the essence of the disk instability, should cause other viscous disks to make transitions from cool to hot states and hence cause similar outburst phenomena in systems other than dwarf novae. In particular, the disk instability theory is important to study outbursts in binary black holes and gives a consistent account of the observations. The similarity of outbursts in dwarf novae and X-ray novae, including neutron star candidates, is also discussed by van Paradijs & Verbunt (1984).

Many X-ray novae have been discovered in recent years that are, like the prototype A0620-00, black hole candidates. Study of these X-ray novae will give new understanding of black hole accretion, means to discriminate neutron stars from black holes, and new connections to the physics on AGN. Recent observations of A0620-00 (McClintock et al. 1993) show a relatively large mass transfer rate $\gtrsim 3\times10^{-11} M_\odot yr^{-1}$, but a very much smaller mass flow through the inner portion of the disk, $\lesssim 4\times10^{-15} M_\odot yr^{-1}$. Marsh et al. (1993) give an upper limit for accretion onto the compact object, $6\times10^{-14} M_\odot yr^{-1}$, based on the lack of HeII($\lambda 4686$) emission and hence ionizing flux. These results are completely consistent with the non-steady state predicted by the disk instability theory in quiescence. The limits on the accretion rates in the inner disk are much lower than the estimate of $\sim 10^{-11} M_\odot yr^{-1}$ based on a steady state disk assumption (de Kool 1988). The time-dependent disk instability models (Mineshige & Wheeler 1989 and Kim et al. 1993) provide much lower rates than the limits derived by McClintock et al. (1993) and Marsh et al. (1993).

The most obvious aspect of X-ray novae is the strong correlation between the optical and soft X-rays (e.g., Chen et al. 1993 and Tanaka 1990). The disk instability theory can naturally produce soft X-rays ($\sim$ keV) and a strong correlation between the optical and soft X-rays in the outburst evolution (Mineshige et al. 1990 and Kim et al. 1993) since both optical and X-ray fluxes are enhanced with the high mass flow rates during outburst. The correlation can be enhanced if the outer portions of the disk reprocess X-rays from the inner disk. We consider the problem of irradiation of the disk below.

2. IRRADIATION IN X-RAY NOVA OUTBURSTS

Typical X-ray novae (A0620-00, GS2000+25, Nova Muscae and GRO J0422+32) show three maxima in both X-rays and optical: the primary outburst, a secondary "reflare" and a final broad bump (e.g., see Figure 1 in Chen et al. 1993). The primary outburst is plausibly due to the disk instability (Huang & Wheeler 1989 and Mineshige & Wheeler 1989). Chen et al. (1993) suggest that the secondary reflare is caused by evaporation of matter from the companion star near the inner Lagrangian region due to X-ray irradiation. We suggest coupling of realistic irradiation effects (coronal scattering, "shadowing", etc.) with the time-dependent vertical structure of the disk instability may give an alternative possible explanation of the secondary reflare. Here we present some simple models relevant to this issue.

We consider two types of irradiation: direct irradiation from the innermost hot disk and reflected irradiation by a corona or chromosphere above the disk. In this study, several model assumptions are made. First, we employ the following viscosity parameter (α) prescription:

$$\alpha(t) = \alpha_o \left(\frac{H(t)}{R}\right)^N, \qquad (1)$$

where, as indicated, both α and the disk height H are time-dependent. This, in turn, means that $\alpha \propto T_c(t)^{N/2}$, a function of the central temperature, because $H = C_s/\Omega_K$ (C_s is the sound velocity and Ω_K is the Keplerian angular velocity). In our preliminary models, we choose the constants $\log \alpha_o = 1.0$, 1.5 and 2.0, and $N = 1.5$, based on the previous studies of Mineshige & Wheeler (1989) and Mineshige & Wood (1989). Secondly, we assume that the principle contribution to the irradiation flux comes from near the innermost disk radius R_{in}, where we set $R_{in} \gtrsim 3R_g$, the last stable orbit for a non-rotating Schwarzschild black hole. Ginga X-ray observations of outbursts (e.g., GS2000+25 and Nova Muscae) imply that the inner disk seems to be almost constant (Ebisawa 1991 and Tanaka 1993). In these observations, $R_{in} \gtrsim 3R_g$ is consistent with a $3M_\odot$ black hole in fits to X-ray spectra in typical X-ray novae such as GS2000+25 and Nova Muscae. Observations of A0620-00 have the same implication (Tanaka 1990 and White et al. 1984). A reasonable size of the outer radius ($R_{out} \gtrsim 0.55R_L$) is obtained by taking the tidally stable circular orbit (Paczyński 1977), while previous studies employed values either larger than the inner Lagrangian point, R_L (Huang & Wheeler 1989), or only $\lesssim 0.3R_L$ due to numerical instability (Mineshige & Wheeler 1989). Our R_{out} is consistent with observations of A0620-00 in quiescence: $0.5 - 0.85R_L$ (Marsh et al. 1993). Our choice of R_{in} results in the following expression for the X-ray luminosity of the irradiation:

$$L_X(t) = \eta \left(\frac{GM}{R_{in}}\right) \dot{M}_{in}(t), \qquad (2)$$

where η is the ratio of X-ray to the accretion luminosity and $\dot{M}_{in}$ is mass accretion rate at R_{in}.

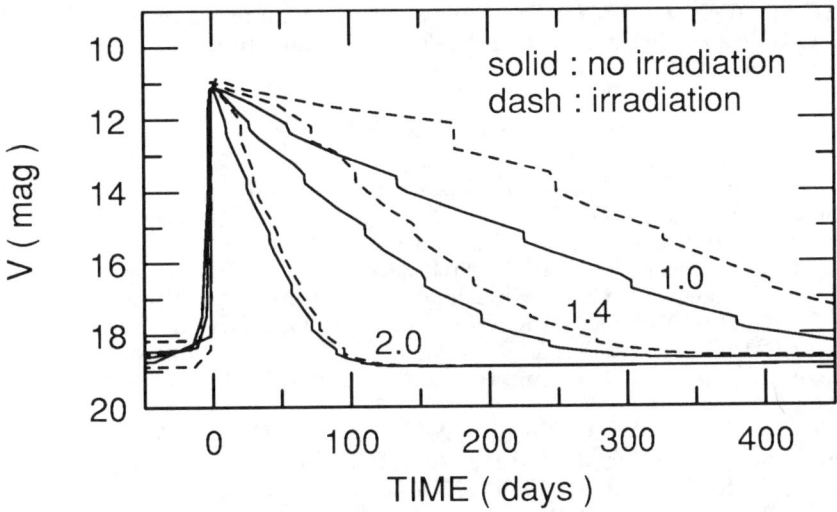

Fig. 1. — Light curves for models with irradiation (dashed lines) and without irradiation (solid lines) are shown for three values of the viscosity parameter, $\log \alpha_o = 1.0$, 1.4 and 2.0, and N = 1.5. In these preliminary models, we take η to be unity and a mass transfer rate from the companion of $\dot{M} = 3.16 \times 10^{16}$ gs^{-1}. We take $4M_\odot$ for a black hole mass. For both irradiated and non-irradiated models, for decreasing α_o, and hence α, the outburst duration is lengthened and the slope of the decay becomes less steep. For illustration purpose, the non-irradiated models have been normalized to maximum light and the corresponding irradiated models adjusted accordingly. The normalization is reflected in small differences in the model distances, 610−660 pc. In the irradiated models, the enhancement by the additional heating in the disk results in longer duration of outburst, about a factor of 1.5.

The irradiated flux is given by:

$$F_{irr}(t) = (1 - A)\left(\frac{L_X(t)}{2\pi R^2}\right)\frac{d}{d(\ln R)}\left(\frac{H(t)}{R}\right)^2, \qquad (3)$$

(Tuchman, Mineshige & Wheeler 1990 and Shakura & Sunyaev 1973), where, for simplicity, we take the X-ray albedo, A=0.5.

After outburst maximum, the additional heating of irradiation causes an extension of the hot state in time. As a result, irradiation will cause the slope of the light curve in the decay to become less steep and the decay time scales to become longer (see Fig. 1). The irradiation also causes an increase of the peak luminosity of outbursts and of the temperature of the inner disk regions. Preliminary results show that the inner hot disk can reach around $\lesssim 1$ keV ($\sim 10^7$ K), which is consistent with observations of the 1991 outburst in Nova Muscae (Mineshige et al. 1993). These increases in temperature and outburst duration result in a longer recurrence timescale because the disk loses more mass after a cycle of outburst. Study with different values of N, α_o and $\dot{M}$, and the indirect irradiation is required (Kim et al. 1993).

ACKNOWLEDGEMENTS: This work was supported in parts by NASA Grant NAGW 2975, NSF Grant 9115143 and a grant from the Texas Advanced Research Program.

REFERENCES

Chen, W., Livio, M., & Gehrels, N. 1993, ApJ, 408, L5
de Kool, M. 1988, ApJ, 334, 336
Ebisawa, K. 1991, Ph.D. Thesis, University of Tokyo
Huang, M., & Wheeler, J. C. 1989, ApJ, 343, 229
Kim, S.-W., Wheeler, J. C., & Mineshige, S. 1993, in preparation
Marsh, T. R., Robinson, E. L., & Wood, J. H. 1993, MNRAS, in press
McClintock, J. E., Horne, K., & Remillard, R. A. 1993, submitted to ApJ
Mineshige, S., Hirano, A., Kitamoto, S., & Yamada, T. Y. 1993, submitted to ApJ
Mineshige S., Kim S.-W., & Wheeler, J. C. 1990, ApJ, 358, L5
Mineshige, S., & Wheeler, J. C. 1989, ApJ, 343, 241
Mineshige, S., & Wood, J. H. 1989, MNRAS, 241, 259
Osaki, Y. 1974, PASJ, 26, 429
Osaki, Y. 1993, in Cataclysmic Variables and Related Physics, eds. O. Regev & G. Shaviv (Bristol: Institue of Physics Publishing), p152
Paczyński, B. 1977, ApJ, 216, 822
Shakura, N. I., & Sunyaev, R. A. 1973, A&A, 24, 337
Tanaka, Y. 1990, in the 23rd ESLAB Symposium on Two Topics in X-ray Astronomy, ed. N. E. White (ESA SP-296), p2
Tanaka, Y. 1993, in From Ginga to ASTRO-D and Further to Duet, Ginga Memorial Symposium, eds. F. Makino & F. Nagase, p19
Tuchman, Y., Mineshige, S. & Wheeler, J. C. 1990, ApJ, 359, 164
van Paradijs, J. & Verbunt, F. 1984, in High Energy Transient in astrophysics, ed. S. E. Woosley (New York: AIP), p49
White, N. E., Kaluzienski, J. L. & Swank, J. H. 1984, in High Energy Transients in Astrophysics, ed. S. E. Woosley (New York: AIP), p31

LOW FREQUENCY OSCILLATIONS FROM ACCRETION DISKS IN X-RAY BINARIES

Xingming Chen and Ronald E. Taam
Department of Physics and Astronomy
Northwestern University, Evanston, IL 60208

ABSTRACT

The global nonlinear time dependent evolution of thermal and viscous instabilities in Keplerian accretion disks in X-ray binary sources has been investigated to understand the low frequency (~ 0.04 Hz) quasi-periodic oscillations (QPOs) seen recently in some black hole candidates (Cyg X–1 and GRO J0422+32) and the Rapid Burster MXB 1730–335. Within the framework of the α-viscosity model, we assume the viscous stress scales with gas pressure only but that the α parameter is formulated as a function of the local scale height, h. Specifically, $\alpha = \min[\alpha_0(h/r)^n, \alpha_{\max}]$, where r is the distance from the compact object, and n, α_0 and $\alpha_{\max}$ are constants. It is found that nonsteady behavior may arise for sufficiently large n (~ 1.2). We show that, the low frequency QPOs may be explicable in terms of a thermal-viscous instability in accretion disks, which results in periodic or quasi-periodic oscillations in the mass accretion rate. These oscillations are globally coherent in the unstable regions of the disk and the disk luminosity is modulated at the same time scale. The observations of QPOs place constraints on the viscosity parameters and indicate that $(n, \alpha_0) \sim (1.6, 30)$ for the Rapid Burster and $\sim (1.3, 30)$ for a $5 M_\odot$ black hole.

1. INTRODUCTION

The quasi-periodic oscillation (QPO) phenomenon in X-ray binaries is believed to be very important to our understanding of the physics of accretion processes in these systems. The well known QPOs are the so called horizontal-branch and normal-branch QPOs with frequencies of $\sim 20 - 50$ Hz and $5 - 10$ Hz respectively. Models proposed for these QPOs require that the central objects be neutron stars (for a review see van der Klis 1989). More recently, low frequency QPOs with characteristic frequencies of ~ 0.04 Hz have been observed from X-ray binary systems, for example, the Rapid Burster MXB 1730–335 (Lubin et al. 1992), Cyg X–1 (Vikhlinin et al. 1993), and GRO J0422+32 (Pietsch et al. 1993). The mechanism responsible for the low frequency QPOs may be different from those proposed for the other classes of QPOs since such oscillations have been observed from both neutron star and black hole candidate systems. Abramowicz, Szuszkiewicz, & Wallinder (1989) proposed that thermal and viscous instabilities in accretion disks could be important for QPOs with frequencies of ~ 1 Hz for galactic X-ray sources. Such types of instabilities are attractive since it is well known that the accretion disk models based upon an α-viscosity prescription (Shakura & Sunyaev 1973) are thermally and secularly unstable in the radiation pressure dominated regime (Lightman & Eardley 1974). However, nonlinear time-dependent calculations by Taam & Lin (1984) reveal that the instability is, indeed, global, but that large-amplitude, burst-like fluctuations (instead of oscillatory variations) of the disk luminosity result. Accordingly, other forms for the viscous stress have been suggested which may be more appropriate. In particular,

Meyer (1986) demonstrated that accretion disk models based on a modified form of the viscous stress, $\tau = -\alpha p_g$ where p_g is the gas pressure and $\alpha = \alpha_0 (h/r)^n$, would be thermally and viscously unstable in the local approximation in the radiation pressure limit with $n > 0.75$. Here h is the local scale height, r is the distance from the compact object, and n and α_0 are constants. In a recent study Milsom, Chen, & Taam (1994; hereafter MCT) generalized the results of Meyer (1986) and, furthermore, pointed out that non-burst-like, low frequency fluctuations in the X-ray luminosity could be produced with this viscous stress formulation.

In this paper, we report on the time-dependent evolutions of the accretion disk based upon the above modified form of the viscous stress and investigate the possible relevance of the disk instability to the low frequency QPOs.

2. RESULTS

We consider an axisymmetric, non-self-gravitating, optically thick Keplerian disk. We assume the disk is sufficiently thin so that it can be described by the vertically integrated equations. In this approximation, the surface density and the mid-plane temperature of the disk, at a given cylindrical radius and time, are governed by the standard time-dependent mass diffusion equation and the energy conservation equation. Here, the effective kinematic viscosity is parameterized in terms of the α model and the heating and cooling rates are determined by the detailed vertical structure calculations (see MCT). The two equations are solved via an explicit method with 41 grid points distributed equally on a logarithmic scale ranging from an inner boundary at $4r_g$ to an outer boundary at $300r_g$, where r_g is the Schwarzschild radius ($r_g = 2GM/c^2$).

As a standard model (sequence 1), we have followed the evolution of the thermal instability in a disk surrounding a neutron star of $1.4 M_\odot$ accreting at $\dot{M} = 0.14 \dot{M}_{Edd}$. Here, $\dot{M}_{Edd}$ is the Eddington value defined as $\dot{M}_{Edd} = \frac{4\pi GM}{\kappa_e c \epsilon}$ where ϵ is the efficiency for the conversion of rest mass energy into radiation taken to be 1/6 for a neutron star and 1/16 for a black hole and κ_e is the electron scattering opacity. We found that the disk luminosity manifests itself as periodic oscillations which have a period of ~ 21.5 s and an amplitude of $\frac{\Delta L}{L} \sim 67\%$ (see Fig. 1a). The unstable region of the disk is spatially confined inside $40 \ r_g$.

To see the sensitivity of the results to the mass accretion rate, the mass accretion rate was decreased to $\dot{M}/\dot{M}_{Edd} = 0.12$ and 0.125 in sequences 2 and 3, with all other input parameters held fixed as in the standard sequence. The disk is unstable as shown by the luminosity variations illustrated in Figures 1b and 1c. In sequence 3, there exist primary and secondary luminosity peaks, whereas in sequence 2 the fluctuations are manifested as oscillations. It is found that the period has decreased to 11 s and the relative luminosity amplitude has decreased to $\sim 20\%$ for sequence 2. Hence, the frequency of the oscillation increases whereas the relative luminosity fluctuation amplitude decreases as the mass accretion rate is decreased. This relation holds only over a limited range in mass accretion rates for which the disk is unstable. For higher mass accretion rates, the amplitude of the oscillations may also decrease when the disk becomes stable due to the saturation of the viscosity parameter, α. For a set of viscosity parameters $(n, \alpha_0, \alpha_{max}) \sim (1.6, 30, 0.2)$, it is found that the accretion disk is stable for $\dot{M} \lesssim 0.11 \dot{M}_{Edd}$ and for $\dot{M} \gtrsim 0.47 \dot{M}_{Edd}$.

The unstable behavior of the accretion disk is sensitive to the viscosity parameter index, n. For example, if we decrease n in the standard sequence to 1.1, the disk is found to be stable. This result does not confirm local analysis

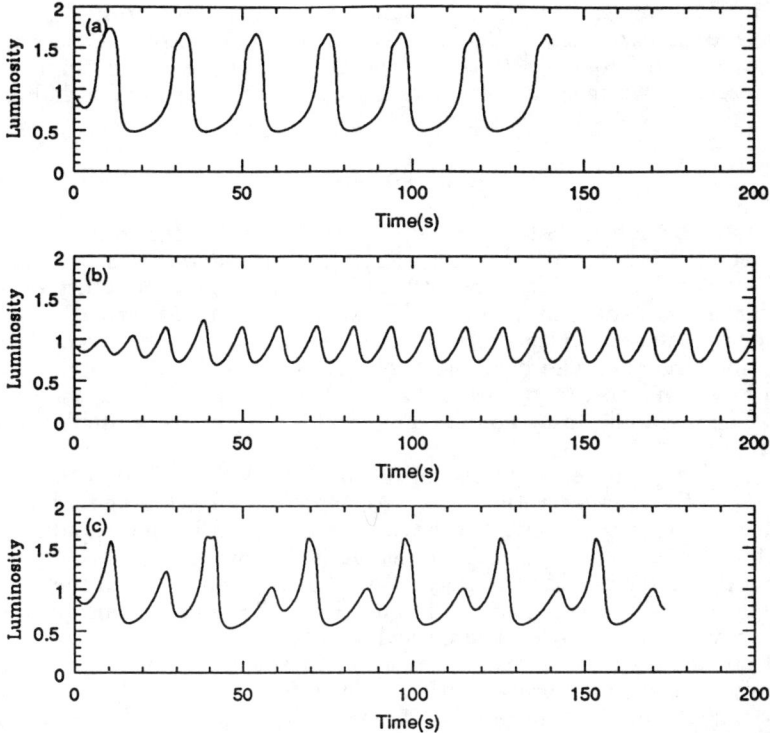

Fig. 1. The variation of the disk luminosity in terms of the steady state value as a function of time for an accretion disk surrounding a neutron star of $1.4 M_\odot$. The mass accretion rates in units of $\dot{M}_{Edd}$ are 0.14, 0.12 and 0.125 for plots (a), (b) and (c) respectively. Note that the period and amplitude of the oscillations decrease as the mass accretion rate is reduced.

which predicts instability for $n \gtrsim 0.75$ (Meyer 1986; MCT). The time-dependent nonlinear calculations reveal that stabilization for $n \lesssim 1.1$ results from the global effects of non-local energy transport (see also Taam & Lin 1984).

The properties of the oscillations are also sensitive to the viscosity parameters α_0 and α_{max}. For an increase in α_0, the surface density decreases and, for a given $\dot{M}$, radiation pressure becomes more important in the disk. Instability can, therefore, occur at lower mass accretion rates. Furthermore, the viscous time scales are correspondingly reduced and the frequencies of the oscillations are increased. The parameter α_{max} determines the range of mass accretion rates over which the disk is unstable. For smaller α_{max}, the range of mass accretion rates is smaller.

To determine the dependence of the oscillations on the mass of the compact object we increased M to $5\ M_\odot$ to model the evolution of an accretion disk surrounding a black hole. In order to produce oscillations at frequencies ~ 0.04 Hz, the viscosity parameter, α, must be larger than that of the neutron star case since the absolute size of the inner disk is larger. We keep $\alpha_0 = 30$ unchanged, but decrease n. The dependence of the results on the mass accretion rate is similar to the case for neutron stars. Specifically, for an increase in the mass accretion rate the oscillation frequency decreases and the amplitude of the luminosity variation increases. The mass accretion rates for which instability is indicated are $0.045 - 0.38\,\dot{M}_{Edd}$ for $n = 1.3$.

3. DISCUSSION

We have demonstrated by time-dependent calculations that the mass flow in a geometrically thin, optically thick disk surrounding either a neutron star or a black hole can be unstable. The instability is due to the inability of the disk to maintain a local thermal balance. The strength of this thermal instability is determined by the sensitivity of the viscous heating rate to the temperature, and is greater for larger n. The unstable behavior is found to lie in a range of mass accretion rates and also to be restricted to a relatively narrow spatial extent in the disk. The instability may be mild and results in luminosity oscillations rather than bursts.

It is a general property of these instabilities that the period of the oscillation is a function of the mass accretion rate, with the period decreasing for lower source intensity. Such behavior appears to have been observed in the Rapid Burster by Lubin et al. (1992) in the hump immediately following the post dip phase of some Type II bursts. In addition, they found that the amplitude of the oscillations ($\lesssim 60\%$) decreased as the persistent flux declined. This is also consistent with the general trends exhibited by our numerical results. Furthermore, the observations suggest that the level of persistent emission in the Rapid Burster corresponds very closely to the minimum accretion rate necessary for instability, since the oscillations disappear as the persistent emission decreases. If we identify the persistent luminosity level to correspond to $\dot{M} \sim 0.15 \dot{M}_{Edd}$, the low frequency (0.04 Hz) and large amplitude of luminosity oscillations imply a large n (~ 1.6) and constrain α_0 to be ~ 30. The limit on α may be constrained by the absence of these oscillations during the bump phase observed on the decline from a Type II burst. This suggests that $\alpha_{max} \lesssim 0.2$.

Finally, the low frequency (~ 0.04 Hz) oscillations from the black hole candidate sources may suggest that $n \sim 1.3$ and $\alpha_0 \sim 30$ for a $5 M_\odot$ black hole.

This research was supported in part by NASA under grant NAGW-2526.

REFERENCES

Abramowicz, M. A., Szuszkiewicz, E., & Wallinder, F. 1989, in Theory of Accretion Disks, ed. F. Meyer, et al. (Dordrecht: Kluwer), 141
Lightman, A. P., & Eardley, D. N. 1974, ApJ, 187, L1
Lubin, L. M., et al. 1992, MNRAS, 258, 759
Meyer, F. 1986, in Radiation Hydrodynamics in Stars and Compact Objects, ed. D. Mihalas & K. -H. A. Winkler (Berlin: Springer-Verlag), 249
Milsom, J. A., Chen, X., & Taam, R. E. 1994, ApJ, 421, in press (MCT)
Pietsch, W. et al. 1993, A&A, 273, L11
Shakura, N. I., & Sunyaev, R. A. 1973, A&A, 24, 337
Taam, R. E., & Lin, D. N. C. 1984, ApJ, 287, 761
van der Klis, M. 1989, ARA&A, 27, 517
Vikhlinin, A., et al. 1993, preprint

ON TURBULENT VISCOSITY IN THIN ACCRETION DISKS

G. Lee
Center for Astrophysical Sciences
Department of Physics and Astronomy
The Johns Hopkins University
Baltimore, MD 21218
lgh@hut4.pha.jhu.edu

ABSTRACT

We estimate the Shakura-Sunyaev α-parameter using the turbulent viscosity model of Canuto et al. (1988). We assume that the over-reflection instability in a differentially rotating accretion disk of the shearing sheet (Narayan et al. 1987) can develop hydrodynamic turbulence. Given a growth rate of the over-reflection instability we can calculate the turbulent viscosity ν_T which is a function of the growth rate. We estimate the α-parameter using the relation $\nu_T = \alpha v_s H$ where v_s is the local sound speed and H is a scale height of the disk. We find $\alpha < 0.6 \times 10^{-2}$ which is similar to the result of Canuto et al. (1988: $\alpha < 10^{-2}$) based on turbulent convective instability.

1. INTRODUCTION

Accreting binary systems containing a neutron star or a black hole are usually defined as X-ray binaries. Formation of an accretion disk around the compact object in which matter slowly spirals inward is proposed to play an important role in the evolution of the X-ray binary systems. The main problem of the accretion disk model is that of removing angular momentum which make the matter drift toward the central object. Molecular viscosity is too weak to cause a timely diffusion of the gas, and an turbulent viscosity is invoked to perturb the otherwise stable Keplerian motion. Shakura and Sunyaev (1973) introduced α parameter to calculate the turbulent viscosity based on purely dimensional analysis, and Stewart's (1976) calculation of the turbulent viscosity by a direct modeling of Reynolds stress $\tau_{ij} = - <v_i v_j> = \nu_T S_{ij}$, where v is the velocity fluctuation, $<...>$ is the ensemble average, and S_{ij} is the mean shear, was not successful. In this paper we assume that an acoustic instability driven by supersonic shear in the Keplerian disk (over-reflection instability: Narayan et al. 1987) can make transition to turbulent flow. Using the linear growth rate of the over-reflection instability, we then calculate the turbulent viscosity from the turbulence model of Canuto et al. (1988).

2. OVER-REFLECTION INSTABILITY

Laminar-turbulent flow transition is caused by some instabilities in the fluid. The nature of the flow in the Keplerian disk (strongly shearing medium with a large Reynolds number) leads one to conclude that the accretion flow must be turbulent. A possible candidate for the instability in the thin disk is the resonant amplification of acoustic waves between the corotation radius (at which wavelike perturbation corotate with the flow) and the reflecting boundary (which provide feedback to the corotation amplifier), known as the over-reflection instability [1].

---

[1] Since the Keplerian flow in thin disks satisfy the Rayleigh criterion for stability of rotating fluid, the flow is unstable only for nonaxisymmetric global instability.

Narayan et al. (1987) did a detail study of the over-reflection instability in the accretion disk using the shearing sheet approximation of the two dimensional Euler equation [2]. Using the normal modes analysis of the linear perturbation theory, they found that the governing equation for the velocity fluctuation is given by the Parabolic Cylinder differential equation (Abramowitz and Stegun 1972). Some important results from their solution of the eigenvalue problem are as follows. 1. Each mode has an associated corotation radius and a forbidden region (where the wave-like disturbances have exponential variation) around corotation. 2. Waves have an associated negative perturbed angular momentum (conserved action) at radii less than the corotation radius and positive outside corotation. 3. Because of the sign change of the conserved action a wave incident from the inner region produces a transmitted wave through the corotation barrier (forbidden region) and a reflected wave with increased amplitude (so called corotation amplifier). 4. Unstable models are obtained when the corotation radius is within the fluid and when there is an inner reflecting wall which can form a resonant cavity (but requires a fine tuning of phase relations). In the case that the shearing sheet is terminated by a wall in one side of the corotation radius and extends to infinity in the other side, all modes are growing mode. The growth rate is given by (Narayan et al. 1987)

$$\Gamma_k = \frac{v_s}{d}\ln(1 + \tau^2) ,\qquad(1)$$

where d is the distance between corotation, and the WKB tunneling probability τ is given by

$$\tau = \exp[-(2\pi/3)(Xk + \frac{1}{Xk})] ,\qquad(2)$$

where $X = v_s/\Omega$ and k is the wave number.

3. TURBULENT VISCOSITY

Analytic modeling for fluid turbulence is very difficult. There is no successful model for compressible turbulence (see Lesieur 1990 p72), thus most of the studies of compressible turbulence were done by numerical simulations (Sarkar et al. 1991 and references therein). Theory of fully developed homogeneous turbulence (spatial and temporal chaos) is one of a few successful models for incompressible fluid (Batchelor 1973). Heisenberg-Kolmogoroff (HK) model for small-scale (isotropic) turbulence (SST) was the most famous theory for incompressible turbulence. Canuto et al. (1988), however, pointed out that the extension of HK model to large-scale turbulence (LST), which may be responsible for turbulence in most astrophysical flow, leads to unphysical results. They solved the closure problem (which means how to calculate the transfer of energy absorbed by a given group of eddies, which represent irregular motion of the fluctuating part of the velocity, to all other eddies via nonlinear interactions) by calculating a correlation time scale subject to the constraint that the energy spectral function must reproduce the HK spectrum ($\sim k^{-5/3}$) in the inertial range. They then model the turbulent viscosity completely in terms of the energy spectral function. The energy spectral function is a function of the linear growth rate of the instability

---

[2]They made a local approximation in the neighborhood of a point (r_o, Ωt) where Ω is the Keplerian angular speed at the radius r_o. Thus the flow in the comoving frame is linear with a constant shear $-3\Omega/2$

that generate the turbulence (see Canuto et al. 1988 for details). The maximum value of the turbulent viscosity in their model is given by

$$\nu_T = \frac{\Gamma_{k_o}}{k_o^2}, \tag{3}$$

where the wave number k_o is defined so that the energy spectral function is zero, and can be calculated by

$$\frac{d}{dk}\left(\frac{\Gamma_k}{k^2}\right) = 0 \text{ at } k = k_o. \tag{4}$$

Attractive features in the model of Canuto et al. (1988) is that their simple calculations can be in good agreement with more sophisticated models such as in the Direct Interaction Approximation (Kraichnan 1987) and in the Renormalization Group Method (Yakhot and Orszag 1986).

4. RESULTS

From eq. (1) the linear growth rate is given by

$$\Gamma_k = \frac{v_s}{d}\tau^2, \tag{5}$$

for small τ. This growth rate in eq. (4) gives $k_o = 0.6\Omega/v_s$. We can then calculate ν_T from eq. (3) and eq. (5). Finally using the relation $\nu_T = \alpha v_s H$ with the isothermal sound speed $v_s = \Omega H$, we have

$$\alpha = 0.6 \times 10^{-2}\frac{H}{d}. \tag{6}$$

Since d is in the order of magnitude of H

$$\alpha < 0.6 \times 10^{-2}, \tag{7}$$

which is similar to the result of Canuto et al. (1988: $\alpha < 10^{-2}$) based on turbulent convective instability model.

5. DISCUSSION

In this paper we calculate the Shakura-Sunyaev α-parameter based on crucial assumption that the over-reflection instability can develop hydrodynamic turbulence in accretion disks. Kaisig (1989) did two dimensional simulations of the nonlinear evolution of the over-reflection instability in accretion disks. The simulations showed no transition to turbulence. However, possibility of transition to turbulence in 3-dimensional flow could not be excluded. Unstable modes in the nonaxisymmetric convection model (Ryu and Goodman 1992) may be one possibility. It may be possible that even for the two dimensional case, 50×50 grid points in second-order finite difference method of Kaisig (1989) could not resolve the subgrid-scale turbulence. To verify this, further investigations using high-resolution numerical schemes such as spectral methods (Marcus 1986) or sixth-order modified Pade scheme (Lee 1992) may be desirable [3]. For further investigations of possible angular momentum transfer via hydrodynamic turbulence, extension of the simulations of Dolez and Leorat (1991), in which set the Reynolds number of 160 (smallist critical number for instability at Mach number of 4.9) for recently found sonic and/or viscous instabilities in the supersonic plane Couette flow (Glatzel 1989), to larger Reynolds numbers are also highly encouraging. Recently many people investigated a possible angular momentum transfer via magnetic viscosity (for example, see Hawley and Balbus 1991). One reason for this activity may be due to current difficulty with pure hydrodynamic turbulence simulation [4]. However, it may be true that the magnetic field effects in astrophysical fluid are unavoidable, we should not discard to investigate possibilities in pure hydrodynamic processes until they are proved to be not significant.

ACKNOWLEDGEMENTS

The author thanks W.-L. Chen for helpful comments.

REFERENCES

Abramowitz, M. and Stegun, I.E. 1972, Handbook of Mathematical Functions
    (Washington D.C. : US Gov. Printing Office)
Batchelor, G.K. 1973, The Theory of Homogeneous Turbulence
    (Cambridge : Cambridge Univ. Press)
Canuto, V.M., Goldman, I., and Chasnov, J. 1988, A&A, 200, 291
Dolez, N. and Leorat, J. 1991, in Turbulence 89
    eds. O. Metais and M. Lesieur (Dordrecht : Kluwer Academic Pub.)
Glatzel, W. 1989, J. Fluid Mech., 202, 515
Hawley, J.F., and Balbus, S.A. 1991, ApJ, 376, 223
Kaisig, M. 1989, A&A, 218, 89
Karniadakis, G.E. and Orszag, S.A. 1993, Phys. Today, 3, 34
Kraichnan, R.H. 1987, Phys. Fluids, 30, 2400
Lee, S. 1992, Ph.D. dissertation, Stanford University
Lesieur, M. 1990, Turbulence in Fluids
    (Dordrecht : Kluwer Academic Pub.)
Marcus, P.S. 1986, in Astrophysical Radiation Hydrodynamics,
    eds. K.A. Winkler and M.L. Norman (Boston : D.Reidel Pub. Co.)
Narayan, R., Goldreich, P., and Goodman, J. 1987, MNRAS, 228, 1
Pringle, J.E. 1981, ARA&A, 19, 137
Reynolds, W.C. 1990, in Whither Turbulence ?,
    ed. J.L. Lumley (Berlin : Springer-Verlag)
Ryu, D. and Goodman, J. 1992, ApJ, 388, 438
Sarkar, S., Erlebacher, G., and Hussaini, M.Y. 1991,
    Theor. Comp. Fluid Dynamics, 2, 291
Shakura, N.I. and Sunyaev, R.A. 1973, A&A, 24, 337
Stewart, J.M. 1976, A&A, 49, 39
Yakhot, V. and Orszag, S.A. 1986, J. Sci. Comp., 1, 3

---

[3] The smallest scales that we must resolve in each direction is $\sim 0.15 Re^{3/4}$ where Re is the Reynolds number for LST (Reynolds 1990), and $Re \sim 10^3$ for accretion flow (Pringle 1981).

[4] It requires at least 10^{12} floating-point operations per second computer which is about 10^3 times higher than the capability of Cray-Y/MP, and the present massively parallel process cannot reach the required the speed (Karniadakis and Orszag 1993).

THE VERTICAL STRUCTURE AND STABILITY OF ACCRETION DISKS SURROUNDING BLACK HOLES AND NEUTRON STARS

John A. Milsom, Xingming Chen, and Ronald E. Taam
Department of Physics and Astronomy
Northwestern University, Evanston, IL 60208

ABSTRACT

The structure and stability of the inner regions of accretion disks surrounding neutron stars and black holes have been studied. Within the framework of the α viscosity prescription for optically thick disks, we assume the viscous stress scales with gas pressure only, and the α parameter, which is less than or equal to unity, is formulated as $\alpha_0 (h/r)^n$, where h is the local scale height and n and α_0 are constants. We neglect advective energy transport associated with radial motions and construct the vertical disk structures by assuming a Keplerian rotation law and local hydrostatic and thermal equilibrium.

The vertical structures have been calculated with and without convective energy transport, and it is found that convection is important especially for mass accretion rates greater than about 0.1 times the Eddington value, $\dot{M}_{Edd}$. Although convective efficiency is low, convection does help to stabilize the disk.

The results show that the disk can be locally unstable and that for $n \gtrsim 0.75$, an S-shaped relation can exist between $\dot{M}$ and the column density, Σ, at a given radius. The upper stable branch exists because of the saturation of α and the disk can be stabilized by this saturation for $\dot{M} \lesssim \dot{M}_{Edd}$.

1. INTRODUCTION

Accretion disks have been applied to explain many astronomical phenomena. In standard theoretical models, the viscosity is responsible for energy generation and angular momentum transfer. Since no fundamental theory for viscosity exists, two phenomenological viscosity prescriptions have been suggested. One takes the form $\tau = -\alpha p$ (Shakura & Sunyaev 1973) and the other is $\tau = -\alpha p_g$ (Lightman & Eardley 1974), where τ, p and p_g are the viscous stress, the total pressure and the gas pressure respectively, and α is a constant. These two prescriptions are essentially the same if the disk is gas pressure dominated. However, for the high temperature disks surrounding neutron stars or black holes, the pressure can be radiation dominated in the inner regions. In such cases, disks calculated with $\tau = -\alpha p_g$ are always thermally and viscously stable for an electron scattering opacity (Lightman & Eardley 1974) while disks with the $\tau = -\alpha p$ viscosity are thermally and viscously unstable (Lightman & Eardley 1974), and periodic burst-like fluctuations of the disk luminosity can occur (Taam & Lin 1984).

Constraints on the α parameter have been derived from observations of dwarf novae outbursts and x-ray novae. To permit understanding of the recurrence time and the duration of the outbursts, it was necessary to relax the assumption that the α parameter remain constant. Specifically, α needed to be larger during outbursts. A widely applied formula is $\alpha = \alpha_0 (h/r)^n$ where h is the local scale height of the disk, r is the radial distance from the star, and the power law index, n, is expected to lie between 1 and 1.5. In the thin disk approximation for a disk with viscous stress scaling as gas pressure and this variable α law, the disk is locally unstable when radiation pressure dominates gas pressure for $n > 0.75$ (Meyer 1986).

Observational evidence may exist for these instabilities. Several black hole candidates have shown small amplitude low frequency quasi-periodic oscillations (QPO's). For example, observations of Cyg X-1 have shown 0.04 Hz oscillations (Angelini & White 1992), while the source GRO J0422+32 has shown oscillations at 0.28 Hz and 0.035 Hz (Vikhlinin et al. 1992). Since observational support for large amplitude outbursts is lacking, the natural choice is to try and model these oscillations using $\tau = -\alpha p_g$ with $\alpha = \alpha_0 (h/r)^n$. Thus, we investigate the circumstances under which instabilities exist in accretion disks surrounding neutron stars and black holes. In particular, we report on calculations of the detailed vertical structure of such disks and discuss their stability.

2. RESULTS

We have calculated vertical structures for accretion disks surrounding compact objects for stellar masses ranging from $1.4 - 10 M_\odot$, mass accretion rates ranging from $0.001 - 5 \dot{M}_{Edd}$, and radii from $5 - 350 r_g$ where r_g is the Schwarzschild radius. Fully radiative structures as well as radiative-convective structures have been studied to determine the sensitivity of the results to the mode of energy transport. For our standard case, we use a black hole mass of $10 M_\odot$ and a viscosity prescription with $n = 1.1$ and $\alpha_0 = 16$.

The vertical structures of these accretion disks admit solutions which exhibit an S-shaped curve in the $\dot{M}$-Σ plane at a given radius (see Fig. 1). Such shapes are characteristic of both radiative and convective models. At low values of $\dot{M}$ gas pressure dominates and the disk is viscously and thermally stable. Energy is transported by radiative diffusion. As $\dot{M}$ increases the radiative contribution to pressure increases while the adiabatic temperature gradients decrease. Eventually, part of the flux is carried by convection. At sufficiently high $\dot{M}$ the disk becomes locally unstable for both models. The convective model is stable until a higher $\dot{M}$ than the radiative model illustrating the stabilizing effects associated with convective energy transfer. At a still higher $\dot{M}$ each model stabilizes again. This occurs because of the saturation of the viscosity parameter α to unity. The stability caused by α saturation is always present in our calculations and can occur for $\dot{M} \lesssim \dot{M}_{Edd}$.

Although convection has significantly affected the solutions in the $\dot{M}$-Σ plane, it is not a highly efficient energy transport process. Convection carries, at most, 40-80% of the flux. The maximum convective velocities are $\lesssim 0.1 C_s$ (where C_s is the adiabatic sound speed) so turbulent pressures are unimportant.

The actual vertical structures for the radiative and convective models naturally depend on $\dot{M}$, M and the parameters describing the viscosity. For the specific case of the models in Fig. 1 at $\dot{M} \sim 0.09 \dot{M}_{Edd}$, the temperature profiles are isothermal close to the midplane while the flux is linearly proportional to height close to the midplane and relatively flat near the surface. Each model has a density inversion since the effective gravity is directed upward in part of the disk. The average density is higher in the convective model since Σ is larger, and the density inversion is smaller because the pressure gradients are smaller. These features are common to most models.

In the following, we focus on the convective models only. First, we discuss

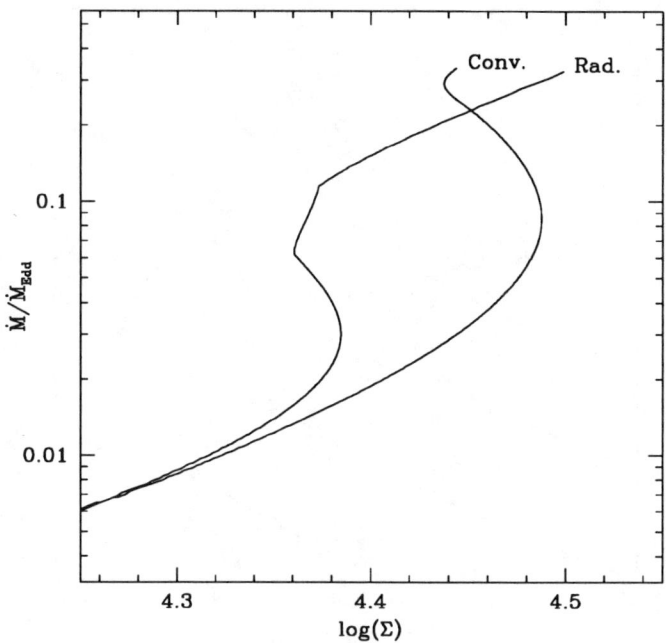

Fig. 1. The relation between mass accretion rate, $\dot{M}$, in units of the Eddington value, $\dot{M}_{Edd}$, and total column density of fully radiative and convective-radiative models for the disk around a $10 M_\odot$ black hole at a distance of $10 r_g$. The viscosity prescription is described by $n = 1.1$ and $\alpha_0 = 16$. Note that, in both cases, there is an upper stable branch which is due to the saturation of the α viscosity parameter.

the sensitivity of our results to the magnitude of the viscosity parameters α_0 and n. These parameters directly affect the energy generation rate. As α_0 is increased, the energy generation rate increases. Our results indicate that (for a fixed n) Σ and the $\dot{M}$ necessary for saturation both decrease when α_0 is increased. For a given n, an S-curve will exist only for a range in α_0 since at high enough α_0 saturation will occur before the instability occurs. An increase in n (at constant α_0) results in an increase in Σ since the energy generation rate has decreased in that case. The instability width (in $\Delta \log(\Sigma)$ or $\Delta \log(\dot{M})$) also increases with n. Our calculations reveal that disks with $n \leq 0.75$ are locally stable while disks with $n > 0.75$ are locally unstable in accord with analytic results from Meyer (1986) and Milsom, Chen & Taam (1994).

Next, we discuss the dependence of our results on the mass of the compact object. The shape of the S-curves is very similar. The major difference is to shift the S-curves to larger Σ with increasing compact object mass. Another difference is that the S-curve for a neutron star is shifted vertically (in terms of $\dot{M}/\dot{M}_{Edd}$) from the S-curve for a black hole reflecting the difference in accretion efficiency between neutron stars and black holes.

The frequencies of any observable oscillations depend on the size of the unstable region. Fig. 2 shows the S-curves for different radii for a $10 M_\odot$ black

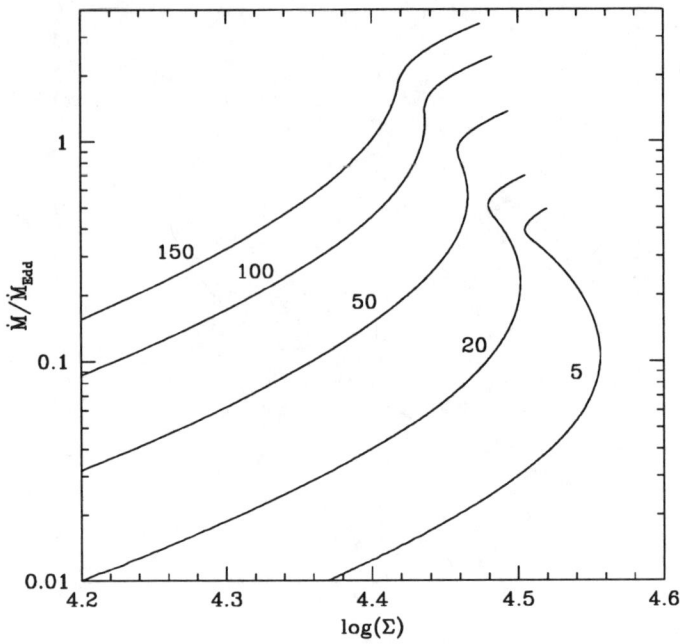

Fig. 2. The variation of $\dot{M}$ with Σ for a range of radii in the disk. A viscosity prescription corresponding to $\alpha_0=10$ and $n = 1$ and a 10 $M_\odot$ black hole was assumed. The numbers on each curve correspond to the radius in the disk (r/r_g). For larger radii the disk is stable.

hole and a viscosity prescription described by $n=1$ and $\alpha_0=10$. At larger radii, the disk becomes less unstable. It is easy to see which radii are unstable for a given $\dot{M}$. Note, however, that this is only a local analysis. A global analysis (see Chen & Taam these proceedings) is needed to determine the nonlinear time dependent behavior of the system. Based upon the small amplitude QPO's seen in black hole candidate systems, our calculations suggest a viscosity prescription with small n and large α_0 in a disk instability interpretation of those QPO's.

This research was supported in part by NASA under grants NAGW-2526, NAGW-2935, and by the NASA National Space Grant College and Fellowship Program.

REFERENCES

Angelini, L., & White, N. 1992, IAU Circ. 5580
Lightman, A. P., & Eardley, D. N. 1974, ApJ, 187, L1
Meyer, F. 1986, in Radiation Hydrodynamics in Stars and Compact Objects, ed. D. Mihalas & K. -H. A. Winkler (Berlin: Springer-Verlag), 249
Milsom, J. A., Chen, X., & Taam, R. E. 1994, ApJ, 421, in press
Taam, R. E., & Lin, D. N. C. 1984, ApJ, 287, 761
Shakura, N. I., & Sunyaev, R. A. 1973, A&A, 24, 337
Vikhlinin, A. et al. 1992, IAU Circ. 5608

THE ULTRAVIOLET SPECTRUM OF 2S 0921-630

Paul C. Schmidtke
Department of Physics and Astronomy, Arizona State University
Box 871504, Tempe, AZ 85287-1504
schmidtke@scorpius.la.asu.edu

ABSTRACT

An ultraviolet spectrum of the long-period low-mass X-ray binary 2S 0921-630 was obtained with IUE near the time of partial eclipse of the accretion disk by its G-giant companion star. The short-wavelength spectrum shows a very weak continuum, with emission lines of N V, Si IV, C IV, and He IV. The strength of these lines during eclipse implies the volume of the line-emitting region must be larger than that which is eclipsed in the optical light curve. The UV spectrum is similar to that of Cyg X-2 (when it is on the X-ray flaring branch of X-ray color-color plots) although the weakness of N V may indicate a lower abundance for that element. Alternatively, the apparent absence of N III and N IV in 2S 0921-630 suggests the upper portion of the accretion disk in 2S 0921-630 may be hotter than that in Cyg X-2.

1. INTRODUCTION

2S 0921-630, optically identified by Li et al. (1978), is a low-mass X-ray binary with an exceptionally long (~9-day) orbital period (Cowley et al. 1982). The optical spectrum is composite, consisting of a G giant that is veiled by light from an accretion disk. Broadband photometry shows a slow decline to minimum followed by a rapid rise to maximum light (e.g., Chevalier & Ilovaisky 1981; Krzeminski & Kubiac 1991), with a deepening of partial eclipse at shorter wavelengths. The light curve varies from cycle to cycle, implying the shape and extent of the disk must change with time.

Mason et al. (1987) found a recurrent X-ray eclipse in $EXOSAT$ data. The apparent stability of this feature (in contrast to the optical behavior) can be explained by the system's orbital inclination. The low value of L_X/L_{opt}=1.6 (Bradt & McClintock 1983) implies the inclination must be close to 90° and the X-ray engine of 2S 0921-630 is hidden from direct view. Hence, the X-rays that we receive are scattered into our line-of-sight by an accretion disk corona. Since the detected X-rays originate in scattering processes in the corona, the X-ray eclipse is more stable than the one present in optical bandpasses, which is subject to short timescale changes in disk geometry. Therefore, to obtain a better understanding of 2S 0921-630, its accretion disk must be examined in more detail. Since the disk radiates primarily in the ultraviolet, IUE observations can provide valuable information on this system's properties.

2. OBSERVATIONS

A low-dispersion, short-wavelength IUE spectrum of 2S 0921-630 was obtained on 1992 July 14, near the predicted time of partial eclipse of the accretion

© 1994 American Institute of Physics

disk by the G-giant star. Using the ephemeris of Mason et al. (1987), the 400-minute exposure spans photometric phases 0.024-0.055, with phase zero corresponding to inferior conjunction of the mass-losing star. The spectrum, shown in Fig. 1, has a very weak continuum and emission lines of N V (1240Å), Si IV (1400Å), C IV (1550Å), and He IV (1640Å).

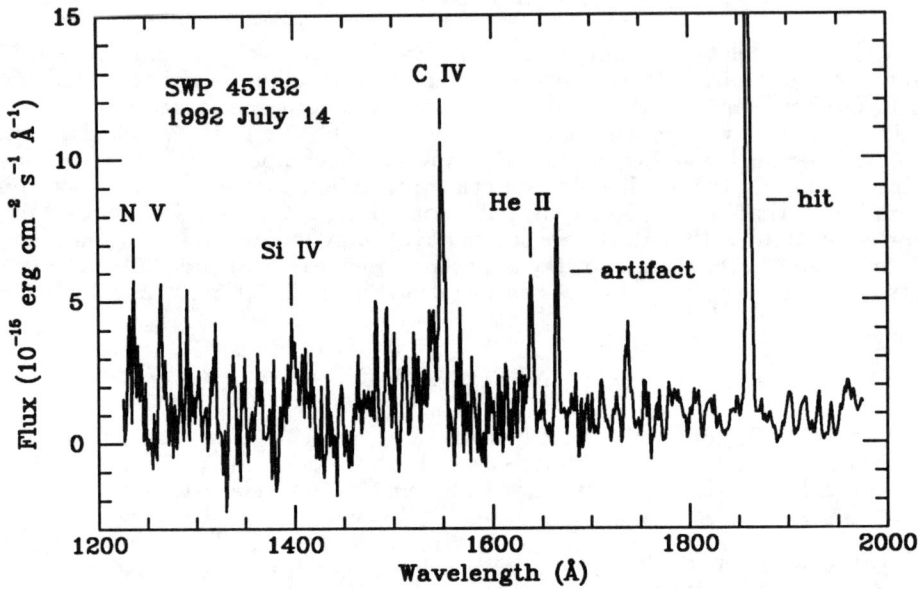

Fig. 1. The Ultraviolet Spectrum of 2S 0921-630.

The equivalent widths of the emission lines are listed in Table 1. They are difficult to measure due to uncertainities in determining the continuum and errors introduced by 'fixed pattern' noise of the detector. Average values of the equivalent widths for the same lines in Sco X-1 (Kallman et al. 1991) are included in the table for comparison. The presence of such strong UV emission lines in 2S 0921-630 during eclipse suggests the volume of the line-emitting region is larger than that which is eclipsed in the optical light curve - although phase-related variations in emission-line strengths or equivalent widths have yet to be measured.

Table 1. Equivalent Widths of Ultraviolet Lines

Source	N V (1240 Å) (Å)	Si IV (1400 Å) (Å)	C IV (1550 Å) (Å)	He IV (1640 Å) (Å)
2S 0921-630	30	10	43–59[a]	15
Sco X-1	8.15	6.3	26.6	3.9

[a] The value depends on the exclusion/inclusion of flux in the violet wing of the profile, see Fig. 1.

The energy distribution of the continuum can be modeled by a blackbody with a temperature of $\sim 28,000$K, assuming $E(B-V)=0.3$. The data also can be fitted using a power-law function, but the low signal-to-noise ratio does not allow us to discriminate between these models. Therefore, at present we cannot say with certainty if the UV flux comes directly from the tail of the X-ray energy distribution or arises from the absorption and redistribution of X-rays within the accretion disk.

3. DISCUSSION

The closest analog to 2S 0921-630 with a history of UV observations is Cyg X-2 (e.g., Chiappetti et al. (1983); McClintock et al. (1984)). The most extensive work comes from a multiwavelength observing campaign, for which *IUE* results have been presented by Vrtilek et al. (1990). They found a direct relationship between the strength of UV line emission (as well as continuum flux) and the system's position in X-ray color-color or hardness-intensity plots, which form Z-shaped patterns. The strongest emission lines are present when Cyg X-2 is on the flaring branch while the weakest occur during the horizontal branch. Although we make no prediction that 2S 0921-630 is a Z source, we note that the observed *IUE* spectrum of 2S 0921-630 is most similar to the spectrum of Cyg X-2 on the flaring branch (see Fig. 3 of Vrtilek et al. (1990)). An exception is the relative weakness of nitrogen lines in 2S 0921-630. Only N V is prominent, albeit at a reduced level compared with Cyg X-2. Whether or not this is a true abundance effect is difficult to resolve. Possibly, the difference is related to differing geometries: 2S 0921-630 eclipses while Cyg X-2 does not. Therefore, the UV spectrum of 2S 0921-630 taken during eclipse may be preferentially sampling a region much further above the accretion disk than that in the spectrum of Cyg X-2. If this is correct, then upper portions of the disk in 2S 0921-630 also may be hotter to account for the suppression of N III and N IV. Again, phase-resolved *IUE* observations may resolve this problem.

ACKNOWLEDGEMENTS

This investigation has been supported by NASA grant NAG5-2101.

REFERENCES

Bradt, H.V.D., & McClintock, J.E. 1983, ARA&A, 21, 13
Chevalier, C., & Ilovaisky, S.A. 1981, A&A, 94, L3
Chiappetti, L., Maraschi, L., Tanzi, E.G., & Treves, A. 1983, ApJ, 265, 354
Cowley, A.P., Crampton, D., & Hutchings, J.B. 1982, ApJ, 256, 605
Kallman, T.R., Raymond, J.C., & Vrtilek, S.D. 1991, ApJ, 370, 717
Krzeminski, W., & Kubiak, M. 1991, Acta Astro., 41, 117
Li, F.K., van Paradijs, J.A., Clark, G.W., Jernigan, J.G, Laustsen, S., & Zuiderwijk, E.J. 1978, Nature, 276, 799
Mason, K.O., Branduardi-Raymont, G., Córdova, F.A., & Corbet, R.H.D. 1987, MNRAS, 226, 423
McClintock, J.E., Petro, L.D., Hammerschlag-Hensberge, C.R., Proffitt, C.R., & Remillard, R.A. 1984, ApJ, 283, 794
Vrtilek, S.D., Raymond, J.C., Garcia, M.R., Verbunt, F., Hasinger, G., & Kürster, M. 1990, A&A, 235, 162

Binary
Interactions

OBSERVATIONS OF ACCRETING PULSARS

Thomas A. Prince, Lars Bildsten, and Deepto Chakrabarty
Division of Physics, Mathematics, and Astronomy
California Institute of Technology
prince@caltech.edu

Robert B. Wilson
NASA/Marshall Space Flight Center - ES66

Mark H. Finger
Compton Observatory Science Support Center

ABSTRACT

We discuss recent observations of accreting binary pulsars with the all-sky BATSE instrument on the *Compton Gamma Ray Observatory*. BATSE has detected and studied nearly half of the known accreting pulsar systems. Continuous timing studies over a two-year period have yielded accurate orbital parameters for 9 of these systems, as well as new insights into long-term accretion torque histories.

1. INTRODUCTION

There are over 30 known accreting pulsar systems. These contain a rotating high-magnetic-field (B $\geq 10^{11}$G) neutron star in orbit with a stellar companion which transfers mass to the neutron star via Roche-lobe overflow or a stellar wind. The gravitational energy released by mass accretion yields thermal and non-thermal radiation, most prominently at X-ray and gamma-ray energies. Neutron stars in accreting pulsar systems are often called "X-ray pulsars" or "accretion-powered" pulsars. An earlier comprehensive review of accreting pulsars is that of Nagase (1989).

In this paper, we discuss recent observational results on 14 of the accreting pulsar systems, using data from the Burst and Transient Source Experiment (BATSE) on the *Compton Gamma Ray Observatory* (*GRO*). We list these systems in Table 1, together with current information on position, pulse period (P_{spin}), orbital period (P_{orb}), and stellar type of the companion. Figure 1 indicates the location of the BATSE-studied pulsars and other pulsars in the ($P_{spin} - P_{orb}$) plane, commonly known as the "Corbet-diagram" (see Corbet 1986; Stella, White, and Rosner 1986; and Waters and van Kerkwijk 1989). The three types of high-mass (≥ 3 M$_\odot$) systems (wind-fed, disk-fed, and Be) populate relatively distinct regions of the diagram. The continuous BATSE spin-period measurements allow tests of models of the long-term spin evolution of pulsars in all three classes of systems.

TABLE 1
ACCRETING PULSARS OBSERVED WITH BATSE AS OF 1993 DECEMBER

System	RA (2000) [hh mm ss.s]	Dec. (2000) [° ′ ″]	Period[a] Pulse [s]	Period[a] Orbital [d]	Companion (MK Type)
Low-mass systems					
Her X-1	16 57 49.7	+35 20 32	1.24	1.7	HZ Her (A9-B)
4U 1626-67	16 32 16.7	-67 27 42	7.66	0.02?	KZ TrA (low mass dwarf)
GX 1+4	17 32 02.1	-24 44 46	120	?	V2116 Oph?[b] (M6III)
High-mass supergiant systems					
Cen X-3	11 21 15.2	-60 37 24	4.8	2.09	V779 Cen (O6-8f)
OAO 1657-415	17 00 47.6	-41 39 14	37.7	10.4	? (B0-6Iab?)[c]
Vela X-1	09 02 06.8	-40 33 18	283	8.96	HD77581 (B0.5Ib)
4U 1538-52	15 42 23.3	-52 23 10	530	3.73	QV Nor (B0I)
GX 301-2	12 26 37.6	-62 46 13	681	41.5	Wray 977 (B1.5Ia)
Be-binary systems					
4U 0115+63	01 18 31.9	+63 44 24	3.6	24.31	V635 Cas (Be)
EXO 2030+375	20 32 15.3	+37 38 15	41.8	46.03	(Be)
A 0535+26	05 38 54.6	+26 18 57	105	110.58	HDE245770 (O9.7IIIe)
A 1118-616	11 20 57.2	-61 55 00	405	?	He 3-640 (O9.5III-Ve)
Systems with an undetermined companion					
GS 0834-430	08 35 55.1	-43 11 22	12.3	111.6	?
GRO J1008-57	10 09 46	-58 17 32	93.5	?	?

[a] For epoch TJD 8500.
[b] Suggested companion (Glass & Feast 1973; Davidsen et al. 1977; Chakrabarty et al. 1994)
[c] Companion not yet identified, but spectral type inferred from orbit and eclipse measurements (Chakrabarty et al. 1993).

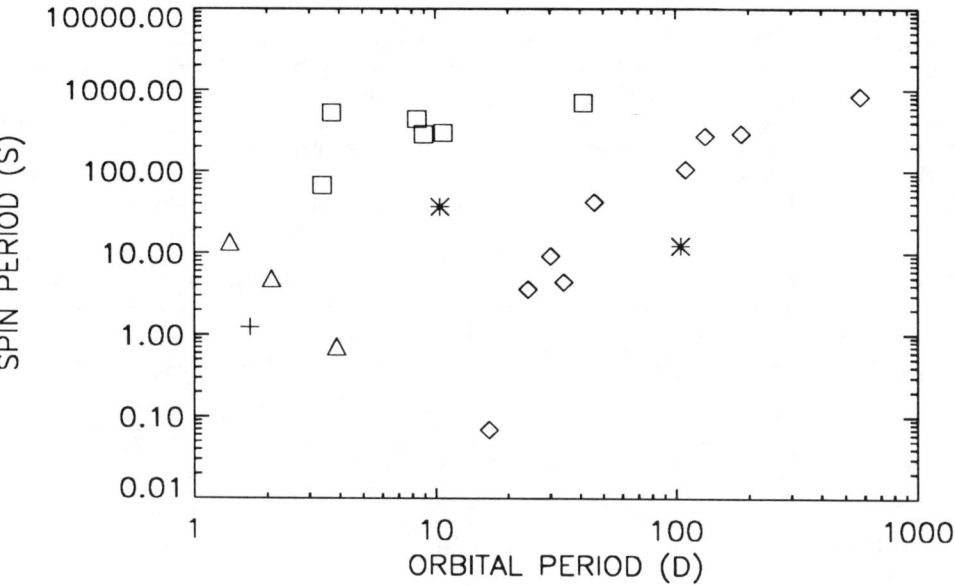

Figure 1. Corbet Diagram. Wind-fed systems are indicated by squares, high-mass disk-fed systems by triangles, Be-binaries by diamonds, low-mass systems by crosses, and systems with an unidentified stellar companion by asterisks

2. OBSERVATIONS

Observations of accreting binary pulsars have been carried out by BATSE since the launch of *GRO* in April 1991. The observations used the eight large area detectors (LADs) of the BATSE instrument, which have overlapping fields of view covering a total of 4π steradian. Each detector has an effective area of about 1500 cm^2 at 40 keV and an energy resolution of about 35% FWHM (Fishman et al. 1989). The LADs are sensitive from about 20 keV to 2 MeV and pulsar timing observations are primarily carried out from 20 keV to 60 keV. Several data types were used for pulsar analysis:

DISCLA data: Count rate samples of all 8 detectors at 1.024 s intervals in 4 energy channels.

CONT data: Count rate samples of all 8 detectors at 2.048 s intervals in 16 energy channels.

PSR data: Count rate samples folded into 64 phase bins with a programmable folding period and 16 energy channels (4 for periods shorter than about 20 ms).

DISCLA and CONT data have been used for timing and spectral studies of pulsars with pulse periods greater than about 2 s and 4 s respectively. PSR data have been used for study of fast accreting pulsars, in particular Her X-1. The DISCLA and CONT data provide continuous flux information on all pulsars not occulted by the earth, while the PSR data have typically been programmed to study specific pulsars with known pulse periods.

To detect a pulsar, the count rates from detectors with significant projected area towards the pulsar are weighted and summed. Systematic orbital background is subtracted either by filtering or by use of a background model. The power spectrum of the resulting residual counting rate is calculated and searched for significant peaks. Once a pulsar is detected, timing measurements are carried out in the usual fashion by constructing a pulse-phase or pulse-arrival time model which accounts for barycenter corrections, the orbit of the pulsar, and torque-induced spin-frequency changes. The best-fit model is used to determine the orbital parameters of the pulsar system.

The measured sensitivity of the BATSE detectors for pulsed flux detection is shown in Figure 2. Searches up to this time (Jan 1994) have used 1-4 day observation periods. The sensitivity is affected by earth-orbital noise for periods greater than about 200 s. The sensitivities for individual CONT energy channels are indicated in Figure 2, as well as sensitivities obtained by combining several energy channels. Typically, the lowest three CONT channels shown in Figure 2 are used for pulsar detection. Also shown for comparison is the approximate *pulsed* Crab flux; the total Crab flux is a factor of about 5-10 higher depending on energy.

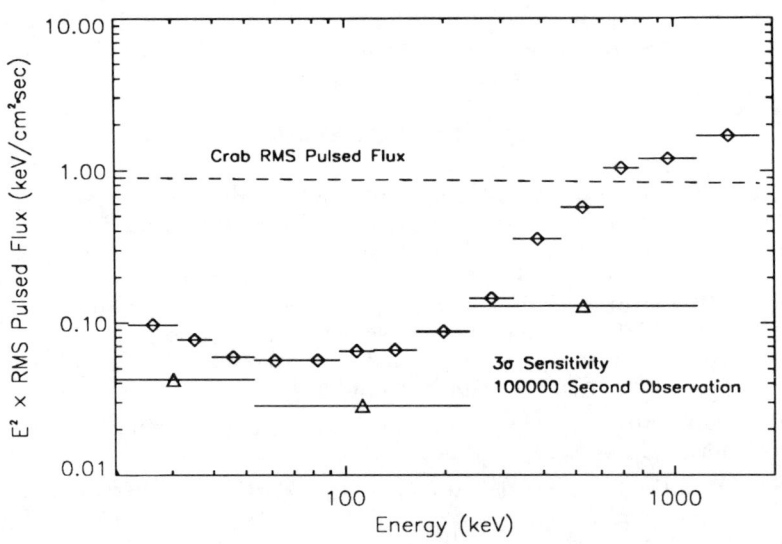

Figure 2. BATSE Pulsed-Flux Sensitivity.

3. RESULTS

Dynamical Studies

BATSE has determined the orbits of OAO 1657-415 and GS 0834-430 and has provided improved or additional orbital solutions for 7 other systems: 3 Be-binary systems (EXO 2030+375 A0535+26, and 4U 0115+63), 3 high-mass supergiant systems (Cen X-3, Vela X-1, and 4U 1538-52), and one low-mass system (Her X-1). In addition, one new accreting pulsar has been discovered, GRO J1008-57. Because many of these sources are transient in nature, the continuous monitoring capability of BATSE allowed timing measurements to be performed whenever the systems were active, thus providing the necessary coverage to determine or improve the orbital parameters.

Table 2 provides the current best-fit orbital parameters for 15 accreting pulsar systems. We have used BATSE data to derive orbital parameters for 9 of these systems. We discuss a few of these briefly below.

OAO 1657-415. This system was initially discovered as an X-ray source by the *Copernicus* satellite (Polidan et al. 1978). The source was observed by several other X-ray satellites and showed both spin-up and spin-down behavior but no definite indication of orbital modulation of the pulse period. Observations with BATSE (Chakrabarty et al. 1993) detected the 10.4 day orbital period of this system and determined that OAO 1657-415 was an eclipsing high-mass system. The optical companion has not yet been identified, but the spectral type is likely B0-6Iab, inferred from the orbit and eclipse measurements. Identification of the companion and measurement of its orbital Doppler curve would constrain the mass of the neutron star in this system.

GS 0834-430. This system was initially discovered in an outburst during February, 1990 by the WATCH detectors on the *Granat* spacecraft (Lapshov et al. 1992). Pulsations at 12.3 s were discovered by *Ginga* in November, 1990 and an accurate position determined (Aoki et al. 1992). A 114 day periodicity in the outbursts was later reported by *Granat*/WATCH, indicating a likely orbital period (Lapshov et al. 1992). BATSE observations were carried out using both occultation analysis for unpulsed flux, and standard timing analysis for the pulsed component and has been reported in Wilson et al. (1994a). BATSE has observed seven outbursts of GS 0834-430 since *GRO* began observations, allowing measurement of the orbital parameters of the system (see Table 2).

Accurate measurement of the orbital eccentricity is important, but is complicated by the uncertainty in the decoupling of the accretion and Doppler induced pulse-frequency changes. If the eccentricity is very close to zero, the companion is probably a 2-5 $M_\odot$ giant that has circularized the orbit due to dissipation in its envelope. On the other hand, if the eccentricity is small but finite (e.g. 0.1-0.2) the companion is almost certainly not a giant, but might possibly be a B or perhaps a Be star. An optical/IR companion has not yet been identified.

TABLE 2
Orbital Parameters of Accreting Pulsar Systems

	Orbital epoch (TJD)	P_{orb} (days)	$a_x \sin i$ (light seconds)	e	ω (degrees)	$f_x(M)$ (solar masses)	Refs.[a]
Low-mass system							
• Her X-1	8799.61235 ± 0.00001^c	$1.700167412 \pm (4.0 \times 10^{-8})^i$	13.1853 ± 0.0002	$< 1.3 \times 10^{-4 f}$	...	0.8517 ± 0.0001	1,2
High-mass supergiant systems							
LMC X-4	7741.9904 ± 0.0002^c	1.40839 ± 0.00001	26.31 ± 0.03	0.006 ± 0.002	...	9.86 ± 0.03	3
Cen X-3	$8561.656702 \pm (7.1 \times 10^{-5})^c$	$2.08706533 \pm (4.9 \times 10^{-7})$	39.627 ± 0.018	$< 1.6 \times 10^{-3 g}$	...	15.343 ± 0.021	4
4U 1538-52	5278.979 ± 0.020^c	3.72840 ± 0.00003^j	52.8 ± 1.8^h	...	...	11.4 ± 1.2	5,6,7
SMC X-1	7740.35906 ± 0.00003^c	$3.89229118 \pm (4.8 \times 10^{-7})^e$	53.4876 ± 0.0004	$< 0.00004^g$	...	10.8481 ± 0.0002	8
4U 1907+09	5575.465 ± 0.35^k	8.3745 ± 0.0042	80.2 ± 7.2	$0.16^{+0.14}_{-0.11}$	330^{+18}_{-56}	7.9 ± 2.1	9,10
• Vela X-1	8563.5364 ± 0.0033^c	8.964416 ± 0.000049^h	113.61 ± 0.30	0.0883 ± 0.0023	153.2 ± 1.7	19.60 ± 0.16	11,12
• OAO 1657-415[b]	8515.99 ± 0.05^c	10.4436 ± 0.0038	106.0 ± 0.05	0.104 ± 0.005	93 ± 5	11.7 ± 0.2	13
GX 301-2	3906.06 ± 0.16^d	41.508 ± 0.007	371.2 ± 3.3	0.472 ± 0.011	309.9 ± 2.6	31.9 ± 0.8	14
Be-binary systems							
4U 0115+63	8355.206 ± 0.004^d	24.309 ± 0.010^h	140.13 ± 0.08^h	0.3402 ± 0.0002^h	47.66 ± 0.03	5.00 ± 0.01	15,16
2S 1553-54	2596.67 ± 0.03^c	30.2 ± 0.1	162.7 ± 1.0	...	...	5.0 ± 0.1	17
V 0332+53	5651.5 ± 1^d	34.25 ± 0.10	48 ± 4	0.31 ± 0.03	313 ± 10	0.10 ± 0.02	18
EXO 2030+375	8798.2 ± 0.7^d	46.03 ± 0.01	268 ± 25	0.33 ± 0.03	228.2 ± 5.7	9.8 ± 2.7	19
A 0535+26	9058.7 ± 0.6^d	110.3 ± 0.3	267 ± 13	0.47 ± 0.02	130 ± 5	1.64 ± 0.23	20
System with undetermined companion							
• GS 0834-430	8591.70 ± 0.51^c	111.64 ± 0.18	205.7 ± 5.0	0.128 ± 0.063	275 ± 30	0.75 ± 0.05	21

Orbital elements for sources marked with bullets (•) have been measured with *GRO*/BATSE.

[a] References: (1) Deeter et al. 1991; (2) Wilson et al. 1994b; (3) Levine et al. 1991; (4) Finger et al. 1993; (5) Makishima et al. 1987; (6) Corbet et al. 1993; (7) Rubin et al. 1994; (8) Levine et al. 1993; (9) Makishima et al. 1984; (10) Cook & Page 1987; (11) Deeter et al. 1987; (12) Finger 1993; (13) Chakrabarty et al. 1993; (14) Sato et al. 1986; (15) Rappaport et al. 1978; (16) Cominsky et al. 1994; (17) Kelley et al. 1983; (18) Stella et al. 1985; (19) Stollberg et al. 1994; (20) Finger et al. 1994b; (21) Wilson et al. 1994a.
[b] Companion not yet identified, but inferred to be a B-supergiant from orbit and eclipse measurements (Chakrabarty et al. 1993).
[c] $T_{\pi/2}$ = epoch of 90° mean orbital longitude.
[d] T_{peri} = epoch of periastron passage.
[e] Epoch TJD 2836.18277 ± 0.00020. $\dot{P}_{orb}/P_{orb} = (-3.36 \pm 0.02) \times 10^{-6}$ yr^{-1}.
[f] 2σ upper limit.
[g] 3σ upper limit.
[h] This element held fixed at this value in fitting other elements.
[i] Orbital period for specified orbital epoch, computed using P_{orb} and $\dot{P}_{orb}$ from Deeter et al. 1991. Held fixed in fitting other elements.
[j] Orbital period for specified orbital epoch, computed using P_{orb} and $\dot{P}_{orb}$ from Rubin et al. (1994).
[k] Orbital epoch for longitude 309° ± 15°. Held fixed in fitting other elements.

EXO 2030+375. This system was discovered by *EXOSAT* in May, 1985 (Parmar *et al.* 1989). The orbital period was determined to be ∼ 46 days, with some ambiguity in the orbital determination due to the finite extent of the *EXOSAT* observations. BATSE has observed EXO 2030+375 in a dozen consecutive orbits, allowing a very precise orbit to be determined, particularly for orbital phases near periastron (see Stollberg *et al.* 1994 and Table 2).

A 0535+26. Although this system has been observed in outburst many times and at many wavelengths (for a review, see Giovannelli and Graziati, 1992), the orbit of the system has never been accurately determined. Recent activity in 1993 of the A 0535+26 system has allowed observations over three consecutive orbits with BATSE leading to the first definitive published orbital parameters (see Finger *et al.* 1994b and Table 2).

GRO J1008-57. This 93.5 s pulsar was discovered by BATSE on 14 July 1993, reached maximum intensity approximately 10 days later, and remained active for a total of about one month (Wilson *et al.* 1994c). The pulsar showed significant frequency evolution with spin-down and spin-up behavior roughly correlated with the hard X-ray flux increase and decrease respectively, similar to the Doppler-induced behavior observed in outbursts of EXO 2030+375. The spin evolution of the system, its transient nature, the fact that the luminosity is a significant fraction of the Eddington luminosity for a fiducial distance of 5 kpc, and the low galactic latitude of 1° are all strongly suggestive of a Be-binary pulsar. From the Corbet-diagram (Figure 1) a 93.5 s Be-binary pulsar might be expected to have an orbital period of 100-200 d. However, no additional emission has been detected from this source through the end of 1993.

Her X-1. This system is perhaps the best studied of all accreting binary pulsars. New results from BATSE (Wilson *et al.* 1994b) show a correlation between the hard X-ray flux at the peak of the main-on portion of the 35d cycle and the spin-frequency derivative, i.e. spin-down is correlated with low-flux as predicted by models such as those of Ghosh and Lamb (1979). Also, BATSE observations of the turn-on times of the 35d cycle indicate a possible correlation of early turn-on with decreased mass-transfer.

4U 1538-52. Analysis of this system using BATSE data has only recently begun. Preliminary results given by Rubin *et al.* (1994) show a secular spin-up trend after almost 10 years of spin-down behavior. Further BATSE analysis will yield a significantly improved orbital solution for this system.

Torque Studies

Accreting pulsars provide a laboratory for the studies of the torque exhibited during magnetic accretion, allowing for a comparison to existing theory (see Ghosh & Lamb 1979). The continuous monitoring capability of BATSE has allowed long-term studies of accretion torques in several systems. These include Her X-1, Vela X-1, Cen X-3, GX 1+4, 4U 1626-67, and OAO 1657-415. The frequency histories for these sources for a period of over two years are shown in Figures 3a&b. Figure 3a shows spin-frequency histories for sources whose accretion is thought to be disk-fed. These are discussed in more detail below. For comparison, Figure 3b shows measurements of Vela X-1, which is a wind-fed accretor, and OAO 1657-415, which may be either disk-fed or wind-fed and sits in the Corbet diagram intermediate between the two types of systems.

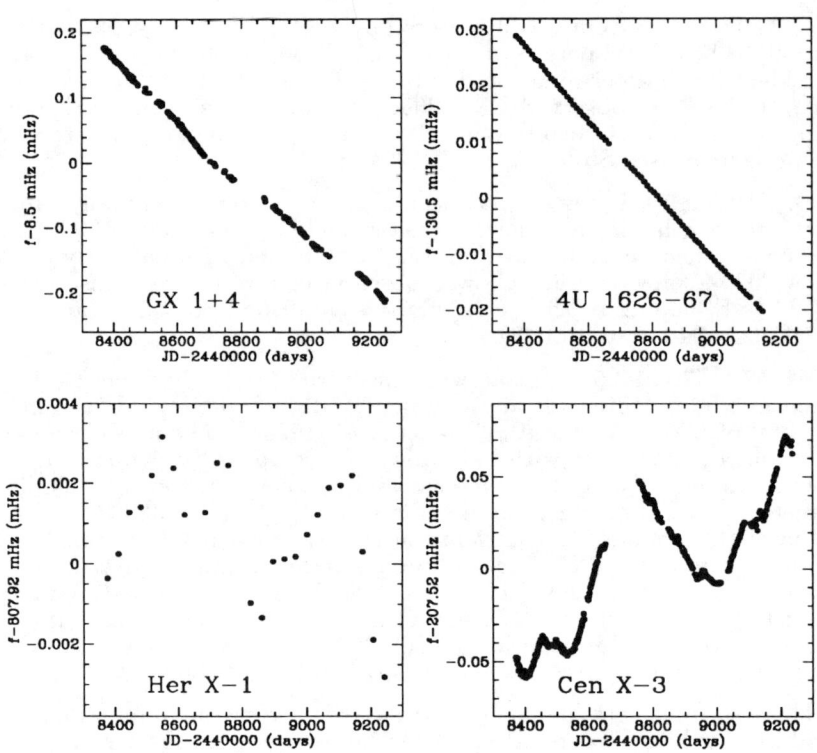

Figure 3a. Spin-frequency history for probable disk-fed systems. (For Her X-1, mean frequencies for each main-on portion of the 35d cycle.)

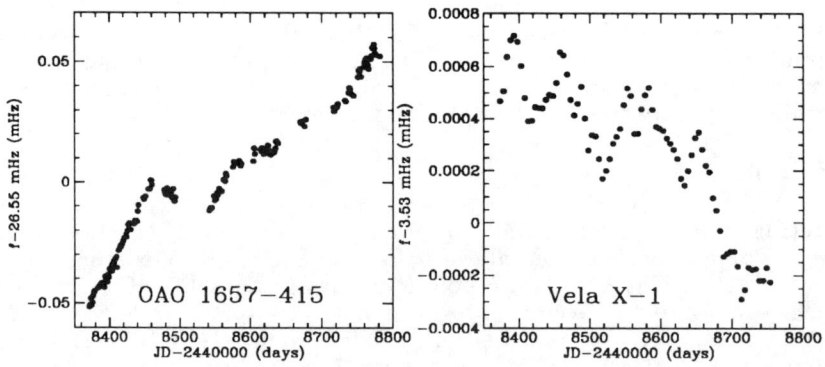

Figure 3b. Spin-frequency history for probable wind-fed systems.

The very different qualitative behavior of the various systems is immediately apparent from Figure 3. In particular, GX 1+4 and 4U 1626-67 both show a monotonic spin-down with nearly constant average rate while other disk-fed and wind-fed systems exhibit frequent changes in sign of the frequency derivative.

The torque however does switch sign on decade time scales in GX 1+4 and 4U 1626-67. From its discovery until the late 1980s, GX 1+4 was observed to be in a state of monotonic spin-up. Given its short spin-up time ($t_{su} \approx 40$ yr) compared to the system evolution time ($t_{evol} \gtrsim 10^6$ yr), this was clearly not a permanent state, and *GINGA* and BATSE observations of a switch to spin-down in GX 1+4 (Makishima et al. 1988 Chakrabarty et al. 1994) showed that this was indeed the case. BATSE has found that 4U 1626-67 ($t_{su} \approx 4000$ yr) has also changed from spin-up to spin-down (Bildsten et al. 1994), and surprisingly, with the same average value of the torque but opposite sign. Since a disk reversal in this Roche-lobe overflow system is hard to imagine, we interpret the steady spin-down as a sign that the pulsar is spinning near its equilibrium period, where the magnetospheric radius equals the co-rotation radius (see Ghosh & Lamb 1979), requiring a magnetic field strength of order $\approx 3 \times 10^{12}$ G for 4U 1626-67 (and 10^{14} G for GX 1+4).

The simplest working hypotheses given the observations and the accretion torque theories of Ghosh and Lamb are then: (1) since $t_{su} \ll t_{evol}$, X-ray pulsars quickly reach the equilibrium period, as determined from the accretion rate and magnetic field strength, (2) once there, the X-ray pulsar oscillates about the equilibrium period by having counteracting periods of spin-up and spin-down. This hypothesis was checked with pre-BATSE data by comparing the observed long-term ($\gtrsim$ yrs) torque to the fiducial value, $N_f = \dot{M}(GMr_{co})^{1/2}$, where $\dot{M}$ is the maximum inferred mass accretion rate, M is the neutron star mass, and $r_{co} = (GM/\omega_s^2)^{1/3}$ is the co-rotation radius and ω_s is the angular spin frequency. For both GX 1+4 and 4U 1626-67 (and also SMC X-1 and 1E 2259+586) the observed long-term average torque, N_o, always satisfies $N_o \gtrsim 0.2 N_f$, which, within the uncertainties, is consistent with accretion from matter near the co-rotation radius. These sources, which might be called "steady-staters", show monotonic spin-up or spin-down over a few year time scale.

As can be seen from Figure 3, other disk-fed sources such as Her X-1 and Cen X-3 show quite different behavior. The long-term ($\gtrsim$ yrs) torque, N_o, measured for these sources (as well as LMC X-4) was nearly a factor of 100 smaller than N_f, implying different torquing behavior than in the "steady-staters". These "wanderers" have frequency histories that are more consistent with a random-walk (Baykal & Ogelman 1993), which could arise if the torque had a value near N_f, but changed sign on a shorter timescale. The measured long term torque would thus be less than N_f. Recent short timescale ($\lesssim 10-20$ d) torque measurements for Her X-1 and Cen X-3 by BATSE always find a torque larger than N_o, but never in excess of N_f (Wilson et al. 1994b, Finger et al. 1994a).

This classification scheme suggests that all disk-accreting pulsars show torques with magnitude $\lesssim N_f$ on the shortest timescales and differentiate themselves by the torque switching time. The wanderers switch within ~ 60 days, whereas the steady-staters switch once in 10-20 years. The primary issue for these systems is identifying the physics that sets this timescale.

ACKNOWLEDGEMENTS

We acknowledge the important contributions of the entire BATSE team at NASA/ Marshall Space Flight Center to this work. We recognize in particular the individual contributions of M. Briggs, J. Chiu, G. J. Fishman, L. Gibby, J. M. Grunsfeld, B. A. Harmon, T. Koh, C. A. Meegan, W. S. Paciesas, G. N. Pendleton, B. C. Rubin, M. T. Stollberg, C. A. Wilson, and N. S. Zhang. This work is funded in part by NASA grants NAGW-1919, NAG 5-1458, NGT-51184 and a Lee A. DuBridge fellowship to L.B. funded by the Weingart Foundation.

REFERENCES

Aoki, T. et al. 1992, PASJ, 44, 641.
Baykal, A. and Ogelman, H. 1993, A&A, 267, 119.
Bildsten, L. et al. 1994, in Proc. of the Second Compton Symp., in press.
Chakrabarty, D. et al. 1993, ApJ, 403, L33.
Chakrabarty, D. et al. 1994, in Proc. of the Second Compton Symp., in press.
Cominsky, L. et al. 1994, in Proc. of the Second Compton Symp., in press.
Cook, M. C. and Page, C. G. 1987, MNRAS, 225, 381.
Corbet, R. H. D. 1986, MNRAS, 220, 1047.
Corbet, R. H. D. et al. 1993, A&A, 276, 52.
Davidsen, A., Malina, R., and Bowyer, S. 1977, ApJ, 211, 866.
Deeter, J. E. et al. 1987, AJ, 93, 877.
Deeter, J. E. et al. 1991, ApJ, 383, 324.
Finger, M. H. 1993, private communication.
Finger, M. H. et al. 1993, in Compton Gamma Ray Observatory, ed. M. Friedlander et al., (New York: AIP), 386.
Finger, M. H., Wilson, R. B., and Fishman, G. J. 1994a, Proc. of Second Compton Symp., in press.
Finger, M. H. et al. 1994b, these proceedings.
Fishman, G. J. et al. 1989, in Proc. of the GRO Science Workshop, ed. W. N. Johnson, (Greenbelt: NASA/GSFC), 2-39.
Ghosh, P. and Lamb, F. K. 1979, ApJ, 234, 296.
Giovannelli, F. and Graziati, L. S. 1992, Space Sci.Rev., 59, 1.
Glass, I. S. and Feast, M. W. 1973, Nature Phys. Sci., 245, 39.
Kelley, R. L. et al. 1983, ApJ, 274, 765.
Lapshov, I. Y. et al. 1992, Soviet Ast. Lett., 18, 12.
Levine, A. et al. 1991, ApJ, 381, 101.
Levine, A. et al. 1993, ApJ, 410, 328.
Makishima, K. et al. 1984, PASJ, 36, 679.
Makishima, K. et al. 1987, ApJ, 314, 619.
Makishima, K. et al. 1988, Nature, 333, 746.
Nagase, F. 1989, PASJ, 41, 1.
Parmar, A. N. et al. 1989, ApJ, 338, 359.
Polidan, R. S. et al. 1978, Nature, 275, 296.
Rappaport, S. et al. 1978, ApJ, 224, L1.
Rubin, B. C. et al. 1994, these proceedings.
Sato, N. et al. 1986, ApJ, 304, 241.
Stella, L., White, N. E., and Rosner, R. 1986, ApJ, 308, 669.
Stella, L. et al. 1985, ApJ, 288, L45.
Stollberg, M. H. et al. 1994, these proceedings.
Waters, L. B. F. M. and van Kerkwijk, M. H. 1989, A&A, 223, 196.
Wilson, C. A. et al. 1994a, these proceedings.
Wilson, R. B. et al. 1994b&c, these proceedings.

COALESCENCE OF NEUTRON STAR BINARIES

Omer Blaes
Dept. of Physics, University of California,
Santa Barbara, CA 93106-9530.
blaes@vela.physics.ucsb.edu

ABSTRACT

I review the current theoretical views on the formation, evolution, and fates of merging double neutron star binaries, with particular emphasis on the possibility that these are detectable sources of gravitational waves and cosmic gamma-ray bursts.

1. INTRODUCTION

In the last few years there has been an explosion of interest in the coalescence of double neutron star binaries.[†] That such events actually occur in the universe is known from the fact that we observe three such binaries which will merge in less than a Hubble time under the action of gravitational radiation reaction (Table 1). Two of these (1913+16 and 1534+12) are in the Galaxy while the other (2127+11C) is in the globular cluster M15. In addition coalescing compact binaries are the systems which we best understand among the possible sources of gravitational radiation for the forthcoming interferometer detectors LIGO and VIRGO. Finally, coalescing neutron stars are perhaps the least fanciful of the proposed cosmological sources of gamma-ray bursts.

In this paper I will review the current theoretical research in this area. The reader is warned that, because we are not certain that we have ever observed a merger event, much of this material is necessarily more speculative than many of the other topics discussed at this conference.

TABLE 1. OBSERVED BINARY NEUTRON STARS WITH $\tau_{merge} < H_0^{-1}$

Pulsar	$m_{PSR}(M_\odot)$	$m_{comp}(M_\odot)$	P_b(hr)	e	τ_{merge}(yr)	$P/2\dot{P}$(yr)
2127+11C[a]	1.3 ± 0.2[b]	1.4 ± 0.2[b]	8.0	0.68	2.2×10^8	9.7×10^7
1913+16[c]	1.442 ± 0.003	1.386 ± 0.003	7.8	0.62	3.0×10^8	1.1×10^8
1534+12[d]	1.32 ± 0.03	1.36 ± 0.03	10.1	0.27	2.7×10^9	2.5×10^8

[a] Prince et al. (1991), [b] Taylor (1992), [c] Taylor & Weisberg (1989), [d] Wolszczan (1991)

2. FORMATION OF TIGHT DOUBLE NEUTRON STAR BINARIES

The main problem in understanding the formation of double neutron star

---

[†] Coalescing double neutron star binaries were first considered as a subject of interest by Dyson (1963), four years *before* the discovery of pulsars.

binaries is how the progenitor system survives a second supernova explosion. The most popular solution comes from considering the final evolution of high mass X-ray binaries (see e.g. Bhattacharya & van den Heuvel 1991; van den Heuvel 1993, these proceedings). Because mass is transferred from the high mass star to the low mass star, the binary orbit shrinks and the neutron star moves into the envelope of its companion. During the subsequent common envelope evolution most of the mass is ejected from the binary as the neutron star spirals in towards the core. When the core finally collapses to a new neutron star, little mass is lost from the system and the binary survives in a tight, albeit eccentric, orbit. Note that in this formation scenario the observed pulsar is presumably the older neutron star which has been recycled during the accretion phase (cf. the long spin down ages $P/2\dot{P}$ in Table 1).

In addition to this process, stellar encounters in globular clusters and galactic nuclei may also play a role in the genesis of tight double neutron star binaries. This is strongly indicated for the case of PSR 2127+11C which, unlike all the other pulsars and X-ray binaries in M15 which are concentrated in the core, is > 20 core radii out from the cluster center (Prince et al. 1991). This suggests that this binary may have been formed in a collision between, say, a single neutron star and a binary containing a neutron star and another star. During the collision the field neutron star and the ordinary star changed places and the binary was ejected from the cluster core (Phinney & Sigurdsson 1991). More complex processes may also be important. Sigurdsson & Hernquist (1992) have shown that double neutron star binaries can be formed in the collision between two binaries, each containing a neutron star and a main sequence star. The end result is a neutron star binary surrounded by a common envelope of disrupted main sequence stellar material.

3. ORBITAL EVOLUTION

Gravitational radiation emission dominates the evolution of a tight neutron star binary. Approximating the system as two point masses in a Newtonian orbit (Peters 1964), the stars spiral in and merge after a time

$$\tau_{merge} = 10^{10} \text{yr} (1-e^2)^{7/2} f(e) \left(\frac{P_b}{15.2\text{hr}}\right)^{8/3} \left(\frac{\mu}{0.7 M_\odot}\right)^{-1} \left(\frac{M}{2.8 M_\odot}\right)^{-2/3}, \quad (1)$$

where e is the eccentricity, $f(e)$ is a factor of order unity which is weakly dependent on e and illustrated in figure 1, P_b is the orbital period, μ is the reduced mass, and M is the total mass. Eccentric orbits can merge much faster than circular orbits with the same semimajor axis (or orbital period) because most of the wave emission occurs at the time of the greatest accelerations, i.e. at periastron (Peters & Matthews 1963). For the same reason the orbit will circularize on the time scale τ_{merge}, with $e \propto P_b^{19/18}$ at late times (Peters 1964). Immediately prior to coalescence the orbit will therefore be almost exactly circular.

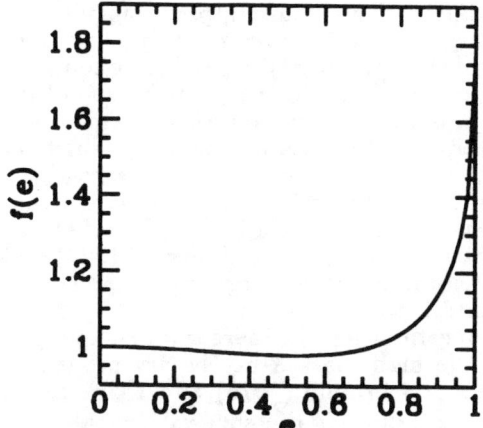

Fig. 1. Weak eccentricity factor which enters merger time, eq. (1).

4. MERGER RATES

In order to see whether binary neutron star coalescence has any observable consequences, we can try to estimate how often it occurs. There are two ways of doing this. The first is to extrapolate from the observed binary pulsars in Table 1 to see how many similar systems there are in the Galaxy and the universe. The first people to try this were Clark, van den Heuvel, & Sutantyo (1979). At the time there was only one binary pulsar known (1913+16), along with 315 radio pulsars. They therefore assumed that the birth rate of double neutron star binaries relative to that of radio pulsars is $\sim 1/315$. Adopting a pulsar birth rate of $\simeq 0.1$ yr^{-1}, they deduced that the Galactic birth rate of binaries similar to 1913+16 is $\simeq 3 \times 10^{-4}$ yr^{-1}. After estimating the birth rate of high mass X-ray binaries to be $\sim 10^{-3}$ yr^{-1}, they concluded that the probability that these systems evolve into double neutron star binaries is ~ 20 percent. Hils, Bender, & Webbink (1990) have made more elaborate calculations based on the same basic argument. This analysis is too naive partly because it neglects the differences in lifetimes and luminosities of the (recycled) binary pulsars and most isolated radio pulsars.

More careful calculations which take these into account as well as the various sampling volumes of the different pulsar surveys have been done by Narayan, Piran, & Shemi (1991) and Phinney (1991). For the birth rate of systems like 1913+16 these authors find a best estimate of $1 - 3 \times 10^{-7}$ yr^{-1}, roughly 10^3 times smaller than the estimate of Clark, van den Heuvel, & Sutantyo (1979). The other Galactic merging neutron star binary, 1534+12, which is much closer and intrinsically much fainter than 1913+16, is estimated to be much more common with a Galactic birth rate of $1 - 2 \times 10^{-6}$ yr^{-1} (Narayan, Piran, & Shemi 1991; Phinney 1991). Depending on the treatment of the pulsar beaming angles and radial distribution in the Galaxy, these authors conclude that the total Galactic birth rate of all binary pulsars which will merge in less than a Hubble time is $\simeq 0.3 - 1 \times 10^{-5}$ yr^{-1}. Note that although this rate is based on two observed systems, the existence of 1913+16 is really irrelevant because 1534+12 dominates the extrapolation. The resulting probability that a high mass X-ray binary evolves to a merging double neutron star binary is a few percent. Presumably the low value of this probability is partly due to the fact that sometimes a second supernova does unbind the binary. Because collapse to a black hole need not lead to ejection of much mass from the system, the survival probability of very massive X-ray binaries to black hole-neutron star binaries is likely to be much higher. Hence even though the number of massive, black hole progenitors is smaller than the number of stars which give rise to neutron stars, the number of black hole-neutron star binaries might be comparable to double neutron star binaries (Narayan, Piran, & Shemi 1991; Phinney 1991). Phinney (1991) has also estimated the contribution to the merger rate from globular cluster binaries like 2127+11C and found it to be small.

The second way of estimating the merger rate is to try and model the final stages of evolution of the high mass X-ray binary progenitors. This was first attempted by Lattimer & Schramm (1976) and Clark & Eardley (1977) who however simply guessed a value for the survival probability of 10^{-2}, therefore ending up with a Galactic merger rate of 6×10^{-6} yr^{-1}, consistent with the above authors. More recently Tutukov & Yungelson (1993) have brought their evolutionary calculations to bear on the problem and estimate a double neutron star merger rate of 3.2×10^{-4} yr^{-1}, comparable to the original estimate of Clark, van den Heuvel, & Sutantyo (1979). Most of the binaries contributing to this rate are born with small orbital periods and therefore have very short merger times. Although they contribute to a rapid overall merger rate, these short-lived systems have a small space density and could escape detection. This estimate is therefore not in conflict with estimates based on the much longer-lived observed binary pulsars. It is nevertheless probably very uncertain due to our poor understanding of common envelope evolution, in particular how efficiently orbital energy is used to eject the envelope (see e.g. Livio 1993 for a discussion of this in the context of double white dwarf binaries).

According to the various authors, the double neutron star merger rate within 200 Mpc is $\sim h^3$ yr^{-1} (Narayan, Piran, & Shemi 1991), $\sim 4h$ yr^{-1} (Phinney 1991), and $\sim 100h$ yr^{-1} (Tutukov & Yungelson 1993), where h is the present value of the Hubble parameter H_0 in units of 100 km s^{-1} Mpc^{-1}. It is this reviewer's opinion that the uncertainties are so large that it is not clear whether these numbers even bracket the true value.

5. THE INSPIRAL GRAVITATIONAL WAVE FORM

After the orbit has circularized, the gravitational wave luminosity from the binary is

$$L = 2 \times 10^{52} \text{ergs s}^{-1} \left(\frac{P_b}{10^{-2}\text{s}}\right)^{-10/3} \left(\frac{\mu}{0.7 M_\odot}\right)^2 \left(\frac{M}{2.8 M_\odot}\right)^{4/3} \quad (2)$$

In the last few minutes prior to coalescence this becomes sufficiently high that it is conceivable that the radiation might actually be observable by the upcoming interferometric detectors LIGO (Abramovici et al. 1992) and VIRGO (Bradaschia et al. 1990), at least in the advanced detector systems. The emitted wave frequency $f = 2/P_b$ and therefore increases with time ("chirps") through the detector bandpass.

Many people are calculating inspiral waveforms to serve as detection templates (e.g. Lincoln & Will 1990; Wiseman 1992; Junker & Schäfer 1992; Kidder, Will, & Wiseman 1993; Poisson 1993; and Cutler et al. 1993a). Should the inspiral waveform be observed much can be learned about the parameters of the binary. Three independent detectors are sufficient to measure the direction to the binary and its orientation in the sky. If this is known, then measurement of the amplitude and the rate of change of frequency of the wave give separately the "chirp mass" $\equiv (\mu^3 M^2)^{1/5}(1+z)$, where z is the redshift of the source, and the luminosity distance of the source. Identifying the host galaxy of the source (by searching the gravitational wave error box - probably difficult in practise) would then permit direct measurement of the redshift and in combination with the luminosity distance would measure H_0 (Schutz 1986).

In the advanced detector systems, one can hope to see the inspiral waveform from frequencies $f_1 \sim 10$ Hz to $f_2 \sim 1000$ Hz (e.g. Abramovici et al. 1992). The number of wave periods in this bandpass, neglecting post-Newtonian corrections to the orbit, is $N = 1.6 \times 10^4 (f_1/10\text{Hz})^{-5/3}(\mu/0.7M_\odot)^{-1}(M/2.8M_\odot)^{-2/3}$ for $f_2 \gg f_1$. Such a large number implies that small effects that accelerate or decelerate the coalescence will lead to secular changes in the phase which can be measured extremely accurately, in principle permitting a determination of the individual stellar masses and spins (Cutler et al. 1993b). If the measured masses are not too far from $1.4M_\odot$, one could assume they are neutron stars with masses exactly equal to this value, thereby giving a redshift determination without having to find the host galaxy. This is perhaps a more promising way in which a redshift-distance relation might be produced, at least statistically, thereby providing a way of measuring cosmological parameters (Cutler et al. 1993b).

6. THE MERGER ITSELF

Because neutron stars are so compact, tidal interactions are generally ineffective in synchronizing the spin and orbital periods (Haensel, Paczyński, & Amsterdamski 1991; Mészáros & Rees 1992a; Kochanek 1992; Bildsten & Cutler 1992). A crude calculation of the tidal effects may be made in the following way (Bildsten & Cutler 1992). The maximum possible tidal torque that can be exerted on a neutron star of mass m_{NS} by a companion of mass m_{comp} is $\sim Gm_{comp}^2 R^5 \sin(2\alpha)/a^6$, where R is the neutron star radius, a is the binary semimajor axis, and α is the angle between the tidal bulge and the line of centers. This is far too small to affect the neutron star spin except possibly just prior to coalescence. The spin period will be much longer than the orbital period at this stage. If we parameterize the volume-averaged dynamic viscosity of the neutron star by $\eta = \beta m_{NS} c/R^2$, where β is some number less than unity, then $\alpha \sim \beta (GM/a^3)^{1/2}(cR^2/Gm_{NS})$. The time required for tides to synchronize the spin and orbital periods is therefore

$$\tau_{synch} \gtrsim \frac{1\text{s}}{\beta}\left(\frac{P_b}{10^{-2}\text{s}}\right)^4 \left(\frac{m_{NS}}{1.4M_\odot}\right)\left(\frac{M}{m_{comp}}\right)^2 \left(\frac{I}{10^{45}\text{g cm}^2}\right)\left(\frac{R}{10\text{km}}\right)^{-7}, \quad (3)$$

where I is the moment of inertia of the neutron star. Comparing this to equation (1), we conclude that unless the viscosity is extremely high ($\beta \sim 1$), tidal decreases in the neutron star spin period cannot keep up with the decreasing orbital period. Note, however, that τ_{synch} has a strong dependence on the neutron star radius so that neutron stars much lighter than those in the observed systems listed in Table 1 might achieve synchronization. Tidal distortions still slightly accelerate the coalescence, leading to orbital phase changes which might be measurable towards the end of the inspiral gravitational waveform (Kochanek 1992, Bildsten & Cutler 1992). In addition as the orbital frequency increases, it sweeps through many neutron star oscillation frequencies and resonant excitation of modes occurs (Reisenegger & Goldreich 1993). The resulting phase shift due to this effect is apparently too small to be detectable, however.

Tidal dissipation will also heat the neutron stars prior to coalescence. The heating rate is $\sim GMc\beta m_{comp}^2 R^7/(m_{NS} a^9)$, or

$$\dot{E} \sim 10^{50} \text{ergs s}^{-1} \beta \left(\frac{P_b}{10^{-2}\text{s}}\right)^{-6} \left(\frac{m_{NS}}{1.4 M_\odot}\right)^{-1} \left(\frac{m_{comp}}{M}\right)^2 \left(\frac{R}{10\text{km}}\right)^7. \quad (4)$$

For $\beta \gtrsim 10^{-12}$ this exceeds the Eddington luminosity so significant mass loss may occur prior to the coalescense itself if enough of the energy is deposited in the outer layers of the star (Bildsten & Cutler 1992, see Kochanek 1992 for a discussion of viscous dissipation mechanisms in a neutron star crust). As discussed below, this is important for the dynamics of the resulting fireball.

Early work on coalescing double neutron star binaries by Clark & Eardley (1977) concluded that upon reaching the tidal radius, stable mass transfer from the lighter to the heavier star could act to stop the merger. Kochanek (1992) and Bildsten & Cutler (1992) have reexamined this question in considerably more detail, and find that for $m_{comp} = 1.4 M_\odot$, the neutron star must be very light, $m_{NS} \lesssim 0.5 M_\odot$, for mass transfer to reverse the coalescence. Mass transfer cannot prevent coalescence in black hole-neutron star binaries. This conclusion is strengthened further by the existence of two physically independent dynamical instabilities which come into play at the late stages of the evolution of the binary. The first is that general relativity itself implies that there is an innermost stable orbit for two "point" masses in orbit around each other (Kidder, Will, & Wiseman 1993). The critical orbital separation is $\simeq (6M + 4\mu)G/c^2$ (in Schwarzschild coordinates; see Lai, Rasio, & Shapiro 1993b), or $\simeq 30$ km for two $1.4 M_\odot$ neutron stars. Inside this point the binary members will simply fall toward each other even without gravitational radiation reaction. In addition Lai, Rasio, & Shapiro (1993a) have discovered a purely Newtonian hydrodynamical instability. Inside a certain critical radius, tidal distortions of a sufficiently incompressible star can actually decrease the total binding energy of the binary, again implying the existence of an innermost stable orbital separation of $\sim 2R(M/m_{NS})^{1/3}$ (Lai, Rasio, & Shapiro 1993b). For two $1.4 M_\odot$ neutron stars this happens to have roughly the same dimensions as the general relativistic unstable orbit. Unless synchronous rotation can be achieved, the tidal limit is generally inside the innermost stable orbit and so the issue of mass transfer is irrelevant to the evolution (Lai, Rasio, & Shapiro 1993b).

A number of groups have attempted to simulate the coalescence itself using three dimensional hydrodynamic codes. Rasio & Shapiro (1992) have run SPH simulations of two stars in synchronous rotation (which will never occur, as discussed above) starting at the point where the tidal instability begins. Using purely Newtonian gravity with no gravitational radiation reaction, they find that the two stars nevertheless coalesce within an orbital time. The final remnant consists of a rigidly rotating star surrounded by a hot ($T \sim 10^{11}$ K), differentially rotating ($\Omega \propto r^{-1.8}$) thick disk. More recent SPH simulations by Davies et al. (1993) reach similar conclusions. These simulations are again Newtonian, but they start the stars off at larger radii and extract orbital angular momentum according to the quadrupole formula which then drives the stars to coalescence. The stars are not assumed to be synchronously rotating, but again a rigidly rotating remnant is formed surrounded by a differentially rotating thick disk with the the same rotation law as found by Rasio & Shapiro (1992). Finally there have been many Newtonian finite difference simulations by Oohara & Nakamura (1989, 1990), Nakamura & Oohara (1989, 1991), and Shibata, Nakamura, & Oohara (1992, 1993). In some cases these simulations result in a relatively hollow torus with no compact remnant inside. All of these simulations do not fully include general relativity (a difficult problem!) and whether or not a spinning black hole results is not yet known. The finite difference simulations often produce results where all the matter is within its own Schwarzschild radius suggesting that this outcome is likely.

7. THE MERGER FIREBALL

In addition to gravitational waves, the high ambient temperatures resulting from the merger process will produce up to $\sim 10^{53}$ ergs in thermal neutrinos and antineutrinos. Photons, electron-positron pairs, magnetic fields, and baryons are also likely to be generated and/or ejected from the system. (Ejected neutron-rich matter is a possible site for r-process nucleosynthesis, e.g. Eichler et al. 1989 and references therein.)

Understanding the electromagnetic display is important when searching for counterparts to future LIGO/VIRGO source detections, but most of the recent interest in this comes from the possibility that neutron star-neutron star or neutron star-black hole mergers are a source of cosmic gamma-ray bursts (Paczyński 1986; Goodman 1986;, Eichler et al. 1989; Paczyński 1990, 1991; Mészáros & Rees 1992a, 1992b; Nakamura et al. 1992; Narayan, Paczyński, & Piran 1992; Mochkovitch et al. 1993). These bursts have an angular distribution on the sky which is consistent with isotropy but their brightness distribution indicates spatial inhomogeneity (see e.g. Meegan et al. 1993 for a discussion of the BATSE data from the *Compton Gamma Ray Observatory*). If the source emits isotropically, then the energy released in gamma-rays is $E \sim 10^{50} h^{-2} (F/10^{-7} \mathrm{ergs\, cm}^{-2})$ ergs, where F is the observed burst fluence. This is a small fraction of the total energy available in a binary neutron star merger event. Given a galaxy density of $\sim 10^{-2} h^3$ Mpc^{-3} (e.g. Efstathiou, Ellis, & Peterson 1988), the BATSE all sky burst rate of 800 yr^{-1} corresponds to $\sim 10^{-6}$ bursts yr^{-1} per galaxy, or about 1 burst per million years in each galaxy. This rate lies within the range of binary neutron star merger rate estimates discussed above.

A variety of processes have been suggested whereby a gamma-ray burst might be initiated. If the compact merger remnant survives as a neutron star for as long as 10-100 ms, then strong convection and differential rotation create an $\alpha\Omega$ dynamo that can build up an enormous dipole magnetic field $\sim 10^{16}$ G (Duncan & Thompson 1992). The resulting rapid spin down of this "magnetar" could power a gamma-ray burst through pulsar-like mechanisms (Duncan & Thompson 1992, Usov 1992). It is not clear whether the rigid body rotation produced in the merger simulations discussed above (which do not include cooling and may have high numerical viscosity) presents a problem for this scenario.

A small fraction of the thermal neutrinos and antineutrinos could annihilate to form electron-positron pairs which might power a gamma-ray burst (e.g. Eichler et al. 1989; Dar et al. 1992; Mészáros & Rees 1992a, 1992b). Viscous dissipation in the surrounding torus could also power bursts through neutrino-antineutrino annihilation along the rotation axis (Woosley 1993b, Mochkovitch et al. 1993), flaring in superstrong fields ($\sim 10^{16} - 10^{17}$ G) above the disk (Narayan, Paczyński, & Piran 1992), and/or the Blandford-Znajek process if the compact remnant is a rapidly spinning black hole (Nakamura et al. 1992, Woosley 1993a).

If $\sim 10^{50}$ ergs of gamma-rays are injected isotropically into a region of size ~ 100 km by any of these mechanisms, the optical depth to electron-positron pair production is huge, and photons will not be able to escape. The electron-positron pairs will also be trapped and will in turn annihilate to produce photons. These two processes together with additional non-number conserving reactions will quickly produce a thermal distribution of pairs and photons. The evolution of this thermal "fireball" has been the subject of much recent research. In the simplest case where only photons and pairs are present, the fireball expands and cools adiabatically until the rest frame temperature drops below the electron rest mass energy ~ 1 MeV (Paczyński 1986, Goodman 1986). At this point photons can no longer produce pairs and the pair density drops exponentially. When the Thomson optical depth then drops below unity the photons finally escape. The relativistic outward motion of the fireball at this point turns out to be such that the lab frame energies of the photons are comparable to the energy at which they were injected at the base of the fireball. These high energy photons can still escape, however, because in the lab frame they are moving sufficiently parallel that they cannot produce pairs and conserve linear momentum. The total burst duration is simply the cooling time of the injection source at the base of the fireball (Paczyński 1986) or the light crossing time of the injection region (Goodman 1986), whichever is longer.

Baryon loaded fireballs have been studied by Shemi & Piran (1990) and in more detail by Mészáros, Laguna, & Rees (1993) and Piran, Shemi, & Narayan (1993). The main effect of baryons (protons and neutrons) is that their much larger rest mass implies that they can soak up much of the fireball energy as kinetic energy. The issue then is how much this occurs before the fireball becomes optically thin. An additional twist is that the electrons associated with the protons provide an extra source of scattering opacity which does not disappear when the rest frame temperature drops below the electron rest mass energy.

The main problem with this picture for gamma-ray bursts is that the resulting spectrum is generally thermal, in contrast to the very broad, nonthermal spectra which are observed. The way to solve this problem is perhaps not to inject the energy in the form of gamma-rays in the source region, but instead to transport it outwards in another form and then convert it into photons in an optically thin region. Rees & Mészáros (1992) have proposed an ingenious solution of this type whereby the accelerated baryons themselves carry the energy outwards until they are decelerated by external matter such as the ambient interstellar medium. This deceleration sends a forward shock into the ambient material and a reverse shock into the fireball, randomizing the bulk kinetic energy and radiating gamma-rays to produce the burst. This idea has been further developed in more detail by Mészáros & Rees (1993) and by Mészáros, Laguna, & Rees (1993). These baryon-loaded fireballs can produce short precursor bursts, produced when the fireball becomes optically thin prior to the baryons interacting with the ambient medium, as well as long afterbursts (Mészáros & Rees

1993). Transient radio emission within a few days of the burst may also be detectable (Paczyński & Rhoads 1993). Ordered magnetic fields in a wind may also transport energy outwards and convert it into radiation through reconnection (C. Thompson 1993, private communication). Tangled fields behave just like radiation (Mészáros, Laguna, & Rees 1993).

8. CONCLUSIONS

We know for a fact that double neutron star binaries do exist and that some will merge within a Hubble time. The formation processes need to be better understood, particularly where collisions are involved as in globular clusters and perhaps in galactic nuclei. Due to the paucity of observed binary neutron stars, it would be highly desirable to try and pin down the merger rate from understanding the evolution of the progenitor systems. If gravitational waves from mergers are ever detected, then the inspiral waveform contains much information which is based on simple physics and therefore easily interpreted. The coalescence waveform itself will also contain much information based on more complicated physics, and numerical simulations need to be improved enormously (particularly by including the full general relativistic effects) before this can be extracted. Finally the electromagnetic display that might result from a coalescence is poorly understood at present and indeed highly speculative, based largely on the unproven hypothesis that these events are sources of observed cosmic gamma-ray bursts.

It is a pleasure to thank L. Bildsten, C. Evans, M. Livio, R. Nelson, E. S. Phinney, F. Rasio, M. Rees, A. Reisenegger, S. Sigurdsson, and C. Thompson for enlightening discussions.

REFERENCES

Abramovici, A., et al. 1992, Science, 256, 325
Bhattacharya, D., & van den Heuvel, E. P. J. 1991, Phys. Rep., 203, 1
Bildsten, L., & Cutler, C. 1992, ApJ, 400, 175
Bradaschia, C., et al. 1990, Nuc. Instr. Meth. Phys. Res., A289, 518
Clark, J. P. A., & Eardley, D. M. 1977, ApJ, 215, 311
Clark, J. P. A., van den Heuvel, E. P. J., & Sutantyo, W. 1979, A& A, 72, 120
Cutler, C., Finn, L. S., Poisson, E., & Sussman, G. J. 1993a, Phys. Rev. D, 47, 1511
Cutler, C., et al. 1993b, Phys. Rev. Lett., 70, 2984
Dar, A., Kozlovsky, B. Z., Nussinov, S., & Ramaty, R. 1992, ApJ, 388, 164
Davies, M. B., Benz, W., Piran, T., & Thielemann, F. K. 1993, ApJ, submitted
Duncan, R. C., & Thompson, C. 1992, ApJ, 392, L9
Dyson, F. J. 1963, Interstellar Communication, A. G. W. Cameron, ed. (New York: Benjamin), 115
Eichler, D., Livio, M., Piran, T., & Schramm, D. N. 1989, Nature, 340, 126
Efstathiou, G., Ellis, R. S., & Peterson, B. A. 1988, MNRAS, 232, 431
Goodman, J. 1986, ApJ, 308, L47
Haensel, P., Paczyński, B., & Amsterdamski, P. 1991, ApJ, 375, 209
Hils, D., Bender, P. L., & Webbink, R. F. 1990, ApJ, 360, 75
Junker, W., & Schäfer, G. 1992, MNRAS, 254, 146
Kidder, L. E., Will, C. M., & Wiseman, A. G. 1993, Phys. Rev. D, 47, 3281
Kochanek, C. S. 1992, ApJ, 398, 234

Lai, D., Rasio, F. A., & Shapiro, S. L. 1993a, ApJ, 406, L63
Lai, D., Rasio, F. A., & Shapiro, S. L. 1993b, ApJ, submitted
Lattimer, J. M., & Schramm, D. N. 1976, ApJ, 210, 549
Lincoln, C. W., & Will, C. M. 1990, Phys. Rev. D, 42, 1123
Livio, M. 1993, *Evolutionary Links in the Zoo of Interacting Binaries*, F. D'Antona, ed., in press
Meegan, C., Fishman, G., Wilson, R., Brock, M., Horack, J., Paciesas, W., Pendleton, G., & Kouveliotou, C. 1993, *Compton Gamma Ray Observatory*, M. Friedlander, N. Gehrels, and D. J. Macomb, eds. (New York: AIP), 681
Mészáros, P., & Rees, M. J. 1992a, ApJ, 397, 570
Mészáros, P., & Rees, M. J. 1992b, MNRAS, 257, 29P
Mészáros, P., & Rees, M. J. 1993, ApJ, 405, 278
Mészáros, P., Laguna, P. & Rees, M. J. 1993, ApJ, 415, 181
Motchkovitch, R., Hernanz, M., Isern, J., & Martin, X. 1993, Nature, 361, 236
Nakamura, T., & Oohara, K. 1989, Prog. Theor. Phys., 82, 1066
Nakamura, T., & Oohara, K. 1991, Prog. Theor. Phys., 86, 73
Nakamura, T., Shibazaki, N., Murakami, Y., & Yoshida, A. 1992, Prog. Theor. Phys., 87, 879
Narayan, R., Paczyński, B., & Piran, T. 1992, ApJ, 395, L83
Narayan, R., Piran, T., & Shemi, A. 1991, ApJ, 379, L17
Oohara, K., & Nakamura, T. 1989, Prog. Theor. Phys., 82, 535
Oohara, K., & Nakamura, T. 1990, Prog. Theor. Phys., 83, 906
Paczyński, B. 1986, ApJ, 308, L43
Paczyński, B. 1990, ApJ, 363, 218
Paczyński, B. 1991, Acta Ast., 41, 257
Paczyński, B., & Rhoads, J. 1993, ApJ, submitted
Peters, P. C. 1964, Phys. Rev., 136, B1224
Peters, P. C., & Matthews, J. 1963, Phys. Rev., 131, 435
Phinney, E. S. 1991, ApJ, 380, L17
Phinney, E. S., & Sigurdsson, S. 1991, Nature, 349, 220
Piran, T., Shemi, A., & Narayan, R. 1993, MNRAS, 263, 861
Poisson, E. 1993, Phys. Rev. D, 47, 1497
Prince, T. A., Anderson, S. B., Kulkarni, S. R., & Wolszczan, A. 1991, ApJ, 374, L41
Rasio, F. A., & Shapiro, S. L. 1992, ApJ, 401, 226
Rees, M. J., & Mészáros, P. 1992, MNRAS, 258, 41P
Reisenegger, A., & Goldreich, P. 1993, ApJ, submitted
Schutz, B. F. 1986, Nature, 323, 310
Shemi, A., & Piran, T. 1990, ApJ, 365, L55
Shibata, M., Nakamura, T., & Oohara, K. 1992, Prog. Theor. Phys., 88, 1079
Shibata, M., Nakamura, T., & Oohara, K. 1993, Prog. Theor. Phys., 89, 809
Sigurdsson, S., & Hernquist, L. 1992, ApJ, 401, L93
Taylor, J. H. 1992, Phil. Trans. R. Soc. Lond. A, 341, 117
Taylor, J. H., & Weisberg, J. M. 1989, ApJ, 345, 434
Tutukov, A. V., & Yungelson, L. R. 1993, MNRAS, 260, 675
Usov, V. V. 1992, Nature, 357, 472
Wiseman, A. G. 1992, Phys. Rev. D, 46, 1517
Wolszczan, A. 1991, Nature, 350, 688
Woosley, S. E. 1993a, *Compton Gamma Ray Observatory*, M. Friedlander, N. Gehrels, and D. J. Macomb, eds. (New York: AIP), 995
Woosley, S. E. 1993b, ApJ, 405, 273

RECENT OBSERVATIONS OF EXO 2030+375 WITH BATSE

M. T. Stollberg, W. S. Paciesas
Dept. of Physics, University of Alabama in Huntsville, 35899.
stollberg@gibson.msfc.nasa.gov

M. H. Finger, G. J. Fishman, R. B. Wilson, B. A. Harmon, C. A. Wilson
ES 66, NASA Marshall Space Flight Center, Huntsville, AL 35812.

ABSTRACT

The transient x-ray pulsar EXO 2030+375 has been detected in the Large Area Detectors (LADs) of BATSE during twelve outbursts from February 1992 to July 1993. These data have been fit to a model of the pulsar's rotational phase which assumes an independent torque for each outburst. An intensity history, a pulse period history, a pulsed fraction, and the latest orbit parameters resulting from this fit are presented.

1. INTRODUCTION

EXO 2030+375 has been detected in the BATSE Large Area Detectors (LADs) for twelve consecutive outbursts beginning in February 1992. Each outburst is detected in the first seven of the LADs 16 CONT energy channels (2.048 sec resolution, 20 – 120 keV). The pulse profiles show only slight variations when summed over an entire outburst (typically 10 – 12 days). Details on profiles and spectra have been reported by us previously in Stollberg et al (1993).

This transient has also been seen in Earth occultation data. However, due to its closeness in the sky to Cygnus X–1 and Cygnus X–3, time intervals when the flux from EXO 2030+375 can be extracted are limited. Only one outburst (TJDs 9120 – 9133) was bright enough to be detected by earth occultation at a time when the source flux could be reliably separated from that of the other two sources.

2. OBSERVATIONS

Data from DISCLA channel 1 (20 – 50 keV, 1.024 sec time resolution) were epoch–folded to produce an intensity history for the twelve outbursts seen. Figure 1 shows the intensity history for the twelve outbursts. Each outburst lasts from the day of periastron passage until 10 – 12 days after, with the peak intensity usually occurring five days after periastron passage. The strongest seen by BATSE was the first (February 1992), while the eighth (December 1992) has been the weakest to date.

The gaps between outbursts in Figure 1 have been searched for pulsations from the source: a grid search was performed on the DISCLA channel 1 data over a range of periods ~40 times the variation in period that is seen during a single outburst. The spacing between trial periods was ~3 msec. Figure 2 shows the periods with maximum χ^2 values plotted against the Truncated Julian Day (TJD). Shown are the last four outbursts (February 1993–August 1993). The size of the plotted points is proportional to the significance of the set of tested periods. The random scatter observed between outbursts shows that the source is not detected. The detection limit is $\sim 10^{35}$ ergs-sec^{-1} at 20– 50 keV, based on a 5.3 kpc distance to the pulsar as was reported by Parmar et al (1989).

Our current model for the orbit of EXO 2030+375 assumes an applied torque which is constant throughout the pulsar's orbit but is variable from orbit to orbit. Figure 3 shows the average pulse period derived for each outburst based upon this model. A constant spin–up has been observed since the beginning of the Compton Gamma–Ray Observatory (CGRO) mission.

Using data from the eleventh outburst shown in Figure 1 (TJDs 9120 – 9131), we obtain a pulsed fraction of 0.36 ± 0.05 for the energy range 30 – 70 keV. The value obtained for each subinterval is consistant with this value. This fraction agrees with the fraction obtained from the model of Parmar, White, and Stella (1989), which assumed emission from the poles of the pulsar was a near–pencil beam by the end of EXOSAT's observations of EXO 2030+375 in July 1985.

3. DISCUSSION

For the twelve outbursts seen, the χ^2 per degree of freedom from the current phase model is 3.39 for 20 free parameters. The phase residuals for EXO 2030+375 show a parabolic trend in many of the outbursts once the orbital correction from the current phase model is removed. This indicates that a constant torque applied to the pulsar over an entire orbit is too simple. In the future, we plan to investigate a phase model in which the torque is applied only during each outburst. Nothing further can be concluded about our pulsed fraction until times from other outbursts can be obtained where reliable fluxes for EXO 2030+375 exist.

Finally, the orbital parameters from the latest orbital fit are presented. The errors are an estimate of the total error for each element, adjusted by a factor of ~ 1.84 to account for the observed scatter in the residuals.

$$P_{orb} = 46.03 \pm 0.01 \text{ days}$$
$$e = 0.33 \pm 0.03$$
$$a_x \sin i = 268 \pm 25 \text{ lt-sec}$$
$$\omega = 228.2 \pm 5.7 \text{ degrees}$$
$$T_{\pi/2} = \text{JD } 2448781.0 \pm 0.3$$
$$T_p = \text{JD } 2448798.7 \pm 0.3$$

4. REFERENCES

Harmon, B. A., 2nd Compton Symposium, 1993, to be published.
Parmar, A. N., White, N. E., Stella, L., Izzo, C., & Ferri, P. 1989, ApJ, 338, 359
Parmar, A. N., White, N. E., & Stella, L. 1989, ApJ, 338, 373
Stollberg, M. T., Pendleton, G. N., Paciesas, W. S., Finger, M. H., Fishman, G. J., Wilson, R. B., Meegan, C. A., Harmon, B. A., & Wilson, C. A. 1993, in Compton Gamma-Ray Observatory, eds. M. Friedlander, N. Gehrels, & D. J. Macomb (New York: AIP Press) p.371
Wilson, R. B., Harmon, B. A., Fishman, G. J., Paciesas, W. S., & Prince, T. A. IAU Circular, 1992, 5454, 1

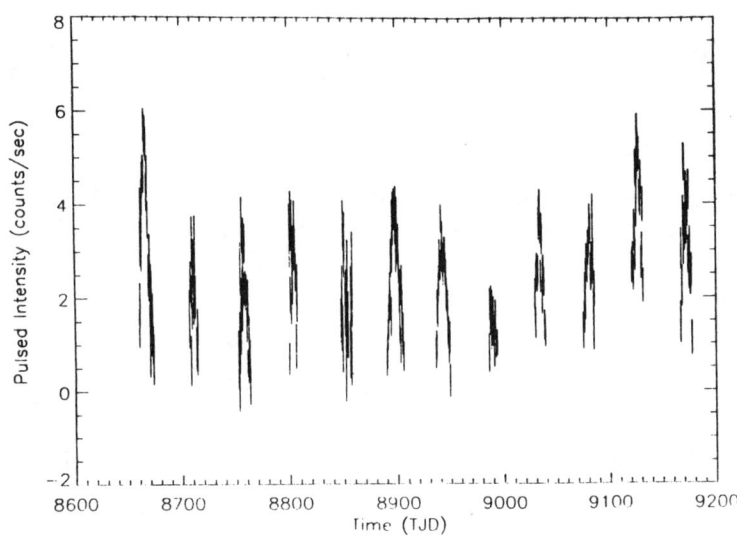

Fig. 1. Intensity history for the twelve outbursts of EXO 2030+375 seen by BATSE from February 1992 to June 1993.

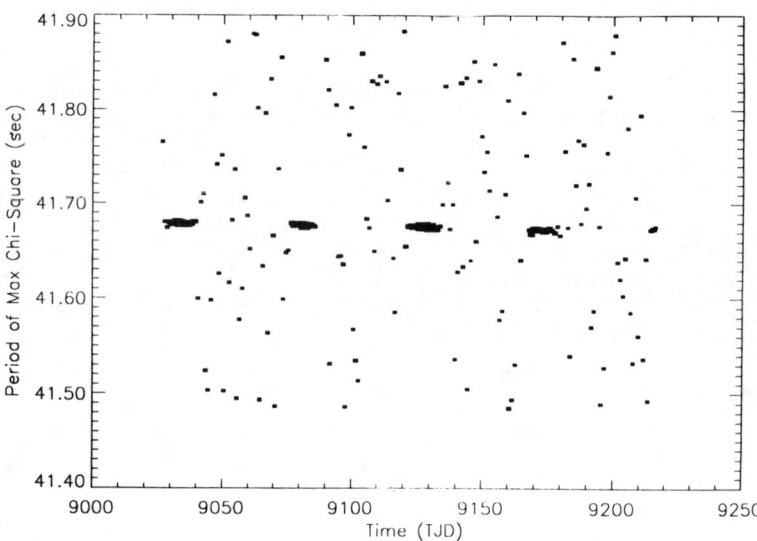

Fig. 2. Epoch folded search for detections of EXO 2030+375 based on the pulsar's pulse period.

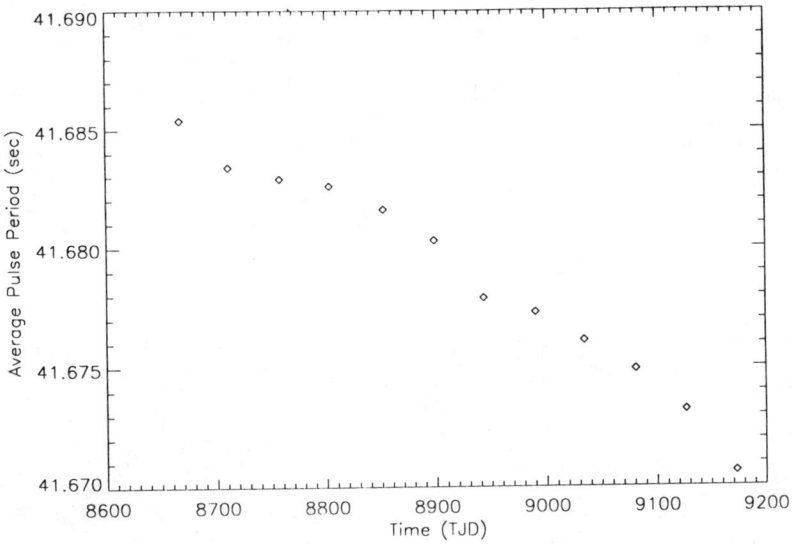

Fig. 3. The average pulse period history of EXO 2030+375 for the twelve outbursts seen by BATSE. A spin-up trend is seen. The errors associated with the pulse periods are smaller than the data points and are not shown.

BATSE OBSERVATIONS OF GS 0834-430

C.A. Wilson, B.A. Harmon, R.B. Wilson, G.J. Fishman
ES-66, NASA Marshall Space Flight Center, Huntsville, Al 35812.
wilsonc@gibson.msfc.nasa.gov

M.H. Finger
Compton GRO Science Support Center, GSFC/CSC

ABSTRACT

GS 0834-430 has been observed in seven outbursts with the Large Area Detectors (LAD's) of the Burst and Transient Source Experiment (BATSE) on the Compton Gamma Ray Observatory. We have fit observed pulse frequencies and total flux with a model that assumes binary motion and a relationship between the total flux and accretion induced torque. In addition, we have fit the first five harmonics of 4 - 6 day summed pulse profiles with a template pulse profile for each channel to detect changes in pulse shape. We present the history of the total flux and the pulsed frequency, pulse profiles, and orbital parameters.

1. INTRODUCTION

The X-ray pulsar GS 0834-430, first seen by GRANAT in 1990 (Sunyaev 1990), and identified as a new transient pulsar by GINGA (Makino 1990) has a 12.3 second spin period and is in a 111 day period binary system. The position of this system on the orbital period versus spin period diagram suggests it is a Be/X-ray binary. GS 0834-430 has been observed continuously by BATSE's LADs since April 1991, using earth occultation, epoch-folding, and fast Fourier transform analyses. The first 5 outbursts observed by BATSE occurred at regular intervals of approximately 110 days; however, the final two outbursts occurred at intervals of over 140 days. As of November 21, 1993, BATSE has seen no significant outburst from GS 0834-430 for about 175 days.

2. PULSE PROFILE EVOLUTION

Apparent variations of pulse profiles with energy have been addressed in earlier publications (Wilson 1992, Aoki 1993). We have also observed changes in pulse profile with time. For this analysis, we compared harmonic representations of 4-6 day summed pulse profiles. Only the first 5 harmonics were considered in the analysis because the sixth harmonic of this 12.3 second pulsar is very close to the 2.048 second width of the data bins. The 4-6 day intervals had a significant signal, while minimizing phase slippage. We chose the time interval from Truncated Julian Day (TJD) 8518-8521 as the template and compared it in each energy channel shown in figure 1 with 41 intervals spanning the 7 outbursts. Intervals with significant signals in less than 3 channels were not included in the comparisons. Of the 41 intervals studied, 8 showed significant deviations from the template in at least 3 channels with chance probabilities less than 5×10^{-4}. The changes appear to persist for 4-14 days, a large fraction of an outburst interval. We observed significant deviations in the first, third, fifth, and seventh outbursts, but the changes did not appear to be correlated with intensity. Possibly, these changes result from differences in detector geometries, but we believe this to be unlikely. Figure 1 shows data from the second and fifth outbursts.

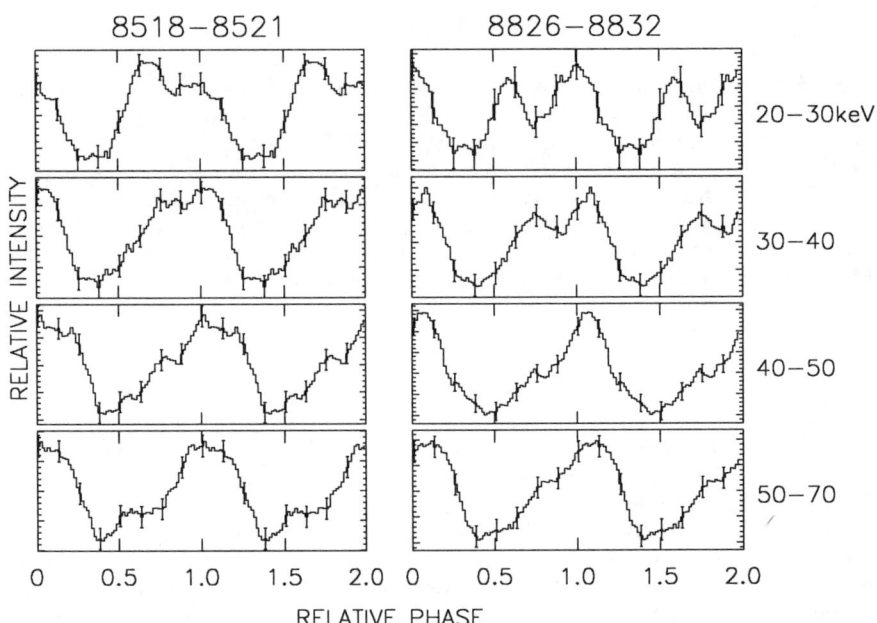

Figure 1. Our most striking example of a change in pulse shape can be seen by comparing our template (left) from September 19–22, 1991, to the shape from July 23–29, 1993. The probability the difference in shape is due to statistical errors is 4×10^{-5} for 20–30 keV, 1×10^{-8} for 30–40 keV, 9×10^{-9} for 40–50 keV, and 2×10^{-6} for 50–70 keV.

3. EARTH OCCULTATION OBSERVATIONS

Using Earth occultation analysis (Harmon 1992), BATSE has been able to observe GS 0834-430 continuously since the start of the mission. A recently developed analysis tool allows us to remove time intervals of source confusion based on the relative geometries of the earth's limb for selected sources. Figure shows the flux history of GS 0834-430 with time intervals removed where the limb separations are less than 2 degrees for Vela X-1, Cen X-3, Cen A, or GX 301-2 and GS 0834-430. Also, the time interval when GRO J0422+32 was bright has been removed. The flux data in figure 2 have been overlaid with the flux model described in section 4.

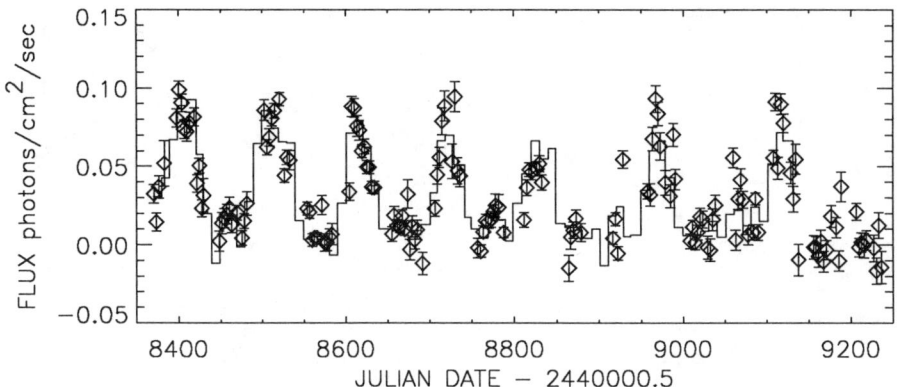

Figure 2. The flux history for GS 0834–430 was calculated by fitting a power law with a fixed index of –3.7 over the energy range 20-120 keV. The flux data is overlaid with the flux model described in section 4. The flux model still includes intervals of source confusion.

4. MODEL FITTING AND ORBITAL PARAMETERS

We simultaneously fit frequency and flux observations using a standard orbital model and

$$\dot{f} = \alpha + \beta F^\gamma$$

where $\dot{f}$ is the frequency derivative (figure 4) and F is the flux model value, computed for each 10 day interval (figure 2). The resulting χ^2 per degree of freedom was 5.06 for 363 degrees of freedom. The model parameters are reported below, with error values inflated for the large χ^2 value. The flux model fit makes a much larger contribution to the χ^2 value than the frequency model fit. Since this fit involved flux data with intervals of source confusion still included, we hope to improve the χ^2 by excluding those intervals. Figures 2 and 3 show the flux and frequency data overlaid with the models. The parameters from our best fit model are:

$P_{orb} = 111.64 \pm 0.18 \text{days}$ $T_{\pi/2} = \text{JD}2448592.20 \pm 0.51$
$e = 0.128 \pm 0.063$ $\alpha = -0.00919 \pm 0.0032 \text{cyc/day}^2$
$a_x \sin i = 205.7 \pm 5.0 \text{lt-sec}$ $\beta = 0.786 \pm 0.12 \text{cm}^2/\text{day}$
$\omega = 257 \pm 30 \text{degrees}$ $\gamma = 0.924 \pm 0.085$

5. DISCUSSION

BATSE has observed GS 0834-430 continuously for 2.5 years. Using this long-term observation, we have modelled the orbit and torque behavior of the system. We have found variations in pulse shape with energy and time. We plan to do spectral studies and further study of the pulse profile evolution. Discovery of an optical counterpart will help our understanding of the source and confirm whether or not the system is a Be/X-ray binary.

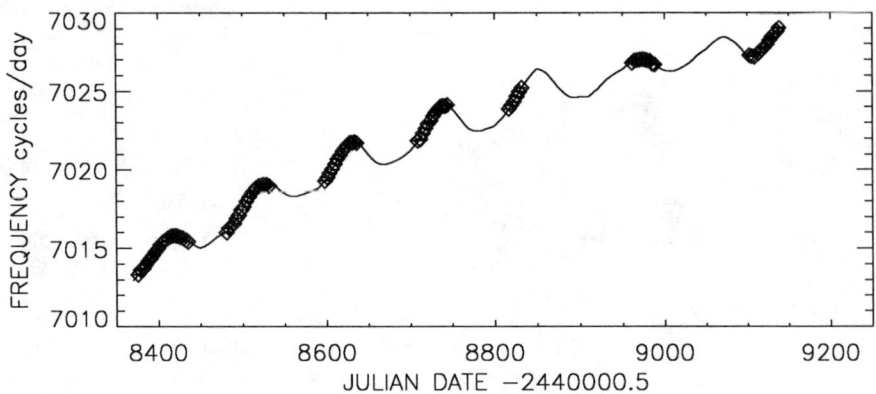

Figure 3. The frequency history generated using epoch-folding analysis has been overlaid with our frequency model.

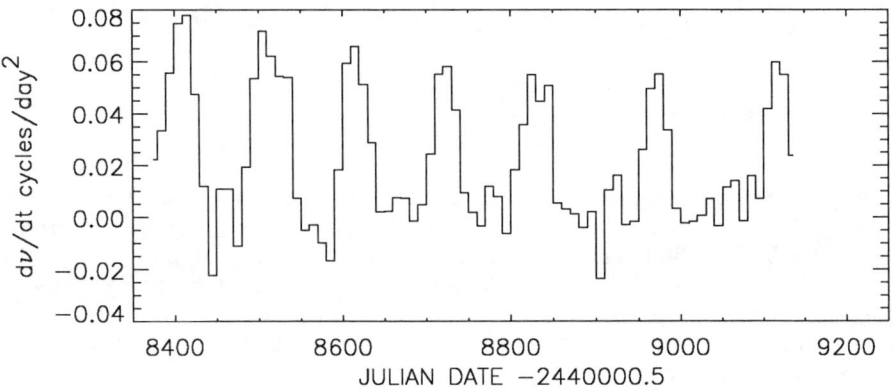

Figure 4. The frequency derivative was calculated from $\dot{f} = \alpha + \beta F^\gamma$, where F is the flux model value.

REFERENCES

Aoki, T. et al. 1993, Pub. Astro. Soc. Jap.,45
Corbet, R.H.D. 1986, Mon. Not. R. Astr. Soc., 220, 1047
Harmon, B.A. et al. 1992, in The Compton Observatory Science Workshop, eds. C.R. Shrader, N. Gehrels, & B. Dennis (NASA CP-3137) p.69
Makino, F. 1990, IAU Circ., No. 5142
Sunyaev, R.A. 1990, IAU Circ., No. 5122
Wilson, C.A. et al. 1992, in Compton Gamma-Ray Observatory, eds. M. Friedlander, N. Gehrels, & D.J. Macomb (New York: AIP Press)p.376

HARD X-RAY VARIABILITY OF CYGNUS X-3

S. M. Matz, D. A. Grabelsky, W. R. Purcell, M. P. Ulmer
Department of Physics and Astronomy
Northwestern University, Evanston, IL 60208

W. N. Johnson, R. L. Kinzer, J. D. Kurfess, M. S. Strickman
E. O. Hulburt Center for Space Research
Naval Research Laboratory, Washington DC 20375

ABSTRACT

The OSSE instrument on the *Compton Gamma-Ray Observatory (GRO)* made 4 observations of Cygnus X−3 between May 1991 and December 1992. OSSE observed a substantial change in the average hard X-ray flux, which has previously been observed to be stable (variations < 20%). Over a 2-3 day period the flux above ~ 50 keV decreased by roughly a factor of 3. This occurred about 15 days prior to a large radio flare from this source. OSSE detected the 4.8 hr period and observed significant orbital modulation of the hard X-ray ($\gtrsim 50$ keV) flux, in conflict with some extrapolations from lower energies. The spectrum in the high state is consistent with a power law, $\propto E^{-3.0}$.

1. INTRODUCTION

Cygnus X-3 is a bright X-ray source located in the plane of the Galaxy behind enough material to make it effectively unobservable at optical wavelengths. Soft X-ray and infrared observations have established that it has a very stable 4.8 hour period and an asymmetric light curve. The period probably corresponds to the orbit of a compact source around a primary star. This period is typical of low-mass X-ray binaries; however, there is evidence that the companion is actually a high-mass Wolf-Rayet star (van der Kerkwijk et al. 1992). Cyg X-3 produces occasional, very intense radio flares, reaching levels up to 20 Jy (e.g., Waltman et al. 1991), with associated jets (Geldzahler et al. 1983; Molnar, Reid, & Grindlay 1988). Bonnet-Bidaud & Chardin (1988) have published a valuable review of Cyg X-3 observations and theories.

2. OBSERVATIONS

The Oriented Scintillation Spectrometer Experiment (OSSE) on *GRO* (Johnson et al. 1993) consists of 4 large, independent NaI/CsI phoswich detectors collimated to $3.8° \times 11.4°$ fields of view and covering the energy range 50 keV to 10 MeV. OSSE observed Cyg X-3 four times: 1991 May 30 – Jun 15 (viewing period 2), Aug 8 – 15 (VP 7), Nov 28 – Dec 11 (VP 15), and 1992 Dec 9 – Dec 15 (VP 203). Spectral data were accumulated in two-minute intervals, alternating between source and background pointings. There was no significant contamination from Cyg X-1 in any of the pointings. Using the standard OSSE analysis techniques, estimated background spectra were subtracted from each source spectrum. OSSE detected significant flux from Cyg X-3 up to ~ 200 keV.

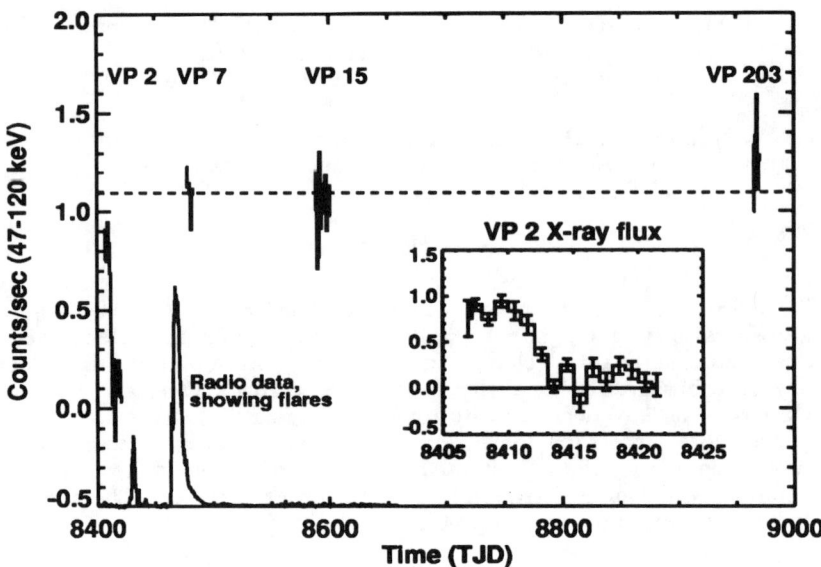

Figure 1: The daily average Cyg X–3 intensity, 47–120 keV, for the four OSSE observations. The inset shows an expanded view of the decrease during VP 2. Also included at the bottom of the plot is the 2.25 GHz radio intensity, showing the flares following the hard X-ray decrease.

3. HARD X-RAY STATE CHANGES

The soft X-rays from Cyg X–3 have two distinct intensity states, with state changes occurring on times scales of $\sim$ 0.5–2.0 years (Smale & Lochner 1992; Priedhorsky & Terrell 1986). Prior to OSSE, all hard X-ray experiments measured approximately the same average flux from Cyg X–3 with no significant variations during their observations, even when the soft X-ray flux changed state (e.g., White & Holt 1982; Hermsen et al. 1987). This suggests that the hard X-rays may represent a separate physical component.

OSSE observed about a factor of 3 decrease in the average hard X-ray intensity of Cyg X–3 over a period of several days during VP 2 (Fig. 1). The flux remained low for the rest of the observation. When OSSE observed this source again the flux had risen to a level higher than that seen prior to the decrease. The flux in the final three observations was roughly constant. The X-ray decrease occurred about 15 days prior to an intense (5 Jy) radio flare (R. Fiedler, private communication).

4. ORBITAL LIGHT CURVES

The two-minute integrated background-subtracted count rates (47–120 keV) during each of the four observation intervals were epoch folded using both the quadratic and cubic X-ray ephemerides of van der Klis & Bonnet-Bidaud (1989) to produce orbital light curves. The EXOSAT template light curve (van der Klis

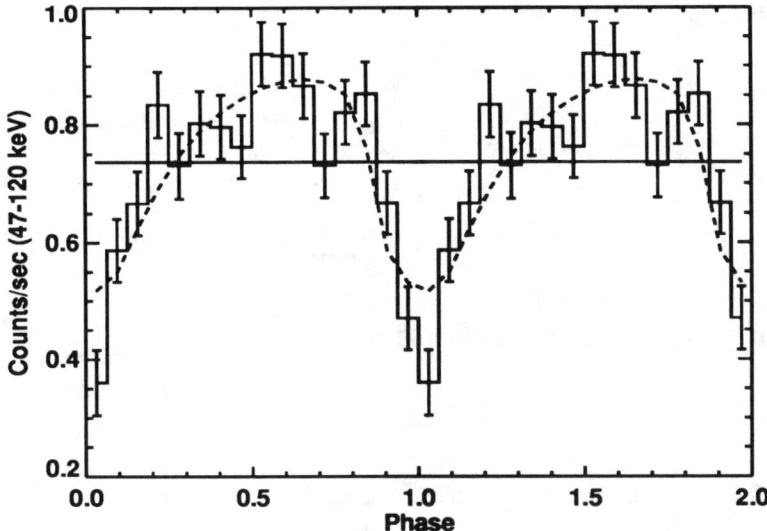

Figure 2: The average Cyg X-3 high-state light curve, 47–120 keV, folded using the cubic X-ray ephemeris of van der Klis & Bonnet-Bidaud (1989).

& Bonnet-Bidaud 1989) was fit to the OSSE light curve for each observation, allowing the phase, amplitude, and background (DC) level to vary. The "zero phase" of the fitted light curve produced using the cubic ephemeris (0.046 ± 0.015) was marginally consistent with 0 but the phase using the quadratic ephemeris (0.879 ± 0.016) was not. The average light curve in the high state is shown in Figure 2, along with the best fit of the soft X-ray light curve from EXOSAT.

EXOSAT measured the modulation ((max - min)/max) in the orbital light curve as a function of energy (Willingale, King, & Pounds 1985) and found only weak modulation in the 20–50 keV band. In contrast, OSSE observed strong modulation of the $\gtrsim 50$ keV flux as seen in the average light curve (Fig. 2). The average value of the modulation is 0.38 ± 0.04 for all the observations. There is no statistically significant variation in this value between observations. The amplitudes of the EXOSAT template fitted to the OSSE light curves are also consistent with a constant value for all the observations. The statistics are not adequate to determine if the light curve amplitude changes with the drop in DC flux during VP 2.

5. SPECTRA

The integrated spectrum from each interval was consistent with a photon power-law, though other models cannot be excluded. The spectral index during all the high state observations (VP 7, 15, and 203) was -3.0 ± 0.1. The index in the low state was -2.2 ± 0.4, consistent with the high state value at the 2σ level. The index for the first part of VP 2 was -2.5 ± 0.1, marginally ($\sim 4\sigma$) harder than that seen in the high state. This may be a third distinct state or a mixture of the high and low states. Figure 3 compares the high and low state spectra.

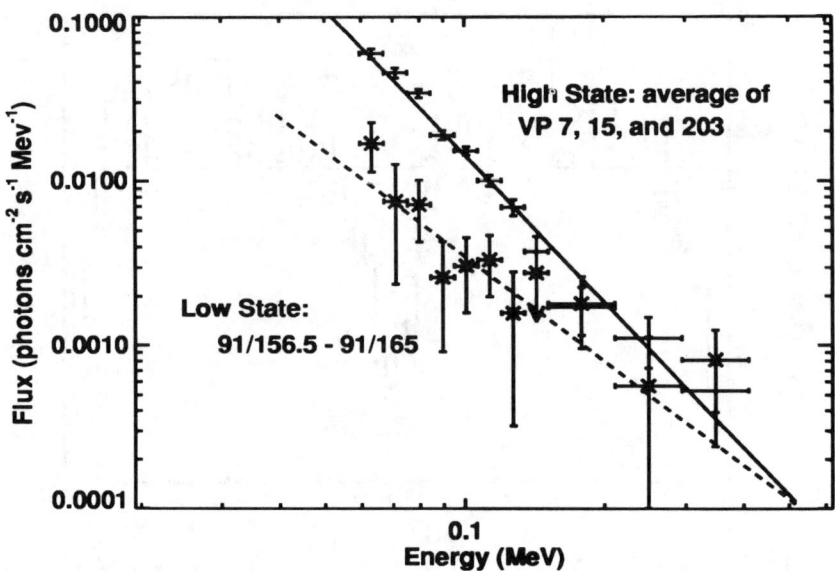

Figure 3: The hard X-ray spectra of Cyg X–3 in both the high and low states.

We accumulated the two-minute background-subtracted spectra for each observation interval into 8 orbital phase bins. The spectra from each interval and phase bin were then fit to a power-law photon spectrum. We saw no statistically significant phase variation of the spectral hardness $\gtrsim 50$ keV in the high state.

ACKNOWLEDGEMENTS

We thank R. Fiedler and E. Waltman for supplying the NRL-GBI radio data for Cyg X-3. This work was supported by NASA grant DPR S–10987C.

REFERENCES

Bonnet-Bidaud, J. M. & Chardin, G. 1988, Physics Reports, 170, 326
Geldzahler, B. J., et al. 1983, ApJ, 273, L65
Hermsen, W., et al. 1987, A&A, 175, 141
Johnson, W. N., et al. 1993, ApJS, 86, 693
Molnar, L., Reid, M. J., & Grindlay, J. E. 1988, in The Impact of VLBI on Astrophysics and Geophysics, ed. M. J. Reid & J. M. Moran (Dordrecht: Reidel), 279
Priedhorsky, W. & Terrell, J. 1986, ApJ, 301, 886
Smale, A. P. & Lochner, J. C. 1992, ApJ, 395, 582
van der Kerkwijk, M. H., et al. 1992, Nature, 355,703
van der Klis, M., & Bonnet-Bidaud, J. M. 1989, A&A, 214, 203
Waltman, E. B., et al. 1991, ApJS, 77, 379
White, N. E. & Holt, S. S. 1982, ApJ, 257, 318
Willingale, R., King, A. R., & Pounds, K. A. 1985, MNRAS, 215, 295

THE PRECESSIONAL PHASE DEPENDENT X-RAY EMISSION OF SS433

W.Yuan[†] N. Kawai and M. Matsuoka
The Institute of Physical and Chemical Research
Wako, Hirosawa 2-1, Saitama, Japan

ABSTRACT

The evolution of X-ray emission of SS433 over its precessional period gave an evidence to the existence of an alternative harder X-ray emission region apart from the jets. The spectral fits with a two continuum component model involving such a hard X-ray component gave further support to this scenario.

1. INTRODUCTION

SS433 was observed with Ginga in 1987, 1988, and 1989, whose results have been presented by Kawai et al.(1989), Kawai (1989), and Brinkmann et al.(1989, 1991). With these observations, considerable progress has been made in understanding the properties of the jets and and the physical conditions of the X-ray emitting regions in the enigmatic object SS433. The Ginga observations confirmed the presence of a Doppler shifted FeXXV Kα emission line (narrow line), emitted from the outer part of the X-ray jet (Watson et al.1986). These observations were explained in terms of emission from one jet only, which was found incorrect by ASCA observation (the ASCA team 1993). However, we have shown that the Ginga observations are in fact consistent with the ASCA results and therefore the above jet emission model derived from the Ginga data is justified (Yuan et al.1993). Further, a broad emission line structure (broad line) with a line width of $\sigma \sim 0.7\text{-}1.3$ keV was seen which does not follow the precessional Doppler shift but always centers at around 7 keV. The X-ray continuum, expected to be emitted mainly from the jets, turned out to be thermal bremsstrahlung with temperatures up to more than kT = 30 keV. The large equivalent width of the broad line requires either a large iron overabundance of 5 to 6 times the cosmic value, or it originates from "leakage" of radiation from the central region near the compact object (Brinkmann et al.1991). New Ginga observations have been performed in 1990 and twice in 1991. We combined all Ginga data covering the more precessional phases (Fig.1) to study the X-ray evolution over the 163-day precession period of SS433.

2. PRECESSIONAL PHASE DEPENDENT X-RAY EMISSION

We fitted the average spectra over longer observation time for each observation run to get better statistics and to reduce the possible effects of short term fluctuations. The acceptable fits for most of the data were in accordance with the results of the previous fits to the individual data segments. We plotted the observed parameters of SS433 as function of the precessional phase to investigate

† On leave from: Beijing Astronomical Observatory, Beijing, 100080, China

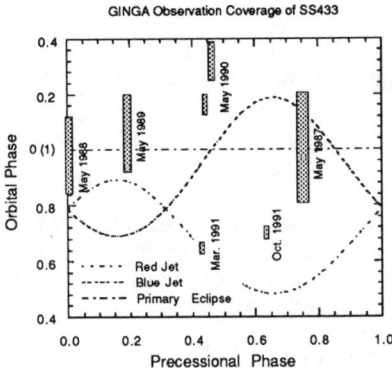

Fig.1
The observation coverage of the precessional and binary phase of SS433 by Ginga, the binary phase refers to the optical minimum. For illustrative purpose, the well known Doppler curve (redshift versus precessional phase) is displayed.

the intrinsic variation of the X-ray emission over the precession period. To obtain the intrinsic physical quantities in the rest frame of the jets, Doppler corrections have been applied to the observed fluxes and temperatures corresponding to the precessional phase of the observations. For the details of Doppler correction for each observation, refer to Yuan et al.(1993). The derived intrinsic intensities for the continuum and the narrow line in the plots are referred to the emission from one jet only and are represented in terms of energy flux (erg s^{-1} cm^{-2}).

The variations of the observed and intrinsic continuum luminosities are shown in Fig.2a with the precessional phase. It shows that at later phases where the jet moves more towards the observer, the intrinsic X-ray flux increases, as expected by relativistic boosting. However, it is noted that the intrinsic luminosity of the jet clearly shows the same tendency of variation, which is rather regular with respect to the precession motion of the jets. This fact suggests that the interpretation in terms of beaming effect cannot account for the X-ray luminosity evolution over the precession period of SS433, *i.e.* the luminosity evolution probably arises from the intrinsic variant radiation of the system, rather than Doppler boosting effects only. Fig.2b shows the variation of intrinsic temperature of thermal bremsstrahlung which shows the similar tendency.

The variation of the narrow line intensity in the rest frame of the jets (Fig.3a) shows that the narrow line intensity is relatively stable (the variation is less than 50%) compared to the significant variation in the continuum ($\sim$ 200%). Fig.3b clearly shows that the broad line intensity varies significantly similar to the behavior of the continuum. The variation of the width of the broad line, shown in Fig.3c, is positively correlated to the broad line intensity.

3. PHYSICAL PICTURE

The intrinsic luminosity of the narrow iron line obtained in observations is relatively constant, suggesting that the X-ray jets are relatively stable. We conclude that the X-ray variation of SS433 is dominated by a precessional phase dependent modulation rather than temporal variability, and is therefore attributed to an alternative emission region following the precession of the jets. Thus We conclude that we have found an evidence for the existence of an additional X-ray emission region apart from the jets, which was suggested by Brinkmann et al.(1991) as a scattering region at the base of the jets or the center of the disc, accounting for the emission of part of the continuum and for the broad line. The increase in X-ray luminosity at later phases of high jet inclination angles is attributed either to the fact that we are seeing deeper into the base of the

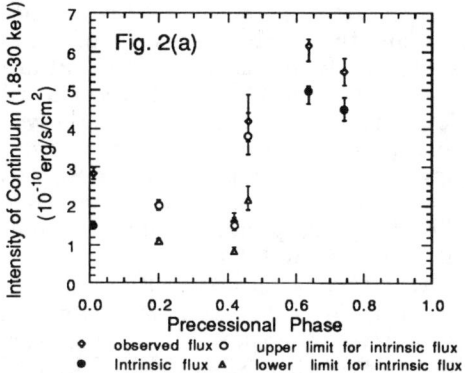
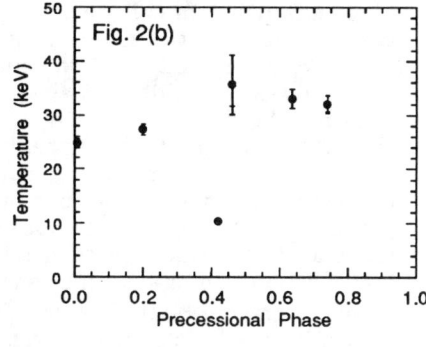

Fig.2 (a) The observed luminosity and the derived intrinsic values, (or upper and lower limit) for the continuum versus the precessional phase. The intrinsic values are the luminosities from one jet only.
(b) The thermal bremsstrahlung temperatures (relativistic correction has been made) fitted with the two-line model versus precessional phase.

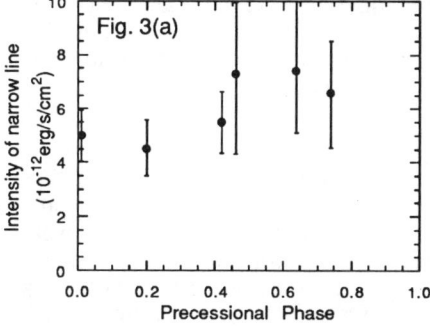
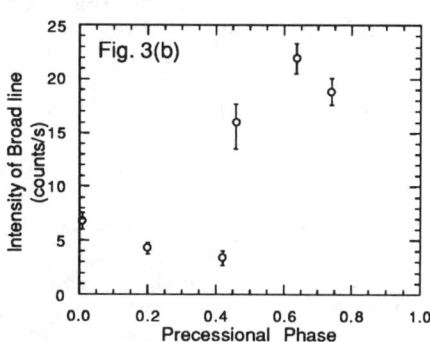

Fig.3 The variation of the line parameters versus the precessional phase.
(a) The Doppler corrected luminosity of the narrow line (from one jet only).
(b) The broad line intensity in counts/s. (c) The broad line width in keV.

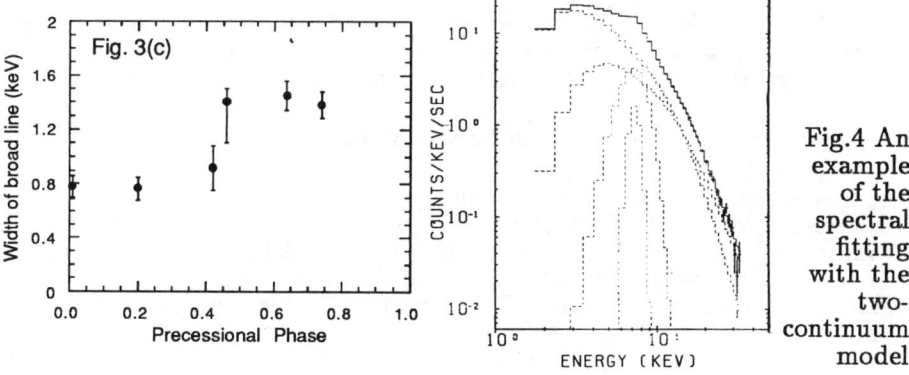

Fig.4 An example of the spectral fitting with the two-continuum model

blue-shifted jet, or we are seeing a larger emission region due to the change of the viewing conditions, *i.e.* subject to a modulation of the precession of the jets.

We have tried to fit the spectra with models consisting of two continuum components with different absorption column densities, and with two Gaussian lines for the line region as in the previous fitting. For the jet emission we fix the thermal bremsstrahlung temperature at around 12 keV which was estimated as the spectral temperature found during the X-ray eclipse when most of the central emission region is occulted. This two-continuum model generally improves the fits for most of the data. The hard component could be fitted as thermal bremsstrahlung with temperature from 50 to 80 keV, or a hard power law. The absorbing column density for the harder component turned out to be in the range $N_H \sim 10^{22} - 10^{23} \mathrm{cm}^{-2}$. An example of the fits with this model is shown in Fig.4. We have searched for a best fit region for this combination of the low and high temperature continuum in a two dimensional grid. The region of acceptable fits with confidence level higher than 90% turned out to be so large (*e.g.* kT~12-24 keV for the jet temperature) that we are not able to constrain the model parameters. And the fitted high temperatures exceeded the limit of Ginga capability. Nevertheless, the fits suggest that more component continuum models are acceptable.

The broad line is regarded originating from such a scattering region. A "leakage" model, in which at all observable viewing angles we can see the reprocessed radiation only and the direct emission from a central object is occulted by the thick disc, can lead to a larger equivalent width of broad line, and thus reduces the required abundance of iron significantly.

4. CONCLUSIONS

We conclude that the X-ray continuum and broad line emission are modulated by the viewing angle (precessional phase), whereas the jets are relatively stable, though one should keep in mind that the temporal fluctuation of the system could never be ruled out. The above discussion suggests consequently that the observed X-rays are in fact composed of two distinct components: thermal emission from the jets, and a harder component from the base of the jets or the central region of the accretion disc.

ACKNOWLEDGEMENTS

We thank W. Brinkmann for helpful discussions. W.Yuan is grateful to the scientific exchange programme of RIKEN and Chinese Academy of Sciences. The authors thank all members of the Ginga team.

REFERENCES

the ASCA team, 1993, in preparation
Brinkmann W., Kawai N., Matsuoka M., Fink H.H.,1991, A&A 241, 112
Brinkmann W., Kawai N. Matsuoka M., 1989, A&A 218, L13
Kawai N., 1989, paper presented at the 23 ESLAB Symp.
Kawai N., Matsuoka M., Pan H.C., Stewart G.C., 1989, PASP 41, 491
Watson M.G., Stewart G.C. Brinkmann W., King A.R., 1986, MNRAS 222,102
Yuan W. et al., 1993, in preparation

IUE LOW AND HIGH DISPERSION SPECTRA OF THE Be-X STAR BD+53°2790 = 4U2206+54

Massimo Teodorani, Adriano Guarnieri, Corrado Bartolini, Adalberto Piccioni
Dipartimento di Astronomia, Universitá di Bologna, Italy
adriano@alma02.bo.astro.it

ABSTRACT

The results of the analysis of two IUE spectra of BD+53° 2790 are presented. Most of the lines were in absorption showing a blue-shifted core indicative of mass ejection from thestar. A wide post-corona zone was put in evidence by the presence of emission lines. P Cygni profiles showed a terminal velocity of the stellar wind in the range 1000 ÷ 2000 km/s.

1. INTRODUCTION

BD+53° 2790 is the optical counterpart of the X-ray source 4U2206+54, a B1Ve star whose V magnitude is variable on medium (days) and long (months) time-scales around $9^m.8$. Some features detected in our optical spectra, particularly the $H\alpha$ emission line, showed evidence of aperiodic mass ejection events affecting the equatorial envelope of the Be star (Teodorani et al 1991). The X-ray luminosity is variable in the range $10^{34} \div 10^{35}$ erg/s (Steiner et al. 1984).

In 1988 BD +53° 2790 showed a strong emission of the lines He I 4121, 4921, 5015 which are normally seen in absorption (Guarnieri et al. 1991): this transient event along with the unusually high X-ray luminosity of 4U2206+54, a factor 10 ÷ 100 over the normal coronal emission of O-B stars, suggests that the Be star is accompanied by a neutron star which might accrete matter from the equatorial envelope of the optical component (van den Heuvel 1987).

We present here the first results of our analysis of the ultraviolet spectra of the star BD+53° 2790, which we observed on June 18, 1990 at the ESA Vilspa tracking station with the IUE satellite.

2. LOW AND HIGH DISPERSION IUE SPECTRA

We acquired two IUE low dispersion spectra using the SWP and LWP facilities of the IUE satellite in the ranges 1200 ÷ 1900 Å and 1900 ÷ 3300 Å respectively. During the same shift we acquired one high dispersion SWP spectrum in the range 1200 - 1900 Å. We analyzed both low and high dispersion spectra by means of the MIDAS package. After merging and processing the two SWP and LWP low dispersion spectra calibrated in wavelength and flux we determined the value of E(B-V) = 0.60 on the basis of the analysis of the interstellar band at 2000 ÷ 2500 Å (Fig. 1). Then, after accomplishing a preliminary line identification we normalized the high dispersion spectrum to the continuum and measured the equivalent width, the terminal velocity and the blue shifts of its superionized lines (Doazan 1987, Snow 1987).
The identified lines in the high dispersion SWP spectrum are shown in Table 1.

Table 1
BD+53° 2790 identified lines in the range 1200 ÷ 2000 Å

Line	Identification	Profile	Origin	e.w. (Å)
1238.80	N V	Absorp.	Stellar	2.48
1242.80	N V	Absorp.	Stellar	1.60
1245	N V ?	Emiss. ?	Stellar ?	
1287	?	Absorp.	Stellar	1.02
1314	?	Absorp.	Stellar ?	0.65
1339	?	Absorp.	Stellar ?	0.53
1343	O IV ?	Absorp.	Stellar ?	1.22
1362	Si III	Absorp.	Stellar ?	1.15
1375	?	Absorp.	Stellar ?	0.83
1393.75	Si IV	Absorp.	Stellar	1.77
1396	Si IV ?	Emiss. ?	Stellar ?	
1402.77	Si IV	Absorp.	Stellar	1.82
1431-34	?	Absorp.	Stellar ?	0.99÷0.72
1533.45	Si II	Absorp.	Stellar	2.20
1548.18	C IV	Absorp.	Stellar	2.21
1550.76	C IV	P Cygni	Stellar	0.81(em. 6.68)
1592	O III ?	Absorp.	Stellar ?	1.42
1617	?	Absorp.	Stellar ?	0.98
1620	N V ?	Absorp.	Stellar ?	1.97
1631	Si IV ?	P Cygni ?	Stellar	1.16(em. 0.86)
1640.53	He II	Absorp.	Stellar	1.85
1647.10	Fe II	Absorp.	Stellar	0.81
1663	O III	P Cygni ?	Stellar	0.82(em. 0.48)
1681	?	Emission	Stellar ?	1.43
1718	?	Absorp.	Stellar ?	1.59
1724	?	Absorp.	Stellar ?	1.72
1746.80	N III	Absorp.	Stellar	0.40
1757	?	Emission	Stellar ?	0.48
1791	?	Absorp.	Stellar ?	0.71

3. SOME CONSIDERATIONS AND CONCLUSIONS

The profile of the C IV doublet 1548-51 is clearly of P Cygni type (Fig. 2). We found some evidence of P Cygni effect also from the profile of N V 1239-43 and Si IV 1394-1403 doublets and, possibly, of Si IV 1631 and O III 1663. From tentative measurements of the blue edges of their (asymmetrical) absorption components we estimated that the terminal velocity of the wind ranged from 1000 to 2000 Km/s. Some unidentified prominent emission features were present at 1681 Å and 1757 Å. No apparent evidence of narrow absorption components (Henrichs 1984) was detected.

Moreover, our SWP high dispersion spectrum showed evidence of mass ejection: we interpret the blue shifts of the center of the absorption profiles as indicative of ejection velocities of the order of 200-300 Km/s. This characteristic of the "lines dynamics" probably originates in the inner part of the UV emitting envelope, from which the ejected gas is accelerated at terminal velocities to

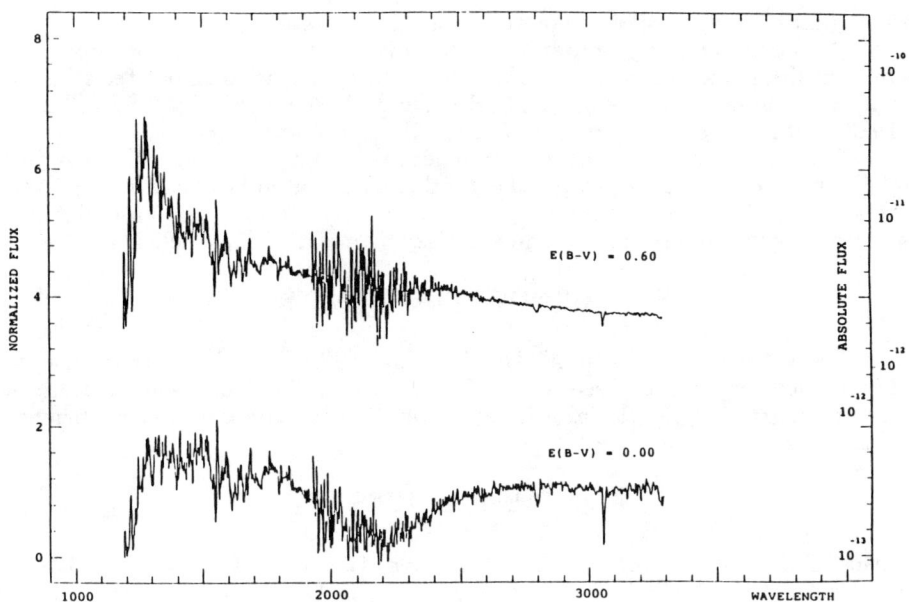

Fig. 1. Determination of the E(B-V) of BD+53° 2790 from a IUE spectrum

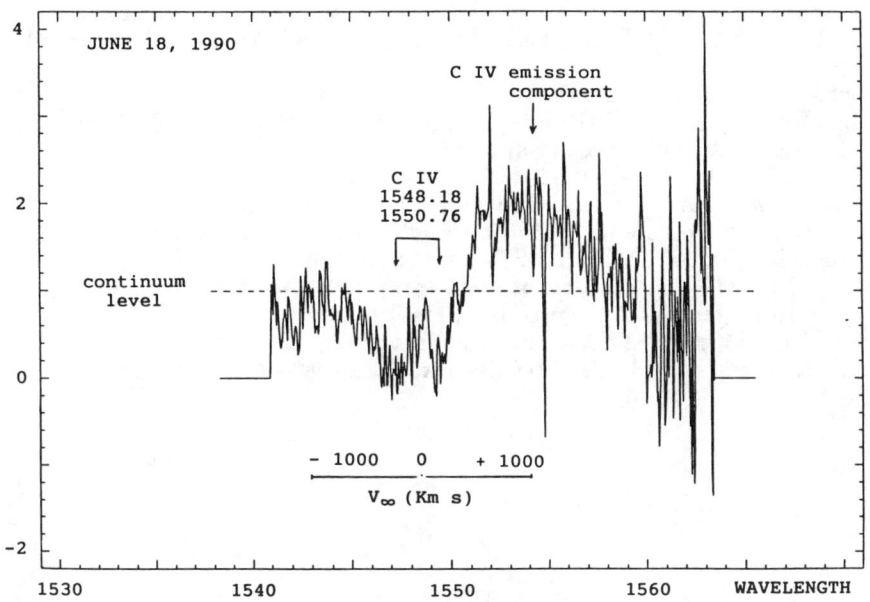

Fig. 2. P Cygni profile of the C IV doublet at 1548-51 Å in BD+53° 2790

higher distances by a piston-like action of the radiation pressure.

The observed morphology and the quantitative analysis of superionized resonance lines, particularly of C IV doublet, are similar to characteristics found in the Be-X binaries LSI +61° 303 (Hutchings 1979) and A0538-66 (Charles et al 1983), which are strong multi-wavelength transient sources.

Finally, the presence of emission features in the UV spectrum indicates that the post-corona zone, where the wind is accelerated (Doazan 1987), extends in this case to larger distances from the central star than in the case of common Be stars, which usually do not show emission profiles (Snow 1987).

ACKNOWLEDGEMENTS

The authors wish to thank Dr. Pierluigi Selvelli for his help in discussing the identification of the lines in the IUE spectra. This work was partially supported by a grant of the Italian Ministero per l'Universitá e la Ricerca Scientifica e Tecnologica.

REFERENCES

Charles, P.A., Booth, L., Densham, R.H., Bath, G.T., Thorstensen, J.R., Howarth, I.D., Willis, A.J., Skinner, G.K., & Olszewski, E. 1983, MNRAS, 202, 657

Doazan, V. 1987, in Physics of Be Stars eds A. Slettebak & T.P. Snow (Cambridge: Cambridge Univ. Press) p.384

Guarnieri, A., Bartolini, C., Civello., R., Piccioni, A., & Teodorani, M. 1990, in Structure and properties of Accretion Disks, eds. C. Bertout, S. Collin, J.P. Lasota & J. Tran Thanh Van (Paris: Ed. Frontières) IAU Coll. 129, p.435

Henrichs, H.F. 1984, ESA SP-218, p.43

van den Heuvel, E.P.J., & Rappaport, S. 1987, in Physics of Be Stars, eds. A. Slettebak & T.P. Snow (Cambridge: Cambridge Univ. Press) p.291

Hutchings, J.B. 1979, PASP 91, 657

Teodorani, M., Bartolini, C., Guarnieri, A., & Piccioni, A. 1990, in Confrontation between stellar pulsation and evolution, eds C. Cacciari & G. Clementini (San Francisco: ASP Conf. Ser. Vol. 11)p. 308

Snow T. P. 1987, in Physics of Be Stars, eds. A. Slettebak & T.P. Snow (Cambridge: Cambridge Univ. Press) p.250

Steiner, J.E., Ferrara, A., Garcia, M., Patterson, J., Schwartz, D.A., Warwick, R.S., Watson, M.G., & McClintock, J.E. 1984, ApJ, 280, 688

ON THE NATURE OF LSI +65° 010, THE OPTICAL COUNTERPART OF THE X-RAY BINARY 2S0114+650

R. Minarini, M. Teodorani, C. Bartolini, A. Guarnieri, A. Piccioni
Dipartimento di Astronomia, Universitá di Bologna, Italy
bartolini@alma02.bo.astro.it

ABSTRACT

Spectroscopic and photoelectric observations taken at the Bologna Observatory show that LSI +65° 010 presents a variability on a time scale of days. We report two optical transient events of this system, which are more consistent with the nature of a Be-X ray binary than that of a supergiant-X ray binary.

1. INTRODUCTION

LSI +65° 010 is the optical counterpart of the massive X-ray binary 2S0114+650, whose orbital period is 11.6 days (Crampton et al. 1985). This system has been observed in a number of spectral regions, ranging from the optical band to X-rays; its quiescent X-ray luminosity is $L_x \simeq 10^{33}$, but aperiodic bursts up to 10^{35} erg/s are also present. There are indications for pulsations with a period P$\simeq 14 \div 15$ min. (Koenigsberger et al. 1983; Yamauchi et al. 1990).

LSI +65° 010 is an eleventh-magnitude reddened star (B-V=1.09, U-B =-0.06) of spectral type B0.2. Hα has been always observed in emission. The determination of luminosity class leads to different results depending on the lines which are used for the classification: H lines suggest a luminosity class III while metal lines indicate Ia (Aab et al. 1983). The broadening of photospheric lines indicates a rotational velocity $v_{rot} \sin i \simeq 100$ Km/s. For complete information on this object see also Finley et al. (1992) and references therein.

It is known that there are two types of massive X-ray binaries (MXRBs), differing in the optical star: in the "standard" MXRBs this is a supergiant star (luminosity class I-II), whereas a Be star (luminosity class III-V) characterizes the Be-X ray binaries. Standard MXRBs are characterized by a high and almost constant X-ray luminosity ($L_x \sim 10^{37} \div 10^{38}$ erg/s), a short orbital period (tipically a few days long), and a relatively long pulse-period of the neutron star ($P_{pul} \sim 10^2 \div 10^3$ s). In these systems the neutron star accretes matter via stellar wind and/or Roche-lobe overflow of the primary star. The Be-X ray binaries show a variable and sometimes transient X-ray luminosity ($L_x \sim 10^{33} \div 10^{39}$ erg/s; but some low luminosity systems have a nearly constant luminosity), a long orbital period (weeks or months), and a short pulse period ($P_{pul} \sim 1 \div 10^2$ s). In these systems the neutron star accretes matter from a disk variable in density and dimensions, which surrounds the Be star.

The system LSI +65°010 = 2S0114+650 presents characteristics of both these subclasses, both in the optical and in the X-ray range. In the Corbet (1986) diagram 2S0114+650 falls in the "standard" MXRBs zone, assuming that the quoted pulse period is correct. Because of this ambiguity, in 1984 we started a programme of optical observations to get information about the variability of

LSI +65°010 and, if possible, to discriminate between the two classes to which this system appears to belong. In this paper we report two transient phenomena that we have observed.

2. TWO OPTICAL TRANSIENT EVENTS

On September 3, 1984, we obtained a spectrum of LSI +65°010 using the 152-cm telescope of the University of Bologna and a Boller & Chivens 26767 grating spectrograph with an image tube intensifier EMI 9914. A 831 lines/mm grating was used, which corresponds to a reciprocal dispersion of 52 Å /mm in the red band. This spectrum (Fig.1, top) is very different from the other spectra we acquired both before and after the time of this observation. FeII lines in emission and some strong absorptions of C, N, O and Si were present. Hα was very weak in emission. These features suggest that a shell of matter was formed near the photosphere of the central star. Four days later, on September 7, these peculiar features disappeared and Hα was strong in emission (Fig. 1, bottom). This could indicate that the emitted shell had arrived in a more external zone of the envelope. Van Kerkwijk & Waters (1989) observed a similar shell event in 1986.

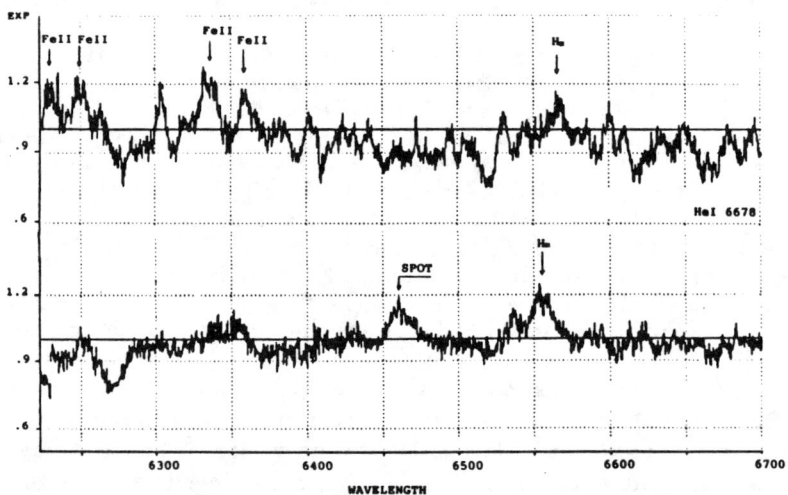

Fig. 1. The red spectrum of LSI +65°010, taken on September 3, 1984, showing a shell-like event (top) compared to a "normal" one(bottom), taken four days later.

We performed UBV photometry at the 60-cm telescope of the Bologna University in Loiano (Fig. 2). The luminosity showed an increase of 0.2 magnitudes in U and 0.1 magnitudes in B and V with respect to the mean values of the night. We estimate that the event lasted approximately 10 minutes. This was the second flare observed on this object after the first one reported by Corral and Koenigsberger (1987). We suggest that these flares could be due to a variation of the f-f emission from a part of the envelope of the Be star, caused by a local increase in density. Actually, the greater increase revealed in the U band is not consistent with the expected proportionality of f-f emission to λ^2 but, because of the temporal sequence of observations (U → B → V) and the short duration of the event, we could have detected the increase near the maximum in the U band.

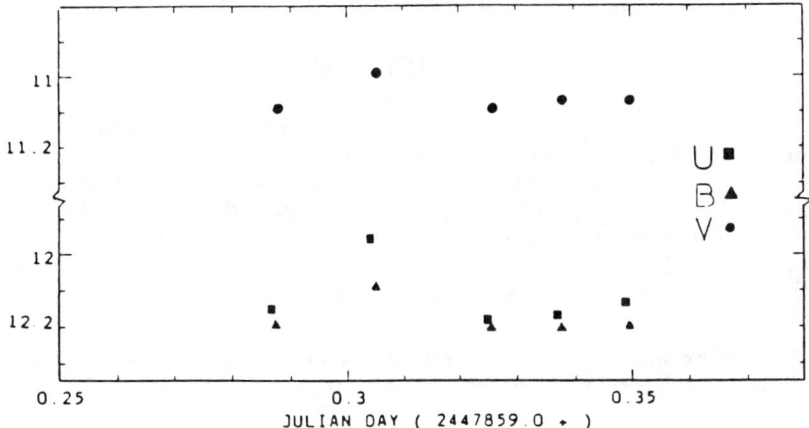

Fig. 2. The optical flare of LSI +65°010 observed on November 28, 1989

3. SPECTROSCOPIC OBSERVATIONS

In the last nine years we collected other spectroscopic data suggesting ideas on the modality of mass emission and crossing the equatorial envelope of the star.

CCD spectra acquired at the Bologna Observatory in the period 1988-1991 show that Hα emission and Hβ and He absorptions are variable on a time scale of days. The values of the EWs of Hα, Hβ and HeI 6678 are reported in Table 1.

Tab. 1. Equivalent widths of Hα, Hβ, HeI 6678

DATE	Hα	Hβ	HeI 6678	DATE	Hα	Hβ	HeI 6678
Nov.26/88	-1.5	1.0	0.80	Mar.01/90	-1.3	–	0.85
Nov.27/88	-1.8	–	0.97	Oct.09/91	-1.6	1.3	1.00
Oct.16/89	-3.0	–	0.58	Nov.02/91	-1.8	1.0	0.89
Nov.07/89	-2.4	–	0.77	Feb.18/92	-1.4	0.77	0.91
Nov.12/89	-1.4	1.2	1.10	Feb.19/92	-1.7	3.0	2.40
Jan.18/90	-1.5	1.0	0.65				

On this time-scale an inverse correlation between the Hα emission and the Hβ and He absorption is evident. It appears that shell effects similar to the one previously reported continuously recur, but on a smaller scale. We notice that when the envelope re-expands and Hα emission lowers, the photosphere seems to be compressed again.

We are not able to find a physical reason for this peculiar behaviour; there are no physical links between the time needed for the shell to travel from the photosphere to the Hα emitting region of the envelope, and the time interval between two compressions of the photosphere. We merely observe that the time scales for the relaxation of the photosphere and for the crossing of the envelope of the star are the same.

4. CONCLUSIONS

Phenomena like those discuted in this paper are frequently observed in most Be stars, which are variable objects in many respects (Slettebak, 1987). Supergiants, on the other hand, are usually more stable objects. Therefore, the observations reported here are more consistent with the hypothesis that the nature of of the optical component of this system is that of a Be star.

ACKNOWLEDGEMENTS

This work was partially funded by a grant from the italian Ministero per l'Universitá e la Ricerca Scientifica e Tecnologica.

REFERENCES

Aab, O.E., Bychova, L.V., & Kopilov, I.M. 1983, Sov. Astron. Lett., 9, 285
Corbet, R.H.D. 1986, in The evolution of galactic X-ray binaries, eds. J. Trumper, W.N.G. Lewin & W. Brinkmann (Dordrecht: Reidel) p. 63
Corral, L., & Koenigsberger, G. 1987, Rev. Mex. Astron. Astrof., 14, 330
Crampton, D., Hutchings, J.B., & Cowley, A.P. 1985, ApJ, 299, 83
Finley, J.P., Belloni, T., Cassinelli, J.P. 1992, A&A, 262, L25
Koenigsberger, G., Swank, J.H., Szimowiak, A.E., & White, N.E. 1983, ApJ, 268, 782
Slettebak, A. 1988, PASP, 100, 770
Yamauchi, S., Asaoka, I., Kawada, M., Koyama, K., & Tawara, Y. 1990, PASJ, 42, L53
Van Kerkwik, M. H., & Waters, L. B. M. F. 1989, Proc. 23rd ESLAB Symp., ESA SP-296, p. 473

OPTICAL BEHAVIOUR OF THE Be-X BINARY V635 CAS = 4U0115+63 BEFORE AND DURING A TRANSIENT PHASE

A. Guarnieri, C. Bartolini, M. Teodorani, R. Silingardi, A. Piccioni
Dipartimento di Astronomia, Università di Bologna - Italy
adriano@alma02.bo.astro.it

ABSTRACT

The main results of the observations of the optical component of the Be-X binary system V635 Cas, carried out at the Bologna Observatory in the period 1985-1990 are presented. Three kinds of variable features were prominent: i) photometric variations on a time-scale of several days; ii) a strong increase in luminosity on a time-scale of about one year; iii) an episodic inversion of the ratio of the equivalent widths of Hα and Fe II 6153.

1. INTRODUCTION

The components of the Be-X binary system V635 Cas = 4U0115+63 are a hard X-ray transient pulsar spinning with a period of 3.61s and a variable B star whose spectrum show variable $H\alpha$ and $H\beta$ emission (compare Mendelson et al. 1991, and Teodorani et al. 1991). The system has undergone several X-ray outbursts since its discovery (Forman et al. 1976; Whitlock et al. 1988). Generally X-ray bursts follow optical flares after 1-2 months, but they do not show a clear periodicity or an apparent relation to the orbital period of 24.3 days. The Be component of the system is variable on different time-scales. The intensity of the optical flares seems to be correlated with X-ray bursts; possibly the strongest accretion events onto the compact companion and the secular maxima of the Be star luminosity are caused by high values of a changing density of the circumstellar envelope.

2. OBSERVATIONAL RESULTS

We present here the main results of our observations of the optical component of this Be-X binary system, carried out with the 152-cm telescope of the Bologna University in the period 1988-90 using CCD BVRI photometry and spectroscopy.

2.1 Photometry: Short time-scale variations

We observed luminosity changes till to 0.15 mag in B and of 0.04-0.07 mag in V, R, I on a time-scale of several days (in some cases hours). This kind of variation seems to be due to a sort of pulsation of the Be star. On a time-scale of 10-20 days the amplitude of the variation increases, possibly due to some kind of magnification effect of the daily pulsation.

We propose that the main cause of this variation could be a "non-conservative pulsation", in which the Be component of the system experiences intermittent mass ejection, possibly driven by shock waves originating in the star interior and compressing the circumstellar envelope at a subsequent stage: as every shock event is supposedly coupled with gas ejections from the central star, the Be envelope is replenished at every expansion phase and the emission coming from the envelope is consequently enhanced. As it is possible to notice in Fig. 1, the increase in the amplitude of the variation is more evident at longer wavelengths, with the exception of the B light (where the contribution of the accretion disk could be prominent). This behaviour is typical of a variable high density and low velocity expanding wind, whose emission properties were described by Waters 1987, and Waters 1988. The longwave excess of radiation with respect to the emission of the central star is probably produced by the free-free and free-bound radiation of the circumstellar envelope, which increases with the square of the wavelength.

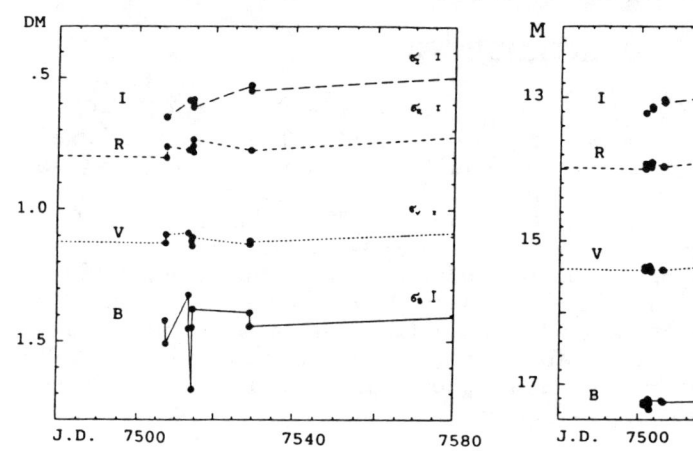
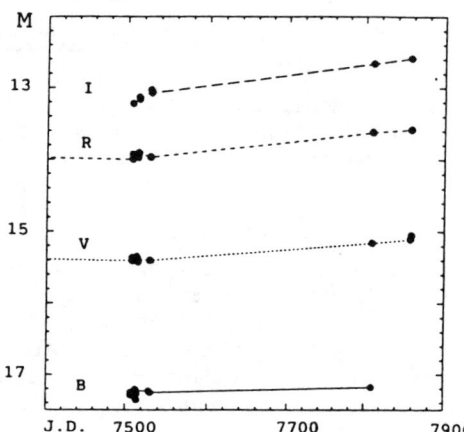

Fig. 1. Short (daily and monthly) time scale photometric variations of V635 Cas (CCD photometry).

Fig. 2. Long (yearly) time-scale photometric variations of V635 Cas (CCD photometry).

2.2 Photometry: Long time-scale variations

During 1989 V635 Cas underwent a slow, steady increase in luminosity. Fig.2 shows increases of 0.20, 0.30, 0.40, 0.60 mag in B, V, R, I respectively up until October 1989. At the end of December, 1989 the luminosity was still increasing. Clearly, the amplitude of the variation increases with the wavelength of observation. These facts seem to indicate a secular increase in the density and dimensions of the Be circumstellar envelope (Waters 1987; Waters 1988).

In the framework of our hypothesis of the presence of a non-conservative pulsation mechanism, we think it is also possible to explain the increase in the optical luminosity on the long time-scale if we assume that during every expansion phase not all the gas of the envelope is lost: if so, during many cycles of pulsation more and more gas is stored up in the Be circumstellar envelope, causing an increase of the emission as it is observed. Even if a time lag of about 60 days between the optical enhancement and the X-ray burst (Makino et al.1990) could be justified as the diffusion time needed by the accreting gas to move from the external to the internal region of the accretion disk, the emissive behaviour we observed excludes the hypothesis that the increase in luminosity is due to radiation produced in the external region of the accretion disk (Kriss et al. 1983).

2.3 Spectroscopy: Short time-scale variations

We obtained two spectra of V635 Cas, one day apart, about three weeks after the optical luminosity reached its maximum (Mendelson et al. 1991). These spectra show strong variations of the emission features, in contrast with the quiescent behaviour observed by other authors over two weeks before (Wagner et al. 1990). This is clearly shown in Fig. 3.

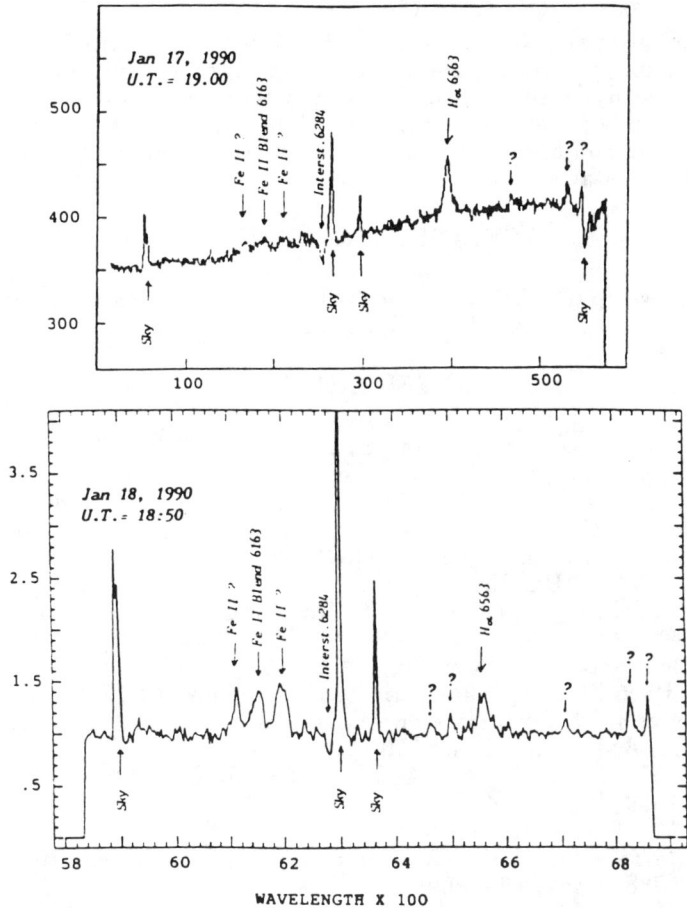

Fig. 3. Short (one day) time-scale variations in the spectrum of V635 Cas

In both spectra we noticed prominent emission lines, particularly in the 6100-6200Å range, which we tentatively identify as due to Fe II. $H\alpha$ is also strong in emission.

On January 17, 1990 the equivalent width of $H\alpha$ in emission was clearly greater than the e. w. of Fe II 6153 in emission. On January 18, 1990 we observed the opposite situation. We think that this kind of variation could also be explained in the framework of a non-conservative pulsation. If the Fe II emission is formed in the hotter and more internal zone of the envelope of Be stars, as many authors think, while the $H\alpha$ emission line is formed in the cooler and thicker external zone (Waters et al. 1988) the strong $H\alpha$ increase observed on January 17 could be due to a compression of the external envelope during the last stage of a previous episode of gas injection, while the strong Fe II increase observed on January 18 could be the result of a new gas injection which was crossing and compressing the internal region of the envelope.

3. CONCLUSIONS

The observational properties of the Be-X binary system V635 Cas like those of other similar systems, show that much work is still needed in order to understand the dynamics of the Be circumstellar envelopes and the mode of mass transfer to the compact companion. This mode may be different from object to object, depending on the orbital parameters and on the density and extension of the Be star envelope.

ACKNOWLEDGEMENTS

This work was partially supported by a grant of the Italian Ministero per l'Università e la Ricerca Scientifica e Tecnologica.

REFERENCES

Forman, W., Jones, C., & Tananbaum, H. 1976, ApJ, 206, L29
Hutchings, J.B., & Crampton, D. 1981, ApJ, 247, 222
Kriss, G.A., Cominsky, L.R., & Remillard, R.A. 1983, ApJ, 266, 806
Makino F. 1990, IAUC 4967
Mendelson, H., & Mazeh, T. 1991, MNRAS, 250, 373
Middleditch, J. 1980, IAUC 3510
Rappaport, S., Clark, G.W., Cominsky, L., Joss, P.C., & Li, F. 1978, ApJ Lett, 224, L1
Ricketts M.J. 1981, Space Sci Rev, 30, 399
Teodorani M., Bartolini, C., Guarnieri, A., & Piccioni, A. 1990, in "Confrontation between Stellar Pulsation and Evolution", eds. C. Cacciari & G. Clementini, ASP Conf. Series Vol. 11, 308
Tsunemi, H., & Kitamoto, S. 1988, ApJ Lett, 334, L21
Wagner, R.M. 1990, IAUC 4942
Waters, L.B.F.M. 1987, Ph.D. Thesis
Waters, L.B.F.M., Taylor, A.R., van den Heuvel, E.P.J., Habets, G.M.H.J., & Persi, P. 1988, A&A, 198, 200
Whitlock, L., Roussel-Duprè, D., & Priedhorsky, W. 1989, ApJ, 338, 381

SIX-YEAR PHOTOMETRY OF BQ CAMELOPARADALIS — THE OPTICAL COUNTERPART OF V0332+53

Tsevi Mazeh[1,2] and Haim Mendelson[1]

[1] The Wise Observatory and School of Physics and Astronomy,
Raymond and Beverly Sackler Faculty of Exact Sciences,
Tel Aviv University, Israel

[2] Laboratory for Astrophysics, National Air and Space Museum,
The Smithsonian Institution

ABSTRACT

We report on the results of V,R,I photometry of BQ Cam — the optical counterpart of the transient X-ray pulsar V0332+53, during the years 1985-91. Throughout this period we have detected a long-term modulation correlated with the last observed X-ray eruption. Similar long-term modulation and correlation were observed in 4U0115+63 — another Be-star X-ray source. We discussed briefly two scenarios for the optical modulation, with and without a possible large mediatory disc around the neutron star.

1. INTRODUCTION

BQ Cameloparadalis, an $m_v \sim 15$ variable red star, was identified as the optical counterpart of the transient X-ray pulsar V0332+53 (Argyle 1983; Kodaira 1983), based on its proximity to the X-ray source. The stellar spectrum exhibits no clear stellar features, except for H_α and probably weak H_β emission (Bernacca, Iijima & Stagni 1984; Honeycutt & Schlegel 1985; Stocke et al. 1985; Iye & Kodaira 1985). The star was therefore suggested to be a redden Be star (e.g. Corbet, Charles & van der Klis 1986). In order to learn more on the optical variability of the system, we have launched a long-term photometric monitoring of BQ Cam at the Wise Observatory. This was done in parallel to the optical monitoring of a few other Be-stars associated with X-ray transient sources (Mendelson & Mazeh, 1989; 1991; 1993). The results for BQ Cam are reported in this poster.

2. OBSERVATIONS AND RESULTS

BQ Cam was observed throughout 93 nights between February 1985 and September 1991, at the Wise Observatory in Mitzpe Ramon, Israel, with the 40-inch telescope and the two star photometer. Differential I, R and V photometry (e.g. Mendelson & Mazeh 1989) was obtained relative to three closeby comparison stars. The observational error of each differential measurement is estimated to be somewhat larger than 0.05 in the I and R filters, and up to 0.2 mag in the V filter. Figure 1 depicts the I apparent magnitudes of BQ Cam, derived by using the observed apparent magnitude of one of the comparison stars. One of the X-ray eruption of the pulsar (Makino et al. 1989a,b) occurred during the

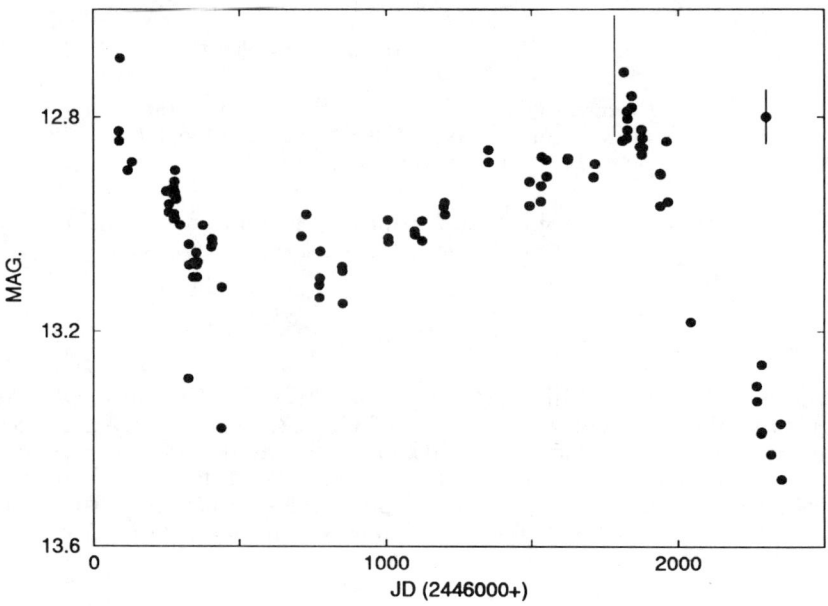

Figure 1: The light curve of BQ Cam in the I band. The error of a typical measurement is indicated. The long vertical bar presents the epoch of the observed X-ray eruption.

period of our observations; this epoch is marked in the Figure by a long bar. The light curve of the R band shows similar features. The stellar V light curve and the $R-I$ colour do not show any clear modulation above their expected noise level.

The red light curve presented in this work reveals a long-term modulation, characterized by gradual decline of ~ 0.3 mag within the first 500 days of the observations, followed by a 1000 day gradual rise to the initial intensity. The last part of the long-term modulation displays a steep decrease of the stellar brightness by ~ 0.8 mag. This modulation is seen at the R and I filters but not in the V band, with no apparent change in the $R-I$ colour.

At the time of the optical maximum, one of the X-ray eruptions were detected. The observed coincidence between the optical maximum and the X-ray outburst establishes for the first time the association of BQ Cam with the X-ray pulsar, which was previously based only on the proximity of the two sources on the sky. Similar long-term variation and correlation with the X-ray behaviour of the source was observed for 4U0114+63 (Kriss et al. 1983; Mendelson & Mazeh 1991), another Be-star X-ray system.

On top of the long-term modulation, the star shows some short-term irregular variations. We note two measurements at around JD 2446400 in particular, which are very different from the general long-term trend of the stellar brightness. In what follows we will ignore the short-term variation and discuss briefly some implications of long-term features of the light curve.

3. DISCUSSION

Be stars commonly show variability with a range of 0.1 − 0.2 mag in the V band and timescales of months and years (e.g. Slettebak 1979). These modulations are believed to arise from an expansion of a circumstellar envelope or a disc around the Be star. This could be the source of the optical rise of BQ Cam too. In this scenario, the modulation of the X-ray output of the system is *triggered* by the variation of the circumstellar circumstances. As suggested by Stella, White and Rosner (1986), the dormant period of the X-ray source could be due to the fact that the density of the circumstellar material was too small for matter to be accreted onto the magnetized neutron star. The production of X-rays started only when the density along the neutron star orbit got high enough to overcome the centrifugal barrier.

Two features of the long-term observed light curve do not reconcile with this simple model. The first one is the observed decrease of 0.8 mag in the stellar brightness in the I and R bands after the X-ray eruption. This decrease is substantially larger than the ones seen in single Be stars. The other feature is the coincidence of the X-ray eruption with the starting point of the long and steep optical decline. If the optical modulation has to do with the Be star and is independent of X-ray activity, this coincidence is accidental. However, the few other coincidences observed for 4U0114+63 (Kriss et al. 1983; Mendelson & Mazeh 1991) make the chance coincidence assumption somewhat less appealing.

A completely different source of the optical brightening might be some matter which has been transferred to the vicinity of the neutron star, accumulated there in a disc, and radiates due to viscous heating. Such a disc could also be the source for the observed near IR excess (Coe et al. 1987). According to this scenario, the X-ray outburst probably occurred when the pressure of matter accumulated in the *disc* around the compact object was strong enough to overcome the centrifugal barrier at the edge of the neutron star magnetosphere (Stella et al. 1986). The long optical rise which preceded the X-ray eruption is consistent with the assumption that the outburst was caused by some accumulating effect.

Provided the optical rise was due to the presence of a transient accretion disc, the optical radiation could be either the blackbody output of a large accretion disc or a reprocessed radiation of far UV radiation produced in a small hot disc around the neutron star. This radiation could be absorbed by the stellar wind and/or the circumstellar equatorial envelope of the Be star, and reprocessed into the optical bands. The appearance of the modulation at the R and I bands and not in the V band indicates that the modulated component of the system has lower temperature than those of the other parts of this source. According to the transient disc model, the long-term optical decrease of the system is due to a decline of the disc around the compact object, or at least the decline of its central part. The disc can decrease by its mass and output either by accretion onto the compact object or by ejection of material outwards. The onset of the long-term decline observed here was probably induced by accretion onto the neutron star, while the final stages of the decline were probably governed by ejection and insufficient supply of new material.

4. CONCLUSIONS

We have presented here a strongly modulated long-term light curve of the optical counterpart of the X-ray pulsar 4U0332+53. The modulation is strongly correlated with the last appearance of an X-ray outburst. Similar long-term modulation and correlation were observed in another Be-star X-ray source. These features were not well known before because no long-term monitoring of the optical components were available. The correlation between the optical and the X-ray modulations shows the potential of long-term monitoring of these systems, to help us understand the nature of the violent irregular eruptions of these sources.

REFERENCES

Argyle, R. W., 1983, IAU Circ., No. 3897
Bernacca, P. L., Iijima, T., Stagni, R., 1984, A&A, 132, L8
Coe, M. J., Longmore, A. J., Payne, B. J., Hanson, C. G., 1987, MNRAS, 226, 455
Corbet, R. H. D., Charles, P. A., van der Klis, M., 1986, A&A, 162, 117
Honeycutt, R. K., Schlegel, E. M., 1985, PASP, 97, 300
Iye, M., Kodaira, K., 1985, PASP, 97, 1186
Kodaira, K., 1983, IAU Circ., No. 3897
Kriss, G. A., Cominsky, L. R., Remillard, R. A., Williams, G. & Thorstensen, J. R., 1983, ApJ, **266**, 806
Makino, F. and *Ginga* team, 1989a, IAU Circ., No. 4858
Makino, F. and *Ginga* team, 1989b, IAU Circ., No. 4872
Mazeh, T., Krymolowski, Y., Latham, D. W., 1987, MNRAS, 263, 775
Mendelson, H., Mazeh, T., 1989, MNRAS, 239, 733
Mendelson, H., Mazeh, T., 1991, MNRAS, 250, 373
Mendelson, H., Mazeh, T., 1993, MNRAS, in press
Slettebak, A., 1979, Space Sci. Rev., 23, 541
Stella, L., White, N. E., Rosner, R., 1986, ApJ, 308, 669
Stocke, J., Silva, D., Black, J. H., Kodaira, K., 1985, PASP, 97, 126

ORBITAL PERIOD CHANGES IN THE ECLIPSING PULSAR BINARY PSR B1957+20

Zaven Arzoumanian[1], Andrew S. Fruchter[2], Joseph H. Taylor[1]

[1] Joseph Henry Laboratories and Department of Physics,
Princeton University, Princeton, NJ 08544
zaven@pulsar.princeton.edu, joe@pulsar.princeton.edu

[2] Astronomy Department and Radio Astronomy Laboratory,
University of California, Berkeley, CA 94720
asf@orestes.berkeley.edu

1. THE SYSTEM

PSR B1957+20 is a 1.6 ms pulsar orbited by a $\sim 0.025\, M_\odot$ companion which eclipses the pulsar's radio signal during approximately ten percent of the 9.2 hour orbit. The eclipses are thought to be caused by an ionized "cometary" wind trailing the companion in its orbit and constantly infused with new matter. The companion has been optically identified — its luminosity is strongly modulated with the orbital period of the system, consistent with synchronous rotation and heating by irradiation from the pulsar. We thus appear to be watching a pulsar in the process of slowly evaporating its companion, the star which presumably spun it up to millisecond periods during an earlier low-mass X-ray binary phase in the system's evolution.

2. ORBITAL VARIABILITY

Ryba & Taylor (1991) first reported a change in the orbital period (P_b) of $\dot{P}_b = (-3.8 \pm 0.9) \times 10^{-11}$ over 2.5 years, implying an unexpectedly short timescale of 30 Myr for orbital decay. Our recent observations extend the overall data span to more than 5 years. During this time, the orbital period of the PSR B1957+20 system has varied typically by $\Delta P_b / P_b \simeq 1.6 \times 10^{-7}$, with $\dot{P}_b$ changing sign in mid-1991. The overall period evolution follows a smooth parabolic path well described by a constant period second derivative $\ddot{P}_b = (1.43 \pm 0.08) \times 10^{-18}\, \text{s}^{-1}$.

Our pulse timing observations, conducted at 430 MHz at the Arecibo Observatory, were fit to a model of the binary pulsar which included as many as 17 spin, astrometric, and orbital parameters. The results of these fits appear in the Table, and the post-fit pulse arrival-time residuals are shown in Figure 1. Figure 2 displays the difference between observed and computed orbital phases over 5 years. Note that this graph differs in sign from an O−C diagram derived from eclipse timings; plotted points are orbital phase deviations at a given time, whereas conventional O−C diagrams plot differences in the time of passage through a specified orbital phase. Thus, a decreasing slope, *e.g.* from positive to negative after 1990, here represents an increase in orbital period. We also divided our data into five non-overlapping subsets, each containing roughly one year of pulse arrival times. These were individually fit for P, $\dot{P}$, higher-order spin derivatives where necessary, and two orbital parameters, T_0 and P_b. The resulting fractional orbital period changes are plotted in Figure 3a.

A varying gravitational acceleration due to a second companion in a distant orbit or large fluctuations in dispersion measure could, in principle, produce changes in the apparent orbital period similar to those seen here. They would also, however, affect our measurement of the pulsar spin period equally well, so that $\Delta P_b/P_b = \Delta P/P$. Figure 3b shows that the spin period displays no such variation at a level nearly five orders of magnitude less than that required.

3. DISCUSSION

Any explanation of orbital variability in the PSR B1957+20 system will likely be based on either the conditions of mass loss through the companion's wind or structural changes in the companion itself. Existing models based on mass loss, which typically ascribe angular momentum changes to the specific geometry of particle injection from the companion's surface, can create orbital period changes of either sign but may be hard-pressed to explain the rapid, well-behaved variation in $\dot{P}_b$ that we see. Alternatively, the PSR B1957+20 system can be likened to other close binaries in which orbital period changes have been observed: Algol-type eclipsing binaries, RS CVn and some cataclysmic variables. The mechanisms thought to underlie orbital variations in these other systems ultimately attribute period changes to magnetic activity in one member of the binary. The orbit then evolves quasi-periodically with the magnetic cycle.

If, as we suspect, PSR B1957+20 undergoes quasi-cyclic period changes, it is reasonable to believe that magnetic activity is responsible. Although Faraday delay measurements have constrained the companion's magnetic field to less than a few gauss parallel to the line of sight at the edges of the eclipse region, a strong, subsurface, toroidal field capable of producing the observed period changes may exist. Fruchter & Goss (1992) argue that the companion just underfills its Roche lobe, making it at least partly convective. The ingredients of a dynamo, convection and rapid rotation, may thus be present in the system.

We believe that rotationally induced magnetic activity plays an important role in the evolution of other close pulsar binaries. The companions to PSRs B1744−24A and B1718−19 both support strong mass outflows despite relatively weak irradiation by the pulsar; a magnetically excited wind would naturally explain the excess mass loss in these systems.

REFERENCES

Fruchter, A. S. & Goss W. M. 1992, ApJ, 384, L47
Ryba, M. F. & Taylor J. H., ApJ, 380, 557

Related poster papers in this volume are
5.01 Callanan & van Paradijs, The Orbital Lightcurve of PSR 1957+20
6.01 McCormick et al., The Evolution of Evaporating Binary Pulsars
6.02 McCormick et al., The Orbital Period Derivative of PSR 1957+20
6.03 Muslimov & Sarna, The Evolutionary Scheme for a Low-Mass Binary with Millisecond Pulsar
6.05 Wijers & Paczyński, PSR B1718−19 and the LMXB-CV connection

Astrometric, Spin, and Orbital Parameters of PSR B1957+20.

Right ascension, α (J2000)[a]	$19^h\ 59^m\ 36\overset{s}{.}76988(5)$
Declination, δ (J2000)	$20°\ 48'\ 15\overset{''}{.}1222(6)$
μ_α (mas yr^{-1})	-16.0 ± 0.5
μ_δ (mas yr^{-1})	-25.8 ± 0.6
Period, P (ms)	1.60740168480632(3)
Period derivative, $\dot{P}$ (10^{-20})	1.68515(9)
$\ddot{P}$ (10^{-31} s^{-1})	1.4 ± 0.4
Epoch (MJD)	48196.0
Dispersion measure, DM (cm^{-3} pc)	29.1168(7)
Projected semi-major axis, x (lt-s)	0.0892253(6)
Eccentricity, e	$< 4 \times 10^{-5}$
Epoch of ascending node, T_0 (MJD)	48196.0635242(6)
Orbital period, P_b (s)	33001.91484(8)
$\dot{P}_b$ (10^{-11})	1.47 ± 0.08
$\ddot{P}_b$ (10^{-18} s^{-1})	1.43 ± 0.08
$\dot{x}$ (10^{-14})	0 ± 3

[a] Coordinates are given in the J2000 reference frame of the DE200 solar system ephemeris. Figures in parentheses are uncertainties in the last digits quoted.

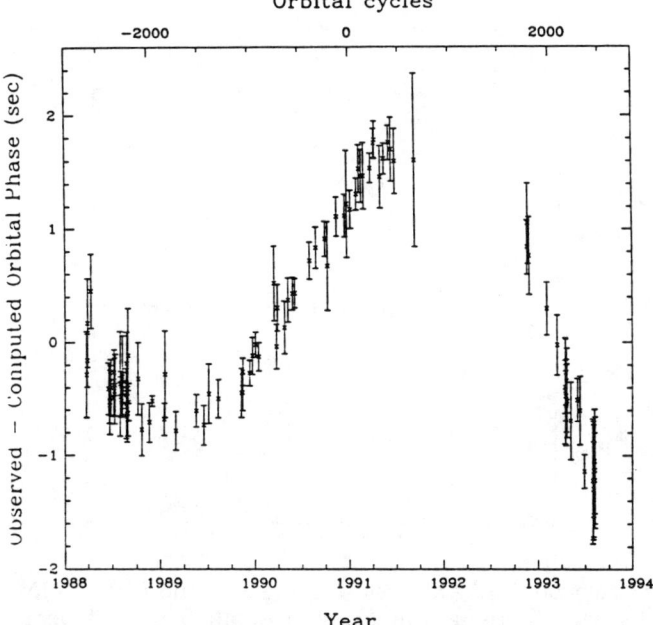

Fig. 1. Post-fit residuals of PSR B1957+20 plotted versus a) date, and b) orbital phase. A typical uncertainty in the pulse arrival times is shown in a).

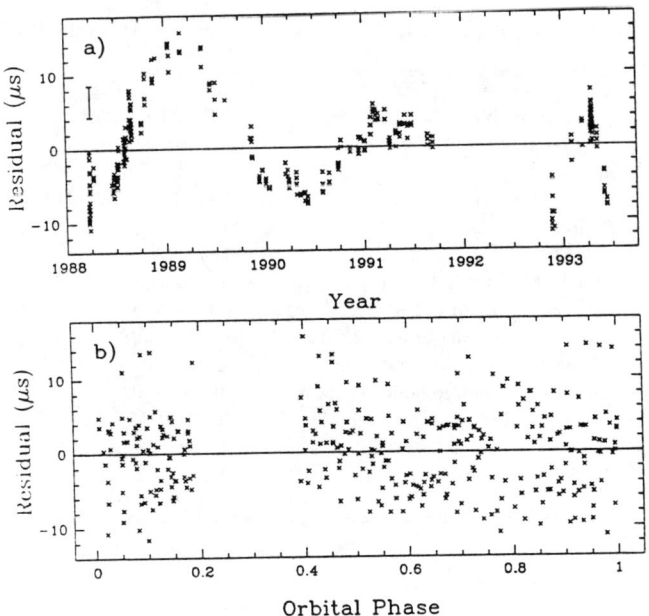

Fig. 2. Orbital phase shifts (observed minus computed) in the PSR B1957+20 system.

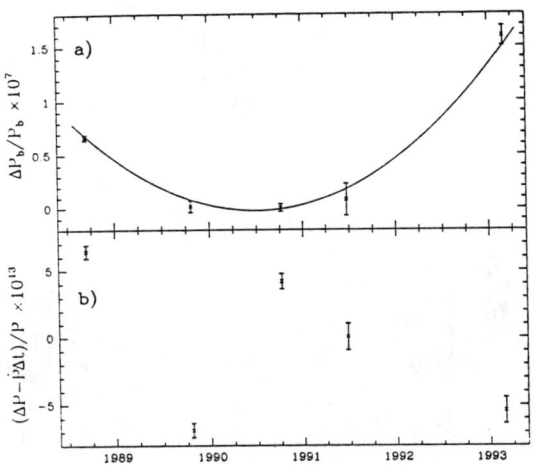

Fig. 3. a) Fractional orbital period changes in the PSR B1957+20 system versus date. The overall changes in P_b span about 5 ms. The solid line curve corresponds to the values of P_b and its derivatives listed in the Table; b) fractional pulse period changes versus date. A correction has been made for the global spin-down rate given by $\dot{P}$ in the Table.

THE EVOLUTIONARY STATUS OF PSR 1259-63

Lynn Cominsky*
Department of Physics and Astronomy, Sonoma State University
Rohnert Park, CA 94928
lynnc@charmian.sonoma.edu

ABSTRACT

The PSR 1259-63 system is unique in that it is the first radio pulsar found to be in a binary system with a massive main sequence companion. As such, it may be the evolutionary missing link connecting the radio pulsars to the X-ray emitting Be-binaries. In this paper, we consider the conditions under which PSR 1259-63 may be the progenitor of the less eccentric, more slowly rotating X-ray Be-binaries such as 4U0115+63, A0535+26, and A0538-66. Scenarios invoking the interaction of the pulsar with the stellar wind of the companion, are proposed to account for the rapid spin down phase which must occur in this system, if it is evolutionarily linked to accreting X-ray binaries. The unexpected X-ray emission observed from PSR 1259-63 may indicate that the interaction between the pulsar and the Be star occurs at greater distances and produces a greater spin down efficiency than has been previously calculated.

1. INTRODUCTION

PSR 1259-63 is a 47.7-msec radio pulsar which is apparently in a 3.4 year, highly eccentric ($\epsilon = 0.87$) binary orbit with the ~ 10 solar mass Be star, SS 2883 (Johnston et al., 1992). The pulsar has a relatively young characteristic age of 3×10^5 y and a moderate magnetic field strength of 3×10^{11} G (Johnston et al., 1993).

Non-pulsed but variable X-ray emission has been detected from the binary system containing the radio pulsar PSR 1259-63 during two pointed ROSAT observations, taken five months apart (Cominsky, Roberts and Johnston, 1993). Detectable X-ray flux has only been observed post-apastron, and has increased with orbital phase. It is significantly greater than what would be expected from the Be-star's corona, and is too variable, yet not sufficiently pulsed, to be due to rotational spin-down losses. Low-level accretion is possible only if the radio pulsar inhibition mechanism can be overcome. However, emission from a shocked pulsar wind remains a viable mechanism. The X-ray observations and the standard scenarios for X-ray emission are briefly reviewed in the sections that follow.

2. OBSERVATIONS

Cominsky, Roberts and Johnston (1993) have obtained $\sim 46 \times 10^3$ s of ROSAT Position Sensitive Proportional Counter data in two multi-day periods, separated by ~ 5 months. The ROSAT PSPC images clearly indicate a significant source at a position consistent with PSR 1259-63. The intensity of the source is $2.38 \pm 0.18 \times 10^{-2}$ counts s^{-1} and $3.15 \pm 0.11 \times 10^{-2}$ counts s^{-1} during the 1992 August 30 - September 4, and the 1993 February 7 - 16 intervals, respectively.

Assuming a distance to the system of 1.5 kpc (Johnston et al., 1993), the X-ray luminosity in the PSPC 0.1 - 2.4 keV band that is required to explain the observed count rates is at least 6×10^{32} ergs s^{-1} and may be as high as 1×10^{34} ergs s^{-1}, depending on the assumed spectral model. Pre-apastron ROSAT PSPC

---

* Visiting Professor, Stanford Linear Accelerator Center

observations of PSR 1259–63 by Bailes and Watson detected less than 10 photons (private communication), yielding a 99% confidence upper limit of $< 0.26 \times 10^{-2}$ counts s^{-1} during a multi-day observation which occurred in 1992 February. The system was also observed prior to apastron in 1991 September using the Ginga LAC. Makino and Aoki (private communication) have found an upper limit for these observations of 0.1 mCrab (2-10 keV), equivalent to an X-ray luminosity of $\sim 6 \times 10^{32}$ ergs s^{-1} (assuming a Crab like spectrum).

The overall flux from PSR 1259–63 has increased by at least a factor of ten between the pre- and post-apastron observations, and by an additional $\sim$ 50% between the first and second post-apastron observation periods, which are separated by about five months. Significant variability on a timescale of hours was also observed during the second observation period. However, no significant pulsed emission has been detected in either observation period, nor in the entire data set. Three sigma upper limits to the pulsed fraction are 21%, 9%, and 9% for the first interval, the second (longer) interval and the entire dataset respectively.

Figure 1 from Cominsky, Roberts and Johnston (1993) depicts the orbital locations of the Ginga observation and the three sets of ROSAT observations.

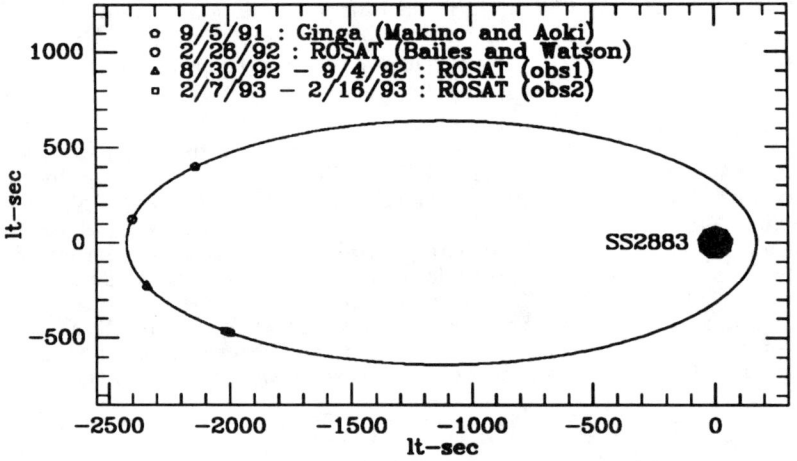

Fig. 1. Schematic of the orbit of PSR 1259–63 indicating observation times. The companion, SS2883, is placed at the center of mass of the system.

3. THEORY

In order to account for the X-ray emission observed from the PSR 1259–63 system post-apastron, Cominsky, Roberts and Johnston (1993) considered four different types of scenarios: coronal emission from the Be star companion, low-level accretion onto the neutron star, rotational spin down energy from the neutron star and emission from interacting winds.

They found that the luminosity observed from PSR 1259–63 was too high by factors ranging from 15 - 700, depending on the spectral model chosen. The X-ray luminosity in the ROSAT band, if consistent with previously observed B and Be stars would only be $\sim 3 \times 10^{31}$ ergs s^{-1} (Pallavicini et al., 1981, Meurs et al., 1992). Rotational spin down was also ruled out as a likely X-ray emission mechanism due to the small percentage of the spin down luminosity which could be present as pulsed flux (less than 0.1%), and the extreme variability on short time scales (more than a factor of 10 in five months.)

Direct accretion onto the neutron star does not seem likely to occur due to the radio pulsar inhibition mechanism (Illarionov and Sunyaev, 1975, Stella, White and Rosner, 1986). This mechanism should operate until the rapidly rotating radio pulsar spins down to a period where the pressure of the infalling material is greater than the pressure of the relativistic pulsar wind. At this time, the system may begin to accrete for low values of the stellar wind velocity and high rates of mass capture onto the neutron star. Otherwise, X-ray emission will be centrifugally inhibited. High mass capture rates and low stellar wind velocities are most likely to occur at periastron. However the lack of any historically observed bright ($\sim 10^{36} - 10^{37}$ ergs s^{-1}) X-ray transient outbursts at previous periastron passages (which could have been observed by X-ray satellites between 1970 and 1991) indicates that these conditions have not yet existed (Cominsky, Roberts and Johnston, 1993.)

As discussed in detail by Kochanek (1993), if the stellar and pulsar winds interact, shocks should occur between the two stars. The shocked stellar wind should radiate X-rays due to thermal bremsstrahlung. The X-ray luminosity expected in this case is a sensitive function of the ratio of the strength of the stellar wind to that of the pulsar (λ). In his model, λ is assumed to be much greater than one and the calculated emission from the stellar wind shock would be a factor of 5-10 less than the expected coronal emission (see above). However, the pulsar wind should also be shocked, and it is not clear how the available energy in this shock (which comes from the pulsar's spin down energy and is potentially much greater) will emerge. For values of $\lambda \geq 20$, it seems possible to release as much energy as was observed (see Figure 6 in Kochanek 1993).

Davies and Pringle (1981) considered four phases in the evolutionary history of a binary neutron star system: radio pulsar spin down phase, very rapid rotator, supersonic (propellar) rotator, and subsonic rotator. Only after the last phase does the neutron star begin to accrete. We wish to determine whether or not the PSR 1259–63 system will evolve into a more circular, more slowly rotating system like the standard X-ray Be-binaries such as 4U0115+63. In order to do this, the neutron star's rotation rate must slow down considerably within the relatively short lifetime of the Be-companion star, SS2883 ($\sim 10^7$ years.) Considering first the radio pulsar spin down phase, and assuming the pulsar's magnetic moment $\mu = 3 \times 10^{29}$ G cm^3, the stellar wind speed, $v_o \sim 300$ km s^{-1}, and a mass capture rate by the neutron star, $\dot{M}_c \sim 10^{15}$ g s^{-1}, the critical spin period for transition from this radio pulsar spin down phase to the next phase is given by:

$$P = 0.8 \left(\frac{\mu}{10^{30} \text{ G cm}^3}\right)^{\frac{1}{3}} \left(\frac{\dot{M}_c}{10^{15} \text{ g s}^{-1}}\right)^{-\frac{1}{6}} \left(\frac{M}{M_\odot}\right)^{\frac{1}{3}} \left(\frac{v_o}{10^3 \text{ km s}^{-1}}\right)^{-\frac{5}{6}} \approx 1 \text{ second}$$

This will happen on a timescale of:

$$\tau = 3.3 \times 10^7 \left(\frac{P}{1 \text{ s}}\right)^2 \left(\frac{\mu}{10^{30} \text{ G cm}^3}\right)^{-2} \left(\frac{I}{10^{45} \text{ g cm}^2}\right)^{-2} \approx 1.5 \times 10^9 \text{ years,}$$

much longer than the life of the Be-companion (and there are still three more phases to consider.) It would therefore appear as though this system will not be able to evolve into a traditional X-ray Be-binary. However, if conditions at periastron act to spin down the system (e.g. by the propellar mechanism), or if the X-ray emission observed post-apastron from the system is indicative of a propellar-type spin down, the time required for accretion to begin could be greatly shortened.

4. CONCLUSIONS

The X-ray emission seen from the PSR 1259–63 system is not likely to be explained by coronal emission from the Be-companion, rotational spin down energy or accretion onto the neutron star's surface. A likely source of energy for the emission is a shocked pulsar wind. This system will not evolve into an accreting X-ray Be-binary within the lifetime of the Be-companion, unless the neutron star can spin down more rapidly than expected. This could occur by a propellar mechanism which operates at periastron, and/or post-apastron, giving rise to the X-ray emission which was observed.

ACKNOWLEDGEMENTS

We are grateful to R. Romani and A. King for useful discussion regarding possible X-ray emission mechanisms, and spin down timescales, and to S. Johnston for helpful comments on this paper. This work has been supported by a grant from the NASA ROSAT Guest investigator program, NAG 5-1684.

REFERENCES

Cominsky, L., Roberts, M., & Johnston, S., 1993, ApJ, submitted
Davies, R.E. & Pringle, J.E., 1981, MNRAS, 196, 209
Illarionov A.F., & Sunyaev R.A., 1975, A&A, 39, 185
Johnston, S., Manchester, R.N., Lyne, A.G., Bailes, M., Kaspi, V.M., Qiao, G., & D'Amico, N., 1992, ApJ, 387, L37
Johnston, S., Manchester, R. N., Lyne, A. G., Nicastro, L., & Spyromilio, J., 1993, MNRAS, submitted
Kochanek, C., 1993, ApJ, 406, 638
Meurs, E. J. A. et al., 1992, A&A, 265 L41
Pallavicini, R., Golub, L., Rosner, R., Vaiana, G. S., Ayres, T., & Linsky, J. L., 1981, ApJ, 248, 279
Stella, L., White, N.E., & Rosner, R., 1986, ApJ, 308, 669

A RADIO PULSAR–B STAR BINARY IN THE SMALL MAGELLANIC CLOUD

V. M. Kaspi
Physics Department, Princeton University, Princeton, NJ 08544
vicky@puppsr.princeton.edu

S. Johnston
Research Centre for Theoretical Astrophysics, University of Sydney, NSW 2006, Australia
simonj@physics.su.oz.au

R. N. Manchester, M. Bailes
Australia Telescope National Facility, CSIRO, PO Box 76, Epping NSW 2121, Australia
rmanches@atnf.csiro.au, mbailes@atnf.csiro.au

J. F. Bell, M. Bessell
Mount Stromlo and Siding Spring Observatories, Private Bag, WestonCreek 2611, Australia
bell@mso.anu.edu.au, bessell@mso.anu.edu.au

A. G. Lyne
University of Manchester, Jodrell Bank, Macclesfield, SK11 9DL, UK
agl@jb.man.ac.uk

N. D'Amico
Istituto di Fisica dell'Università and Istituto di Radioastronomia del CNR, Italy
damico@38167.span%Sdsc.BitNet

ABSTRACT

We report the discovery of regular Doppler shifts of the pulse period of PSR J0045−7319, the only known pulsar in the Small Magellanic Cloud. The pulsar is in a highly eccentric 51-day orbit with a companion star of mass greater than $4M_\odot$. Optical observations in the direction of the pulsar reveal a 16th magnitude B star companion. The PSR J0045−7319 system is likely an X-ray binary progenitor.

1. INTRODUCTION

The discovery of PSR B1259−63 (Johnston et al 1992) provided the first example of a radio pulsar in a binary system with a non-degenerate companion. PSR J0045−7319 (PSR B0042−73) was discovered in a search of the Magellanic Clouds for radio pulsars (McConnell et al 1991). It is the only known pulsar in the Small Magellanic Cloud (SMC), and has a pulse period of 0.926 s. Although it is a faint source, its large distance makes it the most luminous binary radio pulsar known. The pulse profile is a single peak of duty cycle 4%, typical of most pulsars. Its association with the SMC is assured by its dispersion measure $DM \simeq 105$ pc cm^{-3}, since models of the galactic electron distribution account for no more than ~ 25 pc cm^{-3} along that line of sight (Taylor & Cordes 1993).

© 1994 American Institute of Physics

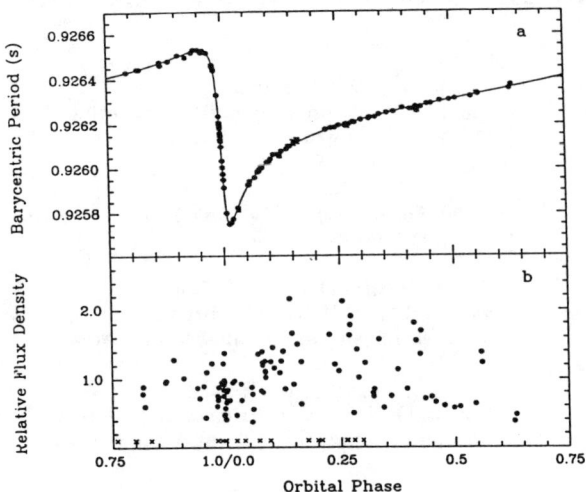

Fig. 1. (a) Period and (b) flux density at 430 MHz vs. orbital phase.

2. OBSERVATIONS AND RESULTS

We have monitored PSR J0045−7319 on a regular basis from 1991 February through 1993 July using the 64-m radio telescope at Parkes. Of the 103 successful observations, 101 were obtained at 430 MHz, with the other two at 660 MHz. Barycentric periods derived from the individual observations are shown in Figure 1a as a function of orbital phase; the curve shown is for the binary orbit parameters given in Table 1. The mass function is given by

$$f(M_p) = \frac{(M_c \sin i)^3}{(M_p + M_c)^2} = \frac{4\pi^2 (a_p \sin i)^3}{G P_b^2} = 2.17 M_\odot, \quad (1)$$

where M_p and M_c are the pulsar and companion masses, i is the orbital inclination angle, and the other quantities are defined in Table 1. This is the largest mass function known for a binary radio pulsar. Assuming $M_p = 1.4 M_\odot$, the minimum companion mass, corresponding to $i=90°$, is $3.97 M_\odot$.

We obtained CCD images of the PSR J0045−7319 field using the Australian National University 2.3-m telescope which reveal a star at the J2000 position $\alpha = 00^h\ 45^m\ 35.3^s \pm 0.4^s$, $\delta = -73°\ 19'\ 01.9'' \pm 1.1''$, consistent within uncertainties with that of the pulsar, having V magnitude 16.19 and B magnitude 16.03. Allowing for the range of possible reddening of the SMC, we obtain $-0.28 < (B-V)_o < -0.22$ for the intrinsic color. At the distance of the SMC, the star has an absolute V-magnitude of $-3.2 < M_v < -2.6$. We also obtained a low-resolution optical spectrum of the candidate under the service spectroscopy program of the Anglo-Australian Telescope and with the ANU 2.3-m. The observed Balmer jump and weak helium lines show that the star is a main sequence star of spectral class B1. Combined with the photometry, this implies a mass for the companion of $10.0 - 12.5\ M_\odot$. No emission lines are evident in the spectrum. Since the probability of chance occurrence of such a star at the pulsar position is small, we conclude that this star is the companion to PSR J0045−7319.

Right Ascension, α (J2000)	$00^h\ 45^m\ 34.9^s \pm 0.2^s$
Declination, δ (J2000)	$-73°\ 19'\ 03.2'' \pm 0.8''$
Dispersion Measure, DM	105.4(7) pc cm^{-3}
R.M.S. timing residual	7.4 ms
Mean flux density at 400 MHz	0.8 mJy
Luminosity at 400 MHz	2800 mJy kpc^2
Period, P	0.9262758349(1) s
Period Derivative, $\dot{P}$	$4.465(7) \times 10^{-15}$
Epoch of Period	MJD 48964.2000
Orbital Period, P_b	51.16926(2) days
Projected semi-major axis, $a_p \sin i$	174.235(2) lt s
Longitude of periastron, ω	115.236(2)°
Eccentricity, e	0.80798(1)
Epoch of Periastron	MJD 49220.3817(1)
Magnetic field strength, B	2.1×10^{12} G
Characteristic age, τ_c	3.3×10^6 yr

Table 1. Observed and Derived Parameters for PSR J0045−7319.

3. DISCUSSION

At periastron, the distance between the pulsar and the companion is $a_p(1-e)(1+M_p/M_c) < 19/\sin i$ R$_\odot$, for $M_p = 1.4$M$_\odot$. For a 10M$_\odot$ companion, the inclination angle $i = 41°$, so that at periastron the pulsar approaches to within ~ 6 stellar radii of the companion. The radio signal from the pulsar might be expected to be dispersed, scattered, or absorbed by the companion's ionized mass outflow, effects that would vary with orbital phase. However, thus far, we have observed no systematic variations in dispersion measure within our measurement uncertainty; our 3σ upper limit is 3.2 pc cm^{-3}. This limit implies that the mass-loss rate is $< 8 \times 10^{-12} v_0$ M$_\odot$ yr^{-1}, where v_0 is the stellar surface wind velocity in units of 100 km s^{-1}. This is somewhat lower than expected for an isolated B star (de Jager et al 1988).

We observe no obvious systematic variation in the intensity of the pulsed emission with orbital phase (Figure 1b). The pulsar's signal was detected at periastron and at superior conjunction. During several observations, the pulsar was not detected at all; the corresponding orbital phases are indicated by crosses in Figure 1. The absence of regular radio eclipses implies that the neutron star does not accrete matter at periastron. The non-detections may indicate occasional B-star mass outflow enhancements on time scales of a few days. Much of the observed variation is likely to be due to scintillation in the inhomogeneous interstellar plasma.

The proximity of the pulsar to the companion at periastron is expected to result in an advance of the line of apsides. We observe a marginally significant value for the advance: $\dot{\omega} = 0.010 \pm 0.003°$ yr^{-1}. From standard expressions for the precession and for apsidal constants (Claret & Giménez 1991; Will 1993) and using standard assumptions, our observed 3σ upper limit of 0.019° yr^{-1} to $\dot{\omega}$ implies that the B star must have mass of less than ~8M$_\odot$. This is somewhat less than the mass estimated from the spectral type; this requires further investigation.

The characteristic age of the pulsar, given by $P/2\dot{P} = 3.3 \times 10^6$ yr, suggests that it has not been spun up by accretion. The system likely formed with the initially more massive star evolving first and explosively collapsing to form the neutron star (van den Heuvel 1992). If we assume the explosion was symmetric and that less than half the total mass was ejected, the pulsar progenitor mass is given by

$$M_{\rm pre} = (1+e)(M_c + M_p) - M_c \qquad (2)$$

(Gott 1972). Equation 2 implies $M_{\rm pre} \approx M_c$, which is large for a pre-supernova mass. If the pre-supernova mass were smaller than this, then the system may have remained bound in spite of mass ejection because of an asymmetric explosion having imparted a velocity kick.

The PSR J0045−7319 system represents the second eccentric radio pulsar–non-degenerate companion binary system after the PSR B1259−63 system, and together, they constitute a new class of young binary radio pulsar systems. These systems are likely X-ray binary progenitors since, as the companion evolves, it will expand and overflow its Roche lobe, transferring matter onto the neutron star. As the mass transfer continues, a common envelope will form, and frictional drag will shrink the orbit. If it becomes sufficiently small, complete spiral-in and a "Thorne-Żytkow" object (Thorne & Żytkow 1977), a red supergiant with a neutron-star core, may be formed. Alternatively, the envelope may be ejected before spiral-in is complete. In that case, the companion will evolve either to a massive white dwarf, or to a second neutron star.

A measurement of orbital Doppler shifts in the B-star's absorption lines would verify the association, and determine the mass ratio of the system components. A velocity curve for the companion to a radio pulsar has never before been detectable.

ACKNOWLEDGEMENTS

VMK received support from an NSERC 1967 Fellowship and from an NSF Grant for US-Australia Collaborative Research. JFB acknowledges an Australian Government Postgraduate Research Award.

REFERENCES

Claret, A. & Gimenez, A. 1991, A&A Suppl. Ser., 87, 507
de Jager, C., Nieuwenhuijzen, H., van der Hucht, K. A. 1988, A&A Suppl. Ser., 72, 259
Gott, J. R. 1972, ApJ, 173, 227
Johnston, S., Manchester, R. N., Lyne, A. G., Bailes, M., Kaspi, V. M., Qiao, G., D'Amico, N. 1992, ApJ, 387, L37
McConnell, D., McCulloch, P. M., Hamilton, P. A., Ables, J. G., Hall, P. J., Jacka, C. E., Hunt, A. J. 1991, MNRAS, 249, 654
Taylor, J. H. & Cordes, J. M. 1993, ApJ, 411, 674
Thorne, K. S., & Zytkow, A. N. 1977, ApJ, 212, 832
van den Heuvel, E. P.J. 1992, in van den Heuvel E. P. J., Rappaport S. A., eds, X-ray Binaries and Recycled Pulsars. Kluwer, Dordrecht, Reidel, p. 233
Will, C. M. 1993, Theory and Experiment in Gravitational Physics, Cambridge University Press, Cambridge

A SEARCH FOR PULSAR COMPANIONS TO RUNAWAY OB STARS

Colin Philp
Department of Physics and Astronomy
University of North Carolina at Chapel Hill
CB#3255 Chapel Hill, N.C. 27599-3255

Dale A. Frail
NRAO-VLA, P.O. Box O, Socorro, N.M. 87801

Charles R. Evans
Department of Physics and Astronomy
University of North Carolina at Chapel Hill
CB#3255 Chapel Hill, N.C. 27599-3255

Peter J. T. Leonard
T-6, MS B288 LANL, Los Alamos, N.M. 87545

ABSTRACT

"Runaway" OB stars are distinguished from the normal population by large peculiar velocities (> 30 km s^{-1}) and/or large heights above the galactic plane (up to several kpc). There are two competing models to explain these properties: the supernova ejection mechanism and the dynamical ejection mechanism. The first predicts a high incidence of compact companions to the OB runaways ($> 50\%$), while the second predicts a low percentage ($< 10\%$). We have initiated a Very Large Array (VLA) search for radio pulsars at the positions of known OB runaways. We will use the High Time Resolution Processor (HTRP) to search for pulsed 20cm emission, while also making continuum images at both 6cm and 20cm to distinguish thermal wind sources from nonthermal emitters. These two aspects of the search are complementary and can be performed simultaneously. As well as testing the two ejection hypotheses, this search might discover an object similar to PSR 1259−63 (a pulsar–Be star binary system) which would shed further light on the link between neutron star formation in massive binaries and massive X-ray binaries. To improve the chances of detecting pulsars, we propose to observe stars later than O5. The winds from these stars, with their lower mass loss rates, are less likely to occult the pulsar's radio beam through free-free absorption or serve as a confusing point source.

1. INTRODUCTION

Surveys of galactic OB Star populations reveal that a significant percentage have large peculiar velocities ($|V_p| > 30$ km s^{-1}) and/or heights above the galactic disc (hundreds of pc). Since the average velocity dispersion of interstellar gas is of order 10 km s^{-1} and OB stars are thought to originate in the galactic plane, these "runaways" must undergo some physical process to explain their large number. Stone (1991), in assuming two distinct populations, derives runaway frequencies of 46% and 4% (respectively) for O and B stars.

The two most widely accepted explanations for OB runaway velocities and heights above the galactic disc are the supernova ejection mechanism and the cluster or dynamical ejection mechanism. In the former, post main-sequence mass loss by the more massive "primary" star in an O or B binary system will reverse the mass ratio and circularize the orbit. Since half the system mass must be lost in order to unbind a circular binary, a supernova will not cause disruption unless other forces play a role. Even including the effects of the expanding supernova shell on the secondary star and admitting the possibility of an asymmetric supernova, Leonard and Dewey (1992) find that most systems with primaries massive enough to form neutron stars should remain bound. A bound OB-neutron star system will acquire a significant eccentricity and speed off relative to the pre-supernova center of mass with a velocity

$$V'_{cm} = \epsilon V_{1,orb} \quad (1)$$

where $V_{1,orb}$ is the pre-supernova orbital velocity of the primary (Dewey & Cordes 1987).

Close dynamical interactions (scatterings) between OB stars in young open clusters can "slingshot" stars out of the association. This mechanism is especially effective at producing runaways when two binaries collide and one of the binaries becomes unbound. This mechanism predicts that fewer than 10% of OB runaways will have companions of any kind, compact or otherwise (Leonard & Duncan 1988, 1990).

2. OBSERVATIONS

Gies and Bolton (1986) conducted a radial velocity survey of 36 bright OB Runaways. They searched for time-variations indicative of a compact companion. They also analyzed existing X-ray data to determine if any of their candidates are X-ray sources (indicative of accretion onto a neutron star or black hole). They found no evidence for compact companions, leading them to accept the dynamical mechanism as being more plausible.

Leonard and Dewey (1992) used a Monte Carlo program to simulate OB runaway production by the supernova ejection mechanism. They found that runaway velocity is anti-correlated with mass of the progenitor system and that fewer than 10% of O stars will have peculiar velocities greater than 50 km s^{-1}. Note, however, that this is not necessarily at odds with Stone's estimate of 46%. Since he fit the runaway population with a gaussian centered on $V_p = 40$ km s^{-1}, $V_p \gtrsim 50$ km s^{-1} represents a high velocity tail.

The case for the supernova mechanism has been outlined by Blauuw (1993). The large rotation velocities, anomalous helium abundances, and the blue-straggler nature of some OB runaways all tend to favor the supernova mechanism.

Johnston *et al.* (1992) discovered PSR 1259−63, a high velocity pulsar-Be binary system. The orbit has a long period and is highly eccentric. Gies and Bolton's search would very likely have missed an object of this type because of the very small variation in the Be orbital velocity except near periastron passage. Binary pulsars such as these may be common, but were missed by previous pulsar surveys because of selection effects. PSR 1259−63 lies far out from the galactic plane, has a short pulse period, has a binary companion, and has not been observed at 400 MHz.

3. OUR PROPOSAL

The discovery of PSR 1259−63 highlights the need for an alternative approach to searching for binary companions to OB runaways, one which is sensitive to a companion near apastron. We have submitted a VLA proposal (June 1993) to conduct a search for radio pulsars at the optical positions of 40 OB runaways. We will search for point sources in 6cm and 20cm images and simultaneously, at 20cm, use the HTRP to search for pulsed emission.

These two aspects of the search are complementary. For a single pulsar, the HTRP can detect a weak pulsed signal that might not be detectable as a point source above the noise in a continuum image. However, in a binary system the OB wind might smear out the pulses (Johnston et al. 1992) without eclipsing the continuum signal. If the pulsar is bright, it may therefore only be detectable as an unpulsed point source.

With 12 minutes of data on each source at each frequency, the imaging $3 - \sigma$ detection limit will be roughly 0.3 mJy (Bridle 1989). The mean pulsar radio luminosity is 10 mJy kpc^{-2} at 20-cm (Taylor et al. 1993). We have chosen all but a few of our candidates to lie within 1 kpc, so for pulsars lying outside the radio photosphere the vast majority should be visible as non-thermal point sources.

A massive stellar wind will present a two-fold problem. The ionized wind could both scatter and absorb the radio beam coming from a pulsar, smearing out the pulses, reducing the pulsar's luminosity, or possibly eclipsing it completely. The wind may also be a source of thermal (bremsstrahlung) radiation (Bieging et al. 1989).

Our method attempts to deal with these problems. If pulses are smeared out, a pulsar may still be visible as a point source in the continuum map. By observing at two different frequencies, it should be possible to distinguish between a non-thermal point source (the pulsar) and a thermal source (the wind) by estimating a spectral index. To minimize these problems, we will observe candidates later than O5.

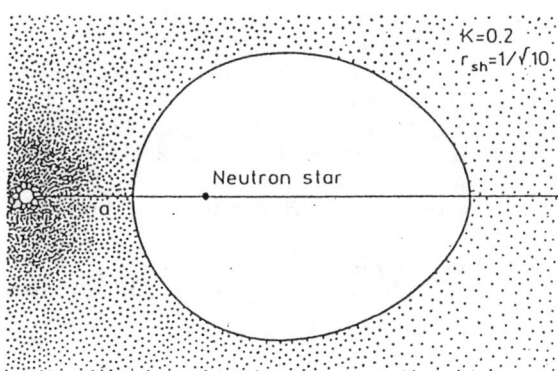

Fig. 1. A pulsar embedded in an ionized wind (Lipunov 1992).

4. SENSITIVITY

The two methods have differing sensitivities. The HTRP's sensitivity is ultimately limited by smearing of the pulse. The pulses are smeared by dispersion across the finite bandpass of the VLA and by scattering. In the absence of significant dispersion and scattering, adding together several harmonics of a well-defined pulse potentially allows pulse searches to have higher sensitivity than the VLA imaging counterpart.

The continuum maps are limited by the thermal noise of the VLA electronics. Confusion due to extragalactic point sources could conceivably be a problem. To minimize this effect, we will use the VLA in its A-array configuration.

5. CONCLUSIONS

With this search, we propose a new method to determine the origin of the OB runaways. Our sizable candidate list should allow us to draw meaningful statistical conclusions about their origin. As well as testing the two ejection mechanisms, our search could produce an object similar to PSR 1259−63 in the northern hemisphere.

ACKNOWLEDGEMENTS

C. P. would like to thank the NRAO and the staff of the VLA for hospitality in summer 1993. This work is supported in part by NSF grant PHY90-57865 and NASA grant NAGW-2934.

REFERENCES

Bieging, J. H., Abbott, D. C. & Churchwell, E, B. 1989, ApJ, 34 0, 518
Blaauw, A. 1993 in Massive Runaway Stars
Bridle, A. H. 1989, in Synthesis Imaging in Radio Astronomy, ed. R. A. Perley
    (Astronomical Society of the Pacific: San Francisco), pp. 443-476
Dewey, R. J. & Cordes, J. M. 1987, ApJ, 321, 780
Gies, D. R., and Bolton, C. T. 1986, ApJS, 61, 419
Johnston, S., Manchester, R. N., Lyne, A. G., Bailes, M., Kaspi, V. M., Guojun,
    Q. & D'Amico, N. 1992, ApJL, 387, L37
Leonard, P. J. T. & Dewey, R. J. 1992, BAAS, 24, 1177
Leonard, P.J.T. & Duncan, M.J. 1988, AJ, 96, 222
Leonard, P.J.T. & Duncan, M.J. 1990, AJ, 99, 608
Lipunov, V. M. 1992, Astrophysics of Neutron Stars (Springer-Verlag: Berlin),
    p. 248
Stone, Ronald C. 1991, AJ, 102, 333
Taylor, J. H., Manchester, R. N. & Lyne, A. G. 1993, submitted to ApJS

HYDRODYNAMICAL INSTABILITY AND ORBITAL EVOLUTION OF CLOSE BINARY SYSTEMS

Dong Lai[a], Frederic A. Rasio[b], & Stuart L. Shapiro[a]

[a] Center for Radiophysics and Space Research, Cornell University, Ithaca, NY 14853
[b] Institute for Advanced Study, Princeton, NJ 08540

1. INTRODUCTION

Essentially all recent theoretical work on close binary systems has been done in the Roche approximation, where the components are modeled as non self-gravitating gas in hydrostatic equilibrium in the effective potential of a point-mass system (e.g., Paczyński 1971). This model applies well to very compressible objects with centrally concentrated mass profiles, such as giants and early-type main-sequence stars. Some theoretical work has also been done in the completely opposite limit of binaries containing a self-gravitating, incompressible fluid (Chandrasekhar 1969). However, many binary systems of astrophysical interest contain stars that are neither very centrally concentrated nor homogeneous. In particular, low-mass white dwarf and main-sequence stars have effective polytropic indices $n \simeq 1.5$, and neutron stars typically have $n \sim 0.5-1$.

In our recent papers (Lai, Rasio & Shapiro 1993a,d), we have presented a comprehensive analytic study of the equilibrium and stability properties of close binary systems containing polytropic components. In addition to providing compressible generalizations for all the classical incompressible configurations as discussed in Chandrasekhar (1969), our method can also be applied to more general binary models where the stellar masses, radii, spins, entropies, and polytropic indices are all allowed to vary over a wide range and independently for each component. As a result, a variety of dynamical behaviors for various types of binary systems can be identified. Most importantly, we find that for sufficiently incompressible systems, both secular and dynamical instabilities can develop before a Roche limit or contact is reached along a sequence of models with decreasing binary separation. These instabilities result from Newtonian tidal interactions between equilibrium stars.

The development of a dynamical instability can have a profound effect on the terminal evolution of coalescing binaries (Lai, Rasio & Shapiro 1993b,c). In particular, it causes binary neutron stars whose orbits decay via gravitational wave emission to undergo rapid merging just prior to contact. The final coalescence can take place on a timescale much shorter than the energy dissipation time scale. For the coalescence of binary neutron stars, the radial infall velocity at contact is comparable to the free-fall velocity. As a result, the imploding stars will experience appreciable shock heating as they come into contact. Some high-mass X-ray binaries are expected to eventually evolve to compact binaries containing neutron stars or white dwarfs (van den Heuvel 1991). Our results are therefore important to determining the final evolution of such systems.

2. COMPRESSIBLE DARWIN-RIEMANN BINARY MODELS

A binary system in steady state is characterized by conserved global quantities such as masses M, M' for the two components, the fluid circulations C, C', and total angular momentum J. The total energy the system E can always be written as a functional of the fluid density and velocity fields $\rho(\mathbf{x})$ and $\mathbf{v}(\mathbf{x})$. In principle, an equilibrium configuration can be determined by extremizing this energy functional with respect to all variations of $\rho(\mathbf{x})$ and $\mathbf{v}(\mathbf{x})$ that leave the conserved quantities unchanged. The essence of our method is to replace the infinite number of degrees of freedom contained in $\rho(\mathbf{x})$ and $\mathbf{v}(\mathbf{x})$ by a limited number of parameters $\alpha_1, \alpha_2, \ldots$, in such a way that the total energy becomes a function of these parameters,

$$E = E(\alpha_1, \alpha_2, \ldots; M, C, J, \cdots). \tag{1}$$

An equilibrium configuration is then determined by extremizing the energy according to

$$\frac{\partial E}{\partial \alpha_i} = 0, \quad i = 1, 2, \ldots \tag{2}$$

where the partial derivatives are taken holding M, J, $\cdots$ constant.

Under the combined effects of centrifugal and tidal forces, the stars in a binary assume nonspherical shapes, which we model as ellipsoids. Moreover, we assume that the surfaces of constant density within each star are self-similar ellipsoids, and the density profile $\rho(m)$, where m is the mass interior to an isodensity surface, is identical to that of a spherical polytrope of the same mass. The independent variables $\{\alpha_i\}$ which specify the structure of our binary models are the binary separation r and the three axes of the two ellipsoids a_i, a'_i ($i = 1, 2, 3$). The velocity field of the fluid is modeled as either uniform rotation or uniform vorticity, with the spin axis perpendicular to the orbital plane. Thus both synchronized and nonsynchronized systems are considered.

3. STABILITY LIMITS AND ROCHE LIMIT

When viscosity is negligible, the fluid circulations of the stars are individually conserved. Figure 1 illustrates three different dynamical behaviors for equilibrium binary sequences with constant circulations. Such sequences are especially relevant for binary systems whose orbits decay due to gravitational radiation. This is because the gravitational radiation reaction forces conserve the fluid circulation (Miller 1974).

The three types of behaviors are (cf. Fig. 1):

(a) For sufficiently compressible systems (large n), the stars behave like two point masses. The energy E decreases monotonically as r decreases, and stable equilibrium solution exists all the way to contact.

(b) For more incompressible systems (smaller n), tidal interaction plays an important role in determining the binary equilibrium. As r decreases, E reaches a minimum before contact. Such a turning point in the equilibrium energy curve exactly coincides with the point of onset of dynamical instability. Beyond this stability limit, all equilibrium solutions become unstable. The physical nature of this instability is common to all binary interaction potentials that are sufficiently steeper than $1/r$. It is analogous to the familiar instability at $r = 6M$ of circular orbits for test particles around a Schwarzschild black hole. Here, however, it is the purely Newtonian tidal effects that are responsible for the steepening of the effective binary interaction potential and for the destabilization of the circular orbit.

(c) When the masses of the two components are different, the binary can encounter a Roche limit before contact. The Roche limit is the point where the binary separation has a minimum value below which no equilibrium solution exists. Typically, both the stability limit and Roche limit occur around orbital separation $r_m \sim 3(M'/M)^{1/3}R$, where R is the stellar radius. But note that the dynamical stability limit is always encountered prior to the Roche limit.

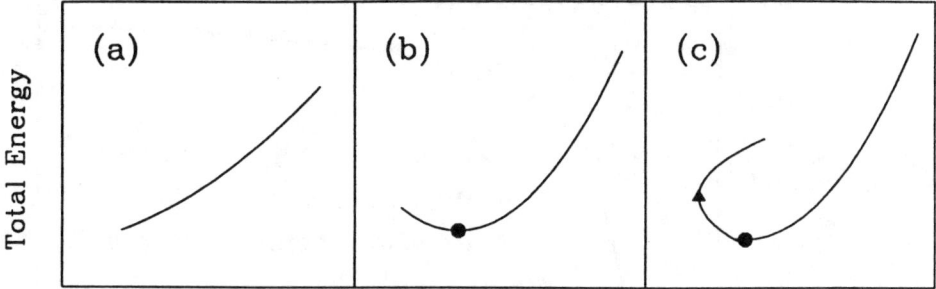

Fig. 1. General classification of equilibrium binary sequences according to terminal configurations and stability limit. The existence and ordering of the stability limit (round dots) and Roche limit (triangle) is shown schematically along equilibrium energy curves $E(r)$. All curves terminate at the point when the two stars contact.

In the opposite limit, when the effective viscosity is sufficiently large to maintain synchronization of spins and orbital motion, corotating sequences are relevant. Depending on the polytropic indices and the mass ratio, different types of equilibrium behaviors similar to Figure 1 can also be identified. However, for a corotating sequence, the minimum in the energy curve corresponds to the secular stability limit, while the true dynamical instability occurs at somewhat smaller orbital separation along the sequence.

4. ORBITAL EVOLUTION AND BINARY COALESCENCE

The importance of the dynamical instability can be easily seen. When the energy loss timescale of the system is much longer than the dynamical timescale, the orbital evolution is quasi-static. The rate of change of the orbital separation is given by $\dot{r} = \dot{E}/(dE/dr)$. As the binary approaches the dynamical stability limit r_m, where $dE/dr \to 0$, we have $\dot{r} \to \infty$. Clearly, the quasi-static description is not valid near r_m.

Figure 2 shows the results of our dynamical calculation for the terminal evolution of binary neutron stars due to gravitational radiation. The development of a dynamical instability causes a rapid acceleration of the coalescence, and the radial infall velocity at contact can be a significant fraction ($\sim 10\%$) of the tangential orbital velocity.

The effects of viscosity can also be considered. Since viscous forces conserve the total angular momentum, a binary system evolving through viscosity only will follow a sequence of configurations with constant J. Such viscous evolution may be responsible for the orbital period changes detected in some high-mass X-ray binaries such as Cen X-3, SMC X-1, LMC X-4 (Levine et al 1993). The orbital evolution of the binary system depends on its initial angular momentum J_i. The binary either evolves toward a stable synchronized state, or is driven to coalescence by viscous dissipation. Again, we find that for sufficiently incompressible systems, the binary can encounter a dynamical instability before the final merger (Lai, Rasio & Shapiro 1993d).

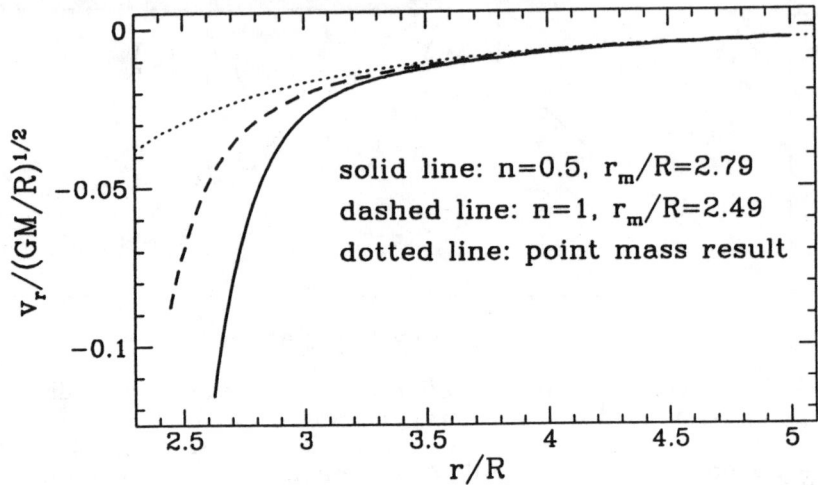

Fig. 2. The infall radial velocity of a coalescing neutron star binary. Results for binary models with different polytropic indices n are shown. The two neutron stars are assumed to be identical, with mass $M = 1.4 M_\odot$, radius $R = 10$ km, and both have zero spin. Here r_m is the dynamical stability limit.

This work has been supported in part by NSF Grant AST 91-19475 and NASA Grant NAGW-2364 to Cornell University, and Grant HF-1037.01-92A from STSI.

REFERENCES

Chandrasekhar, S. 1969, Ellipsoidal Figures of Equilibrium (New Haven: Yale University Press)
Lai, D., Rasio, F. A., & Shapiro, S. L. 1993a, ApJS, 88, 205
———. 1993b, ApJL, 406, L63
———. 1993c, ApJ, in Press
———. 1993d, ApJ, in Press
Levine, A., Rappaport, S., Deeter, J. E., Boynton, P. E., & Nagase, F. 1993, ApJ, 410, 328
Miller, B. D. 1974, ApJ, 187, 609
Paczyński, B. 1971, ARA&A, 9, 183
van den Heuvel, E. P. J. 1991, in "X-Ray Binaries and Recycled Pulsars", ed. E.P.J. van den Heuvel & S.A. Rappaport (Kluwer, Dordrecht)

EVOLUTION of LMBs and ASTEROSEISMOLOGY

Marek J. Sarna
N.Copernicus Astronomical Center, Polish Academy of Sciences, Warsaw 00-716 Poland
Marek_Sarna@camk.edu.pl

Umin Lee and Alexander G. Muslimov
Department of Physics and Astronomy, University of Rochester, Rochester NY 14627 USA
umin@callisto.pas.rochester.edu, muslimov@callisto.pas.rochester.edu

ABSTRACT

We investigate pulsations of the red–dwarf companion of the neutron star (NS) in a low–mass binary (LMB). The illumination of the donor star by the pulsar's high–energy, non–thermal radiation and relativistic wind may substantially affect its structure. We present a quantitative analysis of the oscillation spectrum of a red dwarf which evolved in an LMB and underwent a stage of evaporation. We calculate the $p-$ and $g-$modes for red dwarfs of masses within the interval $0.2-0.6\ M_\odot$. For comparison, we present similar calculations for ZAMS stars of the same masses. In the case of less massive donor stars ($\sim 0.2\ M_\odot$) the oscillation spectrum becomes qualitatively different from that of their ZAMS counterparts. We also consider tidally $forced$ $g-$modes and perform a linear analysis of these oscillations for different degrees of the non–synchronization between the orbital rotation and spin rotation of the red-dwarf component. We discuss the triggering by these oscillations of the Roche–lobe overflow and sudden mass loss by the donor star. Further implications of this effect for the $\gamma-$ and X-ray burst phenomena are outlined.

1. INTRODUCTION

Here we propose the systematic study of the oscillations of the donor star in LMBs with NSs and focus our attention on the global low–order $p-$ and $g-$mode oscillations. The acoustic $p-$modes we consider include both radial and non–radial oscillations. Also, within the framework of linear analysis, we investigate the resonant $g-$modes excited by tides from the NS, provided that there is some non–synchronization between the spin and orbital motion of the red–dwarf component. Since the red dwarf in an LMB loses mass on a relatively short timescale ($\lesssim 10^8\ yr$) and very likely undergoes a stage of illumination by the hard radiation and/or relativistic particles of a NS, its internal structure will differ from that of a main–sequence star of the same mass. This difference, being undetectable in terms of the mass or even the radius of the star, can strongly affect the spectrum of global oscillation modes. For example, an isolated red dwarf of mass $0.2 - 0.3\ M_\odot$ is thought to be fully convective, while a red dwarf of the same mass evolved in an LMB with a NS will be only partially convective. As a result, $g-$mode oscillations can only exist in the latter case. This fact motivates our particular interest in these modes. The detection of red–dwarf oscillations in LMBs would provide astrophysicists with valuable information about their masses and radii.

For our purpose we will use a particular evolutionary sequence for the system consisting initially of a 1.4 $M_\odot$ NS and a 1 $M_\odot$ main–sequence star. At the moment of filling of the Roche lobe by the red dwarf, the orbital period of the system is $9.^h4$ (see for details Muslimov & Sarna 1992). This sequence is characterized by a relatively high mass–loss rate ($\sim 8 \times 10^{-9}$ $M_\odot/yr$) by the donor star and a short ($\sim 7 \times 10^7$ yr) duration of the evaporation stage before the radius of the red dwarf shrinks (at $M_1 = 0.29$ $M_\odot$) within its Roche lobe. Because of the heating (due to illumination from pulsar), the red dwarf does not contract to the main–sequence equilibrium radius, and at $M_1 = 0.29$ $M_\odot$ it is about 1.7 times larger than a normal main–sequence star of the same mass.

The oscillation properties of the models from our evolutionary computations will be compared with the ZAMS (equilibrium) models of the same masses computed by using a stellar evolution code based on the standard Henyey–type code of Paczyński (1970).

2. RESULTS

Here we sketch briefly the main results of our analysis, with the detailed paper will be published elsewhere. In the study of close interacting binary systems, such as the cataclysmic systems and LMXBs, it is usually assumed that the mass–lossing component is in a circular orbit and in a state of synchronous rotation.

Note that in the case of non–conservative evolution of an LMB with the NS, the assumption of complete synchronization is questionable, at least for the late stages of evolution. The reason is that the orbital evolution of the LMB with a rapidly rotating NS proceeds on a relatively short timescale ($\sim 10^7$ yr or less), so that the synchronous rotation of the primary may be violated. Our estimates show that in the LMB at the late stage of its evolution when the mass of the donor star is about 0.2–0.3 $M_\odot$ or less, the effectiveness of synchronization decreases and the system tends to rotate non–synchronously.

In the following analysis of tidally induced oscillations we shall, therefore, assume that the spin rotation of the donor star of mass $\lesssim 0.3$ $M_\odot$ in the LMB is not synchronized with its orbital rotation.

We calculate non–adiabatic and nonradial oscillations of low–mass stars employing a linear theory of stellar oscillations (see e.g. Unno et al 1989). We neglect the effects of rotation on the oscillations. This is justified in the case of *free* radial and nonradial modes, because the typical periods of the oscillations are shorter than the rotation periods. The effects of rotation can be important for low–frequency, tidally forced oscillations, particularly for modes with oscillation periods longer than the rotation period of the star.

Our analysis of the oscillations of 0.3 $M_\odot$ donor star shows that the periods of p–modes for the non–equilibrium model are about 60 per cent longer than those for the equilibrium model. Also, the periods of f–modes and radial oscillations for the non–equilibrium model are, respectively, about 50 and 60 per cent longer than those for the equilibrium model. We have calculated the tidally *forced* oscillations for the non-equilibrium model of mass $0.23 M_\odot$. The results are summarized in Figure 1, in which we have plotted the normalized surface amplitudes of the *forced* oscillations as functions of the forcing period observed in the corotating frame of the primary. Figure 1 clearly shows the resonance between the tidally *forced* and *free* g–mode oscillations.

3. DISCUSSION

We have demonstrated that even insignificant departures from the internal structure from the ZAMS model can affect the oscillation periods. The non-equilibrium models have more extended and rarefied envelopes, so the periods of acoustic (p–) modes, localized in the outer layers, are longer for the non-equilibrium model than for the equilibrium one. In contrast, the gravity (g–) modes are localized in central regions of a star and depend on the superadiabatic temperature gradient. The non–equilibrium models have thiner convection zones and, therefore, shorter periods of g–modes. A ZAMS star of sufficiently low mass ($\lesssim 0.3\ M_\odot$) is fully convective, in which case the gravity modes are absent. We have shown that for the non–equilibrium models g–modes may exist even for star with mass as low as $0.23\ M_\odot$.

We also suggest that at the late stage of the evolution of the LMB with a NS conditions may be favorable for violation of synchronous rotation. If there is non–synchronization between the spin and orbital rotations of the red dwarf, then one can expect the occurence of dynamical tides. We have calculated a series of tidally induced resonances for the red dwarf of mass $0.23\ M_\odot$.

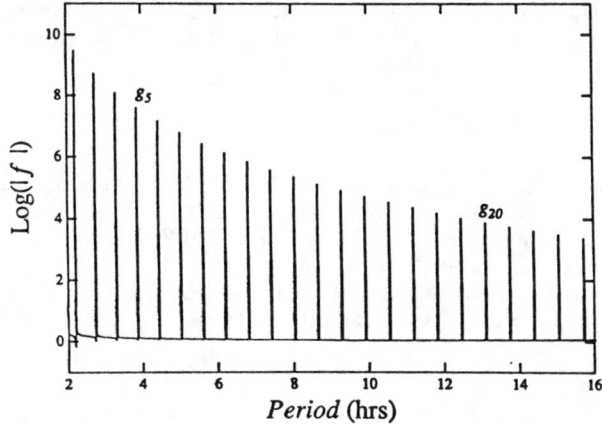

Fig. 1. The normalized surface amplitude f of the tidally forced oscillations as a function of the forcing period P, given in the frame of reference corotating with the red dwarf.

In Figure 1 we have illustrated the resonance amplitudes of gravity modes for different values of the forcing period P, corresponding to different values of the parameter $\Delta\Omega/\Omega_{orb} = P_{orb}/2P$ (where $\Delta\Omega = |\Omega_{orb} - \Omega_s|$; Ω_{orb} and Ω_s are the angular velocities of orbital and spin rotation, respectively) characterizing the degree of non–synchronization. In Figure 1 the parameter $\Delta\Omega/\Omega_{orb}$ ranges from 0.2 to 1. The positions of the resonances shift to shorter forcing periods (larger values of $\Delta\Omega/\Omega_{orb}$) as the mass of the primary decreases. This shift is due to the decreasing eigenperiods of g–modes with decreasing mass of the donor star. It is interesting that the forcing period also decreases if the degree of non–synchronization gradually increases with time, so that the resonance condition may be self–adjusting. The widths of the resonances will also be broadened by turbulent damping. These arguments enable us to suggest that in an LMB with a NS, resonances may occur much more often than once per

$10^6 - 10^7$ yr. The effect of sudden mass loss by the donor star triggered by the dynamical tide may explain some burst phenomena in LMXBs. Also, as has been pointed out recently by Lasota, Frank, & King (1992), some γ-ray burst sources may be identified with the end point of LMXB evolution. In this context, the mechanism of tidally induced Roche–lobe overflows can provide, in addition to the statistical argument suggested by Lasota, Frank, & King (1992), a physical reason for the hypothesis that some γ-ray burst sources (e.g., the so-called soft γ-ray repeaters) may be associated with the NSs in LMBs. The effect of tidally induced Roche–lobe overflow and episodic mass loss by the companion of a NS may be important in binaries with eccentric orbits. In this case one can expect the dynamical tides and accompanying phenomena to occur at periastron and to be modulated by the orbital period.

ACKNOWLEDGEMENTS

We are grateful to Jack Thomas for his reading of the manuscript and Wojtek Dziembowski for his comments and discussion. This work was supported in part by NSF grant AST 91–15132 and in part by NASA grant NAGW–2444, both through the University of Rochester, which we gratefully acknowledge. Also, it was supported in part by the Polish National Committee for Scientific Research under grant 2–2115–92–03.

REFERENCES

Lasota, J.–P., Frank.J. & King, A.R. 1992, in Gamma–Ray Bursts, eds. W.S. Paciesas & C.J. Fishman (New York: AIP) p.126
Muslimov, A.G. & Sarna, M.J. 1993, MNRAS, 262, 164
Paczyński, B. 1970, Acta Astro., 20, 47
Unno, W., Osaki, Y., Ando,H., Saio, H., & Shibahashi, H. 1989, Nonradial Oscillations of Stars, (Tokyo: University of Tokyo Press)

EXCITATION OF NEUTRON STAR OSCILLATION MODES DURING BINARY INSPIRAL

Andreas Reisenegger
Institute for Advanced Study, Princeton, NJ 08540
andreas@guinness.ias.edu

Peter Goldreich
California Institute of Technology, Pasadena, CA 91125
peter@deimos.caltech.edu

ABSTRACT

As a compact binary inspirals due to the emission of gravitational waves, its orbital period decreases continuously down to $\sim$ 1ms, its value at coalescence. During the last part of the inspiral, the two stars are close together and their tidal interactions become strong. Neutron stars have many normal modes whose periods lie in the range swept by the orbital period. Some of these modes are resonantly excited by the tidal force. The amount of energy a mode absorbs is proportional to the square of the overlap integral between its displacement field and the tidal force field. For all modes of interest, this overlap is poor, resulting in relatively weak excitation. The absorbed energy is only a small fraction ($\lesssim 10^{-6}$) of the orbital energy, so the orbital phase shift is too weak to be detected by observations of the gravitational wave signal emitted by the inspiraling binary. However, with displacement amplitudes of excited quadrupole modes ranging up to 0.5% (or more) of the stellar radius, the possibility of a detectable electromagnetic signature cannot be dismissed. Both the periods of the modes and the energy they absorb depend quite strongly on the internal structure of the star. Their observation could shed light on the correct high-density equation of state.

1. MOTIVATION

Inspiraling compact binaries (two neutron stars, two black holes, or one of each) are seen as "safe" sources of gravitational radiation (Schutz 1986; Narayan, Piran, & Shemi 1991; Phinney 1991) to be detected by proposed instruments such as LIGO (Abramovici et al. 1992) and VIRGO (Bradaschia et al. 1990), and also as possible sources of gamma-ray bursts (see Narayan, Paczyński, and Piran 1992 and references therein). Both aspects are reviewed by O. Blaes in the present volume. Because of low signal-to-noise ratios, the detection of gravitational waves from these events relies heavily on accurate prior knowledge of the class of waveforms in the frequency range ($\sim$ 50Hz) to which the detectors are most sensitive (e.g., Thorne 1987; Cutler et al. 1993). Currently, only speculation connects coalescing binaries and gamma-ray bursts. Observational signatures that might distinguish between the "standard model" mentioned above and alternatives proposed or yet to be thought of should be of interest. For both reasons, we examine whether the energy absorbed by neutron star normal modes that are resonantly excited during the inspiral is sufficient to produce detectable changes in the orbital evolution of the binary or some other measurable effect. For reasons of space, here we limit ourselves to present the main results. Their derivation is given in Reisenegger & Goldreich (1993), and some related issues are discussed in Reisenegger (1994).

2. TIDAL EXCITATION OF MODES

We consider a neutron star of mass $M_\star$ and radius $R_\star$ whose normal modes are being tidally excited by a companion star of mass M in a nearly circular orbit. We assume that the rotational frequency of the neutron star (in an inertial reference frame) is much smaller than the orbital (angular) frequency Ω, which increases continuously (on a time scale $\gg \Omega^{-1}$) as the system inspirals due to the emission of gravitational radiation.

It has been shown (McDermott et al. 1985; Finn 1987; McDermott, Van Horn, & Hansen 1988; Reisenegger & Goldreich 1992) that neutron stars have a dense spectrum of normal modes with a variety of restoring mechanisms. The modes are completely described by their displacement fields $\xi_{nlm}(\mathbf{x})$, where l and m are the usual spherical-harmonic indices for the angular dependence, and n is an additional label that distinguishes different radial waveforms. Many modes, particularly some of those associated with the stable stratification of the stellar matter or with the finite shear modulus of the neutron star crust, have frequencies ω_{nl} in the range $\sim 10^3$ s swept by Ω close to coalescence. When $\Omega = \omega_{nl}/m$, the orbital motion will resonate with the mode ξ_{nlm}.

The tidal force exerted by the binary companion, which varies sinusoidally with time due to the orbital motion, will stay in phase with the mode for a time interval $\sim (2\pi/\dot\Omega)^{1/2}$, during which it the mode absorbs from the orbital motion a total amount of energy

$$\Delta E_{nlm} = \frac{5\pi}{384m} \frac{M/M_\star}{(1+M/M_\star)^{(2l+1)/3}} \left(\frac{c^2 R_\star}{GM_\star}\right)^{5/2}$$
$$\times \left(\frac{\omega_{nl}}{m\Omega_d}\right)^{(4l-7)/3} \frac{GM_\star^2}{R_\star} |S_{nlm}|^2 \qquad (1)$$

(Reisenegger & Goldreich 1993). Here, $\Omega_d \equiv (GM_\star/R_\star^3)^{1/2}$ is approximately the inverse dynamical time of the neutron star, and the dimensionless quantity

$$S_{nlm} = 2W_{lm} \frac{\int_{M_\star} \nabla[r^l Y_{lm}^*(\theta,\phi)] \cdot \xi_{nlm}(\mathbf{x}) dM/M_\star}{R_\star^{l-1} \left(\int_{M_\star} |\xi_{nlm}(\mathbf{x})|^2 dM/M_\star\right)^{1/2}} \qquad (2)$$

(of order unity or less) gives the overlap of the modal displacement field $\xi_{nlm}(\mathbf{x})$ with the 2^l-pole component of the tidal force field (Press & Teukolsky 1977). The numerical factor

$$W_{lm} = (-)^{(l+m)/2} \frac{[4\pi(l-m)!(l+m)!/(2l+1)]^{1/2}}{2^l[(l-m)/2]![(l+m)/2]!}, \qquad (3)$$

where the symbol $(-)^k$ is to be interpreted as zero if k is not an integer.

For a constant-density star, $S_{nlm} = 0$ for all modes other than the f-modes (for low-l f-modes, $|S_{0ll}| \sim 1$), and therefore only these can be excited by the tidal potential. This is not exactly true for more general stellar models, but for "realistic" neutron stars $|S_{nlm}|$ is still small ($\ll 1$) for all modes other than the f-modes, and especially for the relatively long-period modes of interest here (Reisenegger 1994).

Evaluating the absorbed energy for $l = m = 2$ (clearly the most strongly excited modes), with "typical" parameters $M = M_\star = 1.4 M_\odot$ and $R_\star = 10\,\text{km}$, one obtains

$$\Delta E_{n22} \approx 6 \times 10^{46} \left(\frac{P_{n2}}{5\text{ms}}\right)^{-1/3} \left(\frac{|S_{n22}|}{10^{-3}}\right)^2 \text{erg}, \qquad (4)$$

where P_{nl} is the period of a mode. (The reference numbers, $P_{n2} = 5\text{ms}$ and $|S_{n22}| = 10^{-3}$, correspond to the lowest-order quadrupole g-mode, as calculated for Model 2 in Reisenegger & Goldreich 1992.) Since $|S_{nlm}|$ decreases rapidly with increasing n (number of radial nodes, say), the energy absorbed by the lowest order quadrupole mode is a large fraction of the total energy absorbed by all modes (Reisenegger & Goldreich 1993). It is, however, only a small fraction, $\approx 2 \times 10^{-6} (P_{n2}/5\text{ms})^{1/3}(|S_{n22}|/10^{-3})^2$, of the orbital energy at resonance, so that the effect on the orbital evolution of the system is negligible. Nevertheless, the typical (mass-weighted rms) displacement in the oscillations can be significant,

$$\xi \approx 5 \times 10^{-3} \left(\frac{P_{n2}}{5\text{ms}}\right)^{5/6} \left(\frac{|S_{n22}|}{10^{-3}}\right) R_\star. \qquad (5)$$

3. PHASE ERROR

It is also interesting to estimate the phase error made by calculating the orbital evolution of the binary system neglecting the effect of the resonant excitation of normal modes. If the phase of the true evolution is set equal to that of the evolution in the absence of mode excitation at any time before the resonance, then their difference at any time t after the resonance is

$$\Delta\Phi(t) \approx 2 \times 10^{-4} \left[\frac{\Omega(t)}{\omega_{n2}/2} - 1\right] \left(\frac{P_{n2}}{5\text{ms}}\right)^2 \left(\frac{|S_{n22}|}{10^{-3}}\right)^2 \qquad (6)$$

(Reisenegger & Goldreich 1993), for the same parameter values of the previous section. Again, the effect of modes with $l > 2$ is negligible compared to that from modes $l = 2$, and a large fraction of the total phase shift comes from the lowest-order, quadrupole g-mode.

We emphasize that all these numerical results involve relatively uncertain parameters such as the neutron star radius, the mode period, and especially the overlap parameter S_{nlm}, all of which depend upon the state of matter above nuclear density.

4. CONCLUSIONS

Neutron star oscillation modes excited by resonant tidal forces in an inspiraling binary system absorb a negligible fraction of the total orbital energy, and their effect on the orbit is too weak to be detected by instruments such as LIGO and VIRGO. Nevertheless, the absorbed energy is quite high compared to the energies involved in phenomena observed electromagnetically, such as radio pulses and x-ray bursts. Moreover, the oscillation amplitudes of excited modes can be as large as $\sim 0.5\%$ of the radius for core g-modes, and even larger for modes concentrated in or near the crust. Thus, the possibility that tidally excited oscillation modes might reveal themselves through electromagnetic signals cannot be easily dismissed. Due to the great sensitivity of the periods and amplitudes of the excited modes to the state of matter in the core of the neutron star, such a detection would be an interesting probe of the properties of matter at supernuclear densities.

ACKNOWLEDGEMENTS

We are grateful to Lars Bildsten, Eanna Flanagan, and Yanqin Wu for useful conversations. This work was done while A. R. was a Research Fellow at the California Institute of Technology. It was supported by NSF grant AST 89-13664 and NASA grant NAGW 2372.

REFERENCES

Abramovici, A., et al. 1992, Science, 256, 325

Bradaschia, C., et al. 1990, Nucl. Instrum. Methods Phys. Res. A, 289, 518

Cutler, C. et al. 1993, Phys. Rev. Lett., 70, 2984

Finn, L. S. 1987, MNRAS, 227, 265

McDermott, P. N., Hansen, C. J., Van Horn, H. M., & Buland, R. 1985, ApJ, 297, L37

McDermott, P. N., Van Horn, H. M., & Hansen, C. J. 1988, ApJ, 325, 725

Narayan, R., Paczyński, B., & Piran, T. 1992, ApJ, 395, L83

Narayan, R., Piran, T., & Shemi, A., 1991, ApJ, 379, L17

Peters, P. C., & Mathews, J. 1963, Phys. Rev., 131, 435

Phinney, E. S. 1991, ApJ, 380, L17

Press, W. H., & Teukolsky, S. A. 1977, ApJ, 213, 183

Reisenegger, A. 1994, ApJ, submitted

Reisenegger, A., & Goldreich, P. 1992, ApJ, 395, 240

Reisenegger, A., & Goldreich, P. 1993, ApJ, in press

Schutz, B. F. 1986, Nature, 323, 310

Thorne, K. S. 1987, in 300 Years of Gravitation, ed. S. W. Hawking & W. Israel (Cambridge Univ. Press)

BINARY-BINARY COLLISIONS INVOLVING MAIN-SEQUENCE STARS, WHITE DWARFS AND NEUTRON STARS IN GLOBULAR CLUSTERS

Peter J. T. Leonard
T-6, MS B288, Los Alamos National Laboratory, Los Alamos, NM 87545
pjtl@eagle.lanl.gov

Melvyn B. Davies
130-33, California Institute of Technology, Pasadena, CA 91125
mbd@tapir.caltech.edu

ABSTRACT

We consider collisions between dynamically-evolved primordial binaries consisting of main-sequence stars, white dwarfs and neutron stars in globular clusters. In our four-body binary-binary scattering experiments, we allow stars to "stick" if they pass close enough to each other, which leads to the formation of a wide variety of exotic objects. Most of these objects have binary companions. Also, relatively clean exchange interactions can produce binaries containing neutron stars that eventually receive material from their companions. Such systems will be observable as X-ray binaries.

1. INTRODUCTION

Collisions between dynamically-evolved primordial binaries consisting of main-sequence stars (MSSs), white dwarfs (WDs) and neutron stars (NSs) must occur in globular clusters, and the purpose of this project is to study such interactions. To reduce the parameter space, we consider only binaries containing at least one MSS. In our scattering experiments, we allow stars to "stick" if they pass close enough to each other. Multiple collisions can take place. The various kinds of stellar mergers that can occur directly or following a common envelope phase and/or the emission of gravitational radiation are 1) MSS-MSS, 2) MSS-WD, 3) MSS-NS, 4) WD-WD, 5) WD-NS, and 6) NS-NS. The exotic phenomena that result from such mergers may include blue stragglers, red giants, Thorne-Żytkow objects, accretion disks around compact objects, spun-up pulsars, ms pulsars formed via AIC, Type Ia supernovae, black holes, and gamma-ray bursts.

2. THE SCATTERING EXPERIMENTS

The scattering experiments were carried out using a modified version of the code originally developed by Alexander (1986). The adopted masses are $M_{MSS} = 0.7\ M_\odot$, $M_{WD} = 0.7\ M_\odot$ and $M_{NS} = 1.4\ M_\odot$. The adopted radii are $R_{MSS} = 0.7\ R_\odot$, $R_{WD} = 0.01\ R_\odot$ and $R_{NS} = 0.0\ R_\odot$. Stars are allowed to "stick" together if they pass close enough. The radius of "stickyness" is $R_{stick} = \Sigma\ R_{MSS}$ for MSS-MSS encounters, and $R_{stick} = 2\ R_{MSS}$ for MSS-WD and MSS-NS encounters. Two MSSs that merge become a 1.4 $M_\odot$ MSS with a radius of 1.4 $R_\odot$. For WD-WD encounters, $R_{stick} = \Sigma\ R_{WD}$. However, direct collisions between WDs are very rare, since WDs are much more likely to collide with MSSs first. The semi-major axes and eccentricities of both binaries are $a = 50\ R_\odot$ and $e = 0.6$, respectively. The relative velocity at infinity well before the interaction begins is $V_{rel,\infty} = 10$ km s^{-1}. The impact parameter, b, is generated from the distribution $dN/db \propto b$ for $0 \leq b \leq b_{max}$. The values of b_{max} are $36.29a$, $40.46a$ and $44.24a$ for collisions involving zero, one and two NSs, respectively. In all cases, b_{max} corresponds to a pericenter distance of $6a$ for the relative two-body hyperbolic orbit of the binary centers of mass, which is extremely unlikely to lead to a strong interaction.

3. THE RESULTS

The following tables state the number, n, of the various types of mergers that occurred in 1000 examples of various kinds of binary-binary collisions. The corresponding cross section is $(n/1000)\ \pi\ b_{max}^2$. The number in brackets is the number of merged stars that have binary companions.

A. (MSS+MSS)–(MSS+MSS) Collisions

no merger 673
MSS-MSS merger .... 236(142)
MSS-MSS-MSS merger .... 75(35)
two separate MSS-MSS mergers .... 14(4)
MSS-MSS-MSS-MSS merger .... 2(0)

B. (MSS+MSS)–(MSS+WD) Collisions

no merger 648
MSS-MSS merger .... 174(87)
MSS-WD merger .... 96(85)
MSS-MSS-WD merger .... 47(24)
separate MSS-MSS and MSS-WD mergers .... 21(11)
MSS-MSS-MSS merger .... 14(0)

C. (MSS+MSS)–(MSS+NS) Collisions

no merger 674
MSS-MSS merger .... 165(49)
MSS-NS merger .... 100(91)
MSS-MSS-NS merger .... 30(15)
separate MSS-MSS and MSS-NS mergers .... 18(11)
MSS-MSS-MSS merger .... 11(1)
MSS-MSS-MSS-NS merger .... 2(0)

D. (MSS+WD)–(MSS+WD) Collisions

no merger 658
MSS-WD merger .... 220(141)
MSS-MSS merger .... 49(31)
MSS-MSS-WD merger .... 34(13)
MSS-WD-WD merger .... 21(11)
two separate MSS-WD mergers .... 17(7)
MSS-MSS-WD-WD merger .... 1(0)

E. (MSS+WD)–(MSS+NS) Collisions

no merger 693
MSS-WD merger .... 124(44)
MSS-NS merger ....89(85)
MSS-MSS merger ....47(23)
MSS-WD-NS merger ....18(13)
MSS-MSS-WD merger ....11(0)
separate MSS-WD and MSS-NS mergers ....10(3)
MSS-MSS-NS merger ....7(6)
MSS-MSS-WD-NS merger ....1(0)

F. (MSS+NS)–(MSS+NS) Collisions

no merger 729
MSS-NS merger .... 193(128)
MSS-MSS merger ....31(10)
MSS-MSS-NS merger ....22(6)
two separate MSS-NS mergers ....12(7)
MSS-NS-NS merger ....12(3)
MSS-MSS-NS-NS merger ....1(0)

4. MERGED STARS IN BINARIES

Most of the merged stars have binary companions, and a small fraction of them are in triple systems. The binaries tend to have periods that are comparable with or longer than the binaries that originally collided, and the orbits tend to be quite eccentric. These are the same trends as found for the binary properties of blue stragglers produced via physical stellar collisions during strong binary-binary interactions (Leonard & Fahlman 1991).

5. WORK IN PROGRESS

We have just begun this project, which is the binary-binary analog of the binary-single study of Davies, Benz & Hills (1993). To provide better statistics, we will consider 10^4 runs per kind of collision instead of only 10^3. Point-particle (i.e., zero-radius) experiments corresponding to the cases presented in this paper will be carried out for comparison. We will also carry out hydrodynamical simulations of the most interesting merger cases. Merger rates in real globular clusters will be estimated. Cross sections for non-merger outcomes (e.g., exchanges, triple formation, etc...) will also be calculated.

ACKNOWLEDGEMENTS

This work was supported by the United States Department of Energy, and through an R. C. Tolman Fellowship awarded to MBD at Caltech.

REFERENCES

Alexander, M. E. 1986, Journal of Computational Physics, 64, 195
Davies, M. B., Benz, W., & Hills, J. G. 1993, ApJ, 411, 285
Leonard, P. J. T., & Fahlman, G. G. 1991, AJ, 102, 994

Evolution

ON THE ORIGIN OF LOW-MASS X-RAY BINARIES

Ronald F. Webbink and Vassiliki Kalogera
Dept. of Astronomy, University of Illinois,
1002 W. Green Street, Urbana, IL 61801
astro@sirius.astro.uiuc.edu

ABSTRACT

Several evolutionary scenarios have been proposed for the origin of low-mass X-ray binaries, including accretion-induced collapse, common envelope evolution of a massive binary with an extreme mass ratio, black hole formation in Wolf-Rayet binaries with extreme mass ratios, triple star evolution, and (in globular clusters) tidal capture. In this brief review, we examine some of the factors which determine whether a primordial binary may survive to become a neutron star- or black hole-binary with a low-mass nondegenerate component, and what structural criteria it must satisfy, having reached that state, in order to appear as an X-ray binary. These constraints include: survival of common envelope evolution, survival of the supernova or core-collapse event, duration of the non-interactive phase following formation of the compact component and evolutionary state of the donor star, and finally the critical mass ratios which dictate the time scale for mass transfer in the X-ray state. A population synthesis model for newly-formed LMXBs illustrates a number of these constraints.

1. INTRODUCTION

The origin of the low-mass X-ray binaries (LMXBs) in the galactic field presents an intriguing problem for close binary evolutionary theory. The small number of such systems found in the Galaxy, combined with the long life times implied by the modest mass accretion rates needed to power their X-ray luminosity, implies a very low birth rate for LMXBs, of order 10^{-6} yr^{-1}. The initial conditions which lead to LMXB formation must therefore require meeting very exacting criteria, for this birth rate is many orders of magnitude smaller than the death rate of isolated field stars.

A number of scenarios have been put forward for the origin of LMXBs. The most familiar of these invoke (i) neutron star formation in a short-period helium star-main sequence star binary which is itself a remnant of common envelope evolution, and (ii) accretion-induced collapse of a massive white dwarf in a cataclysmic (white dwarf-main sequence star) binary (Sutantyo 1975; van den Heuvel 1981). However, neither of these scenarios offers a satisfactory explanation for the existence of low-mass binaries, such as A0620-00 (V616 Mon) and GS 2023+338 (V404 Cyg), containing ≥ 3 M$_\odot$ black holes, a circumstance which has led to other scenarios as well, involving triple star evolution (Eggleton & Verbunt 1986) and Wolf-Rayet binaries (Romani 1992).

Despite some considerable differences in the properties posited for LMXB progenitors, and in the sequence of events leading to LMXB formation, these scenarios invoke many similar processes and submit to similar constraints. Substantial uncertainties still attend the quantitative definition of most of these processes and constraints, and these afflict efforts to quantify birth rate estimates, but qualitatively they form important links in any evolutionary chain.

Our purpose here will be to identify many of the most important processes and constraints for the reader, rather than to critique different scenarios (see, e.g., Webbink 1992). In doing so, we follow essentially the sequence of events and constraints appropriate to the first scenario identified above (LMXB formation from a helium star-main sequence binary remnant of common envelope evolution), bearing in mind that most of these ingredients and constraints also apply, for example, to the accretion-induced collapse scenario.

2. EVOLUTIONARY CONSTRAINTS

1. The initial primary must be massive enough to reach core collapse (although not necessarily before mass transfer). The mass spectrum of primary components can be divided into four ranges: (i) At low initial mass ($M_1 \lesssim 8\ M_\odot$), the carbon-oxygen core produced by shell helium burning becomes degenerate. If the primary is massive enough to survive stellar wind mass loss up to the point of carbon ignition, that ignition probably leads to a carbon deflagration supernova and complete disruption of the star. (ii) In a narrow intermediate mass range (contained within the interval $8\ M_\odot \leq M_1 \leq 12\ M_\odot$), a non-degenerate or partially-degenerate carbon-oxygen core reaches carbon ignition before it grows to a Chandrasekhar mass. In this case, carbon ignition is not violent enough to disrupt the star, and an oxygen-neon-magnesium core is produced which can cool to degeneracy before reaching oxygen ignition and the subsequent nuclear burning phases which lead rapidly to core collapse. Systems with primaries in this range caught after carbon ignition, but before core collapse, may produce cataclysmic binaries containing ONeMg white dwarfs which are candidates for accretion-induced collapse. (iii) At higher initial mass ($12\ M_\odot \lesssim M_1 \lesssim 40\ M_\odot$), the primary exhausts a core exceeding a Chandrasekhar mass while still on the main sequence. Such a helium core, even if subsequently stripped of its hydrogen-rich envelope, can evolve to core collapse. This is the mass range of interest for the scenario considered here. (iv) At very high mass ($M_1 \gtrsim 40\ M_\odot$), the primary exhausts a core so massive that it remains non-degenerate through the entire chain of nuclear burning phases up to the formation of a massive iron core. In this case, the core mass is not limited by the Chandrasekhar mass, and so may collapse directly to a black hole. This is the mass range presumed relevant to the origin of A0620–00 and GS 2023+338.

2. The progenitor binary must be close enough to reach interaction prior to the supernova stage. Generally speaking, if a binary was not close enough to produce tidal mass transfer during the lifetime of its primary (initially more massive) component, mass loss and the consequent expansion of the binary orbit will preclude it from doing so during the evolution of the secondary. In practical terms, this means that the Roche lobe equivalent radius of the primary cannot exceed $\sim 1000\ R_\odot$ during the post-main-sequence evolution of the primary, even allowing for orbital expansion due to stellar winds, if the binary is to be considered "close." However, if the primary component is very massive, it may be able to shed its hydrogen-rich envelope in a stellar wind while still on the main sequence, and evolve directly to a Wolf-Rayet stage without ever developing an extended quasistatic envelope (Chiosi & Maeder 1986); in this case, the secondary star may be the first to fill its Roche lobe.

3. The progenitor binary must be wide enough to permit formation of a massive core. It is essential that the initial primary develop a massive core before the binary first interacts, because rapid mass transfer will strip away its entire hydrogen-rich envelope, aborting further growth of its hydrogen-depleted core.

The exposed core must thus be capable of evolving in isolation to core collapse (due to its own internal evolution), or at any rate be near the threshold for collapse driven by accretion from the companion star. Among stars of moderate mass ($M_1 \lesssim 12$ M$_\odot$), core masses near the Chandrasekhar limit develop only late in asymptotic giant branch evolution, requiring Roche lobe dimensions $R_{L_1} \gtrsim 500$ R$_\odot$. Among more massive stars ($M_1 \gtrsim 12$ M$_\odot$), this condition if fulfilled when the star leaves the main sequence.

4. *The progenitor binary must be wide enough to survive common envelope evolution.* The pivotal role played by common envelope evolution in creating LMXBs can be understood from two perspectives: Empirically, the compactness of observed LMXBs requires a very efficient angular momentum loss mechanism if they arose from very much longer period binaries, as required above if the primaries were originally stars of moderate mass. Theoretically, one is inevitably forced by stability constraints (see item 9 below) to consider binaries in which the original secondary was already of low mass ($M_2 \lesssim 1.5$ M$_\odot$), so that the progenitor system must have had an extreme component mass ratio a circumstance which is expected in any case to lead to common envelope evolution.

Because of the extreme difficulty in physically modeling common envelope evolution (see, *e.g.*, Livio & Soker 1988; Taam & Bodenheimer 1989), the theory which might quantify the relationship between initial and final state of the binary is still seriously incomplete. As a matter of expedience in evaluating evolutionary scenarios, it is therefore convenient to cast the process in strictly energetic terms and bury the theoretical uncertainties in a few dimensionless parameters. One supposes that the energy initially binding the common envelope to the binary is provided by some fraction, α, of the difference between final and initial orbital energies of the binary, *i.e.*, that it is the dissipation of orbital energy with average efficiency α within the common envelope which ultimately unbinds that envelope from the embedded binary. One therefore writes:

$$\alpha\left(-\frac{GM_{1c}M_2}{2A_f} + \frac{GM_1 M_2}{2A_i}\right) = -\frac{GM_1 M_{1e}}{\lambda r_{L_1} A_i},$$

where $M_1 = M_{1c} + M_{1e}$ is the initial mass of the primary component (M_{1c} and M_{1e} its core and envelope masses, respectively), M_2 that of the secondary component (assumed unchanged – see Webbink 1988 and Hjellming & Taam 1991), and A_i and A_f are the initial and final orbital separations, respectively. In this expression, the initial energy of the envelope (the right-hand term) has been identified with its energy at the instant at which the primary first fills its Roche lobe (at radius $R_1 = r_{L_1} A_i$), because the energy source which subsequently inflates it to engulf both cores is gravitational (accretion), and thus must ultimately come from the binary orbit (Webbink 1984). The dimensionless parameter λ is a measure of the degree of central concentration of the envelope at this stage, and is typically of order unity in asymptotic giant branch models. (Note that the ionization energy of the envelope has not been included in the right-hand side of this energy equation, although for luminous asymptotic branch giants it can exceed the binding energy: if converted into mechanical work, rather than lost to radiation, the recombination energy of the envelope could profoundly change the complexion of this problem.) The above expression is easily rearranged into the form:

$$\frac{A_f}{A_i} = \frac{\alpha\lambda r_{L_1}}{2}\left(\frac{M_2}{M_1}\right)\left(\frac{M_{1c}}{M_{1e} + \frac{1}{2}\alpha\lambda r_{L_1} M_2}\right).$$

Each of the three terms on the right-hand side of this equation is usually smaller than unity, the two in brackets typically very much so. The resulting orbital contraction is quite spectacular: for a (typical) 15 $M_\odot$ initial primary (3.4 $M_\odot$ helium core) with a 1 $M_\odot$ secondary, one finds $A_f/A_i = 0.0029$ assuming $\alpha = 0.5$ and $\lambda = 1$. It is therefore necessary to appeal to the long-period extreme of candidate progenitor systems in order to ensure survival.

5. *The post-common-envelope binary must be wide enough to accommodate the secondary within its Roche lobe.* Because of the extreme initial mass ratio of the binary, the secondary will scarcely have evolved when the system completes common envelope evolution, so this condition in effect reduces to one that the Roche lobe radius of the secondary exceed its main sequence stellar radius.

6. *The post-common-envelope binary must be wide enough to allow the primary to evolve to core collapse.* If the neutron stars in LMXBs form by core collapse in helium star-main sequence binaries, as considered here, rather than by accretion-induced collapse, the evolution of that helium star to incipient core collapse must not be terminated by a second round of mass transfer. In the mass range immediately above the Chandrasekhar limit, helium stars undergo dramatic expansion following core helium exhaustion (Paczyński 1971; Habets 1986), and this growth must then be accommodated within the binary in much the same way as the initial growth of the primary was accommodated following core hydrogen burning (item 3 above), in order to assure that the core of this star, stripped now of its *helium* envelope will proceed to core collapse. In Figure 1, we

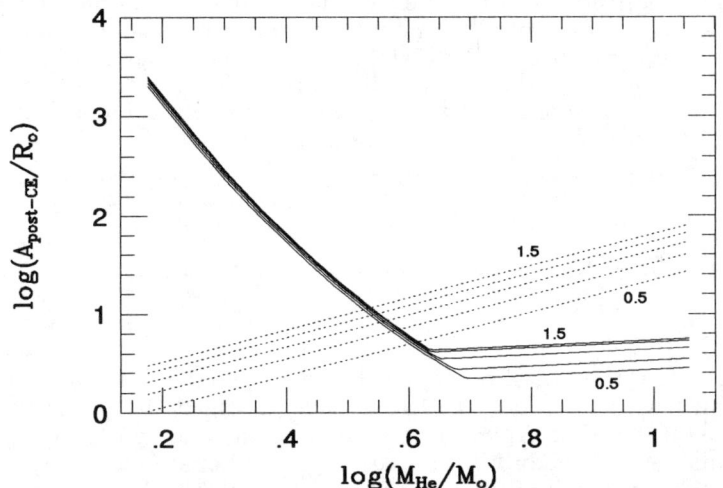

Fig.1. Limits to the orbital separation of post-common-envelope LMXB progenitors, as functions of the post-common-envelope mass of the primary. The dashed curves, labeled by secondary mass, are *upper* limits (for $\alpha = 1$) corresponding to the primary first filling its Roche lobe at the instant it reaches core collapse. The solid curve, also labeled by secondary mass, are *lower* limits corresponding (for $M_{He} \gtrsim 4.5\ M_\odot$) to the secondary just filling its Roche lobe at the end of common envelope evolution, or (for $M_{He} \lesssim 4.5\ M_\odot$) to the post-common-envelope primary reaching core collapse just as it fills its Roche lobe.

have identified this point of no return as the radius of the helium star at the onset of convective core carbon burning.

7. *The binary must survive the core collapse event.* The high space velocities typical of single pulsars provides strong evidence that they are imparted a substantial "kick" at their creation (*e.g.*, Harrison, Lyne & Anderson 1993), large enough to influence very strongly the survival probability and characteristics of a binary in which such an event occurs. In the absence of a kick, a binary undergoing essentially instantaneous mass loss in a supernova explosion will remain bound only if it retains at least half its initial mass. Scrutiny of Figure 1 shows that this condition is marginally fulfilled only in a vanishingly small region of parameter space.

The presence of kicks can both unbind binaries which would otherwise remain bound, or (of greater interest here) bind systems which would otherwise be disrupted. It is important to recognize that if the amount of mass lost in a supernova explosion significantly exceeds the combined mass of the compact remnant and its companion – a situation prevalent among the binaries satisfying the constraints illustrated in Figure 1 – then a kick will tend to leave a bound system only if it imparts a velocity to the compact component comparable in direction and magnitude to that of its stellar companion in the pre-supernova orbit. The magnitude typical of kick velocities thus exercises a strong influence on the orbital separations most likely to occur among binary survivors: In binaries which are initially strongly bound, a kick is unlikely to be large enough to compensate for the large relative orbital velocity of the binary components. Conversely, in binaries which are initially weakly bound, a kick small enough to produce a bound outcome becomes unlikely. This selection effect is clearly

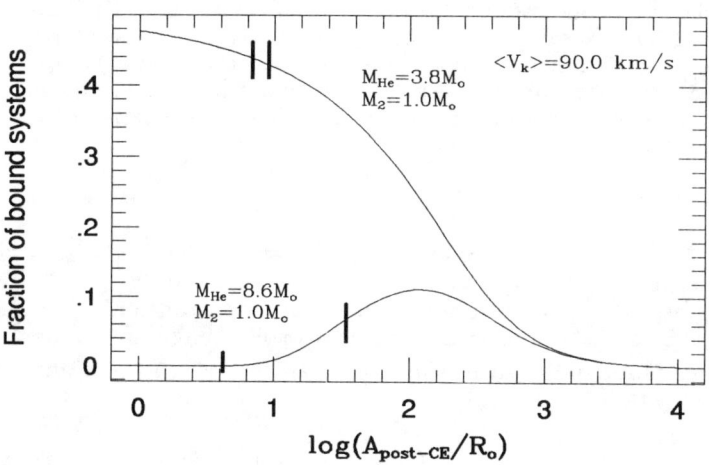

Fig.2. Survival probabilities, as a function of pre-supernova binary separation, for binary systems losing 50% and 75% of their initial mass (3.8 $M_\odot$ + 1.0 $M_\odot$ and 8.6 $M_\odot$ + 1.0 $M_\odot$, respectively), corresponding, in the case of negligible kick velocities, to marginally bound and strongly unbound post-supernova systems. Only a small portion of the range of binary separations illustrated here is actually populated by LMXB progenitors (see Figure 1).

evident in Figure 2. Of lesser importance, but still significant, is the requirement that periastron in the post-supernova orbit not lead to an actual collision between binary components.

8. *The post-supernova binary must be close enough to reach interaction within a Hubble time.* As a rule, those binaries which survive the supernova event are left with eccentric orbits, and with secondary components residing well inside their Roche lobes. Stellar evolution and orbital angular momentum loss must then combine to drive the binary into a further mass-transfer phase. Secondary stars of Population I composition with masses $\gtrsim 0.9$ $M_\odot$ may expand enough in 10^{10} years to fill their Roche lobes by stellar evolution alone. However, binaries with less massive secondaries must be driven to interaction by orbital angular momentum loss. Those losses are widely attributed to a combination of magnetically-coupled stellar winds and (in the short-orbital-period extreme) gravitational radiation. General relativity provides a firm theoretical basis for evaluating gravitational radiation losses, but estimates of magnetic stellar wind losses rest completely on empirical arguments (Verbunt & Zwaan 1981, *et seq.*). It is worth noting that tidal dissipation will in general circularize the orbits of potential LMXBs before the secondary can fill its Roche lobe.

9. *The time scale for mass transfer once the binary reaches interaction must be appropriate to power a luminous X-ray source.* Having survived as a binary with a collapsed component, and having reached an interactive state, a candidate LMXB must satisfy yet another condition if it is to have an appreciable lifetime as an X-ray source. An individual star can be characterized by three time scales: dynamical, thermal, and nuclear. The time scale on which such a star loses mass as the donor in an interactive binary depends upon the relationship between the structural properties of the star and the orbital (and Roche lobe) evolution of the binary in response to mass transfer (see, *e.g.*, Hjellming 1989). To sustain a long-lived X-ray phase, a binary should transfer mass at a rate which will support an accretion rate onto the compact object which is comparable to, or slightly smaller than, one which produces an Eddington luminosity, L_{Edd}. Expressed in terms of the mass transfer time scale of the donor secondary, this condition becomes:

$$\tau_{\dot{M}} \equiv \left| \frac{M_2}{\dot{M}_2} \right| \gtrsim \frac{\beta G M_1 M_2}{R_1 L_{Edd}},$$

where $\beta \equiv -\dot{M}_1/\dot{M}_2$ is the fraction of mass lost by the secondary which is accreted by the primary (compact) star.

The longest-lived X-ray sources occur when mass transfer is driven by the nuclear evolution of the donor star, or by an orbital angular momentum loss rate, $\dot{J}_{orb}$, which is independent of the mass transfer rate in the binary. In this case, the appropriate mass transfer time scale is

$$\tau_{\dot{M}} \approx \tau_{ev} \approx \left(\frac{\dot{R}_2}{R_2} - 2 \frac{\dot{J}_{orb}}{J_{orb}} \right)^{-1}.$$

However, this time scale is only appropriate if the binary is stable against both thermal and dynamical time scale mass transfer, a condition which requires that the donor be no more massive (and possibly significantly less massive) than the compact component. Lower main sequence stars ($M_2 \lesssim 1.4$ $M_\odot$) and low-mass stars ($M_2 \lesssim 1$ $M_\odot$) on the lower giant branch may satisfy these constraints if the accretor is a canonical 1.4 $M_\odot$ neutron star and if mass transfer is essentially conservative (negligible mass and orbital angular momentum losses from the transfer stream).

Thermal time scale mass transfer may occur if the donor star cannot simultaneously remain inside its Roche lobe and in thermal equilibrium, that is, if the Roche lobe of the secondary contracts more rapidly (expands more slowly) in response to mass loss than does stellar radius:

$$\frac{dR_{L_2}}{dM_2} > \left(\frac{\partial R_2}{\partial M_2}\right)_{thermal\ eq}.$$

In this case, the corresponding mass transfer time scale is

$$\tau_{\dot{M}} \approx \tau_{th} \approx \frac{GM_2^2}{R_2 L_2},$$

but this will occur only if the donor is *stable* against dynamical time scale mass transfer. For conservative mass transfer conditions, the thermal time scale is appropriate to moderate-mass main sequence stars ($\sim 1.4 - 4$ $M_\odot$) and to stars of similar or lower mass in the Hertzsprung gap. More massive donor stars on the main sequence or in the Hertzsprung gap may also lose mass initially on their thermal time scales, but evolve to dynamically unstable mass transfer. In all of these cases, however, the characteristic mass transfer rate is super-Eddington onto a neutron star, and will probably quench X-ray emission.

Dynamical time scale mass transfer occurs if the donor star cannot remain in hydrostatic equilibrium within its Roche lobe for *any* mass loss rate. This condition is set by the adiabatic response of the star to mass loss:

$$\frac{dR_{L_2}}{dM_2} > \left(\frac{\partial R_2}{\partial M_2}\right)_{adiabatic}.$$

The *growth* time scale for dynamical instability is typically of order

$$\tau_{\dot{M}} \approx \tau_{dyn} \approx (P_{orb}\tau_{ev}^2)^{1/3},$$

and leads to peak mass transfer rates of order M_2/P_{orb}, instigating common envelope evolution. This circumstance arises, even for conservative mass transfer, for giant branch donors more massive than about two-thirds the mass of the accretor, or (developing out of thermal time scale mass transfer) for main sequence and Hertzsprung gap donors more than two or three times as massive as the accretor.

The discussion above of mass transfer time scales is predicated on the assumption that transfer occurs via Roche lobe overflow. Another possibility is mass transfer in a stellar wind, as occurs among massive X-ray binaries. In this case, the efficiency for classical Bondi-Hoyle accretion is of order

$$-\frac{\dot{M}_1}{\dot{M}_2} \approx \frac{1}{4}\left(\frac{M_1}{M_1+M_2}\right)^2 \left(\frac{v_{orb}}{v_w}\right)\left[1+\left(\frac{v_w}{v_{orb}}\right)^2\right]^{-3/2},$$

and is invariably small. Among low-mass donor stars, mass loss rates in a stellar wind high enough to power a bright X-ray source are likely only for late in giant branch evolution or on the asymptotic giant branch, and should therefore have luminous, late-type optical counterparts. The X-ray lifetime of such systems is short, $\lesssim 10^7$ years. However, the formation scenario for low-mass X-ray binaries with black hole accretors proposed by Romani (1992) produces relatively long-period systems of just this type, with a high estimated birth rate.

3. CONCLUSIONS

Despite the concatenation of a rather long series of uncertain steps, it is still possible to model the birth rate distribution of LMXBs with respect to their system parameters upon entering the X-ray phase. Figure 3 contains three such model distributions, illustrating the influence of supernova kick velocities upon those distributions. The absolute normalization of the birth rates is extremely uncertain, and likely to remain so in view of the observational difficulty in determining the frequency of low-mass, distant companions to the O and B stars which ultimately give birth to the compact components. Nevertheless, the salient features of these distributions can be readily interpreted in terms of the constraints outlined above.

Each distribution shows a ridge line extending to low secondary masses and orbital separations, representing LMXBs born with essentially zero-age donor stars. It is important to recognize that the relative prominence of this feature is determined by the magnitude of orbital angular momentum loss in magnetic stellar winds during post-supernova evolution; it would be very much weaker or absent altogether at the low-mass extreme if gravitational radiation alone were responsible for instigating mass transfer.

Above ~ 1 $M_\odot$, a second ridge line appears at larger orbital separation. This ridge is an artifact of binaries reaching interaction because of the evolutionary expansion of the donor, with that star filling its Roche lobe just prior to reaching the base of the giant branch. At larger separations, the distribution is cut off by a rapid decrease in the threshold mass ratio for instability against dynamical time scale mass transfer, which restricts potential LMXB donors to masses less than two-thirds the mass of the accretor. At smaller separations, the post-supernova distribution is depleted toward smaller separations by the rapid increase in the angular momentum loss rate from magnetic stellar winds with increasing orbital angular frequency (decreasing separation). This depletion creates a minimum, or valley, between large- and small-separation systems.

The entire distribution is cut off at large donor star mass by instability against thermal time scale mass transfer when the mass of that star exceeds $\sim$ 1.4 $M_\odot$. The ridge line to lower secondary star mass disappears where that star is no longer massive enough to provide the orbital energy needed in the primordial binary to eject the common envelope; lower mass companions are tidally destroyed by the core of the primary during the course of common envelope evolution.

It remains to address the secular evolution of the nascent LMXBs which these distributions represent, to relate those evolving models to the distribution of structural properties of LMXBs actually observed (to the extent that they are known), and to repeat the exercise for other possible channels for LMXB creation. Despite the uncertainties involved, however, these preliminary results hold the promise of offering new insights into many of the evolutionary constraints which shape the LMXB population.

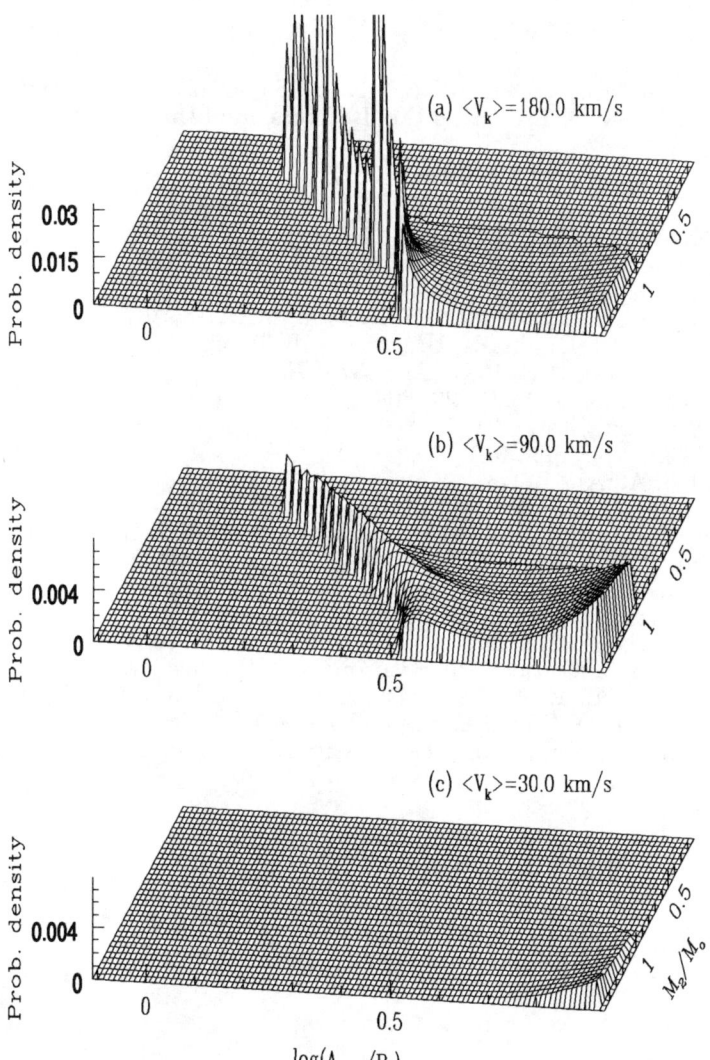

Fig. 3. Population synthesis models for the distribution of newly-formed LMXBs with respect to donor star mass and orbital separation. The three frames illustrate the systematic bias introduced by varying the mean supernova kick velocity imparted to the compact component, as reflected in the distribution of orbital separations at different masses. The vertical scale of frame (a) has been reduced by a factor of 4 for the sake of clarity. The principal features of these distributions are interpreted in the text.

ACKNOWLEDGEMENTS

This research was supported by NSF grant AST 92-18074. We are grateful to Ms. Deana Pettigrew for preparing the TeX version of this review.

REFERENCES

Chiosi, C., & Maeder, A. 1986, ARA&A, 24, 329
Eggleton, P.P., & Verbunt, F. 1986, MNRAS, 220, 13P
Habets, G.M.H.J. 1986, A&A, 167, 61
Harrison, P.A., Lyne, A.G., & Anderson, B. 1993, MNRAS, 261, 113
Hjellming, M.S. 1989, Ph.D. Thesis, University of Illinois
Hjellming, M.S., & Taam, R.E. 1991, ApJ, 370, 709
Livio, M., & Soker, N. 1988, ApJ, 329, 764
Paczyński, B. 1971, Acta Astr., 21, 1
Romani, R.W. 1992, ApJ, 399, 621
Sutantyo, W. 1975, A&A, 41, 47
Taam, R.E., & Bodenheimer, P. 1989, ApJ, 337, 849
van den Heuvel, E.P.J. 1981, in IAU Symposium 93, Fundamental Problems in the Theory of Stellar Evolution, ed. D. Sugimoto, D.Q. Lamb, & D.N. Schramm (Dordrecht: Reidel), p. 137
Verbunt, F., & Zwaan, C. 1981, A&A, 100, L7
Webbink, R.F. 1984, ApJ, 277, 355
Webbink, R.F. 1988, in Critical Observations vs. Physical Models for Close Binary Systems, ed. K.-C. Leung (New York: Gordon& Breach), p. 403
Webbink, R.F. 1992, in X-Ray Binaries and Recycled Pulsars, ed. E.P.J. van den Heuvel & S.A. Rappaport (Dordrecht: Kluwer), p. 269

RECYCLED RADIO PULSARS

A. G. Lyne
University of Manchester, Jodrell Bank, Macclesfield, Cheshire SK11 9DL, UK
agl@jb.man.ac.uk

ABSTRACT

We review the various radio surveys which have been designed primarily for discovering recycled pulsars. In the 11 years since the first millisecond pulsar, PSR 1937+21, was found, a total of about 55 have been discovered, although most of these have only been found since 1988. Around that time, the first millisecond pulsar was found in a globular cluster and by now a total of 34 pulsars have been found in these objects. These include a total of 29 pulsars with period less than 50ms and 14 are in binary systems. New galactic surveys have now discovered 21 millisecond pulsars, mostly discovered in the past 2 years. These constitute a very nearby population which has a space density similar to that of normal pulsars. However, their formation rate is only one every 10^4 - 10^5 years.

1. INTRODUCTION

Arguably, the era of the Recycled Radio Pulsars began in 1982 with the discovery of PSR 1937+21 (Backer et al 1982). While there were some indications that some processing of old neutron stars had occurred in some binary pulsar systems, the 1.56 millisecond period and low magnetic field of PSR 1937+21 made it quite different from anything else and, despite the absence of any binary companion, many authors made a strong case for recycling as the source of the rapid rotation rate (Alpar et al. 1982; Fabian et al. 1983; Webbink, Rappaport & Savonije 1983). In this process, old neutron stars are spun up by accretion of material from a companion star as it overflows its Roche lobe during a giant phase (Smarr & Blandford 1976) and transfers orbital angular momentum to the neutron star. This phase is possibly witnessed as a low-mass X-ray binary system (LMXB) by virtue of the hot, infalling gas.

Following this first discovery, the hunt for more began. This paper provides a more or less historical account of these searches and of some of the striking astrophysical phenomena seen in this relatively new class of object.

2. OBSERVATIONAL ASPECTS

The search for recycled radio pulsars is essentially a search for objects with short rotational period, possibly complicated by the Doppler effects of their motion in binary systems. The sensitivity and speed of these searches has been limited primarily by the computational resources available to pulsar astronomers. This is because, unfortunately, the quest is not simply a search for periodicity in a time sequence of data: dispersion in the ionised component of the interstellar medium results in the broadening of pulses by an amount equal to the time it takes the dispersed signal to traverse the receiver passband, causing a corresponding reduction in detectability. Since large bandwidth is required for high sensitivity, the solution is to record many narrow frequency channels within the receiver band and, offline, to 'de-disperse' the data for a large number of trial values of the dispersion, each producing a 1-dimensional time sequence in which periodicity can be sought. A modern search consists of recording typically 256 frequency channels every 300 microseconds, i.e. about 900,000 samples/second. In order to retain the same sensitivity to shorter periods, P_{min}, at a given dispersion measure, both the number of frequency channels as well as the sample rate have to be increased appropriately so that the computing resources required increase roughly as P_{min}^{-2} (Lyne & Smith 1990).

A 2.5 minute observation with the system above would result in a data set of 128 million data samples, the processing of which would have taken about 1 day on a Vax 780, the typical resource available to astronomers 10 years ago. It is therefore easy to understand the slow discovery rate of further millisecond pulsars in the years following the discovery of PSR 1937+21. In the next 5 years, despite the intense efforts of astonomers at the world's major observatories, only two more were discovered, PSR 1953+29 (Boriakoff, Buccheri & Fauci 1983) and PSR 1855+09 (Segelstein et al. 1986). Both of these pulsars were found to be in binary systems with low-mass companions, reinforcing the hypothesis that recycling was responsible for the high spin rate.

3. GLOBULAR CLUSTER SEARCHES

As described above, shortly after the discovery of PSR 1937+21, it was recognised that the spin-up was probably being witnessed in LMXBs. Present statistics indicate that the probability of LMXB occurrence is a few orders of magnitude higher for stars in globular clusters than in the galactic disk (Verbunt & Hut 1987). This is underlined by the fact that globular clusters contain only $\sim$ 0.05% of the mass of the Galaxy, but almost 20% of the LMXBs (Verbunt 1987). Globular clusters were thought make particularly suitable birthplaces for LMXBs because of their high stellar concentrations and attendant higher rate of binary formation (Fabian, Pringle & Rees 1975). The formation of an appropriate system would result in accretion of material from an evolving star onto a compact companion. The natural implication was that there may be numerous recycled pulsars in globular clusters. Hamilton, Helfand & Becker (1985) therefore carried out a search on a dozen clusters with the VLA, seeking small diameter radio sources which may be pulsars. One weak source was found near the core of M28. Subsequent observations showed this to be highly polarised and to have a steep spectrum, both characteristic properties of pulsars. Finally, in 1987, pulsations with a periodicity of 3.1 ms were discovered using the 76m telescope at Jodrell Bank Lyne et al. 1987). Although, like PSR 1937+21, this was not in a binary system, its presence in the cluster did point to accretion spin-up as the origin of the MSPs. Within 3 months, a second pulsar, PSR 1620-26, was found in M4 (Lyne et al 1988). This time it was a member of a binary system, the companion presumably being the remains of the star responsible for the spin-up.

These discoveries presented pulsar astronomers with a potentially rich source of millisecond pulsars, since now only a hundred or so telescope pointings would be required to survey most of the galactic globular cluster population. During the next 4 years many pulsars were discovered by two main groups, one utilising the Arecibo antenna, the other using the Jodrell Bank and Parkes

TABLE 1: Globular Cluster Pulsars Discovered
in the Arecibo Surveys

Cluster	Pulsar	P_{bc} (ms)	DM	P_{orb}	Reference
M53	1310+18	33.163	25	255 days	Anderson et al. (1989b)
M5	1516+02A	5.553	29.5	Single	Wolszczan et al. (1989a)
	1516+02B	7.947	29.5	6.9 days	"
M13	1639+36A	10.378	30.4	Single	Anderson et al. (1989a)
	1639+36B	3.528	29.5	1.26 days	Anderson et al. (1991)
NGC 6760	1908+00	3.6	200	3.4 hrs	Deich et al. (1993)
M15	2127+11A	110.665	67.25	Single	Wolszczan et al. (1989b)
	2127+11B	56.13	67.3	Single	Anderson et al. (1990)
	2127+11C	30.53	67.1	8.0 hrs	"
	2127+11D	4.65	67.3	Single	Anderson et al. (1991)
	2127+11E	4.80	67.3	Single	"
	2127+11F	4.03	67.3	Single	Anderson et al. (1993)
	2127+11G	37.66	67.3	Single	"
	2127+11H	6.74	67.3	Single	"

antennae. In total, 34 pulsars have been found in globular clusters, of which 14 are members of binary systems and 30 have periods of less than 60 milliseconds.

The 14 pulsars found in the Arecibo searches are presented in Table 1. The most notable discoveries are the 8 pulsars found in the one cluster M15. The first of these, PSR 2127+11A, was remarkable in that it was the first known pulsar to have a negative period derivative (Wolszczan et al. 1989). This apparent spin-up was recognised as being due to the acceleration of the pulsar in the gravitational potential well of the cluster. The period derivative, $\dot{P}$, arising from an acceleration component along the line-of-sight, a, is given by (Phinney 1992) $P/\dot{P}$ = a/c. Of course, the position along the line-of-sight relative to the cluster centre is unknown. Assuming that the intrinsic period derivative of the pulsar is zero, the observed value of $\dot{P}$ then provides a lower limit to the space density of the cluster within a given radius. Both PSR 2127+11A and PSR 2127+11D both lie 1.1 arcsec from the cluster core and have negative period derivatives, giving a lower limit to the density of $2.7 \times 10^6 M_\odot pc^{-3}$ within 0.055 pc of the cluster centre.

Once the dispersion measure of a cluster is determined by the detection of the first pulsar, the search in dispersion measure is no longer necessary and the resultant savings in CPU resources by a factor of 100-1000 can be expended upon searches for pulsars which are accelerated in their binary motion. It was in such a search that PSR 2127+11C was discovered. This notable object is in an eccentric relativistic binary system with another neutron star, very similar to that containing PSR 1913+16. This system is many core radii from the cluster centre and is only just gravitationally bound to it. Unfortunately, this binary will not provide the same test-bed for general relativity and gravitational radiation that PSR 1913+16 has, because it is not in an inertial frame. However, general relativistic effects have allowed the measurement of the two neutron star

TABLE 2: Globular Cluster Pulsars Discovered
in the Jodrell Bank/Parkes Surveys

Cluster	Pulsar	P_{bc} (ms)	DM	P_{orb}	Reference
47 Tuc	0021–72C	5.757	24.4	Single	Manchester et al. (1990)
	0021–72D	5.358	24.7	Single	Manchester et al. (1991)
	0021–72E	3.536	24.2	2.2 days	"
	0021–72F	2.624	24.4	Single	"
	0021–72G	4.040	24.2	Single	"
	0021–72H	3.210	24.3	~ days ?	"
	0021–72I	3.485	23.7	~ days ?	"
	0021–72J*	2.101	24.6	2.9 hours	"
	0021–72L	4.346	24.5	?	"
	0021–72M	3.677	24.4	Binary	"
	0021–72N	3.075	24.4	Single	"
M4	1620–26	11.08	62.9	191 days	Lyne et al. (1988)
NGC 6342	1718–19	1004.0	70	6 hours	Biggs & Lyne (1991)
NGC 6440	1745–20	288.6	220	Single	Manchester et al. (1989)
Terzan 5	1744–24A	11.56	242	1.8 hrs	Lyne et al. (1990)
	1744–24B**	442.8	210	Single	"
NGC 6539	1802–07	23.10	187	2.6 days	D'Amico et al. (1990)
NGC 6624	1820–30A	5.440	86.8	Single	Biggs et al. (1990)
	1820–30B	378.6	86.7	Single	"
M28	1821–24	3.054	120	Single	Lyne et al. (1987)

* *Possibly 4.201 ms pulsar with strong interpulse.*
** *Possibly a foregound object not associated with the cluster.*
Note that PSR 0021–72A (Ables et al. 1989) and PSR 0021–72B (Ables et al. 1988)
not thought to be located within 47 Tuc and efforts to detect them have proved unsucces
(Manchester et al. 1990).

masses, 1.34 $M_\odot$ and 1.37 $M_\odot$ and future determination of the orbital period derivative should provide a direct measurement of the system's radial acceleration.

While only half of the pulsars in M15 are true millisecond pulsars and only 2 are in binary systems, their short periods and small period derivatives indicate that they are probably all recycled to some extent, the solitary pulsars having lost their erstwhile companions through collision or ablation by radiation from the pulsar.

M15 is a relatively dense, massive cluster with a high collision rate in the core, which might be expected to provide the binary progenitors of the recycled pulsars. However, something of a surprise is the number of such pulsars found at Arecibo in the other clusters, all of which have relatively low central density and hence small collision rates. This suggests that many of the progenitor binaries may be primordial.

The pulsars discovered in the Jodrell Bank/Parkes survey are listed in Table 2. The most remarkable observation here is the detection with the Parkes radiotelescope of no fewer than 11 millisecond pulsars. Initially, a 5.75 millisecond pulsar was discovered with a dispersion measure of 24.5 pc cm^{-3} (Manchester et al. 1990). Following a local search around this dispersion measure, a further 10 millisecond pulsars were found in the cluster, all with period of less than 6 milliseconds (Manchester et al. 1991). No longer period objects were detected. More than half of the 11 are in binary systems, one with an orbital period of only about 3 hours and which is probably eclipsed by its companion. More than half the known millisecond pulsars with such short period and more than a quarter of the known binary pulsars lie within this one globular cluster. This preponderance of millisecond pulsars contrasts strongly with M15 in which three of the 5 pulsars have period in excess of 30 milliseconds. What is so special about this cluster to produce such an abundance of millisecond pulsars is not clear, but the discovery does reveal that such clusters must have had many massive stars in their youth in order to produce such an abundance of neutron stars. There is clearly much to be learned about these pulsars and their binary systems in the future. Moreover, with such a large number in this one cluster, measurement of their positions and period derivatives will give an interesting insight into the distribution of mass through the cluster.

Three of the pulsars are in eclipsing binary systems (see section 5). Apart from these, the most notable pulsar is PSR 1802–07 which is in the globular cluster NGC 6539. This has an eccentricity of 0.21, much larger than any of the other cluster binary pulsars which typically have eccentricities of 10^{-4}. The circularity of these orbits probably results from the accretion process. In the case of PSR 1802–07, it seems that since accretion finished the binary system has suffered a close encounter with another body in the core of the cluster.

4. GALACTIC PLANE SURVEYS

While the globular cluster surveys provided a rich source of recycled pulsars, and have given information on the dynamics of clusters and binary evolution, we still have little knowledge of the overall galactic population of recycled pulsars. Computer technology continued to develop during this time, and by the early nineties it became feasible to conduct ambitious surveys with the object of covering the whole of the sky between them in only a matter of a few years. This involves about 10^6 telescope pointings of between one and a few minutes each. In the 10 years to 1992.0 only 5 millisecond pulsars were known outside the globular cluster system. The new surveys are well under way now and by now 1993.7, a total of about 21 galactic millisecond pulsars have been discovered.

5 new millisecond pulsars have been found at Arecibo (Nice, Taylor & Fruchter 1993; Camilo, Nice & Taylor 1993; Foster, Wolszczan & Camilo 1993), 2 at Jodrell Bank, and 9 in the new survey at Parkes (Johnston et al 1993; Bailes et al 1993). 16 of the 21 pulsars are in circular orbits with low-mass companions, mostly between 0.1 and $0.3M_\odot$. However, the most important aspect of these discoveries is their proximity to the Sun. Not only do they have an almost isotropic distribution around the sky, but their dispersion measures are all small.

One most striking discovery is that of PSRJ 0437–4715, which was found in the Parkes survey and at a distance of about 100pc, is the closest of all 600 known pulsars. It is a 5.75ms pulsar in a 5.7 day circular orbit with a companion of mass $0.14/\sin i M_\odot$. It is also by far the brightest of all known millisecond pulsars so that single pulses are clearly observed and it promises to be most valuable in studying the radiation mechanism of the objects. The spin-down energy flux density at the Earth, $S = \dot{E}/d^2 = I\Omega\dot{\Omega}/d^2$, is the third largest behind the Crab and Vela pulsars and greater than that of PSR 1706–44. Already the observation of pulsed X-rays by ROSAT makes it the first millisecond pulsar to be detected outside the radio and, since the 3 pulsars above all have clear γ-ray detections, it may also be detectable at these energies. Its proximity to the Earth has also resulted in the identification of the companion star with a 22^m white dwarf (Johnston et al. 1993).

The growing number of these discoveries and their closeness imply a space density which is not very dissimilar to that of the normal pulsar population, around 10^5 active ones in the whole Galaxy. However, their ages are typically 100-1000 times those of normal pulsars so that their formation rate is only $10^{-2} - 10^{-3}$ as great. Since the galactic birthrate of normal pulsars is about 1 every 100 years, that of millisecond pulsars is one every $10^4 - 10^5$ years, making their births rather rare events.

5. ECLIPSING SYSTEMS

In 1988, Fruchter, Stinebring & Taylor (1988) discovered PSR 1957+20, a 1.6ms pulsar in an 8-hour binary system with a very low-mass companion which occults the pulsar each orbit. It was clear that the pulsar radiation was causing evaporation from the surface of the companion star and driving a wind of material away from the system. This process is strongly supported by the observation of an orbital modulation of the brightness of the companion star and by the detection of a nebula where the wind strikes the interstellar medium. It seems that this process will result in the complete evaporation of the companion star in about 10^7 years. This process will thus result in a solitary pulsar like PSR 1937+21 and may account for the other rapidly rotating solitary pulsars. In globular clusters it is possible that the solitary pulsars result from collisions.

Three of the pulsars in Table 2, PSRs 0021-72J, 1718-19 and 1744-24A display eclipses by the atmospheres of their companions, similar to those observed in PSRB 1957+20. In particular, PSRB 1744–24A in the cluster Terzan 5, is an 11ms pulsar in a 1.8 hour orbit with a low-mass companion (Lyne et al. 1990). The eclipses are very variable, so that on some occasions no pulses are seen anywhere around the orbit and the system is completely engulfed in the atmosphere of the companion star. On other occasions, the pulses may be seen for the whole orbit, if rather attenuated when the pulsar is on the far side of the companion.

PSR 1718–19 is a remarkable member of this group of eclipsing pulsars (Lyne et al 1993) in that it has a period of about 1 second and the spin-down energy density at the companion is extremely small, compared with these other systems. It is not clear whether in this case the mass outflow is driven by irradiation of the companion by the pulsar or is intrinsic to the companion star. At low frequency the whole system is engulfed in the outflowing material from the companion star so that the pulsar is only seen for perhaps 1 hr out of the 6.2 hr orbital period when the pulsar is on the near side of the companion, suggesting that in more extreme cases the pulsar will not be visible at all.

Another source detected by the VLA lies in the cluster(Fruchter & Goss 1990), very close to the core. This has the steep spectrum and variability associated with pulsars, but so far no pulsed emission has been detected from it. If it is a pulsar, the pulsations may be rendered undetectable by strong scattering in a surrounding mass outflow. Alternatively it may have either a very short period or be highly accelerated.

These systems allow us to study the physics of the interactions between pulsars and their companion stars.

6. CONCLUSIONS

Although the globular cluster and galactic surveys for millisecond pulsars have been very successful, we must remember that these surveys are extremely incomplete, not only because the pulsars are very weak and we only see the closest, but because of other selection effects as well. For instance, some Arecibo pulsars in M15 were found by stacking spectra over many weeks of observation. While this process may have good sensitivity to solitary pulsars, the sensitivity is likely to be significantly reduced for pulsars with varying velocity which may cause the pulsar frequency to move between bins during the stacking. This may account for the relatively small number of binaries detected in this cluster.

Most searches use sampling intervals between 0.3 and 0.5ms so that the sensitivity is reduced severely for typical pulsars with period of 6ms or less. For a period of 1ms, the reduction is likely to be at least a factor of 4 compared with longer period objects. In spite of the short periods of the pulsars in 47 Tucanae, there can be little doubt that the intrinsic distribution will peak at even shorter period. The observed binary frequency and the observed spin period distributions may therefore be very poor indications of the true distributions.

It seems that there may be many more inherently detectable millisecond pulsars in the sky and there may be even more exotic objects not yet detected. We just need superior data acquisition and CPU resources to open up the parameter space even further than these surveys have done.

REFERENCES

Alpar M. A., Cheng A. F., Ruderman M. A., Shaham J., 1982, Nature, 300, 728
Backer D. C., Kulkarni S. R., Heiles C., Davis M. M., Goss W. M., 1982, Nature, 300, 615
Bailes M. et al., 1993, Astrophys. J., submitted
Boriakoff V., Buccheri R., Fauci F., 1983, Nature, 304, 417
Camilo F., Nice D. J., Taylor J. H., 1993, Astrophys. J., 412, L37
Fabian A. C., Pringle J. E., Verbunt F., Wade R. A., 1983, Nature, 301, 222
Fabian A. C., Pringle J. E., Rees M. J., 1975, Mon. Not. R. astr. Soc., 172, 15P
Foster R. S., Wolszczan A., Camilo F., 1993, Astrophys. J., 410, L91
Fruchter A. S., Goss W. M., 1990, Astrophys. J., 365, L63
Fruchter A. S., Stinebring D. R., Taylor J. H., 1988, Nature, 333, 237
Hamilton T. T., Helfand D. J., Becker R. H., 1985, Astron. J., 90, 606
Johnston S. et al., 1993, Nature, 361, 613
Lyne A. G., Smith F. G., 1990, Pulsar Astronomy. Cambridge University Press
Lyne A. G., Brinklow A., Middleditch J., Kulkarni S. R., Backer D. C., Clifton T. R., 1987, Nature, 328, 399
Lyne A. G., Biggs J. D., Brinklow A., Ashworth M., McKenna J., 1988, Nature, 332, 45
Lyne A. G. et al., 1990, Nature, 347, 650
Lyne A. G., Biggs J. D., Harrison P. A., Bailes M., 1993, Nature, 361, 47

Manchester R. N., Lyne A. G., D'Amico N., Johnston S., Lim J., Kniffen D. A., 1990, Nature, 345, 598
Manchester R. N., Lyne A. G., Robinson C., D'Amico N. D., Bailes M., Lim J., 1991, Nature, 352, 219
Nice D. J., Taylor J. H., Fruchter A. S., 1993, Astrophys. J., 402, L49
Phinney E. S., 1992, Phil. Trans R Soc. Lond. A, 341, 39
Segelstein D. J., Rawley L. A., Stinebring D. R., Fruchter A. S., Taylor J. H., 1986, Nature, 322, 714
Smarr L. L., Blandford R., 1976, Astrophys. J., 207, 574
Verbunt F., Hut P., 1987, in Helfand D. J., Huang J., eds, The Origin and Evolution of Neutron Stars, IAU Symposium No. 125, Reidel, Dordrecht, p. 187
Verbunt F., 1987, Astrophys. J., 312, L23
Webbink R. F., Rappaport S., Savonije G. J., 1983, Astrophys. J., 270, 678
Wolszczan A., Kulkarni S. R., Middleditch J., Backer D. C., Fruchter A. S., Dewey R. J., 1989, Nature, 337, 531

DIM X-RAY SOURCES IN GLOBULAR CLUSTERS

Jonathan E. Grindlay
Harvard-Smithsonian CfA, 60 Garden St., Cambridge, MA 02138

ABSTRACT

Recent results are reviewed from sensitive X-ray observations of globular clusters with ROSAT. It is clear that virtually all globulars contain a "core source" with soft X-ray luminosity of $\sim 10^{32.5}$ erg/s, thus confirming the orginal Einstein Observatory discovery of such "dim sources" within globular clusters. High Resolution X-ray images (HRI) with ROSAT show that these sources can be resolved into $\gtrsim$ 3-5 sources with correspondingly reduced luminosities. High resolution optical images with HST in Hα vs. R are reviewed and show that emission line objects with uv excesses are found within the $\lesssim$ 6″positional uncertainties of the dim X-ray sources in NGC 6397 and possibly also NGC 6752, strongly suggesting these are accreting white dwarfs or cataclysmic variables. The distinction between CVs and quiescent LMXBs is discussed, as well as additional constraints imposed by the dim sources in globulars. The overall results now becoming available from high resolution imaging with ROSAT (HRI) and HST argue that CVs have finally been discovered in globular clusters.

1. INTRODUCTION

The Einstein X-ray survey of globular clusters (Hertz and Grindlay 1983, hereafter HG) discovered two apparently distinct classes of X-ray sources in globular clusters. Whereas the bright sources have long been identified with moderately high accretion rate low mass X-ray binaries (LMXBs) containing neutron stars (cf. Grindlay 1994a for review and references), it has been debated whether the dim sources are merely very low accretion rate (or quiescent transient) LMXBs (cf. Verbunt et al 1984) or are instead cataclysmic variables (CVs), as originally suggested by HG and Hertz and Wood (1985, hereafter HW). Other explanations for the dim sources have also been proposed, including RS CVn stars (Bailyn, Grindlay and Garcia 1992, Belloni et al 1993), millisecond pulsars (MSPs) (Johnston et al 1994), or perhaps "buried" MSPs resembling blue stragglers (Tavani, these proceedings).

We summarize the recent results of our investigations with ROSAT and HST and consider the possible signatures of CVs vs. quiescent LMXBs or the other sources. This review differs from and extends our recent reviews of prelimary ROSAT results of globular cluster observations (Grindlay 1993a), compact objects in globulars (Grindlay 1993b) and the properties of the their host globulars, and multiwavelength studies of cluster binaries (Grindlay (1993c). We shall also not repeat the extensive discussion of the general properties of the bright and dim sources themselves (i.e. their X-ray and optical characteristics) which was discussed in Grindlay (1994a). Instead, in this review we provide more complete details of the properties of the dim sources in both NGC 6397 and NGC 6752, which we have observed extensively with ROSAT and HST. Although we presented the first evidence for the discovery of CVs as the optical counterparts for the NGC 6397 dim sources in Grindlay (1993c), in this review we draw on the more complete results of Cool et al (1994) for NGC 6397 and of Cool (1993) and Grindlay and Cool (1994) for NGC 6752 to conclude that

© 1994 American Institute of Physics

indeed CVs have at long last been identified as the counterparts of at least the fainter ($\lesssim 10^{32}$ erg/s) dim sources and quite possibly the most luminous ones (e.g. in 47 Tuc, Paresce et al 1992; and in M3, Hertz et al 1993) as well. Our emphasis here is on the nature of the dim sources and the implications of their probable identification with CVs for LMXB and cluster evolution.

We also discuss the problem of how CVs might differ in their X-ray and optical properties from quiescent LMXBs, and we compare the properties of our cluster CV candidates with those in the field. It appears that the cluster sample may differ (e.g. be weaker lined) from field CVs or else that the cluster CV candidates may be either magnetic (DQ-Her type) or nova-like objects. Finally, we compare our results, both X-ray and optical, to other on-going searches for CVs in globulars and conclude there is no conflict.

2. THE LMXB-CV FORMATION PROBLEM AND GLOBULAR CLUSTERS

Although it has been thought for some time that the formation of LMXBs in globulars was relatively straightforward by the process of tidal capture, recent results point to the difficulty of keeping the resultant binary from merging (see Hut et al 1992 for a complete review). Tidal capture is still possible, but only for a much more restricted range of impact parameters than originally thought. Thus it now appears that exchange collisions, where an isolated compact object (neutron star or white dwarf,) encounters a binary and is exchanged for one of the binary components, must be dominant. Such exchange collisions require, of course, a significant population of primordial binaries. Evidence for a significant fraction ($\sim 10\%$) of binaries, which may be largely primordial, has been growing in recent years (cf. Hut et al 1992). However, uncertainties still abound in the long-term evolution and circularization of tidal capture binaries, and indeed recent calculations by Mardling (1993) indicate that merger is less likely.

These complications for the formation of cluster LMXBs should operate equally on systems formed from neutron stars (NSs) or white dwarfs (WDs) as the primary star which captures a low mass cluster secondary star. However, mass segregation of the heavier NSs into the core (Verbunt and Meylan 1988) as well as restrictions on WD and/or secondary masses for stable mass transfer (requiring mass ratio $\lesssim 0.8$ – cf. Bailyn et al (1990)) will favor LMXBs – particularly for "reasonable" initial mass functions (IMFs). The NS/WD ratio, and the fundamental question of what can be inferred about primordial IMFs and the formation of globulars, is discussed further in Grindlay (1994b) in which we also discuss the connections between LMXBs, CVs and millisecond pulsars (MSPs) in cluster cores. Here we focus on the nature of the dim sources and what they can tell us about the formation and evolution of X-ray binaries in clusters. If they are indeed CVs, can we constrain the type of CV (e.g. dwarf nova, magnetic, nova-like) ? If they are indeed CVs, is their number per unit mass or per unit (primordial) binary comparable to that found in the Galaxy or are they (as expected) nearly as overabundant as the LMXBs ?

Questions such as these suggest that the nature of the dim sources can constrain aspects of LMXB (and CV) origin and evolution. For example, if the dim sources are dominated by quiescent LMXBs, their numbers and properties must constrain LMXB lifetimes and birthrates. If the dim sources are instead dominated by CVs, they may be sufficient to account for at least some fraction of the cluster MSPs (as discussed in Grindlay 1994b) and possibly also LMXBs. Thus the nature of the dim sources holds clues to a variety of key problems.

3. ROSAT DISCOVERY OF MULTIPLE DIM SOURCES

We have described our HRI observations of NGC 6397 and NGC 6752 in preliminary form (Grindlay 1993a) and in detail (Cool 1993, Cool et al 1993a, Grindlay and Cool 1994). These were the first to reveal that globular cluster dim sources have X-ray luminosities as low as $10^{31.5}$ erg/s in the ROSAT band (0.2-2.4 keV) and thus are fully consistent with being CVs. At least five dim sources were detected in NGC 6397 in a 18 ksec exposure and at least three were detected in NGC 6752 in a 30 ksec exposure, which only partially compensated for the factor of 2 increased distance (4.2 vs. 2.2 kpc) for NGC 6752. Thus although the faintest source in NGC 6752 is $\sim$3 times more luminous than that detected in NGC 6397, it is likely that NGC 6752 also contains sources as faint so that its total dim source content is at least comparable to (or likely in excess of) that in NGC 6397. We compare these results with the results of PSPC pointed observations for 9 globulars recently reported by Johnston et al (1994).

An important diagnostic is the radial distribution of the dim sources. This can be used to constrain the source mass, as we have done (Grindlay et al 1984) using the Einstein data for seven of the (now 12) bright sources plus the 47 Tuc source (which is the brightest of the dim sources), or for the "pure" bright source distribution (7 LMXBs) alone (Grindlay 1985). However, this statistical measure is complicated for NGC 6397 and 6752 since they are both likely to be post core collapse (PCC) clusters, and thus not "simple" isothermal models as assumed for the LMXB analysis (which in retrospect is also probably compromised, since at least 3 of the 7 LMXB clusters are PCC). We defer a discussion of this problem to our forthcoming comparison of dim source distributions to MSP (and LMXB) distributions (cf. Grindlay 1994b) and instead consider here only the crowding and resolution effects imposed by the observed radial offsets of the dim sources. This is important in deriving the total number of such sources, and for comparison with lower spatial resolution PSPC surveys and pointed observations from ROSAT.

The relevant quantity, then, is the number of dim sources detected within the $\sim$25" diameter (FWHM) resolution of the PSPC. Three (of the 5) sources in NGC 6397 and two (of the 3) in NGC 6752 are located within this effective PSPC resolution element, and indeed only one source (with $L_x \sim$2-3 $\times 10^{32}$ erg/s) is detected in each cluster with the PSPC (cf. Johnston et al 1994). These results may be extended to other globulars by comparing the approximate core radii of the clusters. Even though both of the clusters discussed here are PCC clusters, and the core radii are poorly determined, an approximate value for the underlying faint stars in each is 12" (cf. Cool 1993 and references therein). Thus the radial distribution of the dim sources in both clusters is such that $\sim$60% of the dim sources detected are within $\sim$1 core radius, which is (coincidentally) about the PSPC resolution radius. It is therefore not surprising that the PSPC pointed observations of 9 clusters (Johnston et al 1994) reveal but a single "core source" (in two cases elongated) in 7 of the clusters with luminosity $\sim$1.5-6 $\times 10^{32}$ erg/s.

It is also important to consider the limiting sensitivity of the PSPC vs. HRI results. With the much longer exposures for the HRI observations (18 and 30 ksec for NGC 6397 and 6752, respectively), as well as the relative proximity of these closest PCC globulars (2.2 and 4.2 kpc), these observations reached luminosity limits of about 4 and 8 $\times 10^{31}$ erg/s in the ROSAT band. The PSPC observations of Johnston et al reached the upper end of this range for two clusters (NGC 6544, with no core source detected, and NGC 6656, with one such source detected) and factors of ~2-6 higher luminosities for the remaining 7 clusters. The typical limiting sensitivity of these PSPC observations was thus $\gtrsim$ 1-3 × 10^{32} erg/s and was therefore dominated by the few brightest sources. This is also true, though less so, for our HRI observations so that only very much deeper HRI observations (particularly for clusters with relatively large core radii) can probe the total dim source population. Until AXAF, it will be essentially impossible to push this below $L_x \sim 10^{31}$ erg/s, so that most CVs (with median luminosity below this limit – cf. Patterson and Raymond 1985; hereafter PR) will be individually inaccessible if they dominate the dim sources.

4. HST DISCOVERY OF CV CANDIDATES

We now turn to the likely identification of the dim sources. With HST in its initially launched condition, it is important to recall that spatial resolution vastly exceeding the best ground based seeing was obtained and thus optical counterparts for the dim sources could be found even in the centers of PCC clusters. The first example was the discovery of the counterpart for the brightest of the dim sources, the core source in 47 Tuc (HG), with X-ray luminosity in the Einstein band $\sim 10^{34.5}$ erg/s. A uv-excess and variable object was discovered by Paresce et al 1992 in the ~2″Einstein error circle (Grindlay et al 1984) with the FOC on HST. Paresce et al derived an approximate X-ray to optical flux ratio $f_x/f_{opt} \sim$ 10-50, but this assumed the Einstein value for the X-ray flux. The ROSAT HRI observations of the core of 47 Tuc (cf. Verbunt et al 1993) make it apparent that the source had dimmed by a factor of ~10 (and was joined by 2-3 other core sources of comparable luminosity, $\sim 10^{33}$ erg/s, which would have been invisible to Einstein due to the difficulty of resolving these fainter sources in the wings of the bright Einstein source). Thus the "quiescent" value of the X-ray/optical flux or luminosity ratio for this core source in 47 Tuc is probably ~1, and thus even more like a typical CV (cf. PR) than supposed by Paresce et al, who suggested that the high ratio might instead indicate it was a DQ Her type CV. This may still be true, and the lack of simultaneity in the X-ray and optical images is obviously a complication, but the more nearly contemporaneous ROSAT X-ray image could remove this requirement.

As first mentioned in Grindlay (1993c), and described in detail in Cool et al (1994), we have carried out extensive HST observations of the core of NGC 6397 and have found the first CV candidates for faint dim sources. We have also performed similar studies for NGC 6752, though with (much) reduced sensitivity (cf. Grindlay 1993c, Grindlay and Cool 1994). Our strategy was to search for Hα emission objects since such emission is a ubiquitous signature of all but the brightest CVs (e.g. dwarf novae in outburst) and is thus a more reliable indicator than mere uv excess or variability, both of which can be masked by other globular cluster stars (Grindlay 1992). Our program was therefore designed to use the narrow band (20 Å) Hα filter, F656N, and the broad band (913 Å) R filter, F675W, and the Planetary Camera (PC) of the WFPC-1 on HST for highest spatial resolution (0.042 arcsec pixels) over the broadest possible field (four adjacent frames, each with dimension 38″× 35″). All our analysis to date has been of just the one PC frame (PC6) centered on the cluster center. In order to confirm CVs and reject unlikely foregound dMe stars, we have also examined archival uv images from the FOC on HST (used at both f/48 and f/96) with several uv filters (F140W, F220W, and F346M); we have independently discovered

(Cool et al 1993b) the uv-candidates reported by DiMarchi and Paresce (1994; hereafter DP). Here we describe results for both NGC 6397 and NGC 6752 which have yielded the first CV emission line candidates in globular cluster cores.

NGC 6397

This cluster was observed during our HST Cycle 2 program on August 22-23, 1992 (and thus after our trial Cycle 1 observations of NGC 6752). Approximately 5 hours of total exposure in Hα and (for balance) 25 min in R were obtained in a series of 15-18 exposures at three slightly different offset positions (to minimize systematic variations across the CCD). The images were shifted and co-added with a sigma-clipping algorithm to eliminate cosmic rays. The final image obtained is shown in Figure 1, in which are marked the ROSAT error circles ((4-6"; 90% confidence) for the central sources C1-C3 (from Cool et al 1993a) and the three emission line objects found in the Hα vs. R images.

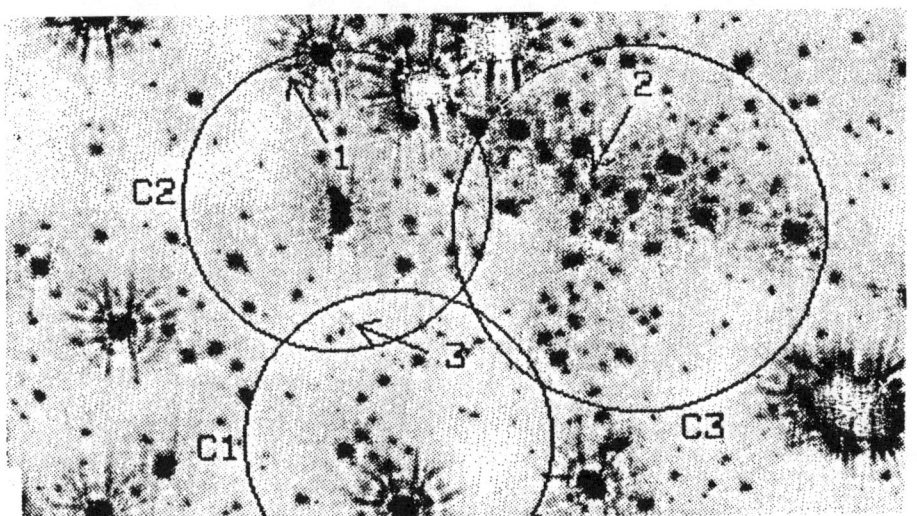

Fig. 1. Probable identification of CVs in NGC 6397 (from Cool et al 1994). North is approximately left, and cluster center is near bright star to West (up) of bright star near candidate 2.

The photometry was carried out using DAOPHOT II (Stetson 1991) with additional modifications (cf. Cool et al 1994). The summed R frame, deeper than the Hα sum, was used as a template for finding and measuring the ~900 stars in the PC6 image. Calibration of the image was done by using ground-based (CTIO 4-m) CCD frames in which several relatively isolated cluster stars could be accurately measured and compared to nearby standards. Results for the color magnitude diagram (CMD) of R vs. (Hα -R) are shown in Figure 2.

The three Hα objects are found to have apparent R magnitudes of 17.8, 18.8 and 19.5 which imply (for assumed colors (V-R) = 0.5) that the absolute magnitudes of these objects are M_V = 6.2, 7.1, and 7.9. The (Hα -R) colors, calibrated by the observed (and expected !) bifurcation of the giant branch from the horizontal branch and blue straggler stars (the squares and circles, respectively, in Figure 2) with their known Hα absorption, show that the emission objects have equivalent widths EW(Hα) $\gtrsim$ 21, 17, and 30Å. All three objects are also found to be uv-bright: they are fortuitously included in several FOC frames, as mentioned above, and are among the bluest objects in the field. The CMDs formed by comparing these uv images with the R image are given in Cool et al (1994) and show that the three objects have approximate (U-R) $\sim$ -0.5 - +0.7, or in the range found for CVs and certainly not red Hα emitters such as dMe stars. In Figure 3 below we reproduce the CMD for the FOC field taken with the F346M filter (effectively, U) vs. our PC field with the F675W filter (effectively, R) which shows how the object is cleanly separated from the main sequence of the cluster and in just the range of (U-R) colors expected for CVs. This same object was independently discovered as uv bright (and possibly variable) in the FOC observations of DP, which are discussed in our section below on completeness and other searches.

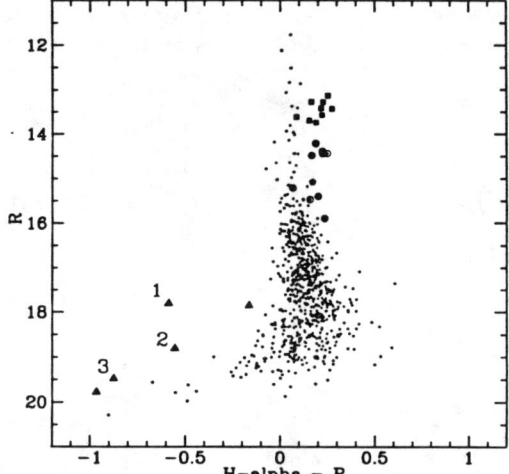

Fig. 2. CMD for our Hα vs. R HST observations of NGC 6397 showing the 3 significant CV candidates (solid triangles) and two possible ones (open triangles).

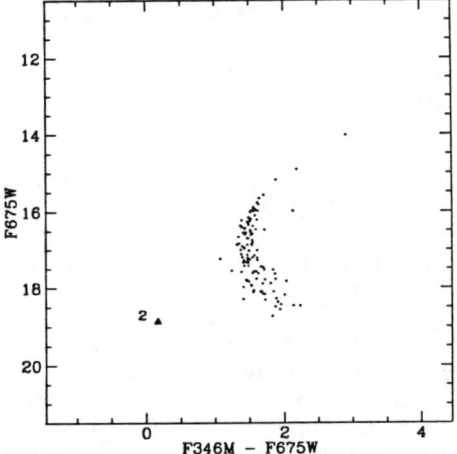

Fig. 3. CMD for the FOC (U or F346M filter) vs. PC6 (R or F675W) frame in HST observations of NGC 6397 showing the clear uv excess of Hα object no. 2.

Several other ($\gtrsim$ 1-2) Hα candidates (open triangles in Fig. 2) have also been found in the central PC field analyzed thus far. However, the most promising of these is not included in the smaller FOC field and so cannot (yet) be confirmed as a bonafide CV candidate with uv excess.

We thus conclude that these three objects are the likely optical counterparts for the ROSAT sources C1-C3 (cf. Cool et al 1993a). In this case, the derived X-ray to optical flux ratios, f_x/f_{opt}, may be compared with the correlation of this quantity with EW(Hβ) found by PR (also for non-simultaneous X-ray and optical data) for CVs in the field. In deriving this flux ratio, we first convert the X-ray flux from the ROSAT band (0.1-2.4 keV) to Einstein band (0.5-4 keV), assuming a 5 keV brems spectrum (as in HG) since the Einstein band is used by PR. We also assume EW(Hβ) $\sim$ EW(Hα), since galactic CVs have EW(Hα)/EW(Hβ) $\sim$ 0.5-2.0 (cf. Cool 1993). The result is shown in Figure 4.

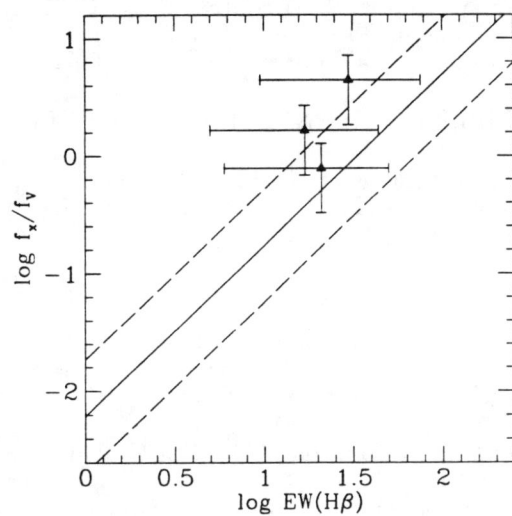

Fig. 4. Approximate X-ray/optical flux ratio vs. Hα strength for the three CV candidates in NGC 6397 compared with the field CV correlation found by PR.

The candidates are roughly consistent with the trend for field CVs, which would be found within the dashed line envelope surrounding the mean correlation line of PR, though they are all three above the envelope. The error bars plotted are approximate and described in Cool et al (1994). Thus it appears that these first CV candidates for the faintest dim sources may be systematically weak in their Hα or (conversely) systematically faint in their optical continuum strength for their given X-ray luminosity. This is discussed further below.

NGC 6752

We have carried out a similar CV search in the core of NGC 6752. Again, Hα vs. R frames were taken with the PC on HST in a program totalling some 3 hours of telescope time in HST Cycle 1. Given the factor of 2 increase in distance, these observations are about a factor of 10 less sensitive than for NGC 6397 and reach limiting absolute magnitude of about $M_V \sim$ 5.5 - 6 in the summed R frames. These observations have thus far been analyzed only for the two ROSAT error circles contained in the central frame; the results are given in Cool (1993) and

Grindlay and Cool (1994). Two marginal Hα excess candidates are found, with $M_R \sim 4.8$ and 5.4 (and thus ~ 0.5 mag fainter still for M_V, assuming a typical (V-R) color of 0.5), and so much more luminous than the NGC 6397 objects. These objects are found one in each of the two HRI source error circles (sources A and B in Grindlay and Cool 1994).

The Hα equivalent widths for such luminous CVs would be expected to be small, since at $M_V \sim 5.3$, these are nearly as bright as dwarf novae in outburst (when in fact the Hα line is driven into absorption). Indeed the Hα equivalent widths of each appear to be ~ 10Å. If these are each associated with the ROSAT source in whose HRI error circle each is found, their f_x/f_{opt} values (which are 0.8 and 1.8) vs. EW(Hα) are once again (even farther) above the PR correlation shown in Figure 4. Given the completeness results in the next section, it is entirely possible that these two marginal Hα objects found are not the optical counterparts for the two central HRI sources in NGC 6752. The true optical counterparts may be a magnitude or more fainter (as in NGC 6397, and as more typical for CVs) and thus beyond our present limits for NGC 6752.

5. COMPLETENESS AND TOTAL POPULATIONS

We now turn to estimates of the total dim source population in the two clusters observed in our ROSAT/HST studies. We consider the ROSAT results alone, our HST Hα surveys, and finally other CV searches – both with and without HST.

ROSAT Results
Many of the arguments for incompleteness have been given in the section on ROSAT results above. Once again, the main limitations are the limited sensitivity and limited resolution. The total core luminosity of a globular as measured by the PSPC (e.g. Johnston et al 1994) is typically 3×10^{32} erg/s for the central 25"(FWHM) approximate resolution of the PSPC. This is comparable to, but somewhat larger than, the integrated luminosity of the three central core sources (C1-C3) in NGC 6397 (Cool et al 1993a); the fourth source very near the cluster center (source B) would be just resolved by the PSPC but it is too faint to be detectable in the relatively short observations of Johnston et al. Thus taking NGC 6397 as "typical" (although a PCC cluster, it is of only moderate mass), the total number of *observed* dim sources in the cluster core is $\gtrsim 3$ and the actual total must be higher: Cool et al noted that $\sim 10\%$ of the detected flux in the C1-C3 complex was not fit by the three sources. Thus at least a fourth core source, with $L_x \sim 2 \times 10^{31}$ erg/s, but more probably even additional weaker sources, are likely.

We conclude that the total number of core sources (i.e. within the ~ 12"core radius of NGC 6397 or also the effective PSPC resolution radius) is at least 4 and possibly ~ 10. If, as in the discussion of dim source radial distributions above, we then extrapolate to the entire cluster, this leads to a total source population of $\gtrsim 6$-15, *independent* of the nature of the dim sources. Given our HST results, if the dim sources are (dominated by) CVs, then the comparison of the X-ray luminosities of the $\gtrsim 5$-6 sources detected in the cluster with the distribution for CVs given by PR leads to a total estimate for the CV population in NGC 6397 of 10-20, as given in Cool et al (1994). This number is obviously only a crude estimate and is based on the small number statistics of our HRI and HST observations. However, it is approximately consistent with the total number of

CVs expected in NGC 6397 if the predictions for 47 Tuc (cf. di Stefano and Rappaport 1993), in which $\gtrsim 100$ CVs are likely, are scaled by cluster mass (Cool et al 1994).

Our results for NGC 6752 are much more uncertain, given the lower limiting sensitivity. With only two sources in the core (and at least a third and possibly fourth source within $\sim 1'$ of the core – cf. Grindlay (1993a), Cool (1993) and Grindlay and Cool 1994), the total numbers of sources actually detected are not that different from NGC 6397. However they are all factors of 2-3 more luminous (the difference in threshold sensitivity), leading to the obvious conclusion that sources like those in NGC 6397 would not be detected. The luminosity distribution observed in NGC 6397, if applied to NGC 6752, would then imply $\gtrsim 20$ dim sources in the cluster (with individual luminosities $\gtrsim 2 \times 10^{30}$ erg/s. These could easily be "hidden" within the integrated luminosity of the cluster, since they would collectively account for only $\sim 1/3$ of the total luminosity of 3×10^{32} erg/s seen with the PSPC (cf. Johnston et al 1994). Alternatively, if we estimate the total dim source population of NGC 6752 by scaling from the HRI point sources and again assume they are (dominated by) CVs with the PR luminosity distribution, we estimate ~ 30 CVs in the cluster.

HST Hα Survey Results
The three excellent CV candidates we have found in the core of NGC 6397 from their apparent Hα emission and uv excess allow us to estimate completeness factors. We have considered the limiting sensitivities for our search: approximately $M_V \sim 8$ in the equivalent R exposure and EW(Hα) ~ 10Å in identifying an Hα excess. We have compared these limits to tabulations of M_V (using most recent distance estimates) and EW(Hα) values to determine the fractions of CVs of each type (e.g. dwarf novae, DQ-Her, etc.) that could have been detected by CV object class. The complete discussion of this procedure is given in Cool et al (1994), in which we conclude that we were sensitive to perhaps 1/3-2/3 of all CVs *excluding* AM-Her systems (which are typically too faint optically, at $M_V \sim 10$-11) to be detectable. Thus our 3 CV candidates in the core region would imply a total of $\gtrsim$ 4-9 in the core and (again, applying the same radial distribution arguements as above) a total of perhaps 6-13 in the cluster. If AM-Her (i.e. strongly magnetic) CVs are present in globulars, as suggested might be the case by Chanmugam et al (1991), then the numbers could be much higher. This may be required to bring the Hα estimates into agreement with the ROSAT estimates above, although the uncertainties in each are obviously large enough that this is not required.

Comparison With Other CV Searches and Surveys
Our results are consistent with other searches. Our results (both ROSAT and HST) demonstrate the requirement for and power of the highest resolution possible. Ground-based searches (e.g. Shara et al 1994) for dwarf novae can set interesting limits but are not useful for the central 1-2r_c, where our results show that most of the dim sources and CV candidates are located. However when candidates are located, then with sub-arcsec seeing and HST templates it may be possible to conduct meaningful ground-based variability searches (see below).

The most significant other searches are the nova searches (e.g. the search for variable Hα of Tomaney et al (1993) in M31), which should indeed be sensitive into the cluster centers, and the HST/FOC uv observations of DP. The nova searches suggest an underabundance relative to the LMXB excess by a factor of 10 (see also Grindlay 1992), but may be affected by cluster Hα absorption and variable sky background as well as completeness limits in the bulge of M31. The four uv-excess objects found by DP in NGC 6397 are much more promising. All are within just the 6″(radius) error circle of ROSAT source C3, and one is our Hα object no. 2. Two others were independently discovered by us as uv bright (cf. Cool et al 1993b) and yet are not particularly Hα bright, as expected for novalikes (the fourth DP object is too faint to have been detected in our R images). Since these objects appear to be variable (DP), they are excellent CV candidates. Since at least two of them are possible novalikes, and we estimated (Cool et al 1994) we were $\sim$2/3 complete for this class, the DP objects might be the remainder (in just the core) and thus roughly consistent with our total CV completeness estimates.

6. DISTINGUISHING CVs FROM QUIESCENT LMXBs

How can we "prove" the dim sources are dominated by CVs ? In this final section we consider several tests.

Spectra
The most obvious test is to obtain optical spectra of the dim source candidates. We have FOS time approved to do this on our 3 primary candidates with the restored HST in the current Cycle 4. But are spectra alone unambiguous ? LMXBs in quiescence also show strong Hα emission: the LMXB Cen X-4 shows EW(Hα) = 43Å some two years after its 1979 outburst (van Paradijs et al 1987). This is comparable to our values (cf. Fig. 4) for the dim source candidates in NGC 6397. However there are significant differences: the dim source candidates have significant uv excesses (cf. Fig. 3), whereas quiescent LMXBs may not (e.g. the continuum of the Cen X-4 spectrum mentioned is well fit by the K3V secondary, at least at $\lambda \gtrsim$ 4000Å). In the BH transient A0620-00 in quiescence, HST FOS spectra (McClintock, Horne and Remillard 1993) also show a lack of far uv continuum despite a possible ROSAT detection at $\sim 10^{31}$ erg/s; similar uv studies in quiescence are needed for the NS transients Cen X-4 and Aql X-1. It thus appears that at the very low accretion rates of NS (or BH) transients in quiescence, that the inner disks are much cooler and may be more optically thin. In contrast, a CV with similar soft X-ray luminosity (e.g. 10^{31-32} erg/s) has an accretion rate an accretion rate $\gtrsim 10^3$ times higher through the (inner) disk and an optically thick, uv-bright disk.

The X-ray spectra may also reveal clues, but are likely to still be ambiguous. Verbunt et al (1994) have obtained PSPC spectra of Aql X-1 and Cen X-4 in quiescence and find reasonable fits to blackbodies with kT $\sim$0.3-0.5 keV and luminosities $\sim 10^{32.5-33.5}$ erg/s, consistent with blackbody emission from the boundary layer at the inner edge of a thin disk. Johnston et al (1994) find some of their PSPC detections of cluster dim sources (integrated over multiple sources, as described above) similar to this but in other cases (e.g. NGC 6752) apparently harder. Unfortunately, CVs have both hard (accretion shock) and soft components, and the spectral resolution of the PSPC is not adequate to disentangle these. However, ASCA spectra of spatially resolvable single dim sources (e.g. possibly source A in NGC 6752 (cf. Grindlay and Cool 1994), at 1.4′ from the core), may allow this.

A potentially important clue is the X-ray to optical flux ratio vs. emission line strength (cf. Fig. 4). For proper comparison with either CVs or quiescent LMXBs, both better optical and X-ray spectra are needed (e.g. the EW values in Fig. 4 are likely to be underestimates since the Hα filter width is insensitive to broadened lines). Although the trend for the 3 candidates to be above the PR correlation might suggest, for example, that if the objects are CVs, they may be magnetic (DQ Her type) with systematically larger f_x/f_{opt} values (as suggested for the brightest dim source in 47 Tuc by Paresce et al (1992)), this will require optical spectra to confirm. If they are DQ Her types, their spectra will likely show moderately strong He II ($\lambda4686$) emission, whereas this is not detectable in quiescent LMXBs (e.g. Cen X-4; Van Paradijs et al 1987).

Outbursts
Clearly if optical outbursts, as from dwarf novae, are seen from the optical counterparts of the dim sources at least these will be established as CVs. This can now be done, or searches begun, using the known optical postions of our NGC 6397 candidates. Both candidate objects 1 and 3 (Fig. 1) are far enough from the cluster center ($\sim12''$) that even ground-based searches under good seeing could be attempted for monitoring: a DN brightening to a typical outburst magnitude of $M_V \sim 4.5$ would represent a ~0.1 mag brightening of object 1 and its (typically unresolved) bright neighbor star. We are examining our historical CTIO data for this purpose. Of course, if these (first) candidates are novalikes or DQ Her type CVs, as their line strengths might suggest, outbursts will not be expected.

7. CONCLUSIONS

We have summarized the current state of observations of dim sources in globular clusters, with emphasis on the high resolution results from the ROSAT HRI and HST. These results, taken together, are fully consistent with the simplest interpretation (HG, Grindlay 1993a, Cool 1993) that the dim sources are dominated by CVs. Upcoming HST spectra, and future high resolution imaging (ROSAT, HST, AXAF), should provide crucial tests.

If our "typical" clusters, NGC 6397 and NGC 6752, indeed each contain $\gtrsim$ 10-30 CVs, these are roughly consistent with scaling from models of diStefano and Rappaport (1993) and do *not* require any significant CV deficit, as possibly suggested by the M31 novae observations. We note that an absolute minimum CV population, scaling just for cluster mass or primordial binary fraction from the (still uncertain) CV mass fraction in the disk, would lead to $\gtrsim 0.1$-1 CVs in each cluster. Thus our results already suggest that, as expected, CVs are overabundant in globulars by at least a factor of 10. Whether this factor is significantly less than that for LMXBs is not yet clear. However, the likely ratio of $N_{CVs}/N_{LMXBs} \gtrsim$ 10-100 (cf. also Grindlay 1994a) already suggests CVs may play a significant role in the formation of LMXBs and MSPs in globular clusters.

ACKNOWLEDGEMENTS

I am indebted to Adrienne Cool for the data reduction and analysis summarized in Cool (1993), her excellent Ph.D. thesis. This work was supported in part by NASA grants NAGW-3280 and NAG5-1624 and HST grants GO-2555 and GO-3851. Some of the results discussed here are based on observations with the NASA/ESA Hubble Space Telescope, obtained at the Space Telescope Science Institute, which is operated by AURA, Inc., under NASA contract NAS 5-26555. I am pleased to thank the John Simon Guggenheim Foundation for support in the form of a Guggenheim Fellowship.

REFERENCES

Bailyn, C., Grindlay, J. and Garcia, M. 1990, *Ap. J. (Letters)*, 357, L35.
Belloni, T., Verbunt, F. and Schmitt, J. 1993, *Astr. Ap.*, 269, 175.
Chanmugam, G., Ray, A. and Singh, K. 1991, *Ap. J.*, 375, 600.
Cool, A. 1993, *Ph.D. Thesis*, Harvard University.
Cool, A. et al 1993a, *Ap. J. (Letters)*, 410, L103.
Cool, A. et al 1993b, in *Dynamics of Globular Clusters* (S. Djorgovski and G. Meylan, eds.), ASP Conf. Series, Vol. 50, p. 307.
Cool, A. et al 1994, *Ap. J.*, submitted.
de Marchi, G. and Paresce, F. 1994, *Astr. Ap.*, 281, L13.
diStefano, R. and Rappaport, S. 1994 *Ap. J.*, in press.
Grindlay, J. et al 1984, *Ap. J. (Letters)*, bf 282, L13.
Grindlay, J.E. 1985, in *Dynamics of Star Clusters*, Proc. IAU Symp. 113, (J. Goodman and P. Hut, eds.), Reidel, p. 43.
Grindlay, J. 1992, in *X-ray Binaries and Recycled Pulsars*, (E. Van den Heuvel and S. Rappaport, eds.), Kluwer, NATO ASI Series, 377, 365.
Grindlay, J.E. 1993a, in *Dynamics of Globular Clusters* (S. Djorgovski and G. Meylan, eds.), ASP Conf. Series, Vol. 50, p. 285.
Grindlay, J.E. 1993b, in *The Globular Cluster-Galaxy Connection*, (G. Smith and J. Brodie, eds.), ASP Conf. Series, Vol. 48, p. 156.
Grindlay, J.E. 1993c, *Adv. Sp. Res.*, Vol. 13, No. 12, p. 597.
Grindlay, J.E. 1994a, in *Evolutionary Links in the Zoo of Interacting Binaries*. (F. D'Antona, ed.), Mem. della Soc. Astron. Ital., in press.
Grindlay, J.E. 1994b, in *Millisecond Pulsars: A Decade of Surprise*, (A. Fruchter. M. Tavani, and D. Backer, eds.), ASP Conf. Series, in preparation.
Grindlay, J. and Cool, A. 1994, in preparation.
Hertz, P. and Grindlay, J. 1983, *Ap. J.*, 275, 105 (HG).
Hertz, P. and Wood, K. 1985, *Ap. J.*, 290, 171 (HW).
Hertz, P., Grindlay, J. and Bailyn, C. 1993, *Ap. J. (Letters)*, 410, L87.
Hut, P. et al 1992, *PASP*, 104, 981.
Johnston, H.M., Verbunt, F. and Hasinger, G.R. 1994, *Astr. Ap.*, submitted.
Mardling, R.A. 1993, preprint.
McClintock, J., Horne, K. and Remilard, R. 1993, *Ap. J.*, submitted.
Paresce, F., DeMarchi, G. and Ferraro, F. 1992, *Nature*, 360, 46.
Patterson, J. and Raymond, J. 1985, *Ap. J.*, 292, 535 (PR)
Shara, M., Bergeron, L. and Moffat, A. 1994, *Ap. J.*, in press.
Stetson, P. 1991, in *ESO/ST-ECF Data Analysis Workshop*, (P. Grosboel and R Warmels, eds.), ESO Conf. and Workshop Proc., No. 38, p. 187.
Tomaney, A.B., Crotts, A.P., Shafter, A.W. 1993, *Ap. J.*, submitted.
van Paradijs, J. et al 1987 *Astr. Ap.*, 182, 47.
Verbunt, F., van Paradijs, J. and Elson, R. 1984, *M.N.R.A.S.*, 210, 899.
Verbunt, F. and Meylan, G. 1988, *Astr. Ap.*, 203, 297.
Verbunt, F., Hasinger, G., Johnston, H. and Bunk, W. 1993, *Adv. Sp. Res.*, Vol. 13, No. 12, p. 151.
Verbunt, F., Belloni, T., Johnston, H., van der Klis, M. and Lewin, W. 1994, *Astr. Ap.*, in press.

DO MAGNETIC FIELDS OF NEUTRON STARS DECAY?

Frank Verbunt
Astronomical Institute, Utrecht University
P.O. Box 80 000, 3508 TA Utrecht, The Netherlands

ABSTRACT

Neutron stars in young X-ray binaries have strong magnetic fields, those in old binaries usually weak magnetic fields, but exceptions confute simple explanations. Observational evidence claimed to prove spontaneous field decay in single radio pulsars is found wanting on closer scrutiny. Some detailed properties of the z-distributions of radio pulsars are described; these may explain the contradictory results of apparently similar population synthesis studies.

1. INTRODUCTION

Magnetic fields play an important role in our thinking about neutron stars. They are invoked for explanations of the lengthening of pulse periods of single radio pulsars, and of the occurrence of periods and quasi-periods in the emission of X-ray binaries. It is therefore somewhat unsettling that, 25 years after the first discoveries of neutron stars, debate persists on a number of rather basic questions about the magnetic fields of neutron stars. This review is concerned with one of these questions, $viz.$ whether the magnetic field of a neutron star decays with time.

The rapid development during recent years in our thinking about the evolution of the magnetic fields of neutron stars has been described in various reviews, among which I mention Bhattacharya (1989), Lamb (1992), and Bhattacharya & Srinivasan (1994). What do we mean when we refer to the magnetic field of a neutron star?

For **radio pulsars**, it is assumed that they are rotating magnetic dipoles that lose rotational energy via the emission of Poynting flux. This provides a relation between the magnetic field strength B at the neutron star surface, and the pulse (i.e. rotation) period P of the neutron star and its time derivative $\dot{P}$ (Gunn & Ostriker 1970).

$$B^2 = \frac{3Ic^3}{8\pi^2 R^6} P\dot{P} \simeq 10^{39} P\dot{P} \tag{1}$$

where $I \simeq 0.4MR^2$ and R are the moment of inertia and radius of the neutron star. M is the mass of the neutron star. The numerical value is found taking $M = 1.4 M_\odot$ and $R = 10$ km. In this equation the dependence of the Poynting flux on the angle α between the rotation and magnetic axes has been silently dropped. The justification for this is that, although the Poynting flux decreases when this angle becomes smaller, other losses — in particular in the form of particles — increase; so that the overall energy loss may not change much (Goldreich & Julian 1969). Support for this assumption is the observation that the measured product $P\dot{P}$ does not depend on α (Bhattacharya 1989).

Figure 1 shows the values of P and B (calculated with Eq. 1) for 416 radio pulsars, mostly from the catalogue of Manchester & Taylor (1981). It is notable that radio pulsars in binaries have much shorter pulse periods and weaker magnetic fields, on average, than single radio pulsars. From this it is inferred that an isolated

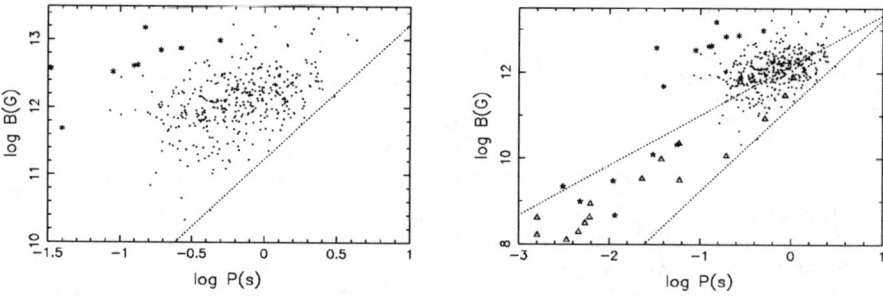

Figure 1: The magnetic field B, according to Eq. 1, as a function of period P for single radio pulsars (•), including those in supernova remnants (*) (left). Recycled pulsars ($\triangle$) and pulsars in globular clusters ($\star$) are added in the right frame. The dashed lines are the death–line, to the right of which no pulsars are observed, and the spin–up line given by Eq. 6 (for the Eddington limit to $\dot{M}$).

pulsar with short period and weak magnetic field may have been in a binary before. The fields of radio pulsars found in the cores of globular clusters have intermediate values. Radio pulsars that are thought to have accreted matter in the past are called recycled pulsars. It should be kept in mind that the magnetic field strength derived for a pulsar in a cluster may be affected by a contribution to the measured value of $\dot{P}$ due to acceleration of the pulsar in the gravitational potential of the cluster (Phinney 1992); and that for a nearby millisecond pulsar in the Galactic Disk by a contribution to the measured $\dot{P}$ due to the proper motion of the pulsar (the 'Shklovskij' effect, Camilo et al. 1993).

By integrating Eq. 1, we obtain an estimate of the age of the radio pulsar

$$\tau_c = \frac{P}{2\dot{P}} \qquad (2)$$

τ_c is called the *characteristic age* of a radio pulsar; it is equal to the actual age of the pulsar if the current pulse period is much longer than the initial pulse period, i.e. $P \gg P_i$, and if the magnetic field has a constant strength.

In **X-ray binaries** strong magnetic fields were first inferred from the pulsed emission of X-rays, which suggest that a magnetic field channels the inflowing matter onto two poles, that appear and disappear as the neutron star rotates. A value for the field strength of Her X-1 has been found by identifying a feature in its hard X-ray spectrum with a cyclotron line (Trümper et al. 1978). The field strength is related to the energy E_c at which the feature appears according to

$$B = \left[\left(\frac{E_c}{mc^2}+1\right)^2 - 1\right]\frac{m^2c^3}{2e\hbar} \simeq \left[\left(\frac{E_c}{511\,\text{keV}}+1\right)^2 - 1\right] 22 \times 10^{12}\,\text{G} \qquad (3)$$

where m is the electron mass. Cyclotron features at energies between 7 and 40 keV have been measured for nine X-ray pulsars, in particular by Ginga, which leads to field strengths comparable to those of the single radio pulsars, but in a surprisingly narrow range of 0.8–3.5×10^{12} G (Makishima et al. 1992).

Values for the magnetic field strength may also be derived from the observed time scales of spin-up/down of X-ray pulsars, as it is the magnetic field that provides a lever arm that allows the accreted matter to exert a torque. The lever arm is given by the radius r_m of the magnetosphere, which is the radius where magnetic pressure equals the ram pressure of infalling matter. For a dipole field and a spherically symmetric accretion at a rate $\dot M$ with free fall velocity one derives

$$r_m = 2^{-3/7} B^{4/7} R^{12/7} \dot M^{-2/7} (GM)^{-1/7} \qquad (4)$$

By dividing the angular momentum of the neutron star by the exerted torque, the timescale of spin-up far from equilibrium is found to be

$$\frac{P}{\dot P} = \frac{I\omega}{\dot M \sqrt{GMr_m}} \simeq 1.4 \times 10^4 \,\mathrm{yr} \left(\frac{B}{10^{12}\,\mathrm{G}}\right)^{-2/7} \left(\frac{\dot M}{10^{-9}\,M_\odot\,\mathrm{yr}^{-1}}\right)^{-6/7} \frac{1}{P} \qquad (5)$$

where is is assumed that the matter has the angular momentum of a circular orbit at r_m when accreting.

The spin-up may lead to an equilibrium pulse period P_{eq}, once the matter at r_m corotates with the neutron star. With Eq. 4 one derives

$$P_{eq} = 2\pi \left(\frac{B^2 R^6}{2\sqrt{2}\dot M}\right)^{3/7} \left(\frac{1}{GM}\right)^{5/7} \simeq 2.3\,\mathrm{s} \left(\frac{B}{10^{12}\,\mathrm{G}}\right)^{6/7} \left(\frac{\dot M}{10^{-9}\,M_\odot\,\mathrm{yr}^{-1}}\right)^{-3/7} \qquad (6)$$

Ghosh & Lamb (1978) interpolate Eqs. 5, 6 (with some refinement) and show that the measured values of P, $\dot P$, and $\dot M \simeq L_x R/(GM)$ for X-ray pulsars imply $B \simeq 10^{12}$ G.

From the absence of X-ray pulsations in many low-mass X-ray binaries, an (uncertain) upper limit to the field strength is estimated at $\sim 10^9$–10^{10} G.

2. X-RAY BINARIES AND IDEAS ON FIELD DECAY

Soon after the discovery of the first X-ray pulsars, it was found that they are usually accompanied by relatively massive, O or B stars. Neutron stars with such companions must be younger than the maximum life time of the massive star, i.e. younger than $\sim 10^7$ yr. On the other hand, it was found that non–pulsing neutron stars usually have low–mass donors, that may have ages of up to 10^{10} yr. The average age of these neutron stars can thus be in excess of 10^9 yr.

Some X-ray pulsars must be relatively old, as they are accompanied by evolved low–mass companions. The best case is Her X-1, whose $\sim 2\,M_\odot$ companion has evolved from the main–sequence, and must therefore be older than $\sim 6 \times 10^8$ yr (Lamb 1981, Verbunt et al. 1990). GX1+4 is another example.

Thus, the X-ray binaries indicate that young neutron stars have strong magnetic fields, and that old neutron stars have weak fields, but that there are exceptions.

The study of radio pulsars originally led to the idea, as discussed in the next section, that a neutron star loses its field on a time scale of $\lesssim 10^7$ yr. Application of this idea to the X-ray binaries nicely explained the dichotomy in field strengths between the neutron stars with high–mass and low–mass companions. To explain the exceptions, a way has been devised to produce a young neutron star in an old binary: accretion–induced collapse. An old white dwarf, with a mass close to the Chandrasekhar limit, can surpass this limit while accreting matter from a companion star, and then implode to form a neutron star. In the case of one system, Her X-1, it could be shown that accretion–induced collapse is not viable;

a similar, perhaps less convincing case can be made for 4U1626−67 (Verbunt et al. 1990).

New studies of radio pulsars, described in Section 4, indicate that the magnetic field of a single neutron star does not decay on a time scale shorter than $\sim 10^8$ yr. One may thus be tempted to assume that the magnetic field of neutron stars decays on a time scale of $\sim 10^9$ yr, long enough for a system like Her X-1 to retain a strong field, and short enough to explain the low field strengths of neutron stars in low-mass X-ray binaries. A problem with this idea (Kulkarni, private communication) is that the radio pulsars in the cores of globular clusters, that must be $\sim 10^{10}$ yr old, have field strengths up to $\sim 10^{10}$ G (see Figure 1).

A third possibility is suggested by the observation that all neutron stars with low magnetic field strengths either are, or presumably have been, in binaries. Perhaps it is the mass accretion onto the neutron star that causes the field to decay (Taam & Van den Heuvel 1986). Alternatively, interaction of the accreting matter with the magnetosphere may cause the neutron star to slow down to a period of ~ 100 s, by which period most magnetic flux tubes have been dragged with the rotation vortices to the neutron star crust, where they may decay (Srinivasan et al. 1990). A problem with the accretion–induced decay may be formed by the system 4U1627−67, which has a donor of extremely low mass (Levine et al. 1988), which suggests that the neutron star has already accreted a fair amount of mass, its high field notwithstanding. A problem with both binary scenarios is the absence of massive X-ray binaries in which the field has already decayed to values comparable with those found in the presumed descendent systems. Both scenarios require further study.

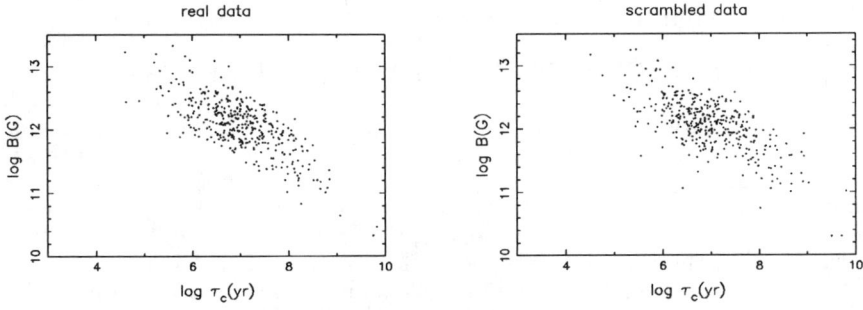

Figure 2: The magnetic field as a function of the characteristic age of radio pulsars, determined from P and $\dot{P}$ with Eqs. 1,2 (left). To show that this plot does not imply field decay, the same plot has made for scrambled $\dot{P}$-values (right).

3. WRONG ARGUMENTS FOR FIELD DECAY IN RADIO PULSARS

A plot of magnetic field strength as a function of characteristic age shows weaker magnetic fields for older radio pulsars (Figure 2). This was taken to imply that the magnetic field of radio pulsars decays on a time scale of $\sim 10^7$ yr (Gunn & Ostriker 1970) The plot is convincing but the argument is wrong! As pointed out by Lyne et al. (1975), the periods of ordinary radio pulsars show much less spread than their period derivatives, so that Figure 2 effectively represents a plot of $\dot{P}$

vs. $1/\dot{P}$ (see Eqs. 1,2). That this indeed explains the anti-correlation between B and τ_c may be shown by the 'scramble test': make a list of periods and period-derivatives, scramble the period derivatives, and make a plot of magnetic field vs. age calculated with these scrambled values. The resulting plot, also shown in Figure 2, indeed shows the same anticorrelation. No conclusions on the relation between field strength and age can be derived from the B–τ_c plot.

A second argument derives from pulsar kinematics. The progenitors of neutron stars, massive stars, are close to the Galactic Plane, and we may estimate the *kinematic age* of a pulsar by dividing its distance z to the plane by its velocity in the z-direction:

$$\tau_k = \frac{z}{v_z} \quad (7)$$

Lyne et al. (1982), and more recently Harrison et al. (1993), show that the kinematic age of radio pulsars tends to be smaller than the characteristic age, and argue that this is caused by decay of the magnetic field: for a decaying field, τ_c underestimates the actual age of the pulsar (see Eq. 1). I will return to this in the next Section, after the discussion of population synthesis of radio pulsars.

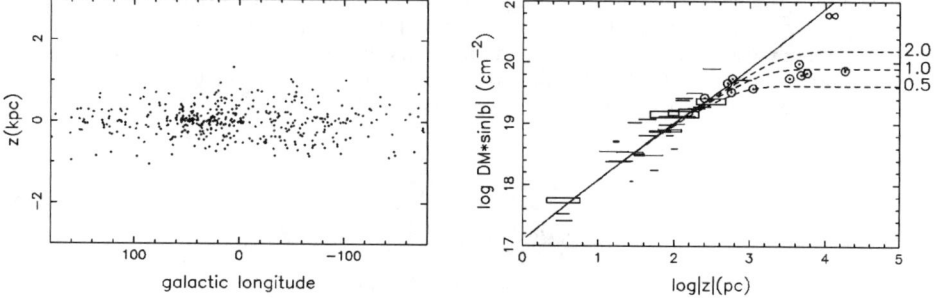

Figure 3: If the distance to a radio pulsar is determined from the dispersion measure, no pulsars are found at distances $|z| \gtrsim 1$ kpc from the Galactic Plane (left). A plot of the (vertical component of) the dispersion measure against $|z|$ for pulsars whose distance is known independently (shown as oblong error boxes), including those in globular clusters ($\odot$), shows that this is due to the absence of electrons at $|z| \gtrsim 1$ kpc (right). The curves give the expected DM for various n_e scale heights. (From Bhattacharya & Verbunt 1991.)

A second version of the kinematic argument is as follows. With a velocity $v_z \sim 100$ km/s, a radio pulsar born at $z \simeq 0$ reaches 1 kpc after 10^7 yr, and 2 kpc after 2×10^7 yr. The absence of observed pulsars at 2 kpc (see Figure 3), even though they have high velocities, has been taken to imply that pulsars are no longer detectable when they are 20 Myr old, presumably because their field decayed. This apparently solid argument collapsed when it was shown that the maximum distance of 1 kpc is an artefact of the method used to determine distances of most radio pulsars (Bhattacharya & Verbunt 1991). These distances are determined from the variation with wavelength of the arrival time of the pulse peaks, which is proportional to the dispersion measure, i.e. the number of electrons between Earth and the pulsar. If no electrons are present at distances more than 1 kpc from the Galactic Plane, pulsars at larger distances will have the same dispersion measure — hence derived distance from the plane — as pulsars at ~1 kpc. That this is indeed the case can be shown by considering the dispersion measures of radio pulsars in globular clusters, whose distance to the Galactic Plane is known (Reynolds 1989, Bhattacharya & Verbunt 1991, see Figure 3).

4. THE B–P DIAGRAM AND POPULATION SYNTHESIS

To study the evolution of radio pulsars, one often considers the B–P diagram (see Figure 1). One method is the investigation of the *pulsar current*. To illustrate this method, we consider the case of no field decay. In this case, a pulsar born at a short period moves to the right at constant B in Figure 1. The investigation of the pulsar density along the line, or its derivative, the pulsar current, thus informs us about $\dot{P}$ as a function of P. One can test whether Eq. 1 holds. In a similar fashion, one may study flow lines followed by the pulsar when its field decays on a given time scale. Eq. 1 implies that pulsars spend more time at long periods, so that we expect them to pile up at the death line. The fact that this is not the case immediately suggests that selection effects may cause a less efficient detection of pulsars at longer periods.

The distribution of pulsars in the B–P diagram can also be studied with population synthesis. Narayan & Ostriker (1990) carefully modelled various selection effects, and used their results to simulate the population of radio pulsars and its observation with radio telescopes. A similar study was performed by Bhattacharya et al. (1992), who based their work strongly on the previous study by Narayan & Ostriker. The structure of these population syntheses is shown schematically in Eqs. 8–12.

$$\vec{r}_i, \vec{v}_i \xrightarrow{\Phi_G} \vec{r}, \vec{v} \tag{8}$$

$$P_i, \dot{P}_i \xrightarrow{B^2 \propto P\dot{P}} P, \dot{P} \longrightarrow B, \tau_c \tag{9}$$

$$\text{beaming} \quad f(P) \tag{10}$$

$$\text{luminosity} \quad L(P, \dot{P}) \xrightarrow{\vec{r}} S \tag{11}$$

$$\vec{r} \xrightarrow{n_e(\vec{r})} DM, \tau_{\text{scatt}} \tag{12}$$

For each pulsar one assumes an initial position $\vec{r}_i$ and velocity $\vec{v}_i$ and then uses the galactic potential Φ_G to calculate its current position $\vec{r}$ and velocity $\vec{v}$, for an age arbitrarily selected between zero and the age of the Universe (Eq. 8). One also assumes the initial period P_i and period derivative $\dot{P}_i$, or equivalently the initial magnetic field B_i, and uses Eq. 1 to calculate the current period P, period derivative $\dot{P}$, and field strength B (Eq. 9). The field strength is assumed to decay on a time scale τ according to $B(t) = B_i \exp(-t/\tau)$. The characteristic age is found from P and $\dot{P}$ with Eq. 2.

Radio pulsars emit their radiation in a beamed pattern, and the opening angle of the beam is thought to be smaller for pulsars with longer periods. As a result, the sweep of the beam of pulsars with long periods has a smaller probability of hitting the Earth. The dependence on P of this probability is called the beaming factor $f(P)$, and has been suggested to be $f(P) \propto P^{-1/3}$ (Lyne and Manchester 1988) or $f(P) \propto P^{-1/2}$ (Narayan and Vivekanand 1983) (Eq. 10). From the current period and period derivative, one may also estimate the luminosity L of the radio pulsar, according to various luminosity laws that have been proposed. Narayan & Ostriker (1990) show that it makes little difference which of these laws is selected, and that it is much more important to correctly model the spread around the average value of $L(P, \dot{P})$ provided by these laws. From the luminosity and the position of the pulsar, the flux S received at earth may then be estimated (Eq. 11). In each case, it is taken into account that pulsars to the right of the 'death–line' in Figure 1 are not detected. Finally, with the use of a model for the galactic distribution of electrons $n_e(\vec{r})$, one may calculate for the pulsar position $\vec{r}$ the dispersion measure DM and the scattering τ_{scatt} of the pulse signal.

To check whether the pulsar was in an observed part of the sky, its position is expressed in right ascension and declination, and if it is, it may be estimated from S, DM, and τ_{scatt} whether it would have been detected in any of the radio surveys that have been made. This whole procedure from Eq. 8 onward is repeated until a sufficient number of simulated detections is obtained for comparison with the real pulsars detected in the surveys.

Bhattacharya et al. (1992) investigate various decay times, and find that simulations in which the field decays on a long time scale, i.e. not perceptibly during the time that pulsars spend to the left of the death–line, give a much better description of the observations than simulations in which the field decays rapidly. Interestingly, the dispersion measure and luminosity are well described by all models; the discrimination between the short and long decay times is provided by the distribution of the pulsars in the B–P diagram. In particular, simulations with a short decay time produce too few pulsars with long periods and strong fields. The characteristic ages of the observed pulsars also indicate a long decay time.

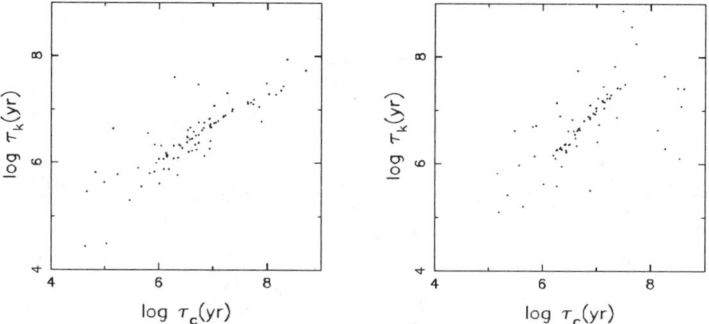

Figure 4: Kinematic age as a function of characteristic age for 100 radio pulsars 'detected' in simulations by Bhattacharya et al.(1992) with decay times for the magnetic field of 10 Myr (left) and 100 Myr (right). 14 and 19 pulsars are moving towards the plane, in the 10 and 100 Myr simulations, respectively, so that their τ_k is not defined.

We briefly return to the argument for field decay derived from the comparison of the kinematic and characteristic ages. In Figure 4 we show these ages for simulations with assumed decay times of 10 and 100 Myr. In the simulation with long decay times, several pulsars with characteristic ages longer than the kinematic age are found at small τ_c. These could be pulsars born some distance from the Galactic Plane with a velocity towards the plane, which have already passed the plane. In the same simulation, a group of pulsars at large τ_c has small τ_k. These are old pulsars, which first moved away from the disk, then returned to the plane, and now are moving away from it a second time (see also Figure 5).

358 Do Magnetic Fields of Neutron Stars Decay?

Thus, the existence of pulsars with $\tau_k < \tau_c$ can not be used as an argument for field decay.

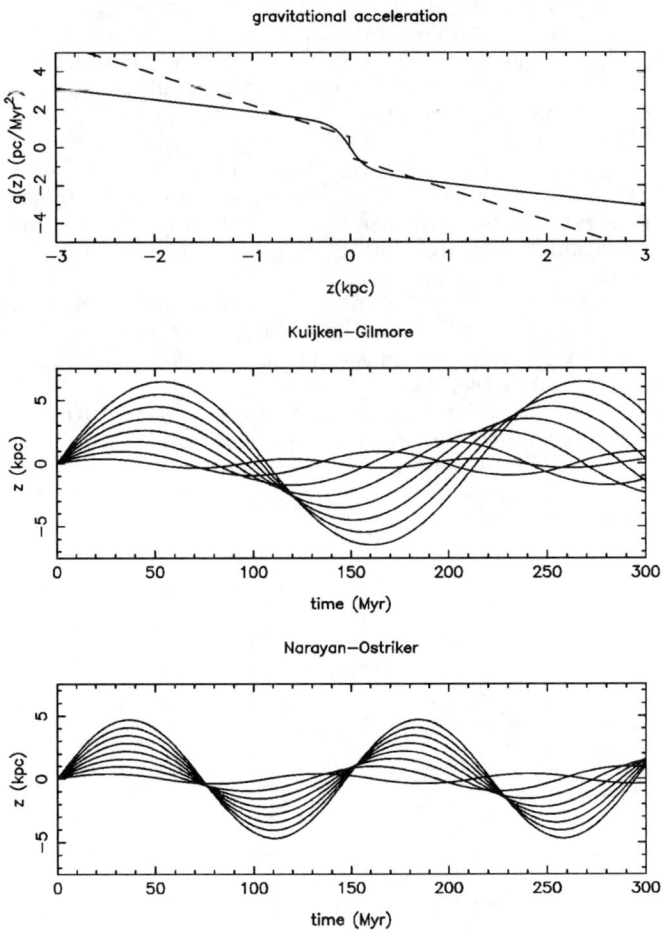

Figure 5: (top:) Models for the z-component of the gravitational acceleration near the Sun used by Bhattacharya et al.(1992), from Kuijken & Gilmore (1989, solid line), and by Narayan & Ostriker (1990, dashed line). (middle and below:) Numerically calculated orbits of radio pulsars with velocities of 25, 50, 75, ...200 km/s in these models.

Since the work of Bhattacharya et al. closely follows that of Narayan & Ostriker (1990), it is rather surprising that the conclusions are opposite. The model favoured by Bhattacharya et al., a single population of radio pulsars with slow field decay, is not only the worst of the 19 models calculated by Narayan & Ostriker, it is worse than the next worst model by a factor 10^{13}... This is all the more striking as the model favoured by Narayan & Ostriker adds a second population of pulsars, formed at some distance from the plane with long periods and high fields, i.e. with exactly the properties of old neutron stars whose fields don't decay.

Let us therefore compare the two simulations somewhat more closely. Eqs. 9–11 are identical in both. Eq. 12 differs in that Bhattacharya et al. use an $n_e(\vec{r})$ model with finite scale height (see Figure 3); however, they show that this alone cannot explain the difference. Narayan & Ostriker compare their simulations with eight radio surveys, with 301 pulsars, whereas Bhattacharya et al. only use four, with 130 pulsars. The advantage that the use of four extra Surveys would bring is in my opinion offset by the need to extrapolate luminosity from one frequency to another in two of them; by the uncertainty in the galactic potential at the large distances covered by a third; and by the fact that the exact positions of the pointings of two NRAO surveys are not available, forcing Narayan & Ostriker to make a rather crude description.

Narayan & Ostriker include a correlation between pulsar velocity and magnetic field strength, which is ignored by Bhattacharya et al. They also use an analytic description of the z- and v_z-distributions, whereas Bhattacharya et al. solve Eq. 8 numerically. The consequences of these differences are discussed in the next Section, which describes recent work by Hartman & Verbunt (1994).

5. THE z-DISTRIBUTION OF RADIO PULSARS

Bhattacharya et al., like Narayan & Ostriker (1990), solve the equation of motion only in the z-direction, using the locally determined galactic acceleration. The acceleration models used are slightly different, as shown in Figure 5, where we also show the numerically solved orbits of radio pulsars shot from the Galactic Plane at different velocities. These solutions show a striking effect: at first, all pulsars move away from the disk, but deceleration by the Galactic gravity returns them to the plane again, first those with low velocities, and subsequently those with higher initial velocities. The effect of the returning pulsars is twofold. First, the low-velocity pulsars cause a maximum in the pulsar density at low $|z|$. Secondly, crossing orbits cause maxima in the density some distance from the plane. These effects are illustrated in Figure 6 for the Kuijken–Gilmore potential; they are also present in the potential used by Narayan & Ostriker.

Narayan and Ostriker use an analytical solution for the equation of motion, which assumes that the z-distribution of the radio pulsars is always Gaussian. As illustrated in Figure 6 this is a reasonable solution for young pulsars, with $t \lesssim 10$ Myr, and thus for the calculations performed by Narayan & Ostriker with short decay times. For pulsars that are much older, Figure 6 shows that the z-distribution is not described by a Gaussian distribution (as also noted by Blaes & Madau 1993). This invalidates the calculations of Narayan & Ostriker with longer decay times, and in particular the one relating to the model favoured by Bhattacharya et al., i.e. a single population with a decay time $\tau \gtrsim 100$ Myr.

There is also a problem at short decay times, however. As shown by Figure 5, for each age of a pulsar there is a clear relation between its height z and its velocity v_z. Narayan and Ostriker appear to select v_z from the distribution of *all* pulsars at a given age, and thus to ignore this relation. In my opinion, this makes any correlation found in their simulations between v_z and other quantities, such as B, suspect.

Do we now understand the different results of Bhattacharya et al. and Narayan & Ostriker? To investigate this, I have performed simulations in which the analytical solution for the z-distribution derived by Narayan & Ostriker is used, but otherwise identical to those of Bhattacharya et al. (1992). I still find that long decay times give a better description than short decay times, even though the difference is smaller. The underlying reason is presumably the very

Do Magnetic Fields of Neutron Stars Decay?

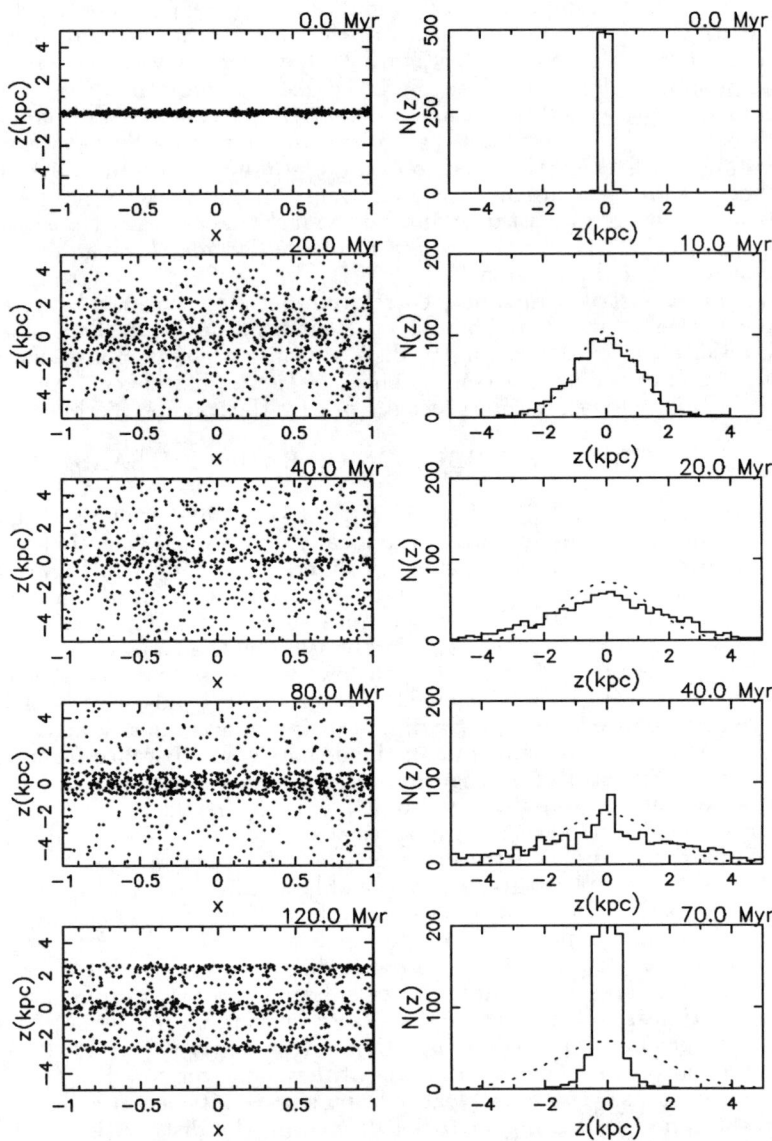

Figure 6: z-distributions of radio pulsars of the same age, as a function of age. No observational or 'death-line' selection effects have been applied. The initial velocity distribution is assumed to be Gaussian, with $\sigma_v = 110$ km/s. (left:) z of thousand pulsars, calculated numerically using the Kuijken–Gilmore (1990) potential. The x-values are randomly distributed. (right:) z-distributions of thousand neutron stars, calculated numerically using the Narayan–Ostriker (1990) potential (histograms) and the analytical solution proposed by these authors (dotted lines).

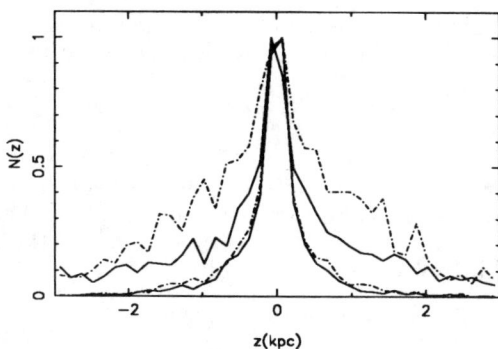

Figure 7: Simulated z-distributions of radio pulsars of all ages in the interval 0–300 Myr, normalized to the same maximum. The solid lines use the numerical solutions by Bhattacharya et al.(1992), the dashed lines the analytic solutions by Narayan & Ostriker (1990). Before detection, but after application of the 'death-line' criterion, (top lines) the distributions are very different; for the pulsars 'detected' in four simulated surveys (lower lines) the difference has largely disappeared.

strong selection against radio pulsars at large distances, which tends to reduce the difference in the z-distributions, as illustrated in Figure 7. Clearly, selection effects have an enormous impact on the simulations.

It is not at all obvious, therefore, that the difference between the results of Narayan & Ostriker and Bhattacharya et al. is due to the different treatment of the z-distributions. However, such an explanation cannot be excluded, as my calculations do not include the four extra surveys used by Narayan & Ostriker, and do not test for a v_z–B correlation. As already noted, use of such a correlation may be subject to error if the relation between z and v_z of a pulsar at a given age is not taken into account.

6. SUMMARY

The original arguments put forward to prove that the magnetic fields of single radio pulsars decay are not correct. The populations syntheses of Narayan & Ostriker (1990) and Bhattacharya et al. (1992) show that the observed properties of radio pulsars cannot be described in models with a single population of radio pulsars with magnetic fields decaying on a time scale $\tau \lesssim 10^7$ yr. Bhattacharya et al. find that a model with a single population and a long ($\tau \gtrsim 10^8$ yr) decay time scale does give an acceptable description. The calculations of Narayan and Ostriker are not valid for such long times scales, due to a simplified description of the z-distribution of radio pulsars.

Narayan & Ostriker (1990) argue that a model with a short decay time for the magnetic field also gives an acceptable description of the observations, if a second population of radio pulsars, born with longer periods, is added. However, Lorimer et al. (1993) find no evidence for such a second population in a pulsar current study. This study, limited to the brightest radio pulsars only, to eliminate selection effects, includes all eight surveys used by Narayan & Ostriker.

The observations show that the magnetic field of radio pulsars in binaries is lower on average than that of single radio pulsars. At the moment, we have no good understanding why this is the case, but several possibilities, such as accretion–induced decay or decay following dramatic slow–down, are being investigated.

ACKNOWLEDGEMENTS

I am very grateful to my co-workers Dipankar Bhattacharya, Ralph Wijers and Jan–Willem Hartman for many discussions on field decay. Discussions with Shrinivas Kulkarni and Sterl Phinney have also helped in sharpening my understanding of this topic. This research is supported by the Netherlands Organization for Scientific Research (NWO) under grant PGS 78-277.

REFERENCES

Bhattacharya, D. 1989, in 23rd ESLAB Symposium: X-ray binaries, ed. B. Battrick, ESA-SP 296, p.179
Bhattacharya, D. & Srinivasan, G. 1994, in X-ray binaries, eds. W. H. G. Lewin, J. van Paradijs & E. P. J. van den Heuvel, in press (Cambridge University Press)
Bhattacharya, D. & Verbunt, F. 1991, A&A, 242, 128
Bhattacharya, D., Wijers, R.A.M.J., Hartman, J.W. & Verbunt, F. 1992, A&A, 254, 198
Blaes, O. & Madau, P. 1993, ApJ, 403, 690
Camilo, F., Thorsett, S.E. & Kulkarni, S.R. 1993, ApJ (Letters), in press
Ghosh, P. & Lamb, F.K. 1978, ApJ, 223, L83
Goldreich, P. & Julian, W.H. 1969, ApJ 157, 869
Gunn, J.E. & Ostriker, J.P. 1970, ApJ, 160, 979
Harrison, P.A., Lyne, A.G. & Anderson, B. 1993, MNRAS, 261, 113
Kuijken, K. & Gilmore, G. 1989, MNRAS, 239, 571 & 605
Lamb, D.Q. 1992, in Frontiers of X-ray astronomy, eds. Y. Tanaka & K. Koyama, p.33 (Universal Academy Press, Tokyo)
Lamb, F.K. 1981, in Pulsars, eds. W. Sieber & R. Wielebinski, IAU Symp. 95, p.309
Levine, A., Ma, C.P., McClintock, J., Rappaport, S., van der Klis, M. & Verbunt, F. 1988, ApJ, 327, 732
Lorimer, D.R., Bailes, M., Dewey, R.J. & Harrison, P.A. 1993, MNRAS, 263, 403
Lyne, A.G. & Manchester, R.N. 1988, MNRAS, 234, 477
Lyne, A.G., Ritchings, R.T. & Smith, F.G. 1975, MNRAS, 171, 579
Lyne, A.G., Anderson, B. & Salter, M.J. 1982, MNRAS, 201, 503
Makishima, K., Mihara, T., Nagase, F. & Murakami, T. 1992, in Frontiers of X-ray astronomy, eds. Y. Tanaka & K. Koyama, p.23 (Universal Academy Press, Tokyo)
Manchester, R.N. & Taylor, J.H. 1981, Astron J, 86, 1953
Narayan, R. & Ostriker, J.P. 1990, ApJ, 352, 222
Narayan, R. & Vivekanand, M. 1983, A&A, 122, 45
Phinney, E.S. 1992, Phil Trans R Soc London A, 341, 39
Reynolds, R.J. 1989, ApJ, 339, L29
Srinivasan, G., Bhattacharya, D., Muslimov, A.G. & Tsygan, A.I. 1990, Curr Sci, 51, 596
Taam, R.E. & van den Heuvel, E.P.J. 1986, ApJ, 305, 235
Trümper, J., Pietsch, W., Reppin, C., Voges, W., Staubert, R., Kendziorra, E. 1978, ApJ, 219, L105
Verbunt, F., Wijers, R.A.M.J. & Burm, H. 1990, A&A, 234, 195

ROSAT MEASUREMENT OF THE EVOLVING ORBITAL PERIOD IN THE LMXB EXO0748-676

P. Hertz[†], Y. Ly[‡], and K. S. Wood[†]
Code 7621, Naval Research Laboratory, Washington, DC 20375

L. R. Cominsky[†]
Phys. and Astron. Dept., Sonoma State University, Rohnert Park, CA 94928

ABSTRACT

The sign and magnitude of the orbital period derivative are perhaps the most critical diagnostics of the evolution of low mass X-ray binaries (LMXBs). However $\lesssim 4$ LMXBs have measured orbital period derivatives. One of these is the eclipsing X-ray binary EXO0748-676, whose sharp edged eclipse transitions provide a good fiducial marker.

We have observed a single eclipse egress with ROSAT. We have combined our data with EXOSAT and Ginga timings to fit the ephemeris of EXO0748-676. Assuming that the observed eclipse times are determined solely by a deterministic period plus measurement error, we confirm Asai et al. — a constant period and a constant period derivative are both ruled out. A constant second period derivative and a 12 yr sinusoidal variability provide equally good fits.

The sinusoidal variation in orbital period is reminiscent of several cataclysmic variables. The period could be interpreted as evidence of a third body or of spin-orbit coupling. However no CV has been observed for more than 1.5 long cycles, and many show orbital period timings that deviate significantly from the predictions of the sinusoidal ephemeris.

We wonder if there might not be intrinsic, stochastic jitter in the time of individual eclipses. If so, the O-C diagram represents a random walk in orbital phase, and it will usually be well fitted by a sinusoid with a period 1-2 times the duration of the observational history. We present statistical tests in support of this interpretation.

1. ROSAT DATA

EXO0748-676 was observed three times during ROSAT AO1 and AO2. The purpose was to time an eclipse and extend the data base of EXO0748-676 eclipses. Although 23 OBIs of data were obtained, only one eclipse egress was observed. The time of egress was estimated to the nearest second and corrected to the solar system barycenter. The source was observed on axis, so the ROSAT wobble is clearly visible in the data. We adopted an uncertainty of 5 seconds until we correct for the wobble.

[†] ROSAT Observatory Guest Investigator.
[‡] also Thomas Jefferson High School, Alexandria, VA.

2. THE O-C PLOT

The eclipse duration for EXO0748-676 has been shown to be variable (Parmar et al. 1991) so we have timed eclipse egress rather than correct the ROSAT measurement to mid-eclipse using an unknown eclipse duration. We have fitted all measured eclipse egress times in TJD at the solar system barycenter. EXOSAT and Ginga data are taken directly from Parmar et al. (1991) and Asai et al. (1992). In Figure 1 we plot the residuals of the observed eclipse egress timing from the best fitted constant period model as a function of TJD. This is the standard O-C plot. Deviations from a horizontal line indicate a changing period.

We have fitted linear (constant period), quadratic (period derivative), cubic (period acceleration), and sinusoidal (long period, third body?) models to the O-C data. The results are also shown in Figure 1, and the best fitted coefficients are given in Table 1. The cubic and sinusoidal models give fits of comparable quality. An extrapolation of both models shows that observations taken after TJD 8900 (October 1992) will be sufficient to discriminate between these models. ASCA observations during the PV phase obtained during 1993 this will be sufficient to distinguish between the cubic and sinusoidal models.

TABLE 1: Models Fitted to O-C Data

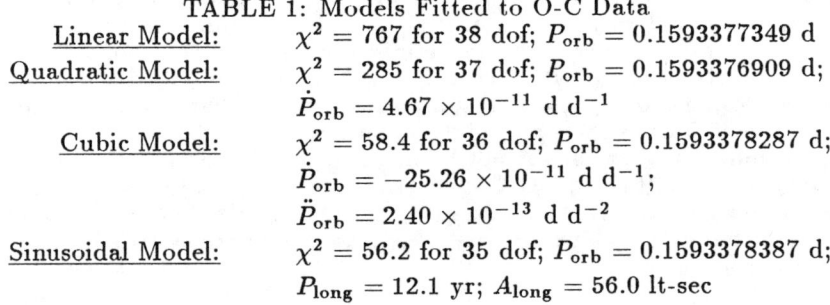

Linear Model: $\chi^2 = 767$ for 38 dof; $P_{\rm orb} = 0.1593377349$ d

Quadratic Model: $\chi^2 = 285$ for 37 dof; $P_{\rm orb} = 0.1593376909$ d; $\dot{P}_{\rm orb} = 4.67 \times 10^{-11}$ d d^{-1}

Cubic Model: $\chi^2 = 58.4$ for 36 dof; $P_{\rm orb} = 0.1593378287$ d; $\dot{P}_{\rm orb} = -25.26 \times 10^{-11}$ d d^{-1}; $\ddot{P}_{\rm orb} = 2.40 \times 10^{-13}$ d d^{-2}

Sinusoidal Model: $\chi^2 = 56.2$ for 35 dof; $P_{\rm orb} = 0.1593378387$ d; $P_{\rm long} = 12.1$ yr; $A_{\rm long} = 56.0$ lt-sec

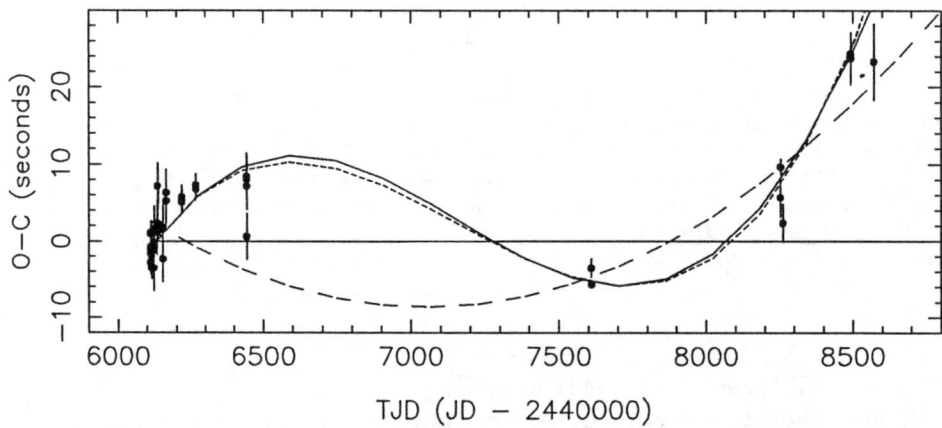

Figure 1: The O-C Plot. The residuals of measured eclipse egresses from a constant period ephemeris are plotted as a function of time. The best fitted linear (light solid), quadratic (dashed), cubic (dotted), and sinusoidal (heavy solid) models are shown.

3. DISCUSSION

It is usually assumed that the time of eclipse is determined solely by the deterministic ephemeris plus measurement error. Under this assumption, the orbital period, as determined by the time between eclipses, has varied over the last 6.5 years by up to 20 seconds. A constant period derivative is inconsistent with the data. Asai et al. (1991) have suggested that the period is varying sinusoidally with a period of approximately 12 years, almost twice the time that the source has been observed. We confirm their fit, and obtain a best fitted long period of 12.1 years and an amplitude of 56 lt-sec.

A number of cataclysmic variables have been observed with variations in orbital period that are also interpreted as sinusoidal variations, e.g., U Gem (18 yr, 1 cycle, Eason et al. 1983, Beuermann & Pakull 1984), IP Peg (5 yr, 1.5 cycles, Wolf et al. 1993), RW Tri (8 or 14 yr, 0.5 or 1 cycle, Africano et al. 1978), DQ Her (14 yr, 1.7 cycles, Patterson et al. 1978), UX UMa (29 yr, 1.5 cycles, Mandel 1965, Quigley & Africano 1978), EX Hyd (20 yr, 1.3 cycles, Bond & Freeth 1988). None of these systems have been observed for more than 1.7 cycles. The sinusoidal variations in cataclysmic variables are variously interpreted as due to the presence of a third body, loss of angular momentum from the system, exchange of orbital and spin angular momentum, apsidal motion, and motion of the "hot spot" relative to the two stars. Only the last can be safely eliminated for EXO0748-676.

4. STATISTICS OF THE O-C PLOT

An examination of the closely spaced EXOSAT eclipse timings shows that there is an rms residual of 3 seconds about any of the model ephemerides. This indicates that there may be intrinsic variability in the eclipse timings. The variable eclipse durations give further evidence of this. If there is some "phase jitter" in the eclipse timings, then the O-C residuals contain serially correlated errors and the statistics used to show that Porb is changing are invalid. The following comments are based on Lombard (1993) and Sterne (1934).

Let T_n be the time of eclipse egress for cycle number n. If the only uncertainty in the measurement of T_n is measurement error, then $T_n = F(n) + e_n$, where $F(n)$ is the ephemeris and e_n are independent identically distributed measurement errors. If this assumption is true, then the analysis in §2-3 is correct.

If there is intrinsic jitter in every cycle, then the instantaneous period during cycle n is $D_n = T_n - T_{n-1} = F(n) - F(n-1) + \varepsilon_n$, where ε_n is the jitter in cycle n. Now we have $T_n = T_0 + \sum_{i=1}^{n} D_n + e_n = F(n) + \sum_{i=1}^{n} \varepsilon_i + e_n$. The T_n are no longer independent, and they are no longer distributed about $F(n)$. They execute a random walk in O-C space, with the random walk controlled by the stochastic process $\{\varepsilon_i\}$. The random walk will move away from a constant period ephemeris and then back. If you fit an ephemeris to an O-C plot when the source is random walking, you will fit a sinusoid with a period approximately the length of the data base.

Is this reasonable? It is statistically. Lombard (1993) has worked out the cumulative statistics for the case when every cycle is observed and shown that several variable stars (RR Sco, T Cen, DY Peg) exhibit this behavior. The complete statistical description of the case where cycles are observed randomly, as for catalysmic variables and EXO0748-676, has not been worked out in full (Lombard & Koen 1993). The fact that the number of cycles observed is always

1-1.5, as well as the fact there are always rms residuals of 5-10 s over any model, support the intrinsic jitter hypothesis.

For EXO0748-676, we show an indicative plot. If the eclipse timing yields T_n at cycle n and T_m at cycle m, $m > n$, then the period residual is $R_m = (T_m - T_n) - (m - n)\overline{P}$. Here $\overline{P}$ is the mean period over the entire data base. In Figure 2 we plot R_m vs. m. Note that R_m is larger when $m - n$ is larger (long gaps between timings), and that the largest values of R_m increase as m increases. This behavior is expected for a random walk process. If there truly is a period derivative, then a regression of $(T_m - T_n)/(m-n) - \overline{P}$ vs. $(m+n+1)/2$ should have non-zero slope even in the presence of random walk noise. However the errors on the data points are highly correlated, and a standard least squares analysis can not be performed. The data are not good enough (yet) to detect a period derivative in the presence of intrinsic jitter.

REFERENCES

Africano, J. L., et al. 1978, PASP, 90, 568
Asai, K., et al. 1992, PASJ, 44, 633
Beuermann, K., & Pakull, M. W. 1984, A&A, 136, 250
Bond, I. A., & Freeth, R. V. 1988, MNRAS, 232, 753
Eason, E. L. E., et al. 1983, PASP, 95, 58
Lombard, F. 1993, presented at Applications of Time Series Analysis in Astronomy and Meteorology, Universita di Padova, Padua, Italy, 6-10 September 1993
Lombard, F., & Koen, C. 1993, in preparation
Mandel, O. 1965, Perem. Zvesdy, 15, 474
Parmar, A. N., et al. 1991, ApJ, 366, 253
Patterson, J., Robinson, E. L., & Nather, R. E. 1978, ApJ, 224, 570
Quigley, R., & Africano, J. 1978, PASP, 90, 445
Sterne 1934, Harvard College Observatory Circular, 366, 1
Wolf, S., et al. 1993, A&A, 273, 160

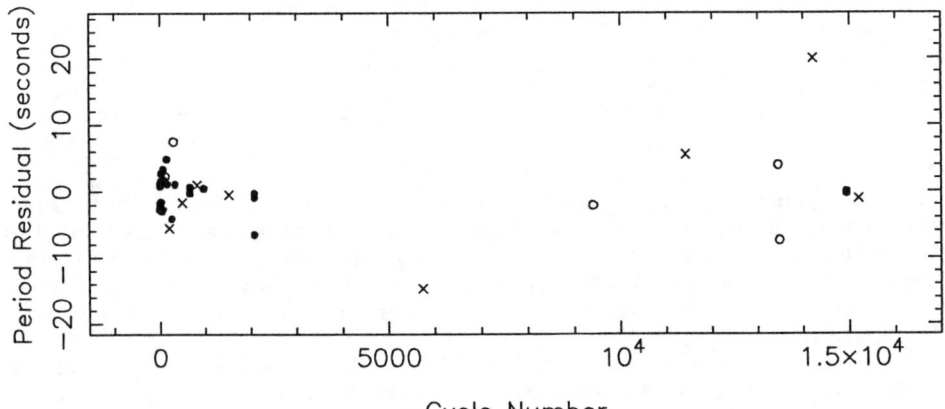

Figure 2: Period residuals vs. cycle number. Single cycle gaps ($m - n = 1$) are plotted with filled circles, short gaps ($m - n < 100$) are plotted with open circles, and long gaps ($m - n > 100$) are plotted with crosses.

UV Polarimetry of the X-Ray Binary Systems 4U1700-37, Vela XR-1, & Cygnus XR-1

Karen G. Wolinski[1,2], Joseph F. Dolan[1], Patricia T. Boyd[1,6],
James L. Elliot[4], Matthew J. Nelson[3], Jeffrey W. Percival[3],
L. Colleen Townsley[3], G. Wayne van Citters[5]

[1] Laboratory for Astronomy & Solar Physics,
NASA Goddard Space Flight Center, Greenbelt, MD 20771.
[2] Dept. of Physics, Purdue University, West Lafayette, IN 47907.
[3] Dept. of Astronomy & Space Astronomy Laboratory,
University of Wisconsin, Madison, WI 53706.
[4] Dept. of Earth, Atmospheric & Planetary Sciences,
Massachusetts Institute of Technology, Cambridge, MA 02139.
[5] Division of Astronomical Sciences, National Science Foundation,
Washington, D.C. 20550.
[6] Universities Space Research Association, Code 610.3, NASA/GSFC

ABSTRACT

Variable linear polarization has previously been observed in the visible from 4U1700-37 (Dolan & Tapia, 1984), Vela XR-1 (Dolan & Tapia, 1988), and Cyg XR-1 (Kemp et al., 1976, 1978). We present the first linear polarization observations of these systems in the ultraviolet (UV), using the High Speed Photometer (HSP) aboard the Hubble Space Telescope. The UV polarimetric light curves obtained from Vela XR-1 and 4U1700-37 display an added component of polarization at first quadrature inconsistent with the standard model of Brown et al., (1978) for scattering in binary systems. However, the UV polarimetric light curves toward Cyg XR-1 are well fit by the standard model. Variations attributable to changes in scattering structure from one orbit to the next are also detected in two of the systems. Possible origins of the additional polarization detected at first quadrature in Vela XR-1 and 4U1700-37 are evaluated.

1. INTRODUCTION

It has already been shown that linear polarization in many X-ray binaries varies with orbital phase, ϕ. These variations distinguish polarization which is intrinsic to the system from the non-varying interstellar component. The mechanism producing the largest intrinsic polarization is scattering of light from the primary by gas streams between the stars. Other factors such as stellar winds and the presence of an accretion disk around the secondary may also contribute to the intrinsic polarization detected in these systems. The tidal distortion of the primary contributes a negligible amount to the total polarization (Bochkarev et al., 1986, Dolan 1992). The three systems whose polarization in the UV we report here are 4U1700-37, Vela XR-1, and Cyg XR-1.

2. OBSERVATIONS

4U1700-37 was observed by the HSP on August 24 to 27, 1993. A description of the instrument can be found in Bless et al. (1992). The eight observations obtained in a bandpass centered at 2770 Å were made approximately equally spaced throughout the 3.411 d orbital period. The error bars associated with the polarimetric and photometric data represent $\pm 1\sigma$ uncertainties derived from photon statistics plus systematic errors associated with HST operations (see Dolan et al. 1993). Vela XR-1 was observed between January 4 and 13, 1993. Simultaneous photometry (at 2480 Å) and linear polarization measurements (at 2770 & 3270 Å) were made twice each day, covering ~ 1.10 orbits of the 8.96 d period. Cyg XR-1 (5.6 d period) was observed polarimetrically at 2770 Å between Aug. 5 and 11, 1993.

3. ANALYSIS

The Stokes parameter light curves were analyzed by the method of Brown et al. (1978). For binary systems with small eccentricities ($\epsilon \leq 0.1$), and linear polarization resulting from single scattering of radiation from the primary, the Stokes curves may be represented by a Fourier series with no terms beyond the second order in orbital longitude, λ (where $\lambda = 2\pi\phi$) (Brown et al., 1978; Rudy & Kemp, 1978). Previous studies of X-ray binaries with OB supergiant primaries show evidence for changes in the scattering structure surrounding the stars on time scales as short as 10 days (Dolan & Tapia, 1984, 1988). Since the distribution of scattering material must remain fixed in the coordinate system corotating with the stars under the assumptions of the standard model, polarimetric measurements covering more than one orbit cannot necessarily be averaged together when deriving the Fourier series fit to the observed Stokes (q, u) curves.

The best fit truncated Fourier series which yielded an acceptable value for χ^2 is plotted along with the observed q, u in Fig.1 for 4U1700-37. An acceptable fit was obtained only when the set of observed data did not contain the observation done at first quadrature ($\phi \sim .25$). Since the eight observations form a contiguous set over one orbit, it is unlikely that the large deviation from the fit at first quadrature is due to a very large and rapid change in the scattering geometry at $\phi \sim .25$. An added component of polarization (not accounted for in the standard model for scattering in binary systems) is present at this phase. The results for Vela XR-1 are shown in Figs.2 & 3. The acceptable fit to these data was obtained when the first two observations (at phases $\phi \sim .70, .79$) and those at $\phi \sim .24$ were removed. Since the set of observations spanned slightly more than one orbit (beginning at $\phi = .70$, ending at $\phi = .76$) it is apparent that the scattering geometry changed over the 10 days. The excursion from the fit at $\phi = .24$ is similar to the results seen in 4U1700-37. An additional component of polarization is present at first quadrature in this system as well. The Stokes q, u curves for Cyg XR-1 are seen in Fig.4. The amplitude of polarimetric variability is not as large as seen in the previous two systems. These data are well fit by the standard model if the point at $\phi \sim .06$ is not included in the data set. This observation was taken last in the series, and probably represents a change in scattering geometry.

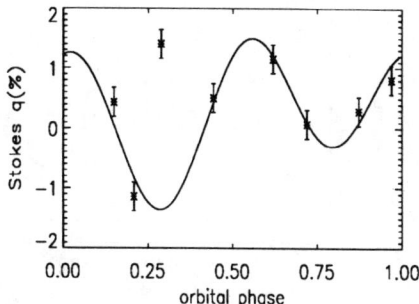
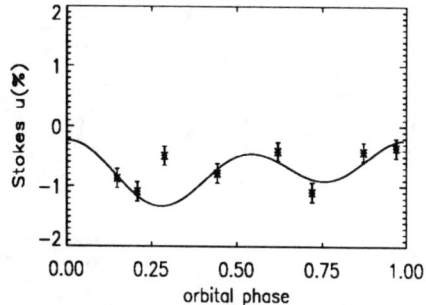

Fig. 1 Stokes (q, u) (equatorial coordinate system) at 2770 Å for 4U1700-37.

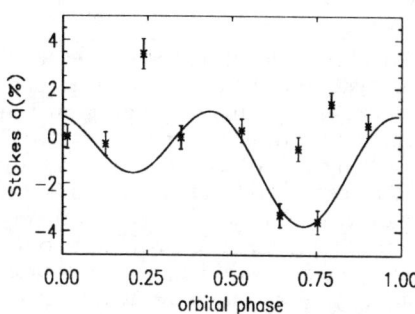
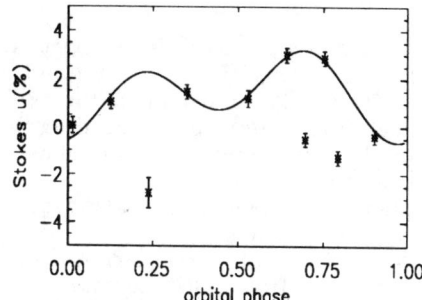

Fig. 2. Stokes (q, u) (equatorial system) at 2770 Å for Vela XR-1.

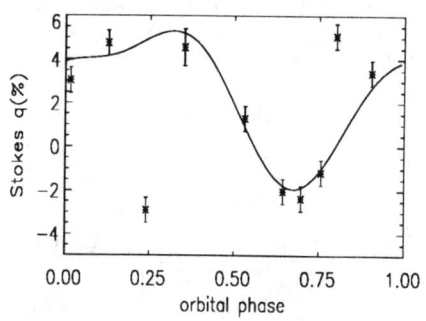
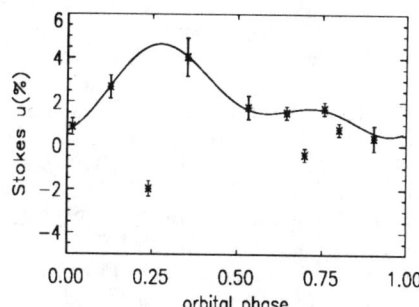

Fig. 3. Stokes (q, u) (equatorial system) at 3270 Å for Vela XR-1.

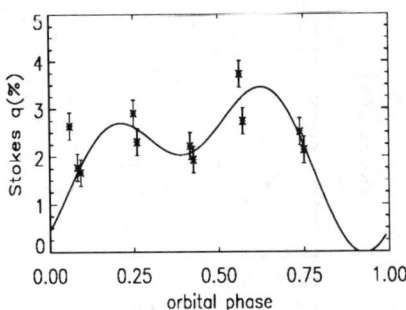 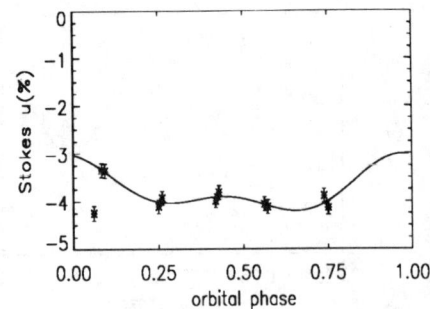

Fig. 4. Stokes (q, u) (equatorial system) at 2770 Å for Cyg XR-1.

4. DISCUSSION

Several possibilities exist which could produce the observed increase in linear polarization at first quadrature for the systems with neutron star secondaries. The most plausible explanation is that gas streams impinging on the accretion structure surrounding the neutron star may produce a small "hot spot" in the outer regions of the accretion disk, resulting in polarized bremstralung radiation. Other possibilities include a polarization "flare", or some transient phenomena evidenced by an increase in UV polarization. These seem unlikely since the increase is seen for two different systems. Cyg XR-1 & LMC X-3, both with black hole secondaries, do not show evidence of an additional component of polarization. The accretion structures in these systems may be different from those for Vela XR-1 and 4U1700-37.

REFERENCES

Bless, R.,C., Percival, J.W., Walter, L.E., & White, R.L., 1992, HSP Instrument Handbook, Space Telescope Science Institute, Baltimore, MD
Bochkarev, N.G., Karitskaya, E.A., Loskutov, V.M., & Sokolov, V.V., 1986, Sov. Ast., 30, 43.
Brown, J.C., McLean, I.S., & Emslie, A.G., 1978, A&A, 68, 415
Dolan, J.F., et al., 1993, in preparation
Dolan, J.F., 1992, ApJ, 394, 249
Dolan, J.F., & Tapia, S., 1984, A&A, 139, 249
Dolan, J.F., & Tapia, S., 1988, A&A, 202, 124
Kemp, J.C., Barbour, M.S., Herman, L.C., & Rudy, R.J., 1978, ApJ, 220, L123
Kemp, J.C., Southwick, R.G., & Rudy, R.J., 1976, ApJ, 210, 239
Rudy, R.J., & Kemp, J.C., 1978, 221, 200

LITHIUM IN QUIESCENT X-RAY NOVAE

P.A. Charles[1], J. Casares[1,2] E.L. Martín[2] & R. Rebolo[2]
[1] Department of Astrophysics, Nuclear Physics Lab., Oxford, OX1 3RH, UK
[2] Instituto de Astrofísica de Canarias, E-38200 La Laguna (Tenerife), Spain
pac@ox3500.astro.ox.ac.uk; jcv@ox3500.astro.ox.ac.uk; 28844::ege; 28844::rrl

1. INTRODUCTION

The soft X-ray transients (SXTs; often referred to as X-ray Novae) are a sub-class of low-mass X-ray binary (LMXB) which are discovered during their bright X-ray outbursts (typically $L_x \sim 10^{38-39}$ erg s^{-1}). These last only a few months and are followed by quiescent intervals of 10–50 years. There are ~10 known and optically identified, all of them fainter than V=18 in quiescence (but 4–7 mags brighter at peak of outburst) and all with spectral types around K0–K5 (see van Paradijs & McClintock, 1993 and references therein). The prototype system is A0620–00 (V616 Mon), and several SXTs have dynamical masses resulting from radial velocity studies of the secondary star indicating they are likely to be black-hole systems, the best of which is GS2023+338 (V404 Cyg) with a mass function of $6.1 M_\odot$ (see Casares & Charles, 1993).

A by-product of these radial velocity studies has been the surprising discovery of strong LiIλ6707 absorption in the SXTs V404 Cyg (Martín et al, 1992) and V616 Mon (Marsh et al, 1993). Normally stars with the spectral types of our secondaries only show Li if they are pre-main sequence or T Tau stars. Li is important because of its production in standard Big Bang nucleosynthesis. Halo dwarfs have been used to set the primordial abundance level of $\log N(Li)$=2.05, whereas studies of the solar neighbourhood (ISM, meteorites) and T Tau stars yield $\log N(Li)$=3.1. *i.e.* the galactic gas is enriched in Li, and has been for $>5.10^9$ years. Spallation processes (*e.g.* cosmic rays in the ISM) have been proposed to explain this enrichment (*e.g.* Walker et al, 1985), but the models fall short of the required quantities of Li. There must be other sites of Li production in the Galaxy, and the observations summarised here suggest that SXTs and X-ray binaries could be ideal candidates.

2. OBSERVATIONS AND ANALYSIS

High resolution (1.5Å) spectra around Hα have been obtained from La Palma (with the 4.2m WHT) and the AAT of V404 Cyg, V616 Mon and V822 Cen (Cen X-4). These are shown in figure 1 with the scale expanded so as to bring out the late-type features of the secondary stars. The LiIλ6707 absorption is clearly visible in each. The equivalent widths are summarised below in table 1 together with the abundances derived from LTE and non-LTE analyses (for full details see Martín et al, 1994). The T_{eff} values are derived from the spectral types, and the veiling factors (where the disc continuum dilutes the absorption spectrum) are negligible for V404 Cyg and V616 Mon, but not for V822 Cen.

The results show that V822 Cen and V404 Cyg have similar Li abundances, whereas that in V616 Mon is significantly less.

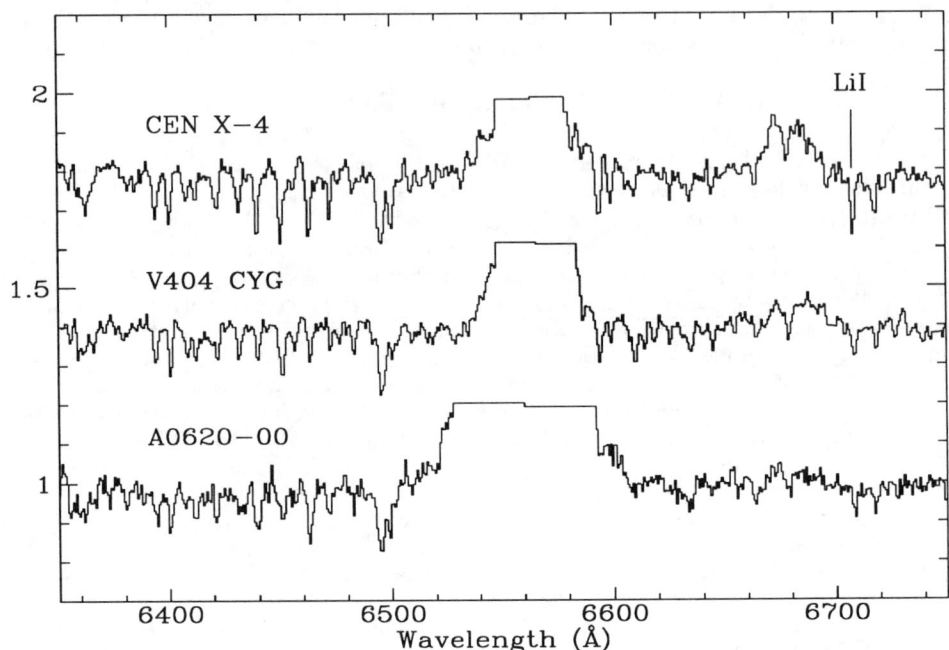

Fig. 1. High resolution spectra of V0822 Cen (Cen X-4), V404 Cyg and V616 Mon (A0620-00) showing LiIλ6707 in each.

Table 1: Surface abundances of 3 SXTs

Name	Sp. Type	T_{eff} K	Veiling	W_λ (LiI) mÅ	log N_{Li} LTE	log N_{Li} NLTE	[Ca/H]
V822 Cen	K5-K7 V	4250	25%	480±30	3.3	3.1	0.0
V404 Cyg	G9-K1 IV	4750	5%	290±40	2.6	2.7	+0.1
V616 Mon	K3-K5 V	4500	6%	250±50	2.0	2.1	-0.1

3. FORMATION AND EVOLUTION

It is well known that low mass stars deplete Li during the early stages of their evolution, and we find that the Li abundance in V404 Cyg and V606 Mon is comparable to that of α Per and the Pleiades, whereas V822 Cen is similar to T Tau stars (see García López, Rebolo & Martín 1993). By the age of the Hyades the Li abundance is much lower, which implies that either SXTs are young ($< 10^7$ years for V822 Cen!) or we need a mechanism for producing the

Li. Such a young age is unlikely given current ideas on formation mechanisms for LMXBs (massive primary in a wide binary; accretion-induced collapse of a white dwarf in a cataclysmic variable; capture of the secondary star by the compact object; see e.g.Verbunt 1993).

It had been suggested by Chevalier et al (1989) that V822 Cen underfills its Roche lobe and transfers material to the compact object by means of a stellar wind. However, ellipsoidal modulations (McClintock & Remillard, 1990; Shahbaz et al, 1993) show it is a sub-giant which is (at least close to) filling its Roche lobe (and there is similar evidence for V404 Cyg and V616 Mon). Also, Marsh et al (1993) derive a companion mass of $M_c \sim 0.2$–0.4 $M_\odot$, as expected for an age of $\sim 10^9$ years (a young K5 star would have a mass of 0.7 $M_\odot$). Hence it is very unlikely that all three SXTs have age $\leq 10^7$ years, and the high Li abundances are therefore surprising. In late spectral types Li is rapidly burnt by proton bombardment as convection takes the surface Li into high temperature regions. And of course mass transfer should deplete it even faster.

We consider three possible sites for the creation of Li. (1) in X-ray bursts on the surface of the neutron star (as the thermonuclear runaway can create Li too, as pointed out by Starrfield et al, 1978). But V404 Cyg and V616 Mon are almost certainly black holes, and so this could only apply to V822 Cen (as X-ray bursts have been seen from Cen X-4). (2) in the supernova explosion when the compact object was initially created (Woosley et al, 1990). But how then is it preserved against destruction by the normal processes described above? (3) during the X-ray outbursts themselves, as this is the only property that the three SXTs have in common (their periods cover a wide range from 0.32 to 6.5 days; two are black holes, one a neutron star).

4. X-RAY OUTBURSTS

The X-ray active phase of LMXBs is estimated to be $\sim 10^7$ years (van den Heuvel, 1992), with $\sim 0.1 M_\odot$ transferred in that time. In fact, the outbursts are $\sim L_{Edd}$ which leads to substantial mass outflow from the system as a whole (as seen in V404 Cyg by Casares et al, 1991). Now if the mean $\dot{M}=10^{-8} M_\odot$ yr^{-1} and 10% of this is ejected, and $\sim (R_s/a)^2$ is accreted by the secondary, then $\sim 10^{-4} M_\odot$ of the expelled material is ultimately transferred to the secondary star. Using Ryter et al's (1970) expression relating the Li abundance to L_X (taken to be $L_X=10^{38}$ erg s^{-1}) we calculate that the transfer and mixing of the Li in the secondary will lead to an abundance of log N(Li)=2.8, very close to that observed. If this process works in SXTs, then it should also work in massive X-ray binaries, such as Cyg X-1, which have a birthrate one hundred times higher than LMXBs.

5. TEST OF LITHIUM PRODUCTION MECHANISM

During the spallation processes (presumably occuring in the inner accretion disc, and then driven off by radiation pressure) we would expect the ^{7}Be and ^{7}Li lines at 431 and 478 keV to be produced (as pointed out by Martín et al, 1992). Now, during the outburst of Nova Mus 1991, a γ-ray line at 476±15 keV was observed and interpreted as a gravitationally-redshifted e$^-$-e$^+$ annihilation line, double-peaked due to the Keplerian rotation of the disc (Sunyaev et al, 1992; Chen et al, 1993). The narrow width is explained by the annihilation taking place in a cool disc. However, an alternative explanation is that the line could be partly due to lithium and beryllium de-excitation if $\alpha + \alpha$ spallations did take place in the vicinity of the compact object.

REFERENCES

Casares, J. & Charles, P.A. 1993 (these proceedings)
Casares, J., Charles, P.A., Jones, D.H.P., Rutten, R.G.M. & Callanan, P.J. 1991, MNRAS, 250, 712
Chen, W., Livio, M., & Gehrels, N. 1993, ApJ, 408, L5
Chevalier, C., Ilovaisky, S.A., van Paradijs, J., Pedersen, H. & van der Klis, M. 1989, A&A, 210, 114
García López, R.L., Rebolo, R. & Martín, E.L. 1993, A&A, in press
Marsh, T.R., Robinson, E.L. & Wood, J.H., 1993 MNRAS, in press
Martín, E.L., Rebolo, R., Casares, J. & Charles, P.A., 1992, Nature, 358, 129
Martín, E.L., Rebolo, R., Casares, J. & Charles, P.A., 1994, ApJ, submitted
McClintock, J.E. & Remillard, R.A. 1990, ApJ, 350, 386
Ryter, C., Reeves, H., Gradsztajn, E. & Audouze, J. 1970 A&A 8, 389
Shahbaz, T., Naylor, T. & Charles, P.A. 1993, MNRAS, in press
Starrfield, S., Truran, J.W., Sparks, W.M. & Arnould, M. 1978, ApJ, 222, 600
Sunyaev, R. et al 1992, ApJ, 389, L75
van den Heuvel, E.P.J. 1992, in X-ray binaries and recycled pulsars, eds. E.P.J. van den Heuvel & S.A. Rappaport, p. 233
van Paradijs, J. & McClintock, J.E. 1993, in X-ray Binaries, eds. W.H.G. Lewin, J. van Paradijs & E.P.J. van den Heuvel, C.U.P.
Verbunt, F. 1993, ARA&A, 31, 93
Walker, T.P., Mathews, G.J. & Viola, V.E. 1985, ApJ, 299, 745
Woosley, S.E., Hartmann, D.H., Hoffman, R.D. & Haxton, W.C. 1990, ApJ, 356, 272

THE ORBITAL LIGHTCURVE OF PSR1957+20

Paul J Callanan
Center for Astrophysics, 60 Garden Street, Cambridge

Jan van Paradijs and Roeland Rengelink
Astronomical Institute 'Anton Pannekoek', Amsterdam, The Netherlands

ABSTRACT

We make a detailed study of the optical light curve of PSR1957+20. We show that the deep, smooth and symmetrical modulation can be successfully modelled by a highly irradiated secondary. Two types of models are consistent with the data *(1)* a secondary close to filling its Roche lobe with 10-20% of the incident flux converted to optical emission, and *(2)* a secondary considerably underfilling its Roche lobe, and a high degree of beaming of the neutron star flux. However, we exclude the highly beamed model on the basis of current estimates of the extinction towards 1957+20. The proximity of the photosphere of the secondary to its Roche lobe facilitates mass loss by irradiation. An inclination as low as 55 degrees is compatible with our data: however, our ignorance of the (non-irradiated) luminosity of the secondary allows us to place only weak constrains on the inclination – 55-80 degrees. There is no evidence in the lightcurve for optical emission from any source in PSR1957+20 other than the secondary itself.

1. INTRODUCTION

PSR1957+20 occupies a unique place in our understanding of the evolution of binary to isolated millisecond pulsars (Fruchter, Stinebring and Taylor 1988a). Is is the only known binary where the effect of irradiation by the radio pulsar on the secondary can be directly observed: the optical counterpart varies by ≥ 3 magnitudes in R throughout the 9.17 hour orbital cycle (Fruchter et al. 1988b, van Paradijs et al. 1988). At radio wavelengths, emission from the pulsar is eclipsed for ~ 10 per cent of the orbital cycle: the eclipsing region is much larger than even the Roche lobe of the secondary, implying that a wind of material is being driven from the secondary by irradiation from the neutron star. Although estimates for the exact timescale vary, it seems plausible that the flux from the neutron star will eventually evaporate the secondary entirely, leaving an isolated millisecond pulsar (e.g. van Paradijs and van den Heuvel 1988).

Efforts to understand the detailed physics of this evaporation mechanism have been hampered by the considerable uncertainties which still exist in many of the basic parameters of this system. To better determine the lightcurve, and constrain the parameters of PSR1957+20 (e.g. the orbital inclination, the flux incident on the secondary, the fraction of Roche lobe filled by the secondary) we obtained 4 nights of photometry using the 4.2m William Herschel Telescope (WHT) on La Palma.

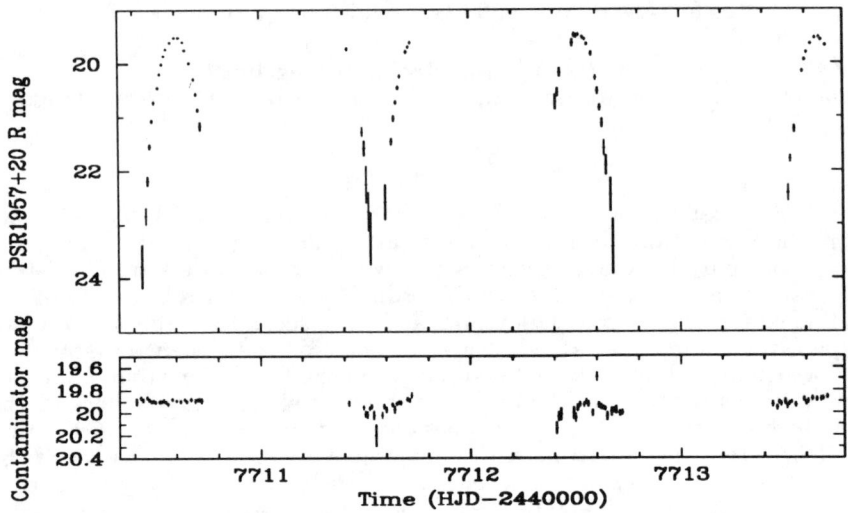

Figure 1. The Orbital Light curve of PSR1957+20

2. OBSERVATIONS

We observed PSR1957+20 with the WHT on the nights of 1989 July 2,3,4 and 5 with a GEC (P8603) chip at the Cassegrain focus of the telescope. Using Taurus-2 as a focal reducer yielded a plate scale of 0.27 arc sec/pixel. We cycled through B, V and R (Johnson) filters at maximum light, and through V and R only through minimum. The data were debiased and flatfielded, and DAOPHOT (Stetson 1987) was used to perform photometry of the pulsar relative to a series of nearby stars. The B, V and R band magnitudes of the PSR1957+20 at maximum were found to be 21.08, 20.16 and 19.53 respectively (±0.05).

In Fig 1 (upper panel) we show the total R band lightcurve of PSR1957+20. For comparison, we plot in the lower panel the lightcurve of the nearby "contaminator star" (a line-of-sight interloper only 0.7 arc sec away from PSR1957+20). In Fig 2 we present the R band lightcurve of the pulsar obtained on the nights of July 2 and 5, folded on the radio ephemeris of Ryba and Taylor (1992): for these nights ~0.8 arc second seeing and especially stable atmospheric conditions allowed us to easily resolve the pulsar from its contaminator. Superimposed on the data is the best fit model discussed in Section 3. We do not detect the pulsar at minimum, to a limiting magnitude of ~24 in R.

3. DISCUSSION

It is clear from our data that the pulsar exhibits negligible intrinsic short-term variability either within an orbital cycle, or from one to another, to within our errors of measurement. This is in stark contrast to most other interacting binary systems (e.g. cataclysmic variables and X-ray binaries), where such vari-

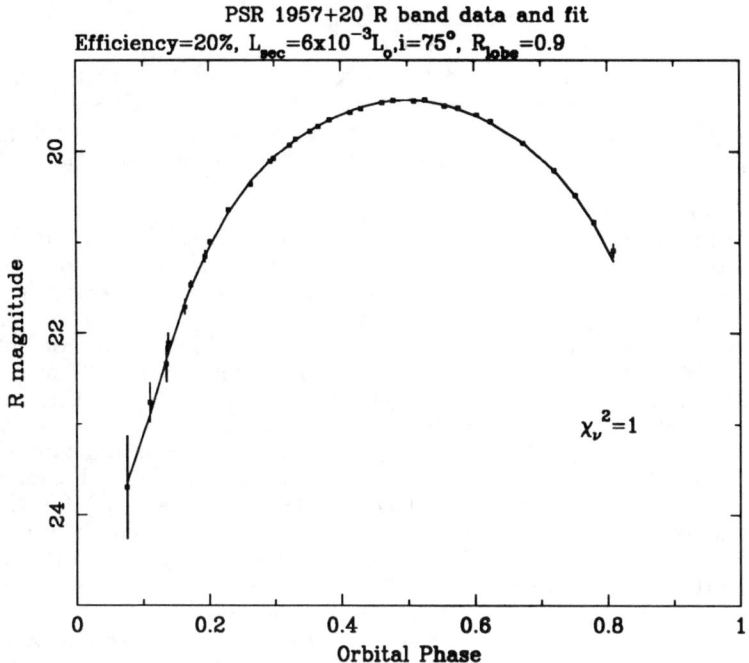

Figure 2. The 1989 July 2 & 5 lightcurve of PSR1957+20 folded on the radio ephemeris (and shifted by 0.25 in phase for consistency with standard nomenclature). The best fit model is superimposed.

ability is ubiquitous. This is consistent with the scenario of simple irradiation of the secondary by the spindown luminosity of the pulsar. The stability of the lightcurve provided us with a strong motivation to model our highest quality data in detail.

The data were fitted to a simple model of an irradiated secondary. We first fitted the R and V-R data from the July 2 and 5 observation, and then checked for consistency with the entire dataset. Four parameters were varied during the fit as follows:

- the fraction of its Roche lobe filled by the secondary: $0.5 < R_{sec} < 1.0$
- the (un-illuminated) luminosity of the secondary: $2.5 \times 10^{-5} L_\odot < L_{sec} < 7.5 \times 10^{-3} L_\odot$
- the orbital inclination of the binary (i): $45 < i < 90$ degrees, and
- the fraction of the neutron star spindown luminosity (1.5×10^{35} ergs s^{-1}) converted into optical emission (the efficiency): $0.1 < \eta < 5.0$

The maximum value of L_{sec} was determined by the amplitude of the optical variability (for $R_{sec}=1$ and i=90). Values of $\eta > 1$ were used to simulate the effect of beamed emission from the neutron star.

For each angle of inclination, the mass of the secondary and the size of its Roche lobe was recomputed using the radio data of Fruchter et al (1988). See Callanan et al (1993) for further details.

3.2 RESULTS

We find that the best fit models (with χ^2_ν =1.2 or better) fall into two categories: *(a)* relatively low values of η ($\sim$10-20 %) and high values of R_{sec} ($\lesssim$1.0) and *(b)* high values of η (3 or more) and low values of R_{sec} ($\sim$0.5). The case *(a)* fits imply that the secondary is close to, but not quite filling, its Roche lobe.

We further constrain the binary parameters as follows:

The Efficiency: The case *(b)* fits imply a high degree of beaming. Indeed, Manchester and Taylor (1977) show that the emission from only one out of every 5 pulsars is beamed towards the Earth: however, subsequent work has shown that for the fastest spinning pulsars, this ratio approaches unity (e.g. Narayan and Ostriker 1990). We believe we can exclude such a high degree of beaming in the case of PSR1957+20: values of $\eta > 3$ imply an absolute R magnitude (M_R) $\sim$6. For a distance of 1.6 kpc (Fruchter and Goss 1991), this requires an R band extinction of $\sim$2.5 magnitudes. Such a large visual extinction is inconsistent with the estimate of A_v <1.0 by Aldcroft, Romani and Cordes (1992), made on the basis of optical spectroscopy of the bow shock nebula. Hence we will only consider the case *(a)* fits (low η, large R_{sec}) for the rest of the discussion. In Fig 2 we have superimposed the lightcurve and colour variation predicted by the model (for a case *(a)* fit and L_{sec} =6.4x10$^{-3}L_\odot$).

The Distance: For a given η, A_v is determined from the observed value of $V - R$ at maximum. η also determines M_R (again at maximum), and hence the distance (D) to the pulsar can also be constrained. These estimates are essentially independent of L_{sec} (because of the high degree of heating at maximum light) and i (because of the increase in size of the Roche lobe with decrease in inclination). Taking E_{V-R}=0.26 A_v (Allen 1973), we find that the observed $V - R$ colour of 0.7±0.1 implies 10%< η <20%. Corresponding values of A_v are 0<A_v <0.8. The absolute R band magnitude predicted by the models for these efficiencies then yield a distance range of 1.6< D <3 kpc. This distance estimate is consistent with that of Fruchter and Goss (1991).

The Inclination: Unfortunately, our limits on the orbital inclination (i) are strongly dependent on the assumed value of L_{sec}. Decreasing L_{sec} lowers the minimum acceptable inclination. For L_{sec} =6.4x10$^{-3}L_\odot$ (the maximum allowed for an acceptable χ^2), i=75±5 (90% confidence) degrees. Acceptable fits are also obtained for L_{sec} =0, in which case i=55±5 degrees.

REFERENCES

Allen C.W., 1973, Astrophysical Quantities, The Athlone Press.
Aldcroft T.L., Romani R.W., Cordes J.M., 1992, ApJ, 400, 638
Callanan P.J., van Paradijs, J., Rengelink, R., 1993, MNRAS, submitted
Fruchter A.S., Stinebring D.R., Taylor, J.H., 1988a, Nat, 333, 237
Fruchter A.F., Gunn J.E., Lauer T.R., Dressler A., 1988b, Nat, 334, 686
Fruchter A.S., Goss W.M., 1991, ApJ, 384, L47
Manchester R.N., Taylor, J.H., 1977, in Pulsars, (San Francisco: Freeman), p18
Narayan R., Ostriker J.P., 1990, ApJ, 352, 222
Ryba M.F., Taylor J.H., 1992, ApJ, 380, 560
Stetson P.B., 1987, PASP, 99, 443
van den Heuvel E.P.J., van Paradijs J., 1988, Nat, 334, 227
van Paradijs et al., 1988, Nat, 334, 684

THE ORBITAL PERIOD DERIVATIVE OF PSR 1957+20

P. McCormick and J. Frank
Department of Physics and Astronomy,
Louisiana State University, Baton Rouge, LA 70803
cormick@rouge.phys.lsu.edu

A. R. King
Astronomy Group, The University Leicester LE1 7RH, UK

and

A. Rajasekhar
Department of Physics and Astronomy,
Louisiana State University, Baton Rouge, LA 70803

ABSTRACT

In light of recent observations of PSR 1957+20, we suggest a mechanism in which the mass loss by the secondary carries away varying degrees of specific angular momentum. We present a model which incorporates Roche–lobe overflow and wind driven mass loss. In our model, Roche-lobe overflow of the companion produces an evaporating disk or ring. An external disk around the binary may also be created by the mass loss. Changes in the structure of the accretion disk and external disk produce variations in the specific angular momentum carried away by the mass loss. These variations are likely to occur over a viscous timescale. This mechanism could account for the dramatic changes of $\dot{P}_b$ observed and predicts alternating cycles of expansion and contraction of the binary orbit.

1. INTRODUCTION

PSR 1957+20 is a member of a particularly interesting subgroup of 4 binary pulsars with orbital periods $P \lesssim 12$ h, very low mass functions and circular orbits. In two cases (1957+20 and 1744-24A) the occurrence of eclipses and accompanying changes in dispersion measure suggest that the pulsar is evaporating mass from the companion star (Fruchter, Stinebring & Taylor 1988; Ryba & Taylor 1991; Lyne et al. 1990; Thorsett & Nice 1991), and there is good reason to believe that this is occurring in 0021-72J also (Manchester et al 1991). The recently discovered fourth member of this class (1908+00) is very similar to 0021-72J and may have eclipses shorter than the integration time (Deich et al. 1993).

An estimate of the evolutionary timescale of PSR 1957+20 is provided by the observation (Ryba & Taylor, 1991) that between 1988 and 1991 its orbital period decreased on a timescale $t_P = P_b / - \dot{P}_b \simeq 2.7 \times 10^7$ yrs. Recently, Arzoumanian, Fruchter & Taylor (1993) have anounced that the orbital period is now *increasing* on a similar timescale. We consider here the implications of these results for the orbital evolution of PSR 1957+20, and we give a more general discussion of the origin and evolution of these systems in another poster (McCormick et al. 1993).

2. ROCHE–LOBE OVERFLOW AND ORBITAL EVOLUTION

We consider a detached companion of small mass, not far from filling its Roche lobe but losing mass through evaporation. We assume that the specific binding energy and angular momentum carried away by the evaporating wind are such that the orbit shrinks and remains circular. The evolution of the orbital period for arbitrary eccentricity depends solely on energy losses but for circular orbits it is convenient to parametrize the evolution in terms of β, the specific angular momentum of the wind in units of the specific angular momentum of the companion, as in Czerny & King (1988). Thus, if no Roche–lobe overflow occurs, the rate of change of the orbital period in a binary driven by evaporation and gravitational radiation is

$$\frac{\dot{P_b}}{3P_b} = \frac{\dot{M_2}M_1}{M_2 M}(\beta - \beta_0) + \left(\frac{\dot{J}}{J}\right)_{GR}. \qquad (1)$$

where M_1, M_2, and M are the masses of the pulsar, its companion, and the whole system respectively, and $\beta_0 = 1 + 2M_2/3M_1$. It is easy to see that in PSR 1957+20 wind losses must dominate to produce period changes on the observed timescale, and therefore $\beta > \beta_0$ is required for the orbit to shrink (see also Banit & Shaham 1992).

Since a low–mass companion will expand under near–adiabatic mass loss while the system shrinks, contact is inevitable. It is therefore natural to consider the possibility of wind–driven Roche–lobe overflow for PSR 1957+20 and related systems. We assume a partially degenerate low–mass secondary filling its Roche–lobe and split the mass loss into wind losses at a given rate $\dot{M}_w$, and Roche–lobe overflow losses at a rate $\dot{M}_{RL}$. We assume that the pulsar is capable of evaporating and expelling from the system both wind and overflow plasma in such a way that $\dot{M} = \dot{M_2} = \dot{M}_w + \dot{M}_{RL}$. The material that flows through the inner Lagrangian point (L_1) will already have a lower specific angular momentum than the secondary as a whole. We parametrize the angular momentum of the expelled overflow material by β_{RL}, while the wind material will be assumed to possess a specific angular momentum β_w. If an accretion disk or annulus forms and spreads by viscosity down to a radius r_{in} before it is blown away, tides will return the excess angular momentum to the orbit. The further in the overflow material spirals before it is expelled by the pulsar wind, the lower β_{RL} will be. If r_{in} is well inside the primary Roche lobe, then $\beta_{RL} \simeq (r_{in}/a)^{1/2}$, where a is the binary separation. Under these assumptions, no accretion occurs and the pulsar is not quenched provided that the overflow material never reaches the light cylinder.

3. SHORT–TERM FLUCTUATIONS OF $\dot{P_b}$

The timescale for changes in overflow rate is the time it would take $|R_2 - R_L|$ to change by an amount comparable to the pressure scale height in the atmosphere of the companion. Any variations in wind losses caused by changes in the degree of lobe filling would also take place on that timescale. For the parameters of PSR1957+20 this timescale is $t_{RL} \sim 10^3$ yrs. To change $\dot{P_b}$ on the observed timescale $t_{obs} \ll t_{RL}$ one therefore either has to change the properties of the wind by modulating β_w and/or $\dot{M}_w$, or to change β_{RL}, on the timescale t_{obs}. In both cases the overflow rate will remain constant, but the period derivative of the binary will change observably in response to the changes in the angular momentum balance. We assume that a disk forms and spreads inwards, while material is evaporated mostly from the inner boundary of this disk, carrying away an angular momentum flux characterized by β_{RL}. As $\dot M_{RL}$ increases, we expect r_{in} and β_{RL} to decrease until stable mass transfer is reached with $\beta_{RL} < \beta_{crit}$.

It is straightforward to derive an expression for the rate of change of the orbital period under the above assumptions, generalizing equation (1) for arbitrary wind and overflow mass-loss rates:

$$\frac{\dot P_b}{3P_b} = \left(\frac{\dot J}{J}\right)_{GR} + \frac{M_1}{M}\frac{\dot M_{RL}}{M_2}(\beta_{RL} - \beta_0) + \frac{M_1}{M}\frac{\dot M_w}{M_2}(\beta_w - \beta_0). \quad (2)$$

Let us now consider what happens if β_w and/or β_{RL} change on a timescale of a few years. (Such changes might for example result from instabilities in the outflow or variations of β_{RL} following viscous outbursts in the disk.) Since this time is $\ll t_{RL}$, neither the binary separation nor the degree of Roche-lobe filling can change significantly, leaving $\dot M_2$ and its division into wind and overflow rates essentially unchanged too. However, $\dot P_b$ *can* change, as it responds dynamically to changes in the angular momentum balance.

The effect of variations in β_w and β_{RL} can be seen directly from (2) by substituting the equilibrium mass-loss rates and letting β_{RL}, β_w take "instantaneous" values $\beta_{RL}^i = \beta_{RL} + \Delta\beta_{RL}$, $\beta_w^i = \beta_w + \Delta\beta_w$. After some algebra one obtains

$$\frac{\dot P_b}{P_b} = -\frac{\dot M_2}{M_2} + \frac{3M_1\dot M_2}{MM_2}\left(\frac{\beta_w - \beta_{crit}}{\beta_w - \beta_{RL}}\Delta\beta_{RL} + \frac{\beta_{crit} - \beta_{RL}}{\beta_w - \beta_{RL}}\Delta\beta_w\right)$$
$$+ \frac{3(\Delta\beta_{RL} - \Delta\beta_w)}{\beta_w - \beta_{RL}}\left(\frac{\dot J}{J}\right)_{GR}. \quad (3)$$

For the evaporation rates and period changes observed in PSR 1957+20 the last term in (3) must be negligible and the sign of $\dot P_b$ is determined by the first two terms. Clearly if $\Delta\beta_{RL} = \Delta\beta_w = 0$, the period increases because of the expansion of the secondary, as already discussed.

4. DISCUSSION

This model can give a quantitative explanation of the behavior of the orbital period of PSR 1957+20. Taking the steady mass-loss rate as 5×10^{16} g s^{-1}, the average expansion proceeds with $\dot P_b \approx +3.3 \times 10^{-11}$. Over the past five or six years $\dot P_b$ has reached values of $\pm 7 \times 10^{-11}$, so equation (3) sets limits on the required instantaneous values. Fig. 1 shows an example of how fluctuations of angular momentum losses around an assumed equilibrium could explain the magnitude of the observed fluctuations of orbital period derivative. To attain the values of β implied by Fig. 1, viscous instabilities in the accretion disk and the

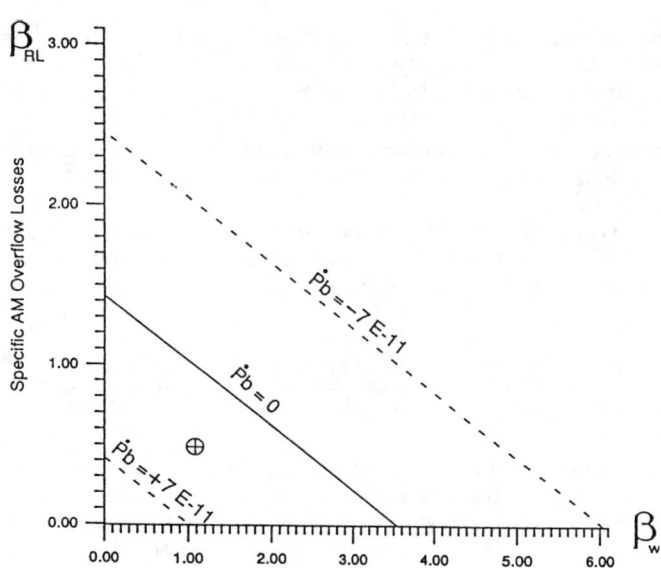

Fig. 1: The allowed ranges of β_w^i and β_{RL}^i is shown for $\dot{M} = 5 \times 10^{16}$ g/s and for a particular choice of equilibrium parameters β_w and β_{RL} indicated by a $\oplus$. The $\dot{P}_b = 0$ locus is indicated by a solid line and the dotted lines show the loci of β_w^i and β_{RL}^i which yield the observed extreme values of $\dot{P}_b$.

external disk may be necessary. A variety of solutions are possible with different mass loss rates 10^{16-17} g/s. Lower rates require specific angular momentum losses too high to be physical, and higher rates conflict with observational limits.

This research was partially supported by NASA grant NAGW-2447 and NSF grant AST-9020855.

REFERENCES

Arzoumanian, Z., Fruchter, A.S. & Taylor, J.H. 1993, in preparation.
Banit, M. & Shaham, J. 1992, ApJ, 388, L19
Czerny, M., & King, A.R. 1988, MNRAS, 235, 33P
Deich, W.T.S., Middleditch, J., Anderson, S.B., Kulkarni, S.R., Prince, T.A. & Wolszczan, A. 1993, ApJ, 410, L95
Fruchter, A.S., Stinebring, D.R., & Taylor, J.H. 1988, Nature, 333, 237
Lyne, A.G., Manchester, R.N., D'Amico, N., Staveley-Smith, L., Johnston, S., Lim, J., Fruchter, A.S., Goss, W.M., & Frail, D. 1990, Nature, 347, 650
Manchester, R.N., Lyne, A.G., Robinson, C., D'Amico, N., Bailes, M., & Lim, J. 1991, Nature, 352, 219
McCormick, P.J., Frank, J. & King, A.R. 1993, ApJ, in preparation
Ryba, M.F., & Taylor, J.H. 1991, ApJ, 380, 557
Thorsett, S.E., & Nice, D.J. 1991, Nature, 353, 731

A STUDY OF ANGULAR MOMENTUM LOSS IN BINARIES USING THE FREE LAGRANGE METHOD

Aruna Rajasekhar, Juhan Frank and Robert Whitehurst
Department of Physics and Astronomy,
Louisiana State University, Baton Rouge, LA 70803-4001
rajase@rouge. phys. lsu. edu

ABSTRACT

Using a 3-dimensional hydrodynamic fluid code based on the Free Lagrange Method, we study the loss of specific angular momentum from a binary system due to an evaporative wind from the companion of a millisecond pulsar. We consider binaries of different mass ratios and winds of different initial velocities and in particular attempt to model the system PSR 1957+20. We are in the process of incorporating the effect of the radiation force from the pulsar and the magnetic field of the companion on the mass outflow. The latter effect would also enable us to study magnetic braking in Cataclysmic Variables and Low Mass X-ray Binaries.

1. INTRODUCTION

The evolution of a binary star system depends greatly on the angular momentum losses in the system brought about by gravitational radiation and mass outflow (e.g. evaporating winds and magnetic braking) from the secondary component of the binary. Hence it is important to calculate the specific angular momentum carried away by the particles leaving the binary system (Huang 1963, Banit & Shaham 1992, Brookshaw & Tavani 1993) for a given set of initial conditions. Many workers have carried out the task of computing numerically the dependence of this quantity on the various parameters such as the mass ratio of the binary, initial velocity of the outflowing mass and the injection geometry. The outflow may be of two types depending on the initial velocity; a Roche lobe overflow through the inner Lagrange point if the initial velocity is almost sonic or a stellar wind from the star surface if the initial velocity is supersonic. In this note we present 3-dimensional simulations of mass outflow from the secondary component of the binary with a pulsar as the primary component. In its present form our code does not include the effect of the radiation emitted by the pulsar on the mass outflow. The two stars are assumed to be in synchronous motion.

2. COMPUTATIONAL METHOD

In this simulation particle motion is carried out by the Free Lagrange Method (Whitehurst 1988). For binary systems with circular orbits and centrally condensed stars the equations of motion are identical to those in the restricted three body problem. A coordinate system (x, y, z), rotating with the binary orbital frequency ω, and with origin at the center of the primary is used. The orbital plane coincides with the xy plane and the axis of rotation lies along the z-axis. The binary separation a is taken as the unit of length and the secondary star is placed at (-1,0). The total mass of the binary $M = M_1 + M_2$, where M_1 and M_2 are the masses of the primary and the secondary star respectively, is chosen as the unit of mass, and the orbital period as the unit of time. In these units the specific angular momentum in the z-direction, (measured with respect to the center of mass of the binary) carried by an escaping particle at position (x, y, z) is

$$h_{esc} = x\frac{dy}{dt} - y\frac{dx}{dt} + (x^2 + y^2)\omega. \qquad (1)$$

The ratio of the average specific angular momentum of the escaping particle to the specific angular momentum of the center of mass of the secondary star is then given as,

$$\beta = \frac{\langle h_{esc} \rangle}{m_1^2 a^2 \omega}; \qquad (2)$$

where $\langle h_{esc} \rangle$ is the average of the values of h_{esc} for all the particles leaving the system and $m_1 = M_1/M$. In most of our simulations we assume that the secondary star is in a state where it completely fills its Roche lobe. In some cases we also consider a detached secondary. Particles are injected all over the secondary surface at random positions with velocities directed either radially outward from the secondary or randomly oriented and with varying Mach numbers. They are then moved in the binary star potential following the equations of motion of a body in a restricted three-body problem. Inter-particle interactions provide the pressure and the viscous forces. The trajectories of the particles under the influence of the binary potential and the fluid properties are studied. The effect of the pressure due to the radiation emitted by the primary will be included later. The time development of the system of particles is done by the Bulirsch - Stoer method of numerical integration for ordinary differential equations (Press et al 1993). The accuracy of the numerical integration is ensured by checking the constancy of Jacobi's integral J in the purely ballistic case for each particle. In these units J at (x, y, z) can be expressed as

$$J = \frac{1}{2}(u^2 + v^2 + w^2) - m_1(\frac{1}{2}r_1^2 + \frac{1}{r_1}) - m_2(\frac{1}{2}r_2^2 + \frac{1}{r_2}); \qquad (3)$$

with $m_2 = M_2/M$, r_1, r_2 the distances of the particle from the primary and the secondary respectively, and u, v, w are the velocities of the particle in the x, y, z directions respectively. It is aimed to achieve an accuracy of one part in a million in the conservation of J. The number of particles injected varies from 5000 per orbital period for low initial velocities (Mach number 3.) to 15,000 per orbital period for high initial velocities (Mach number 100.). The particles are allowed to escape from the binary system at a radius three times the binary separation and the specific angular momentum of each escaped particle is calculated. Particle trajectories for different Mach numbers are studied.

3. DISCUSSION

We present here the distribution of particles for different initial velocities. The figures shown are for a binary system of mass ratio ($q = M_2/M_1$) equal to 0.017 and an orbital period of 33,000 seconds. Each of them is a projection of

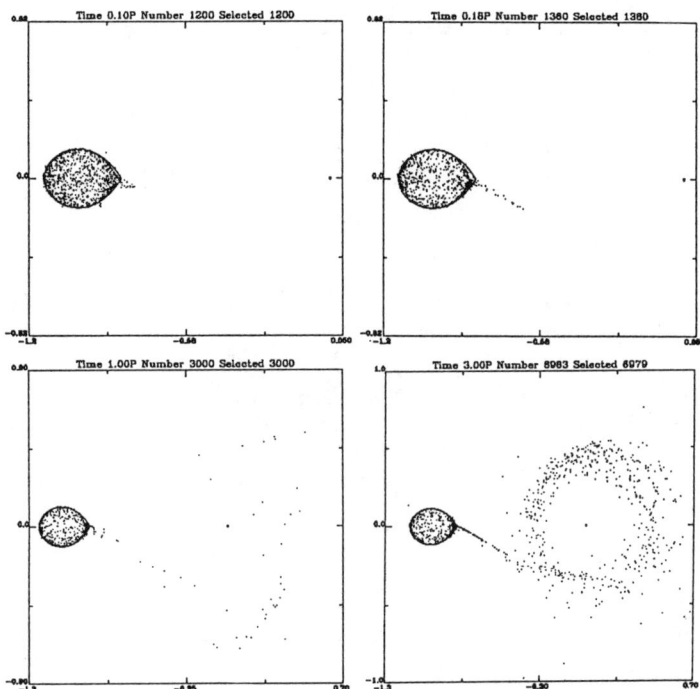

Fig.1 Snapshots of the binary system at different stages of its evolution. The figures are to be read left to right and top to bottom. Particles are injected all over the secondary star surface with Mach number 3. Mass transfer through the inner Lagrange point sets in at about 0.1 orbital period. The last snapshot illustrates the formation of a disk around the primary.

the distribution of particles onto the orbital plane. Evolution time is presented in units of the binary orbital period.

The two extreme cases of Mach number= 3. and Mach number= 100., are chosen to highlight the features. The former case (Fig.1) illustrates the onset of mass transfer from the secondary through the inner Lagrange point. At a later time a snapshot of the simulation shows the formation of a disk around the primary. The inner radius of the disk reduces as the system evolves, followed by the accretion of the particles onto the surface of the primary. A slightly larger initial velocity creates an atmosphere around the secondary before the particles move away from the stars and escape. At a Mach number of 10. the particles are seen to be escaping from the secondary through the L1 and the L2 points. The effect of the tides is noticeable. For the case of very large initial velocity, Mach number = 100. (Fig.2), the particles are injected only on the hemisphere of the secondary that is exposed to the radiation from the primary. For this initial velocity β is calculated in the 2 cases: a) ballistic and b) presence of inter particle collisions. The ballistic case yields a value which is slightly lower than that obtained when the hydrodynamic properties of the system of particles are included. This can be understood as the consequence of the lack of particle collisions.

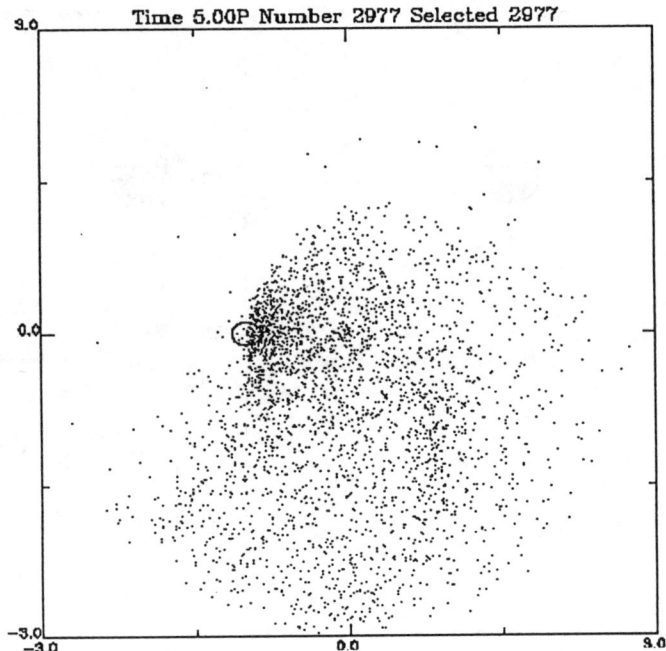

Fig.2. Particles are injected only on the illuminated hemisphere of the secondary star with Mach number 100. This is the ballistic case.

CONCLUSION

For Mach number= 100., the ballistic calculation yields a $\beta = 0.836$, while the non-ballistic case yields $\beta = 0.892$. For the specific case of PSR 1957+20 it is possible to compare the value of β calculated from observed quantities (such as mass transfer rate, orbital period changes etc.) with that from the simulation. This would be useful in the study of orbital period modulations in binary systems such as that observed recently in the above mentioned binary pulsar. The code is in a preliminary stage of development. The aim is to develop it to represent adequately the late stages of evolution of binaries with evaporating secondaries in LMXBs.

This research was partially supported by NASA grant NAGW-2447 and NSF grant AST-9020855.

REFERENCES

Banit, M. & Shaham, J. 1992, Ap.J., 388, L19
Huang, S. 1963, Ap.J, 138, 481
Press, W.H. et al. 1990, 'Numerical Recipes', Cambridge University Press
Brookshaw, L. & Tavani, M., 1993, Ap.J., 410, 719
Whitehurst, R. 1988, MNRAS, 233, 529

DO QUIESCENT SOFT X-RAY TRANSIENTS CONTAIN MILLISECOND RADIO PULSARS ?

L. Stella [1], S. Campana [1], M. Colpi [2], S. Mereghetti [3] & M. Tavani [4]

[1] Osservatorio Astronomico di Brera, Milano, Italy - also affiliated to ICRA
[2] Università degli Studi di Milano, Italy
[3] Istituto di Fisica Cosmica/CNR, Milano, Italy
[4] Princeton University, USA

ABSTRACT

Soft X-Ray Transients (SXRTs) in outburst show properties similar to those of persistent Low Mass X-Ray Binaries (LMXRBs), and therefore likely contain an old weakly magnetic neutron star spun-up by accretion torques. Available X-ray observations can be used to constrain the spin period and magnetic field strength of the neutron stars in SXRTs. The conditions under which a detectable radio pulsar signal can be produced by the rapidly rotating neutron star in the quiescent phase of a SXRT are investigated.

INTRODUCTION

The currently accepted evolutionary scenario for the origin of millisecond pulsars requires that these neutron stars underwent a phase of accretion-induced spin-up as members of LMXRBs (e.g. Bhattacharya & van den Heuvel 1991). Once in these systems accretion stops for evolutionary reasons, the fast rotating neutron star with spin period P in the millisecond range is expected to shine as a radio pulsar. Strong support to this scenario would come from the discovery of fast periodic pulsations in the X-ray emission of LMXRBs.

An alternative approach would be to observe a radio pulsar in a LMXRB; this could be possible only if a severe decrease of the mass inflow rate toward the neutron star takes place (Callanan 1989; Kulkarni et al. 1992). The largest changes of accretion rates are observed in transient LMXRBs and among those in SXRTs (White, Kaluzienski & Swank 1984).

1. THE CENTRIFUGAL BARRIER LIMIT IN THE $B - P$ PLANE

Accretion onto a rotating magnetic neutron star occurs only if the centrifugal drag exerted by the magnetosphere on the accreting material is weaker than gravity (Illarionov & Sunyaev 1975). This requires that the rotation velocity at the magnetospheric boundary is slower than the local Keplerian velocity. Owing to the dependence of the magnetospheric radius on the mass accretion rate, accretion can occur down to a minimum luminosity of

$$L_{\min}(R) = GM\dot{M}/R \simeq 4 \times 10^{36} B_9^2 P_{-2}^{-7/3} M_{1.4}^{-2/3} R_6^5 \quad \text{erg s}^{-1}, \quad (1)$$

where B_9 is the neutron star magnetic field in units of 10^9 G, P_{-2} the spin period in units of 10^{-2} s, $M_{1.4}$ the neutron star mass in units of 1.4 solar masses and R_6 its radius in units of 10^6 cm. Below this minimum accretion rate the neutron star is in the "propeller regime", the accretion flow cannot proceed beyond the magnetospheric boundary, due to the fact that the "centrifugal barrier" is closed. In this case the gravitational energy liberated per unit mass decreases and a sharp reduction of the accretion induced luminosity is produced. In this regime, matter can either accumulate or be ejected. Therefore, for a given source, the minimum observed X-ray luminosity above the "centrifugal barrier" (i.e. that resulting from accretion onto the neutron star surface), $L_{\min}^{\text{obs}}$, allows to constrain the spin period and/or the magnetic field of the neutron star. This can be illustrated in the usual $B-P$ diagram for radio pulsars (Fig. 1), where Eq. (1) is plotted for different values of $L_{\min}$: in order to accrete at the observed rates, a neutron star must possess a spin period and a magnetic field such as to lie to the right of the line corresponding to its minimum accretion-induced luminosity. No evidence for the sharp luminosity decrease at the end of an outburst that should result from the onset of the "centrifugal barrier" has been observed so far during the decay of a SXRT. Therefore one can be confident that the lowest detected X-ray luminosity during the decay of the outburst $L_{\min}^{\text{obs}}$ is produced by accretion onto the neutron star surface. Currently available measurements range from $L_{\min}^{\text{obs}} \sim 3 \times 10^{35}$ to $\sim 10^{37}$ erg s^{-1}. These luminosities are quite high and the corresponding lines from Eq. (1) define relatively large regions in the $B-P$ diagram in which the neutron stars of SXRTs are potentially radio pulsars.

2. APPLICATION TO CEN X-4

Cen X-4 has been detected during quiescence more than one year after the previous outburst, with a luminosity of about 10^{33} erg s^{-1} in the range 0.5–4.5 keV (van Paradijs et al. 1987). In principle, this emission could result from: (a) coronal activity of the companion star; (b) non-thermal processes powered by the rotational energy loss of a rapidly spinning neutron star; (c) thermal emission from the cooling neutron star; (d) accretion onto the neutron star surface or down to the magnetospheric radius (depending on whether the "centrifugal barrier" is open or closed). For this source, possibility a seems unlikely considering that the late type low-mass companion could provide a luminosity of $< 10^{32}$ erg s^{-1} (Vaiana et al. 1981). The constraints on B and P derived from the lowest luminosity measured during the 1979 outburst ($L_{\min}^{\text{obs}} \sim 3 \times 10^{35}$ erg s^{-1}) and Eq. (1) are strong enough to exclude also possibility b. Indeed these constraints limit the maximum pulsar spin-down luminosity to $\dot{E}_{\text{rot}} \leq 10^{34}$ erg s^{-1}, which is insufficient to power the observed X-ray luminosity, L_X, for $L_X/\dot{E}_{\text{rot}} \leq 10^{-3}$ as measured for PSR 1957+20 (Fruchter et al. 1992) and PSR J0437–4715 (Becker et al. 1993).

The quiescent X-ray flux might originate from the release of thermal energy in the neutron star interior, which has been reheated through accretion (case c). However van Paradijs et al. (1987) discarded this possibility in the case of Cen X-4. The quiescent X-ray luminosity of Cen X-4 is likely powered by accretion (case d). Due to the sparse X-ray coverage it is not possible to ascertain whether the persistent emission level is reached above or below the sharp luminosity decrease characterizing the onset of the "centrifugal barrier". In the former case, accretion proceeds to the neutron star surface and the constraints derived from Eq. (1) and the "death line" (see Fig. 1), virtually exclude the possibility that a radio pulsar might turn on, should the mass inflow rate decrease further. Less tight constraints are obtained if in quiescence the "centrifugal barrier" is closed. In this case the flow of matter cannot proceed beyond the magnetospheric radius, r_m, and an accretion luminosity of $L(r_m) = GM\dot{M}/r_m \simeq 10^{34}\, \dot{M}_{15}^{9/7}\, B_9^{-4/7}\, M_{1.4}^{8/7}\, R_6^{-12/7}$ erg s^{-1} ($\dot{M}_{15}$ is the mass inflow rate in units of 10^{15} g s^{-1}) is liberated, which for an optically thick accretion disk is radiated in the UV/soft X-ray band with a typical temperature of $T \simeq 5 \times 10^5\, \dot{M}_{15}^{13/28}\, B_9^{-3/7}\, M_{1.4}^{5/14}\, R_6^{-9/7}$ K. In this regime, the minimum accretion induced luminosity can be estimated by requiring that the inflow is not disrupted by the radio pulsar pressure; the threshold corresponds to $r_m = r_{lc}$ (r_{lc} is the light cylinder radius) and is given by $L_{min}(r_{lc}) \simeq 10^{32}\, B_9^2\, P_{-2}^{-9/2}\, M_{1.4}^{1/2}\, R_6^6$ erg s^{-1}. The limit derived from this equation is indicated with dotted lines in Fig. 1 for selected values of $L_{min}(r_{lc})$. In this interpretation the neutron star of a SXRT must lie to the right of the line corresponding to the lowest detected luminosity in quiescence. This, together with the other constraints, defines a region of the $B - P$ diagram for which a *radio pulsar* might *turn on* in Cen X-4, should the inflow rate decrease below $\dot{M}_{lc} \simeq 3 \times 10^{13}\, B_9^2\, P_{-2}^{-7/2}\, M_{1.4}^{-1/2}\, R_6^6$ g s^{-1}, the value corresponding to $r_m = r_{lc}$.

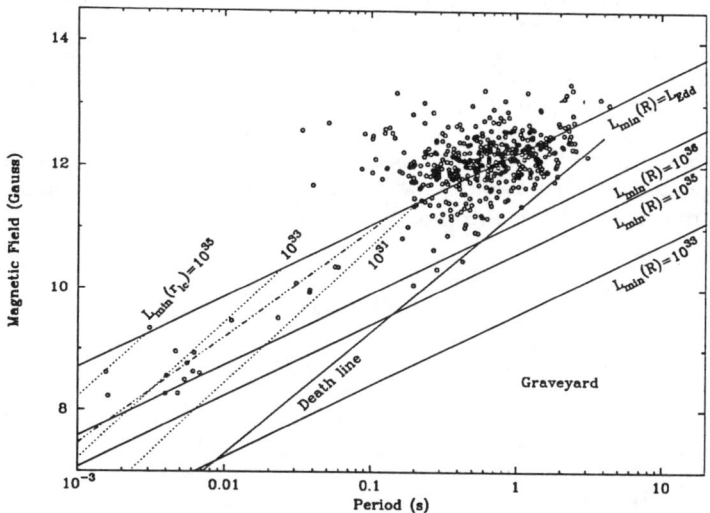

Figure 1. Surface magnetic field versus spin period of known radio pulsars (circles). Solid lines represent the limit given by Eq. (1) for different values of the mass accretion. For $L = L_{Edd} = 1.8 \times 10^{38}\, M_{1.4}$ erg s^{-1}, this corresponds to the so called spin-up line. Below the "death line" no radio pulsars can exist.

The observability of the radio pulsar signal produced depends on the dispersion measure and free-free absorption of the material outside the accretion radius and/or the Roche lobe of the neutron star. Adopting the model of Kochanek (1993), a free-free absorption optical depth of < 1 is obtained for $B_9 < 9 \times 10^{-1}\, P_{-2}^{7/4}\, a_{11}^{1/4}\, M_{1.4}^{-1/4}\, R_6^{-3}\, \nu_9^{1/2}\, f$ G (where a_{11} is the orbital separation in units of 10^{11} cm, ν_9 is the radio frequency in GHz and $f \sim 1$ is a factor that depends on the stellar wind and binary parameters). This equation defines a line in the $B - P$ diagram (shown as dot-dashed line in Fig. 1) below which a detectable radio pulsar signal with dispersion measure ~ 1 pc cm^{-3} or less is produced by a SXRT.

CONCLUSIONS

We have shown that if an *accretion* powered luminosity of $\leq 10^{33} - 10^{34}$ erg s^{-1} is a common characteristic of SXRTs after the end of an outburst or in quiescence, then two very different possibilities present themselves. If the "centrifugal barrier" is open, the accretion flow in quiescence extends to the surface of the neutron star and the energy liberated per unit mass is high. In this case the accretion rates required to produce $10^{33} - 10^{34}$ erg s^{-1} are so low that almost only neutron stars to the right of the "death line" would be able to accrete and, therefore, would not turn on as radio pulsars even if accretion stops completely (cf. Fig. 1). The SXRT stage would then occur very early in the LMXRB phase or belong to a different evolutionary track. If instead in the quiescent state the "centrifugal barrier" prevents the accreting matter from entering the neutron star magnetosphere, then a large region of the $B - P$ diagram would be available for the production of an observable radio pulsar signal, when the accretion flow becomes sufficiently low to be swept away (see Stella et al. 1994 for details). The low-level X-ray luminosity in quiescence does not provide any constraint in the $B - P$ diagram if it results from neutron star cooling.

Monitoring the X-ray luminosity of SXRTs in the decay from the outburst to the quiescent emission level could reveal the onset of the "centrifugal barrier" and provide further clues on the properties of the neutron stars in SXRTs. Prospects for observing a radio pulsar signal would be very promising if SXRTs with a minimum accretion luminosity at the neutron star surface of $\sim 10^{36} - 10^{37}$ erg s^{-1} were detected.

References

Becker, W., Trümper, J., Brazier, K.T.S., & Belloni, T. 1993, IAUC N° 5701
Bhattacharya, D., & van den Heuvel, E.P.J. 1991, Phys. Rep., 203, 1
Callanan, P. private communication quoted in Biggs, J.D., Lyne, A.G., & Johnston, S. 1989, in 23rd ESLAB Symp; eds. J. Hunt, & B. Battrick, ESA SP-296, p 293
Fruchter, A.S., Bookbinder, J., Garcia, M.R., & Bailyn, C.D. 1992, Nature, 359, 303
Kochanek, C.S. 1993, ApJ, 406, 638
Kulkarni, S.R., Navarro, J., Vasisht, G., Tanaka, Y., & Nagase, F. 1992, in X-ray binaries and recycled pulsars, eds. van den Heuvel E.P.J. & Rappaport S.A., Kluwer, p 99
Illarionov, A.F., & Sunyaev, R.A. 1975, A&A, 39, 185
Stella, L., Campana, S., Colpi, M., Mereghetti, S., Tavani, M., 1994, ApJ Lett., in press
Vaiana, G.S. et al. 1981, ApJ, 244, 163
van Paradijs, J., Verbunt, F., Shafer, R.A., & Arnoud, K.A. 1987, A&A, 182, 47
White, N.E., Kaluzienski, J.L., & Swank, J.H. 1984, in High Energy Transients in Astrophysics, ed. S.E. Woosley, AIP Conf. Proc. No. 115, p 31

EVOLUTIONARY SCHEME FOR LOW–MASS BINARY WITH MILLISECOND PULSAR

Alexander G. Muslimov
Department of Physics and Astronomy, University of Rochester, Rochester NY 14627 USA
muslimov@callisto.pas.rochester.edu

Marek J. Sarna
N.Copernicus Astronomical Center, Polish Academy of Sciences, Warsaw 00-716 Poland
Marek_Sarna@camk.edu.pl

ABSTRACT

We present the results of evolutionary computation for the low–mass binary (LMB) with 1.4 $M_\odot$ neutron star (NS) as a compact companion and 1 $M_\odot$ red dwarf as a donor star. At the moment of filling of the Roche lobe by the red dwarf an orbital period of a system is $9.^h4$. We take into account the illumination of the donor star by energetic quanta and particles from the NS. We focus on a regime of "non–catastrophic" heating when at the beginning of mass–transfer phase the pulsar's illumination activates the mass loss by a donor star rather than vaporizes its outer layers. We also incorporate the effect of sweeping of overflowing plasma out of the system by the pulsar's magneto–dipole radiation. One issue is that the evolution of such a system may result in the formation of the LMB with rapidly rotating NS and 0.2 $M_\odot$ red–dwarf companion, and having orbital period $\sim (6-7)^h$. Another issue we investigate is the regime of episodic mass loss (with approximately Eddington rate) by the donor star commencing at the late stage of evolution before the donor star reached the minimum mass $\sim 0.2\ M_\odot$ and shrinked within its Roche lobe. The relevance of this regime to the question of whether the LMXBs may be descendants of weakly magnetized millisecond pulsars is briefly discussed.

1. INTRODUCTION

In the present work we consider evolutionary sequence for a close binary system consisting of a 1.4 $M_\odot$ NS and a main–sequence star of 1 $M_\odot$. We compute evolution of the red–dwarf component using a stellar evolution code based on the standard Henyey-type code of Paczyński (1970).

We consider the LMB with circularized orbit, and assume that the orbital and spin rotation of the donor star are synchronized. Evolution of such a binary is mostly driven (during its *non-conservative* stage) by the loss of orbital angular momentum via magnetic stellar wind from the red–dwarf and mass loss from the system.

In our computations of the sequences of the red–dwarf models we have used the mixing–length algorithm proposed by Paczyński (1969), as well as opacities of Heubner et al. (1977). The processes of evaporation and ablation of the donor star by the hard radiation and relativistic particles from the NS are incorporated in our computations in the same way as described by Muslimov & Sarna (1993).

2. RESULTS

We shall now describe briefly the evolution of a binary system with initially 1.4 $M_\odot$ NS and main–sequence companion of 1.0 $M_\odot$ [see for details Muslimov & Sarna 1993 (evolutionary sequence where the parameter λ, characterizing the effectiveness of magnetic braking, gradually increases from 1 up to 1.7 during the evolution)]. Here we present the case where at the onset of the Roche-lobe overflow by the red–dwarf companion the surface value of the NS magnetic field is equal to 5×10^8 G. In this case the NS spun up to ~ 1.6 ms during $\sim 2 \times 10^7$ yr. Then the condition for the expulsion out of the infalling plasma is satisfied and the stage of evaporation of the donor star sets in. Thus the accretion phase (lasted for 2×10^7 yr) ceases, and the NS (now a radio PSR) resumes the spinning down due to magneto–dipole losses.

During subsequent evolution the orbital period of a system monotonically decreases (due mostly to the magnetic braking) until the mass of the donor star decreases down to $M_1 \sim 0.38$ $M_\odot$. After this stage the system (since the loss of orbital angular momentum at this stage is determined by the mass loss from the binary) tends to force the red dwarf to shrink within its Roche lobe, while still operating magnetic braking of the red dwarf tends to keep it overflowing the Roche lobe and losing mass. As a result, the system evolves non–monotonically. Our computations show that the episodic Roche–lobe overflows by the red dwarf may occur before it shrinks well within its Roche lobe (e.g. at $M_1 \approx 0.23$ $M_\odot$ as in the presented evolutionary scheme). Then evolution of the system substantially slows down, since it is now determined by the loss of orbital angular momentum due to emission of gravitational waves (with the characteristic evolutionary timescale $\sim 10^{10} yr$). When $M_1 \lesssim 0.3$ $M_\odot$ the magnetic braking operates not so efficiently as at larger masses, and the orbital evolution is driven by the mass loss from the system. At this stage the orbital period increases and reachs the value $6.^h 7$ at $M_1 \approx 0.23$ $M_\odot$, thus resulting in a formation of the LMB with orbital period $\sim 7^h$ and consisting of rapidly rotating NS and ~ 0.2 $M_\odot$ red–dwarf companion.

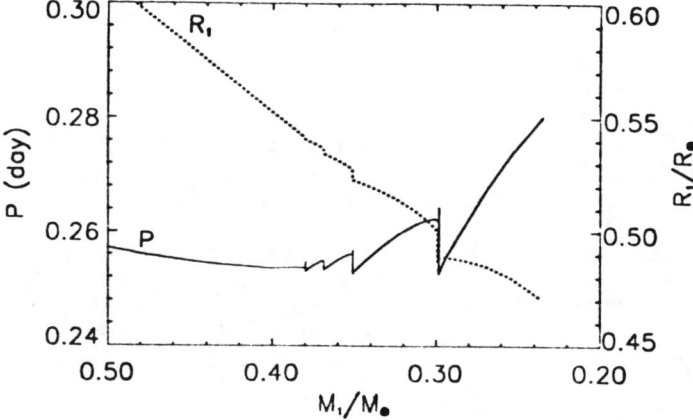

Fig. 1. The orbital period of the system (solid curve) and radius of the donor star (dotted curve) as a function of mass of the red dwarf.

Figure 1 shows the late evolution of the orbital period (solid curve) and radius (dotted curve) of the donor star with its mass (as a time coordinate). Within the mass interval $0.3 - 0.4$ $M_\odot$ the orbital period experiences sawtooth-type oscillations with the maximum relative amplitude about 4 per cent. Because of the heating (due to illumination by PSR), the donor star does not contract to the main-sequence equilibrium radius. For example, as Figure 1 (dotted curve) shows, the non-equilibrium radius of the red dwarf just prior to its final shrinkage (at $M_1 \approx 0.23$ $M_\odot$) within the Roche lobe is about 2 times greater than the radius of main-sequence star in thermal equilibrium of the same mass.

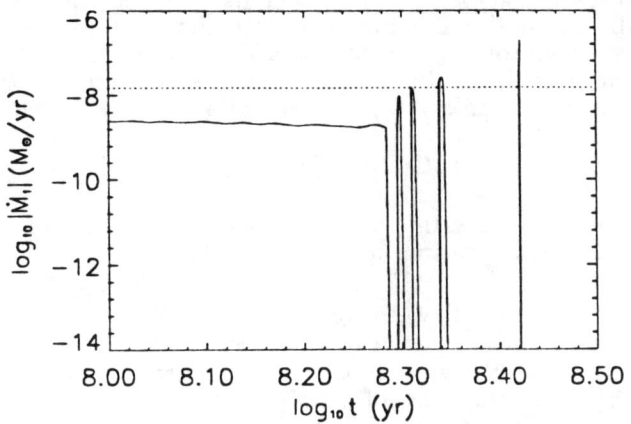

Fig. 2. The rate of mass loss by the donor star as a function of time. The dotted horizontal line corresponds to the value of Eddington accretion rate.

The rate of mass loss by the donor star as a function of time is presented in Figure 2, which clearly shows that the late stage of the LMB evolution is accompanying by alteration of detached and semi-detached phases. The system enters the first detached phase at $M_1 \approx 0.38$ $M_\odot$ and stays there $\sim 5 \times 10^6$ yr. Then the red dwarf fills its Roche lobe and resumes mass loss. The first peak of mass loss (at the rate ~ 0.4 $\dot{M}_{Edd}$) lasts $\sim 1.8 \times 10^6$ yr. Then the mass-loss rate declines and the system enters the second detached phase. The second peak of mass loss is almost identical to the first one. The pause between the second and third peaks is 1.2×10^7 yr. The mass-loss rate in the third peak reachs the value of 1.4 $\dot{M}_{Edd}$. The duration of this peak is $\sim 2.6 \times 10^6$ yr. After the third peak the system enters long ($\sim 4 \times 10^7$ yr) detached phase followed by the last relatively short ($\sim 6 \times 10^5$ yr) episod of mass loss with the super-Eddington (~ 7 $\dot{M}_{Edd}$) rate at the peak.

Evolutionary scheme considered in this paper implies the occurence in the life of the binary of one relatively long ($\sim 2 \times 10^7$ yr) and four episodic (lasting from 6×10^5 to 2.6×10^6 yr) stages, which can be identified with the LMXB stages. The episodic LMXB stages are followed by the stage of the LMB with ms radio PSR. These relatively short LMXB stages are direct consequence of the effect of discrete mass loss by the red dwarf prior to its shrinkage within the Roche lobe at $M_1 \sim 0.2$ $M_\odot$. The total duration of these episodic LMXB stages is about 7×10^6 yr which is more than a factor of 200 less than the characteristic age of the binary system. Our results thus illustrate that the question of whether the LMXBs are progenitors or descendants of binary ms PSRs may be semantic. For example, according to our evolutionary scheme a ms PSR turned on after the first stage of LMXB may be regarded as the descendant of the LMXB. In contrast, at any episodic LMXB stage the binary ms PSR may be considered as progenitor of this LMXB. As our calculation shows the characteristic durations of the stages, which may be identified with the stages of LMXBs, are much shorter than the age of a binary system, and agree well with the statistical argument concerning the relative populations of LMXBs and LMBs with ms PSRs (see e.g. Bhattacharya & van den Heuvel 1991, and references therein). Evolutionary scheme considered in this paper may be regarded as one of the theoretical possibilities not conflicting both with the suggestion that Galactic disc LMXBs are progenitors of binary ms PSRs (Bhattacharya & Srinivasan 1986, and van den Heuvel, van Paradijs & Taam 1986), and that the globular cluster LMXBs are descendants of binary ms PSRs (Chen & Ruderman 1993).

3. CONCLUSIONS

Our principal results are that (1) we proposed a scheme of evolution of the red–dwarf component resulting in the formation of LMB [$M_1 \sim 0.2\ M_\odot$, $P_{orb} \sim (6-7)^h$] with the ms PSR having magnetic field $\lesssim 5 \times 10^8\ G$ and evaporating its companion; (2) we have shown that the duration of the first phase in a LMB evolution which can be identified with the stage of LMXB is about 2×10^7 yr, with the relevant accretion luminosities are $10^{37} - 10^{38}$ erg·s^{-1}. We have found that after the stage of a LMB with ms radio PSR, followed by the first LMXB stage, a sequence of shorter LMXB stages takes place. The total duration of these shorter LMXB stages is comparable with that of first LMXB stage, and the typical accretion luminosities corresponding to these discrete stages are $\sim 10^{38}$ erg · s^{-1}; (3) we have suggested that the episodic mass loss by the donor star (accretion onto a NS with near Eddington rate) followed by the stage of LMB with radio PSR may be important for further discussion of whether the LMXBs are progenitors or descendants of binary ms PSRs.

ACKNOWLEDGEMENTS

This work was supported by NSF grant AST 91–15132 and NASA grant NAGW–2444 through the University of Rochester. Also, it has been supported in part by the Polish National Committee for Scientific Research under grant 2–2115–92–03.

REFERENCES

Bhattacharya, D. & Srinivasan, G. 1986, Curr. Sci., 55, 327
Bhattacharya, D. & van den Heuvel, E.P.J. 1991, Phys. Rep., 203, 1-124
Chen, K. & Ruderman, M. 1993, ApJ, 408, 179
Heubner, W.F., Herts, A.L., Magee, N.H.Jr. & Argo, M.F. 1977, Astrophys. Opacity Library, Los Alamos Scientific Lab. Report No. LA-6760-M
Muslimov, A.G. & Sarna, M.J. 1993, MNRAS, 262, 164
Paczyński, B. 1969, Acta Astro., 19, 1
Paczyński, B. 1970, Acta Astro., 20, 47
van den Heuvel, E.P.J., van Paradijs, J.A. & Taam, R.E. 1986, Nature, 322, 153

THE EVOLUTION OF EVAPORATING BINARY PULSARS

P. McCormick and J. Frank
Department of Physics and Astronomy
Louisiana State University, Baton Rouge, LA 70803
cormick@rouge.phys.lsu.edu

A. R. King
Astronomy Group, The University Leicester LE1 7RH, UK

ABSTRACT

We discuss the origin and evolution of short–period, low–mass binary pulsars with evaporating companions. We suggest that these systems descend from low mass X-ray binaries and that angular momentum loss mainly due to evaporative wind drives their evolution tending toward $d\log P_b/d\log m_2 = -1$ at late stages. We derive constraints on the energy and angular momentum carried away by the wind based on the observed orbital parameters. In our model the companion remains near contact and its quasi-adiabatic expansion causes the binary to expand. Short term oscillations of the orbital period may occur if the Roche–lobe overflow forms an evaporating disk. Possible links to related systems such as 4U16126-67 and 1E2259-586 are briefly discussed.

1. INTRODUCTION

We use a simple numerical Runge Kutta integration routine (Press et al. 1986) to follow the time development of the orbital parameters in compact binaries. We are able to construct evolutionary scenarios for low–mass X-ray binaries (LMXBs) which produce evaporating pulsar binaries. The evolution of compact binaries is primarily determined by the interplay between the mass loss rates and the angular momentum loss driving mechanisms. The most fundamental angular momentum loss driving mechanism is gravitational radiation. Gravitational radiation is present throughout all of the evolution. In addition, magnetic braking and stellar evaporation, under certain conditions, can contribute significantly to the angular momentum losses of these systems. Angular momentum loss is necessary for the secondary to maintain contact with the primary and thus reproduce observed stable long term mass transfer (e.g. King 1988). Magnetic braking is thought to drive mass transfer in both cataclysmic variables (CVs) and in LMXBs with orbital periods above 3 hours. We follow standard practice and cut off magnetic braking for short period binaries when the companion mass drops below some minimum mass. This occurs because magnetic braking ceases to be effective after the star becomes fully convective.

2. METHOD FOR MODELING THE ORBITAL EVOLUTION

Our method for simulating evolution of LMXBs makes use of the stars tendency to approach an equilibrium radius r_{eq}. Low mass, hydrogen burning, stars follow an approximate main sequence relation such that $r_{eq} \propto m_2$. In a binary system, where a secondary is losing mass, hydrogen burning will cease when the secondary reaches about $m_2 = .085 M_\odot$. For smaller masses the equilibrium radius of the secondary is assumed to follow $r_{eq} \propto m_2^{-1/3}$. Semi-detached binaries with thermally relaxed companions would fall along one of the equilibrium sequences shown in Fig. 1. The "bloated sequence" for irradiated companions (Frank, King & Lasota 1992) is shown here for comparison purposes only.

Starting from some initial assumed configuration, we calculate the secondary radius as mass–loss occurs from

$$\frac{\dot{r}_2}{r_2} = -\frac{1}{3}\frac{\dot{m}_2}{m_2} + \frac{1}{t_{kh}}\frac{(r_{eq} - r_2)}{r_2}$$

where r_{eq} is the equilibrium radius as defined above. Adiabatic expansion occurs if the mass transfer time scale is much smaller than the Kelvin-Helmholtz timescale. The time evolution of the radius is thus governed by the mass transfer rate and the tendency of the secondary to maintain an equilibrium radius.

Mass-loss from the companion can occur because of a wind (see below) or because the secondary slightly overflows its Roche lobe or both. The mass transfer rate due to Roche lobe overflow is given by

$$\dot{m}_2 = \dot{m}_{20} \exp\left(500(r_2 - r_{lobe})/r_2\right).$$

where $\dot{m}_{20}$ is the contact mass transfer rate for a main sequence or degenerate companion. The separation between roche lobe and the secondary radius is scaled to avoid numerical oscillations (Hameury 1991).

We use Verbunt & Zwaan's (1981) form of the magnetic braking law, such that the torque is given by

$$G_{mb} = -5 \times 10^{-29} f_{mb}^{-2} k^2 m_2 r_2^4 \omega^3,$$

where k is the gyration radius and f_{mb} is a dimensionless parameter of order unity. Since Verbunt and Zwaan's magnetic braking law was based on observations of main sequence stars $r_2 \propto m_2$, we assume that mass of the secondary is the important factor in determining magnetic braking. That is, we take $G_{mb} \propto m_2^5$. We choose f_{mb} such that it produces the 2-3 hour period gap observed in the orbital period distribution of CVs when magnetic braking is turned off at $m_2 = .22 M_\odot$. We further assume that the same parameters and premises apply to the evolution of LMXBs.

3. LMXB INTO EVAPORATING PULSAR BINARIES

Accretion during the LMXB stage of evolution can produce a spun up neutron star. In the standard model, as the secondary becomes fully convective and magnetic braking becomes an inefficient angular momentum loss mechanism, a temporary suspension of mass transfer will occur. The pulsar may become active ("resurrected") because of the sudden decline of mass transfer. The drop in the mass loss rate allows the magnetostatic radius of the neutron star to reach its light cylinder radius and thus become a pulsar. The mass loss rate at which the pulsar is turned on is thus estimated from $\dot{M}_{psr} \sim 10^{16}(1.9P/1.6\text{ms})^{-3.5}(B/2 \times 10^8 G)^2)$ g/s.

Irradiation from the pulsar may start to evaporate the secondary driving mass and angular momentum losses from the system. The efficiency of the

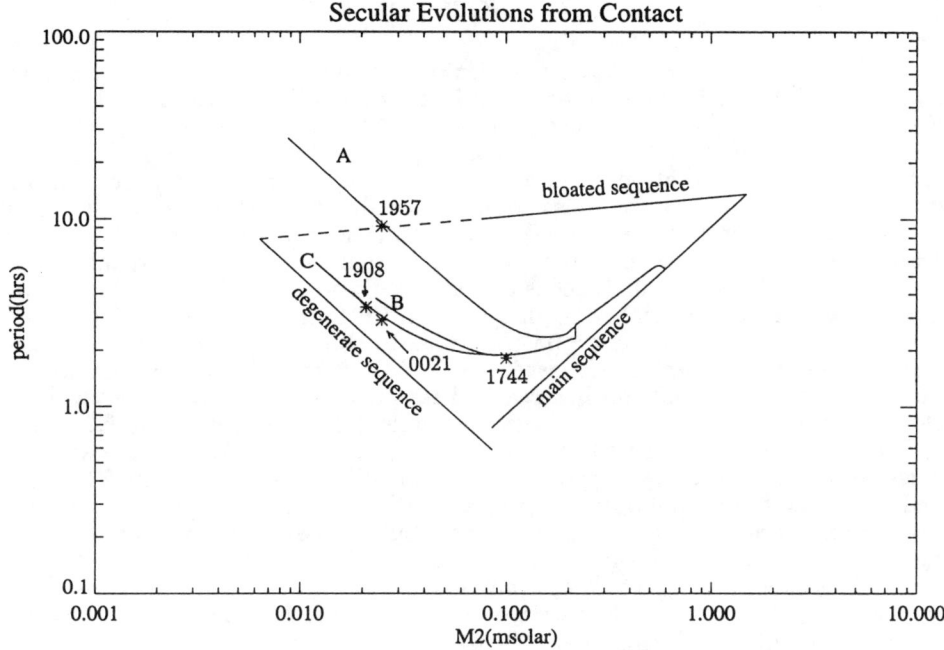

Fig. 1: Possible evolutionary paths for LMXBs which lead to the four known millisecond pulsars with very low mass evaporating companions. Without loss of generality all are assumed to start with a main sequence companion of 0.6 $M_\odot$. Once the pulsar turns on, it is assumed it does not turn off even if Roche lobe overflow occurs. Path A was calculated with $\beta_w = 1.0$, $\beta_{RL} = 0.5$ and shows an example of LMXB evolving into PSR 1957+20. Path B was calculated with the same β values but lower evaporation rate. The pulsar turns on later and this evolution could explain PSR 1744-24A. Path C was calculated with the same β values as the other cases but the evaporation was assumed inefficient resulting in slower mass-loss. The same evolution could explain both PSR 1908+00 and PSR 0021-72J.

evaporating process may vary and the mass-loss due to a pulsar–driven wind is estimated as follows (Deich et al. 1993):

$$\dot{M_2} = f(\pi^2 I/GM_2)(\dot{P}/P^3)(R_2^3/a^2).$$

This evaporating wind can either drive the binary into or out of contact depending on the specific angular momentum loss β (see McCormick et al. in this volume). If Roche–lobe overflow starts and the overflow material is blown away, the pulsar remains active. For the cases shown in Fig. 1 we have assumed that once the pulsar is activated it is not quenched by the overflow even if a disk forms. Thus the secondary loses mass through evaporation and overflow but all this material leaves the system carrying away some angular momentum specified by β_w for wind losses and by β_{RL} for the evaporating disk generated by overflow.

4. DISCUSSION

We have explored a large number of possible scenarios using the "toy" model described above and a more complete acount will be published elsewhere. We present selected evolutionary histories for LMXBs born at some orbital period above 4 hours which evolve to shorter periods driven by magnetic braking and gravitational radiation. The neutron stars are spun up by accretion and turn on as pulsars when the mass transfer rate due to Roche lobe overflow falls below the value estimated above. Depending on the spin period, the efficiency of evaporation, and the assumed magnetic field, their late evolutions diverge after they detach. The main difference is due to the evaporation rate and the concurrent overflow rate: the higher the mass-loss rate the sooner the companion will evolve adiabatically following a path parallel to the degenerate sequence. Thus PSR 1957+20 must be losing mass more rapidly than the other pulsars evaporating their companions and has reached its present mass and period evolving along a path similar to A. Note that the pulsar turns on at a relatively high accretion rate because of its short spin period. Despite posssible differences in mass–loss rates, we expect systems such as PSR 1957+20, PSR 1908+00 and PSR 0021-72J to be similar, with companions near contact and possibly an evaporating disk, undergoing secular expansion but with sort–period oscillations in their orbital periods. AT the present stage PSR 1744-24A may be different because of its slower spin, this delays the pulsar turn on and lowers the subsequent evaporation rate. Nevertheless we would expect PSR 1744-24A also to become more like the other systems in $\sim 10^9$ years.

If the pulsar is quenched by overflow, different evolutionary histories are possible with alternating phases of radio and X-ray pulsar and a complex spin period evolution. The X-ray pulsar 4U 1626-67 (Levine et al. 1988) has a very low mass companion $M_2 \lesssim 0.02 M_\odot$ in an orbit with a period of 42 minutes, a spin period of 7.67 seconds and is spinning up. It is possible that this system was a pulsar similar to those discussed above, which has been quenched when the wind-driven Roche lobe transfer from its He–rich companion increased sufficiently, and it is seen now in an accreting phase. Another related system may be the X-ray pulsar 1E2259+586 (Davies, Wood & Coe 1990) which has a pulse period of 6.98 seconds and is spinning down. It is probably seen in the propeller phase.

This research was partially supported by NASA grant NAGW-2447 and NSF grant AST-9020855.

REFERENCES

Davies, S.R., Wood, K.S. & Coe, M.J. J. 1990, MNRAS, 245, 268
Deich, W.T.S., Middleditch, J., Anderson, S.B., Kulkarni, S.R., Prince, T.A. & Wolszczan, A. 1993, ApJ, 410, L95
Frank, J., King, A.R. & Lasota, J.P. 1992, ApJ, 385, L45
Hameury, J.M. 1991, A& A, 243, 418
King, A. R. 1988, Quarterly Journal, 29, 1
Levine, A. et al. 1988, ApJ, 327, 732
Press et al 1986, Numerical Recipes, p. 547
Verbunt, F. & Zwaan, C. 1981, Astron. Astrophys., 100, L7

EVOLUTION VERSUS VARIABILITY IN NEUTRON STAR BINARIES

Ralph A.M.J. Wijers[*]
Princeton University Observatory, Peyton Hall, Princeton, NJ 08544-1001, USA
rw@astro.princeton.edu

ABSTRACT

In recent years, the discovery of orbital period changes and eclipses in binaries with neutron stars has led to a new wave of excitement over these systems. They have also led to speculation that the traditional views of X-ray binary evolution may be incomplete. Here I briefly discuss the observed orbital period derivatives and possible analogies between these binaries and other types of compact binary. I conclude that stellar activity in a late-type companion is the most likely cause for the anomalous orbital period changes, and that it is also a strong candidate for explaining the eclipse phenomenon in at least some observed systems.

1 INTRODUCTION

The theory of evolution of X-ray binaries and recycled pulsars was developed and successfully applied to a variety of problems in the seventies and eighties (see e.g. Bhattacharya and van den Heuvel 1991). It has become the subject of considerable debate again following discoveries in the past five years of unexpected values of orbital period derivatives, a discrepancy between the birth rates of recycled pulsars and X-ray binaries (calling into question their evolutionary link) and the discovery of four eclipsing binary pulsars. It appears now that these problems are confined to the low-mass short-period systems. The classical scenarios are still perfectly valid for the recycled pulsars in long-period circular orbits around white dwarfs and their progenitors, the low-mass X-ray binaries with giant donors (both for quantitative predictions of parameters of individual systems and for population statistics).

The debate about the short-period systems with main-sequence or degenerate donors concerns both proposed extensions to existing evolutionary scenarios, such as companion evaporation to explain eclipsing and single recycled pulsars (Ruderman et al. 1989), and proposals that new effects, such as radiation-driven winds and irradiation-induced bloating of companions may actually dominate the long-term evolution of many systems. It is with the latter proposals that I take issue here: the observed orbital period changes are demonstrably non-secular or non-intrinsic in most cases (Sect. 2); furthermore, I will appeal to an analogy with other compact binaries with late-type companions to argue that enhanced stellar activity is the common cause of both the orbital period variability and the large mass loss rates (Sect. 3).

2 THE OBSERVED ORBITAL PERIOD DERIVATIVES

The eclipsing pulsars and X-ray binaries with reported values, value ranges, or good limits of the orbital period derivative are the radio pulsars PSR 1957+20 and PSR 1744−24A and the X-ray binaries 4U 1820−30, EXO 0748−676, X 1822−371, Hercules X-1 (X 1656+354), and

---
[*]Compton GRO Fellow

Cygnus X-3 (X 2030+407). The latter two are not short-period and not low-mass, respectively, but I will discuss them anyway to make the census of known orbital period derivatives complete, and because they have occasionally been classified (incorrectly, I think) in this category.

PSR 1957+20 is the archetypical eclipsing pulsar (Fruchter et al. 1988). Its eclipses lent support to the idea of companion evaporation (Ruderman et al. 1989). Its rapid orbital expansion contradicted predictions in both sign and magnitude; it was regarded by proponents as support for radiation-driven accelerated evolution. However, the orbit is now *shrinking* rapidly (Arzoumanian et al. 1993), so the effect is ephemeral and gives no indication of the long-term sense of orbital evolution. The range of period derivatives observed sofar is $\dot{P_b}/P_b = -0.4 - +1.8 \times 10^{-7}\,\mathrm{yr}^{-1}$.

PSR 1744-24A is an eclipsing pulsar in the globular cluster Terzan 5, with a fairly low limit to its orbital period derivative: $|\dot{P_b}/P_b| < 2.4 \times 10^{-8}\,\mathrm{yr}^{-1}$ (Nice and Thorsett 1992).

EXO 0748-676 is an eclipsing X-ray burster; its $\dot{P_b}$ is variable in sign and magnitude, $\dot{P_b}/P_b = -2 - +1 \times 10^{-7}\,\mathrm{yr}^{-1}$ (Asai et al. 1993).

Hercules X-1 is one of the best studied classical X-ray binaries. It is fueled by Roche-lobe overflow from a more massive companion and therefore has a short lifetime. Its orbit is shrinking, as expected, but its $\dot{P_b}/P_b = -1.3 \pm 0.2 \times 10^{-8}\,\mathrm{yr}^{-1}$ (Deeter et al. 1991) is 7 times as fast as expected from its mass transfer rate for standard conservative evolution.

4U 1820-30, an X-ray burster in the globular cluster NGC 6624, is the most compact known binary. Its $\dot{P_b}/P_b = -5.3 \pm 1.1 \times 10^{-8}\,\mathrm{yr}^{-1}$ is of the opposite sign as expected if the companion is a degenerate low-mass star, but the measurement has been questioned recently because the X-ray light curve is not stable (van der Klis et al. 1993). Also, its revised HST position relative to the cluster center (King et al. 1993) implies that such a value of $\dot{P_b}$ is likely due to the gravitational pull of the cluster on the binary; this precludes the use of its $\dot{P_b}$ in discussions of orbital evolution.

X 1822-371 is an accretion disk corona source with $\dot{P_b}/P_b = 3.9 \pm 1.1 \times 10^{-8}\,\mathrm{yr}^{-1}$, again of opposite sign to clasical predictions.

Cygnus X-3 is now again considered most likely a massive binary, the mass donor being a helium star (van Kerkwijk et al. 1992). Its measured $\dot{P_b}/P_b = 1.65 \pm 0.09 \times 10^{-6}\,\mathrm{yr}^{-1}$ is consistent with this for pure Jeans mode mass loss with typical values for mass and mass loss of the helium star. Its $\dot{P_b}$ appears to be somewhat variable as well (van der Klis and Bonnet-Bidaud 1989).

Taking all the above into account, it is fair to say that we have no low-mass short-period neutron star binaries in which the secular value of the orbital period has been measured. The only non-variable $\dot{P_b}$ is that of X 1822-371, but the variability of similar systems makes one wonder how long this constancy will hold up.

3 COMPANION STELLAR ACTIVITY — A COMMON CAUSE?

When studying the literature on all known types of close binary in which one component is a late-type dwarf or subgiant, one quickly finds a few similarities: they all show signs of stellar activity if the cool component can be studied well enough: star spots, X-ray flares and coronal X rays, radio emission, certain optical line profiles, etcetera. If any orbital period changes are measured, $\dot{P_b}$ is seldom constant over a decade. Examples of such systems are CV's and CV-like systems, RS CVn stars, and Algols. The common requirements for these characteristics seem to be a binary member with a convective envelope and strong tidal coupling forcing that binary member to rotate synchronously with the orbit. The link with convection is especially strong in Algols, where period derivatives are erratic if the cooler star has a spectral type

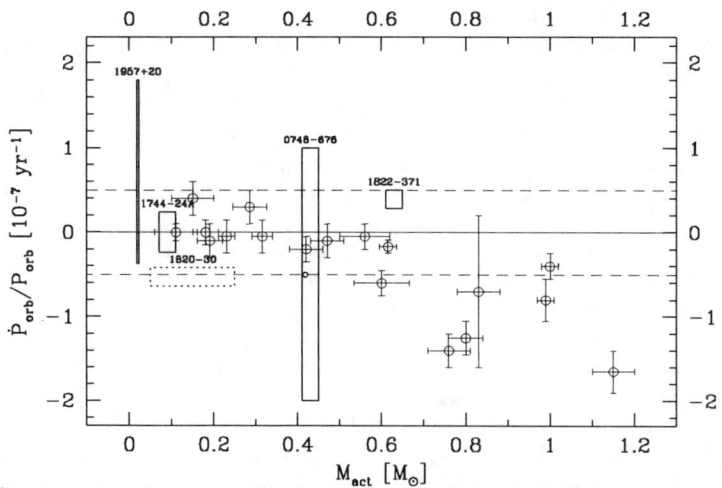

Fig. 1. A comparison between orbital period derivatives of CV's and CV-like systems (open circles) and close neutron star binaries (boxes) as a function of companion mass. The size of each box indicates the donor mass and error and observed range (1957+20 and 0748−676) or measured confidence interval (1-σ) for the orbital period derivative. The box for 1820−30 is dashed because the value is suspect and probably not even intrinsic if correct (see text). The dashed horizontal lines indicate the likely range of erratic period variability in CV's.

later than F8, but smooth and in agreement with simple conservative evolution if its spectral type is earlier (see Hall 1989).

In Fig. 1 I show a comparison between the orbital period derivatives measured in CV's and related systems (taken from Patterson 1984) and those of compact neutron star binaries. The similarity and the fact that the neutron star binaries fulfill the requirements for strong stellar activity listed above strongly suggests that the same mechanism is responsible for the same phenomenon in both types of system. This would argue against irradiation-driven evolution as an explanation for the neutron star binaries, since that cannot operate in most of the others. A wealth of possibilities for explaining short- and medium-term variability in neutron star binaries (in which the companion stars are very hard to study) by analogy with less rare close binaries suggest itself immediately. A case in point may be the eclipses in binary pulsars: there are a few (such as PSR 1718−19) in which evaporation is unlikely to be significant, but the mass loss required to explain the eclipses is well within the range observed in RS CVn stars and CV-like systems such as V 471 Tau (Wijers and Paczyński 1993).

I thank Joe Patterson, Bohdan Paczyński, and Paul Hertz for useful discussions. This work was supported by Compton Fellowship grant GRO/FPF-91-26.

REFERENCES

Arzoumanian, Z., Fruchter, A. S., and Taylor, J. H.: 1993, *private communication*
Asai, K., Dotani, T., Nagase, F., Corbet, R. H. D., and Shaham, J.: 1993, *Publ. Astron. Soc. Jpn. in press*, (ISAS research note 510)

Bhattacharya, D. and van den Heuvel, E. P. J.: 1991, *Phys. Rep.* **203**, 1-124
Deeter, J. E., Boynton, P. E., Miyamoto, S., Kitamoto, S., Nagase, F., and Kawai, N.: 1991, *Astrophys. J.* **383**, 324-329
Fruchter, A. S., Stinebring, D. R., and Taylor, J. H.: 1988, *Nature* **333**, 237-239
Hall, D. S.: 1989, *Space Sci. Rev.* **50**, 219-233
King, I. R., Stanford, S. A., Albrecht, R., Barbieri, C., Blades, J. C., Boksenberg, A., Crane, P., Disney, M. J., Deharveng, J. M., Jakobsen, P., Kamperman, T. M., Machetto, F., Mackay, C. D., Paresce, F., Weigelt, G., Baxter, D., Greenfield, P., Jedrzejewski, R., Nota, A., Sparks, W. B., and Sosin, C.: 1993, *Astrophys. J.* **413**, L117-L120
Nice, D. J. and Thorsett, S. E.: 1992, *Astrophys. J.* **397**, 249-259
Patterson, J.: 1984, *Astrophys. J., Suppl. Ser.* **54**, 443-493
Ruderman, M., Shaham, J., and Tavani, M.: 1989, *Astrophys. J.* **336**, 507-518
van der Klis, M. and Bonnet-Bidaud, J. M.: 1989, *Astron. Astrophys.* **214**, 203-208
van der Klis, M. H. K., Hasinger, G., Verbunt, F., van Paradijs, J., Belloni, T., and Lewin, W. H. G.: 1993, *Astron. Astrophys.* submitted
van Kerkwijk, M. H., Charles, P. A., Geballe, T. R., King, D. L., Miley, G. K., Molnar, L. A., van den Heuvel, E. P. J., van der Klis, M. H. K., and van Paradijs, J.: 1992, *Nature* **355**, 703-705
Wijers, R. A. M. J. and Paczyński, B.: 1993, *Astrophys. J.* **415**, L115-L118

THE FATE OF THORNE-ŻYTKOW OBJECTS

Philipp Podsiadlowski, Robert C. Cannon and Martin J. Rees
Institute of Astronomy, Cambridge CB3 0HA, U.K.

ABSTRACT

We briefly review the processes by which Thorne-Żytkow objects (TŻOs), i.e. red supergiants with neutron cores, are believed to form. The energy source in massive TŻOs is nuclear burning, provided by a modified rapid p process. After $\lesssim 10^6$ yr, this process is expected to break down, leading to a neutrino runaway and the collapse of the TŻO envelope. Part of the envelope will be accreted by the neutron-star core, which will be spun up and may be transformed into a black hole. The rest of the envelope is likely to form a centrifugally supported disk. This disk will ultimately become gravitationally unstable, possibly forming one or more self-gravitating objects (planets or low-mass stars) in the process. The final system may be a spun-up pulsar surrounded by planets, a low-mass X-ray binary, or a low-mass black-hole binary like V404 Cygni.

1. THE FORMATION OF THORNE-ŻYTKOW OBJECTS

Thorne-Żytkow objects (TŻOs) are a class of red supergiants which have degenerate neutron cores at their centers (Thorne & Żytkow 1977; also see Biehle 1991, 1993; Cannon et al. 1992; Cannon 1993). While no such object has yet been identified, they have been predicted to form by a variety of mechanisms:

- the direct disruptive collision of a low-mass main-sequence star with a neutron star in globular clusters,
- the complete coalescence of a neutron star with a massive companion, following a massive X-ray binary phase (Taam, Bodenheimer & Ostriker 1978),
- the disruption of a companion star by a newly formed neutron star, which received a supernova kick in the direction of the companion (Leonard, Hills & Dewey 1993).

In the present study, we are only interested in TŻOs that formed via the second route, i.e., TŻOs that are descendants of massive X-ray binaries (MXBs). In most MXBs, it is unavoidable that the massive component (the mass donor) will, at some point in its evolution, fill its Roche lobe. The resulting mass-transfer rate ($\dot{M} \sim 10^{-3} - 10^{-5}\, M_\odot\, \text{yr}^{-1}$) will exceed the Eddington accretion rate of the neutron star ($\sim 10^{-8}\, M_\odot\, \text{yr}^{-1}$) by several orders of magnitude. As a result, the neutron star is expected to be completely engulfed by the massive star, and the system will enter into a common-envelope phase. If the energy released in the subsequent spiral-in phase is not sufficient to eject the common envelope, the neutron star will settle at the center and the system will become a TŻO. Taam et al. (1978) concluded that a TŻO forms if the initial period of the MXB is less than ~ 100 d. This includes most MXBs with O/B supergiants and many Be X-ray binaries. The birthrate of TŻOs, $\nu_{\text{TŻO}}$, can be estimated from the observed number of MXBs with periods less than 100 d and their expected lifetimes before they fill their Roche lobes. A conservative estimate yields $\nu_{\text{TŻO}} \gtrsim 10^{-4}\, \text{yr}^{-1}$.

This can be compared to the birthrate of low-mass X-ray binaries (LMXBs), $\nu_{\rm LMXB}$. Assuming that there are ~ 100 LMXBs in the Galaxy and that the lifetime of the LMXB phase is longer than 10^7 yr, one obtains a generous estimate for the LMXB birthrate, $\nu_{\rm LMXB} \lesssim 10^{-5}$ yr^{-1}. Biehle (1993) and Cannon (1993) have estimated that a few of the 100 closest red supergiants could be TŻOs. In addition, SS 433 may be a system that in the transition to becoming a TŻO.

2. ENERGY GENERATION IN TŻOS

The outer appearance of a TŻO is that of a more-or-less normal red supergiant (with temperature $T_{\rm eff} \sim 3000$ K, radius $R \sim 10^3 R_\odot$). However, the internal structure is fundamentally different. In general, two types of TŻOs can be distinguished, based on their central energy source. In low-mass models (with envelope masses $\lesssim 8 M_\odot$), the main energy source is gravitational energy, released by the Eddington-limited accretion of matter onto the central neutron core. In massive models (with envelope masses $\gtrsim 14 M_\odot$), the energy source is nuclear energy. However, the region near the core which is hot enough for nuclear reactions to take place contains very little mass ($\sim 10^{-12} M_\odot$), only sufficient to produce the required surface luminosity ($\sim 10^5 L_\odot$) for a few seconds. Therefore, fresh fuel has to be continually injected from the envelope which serves as a fuel reservoir. This implies that the burning region has to be linked with the envelope by convection, i.e., the burning zone has to be at the base of the convective supergiant envelope. A further complication is that the convective turnover time in the burning region is only ~ 0.01 s, much shorter than the β^+-decay times of many weak interactions in the CNO-Ne cycle, which have lifetimes of up to ~ 100 s. As a result, the CNO-Ne reaction chain gets hung up (by the time the β^+-decays have occurred, the matter has moved out of the hot burning region) and cannot provide the necessary energy to support a TŻO envelope.

The problem of the missing energy source has recently been solved by Biehle (1991) and, in most detail, by Cannon (1993). They showed that the energy provided by the rapid proton process (rp process; Wallace & Woosley 1981) is enough to supply the luminosity of a TŻO. The rp process (strictly speaking the irp process, which stands for interrupted rp process; see Cannon 1993) consists of sequences of proton-capture reactions, which are terminated when the timescale for the next proton capture exceeds the β^+-decay time, e.g.,

$$^{23}{\rm Na}\,(p,\gamma)\,^{24}{\rm Mg}\,(p,\gamma)\,^{25}{\rm Al}\,(p,\gamma)\,^{26}{\rm Si}\,(\beta^+\nu)\,^{26}{\rm Al}.$$

The fact that TŻOs generate their luminosities by the exotic irp process has two important implications. One, it is possible that TŻOs provide the site for the generation of proton-rich elements, for which no site has unambiguously been identified in the past. Two, the presence of irp-process elements (e.g., Mo, Br, Rb, Y, Nb) in the photosphere of TŻOs provides a potential signature for distinguishing TŻOs from ordinary red supergiants (Biehle 1993).

3. THE NEUTRINO RUNAWAY

The steady-burning phase of a massive TŻO is terminated by either of two events. One is that the supply of irp-process seed elements is exhausted and that the irp process becomes inefficient (after $\sim 10^6$ yr). The second is that, due to the expected strong stellar wind, the envelope mass decreases below the minimum mass for nuclear burning (for typical red-supergiant wind-mass-loss rates, this will also happen after $\sim 10^6$ yr). At this point, a radiative zone develops somewhere between the burning region and the outer envelope, and the supply of fresh fuel to the burning region is choked off. In either of these two cases, the burning region runs out of fuel. As a result, the region just above the core will heat up until neutrino losses become the dominant energy-loss mechanism (at $\log T \gtrsim 9.4$), and accretion onto the neutron core is no longer Eddington limited. At this point, a neutrino runaway becomes unavoidable (also see Bisnovatyi-Kogan & Lamzin 1984).

In the neutrino runaway, all of the material that has less specific angular momentum than the maximum specific angular momentum allowed for a neutron star will fall onto the neutron star on a dynamical timescale. We can estimate the total amount of accreted mass by assuming that the TŻO envelope was in solid-body rotation (this is justified, since convection is very efficient in redistributing angular momentum) † and that the total angular momentum of the envelope is of order the initial orbital angular momentum of the binary. For an initial binary period P, the mass accreted is $\Delta M_{\rm acc} \sim 10^{-3}\, M_\odot\, P_{\rm 10d}^{-1/3}$, where $P_{\rm 10d} \equiv P/10\,{\rm d}$. The associated angular momentum can spin up a slowly rotating neutron star to a spin period $P_{\rm NS} \sim 30\,{\rm ms}\, P_{\rm 10d}^{2/3}$.

4. THE ENVELOPE COLLAPSE

The fate of the large TŻO envelope is less obvious and is mainly governed by the competition of its cooling timescale and the various (uncertain) viscous timescales (Krolik 1984). The most likely outcome is that most of the envelope will collapse into a centrifugally supported disk with a characteristic radius, $r_{\rm centr}$, of order, but somewhat smaller than, the initial binary separation, $r_{\rm centr} \sim 2 \times 10^{11}\,{\rm cm}\, P_{\rm 10d}^{2/3}$. The further evolution of this massive, disk-like structure depends on when gravitational instabilities start to develop and whether these lead to bar instabilities or fragmentation of the disk. This also depends on the amount of material accreted by the central object, which in turn is governed by the viscous timescale of the inner envelope. As a result, a variety of different outcomes are possible.

The central object may be become either a slightly spun-up neutron star or, perhaps more likely, a stellar-mass black hole. If the massive disk fragments into self-gravitating objects, these will ultimately form one or more low-mass objects. These may be planet-mass objects or, in view of the large available mass in the disk more likely, low-mass stars. Whichever the outcome, the result is bound to be of some interest. (1) The final system may be a (somewhat) spun-up pulsar, probably surrounded by one or more planets. (2) If the orbiting objects are low-mass stars (with a characteristic separation $r_{\rm centr}$), the system would be a potential progenitor for a low-mass X-ray binary. While only a fraction of all TŻOs would be sufficient to produce all LMXBs in the Galaxy, the spatial distribution of LMXBs in the Galaxy suggests that their progenitors are

---

† This implies that the material that is accreted onto the neutron star ($\sim 10^{-5}\, M_\odot$ in total) during the TŻO phase has little specific angular momentum; thus, the neutron star will be spun down rather than spun up in this phase.

either population II systems or population I systems that received a substantial supernova kick (e.g., Naylor & Podsiadlowski 1993). Since this is inconsistent with the distribution of MXBs and hence TŻOs, this possibility is not favoured. (3) The most likely outcome may therefore be a stellar-mass black hole. If, in addition, the disk fragments and forms one or more stellar-mass objects, the resulting system would be an excellent candidate for the progenitor of a soft X-ray transients like V404 Cygni.

5. SOFT X-RAY TRANSIENTS AND V404 CYGNI

V404 Cygni is a member of an apparently relatively common class of X-ray binaries which are widely believed to consist of a stellar-mass black hole and a low-mass companion. In V404 Cygni, the minimum mass of the compact object is 6.3 $M_\odot$ (Casares, Charles & Naylor 1992). Such systems are extremely difficult to form by conventional theories. Not only do conventional (common-envelope) models need an extreme mass ratio for the initial binary, which is not favoured observationally, but they also require that the orbital energy released by a low-mass star is sufficient to eject a very massive common envelope. On the other hand, the formation of such systems appears to be a rather natural outcome of the evolution of TŻOs.

Fortunately, this model allows several possible tests. First, one might expect that, in a significant fraction of systems, more than one companion is formed (i.e., a hierarchical multiple system). Second and most importantly, the companion should be metal-enhanced and show signatures of *irp*-processing (see Section 2). Interestingly enough, in the case of V404 Cygni, the companion has been reported to have an anomalously large ^{7}Li abundance (Martin et al. 1992). This can be immediately explained by our model, since, in a TŻO envelope, ^{7}Li is produced by the ^{7}Be-transport mechanism (Cameron 1955), just as it is in lithium stars (e.g., Sackmann & Boothroyd 1992). Thus, V404 Cygni provides direct support for our suggested scenario and we encourage the search of ^{7}Li and *p*-process elements in soft X-ray transients in general.

REFERENCES

Biehle, G. T. 1991, ApJ, 380, 167
Biehle, G. T. 1993, ApJ, in press
Bisnovatyi-Kogan, G. S. & Lamzin, S. A. 1984, Sov. Astr., 28, 187
Cameron, A. G. W. 1955, ApJ, 121, 144
Cannon, R. C. 1993, MNRAS, 263, 817
Cannon, R. C. et al. 1992, ApJ, 386, 206
Casares, J., Charles, P. A. & Naylor, T. 1992, Nat, 355, 614
Krolik, J. H. 1984, ApJ, 282, 452
Leonard, P. J. T., Hills, J. G. & Dewey, R. J. 1993, preprint
Martin, E. L., Rebolo, R., Casares, J. & Charles, P. A. 1992, Nat, 358, 129
Naylor, T. & Podsiadlowski, Ph. 1993, MNRAS, 262, 929
Sackmann, I.-J. & Boothroyd, A. I. 1992, ApJ, 392, L71
Taam, R. E., Bodenheimer, P. & Ostriker, J. P. 1978, ApJ, 222, 269
Thorne, K. S. & Żytkow, A. N. 1977, ApJ, 212, 832
Wallace, R. K. & Woosley, S. E. 1981, ApJS, 45, 389

IRRADIATION-DRIVEN EVOLUTION OF LOW-MASS X-RAY BINARIES

MARCO TAVANI
Joseph Henry Laboratories and Department of Physics
Princeton University, Princeton, NJ 08544

ABSTRACT

X-ray irradiation in binaries with accreting neutron stars can drive large mass outflows from the surface of companion stars. The resulting mass and angular momentum loss influence binary evolution. We summarize here recent results of detailed calculations of irradiation-driven mass loss rates and binary evolution of low-mass X-ray binaries.

INTRODUCTION

After about 25 years of observations, the evolution of low-mass X-ray binaries (LMXBs) is still somewhat mysterious. LMXBs constitute a heterogeneous class of compact binaries, including bright sources near the Eddington limit for solar mass compact stars, X-ray bursters, QPO sources, steady and transient sources (for a review see, e.g., van Paradijs, 1993). The canonical model of an LMXB is an accreting neutron star (of mass $m_1 = 1.4\, M_\odot$) orbiting around a low-mass ($m_2 \lesssim 1\, M_\odot$) main-sequence or white dwarf companion with orbital periods typically $P_{orb} \lesssim 1^d$ (e.g., White et al., 1993; hereafter WNP93). The evolution of LMXBs has been the subject of a series of theoretical investigations which reached maturity in the early '80. The obvious similarities between LMXBs and cataclysmic variables (CVs) led to consider the loss of binary angular momentum for Roche-lobe filling companions as the main mechanism of evolution (e.g., Rappaport et al., 1983). Binary evolution can be obtained for both CVs and LMXBs from the loss of angular momentum due to gravitational radiation (GR) and magnetic braking (MB). The CV distributions of orbital periods and mass loss rates inferred from accretion luminosities are in qualitative agreement with evolutionary models driven by GR and MB. The only problematic feature of the P_{orb}-distribution of CVs is the existence of a 'period gap' between 2 and 3 hrs which requires to consider when and how the companion star goes out of thermal equilibrium near $m_2 \sim 0.3\, M_\odot$ (e.g., Ritter, 1990).

However, despite obvious similarities, LMXBs are characterized by properties substantially different from CVs. *(1) Mass tranfer rates.* The distribution of X-ray luminosities is restricted in the range 10^{36} erg s^{-1} $\lesssim L_X \lesssim 10^{38}$ erg s^{-1}, suggesting that LMXBs spend most of their active phase with accretion rates of order of the Eddington limit, $\dot{m}_E \sim 10^{18}$ g s^{-1}, without reaching values of stable accretion rates less than $\sim 10^{-2}\dot{m}_E$. *(2) No period gap.* The P_{orb}-distribution of LMXBs does not show any sign of a period gap between 2 and 3 hrs (e.g., WNP93). *(3) Relatively large values of measured $\dot{P}_{orb}$.* Five LMXBs show non-zero (and in some cases, time variable) values of $\dot{P}_{orb}$ with an inferred timescale of order of $10^6 - 10^7$ years (Tavani, 1991b and references therein). On the other hand, the timescale of secular evolution is $10^8 - 10^9$ yrs for GR-MB-driven evolution. The behavior of $\dot{P}_{orb}$ in LMXBs might have the same origin as the one producing the time variable and large values of $\dot{P}_{orb}$ observed in *RS-CVn*-type binaries and in CVs (e.g., Hall, 1990). However, a negative and large value of $\dot{P}_{orb}$ has been determined also for *Her X-1* which has a companion of mass $m_2 = 2.2\,M_\odot$ and a radiative envelope (Deeter et al., 1991). Any explanation of the behavior of $\dot{P}_{orb}$ in RS-CVn-type binaries and in CVs which assumes the existence of a solar-like activity of companionis with convective envelopes (e.g., Hall, 1990) cannot be applied to Her X-1.

For all of these reasons, it is important to consider a mechanism of LMXB evolution that can explain the observed features of $L_X, P_{orb}, \dot{P}_{orb}$ in a coherent way. The feature clearly distinguishing LMXBs from CVs is an X-ray luminosity which is larger by four-five orders of magnitude for LMXBs compared to that one of CVs. The study of irradiation-driven effects in LMXBs has a long history (e.g., London et al., and reference therein) and it was recently reconsidered in the light of a possible connection between progenitor LMXBs and millisecond pulsars (Ruderman et al., 1989a,b; Tavani, 1991a; Shaham, 1992). Further analytical developments of the irradiation-driven (ID) model led to calculations of typical LMXB lifetimes $\tau \sim 10^6 - 10^7$ years and of the binary evolution (Tavani, 1990, 1991b). Here we briefly report results of detailed calculations of ID-mass loss rates, captured fractions of mass losses and outflow angular momentum losses. These calculations are aimed at supporting complete models of LMXB evolution (note that ID-evolution can be applied to both black hole and neutron star binaries). Computational techniques for the study of the hydrodynamics and atomic physics processes influencing ID-outflows make possible today the calculation of ID-LMXB evolution from first principles.

IRRADIATION-DRIVEN MASS OUTFLOW IN X-RAY BINARIES

Irradiating photons in the energy band between 0.3 - 10 keV as well as soft γ-rays of energy $\sim 1/2$ MeV are the most efficient in driving an evaporative wind from an illuminated atmosphere of a companion star in X-ray binaries (London et al., 1981; Ruderman et al., 1989b). In the presence of an external irradiating flux, the ionization-recombination and heating-cooling balance conditions are satisfied at a relatively large depth (new irradiated photosphere) but they cannot be maintained for low values of both density and temperature in the outer and more dilute atmospheric layers. As the density in the outer layers decreases, only a rapid rise in temperature can keep the gas medium in balance with the radiation field. An ID-wind is a consequence of a thermal instability of the irradiated atmosphere developing a high temperature corona. The ID- mass loss rate can be written in spherical geometry as $|\dot{m}_{rad}| = 2\pi R_s^2 \mu n^* z(R_s) v_{w,s}$ where R_s is the sonic point radius, μ the mean molecular weight, $z(r)$ a dimensionless quantity which gives the number density as a function of radial distance r normalized to unity at the base of the corona, and n^* the number density at the base of the corona where heating equals cooling. The quantity n^* depends on the irradiating spectrum, i.e., $n^* \propto \Upsilon S$, with Υ a dimensionless quantity depending on the spectrum and S a dimensionless parameter of order unity obtained from the atomic physics processes determining the cooling at the base of the expanding corona. For typical LMXB spectra, $1 \lesssim \Upsilon S \lesssim 10^2$ (Tavani & London, 1993; hereafter TL93). The effective irradiation of the companion depends also on the solid angle subtended by the companion star and by an attenuation factor χ which takes into account possible absorption/screening and scattering/reprocessing effects in the disk or corona surrounding the compact star (typically χ is assumed to be $\lesssim 0.1$ in the calculations reported here). The calculation of the ID-mass outflow rate in LMXBs requires the study of atomic physics processes at the base of the expanding corona as a function of the irradiating flux and spectrum (London et al., 1981; 1994). The sonic point is determined by a hydrodynamic code in spherical geometry (TL93).

IRRADIATION-DRIVEN EVOLUTION

The evolution of the orbital period is given by (e.g., Tavani, 1991b)

$$\frac{\dot{P}_{orb}}{P_{orb}} = -3\left[1 - (1-\beta)h_{cm} - \beta q - \frac{1}{3}(1-\beta)\frac{q}{1+q}\right]\frac{\dot{m}_2}{m_2} + 3\left[\left(\frac{\dot{j}}{J}\right)_{GR} + \left(\frac{\dot{j}}{J}\right)_{MB}\right] \quad (1)$$

where $q = m_2/m_1$, $\dot{m}_2 = \dot{m}_{2,rad} + \dot{m}_{RL}$, with $\dot{m}_{RL}$ the mass loss rate due to Roche lobe overflow of the companion, $\beta = \dot{m}_{1,acc}/\dot{m}_2$, with $\dot{m}_{1,acc}$ the mass accretion rate onto the surface of the neutron star, and $\dot{J}/J$ the fractional rate of change of the orbital angular momentum. In Eq. (1) h_{cm} is the dimensionless value of the specific angular momentum of the outflow in the center of mass frame, $h_{cm} = <j_w>/j_{cm}$, where $j_{cm} = \omega\, a^2/(1+q)$, with ω the orbital angular frequency and a the orbital radius. Let us define $\Psi \equiv R/R_L$, with R and R_L the companion and its Roche lobe radii, respectively. We obtain (Tavani, 1990)

$$\frac{\dot{\Psi}}{\Psi} = 2D(n, h_{cm}, \beta, q)\frac{\dot{m}_2}{m} - 2\left[\left(\frac{\dot{j}}{J}\right)_{GR} + \left(\frac{\dot{j}}{J}\right)_{MB}\right] + \frac{\dot{K}}{K} \quad (2)$$

with

$$D(n, h_{cm}, \beta, q) = \frac{5}{6} - \frac{1}{2}n - \frac{(1-\beta)}{1+q}\left[\frac{q}{3} + h_{cm}(1+q)\right] - q\beta. \quad (3)$$

Here n is the effective stellar index of the companion star related to the stellar response (expansion/contraction) to its mass loss rate through the relation $R = K\,m^{-n}$. The term $\dot{K}/K$ of Eq. (3) takes into account both possible intrinsic expansion/contraction and ID-'bloating' of the companion (e.g., Hameury et al., 1993) for a constant mass m. Roche lobe overflow of the companion is obtained when $\Psi = 1$; ID-evolution of LMXBs allows $\Psi < 1$.

In the simplest model of *self-sustained* LMXB evolution with $h_{cm} \lesssim 1$, the companion star has a *secular* tendency of evolving inside its Roche lobe with a positive $\dot{P}_{orb}$, and the ID-mass loss rate (corresponding to an accretion luminosity $L_X \propto \beta|\dot{m}|_2$) can reach $|\dot{m}|_{2,rad} \sim 10^{18} - 10^{19}$ g s^{-1}. As orbital radius increases, the ID-mass transfer is more and more difficult to be self-sustained because of the decreasing irradiating flux and solid angle of the companion star as it loses mass (Ruderman et al., 1989a,b). A transition occurs at a value of the companion mass m_c which corresponds to a sudden decrease of the accretion rate or even a permanent accretion turn-off if a millisecond pulsar has been produced (Ruderman et al., 1989a). The characteristics of the self-sustained evolutionary phase are quite different from the canonical orbital evolution that neglects irradiation effects: (1) the mass loss is driven at a relatively large rate which is weakly dependent on orbital radius; (2) the companion may contract inside its Roche lobe and still transfer mass; (3) the orbital evolution can have $\dot{P}_{orb} > 0$ and/or $\dot{P}_{orb} < 0$ depending on the values of β and h_{cm} (Tavani, 1991b).

DETAILED CALCULATIONS

Mass loss rates and sonic points

A calculation of the hydrodynamics of the ID-mass loss has been carried out for a variety of orbital parameters and for the range $1 \lesssim \Upsilon S \lesssim 10^2$ appropriate to LMXBs (TL93). The calculation assumes that only a small fraction of the available X-ray flux irradiates the companion ($\chi = 0.1$) and that the outflow emerges from the half-irradiated emisphere of a Roche lobe underfilling companion. The results show that the irradiating X-ray fluxes and spectra of relatively bright LMXBs with $\Upsilon S \gtrsim 10$ are able to self-sustain mass transfer in the range $|\dot{m}|_{2,rad} \sim 10^{18} - 10^{19}$ g s^{-1}. For bright sources the outflow velocity at the sonic radius

Table 1: Captured fraction of mass loss in ID-LMXBs

m $M_\odot$	a (10^{11} cm)	$\mathcal{M}$	R_c (10^{10} cm)	R_d (10^{11} cm)	h_d (10^{10} cm)	β
0.57	1.4	1	3.5	0.8	2.0	0.312
”	1.8	1	”	”	”	0.231
”	2.0	1	”	”	”	0.176
0.57	1.4	3	”	”	”	0.122
”	1.8	3	”	”	”	0.065
”	2.0	3	”	”	”	0.063

is of order of the orbital velocity and the sonic radius is typically larger than the companion radius by less than 10%.

Captured fraction of the ID-mass loss rate: β
Since the outflow velocity of ID-outflows v_w is comparable with the orbital velocity v_{orb} in typical LMXBs (LT93), only a large-scale three dimensional calculation of the hydrodynamics can determine β for a variety of initial conditions. The calculation of the captured fraction of the mass outflow from the companion surface depends on v_w/v_{orb}, orbital parameters and the size of the accretion disk capturing the outflow. Table 1 summarizes the results obtained by a three-dimensional SPH model of a ID-LMXB (with $m_2 = 0.57\,M_\odot$) for radial injection from the irradiated half-emisphere of the companion (Brookshaw & Tavani, 1994). Table 1 gives an example of the dependence of β on orbital distance a and Mach number $\mathcal{M}$ for a typical outflow temperature $T = 10^7$ K. The mass loss rate has been chosen to be $|\dot{m}|_{rad} \sim 10^{18}\,\mathrm{g\,s^{-1}}$ for a companion that is progressively shrinking inside its Roche lobe as the orbital radius increases. R_d and h_d are the radius and the outer height of the accretion disk assumed to be an ideal absorbing surface (note that the accretion radius $R_G = 2\,G\,m_1/v_w^2$ is of order of R_d).

Average angular momentum loss
The angular momentum loss is crucial in determining the orbital evolution of ID-LMXBs. The average value of the specific angular momentum of the outflow leaving the binary, $<j_w>$, can be substantially different than the angular momentum of the center of mass of the companion, $h_{cm}^* = 1/(1+q)$. A ballistic three dimensional calculation of $<j_w>$ for a variety of initial conditions ($q, v_w/v_{orb}$) confirms that h_{cm} can be even larger than h_{cm}^* (Brookshaw & Tavani, 1993; hereafter BT93. See also Banit & Shaham, 1992). In particular, for $v_w/v_{orb} \sim 1$, the absorption of a component of the outflow by an accretion disk can produce a final h_{cm} systematically different and larger than the value of h_{cm} computed without an accretion disk. Furthermore, h_{cm} depends crucially on the selected area of mass mass injection from the companion. A possible cause for the selection of a specific injection area of the companion surface is a surface magnetic field able to influence the outflow (Tavani, 1994). The injection surface areas on the mass losing star associated to specific values of h_{cm} are shown in BT93. A value of h_{cm} larger than unity can be obtained for a particular choice of binary and outflow parameters. A temporary value $h_{cm} > 1$ can explain the negative (and non-stable) values of $\dot{P}_{orb}$ observed in *4U 1820-30* (van der Klis *et al.*, 1993), *EXO 0748-676* (Asai *et al.*, 1993), and in the eclipsing pulsar system PSR 1957+20 (Arzoumanian *et al.*, 1993). Time dependent effects can alter the outflow injection geometry and the large scale flow producing a time variable $\dot{P}_{orb}$. The interpretation of the observed time variable values of $\dot{P}_{orb}$ in

compact binaries involves a study of angular momentum losses, tidal effects and companion bloating (Tavani, 1994). Since the ID-outflows in LMXBs are characterized by a gas velocity comparable with orbital velocity a three dimensional hydrodynamical calculation of h_{cm} is necessary. This calculation is now in progress (Brookshaw & Tavani, 1994).

Self-sustained orbital evolution

A calculation of the ID-evolution of LMXBs must take into account mass outflows as well as the response of the star to mass loss with a self-consistent determination of the stellar index n of Eqs. (2) and (3). The calculation requires the use of $|\dot{m}|_{rad}$, β, h_{cm} as functions of orbital and outflow parameters. The companion star is initially driven into Roche lobe contact by GR evolution of a detached binary ('Roche-lobe trigger'). The subsequent evolution is followed taking into account mass loss and angular momentum loss from the binary. Preliminary results of LMXB orbital evolution with the use of a stellar evolution code which includes the effects of GR, MB and ID-outflows confirm that ID-LMXBs have relatively short accreting phases ($\tau \sim 10^6 - 10^7$ years) with a few accreting episodes for $h_{cm} < 1$ (Becker *et al.*, 1993). Previous analytic models of ID-evolution (Tavani, 1990; 1991a,b) are therefore supported by more detailed calculations.

The author thanks C. Becker, L. Brookshaw, R. London, S. Rappaport, M. Ruderman, J. Shaham and E.P.J. van den Heuvel for many informative discussions.

REFERENCES

Asai, K., *et al.*, 1993, *Pub. Astr. Soc. Japan*, **44**, 633.
Arzoumanian, Z., Fruchter, A., and Taylor, J.H., 1993, submitted to the *Ap.J.*
Banit, M. & Shaham, J., 1992, in *X-Ray Binaries and Recycled Pulsars*, eds. E.P.J. van den Heuvel and S.A. Rappaport (Dordrecht: Kluwer), p. 401.
Becker, C., Tavani, M., and Rappaport, S., 1993, in preparation.
Brookshaw, L. & Tavani, M., 1993, *Ap.J.*, **410**, 719 (BT93).
Brookshaw, L., & Tavani, M., 1994 to be submitted to the *Ap.J.*
Deeter, J.E., *et al.*, 1991, *Ap.J.*, **383**, 324.
Hall, D.S., 1990, in *Active Close Binaries*, ed. C. Ibanoglu (Dordrecht: Kluwer), p. 95.
Hameury, J.-M., King, A.R., Lasota, J.-P., Raison, F., 1993, *A.&A.*, **277**, 81.
London, R.A., McCray, R., and Auer, L.H., 1981, *Ap.J.*, **243**, 970.
London, R.A., *et al.*, 1994, in preparation.
Patterson, J., 1984, *Ap.J.Suppl.Series*, **54**, 443.
Rappaport, S., Verbunt, F., and Joss, P.C., 1983, *Ap.J.*, **275**, 713.
Ritter, H., 1992, *A.&A.S.*, **85**, 1179.
Ruderman, M., Shaham, J., and Tavani, M., 1989, *Ap.J.*, **336**, 507 (RST).
Ruderman, M., Shaham, J., Tavani, M., and Eichler, D., 1989b, *Ap.J.*, **343**, 292 (RSTE).
Shaham, J., 1992, in *X-Ray Binaries and Recycled Pulsars*, eds. E.P.J. van den Heuvel and S.A. Rappaport (Dordrecht: Kluwer), p. 375.
Tavani, M., 1990, preprint.
Tavani, M., 1991a, *Ap.J.(Letters)*, **366**, L27.
Tavani, M., 1991b, *Nature*, **351**, 39.
Tavani, M., and London, R., 1993, *Ap.J.*, **410**, 281.
Tavani, M., 1994, to be submitted to the *Ap.J.*
van der Klis, M., *et al.*,, *M.N.R.A.S.*, **260**, 686.
van Paradijs, J., 1993, in *X-Ray Binaries*, eds. W.H.G. Lewin, J. van Paradijs and E.P.J. van den Heuvel, Cambridge University Press.
White, N.E., Nagase, F. and Parmar, A.N., 1993, in *X-Ray Binaries*, eds. W.H.G. Lewin, J. van Paradijs and E.P.J. van den Heuvel, Cambridge University Press, (WNP93).

X-Ray Pulsars

THE PROPERTIES OF ACCRETING X-RAY PULSARS

Arvind Parmar
EXOSAT Observatory, Space Science Department of ESA,
ESTEC, 2200 AG Noordwijk, The Netherlands.
aparmar@astro.estec.esa.nl

ABSTRACT

The properties of accreting X-ray pulsars are reviewed paying particular attention to their pulse and orbital periods and high-energy X-ray spectra. The importance of making broad-band X-ray measurements of these systems is demonstrated.

1. INTRODUCTION

An X-ray binary pulsar contains a neutron star accreting material from a companion star. The strong (typically $\sim 10^{12}$ G) magnetic field of the neutron star disrupts the accretion flow at several hundred neutron star radii (Pringle & Rees 1972; Davidsen & Ostriker 1973) and X-ray pulsations are observed as the beamed emission from the magnetic polar caps rotates through the line of sight.

Depending on the nature of the companion, X-ray pulsars can be classified into three categories; (1) those where the companion has evolved away from the main sequence and is a supergiant with spectral type earlier than B2, (2) those where the companion is a Be or Oe star on, or close to, the main sequence, or (3) those in low-mass X-ray Binaries (LMXRBs) where the companion is of spectral type later than A. Since there are only four LMXRB pulsar systems (Her X-1; Tananbaum et al. 1972, X 2259+587; Gregory & Fahlman 1980, X 1626-673; Rappaport et al. 1977 and GX 1+4; Lewin, Ricker & McClintock 1971) it is difficult to generalize about their properties.

Many of the supergiant systems show eclipses by the companion star, strong intensity and absorption variability and in some cases high and low states. The orbital periods are generally $\lesssim$ 10 days. The supergiants lose $\sim 10^{-6} M_\odot$ yr^{-1} in an approximately spherical wind with a terminal velocity of a few $\sim 10^3$ km s^{-1}. Only a small fraction of this wind is captured by the neutron star. Supergiants may also lose mass by incipient Roche lobe overflow (Day & Stevens 1993). X-ray luminosities are in the range $10^{36} - 10^{38}$ erg s^{-1} and examples include SMC X-1 (Lucke et al. 1976) and Cen X-3 (Giacconi et al. 1971).

The Be X-ray systems are often transient sources (Maraschi et al. 1976). Doppler variations in the pulse period give orbital periods of a few tens of days, with a moderate eccentricity ($e \sim$0.3). Such transients include X 0115+63 (Rappaport et al. 1978), X 1553-542 (Kelley, Rappaport & Ayasli 1983), X 0332+53 (Stella et al. 1985) and EXO 2030+375 (Parmar et al. 1989). The outbursts are most likely caused by mass ejection episodes from the Be star, and are probably related to the fact that Be stars rotate close to break up. There are also Be X-ray binaries that are persistent sources. Their X-ray luminosities are generally low, $10^{33} - 10^{35}$ erg s^{-1}. The prototype system in this class is the Be star X Per. The orbital periods are not well known, but the low X-ray luminosity and lack of eclipses suggests that they are long, of order several hundred days.

2. PULSE PERIODS

Table 1 lists the 32 X-ray pulsars currently known, their pulse and orbital periods, and whether the system is an LMXRB or high-mass X-ray binary (HMXRB). The pulse periods lie between 0.069 to 835s, with no evidence for a clustering at any particular period. In Figure 1 the pulse periods are shown versus orbital period for those systems where both are known. The Figure is divided into three depending on whether the underlying system is an LMXRB, or a Be star or supergiant system. For the Be star systems there is a strong correlation between the orbital and pulse periods which was first noticed by Corbet (1984). The Supergiant systems show no obvious dependance. For the LMXRB there are only three points, which makes it hard to draw any firm conclusions.

Table 1. The Pulse Periods of X-ray Binaries

Source	Alternative Name	Pulse Period (s)	Orbital Period (day)	Type	Reference
X 0535−668	A 0538−66	0.069	16.7	HMXRB	1
X 0115−737	SMC X−1	0.71	3.89	HMXRB	2
X 1656+354	Her X−1	1.24	1.7	LMXRB	3
X 0115+634	V635 Cas	3.6	24.3	HMXRB	4
X 0332+530	BQ Cam	4.4	34.25	HMXRB	5
X 1119−603	Cen X−3	4.8	2.1	HMXRB	6
X 1048−594		6.4		?	7
X 2259+587		7.0		LMXRB	8
X 1627−673		7.7	0.029	LMXRB	9
X 1553−542		9.3	30.6	HMXRB	10
X 0834−430	GR 0834−430	12.2	-	?	11
X 0532−664	LMC X−4	13.5	1.4	HMXRB	12
X 1417−624		17.6		HMXRB	13
X 1843+009		29.5		?	14
X 1657−415	OAO 1653-40	37.9	10.4	HMXRB	15
X 2030+375		42	45.6	HMXRB	16
X 2138+568	Cep X−4	66		?	17
X 1836−045		81		?	14
X 1843−024		95		?	14,34
X 0535+262		104	111	HMXRB	18
X 1833−076	Sct X−1	111		?	19
X 1728−247	GX 1+4	114	304?	LMXRB	20,21,22
X 0900−403	Vela X−1	283	8.96	HMXRB	23
X 1258−613	GX 304−1	272	133?	HMXRB	24,25
X 1145−614		298		HMXRB	26,27
X 1145−619		292	187.5	HMXRB	26,27
X 1118−615	A 1118−61	405		HMXRB	28
X 1722−363		413		?	29
X 1907+097		438	8.38	HMXRB	30
X 1538−522	QV Nor	529	3.73	HMXRB	31
X 1223−624	GX 301−2	696	41.5	HMXRB	32
X 0352−309	X Per	835	-	HMXRB	33

References: [1]Skinner et al 1982; [2]Lucke et al 1976; [3]Tananbaum et al 1972; [4]Cominsky et al 1978; [5]Stella et al 1985; [6]Giacconi et al 1971; [7]Corbet & Day 1990; [8]Gregory & Fahlman 1980;

[9]Rappaport et al 1977; [10]Kelley, Rappaport & Ayasli 1983; [11]Grebenev & Sunyaev 1991; [12]Kelley et al 1983; [13]Kelley et al 1981; [14]Koyama et al 1990a; [15]White & Pravdo 1979; [16]Parmar et al 1989; [17]Koyama et al 1991a; [18]Rosenberg et al 1975; [19]Koyama et al 1991b; [20]Lewin, et al 1971; [21]White et al 1976a; [22]Strickman et al 1980; [23]McClintock et al 1976; [24]Huckle et al 1977; [25]McClintock et al 1977; [26]White et al 1978; [27]Lamb et al 1980; [28]Ives et al 1975; [29]Tawara et al 1989; [30]Makishima et al 1984; [31]Davison et al 1977; [32]White et al 1976a; [33]White et al 1976b; [34]Koyama et al 1990b;

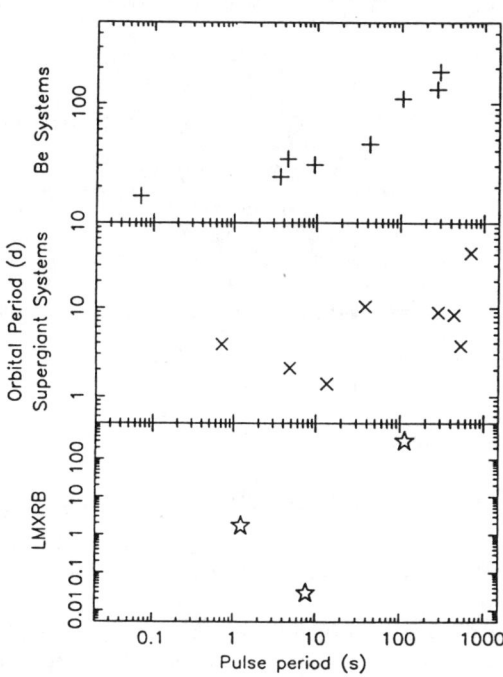

Fig. 1. The distribution of pulse period versus orbital period for the LMXRB and the HMXRB Be and supergiant systems.

Long term-monitoring of X-ray pulsars has revealed that the pulse periods vary with time in two ways. In some systems, the periods show an overall linear decrease with time. In others, the periods show erratic random walk variations with no obvious trend in either direction. The pulse period histories of the two types of behavior are illustrated in Figure 2. The trend to decreasing period, or secular spin-up, is expected because the neutron star will gain the angular momentum of the accretion flow, at the point where it attaches to the magnetic field lines (Pringle & Rees 1972; Lamb, Pethick & Pines 1973; Davidson & Ostriker 1973). The random walk behaviour is harder to understand and suggests that in these cases the accretion flow does not include a stable accretion disk.

The strong spin-up seen in systems such as Her X-1, Cen X-3, SMC X-1 and 4U 1626-67 suggests that the accretion flow is mediated by an accretion

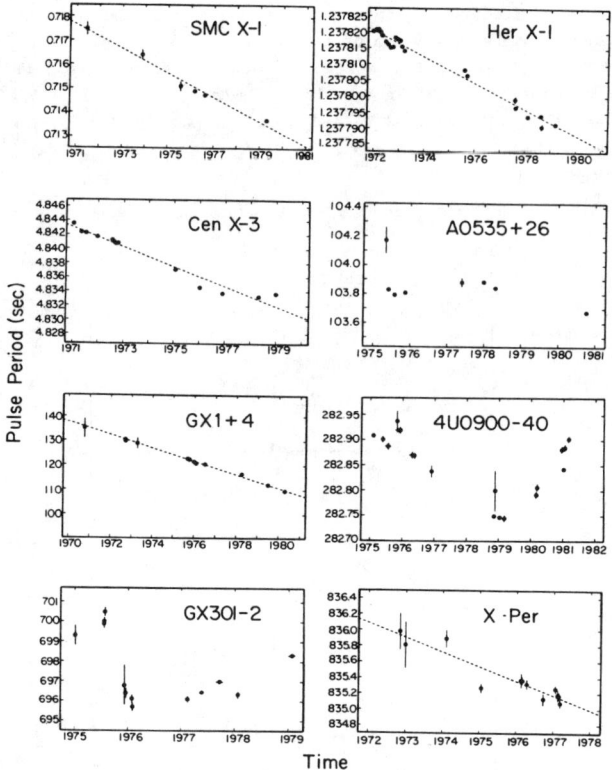

Fig. 2. The spin history of a number of X-ray pulsars

disk. The disk will be disrupted at the Alfvén or magnetospheric radius, r_m. There are several different ways to calculate r_m, but the simplest is to equate the ram pressure of the inflowing material for spherical accretion, to the magnetic pressure (see *e.g.* Henrichs 1983). This gives

$$r_m = 3 \times 10^8 \mu_{30}^{4/7} M_x^{1/7} R_6^{-2/7} L_{37}^{-2/7} \, cm,$$

where $\mu_{30} = \mu/10^{30}$ is the magnetic moment of the neutron star in gauss cm^3; M_x is the mass of the X-ray source in solar units; R_6 is the radius of the neutron star in 10^6 cm and L_{37} is the X-ray source luminosity in units of 10^{37} erg s^{-1}. In the case of an accretion disk, the magnetospheric radius is approximately half that of spherical accretion (Ghosh & Lamb 1978, 1979). This leads to an empirical relationship between the spin-up timescale $P_p/\dot{P}_p$, and the parameter $(P_p L^{6/7})^{-1}$. The constant is primarily determined by the magnetic dipole moment and the mass of the neutron star. The measured values of these parameters for the pulsars that exhibit strong spin-up are in good agreement (Rappaport & Joss 1977; Mason 1977).

The spin-up timescale for the pulsars where a stable secular spin-up is seen ranges from 100–100,000 yrs. This is much shorter than the evolutionary timescale of these systems, suggesting a relatively short duration of rapid spin-up activity. A pulsar will be spun-up to the point where the neutron star magnetosphere corotates with the inner edge of the accretion disk. This critical radius is called the corotation radius, r_c. Elsner & Lamb (1976) divided pulsars into two types: *slow rotators* where $r_c > r_m$ and *fast rotators* where $r_c \simeq r_m$. The boundary layer between the unperturbed disk and the magnetosphere is complex, and there will be a transition region where the magnetic field begins to thread the disk. As the neutron star reaches corotation the spin-up torque diminishes. Ghosh & Lamb (1978, 1979) introduced a dimensionless torque function, called the *fastness parameter* which describes the reduced torque. Close to corotation the field lines in the transition region are swept backwards, and a negative torque exerted on the neutron star. This means that the star can be spun-down, even though accretion continues. In Figure 3 the $\dot{P}_p$ verses $P_p L^{6/7}$ values for a number of pulsars exhibiting strong spin-up are shown, and compared to the theoretical relationship determined by Ghosh & Lamb (1979). The agreement is good. In the X-ray transient system, EXO 2030+375, this relationship has been directly measured over a wide range of luminosity (Parmar *et al.* 1989) and again the agreement is good.

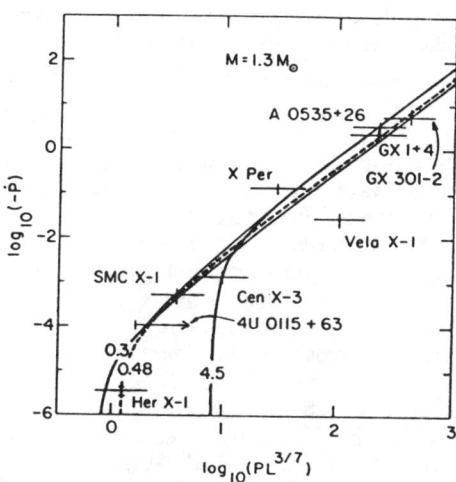

Fig. 3. $\dot{P}_p$ verses $P_p.L^{6/7}$ for a number of X-ray pulsars, superposed on the theoretical relation for the spin-up of an accreting magnetized neutron star (Ghosh & Lamb 1979).

In 1983 March the previously bright pulsar GX 1+4 entered a low state, a factor of 20 below its typical high state intensity (Hall & Davelaar 1983). Subsequent observations by *Ginga* in 1987 and 1988 by Makishima *et al.* (1988) detected pulsations in the low state. The measured pulse period showed that the pulsar was spinning down, even though it was still accreting material. This is a good confirmation of the Ghosh & Lamb accretion torque model. Similar erratic behaviour has also been seen from Her X-1 and Cen X-3, and also seems to be associated with the source entering a low state.

In the lower luminosity systems no overall trend in the pulse period is seen, although there are still large fluctuations in pulse period *e.g.* Vela X-1, GX 301-2 and X Per. These fluctuations persist down to timescales at least as short as a few days and can be described as white noise fluctuations in the angular acceleration of the neutron star (Boynton *et al.* 1984; Deeter *et al.* 1987). In these systems it seems likely the X-ray emission is driven by wind accretion, and that the fluctuations in the pulse period reflect inhomogenieties in the wind or accretion flow.

3. PERIOD EVOLUTION

The evolution of the period of a neutron star throughout its lifetime will depend on the period at birth, the efficacy of spin down and the history of the mass loss from the primary. The spin down efficacy is not well understood, nor is the period distribution at birth (Henrichs 1983). Based on the observations of so many long period X-ray pulsars either spin down is a relatively efficient process, or pulsars are born slow rotators (Lea 1976).

Corbet (1984) was the first to notice that the pulse, P_p, and orbital periods, P_o, of the Be star X-ray binaries are correlated with $P_p \propto P_o^2$. Corbet proposed that this is because these pulsars are *fast rotators* with $r_c \simeq r_m$. The condition $r_c = r_m$ combined with Keplers law and an assumption that the wind is radially expanding at a constant velocity gives $P_p \propto P_o^{4/7}$ (Corbet 1984; Stella, White & Rosner 1986; Van den Heuvel & Rappaport 1987; Waters & van Kerkwijk 1989). This is consistent with the P_P, P_o distribution found for the longer period pulsars in supergiant systems, but not for the Be star systems. However, even for the supergiant systems, the constant requires a mass accretion rate two orders of magnitude lower than observed (Van den Heuvel & Rappaport 1987, Stella, White & Rosner 1986). Assuming $r_c = r_m$ occurred at some point in the spin history of the pulsar, then the supergiant distribution on the P_p, P_o diagram may reflect an earlier evolutionary phase (Waters & van Kerkwijk 1989).

In Be star systems the outflow in the equatorial region is quite different from the poles. An infra-red excess in the spectra of several of the Be X-ray binaries has been shown by Waters *et al.* (1988) to be due to a dense equatorial ring flowing outwards with a much more gradual velocity law than a radially expanding wind. Many of the Be X-ray binaries are much more luminous than predicted for accretion from the Be stars wind (White *et al.* 1982) and enhanced flow in the rotation plane may explain this (Waters *et al.* 1988). Waters & van Kerkwijk (1989) show that for a wind density distribution $\rho(r) = \rho_o(r/R_\star)^{-n}$ then $P_p \propto P_o^{2/7(4n-6)}$. The Be star HMXRB P_p, P_o distribution requires $n \sim 3.5$, comparable to the value found independently from infra-red measurements (Waters & van Kerkwijk 1989).

Stella, White & Rosner (1986) noticed that many Be star transient pulsars turn off abruptly during their decay. They proposed that this results from the centrifugal inhibition of accretion which occurs when the magnetospheric radius exceeds the corotation radius. The magnetospheric radius increases with decreasing mass accretion rate and $r_c > r_m$ may occur during the decay of an outburst, so causing the abrupt turn off. This cut-off point translates to a minimum luminosity given by

$$L_{\min} = 2.5 \times 10^{37} \; R_6^{-1} \; M_x^{-2/3} \; \mu_{30}^2 \; P_p \quad \text{erg s}^{-1},$$

where P_p is the pulse period in seconds. As the pulsar orbits the companion relatively small changes in accretion rate around the eccentric orbit can, for a particular mass loss rate from the Be star, be amplified into a large orbital modulation by this effect. If the luminosity can be measured just before the transient turns off, it provides an estimate of the magnetic dipole moment.

The centrifugal barrier means that an X-ray outburst may not coincide exactly with the optical outburst. This has been seen from 4U 0115+63 where an outburst in December 1980 was preceeded some months earlier by an optical brightening of the Be star, suggesting that the shell episode began much earlier (Kriss et al. 1980). The delay between the X-ray and optical outburst is naturally explained by the centrifugal barrier preventing accretion until late in the mass ejection episode (Stella, White & Rosner 1986). Long term monitoring of V 635 Cas, the optical counterpart to 4U 115+63 by Mendelson & Mazeh (1991) between 1985-1990 revealed three optical outbursts, although only two of them were accompanied by X-ray outbursts. Mendelson & Mazeh (1991) suggest the relatively short 1988 optical outburst did not appear in X-rays because the accretion rate was insufficient to overcome the centrifugal barrier.

The most extreme Be star system found to date is the 69 ms pulsar in the Be X-ray binary A 0538-66 located in the LMC (Skinner et al. 1982). This system sometimes undergoes super Eddington luminosity outbursts. With such a rapid rotation period it must have a magnetic field a factor of ten less than the others to overcome the centrifugal barrier. This suggests that it is an older system where the magnetic field has decayed significantly and that it is nearing the end of the Be X-ray binary phase. This is consistent with the slightly evolved state of the Be star (Pakull & Parmar 1981).

4. PULSE PROFILES

Most X-ray pulse profiles are broad with duty cycles as large as $\sim 50\%$, in contrast to the sharp pulse profiles observed from radio pulsars where typical duty cycles are $\lesssim 3\%$. The pulse shapes are often highly asymmetric and the pulse fractions range from $\sim 10\%$ to 90 %. The X-ray pulse profiles vary dramatically from source to source and may show strong energy dependence (see e.g. White, Swank & Holt 1983). Examples of X-ray pulse profiles are shown in Figure 4 for four bright pulsars: Her X-1; a disk-fed LMXRB, Cen X-3; a luminous HMXRB with a supergiant companion, Vela X-1; a typical wind-fed binary with a relatively low luminosity, and 4U 0115+63; a transient pulsar in a Be-star binary system.

Figure 4 illustrates that all four pulsars show a dependence of profile on energy. In general, there is a tendency for the pulse shape to become simpler and the pulse fraction to become larger at higher energies. This effect can be most clearly seen in the profiles of Vela X-1 and 4U 0115+63. Cen X-3 is a pulsar which normally shows a relatively simple single peaked profile with a weak interpulse, without prominent energy dependence (e.g. White, Swank & Holt 1983; Nagase 1989). However, during *Ginga* observations in March 1989 it showed a double-peak structure at low energies as seen in Figure 4.

It has become clear that X-ray pulse profiles can strongly depend on X-ray luminosity. A good example was obtained from *EXOSAT* observations of the transient X-ray pulsar EXO 2030+375 during the decay phase of its outburst in 1985 (Figure 5). At a high luminosity of 1×10^{38} erg s^{-1} the profile consisted of a smooth asymmetric main pulse separated by $\sim 180°$ in phase from a small interpulse. As the luminosity decreased by a factor of ~ 10 the relative intensity

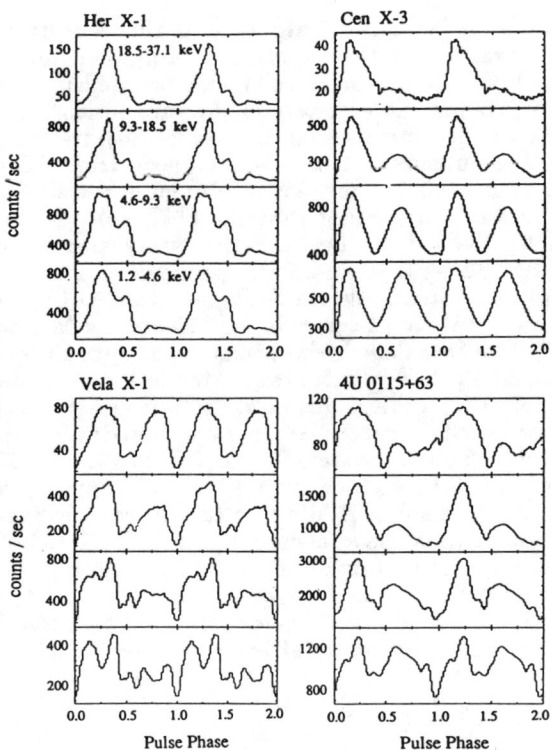

Fig. 4. Folded pulse profiles of four X-ray pulsars in the same four energy bands as shown for Her X-1.

of the two peaks became comparable and the profile became more complex with a narrow dip. As the luminosity decreased by a further factor of ~ 10 to 1×10^{36} erg s^{-1}, the relative intensity of the two pulses reversed, with the interpulse becoming the dominant pulse. This change in profile may be understood as being due to a change in the dominant beaming mechanism from a fan-beam (emission perpendicular to the magnetic field) at high luminosities to a pencil-beam (parallel to the magnetic field) at low luminosities (Parmar, White & Stella 1989).

Sharp dips with widths less than one-tenth of a pulse phase, which are often seen in Vela X-1 and 4U 0115+63 at low energies and EXO 2030+375 at low luminosity, may be a common feature of many pulsars during low intensity states. The galactic center pulsar GX 1+4 showed a simple broad peak structure during the high state in the 1970s, which became a sharp dip-like structure superposed on a broad sinusoidal modulation when observed in 1980s with *Ginga* during the low state (Dotani et al. 1989).

5. PULSAR SPECTRA

The energy spectra of most of the X-ray pulsars are characterized by a power-law continuum at energies below a cut-off energy which is typically at ~ 15 keV. Above this energy, the spectra fall off more rapidly then predicted

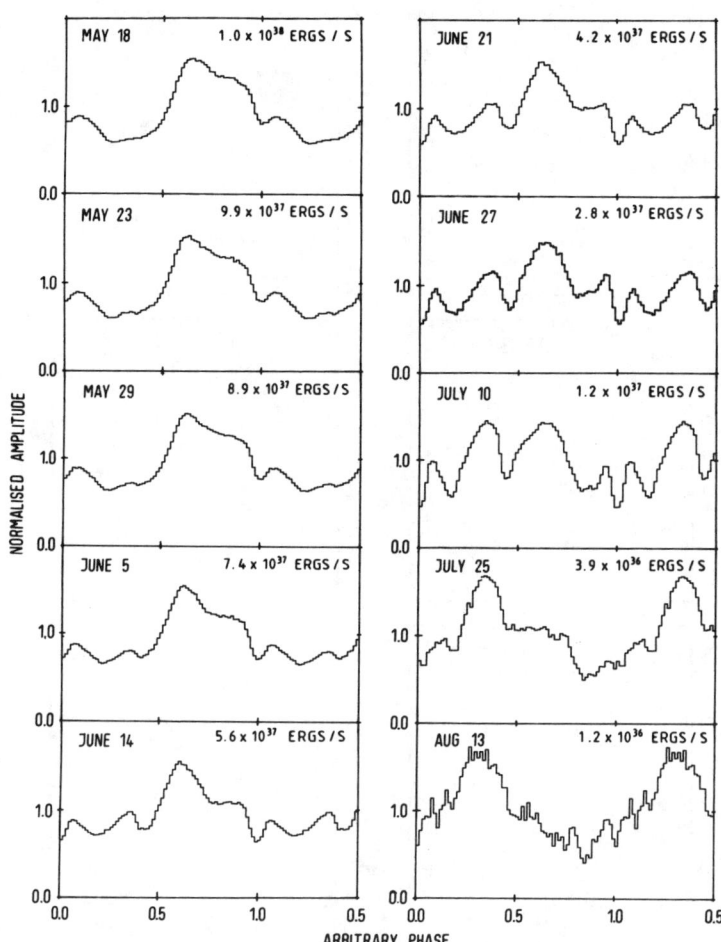

Fig. 5. Folded 1-10 keV pulse profiles of EXO 2030+375 showing the evolution of the pulse profile during the decay of an outburst (Parmar *et al.* 1989).

by an extrapolation of the power-law (*e.g.* White, Swank & Holt 1983). This spectral shape is distinct from that of other X-ray binaries. For example, many non-magnetic LMXRB systems have spectra similar to that of thermal bremsstrahlung, while black hole candidates show bimodal spectra consisting of either a single power-law or an ultra-soft component accompanied by a hard power-law tail. Figure 6 shows phase-averaged spectra of six X-ray pulsars observed by *Ginga*.

The observed spectra of X-ray pulsars are conventionally represented by a power-law with a photon index γ in the range $0.5 - 1.5$, modified at energies above a cut-off, E_c, by $\exp[(E_c - E)/E_f]$. In some pulsars, iron Kα emission and K-edge absorption features, cyclotron resonant scattering features, low energy absorption by cool circumstellar matter and soft excesses below ~ 1 keV are required in order to obtain acceptable fits. However, some low luminosity

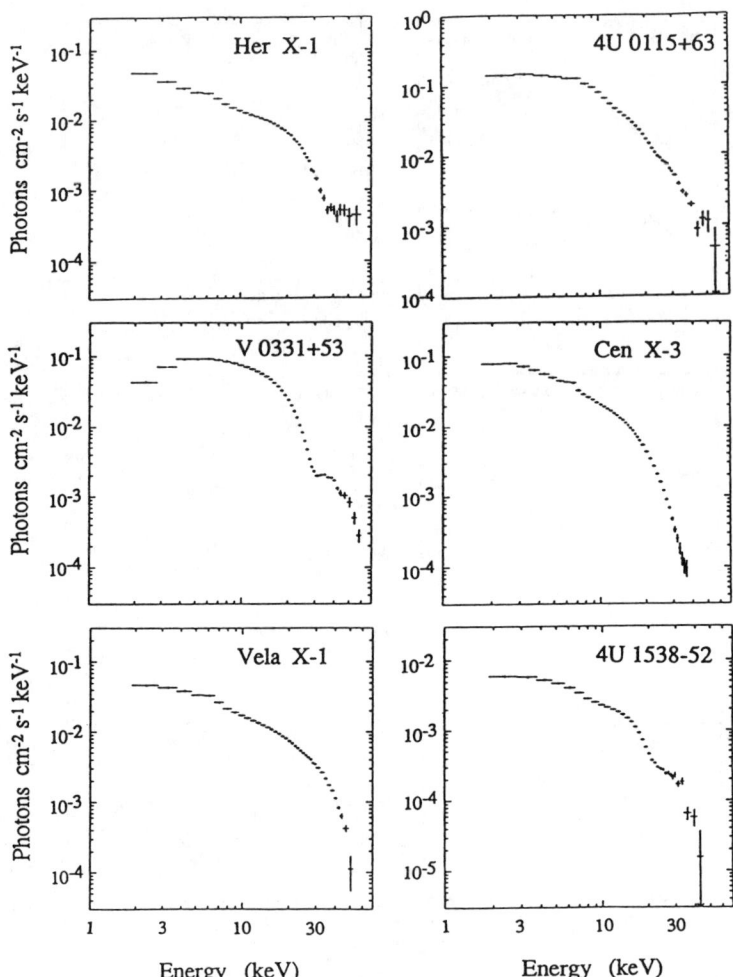

Fig. 6. Phase averaged incident spectra of six pulsars observed with *Ginga*. Cyclotron features are visible in the spectra of Her X-1, X 0115+634, X 0332+530 and X 1538-522.

pulsars, such as X Per (White *et al.* 1982) and Sct X-1 (Koyama *et al.* 1991b) show significantly steeper spectra ($\gamma \simeq 2$). An unusually soft spectrum is also observed from X 2259+586 (*e.g.* Iwasawa, Koyama & Halpern 1992), which is a pulsar located in the supernova remnant G 109.1−1.0. The binary nature of this pulsar is not yet confirmed (see *e.g.* Coe & Jones 1992). One explanation for the the steep spectrum is that the cut-off energy is below the detection threshold of ~ 1 keV.

Improved spectral measurements over a broad energy band and with better statistical precision have shown that a Lorentzian function multiplied by E^2 gives a better fit to the high energy cut-off than the simple exponential cut-off discussed above. In this model the overall continuum is given by:

$$f(E) = AE^{-\gamma} \exp\{-(H(E) + \sigma_a(E)N_H)\}$$

where σ_a are the photoelectric absorption cross sections of material with equivalent hydrogen column density N_H. The function $H(E)$ is given by (Makishima et al. 1990a):

$$H(E) = \frac{\tau_a W_a^2 (E/E_a)^2}{(E - E_a)^2 + W_a^2},$$

where E_a, W_a and τ_a are the energy, width and optical depth of the absorption feature, respectively. This function can simultaneously fit both a high energy cut-off at energy E_c and a cyclotron absorption feature at energy E_a.

Cyclotron scattering resonance features (CSRFs) have now been detected from nine pulsars at energies between 7 keV and 40 keV, corresponding to a range of surface magnetic field strength of $(0.8–4.4) \times 10^{12}$ G. The first detection was by Trümper et al. (1978) from Her X-1 at 35–45 keV. This feature has been extensively studied (see Soong et al. 1990 and references therein) and discussed in terms of both an emission feature at ~ 45 keV or an absorption feature at ~ 35 keV. The absorption line interpretation is more likely, based on modelling the variation of the feature with pulse phase (Voges et al. 1982; Soong et al. 1990; Mihara et al. 1990).

CSRFs at $E_a \sim 11.5$ keV and ~ 23 keV were detected in the spectrum of the transient Be star system X 0115+634 by Wheaton et al. (1979). Both features first appear in absorption during the main-pulse and then in emission during a broad-interpulse. *Ginga* observations of X 0115+634 by Nagase et al. (1991) showed a different behavior, with the same two features in absorption during most pulse phases. The *Ginga* satellite had increased sensitivity above 10 keV compared to earlier X-ray missions and seven more pulsars were discovered to have CSRF: X 1538−522 (Clark et al. 1990), X 0332+530 (Makishima et al. 1990b), Cep X-4 (Mihara et al. 1991), X 2259+586 (Iwasawa, Koyama & Halpern 1992), Vela X-1, X 1907+097 and GX 301−2 (Makishima & Mihara 1992).

Table 2. Cyclotron Features Detected in X-Ray Pulsars

Source	P_p (s)	P_o (d)	E_c (keV)[a]	E_a (keV)[a]
Her X-1	1.24	1.7	17–21	35
X 0115+63	3.6	24.3	7–9	12, 23
X 0331+53	4.4	34.25	14–17	28.5
X 2259+586	7.0	?	≤ 4	$\sim 7?$
Cep X-4	66	?	15–17	32
Vela X-1	283	8.96	15–20	27
X 1907+09	438	8.38	14–16	21
X 1538−52	529	3.73	14–16	21
GX 301−2	696	41.5	19–21	40

[a] E_c and E_a represent the exponential cut-off energy and the cyclotron scattering resonance feature (CSRF) energy, respectively.

For some systems such as Vela X-1, the CSRF is only detected for part of the pulse phase. The absorption features are broad ($\sim 25\%$) and shallow (*e.g.* Tamura et al. 1992). The energy of the fundamental (and also the 2nd harmonic) resonance energies vary by up to 30 % with pulse phase (Voges et al. 1982; Soong et al. 1990; Clark et al. 1990; Nagase et al. 1991; Tamura et al. 1992). There is evidence for a second harmonic in the spectra of X 1538−522 (Clark et al. 1990), Her X-1 (Mihara et al. 1990), X 0332+530 (Makishima et

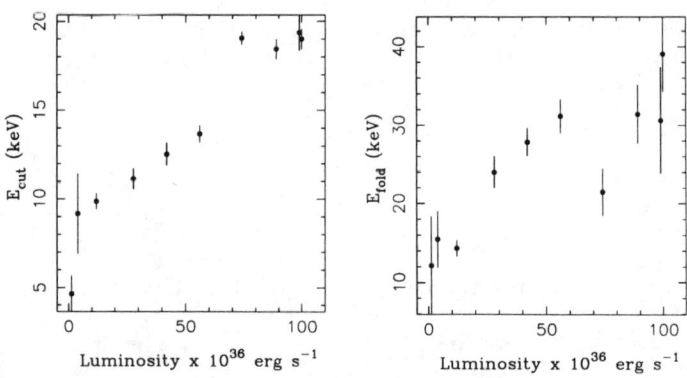

Fig. 7. The variations in cut-off and folding energies observed during the 1985 outburst from EXO 2030+375 (Reynolds, Parmar & White 1993).

al. 1990*b*), Vela X-1 and X 1907+097 (Makishima & Mihara 1992) but their detections are less significant than that in X 0115+634.

The observed cyclotron resonance energy, E_a, and the high-energy cut-off, E_c, (in the old model for describing the high-energy spectrum) are correlated with $E_a \simeq 2 \times E_c$ (Makishima & Mihara 1992). This correlation suggests that the cut-off energy is related to the cyclotron resonant scattering. If this is the case then E_c gives a measure of the magnetic field strength for those pulsars where a CSRF has not been detected. The high-energy cut-off energies are all typically between 10 and 20 keV, giving a narrow scatter in magnetic field strength of $B = (1-4) \times 10^{12}$ G. However, during the 1985 outburst of EXO 2030+375 the cut-off energy decreased from 20 to $\lesssim 10$ keV (Figure 7; Reynolds, Parmar & White 1993). This variation implies that, in this system at least, the cut-off energy does not provide a reliable measure of the surface magnetic field strength.

The luminosities of the nine pulsars for which CSRF have been detected are all $< 4 \times 10^{37}$ erg s^{-1}. A CSRF feature has not been detected from any of the more luminous pulsars such as Cen X-3. The power-law plus Lorentzian model still gives a good fit, but gives a line energy at ~ 30 keV. This is at the upper end of the observable energy range (Makishima & Mihara 1992). Forthcoming missions such as SAX and XTE will allow sensitive searches for CSRFs over a much broader range of energies.

ACKNOWLEDGEMENTS

I thank Nick White and Fumiaki Nagase without whose help this review would not have been possible.

REFERENCES

Boynton, *et al.* 1984, ApJ, 283, L53
Clark, G.W., *et al.* 1990, ApJ, 353, 274
Coe, M. J. & Jones, L. R. 1992, MNRAS, 259, 191
Cominsky, L., Clark, G. W., Li, F., Mayer, W., & Rappaport, S. 1978, Nature, 273, 367

Corbet, R. H. D., 1984, A&A, 141, 91
Corbet, R. H. D. & Day, C. S. R. 1990, MNRAS, 243, 553
Day, C. S. R. & Stevens, I. R. 1993, ApJ 403, 322
Davidsen, K., & Ostriker, J. P. 1973, ApJ, 179, 585
Davison, P. J. N., Watson, M. G., & Pye, J., 1977, MNRAS, 181, 73P
Deeter, J. E., Boynton, P. E., Lamb, F. K., & Zylstra, G. 1987, ApJ, 314, 634
Dotani, T., et al. 1989, PASJ, 41, 427
Elsner, R. F., & Lamb, F. K., 1976, Nature, 262, 356
Ghosh, P., & Lamb, F. K. 1978, ApJ, 223, L83
Ghosh, P., & Lamb, F. K. 1979, ApJ, 232, 259
Giacconi, R., Gursky, H., Kellogg, E., Schreier, E., & Tananbaum, H. 1971, ApJ, 167, L67
Grebenev, S. A., & Sunyaev, R. 1991, IAU Circ, 5294
Gregory, P. C., & Fahlman, G. G. 1980, Nature, 287, 805
Hall, R. & Davelaar, J. 1983, IAU Circ, 3872
Henrichs, H. F. 1983, in *Accretion-driven stellar X-ray sources*, ed. W. H. G. Lewin & E. P. J. van den Heuvel (Cambridge University Press), p. 393
Huckle, H. E., et al. 1977, MNRAS, 180, 21P
Ives, J. C., Sanford, P. W., & Bell-Burnell, S. J. 1975, Nature, 254, 578
Iwasawa, K., Koyama, K. & Halpern, J. P. 1992, PASJ, 44, 9
Kelley, R. L., Apparao, K. M. V., Doxsey, R. E., Jernigan, J. G., Narayan, S., & Rappaport, S., 1981, ApJ, 243, 251
Kelley, R. L., Jernigan, J. G., Levine, A., Petro, L. D., & Rappaport, S., 1983, ApJ, 264, 568
Kelley, R. L., Rappaport, S., & Ayasli, S. 1983, ApJ, 274, 765
Koyama, K. et al. 1991a, ApJ, 366, L19
Koyama, K., Kawada, M., Kunieda, H., Tawara, Y., Takeuchi, Y., & Yamauchi, S., 1990a, Nature 343, 148
Koyama, K., Kunieda, H., Takeuchi, Y., & Tawara, Y., 1990b, PASJ, 42, L59
Koyama, K., Kunieda, H., Takeuchi, Y., & Tawara, Y. 1991b, ApJ, 370, L77
Kriss, G.A., Cominsky, L.R., Remillard, R. A., Williams, G., & Thorstensen, J. R. 1983, ApJ, 266, 806
Lamb, F. K., Pethick, C. J., & Pines, D. 1973, ApJ, 184, 271
Lamb, R. C., Markert, T., Hartman, R., Thompson, D., & Bignami, G. F. 1980, ApJ, 239, 651
Lea, S., 1976, ApJ, 209, L69
Lewin, W. H. G., Ricker, G., & McClintock, J. E. 1971, ApJ, 169, L17
Lucke, R., Yentis, D., Friedman, H., Fritz, G., & Shulman, S. 1976, ApJ, 206, L25
Makishima, K., Kawai, N., Koyama, K., Shibazaki, N., Nagase, F., & Nakagawa, M. 1984, PASJ, 36, 679
Makishima, K. et al. 1988, Nature, 333, 746
Makishima, K., et al. 1990a, PASJ, 42, 295
Makishima, K., et al. 1990b, ApJ, 365, L59
Makishima, K., & Mihara, T. 1992, in *Frontiers of X-Ray Astronomy* (proc. of the 28th Yamada Conference), ed. Y. Tanaka & K. Koyama, (Uni. Acad. Press, Tokyo), p23
Maraschi, L., Huckle, H. E., Ives, J. C., & Sanford, P. W. 1976, Nature, 263, 34
Mason, K. O. 1977, MNRAS, 178, 81P
McClintock, J. E. et al. 1976, ApJ, 206, L99
McClintock, J. E., Rappaport, S., Nugent, J., & Li, F. 1977, ApJ, 216, L15
Mendelson, H. & Mazeh, T. 1991, MNRAS, 250, 373

Mihara, T., et al. 1990, Nature, 346, 250
Mihara, T., et al. 1991, ApJ, 379, L65
Nagase, F. 1989, PASJ, 41, 1
Nagase, F., et al. 1991, ApJ, 375, L49
Pakull, M., & Parmar, A. N. 1981, A&A, 102, L1
Parmar, A. N., White, N. E., & Stella, L. 1989, ApJ, 338, 373
Parmar, A. N., White, N. E., Stella, L., Izzo, C., & Ferri, P. 1989, ApJ, 338, 359
Pringle, J. E., & Rees, M. J. 1972, A&A, 21, 1
Rappaport, S. A., Clark, G. W., Cominsky, L., Joss, P. C., & Li, F. 1978, ApJ, 224, L1
Rappaport, S., & Joss, P. C. 1977, Nature, 266, 683
Rappaport, S. A., Markert, T., Li, F. K., Clark, G., Jernigan, J., & McClintock, J. 1977, ApJ, 217, L29
Reynolds, A., Parmar, A. N., & White, N. E., 1993, ApJ, 414, 302
Rosenberg, F. D., Eyles, C., Skinner, G., & Willmore, A. P. 1975, Nature, 256, 628
Skinner, G. K., Bedford, D. K., Elsner, R. F., Leahy, D., Weisskopf, M. C., & Grindlay, J. E. 1982, Nature, 297, 568
Soong, Y., et al. 1990, ApJ, 348, 641
Stella, L., White, N. E., Davelaar, J., Parmar, A. N., Blissett, R. J. & van der Klis, M. 1985, ApJ, 288, L45
Stella, L., White, N. E., & Rosner, R. 1986, ApJ, 308, 669
Strickman, M. S., Johnson, W. N., & Kurfess, J. 1980, ApJ, 240, L21
Tananbaum, H., Gursky, H., Kellogg, E. M., Levinson, R., Schreier, E., & Giacconi, R. 1972, 174, L143
Tamura, K., Tsunemi, H., Kitamoto, S., Hayashida, K., & Nagase, F. 1992, ApJ, 389, 676
Tawara, Y., Yamauchi, S., Awaki, H., Kii, T., Koyama, K., & Nagase, F. 1989, PASJ, 41, 473
Trümper, J., et al. 1978, ApJ, 219, L105
Van den Heuvel, E. P. J. & Rappaport, S. 1987, in Physics of Be Stars, ed. A. Slettebak & T. D. Snow (Cambridge Univ. Press), 291.
Voges, W., et al. 1982, ApJ, 263, 803
Waters, L. B. F. M., Taylor, A. R., Van den Heuvel, E. P. J., Habets, G. M. H. J., & Persi, P. 1988, A&A, 198, 200
Waters, L. B. F. M., & van Kerkwijk, M. H. 1989, A&A, 223, 196
Wheaton, Wm. A., et al. 1979, Nature, 282, 240
White, N. E., Mason, K. O., Huckle, H., Charles, P., & Sanford, P. W. 1976a, ApJ, 209, L119
White, N. E., Mason, K. O., Sanford, P. W., & Murdin, P. 1976b, MNRAS, 176, 201
White, N. E., Parkes, G., Sanford, P. W., Mason, K., & Murdin, P. G. 1978, Nature, 274, 664
White, N. E., & Pravdo, S. H. 1979, ApJ, 233, L121
White, N. E., Swank, J. H., & Holt, S. S. 1983, ApJ, 270, 711
White, N. E., Swank, J. H., Holt, S. S. & Parmar, A. N. 1982, ApJ, 263, 277

EMISSION PROCESSES IN X-RAY PULSARS

Alice K. Harding
Code 665, NASA Goddard Space Flight Center, Greenbelt, MD 20771.

ABSTRACT

The processes that are important in X-ray pulsar emission are dominated by the strong magnetic fields of the accreting neutron star. These processes include the channeling of accretion flow from the companion via either a disk or a wind onto the neutron star polar caps, the deceleration of the accreting particles and the radiation mechanisms that form the spectrum. I will review X-ray pulsar emission models in both the high-luminosity sources, where the accreting plasma is stopped by radiation pressure, and the low-luminosity sources, where the accretion flow is stopped by either a collisionless shock or Coulomb collisions. The formation of cyclotron lines in pulsar spectra will also be discussed.

1. INTRODUCTION

X-ray pulsars are believed to be strongly-magnetized accreting neutron stars powered by kinetic energy of infalling matter from a companion star. The magnetic fields of $10^{11} - 10^{13}$ G are inferred from the pulsations that require anisotropic infall and radiation, the cyclotron lines observed in a number of pulsar spectra (Nagase 1989), and the observed changes in pulse period (Ghosh & Lamb 1979, Joss & Rappaport 1984). The strength of the magnetic fields are instrumental in the physics of the accretion flow and how the kinetic energy of infall is converted to radiation. We can get a fairly good estimate of the effective temperature observed in X-ray pulsars from $T_{\rm eff} \simeq (L_x/\sigma A_{\rm cap})^{1/4} \simeq 10$ keV, where L_x is the X-ray luminosity and $A_{\rm cap}$ is the heated polar cap area. But clearly, since pulsar spectra are not blackbody, we must have an accurate description of the radiating plasma in order to model the emission.

The conditions in the emission region are closely tied to the dynamics of the accretion flow and infall, and there are two major regimes that depend upon whether or not the radiation pressure of the emitting plasma is capable of decelerating the accretion flow. The critical luminosity $L_{\rm crit}$ where this division occurs is not accurately determined but can be estimated by requiring that the outgoing radiation pressure balance the ram pressure of the infalling matter:

$$L_{\rm crit} = \frac{GMm_p c}{\sigma_\parallel} \frac{A_{\rm cap}}{R_o^2} \simeq 10^{36} \, {\rm ergs}^{-1}, \qquad (1)$$

where $\sigma_\parallel$ is the magnetic Thomson scattering cross section parallel to the field, averaged over the spectrum, polarization and angles of the radiation. The value of L_crit is model dependent and difficult to calculate because the averaged cross section should include the radiation produced throughout the atmosphere in a self-consistent model. Formation of cyclotron lines in the spectrum can also be important and should also be taken into account (Gnedin & Nagel 1984, Zheleznyakov & Litvinchuk 1987). If one takes $\sigma_\| = \sigma_T$ then the reduction of the area, $A_\mathrm{cap} \sim 0.01 R_o^2$ alone gives $L_\mathrm{crit} \sim 10^{36}\,\mathrm{ergs}^{-1} \sim 10^{-2}\,L_\mathrm{Edd}$, where $L_\mathrm{Edd} = 4\pi G M m_p c/\sigma_T$ is the Eddington limit for spherical accretion. The above expression is not really an effective Eddington limit in the sense that such a limit is interpreted for the spherical accretion case as preventing radiation above L_crit. This is because, as we will see below, the channeling of accretion flow along field lines and onto a restricted polar cap area allows radiation to escape from the sides of the accretion column. The high luminosity X-ray pulsars where $L_x > L_\mathrm{crit}$ may radiate several orders of magnitude above L_crit and even above the spherical Eddington limit. In low-luminosity sources where $L_x \ll L_\mathrm{crit}$, radiation pressure is not important in decelerating the accretion flow, in which case it may be decelerated by a collisionless shock above the neutron star surface or by Coulomb and nuclear collisions with atmospheric plasma near the surface. Most (about two thirds) of the observed X-ray pulsars fall into the high-luminosity category.

2. RADIATION PROCESSES IN STRONG MAGNETIC FIELDS

The most important effect of the neutron star magnetic field on the processes that decelerate the accretion flow and that produce the observed radiation is the quantization of particle momentum perpendicular to the field, in Landau states,

$$E_n = (m^2 c^4 + p_\|^2 + 2nm^2 c^4 \frac{B}{B_\mathrm{cr}})^{1/2} \qquad (2)$$

where n is the principal quantum number, $p_\|$ is the momentum parallel to the magnetic field and $B_\mathrm{cr} \equiv m^2 c^3/e\hbar = 4.413 \times 10^{13}$ G is the critical field strength. For $B_n \ll B_\mathrm{cr}$ the energy spacing between Landau states reduces to the cyclotron energy, $\Delta E \simeq \hbar \omega_B = 12\,\mathrm{keV}(B/10^{12})$ G. Since $kT_e \lesssim \hbar \omega_B$ in the atmospheres of X-ray pulsars, electrons will mostly occupy the ground Landau state ($n = 0$) in a one-dimensional distribution. Furthermore, since the rate of collisional excitation to higher Landau states is much less than the cyclotron radiation rate from excited states, the population of the levels may be far from thermal equilibrium, and will be dominated by the radiation field.

The dominant cooling process in X-ray pulsar atmospheres is resonant Bremsstrahlung (or cyclotron cooling), in which an electron in the ground state is collisionally excited and deexcites through spontaneous emission, producing a cyclotron photon. The inverse process of resonant free-free absorption, where an electron is excited by absorption of a cyclotron photon and collisionally deexcited, can in some cases provide significant heating to the atmosphere. So while the vast majority of electron excitation-deexcitation events are resonant

Compton scattering at the cyclotron energy the collisions, although less frequent, control the net production and destruction of cyclotron photons in these atmospheres. Models of X-ray pulsar atmospheres thus usually evaluate the cyclotron cooling contribution by using electron-electron and electron-ion collision rates (Langer & Rappaport 1982), which are second-order in α, coupled with the first-order cyclotron emission rate, with some approximation for line photon escape. The full third-order Bremsstrahlung cross section has been calculated in the non-relativistic limit (e.g. Nagel 1982) and the non-resonant part of this cross section is used to evaluate continuum Bremsstrahlung cooling and photon production.

Another important heating and cooling process is Compton scattering, which is resonant at the cyclotron harmonics in a magnetic field. While scattering does not produce nearly as much heating and cooling as Bremsstrahlung since it does not create and destroy cyclotron photons, it exchanges energy between photons and electrons in the atmosphere. Compton scattering can also produce cyclotron lines in the radiated spectrum (see section 6).

Two-photon emission, where electrons in excited Landau states spontaneously decay to a neighboring state through the emission of two photons, instead of by one photon cyclotron emission, may also be important in X-ray pulsars (Melrose and Kirk 1986). This is a second-order process, because a virtual intermediate state is obviously needed for the electron following its emission of the first photon. Since the two emitted photons share the total energy of the transition, the emission forms a continuum spectrum going down to arbitrarily small photon energies. This process is related to two-photon Compton scattering, where an electron in the ground state is scattered by a photon just above the cyclotron energy to the first excited state, resulting in a soft scattered photon and a cyclotron photon from the spontaneous decay of the electron back to the ground state. Both of these processes may be important sources of soft photons in X-ray pulsars (Bussard et al. 1985).

3. LOW-LUMINOSITY SOURCES

In sources where $L_x \ll L_{\text{crit}}$, radiation pressure is not capable of decelerating the accretion flow and other processes must operate to slow down the plasma. If the infalling matter excites plasma instabilities in the accretion flow then it can convert its kinetic energy to heat at a collisionless shock front. This stopping mechanism was proposed for a non-magnetic neutron star by Shapiro & Salpeter (1975), but whether collisionless shocks can form in strongly magnetized plasmas, where the magnetic energy density is much larger than the particle energy density, is questionable. If such a shock does not occur, then the accreting plasma must decelerate through Coulomb collisions with atmospheric electrons or nuclear collisions with protons.

Langer & Rappaport (1982) investigated the structure of a magnetized, accreting atmosphere assuming that a collisionless shock does form and decelerates the infalling plasma. They solved the time-dependent hydrodynamic equations in a one-dimensional column atmosphere to find the density, velocity and temperature of ions and electrons in the post-shock flow. The upper boundary condition comes from the jump conditions across the shock, assuming that the upstream gas is in free-fall, $v = (2GM/R_o)^{1/2}$. The lower boundary condition assumes the gas has zero velocity at the neutron star surface. Cooling is due to bremsstrahlung and cyclotron emission, with cyclotron emission being dominant, and collisional energy exchange between ions and electrons is included. The protons initially have higher temperatures than the electrons behind the shock and both cool slowly in the subsonic flow region. They find that the shock has a height $r \simeq 1.5 r_*$ above the stellar surface depending on the accretion rate $\dot M$ and the magnetic field strength B. The shock height is larger for lower accretion rates, where cooling is less effective, and also larger for higher magnetic fields, where cyclotron cooling decreases. Cyclotron line emission regulates the cooling of the atmosphere: for $B > 10^{12}$ G, electron temperatures can be maintained at a few times $\hbar\omega_B$, but when $B < 10^{12}$ G, the ion temperature is high enough to collisionally excite the electron Landau levels and the electrons get hot enough ($kT_e \to 10^9$ K) to produce γ-rays. The spectrum from the atmosphere is a Doppler broadened cyclotron line, with almost no continuum contribution from bremsstrahlung.

In the case of low-luminosity sources where the accretion flow is decelerated by collisions, the accreting protons stop by multiple Coulomb scattering with thermal electrons and by nuclear collisions with atmospheric protons. The presence of a strong magnetic field introduces substantial changes in the stopping length due to Coulomb scattering. Since the thermal electrons are mostly in the ground Landau state, their momentum exchange with the infalling protons is highly anisotropic. A proton moving along the magnetic field experiences a much smaller drag force than in the non-magnetic case because the electrons are limited to momentum transfers along the field unless they can be excited to higher states. The infalling protons must diffuse in momentum space through many small-angle collisions, gradually veering away from the magnetic field direction where their drag force will increase and they can decelerate faster. As a result, the stopping lengths are very dependent on the proton momentum diffusion and significantly larger than in an un-magnetized plasma. An example is shown in Table 1 for the case of infalling proton velocity of $v = 0.5c$ and atmospheric electron temperature and density $kT = 10$ keV and $n = 10^{23}$ cm^{-3}. The variation in magnetic stopping lengths from different calculations reflects the increasing sophistication in treating important effects such as the evolution of the proton distribution function, including pitch-angle diffusion, excitation of electron Landau states and the thermal electron velocity. Most importantly, the values of Coulomb stopping length of the most recent calculation are lower than the stopping length of $\sim 50\,\mathrm{g\,cm}^{-2}$ due to nuclear collisions. Therefore, accreting protons will decelerate via Coulomb collisions unless $kT < 10$ keV, where nuclear collisions will become more important (Pakey 1990).

Table 1 - PROTON STOPPING CALCULATIONS

$v = 0.5\,\text{c},\ T = 10\,\text{keV},\ n = 10^{23}\,\text{cm}^{-3}$

	Stopping length (g cm^{-2})
$B = 0$	4.3
$B = 5 \times 10^{12}$ G	
Basko & Sunyaev (1975)	200
Pavlov & Yakovlev (1976)	40
Kirk & Galloway (1982)	15
Miller et al. (1987)	27
Pakey (1990)	38

Since the proton stopping lengths are sensitive to the atmospheric temperature and density which in turn are dependent on the heating from Coulomb collisions, one needs to construct self-consistent models. Meszaros et al. (1983) and Harding et al. (1984) modeled accreting atmospheres where Coulomb and nuclear collision heating was balanced with radiative cooling to determine the atmospheric temperature and density. Using the Coulomb stopping results of Kirk & Galloway (1982), they solved the equations of energy balance between collisional heating and cooling, including bremsstrahlung, Compton scattering and cyclotron line emission, and momentum balance (hydrostatic equilibrium) between the ram pressure of the infalling protons, thermal pressure and gravitational force. This was coupled with radiative transfer in the inhomogeneous atmosphere using polarization dependent magnetic cross section. They found that the Coulomb heated atmospheres are both geometrically and optically thin slabs with optical depth $\tau \simeq \sigma y_o/m_p = 6(\sigma/0.5\sigma_T)(y_o/30\,\text{g cm}^{-2})$, where y_o is the stopping length of the protons. For a typical density of $n_e = 10^{23}\,\text{g cm}^{-3}$, the scale height of the atmosphere is only $h \simeq 200$ cm. Cyclotron emission is the most important cooling process and the magnetic field acts as a thermostat that keeps the temperature below the cyclotron energy. The densities are then determined by the condition of hydrostatic equilibrium. The emergent spectra are primarily thin thermal bremsstrahlung at $B > 10^{13}$ G and closer to blackbody at $B < 10^{13}$ G. Miller, Wasserman & Salpeter (1989) repeated the calculations of Harding et al. (1984) using improved Coulomb deceleration results of Miller, Wasserman & Salpeter (1987), who obtained longer stopping lengths than those of Kirk & Galloway (1982). Consequently their model atmospheres are slightly cooler and denser than those of Harding et al. (1984). In the Coulomb decelerated slab-like atmospheres, the radiation escapes primarily along the magnetic field where the optical depth is minimum, forming a pencil beam (see Figure 1).

4. HIGH-LUMINOSITY SOURCES

Models for sources with $L_x > L_{\rm crit}$ where radiation pressure dominates the deceleration must solve the coupled radiation-hydrodynamic flow to find the structure of the atmosphere. In these solutions, a stationary shock front stands off above the neutron star surface, creating a subsonic sinking zone below which slowing gas cools and radiates. Basko & Sunyaev (1976, see also Inoue 1975) investigated one-dimensional models in which they identified many of the basic problems of these types of solutions. One is the boundary condition assumed at the stellar surface. In order to obtain a physically reasonable solution, i.e. one in which the radiative shock forms for $L_x > L_{\rm crit}$, they had to assume a finite radiative flux into the star to balance the magnetic field pressure. In this case the shock stands off a distance above the surface which increases with $\dot{M}$, as would be expected. Since the atmosphere is a narrow column, the optical depth is much higher along the magnetic field than across it so the radiation escapes from the sides of the column in a fan beam (see Fig. 1). As the accretion rate increases and the shock rises further above the stellar surface, the luminosity increases beyond $L_{\rm crit}$. They derive an ultimate limiting luminosity where either the shock reaches the Alfven surface or the radiation pressure at the base of the column exceeds $B^2/8\pi$. This limit in principal allows luminosities above the spherical Eddington limit for a neutron star and is very sensitive to the accretion geometry. Braun & Yahel (1984) studied one-dimensional radiative shock models which included Coulomb deceleration as well.

Two-dimensional radiation shock deceleration models further elucidated the structure of the atmospheres of high luminosity pulsars. Davidson (1973) studied a simplified model including non-magnetic cross sections and neglected gravity in the subsonic flow. He found that the subsonic flow settles into an accreting mound of gas at the stellar surface. This happens because most of the radiation escapes from the sides of the accretion column, as in the one-dimensional models, but the inclusion of the transverse dimension shows that the radiation density, which varies across the column, is lower at the sides and decelerates the flow more slowly. This causes the nearly stationary gas at the base of the column to built up in the middle. Wang & Frank (1982) refined Davidson's calculation by including magnetic cross sections before ω_B and gravity in the sinking zone. Their numerical solutions also showed a mound of stagnant gas at the base of the accretion column and that the radius of the shock above

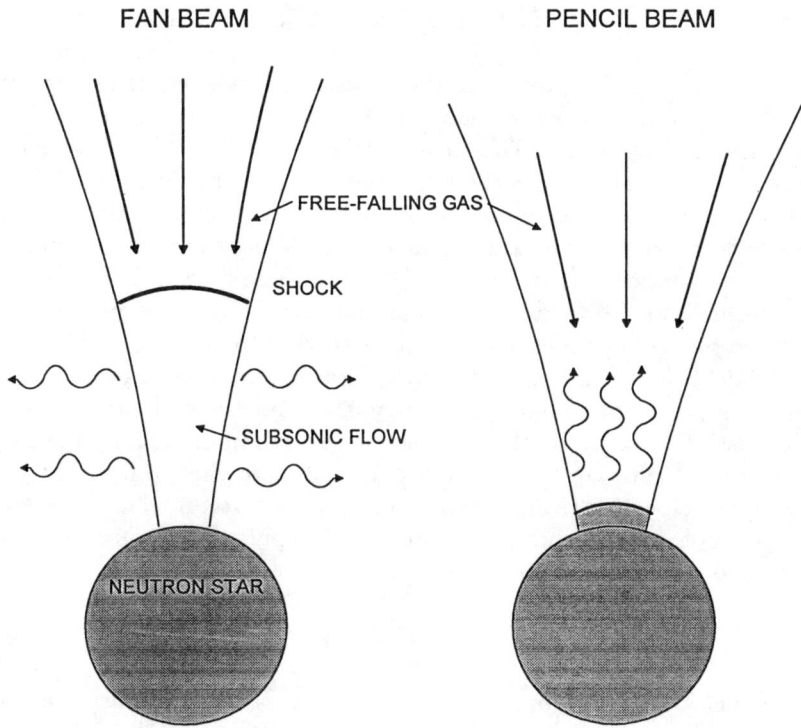

Figure 1. Schematic illustration of the possible geometry of accretion onto the polar cap of a magnetized neutron star, showing radiation into a fan beam from a narrow column atmosphere and a pencil beam from a thin slab atmosphere.

the surface increased with $\dot{M}$, similar to the result of Basko & Sunyaev (1976). Kirk (1985) found analytic solutions to the hydrodynamic flow in the high-luminosity pulsars, assuming a collisionless shock as well as a radiative shock is present to decelerate the accretion flow. Time-dependent, numerical solutions of the radiation-hydrodynamic flow for the high-luminosities were investigated by Arons, Klein & Lea (1985,1987). They treated the cyclotron heating and cooling solving the two-level non-LTE populations of the Landau states, using magnetic resonant cross sections. Their solutions show the formation of "photon bubbles", buoyant cells of radiation rising between columns of accreting gas, but the code could not follow the evolution of these structures for long times. The spectrum of radiation escaping from the accretion mound is a fan beam made up of different temperature blackbody spectra coming from various heights above the surface (Burnard et al. 1991).

5. PULSE PROFILES

The expected pulse profiles in the different models are primarily sensitive to the geometry of the accretion column (i.e. column or slab). In the cases where a shock either radiative or collisionless decelerates the accretion flow a distance above the neutron star surface the atmosphere is a narrow column and radiates in a fan beam. In the case where collisions stop the accretion flow a thin slab atmosphere forms and radiates in a pencil beam along the magnetic field. Since the fan beam radiates perpendicular to the magnetic field and the pencil beam radiates parallel to the field the characteristics of the pulse profiles are very different due to the dependence of the cross sections on energy and angle to the magnetic field (Meszaros & Nagel 1985). Thus the observed pulse profiles are excellent diagnostics of emission geometry and indirectly of the physics of the accretion flow. For example the cross section is a minimum along the magnetic field at energies below ω_B which causes the pulses to change from single above ω_B to double below ω_B (Nagel 1981). This is observed in some pulsars most notably Her X-1. In the column atmospheres the pulses at some observer angles is expected to be double *at* the resonance.

6. CYCLOTRON LINES

The cyclotron lines observed in the spectra of some pulsars are also sensitive diagnostics of the structure of the atmosphere especially the pulse phase spectroscopy of the lines and correlations between the shape of the pulse profiles and the characteristics of the line spectra. The magnetic scattering cross sections and the cyclotron line profiles are strongly dependent on the viewing angle to the magnetic field (Canuto et al. 1971 Harding & Daugherty 1991). Relativistic corrections cause the resonance energy to decrease with increasing angle θ:

$$\omega_n = [(1 + 2nB' \sin^2 \theta)^{1/2} - 1]/\sin^2 \theta \qquad (3)$$

where $\omega = \omega_B$ for $\theta = 0^0$. The strength of the harmonics decreases with increasing harmonic number n faster at smaller angles. In addition the Doppler broadening of the line profiles by a thermal one-dimensional electron distribution is angle dependent and in the non-relativistic limit is

$$\Delta \omega_D = \omega_n (2kT/mc^2)^{1/2} \cos \theta \qquad (4)$$

where kT is the parallel electron temperature. At small viewing angles one sees the full first-order Doppler broadening while at angles near 90^0 one sees no broadening. If relativistic kinematics are included there is a small second-order transverse Doppler broadening which is exclusively a red shift. Thus the line profiles become highly asymmetric at large angles to the field. Because of this angular sensitivity of the lines one might hope to distinguish between a slab and column geometry by comparing models to observed phase dependence of the cyclotron features.

Since the resonant radiative transfer calculations to model cyclotron lines are complex and time consuming results have been obtained only in the case of homogeneous static atmospheres (Meszaros & Nagel 1985 Wang et al. 1988). Even with this simplification the models have been able to show the general dependence of characteristics such as line energy on emission geometry. It was found that the cyclotron line energy varies as a function of pulse phase because the resonance energy is a function of the viewing angle to the magnetic field (Meszaros & Nagel 1985). In a slab atmosphere the line energy is predicted to decrease with phase from 0 (0^0 to B) to $\pi/2$ (90^0 to B) while in the cylinder atmosphere the line energy does the opposite increasing from phase 0 (90^0 to B) to $\pi/2$ (0^0 to B). The decrease in observed cyclotron line energy with increasing phase in pulsars such as Her X-1 (Voges et al. 1982) seems to argue in favor of pencil beam emission from slab-type atmospheres. This is surprising because Her X-1 has $L \sim 10^{37}\,\mathrm{erg\,s^{-1}} > L_{\mathrm{crit}}$ and is in the high-luminosity regime where the formation of a radiative shock should give a fan beam. The discrepancy indicates that either the estimates of L_{crit} are no low or some effects on the emission such as gravitational bending of photon directions (Meszaros & Riffert 1988) are important.

REFERENCES

Arons J. Klein R. & Lea S. 1985 in Plasma Penetration into Magnetospheres ed. N. Kylafis et al. (Crete Univ. Press Crete) p. 141.
Arons J. Klein R. & Lea S. 1987 ApJ 312 666.
Basko M. M. & Sunyaev R. A. 1975 A & A 42 311.
Basko M. M. & Sunyaev R. A. 1976 M.N.R.A.S. 175 395.
Braun A. & Yahel R. Z. 1984 ApJ 78 349.
Burnard D. J. Klein R. I. & Arons J. 1991 ApJ 367 575
Bussard R. W. Meszaros P. & Alexander S. G. 1985 ApJ 297 L21.
Canuto V. Lodenquai J. & Ruderman M. A. 1971 Phys. Rev. D 3 2303.
Davidson K. 1973 Nature 246 1.
Ghosh P. & Lamb F. K. 1979 ApJ 234 296.
Gnedin Yu. N. & Nagel W. 1984 A & A 138 356.
Harding A. K. & Daugherty J. K. 1991 ApJ 374 687
Harding A. K. Meszaros P. Kirk J. G. & Galloway D. J. 1984 ApJ 278 369.
Inoue H. 1975 Publ. Astron. Soc. Japan 27 311.
Joss P.C. & Rappaport S. A. 1984 Ann. Rev. Astron. Ap. 22 537.
Kirk J. G. 1985 A & A 142 430.
Kirk J. G. & Galloway D. J. 1982 Plasma Phys. 24 339.
Langer S. H. & Rappaport S. 1982 ApJ. 257 733.

Melrose D. B. & Kirk. J. G. 1986 A & A 156 268.
Meszaros P. & Nagel W. 1985 ApJ 299 138.
Meszaros P. Harding A. K. Kirk J. G. & Galloway D. J. 1983 ApJ 266 L33.
Meszaros P. & Riffert H. 1988 ApJ 327 712.
Miller G. S. Wasserman I. & Salpeter E. E. 1987 ApJ 314 215.
Miller G. S. Wasserman I. & Salpeter E. E. 1989 ApJ 346 405.
Nagase F. 1989 Publ. Astron. Soc. Japan 41 1.
Nagel W. 1981 ApJ 251 278 (288).
Pakey D. D. 1990 PhD Thesis University of Illinois.
Pavlov G. G. & Yakovlev D. G. 1976 Sov. Phys. JETP 43 389.
Shapiro S. L. & Salpeter E. E. 1975 ApJ 198 671.
Wang J.C.L. Wasserman I. and Salpeter E. E. 1988 ApJ Supp 68 735.
Wang Y. M. & Frank J. 1982 A & A 93 255.
Zheleznyakov V. V. & Litvinchuk A. A. 1987 Sov. Astron. 31 159.

SPIN EVOLUTION OF NEUTRON STARS IN ACCRETION POWERED PULSARS

Pranab Ghosh
Tata Institute of Fundamental Research, Bombay 400 005, INDIA.
pranab@tifrvax.tifr.res.in

ABSTRACT

This is a review of our understanding of the evolution of the spin periods of neutron stars in accretion-powered binary X-ray pulsars during the course of evolution of these binary systems. I discuss how neutron stars born in binaries as rotation-powered pulsars are initially spun down by electromagnetic and plasma torques, until accretion of matter from the companion star begins. I describe how accretion torques spin up and spin down the star during the main X-ray emission phase of neutron stars accreting from disks, winds, or equatorial rings produced by mass loss from their companions. I outline the final phase of spinup of the neutron stars into recycled rotation-powered pulsars.

1. INTRODUCTION

Thirty-three accretion-powered X-ray pulsars are known now (Nagase 1989, 1992; van Paradijs 1993), including the ~ 94 s pulsar J1008-57 recently discovered (Stollberg et al. 1993) by the Burst and Transient Source Experiment (BATSE) on the Compton Gamma Ray Observatory (CGRO), and the first detection of ~ 8.6 s pulsations from a possible thirty-fourth one has just been reported (Israel et al. 1993). The distribution of the pulse periods of the pulsars is shown in Fig.1: an apparent paucity of pulsars in the range ~ 0.1 s to a few seconds is noticeable, and the distribution longwards of a few seconds seems to be consistent with a uniform distribution in $\log P$. One of the ultimate goals of a study of the evolution of spin periods of accreting neutron stars in binary systems is to understand how such a distribution might come about. As also shown in Fig.1, 4 of the 33 pulsars have low-mass ($M \lesssim 2M_\odot$) companions, 9 have massive ($M \gtrsim 20M_\odot$) O/B supergiant companions, and the rest, $\sim 60\%$ of the total, are known or suspected to have massive ($8M_\odot \lesssim M \lesssim 15M_\odot$) Be-type companions (Nagase 1989; van Paradijs 1993).

2. INITIAL SPINDOWN

Scenarios for the formation of neutron stars in high-mass X-ray binaries (HMXBs) have been investigated extensively (van den Heuvel 1992, hereafter vdH92; Bhattacharya & van den Heuvel 1991): the system is not disrupted in the supernova explosion because of the large-scale mass transfer prior to the explosion. In low-mass X-ray binaries (LMXBs), neutron stars can form (Webbink 1992) either (a) in supernova explosion of helium stars produced after the first stage of mass transfer in massive binaries with very disparate original masses ($15M_\odot$ and $2M_\odot$, say, as suggested for the progenitor of Her X-1), or, (b) in accretion-induced collapse of massive white dwarfs driven over the

Chandrasekhar limit by accretion from their low-mass binary companions.

A newborn neutron star functions as a rotation-powered pulsar, until its rapid rotation is braked in the ways described below. The recent discovery of the ~ 48 ms radio pulsar PSR 1259 – 63 (Johnston et al. 1992)) in a highly eccentric ($e \simeq 0.87$, Johnston et al. 1993) binary with a Be star companion (SS 2883) may have provided the first opportunity for observational study of this phase in HMXBs (Cominsky 1993).

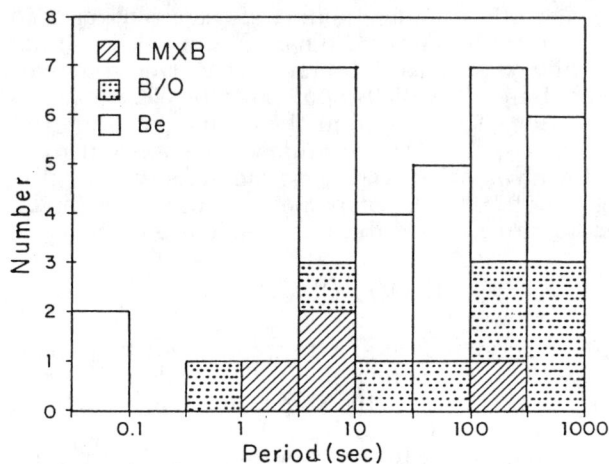

Fig. 1.–Pulse period distribution of the 33 known accretion-powered pulsars, with low-mass, B/O-, and Be-type companions as indicated.

2.1 Relativistic Braking

Electromagnetic braking torques are usually expressed in the form $N \propto \mu^2 (\Omega_s/c)^3$, in analogy with the expression for magnetic dipole radiation in vacuum. Here, μ is the magnetic dipole moment of the star, and Ω_s is its angular velocity. It has been argued that, in the presence of plasma, the braking torque may still scale as above if magnetic stresses dominate plasma stresses at the radius of the light cylinder, but evidence and arguments in favor of such dominance are uncertain. This torque, which spins the pulsar down to a period P in $\sim 3.10^4 (P/50\ ms)^2 I_{45} \mu_{30}^{-2}$ yr, can brake the stellar spin to the critical point where plasma effects first dominate the torque. The critical period depends on the properties of the plasma flow from the companion to the neutron star: for capture at rates $\sim 10^{-11} M_\odot$ yr^{-1} from winds of speed $\sim 10^3$ km s^{-1} from the companion, this period is $\sim 100 \mu_{30}^{1/2}$ ms (Illarionov & Sunyaev 1975, hereafter IS). Here, I_{45} is the moment of inertia of the neutron star in units of 10^{45} gm cm^2, and μ_{30} is μ in units of 10^{30} G cm^3.

2.2 "Propeller" & Related Braking

Even when the spin of the neutron star becomes slower than the above critical value, large-scale accretion is not possible at first. Stresses associated with the rotating magnetic field of the neutron star deposit energy and momentum in matter, causing outflow, in a manner which has been likened to the action of a propeller (IS). The scaling of the propeller braking torque was given in the original formulation (IS) as $N \propto (\mu^2/r_A^3)(\Omega_K(r_A)/\Omega_s)$, where r_A is the Alfvén radius and Ω_K is the Keplerian velocity. Further work (Mineshige et al. 1991; Davies & Pringle 1981; Henrichs 1983) on the propeller mechanism showed that, for subsonic propeller spin rates ($r_A \Omega_s < c_s$, c_s being the sound speed), the braking torque scales as $\mu^2 \Omega_s^2 / GM$, independent of r_A. For supersonic propellers, the torque is smaller by a factor $\sim (c_s/\Omega_s r_A)$, and reduces to the original scaling if $c_s \sim v_{ff}$ at r_A. Here, v_{ff} is the free-fall velocity. Actually, the propeller effect requires a misalignment between the spin and magnetic axes of the neutron star. Even when the two axes are aligned, there may be a significant "frictional" braking torque (Mineshige et al. 1991) on the star, which scales as $N \propto p \, r_A^3$, where p is the total pressure at r_A. Finally, it has been suggested recently (Illarionov & Kompaneets 1990) that the propeller "switches off" when the condition $\Omega_s > \Omega_K(r_A)$ is no longer satisfied. There is, instead, accretion to the star over some parts of the total cross-section, and outflow of a Compton-heated wind over the other parts, the resultant braking torque scaling as $\dot{M}_{out} r_A^2$. Here, $\dot{M}_{out}$ is the mass outflow rate.

For typical values of stellar and binary parameters, the plasma torques described above give timescales for spindown to the point of the onset of accretion which are comparable to or shorter than that given above for the electromagnetic torque. For example, the subsonic propeller torque gives a timescale $\simeq 3.10^3 (P/50 \text{ ms}) \, I_{45} \, \mu_{30}^{-2} (M/M_\odot)$ yr, while the supersonic propeller gives a timescale $\simeq 10^4 \mu_{30}^{-8/7} \, I_{45} \, (M/M_\odot)^{2/7}$ yr for mass capture rates from the stellar wind given above. It has therefore been argued (Henrichs 1983) that the total time taken by the star to reach this point is essentially determined by the "bottleneck" of the elctromagnetic torque, if the different torques act *consecutively*. However, this need not be the case for a highly eccentric binary like PSR 1259 – 63, since, during each orbital cycle, elctromagnetic torques may dominate over those parts of the orbit where the wind density is small, while plasma torques may dominate over those parts where the density is high, particularly near periastron. Then spindown would occur on a timescale which is an appropriate average of those due to the various torques acting over the orbital cycle. Indeed, this seems to be the case for PSR 1259 – 63, since its average spindown rate (Johnston et al. 1993) corresponds to a characteristic time $\approx 3.10^5$ yr. If we assume $\mu_{30} \simeq 0.1$, corresponding to a surface magnetic field $\simeq 10^{11}$ G., the timescales for the electromagnetic, subsonic propeller, and supersonic propeller spindown are $\approx 3.10^6$ yr, 3.10^5 yr, and $1.4.10^5$ yr respectively (see above). Thus the observed timescale lies between the electromagnetic and supersonic propeller (which is probably the correct description for 1259 – 63 near periastron) timescales, and is close to the subsonic propeller timescale.

2.3 Braking by Onset of Disk Accretion

At the onset of accretion to the neutron star from disks, or from stellar winds, the star is spun down further by the accretion torque. Here, I indicate how the braking torque acting at the onset of disk accretion is particularly effective in spinning down the star to $P \sim 100$ s (Elsner, Ghosh, & Lamb 1980, hereafter EGL).

The principal features of the accretion torque for disk accretion by magnetic neutron stars are summarized in §3.2. The torque spins the star down if the stellar spin rate, as measured by the dimensionless fastness parameter ω_s, exceeds a critical value ω_c, i.e., the spin period is shorter than the critical value P_c which corresponds to the point $\omega_s = \omega_c$ where the accretion torque vanishes. Under the action of the braking torque, the star spins down from an initial period P_i to a final period $P_f \approx P_c$ in a time $\simeq 2.10^6 \mu_{30}^{-2}(P_f/100\ s)^2(P_i/1\ s)^{-1}$ yr, if the accretion rate is a constant (EGL). If $\dot{M}$ goes through outbursts or flares, spindown can occur even more rapidly, as the braking torque can then continue to be very effective as the star spins down. In this way, the star can be spun down to a final period P_f from an initial period $P_i \ll P_f$ in a time as short as $\simeq 1.5.10^5 \mu_{30}^{-2}(P_f/100\ s)$ yr. Hence, this mechanism is clearly capable of producing the initial spindown to the long ($\gtrsim 100$ s) periods, which ~ 45 % of the known accretion-powered pulsars have, in a time which is short compared to the main-sequence lifetime of their companions (EGL).

3. SPIN EVOLUTION DURING MAIN X-RAY PHASE

Following the onset of accretion, the system enters the X-ray binary phase which lasts for typically $\sim 2.5.10^4$ yr for HMXBs and $\sim 7.10^7$ yr for LMXBs (vdH92). During this phase, angular momentum exchange between the magnetic neutron star and the matter being accreted from the stellar wind (in HMXBs) or the Roche-lobe-overflow accretion disk (in LMXBs) determines the spin evolution of the neutron star.

3.1 Observation

For about half of the known accretion-powered pulsars, long-term period histories have been compiled from data collected over ~ 2 decades of satellite observation (Nagase 1989; Corbet & Day 1990; Iwasawa et al. 1992; Chakrabarty et al. 1993b; Bildsten et al. 1993). From these compilations, which document period changes on timescales $\sim$ months to years, the following approximate categories of secular trends in spin evolution can be identified. First, there are those pulsars which have shown secular spinup over essentially the entire stretch of observation (except possibly for a brief episode of standstill or slight spindown), e.g., Her X-1, SMC X-1, LMC X-4, and Cen X-3. Second, there are those which have, similarly, shown essentially secular spindown, e.g., 1E 1048.1-5937, 1E 2259+586, E1145.1-614, and 4U 1538-52. Third are those which have shown considerable stretches of both spinup and spindown, e.g., A 0535+26, Vela X-1, GX 301-2, GX 1+4, 4U 1626-67, and OAO 1657-415. Finally, there are those that seem to show "erratic" variation on these timescales, e.g., 4U 1145-619, 4U 0115+63, and X Per.

BATSE observations over the last ~ 3 years are now producing detailed period histories of pulsars over this duration, documenting changes on timescales $\sim$ hours to days to months. Examples are Her X-1 (Wilson et al. 1993), Cen X-3 (Finger et al. 1993), OAO 1657-415 (Chakrabarty et al. 1993a), GS 0834-430 (C. A. Wilson et al. 1993), GX 1+4 (Chakrabarty et al. 1993), and 4U 1626-67 (Bildsten et al. 1993). These observations have given the first indication that 4U 1626-67 (a 7.7 s pulsar with a very low-mass companion), which had been observed to spin up at almost a constant rate since its discovery, is now spinning down (Bildsten et al. 1993).

3.2 Disk Accretion

Well-developed accretion disks are expected to form in LMXBs due to

Roche lobe overflow, and perhaps also in the Be-star systems during accretion from equatorial rings. Interactions between the rotating plasma in the accretion disk and the magnetic field of the rotating neutron star determines the rate of angular momentum transport and the geometry of the accretion flow.

3.2.1 Disk-Magnetosphere Interaction. The stellar magnetic field has a tendency to penetrate into the disk plasma due to a variety of processes, *e.g.*, Kelvin-Helmholz instability, turbulent diffusion, and reconnection to small-scale fields in the disk (Ghosh and Lamb 1979a,b, 1991; hereafter GL). Penetration is actually expected to occur over a considerable part of the disk because the rate of inward radial drift at which the ionized disk plasma can "sweep" the field inward is much slower than the rates of the above processes. Thus, the diamagnetic disk picture, in which the disk is completely excluded from stellar magnetic field (Aly 1980; Kundt & Robnik 1980), is not a self-consistent description of the physical situation.

The precise extent and structure of the disk-magnetosphere interaction region is still a matter of some debate. GL originally described the region in terms of its two conceptually different components: (a) a broad outer transition zone, where magnetic stresses participate in the angular momentum transport, but are not strong enough to cause deviation from Keplerian flow, and, (b) a narrow boundary layer at the inner edge of the disk, where strong magnetic stresses dominate the angular momentum transport, terminate the disk, and channel the accretion flow along the stellar field lines. The location of the inner edge of the disk, r_0, is given by the conservation of angular momentum in the boundary layer,

$$\frac{B_p B_\phi}{4\pi} 4\pi r^2 \delta \simeq \dot{M} r v_K. \tag{1}$$

Here, B_p and B_ϕ are respectively the poloidal and toroidal components of the magnetic field, δ is the width of the boundary layer, and v_K is the Keplerian velocity. The value of r_0 obtained in this way, and its scaling with the essential variables $\dot{M}, \mu,$ and M, depends on the model of the boundary layer, which, in turn, depends on the model of accretion disk appropriate at such radii. For accretion rates and magnetic fields typical of accretion-powered pulsars, the so-called "middle" region of the Shakura-Sunyaev (1973) disk model is appropriate, and the inner radius is given (GL) by

$$r_0 \simeq 1.3 \times 10^8 \dot{M}_{17}^{-46/187} \mu_{30}^{108/187} (M/M_\odot)^{-41/187} \ cm. \tag{2}$$

Here, $\dot{M}_{17}$ is $\dot{M}$ in units of $10^{17} \mathrm{gs}^{-1}$. Note that the above scalings are close, but not identical, to those of the Alfvén radius for spherical accretion, $r_A = \dot{M}^{-2/7} \mu^{4/7} (2GM)^{-1/7}$.

While GL treated the two above regions separately by semi-analytic or numerical methods, it should be possible to treat the entire region numerically in a unified scheme. Preliminary results of such calculations (Daumerie *et al.* 1993) show that the full region does, indeed, have an outer part where the physical variables change on a lengthscale $\sim r$, and an inner, boundary-layer like, part where the variables change on a lengthscale $\sim \delta \ll r$. However, the width of the boundary layer is larger than that calculated by GL.

Possible structures of the disk-magnetosphere interaction region which are related to the GL structure, but differ from it in detail, have been discussed qualitatively in recent years. Arons (1987) and co-authors have suggested, for example, that some stellar field lines in the outer parts of the interaction region may become open if the disk plasma loaded on them can be spun up to super-Keplerian rotation. A centrifugal wind would then be driven along these field lines. In the picture of Aly and Kuijpers (1990), on the other hand, the interaction region is confined to a thin annular zone where the strength of the stellar magnetic field is $\approx$ that of the small-scale magnetic fields in the disk (so that reconnection between the two proceeds the most efficiently), and a highly pinched version of the disk extends far into the magnetosphere.

Understanding the electrodynamics of the interaction region is a complicated and challenging task. The stellar magnetic field is wound up in the azimuthal direction and pinched inward in the radial direction by the differential motion between the disk and the star. While it is reasonable to suppose that these distortions are ultimately limited by a variety of processes, *e.g.*, reconnection (GL), flux escape (Wang 1987), and current-driven instabilities, the expected saturation value of the magnetic pitch B_ϕ/B_p in a steady state, and the dominant process that determines this value, are still uncertain. Indeed, the process of buildup and release of magnetic energy is inherently episodic, and the adequacy of a time-averaged description of it in terms of steady-state models may be in question. GL focused on reconnection-limited pitch, and showed that steady-state electrodynamics can then be formulated in terms of an effective elctrical conductivity, assumed isotropic for simplicity. The idea is generally valid, of course, and Campbell (1987) turned it around to calculate the pitches corresponding to simple, mathematically tractable, models of disk conductivity.

3.2.2 Accretion Torque. The accretion torque on the neutron star consists of two parts. The first comes from stresses associated with matter accreting from the inner edge of the disk, and is given by (GL)

$$N_0 \equiv \dot{M}(GMr_0)^{1/2}. \tag{3}$$

The second comes from the stresses associated with the magnetic field coupling the star with the disk. The total torque N can be conveniently expressed in terms of N_0, and the dimensionless torque, $n \equiv N/N_0$, depends only on the fastness parameter (GL; EGL),

$$\omega_s \equiv \Omega_s/\Omega_K(r_0) \simeq P^{-1}\dot{M}_{17}^{-3/7}\mu_{30}^{6/7}, \tag{4}$$

of the star. Slow rotators ($\omega_s < \omega_c$) are spun up ($n > 0$), while fast rotators ($\omega_s > \omega_c$) are spun down ($n < 0$), by this accretion torque. For $\omega_s > 1$, steady accretion is of course not possible. Here, ω_c is the critical fastness at which the torque changes direction, whose value was found to be $\sim 0.3 - 0.5$ in the original GL calculations. A simple analytic approximation to the torque calculated numerically by GL is

$$n(\omega_s) \simeq 1.4 \left(\frac{1 - \omega_s/\omega_c}{1 - \omega_s}\right). \tag{5}$$

It was suggested by EGL that, due to the balance between stretches of spinup and spindown, the period of a pulsar can be maintained in a range around its critical period, if its luminosity goes through alternate high and low states. This would explain the observability of long-period sources like GX 1+4 and A

0535+26, despite their short spinup timescales in high states. Extensive stretches of spinup and spindown seen in these and other sources since then have borne out this conclusion (also see below).

While the results of subsequent torque calculations using GL-type models (Campbell 1987; Wang 1987; Daumerie et al. 1993) have been qualitatively similar, a major issue has been the value of the critical fastness. This parameter depends on the magnetic spindown torque, and so on the magnetic pitch (see above) in the outer parts of the interaction region. GL undoubtedly overestimated the pitch in their approximate treatment of the electrodynamics, so that a more exact calculation should lead to smaller pitches and therefore to larger values of ω_c. However, the extreme value of $\omega_c \simeq 0.97$ obtained by Wang (1987) seems untenable on both theoretical and observational grounds. Wang (a) made an incorrect algebraic approximation, as a result of which the torque diverges as $\omega_s \to 0$, and, (b) neglected screening of the poloidal field. Observationally (see below), there is little evidence for such a large value of ω_c. Indeed, $\omega_c \simeq 0.97$ would imply a tiny range of accretion rates for each pulsar over which spindown could occur (see eq.[4]), making it a rare phenomenon. This is clearly not the case, as most pulsars show spindown, and many of them have long and repeated intervals of it (see above).

3.2.3 *Comparison With Observation.* Predictions from accretion torque theories were compared (GL and references therein) with the observed spinup rates of pulsars in the 1970s, as few episodes of spindown were known at the time. The rich structure of spinup/spindown patterns revealed by further observations (see above) makes these comparisons more interesting and complex, and provides more stringent tests of accretion torque models. Observation of spinup and spindown during an outburst of the transient Be-star binary source EXO 2030+375 (Parmar et al. 1989) with a pulse period ~ 42 s provided the first opportunity to study the spinup-spindown transition in detail as the luminosity declines and a slow rotator turns into a fast one.

If, as outlined above, both spinup and spindown can be described in terms of a dimensionless torque n which is a function of a dimensionless stellar spin rate ω_s alone, this universal scaling (GL) can be demonstrated in the following way. Since $n \propto -\dot{P}P^{-2}L^{-6/7}\mu^{-2/7}$ roughly (as can be shown by combining eqns.[2] and [3]), and $\omega_s \propto P^{-1}L^{-3/7}\mu^{6/7}$, the n vs. ω_s curve can be mapped out from period-luminosity data. Here, L is luminosity of the pulsar. With data on one individual pulsar (μ = constant), we need only plot $-\dot{P}P^{-2}L^{-6/7}$ vs. $P^{-1}L^{-3/7}$. Data on a collection of pulsars can be combined if we have a knowledge of the relative values of μ of these from, say, cyclotron line energies (Nagase 1992; Mihara 1993). I show below the results for two well-studied pulsars.

The ~ 120 s pulsar GX 1+4 with an M6 III giant companion was observed to spin up at a rate $\dot{P}/P \approx -3.10^{-2}$ yr^{-1} in its high-luminosity state in the 1970s (Nagase 1989). It entered an extended low state in the 1980s, from which it reappeared in 1987 in a low-luminosity spindown state (Makishima et al. 1988), and has continued to spin down at a rate $\dot{P}/P \approx 2.10^{-2}$ yr^{-1} upto the BATSE observations. Observations of this source first resolved the dilemma of the short spinup timescales of long-period pulsars (EGL). Shown in Fig. 2a is the observed torque curve for GX 1+4, using a compilation (Rao et al. 1993) of the long-term period history given in the literature. The uncertainties in the observed values are dominated by the range of variation in the luminosity (typically by a factor ~ 4) during each observation. A theoretical torque curve given by eqn. (5)

with $\omega_c = 0.3$ accounts well for the observations, except perhaps for the highest-fastness point, as Fig.2a shows.

For Her X-1, one of the best-studied pulsars ($P = 1.24$ s), BATSE data with $\dot{P}$ measured $\sim$ every month (Wilson et al. 1993) has been used in Fig. 2b to map out the torque curve for this source from shorter-term period variations, which show spinups and spindowns of comparable maximum strength $|\dot{P}/P| \approx 2.10^{-5}$ yr^{-1}. In this case, an additional assumption about the relation between the pulsed flux (which is what BATSE normally sees) and the total flux is necessary, and I have used a direct proportionality, guided by observations of those sources (e.g., GS 0834-430; C. A. Wilson et al.1993) where the latter flux is also available from Earth occultation measurements. Observational uncertainties are again dominated by the range of variation in L, as given by BATSE data (Wilson et al. 1993). The same theoretical curve as in Fig.2a fits these observations, as shown, although other choices of ω_c (see below) are also possible.

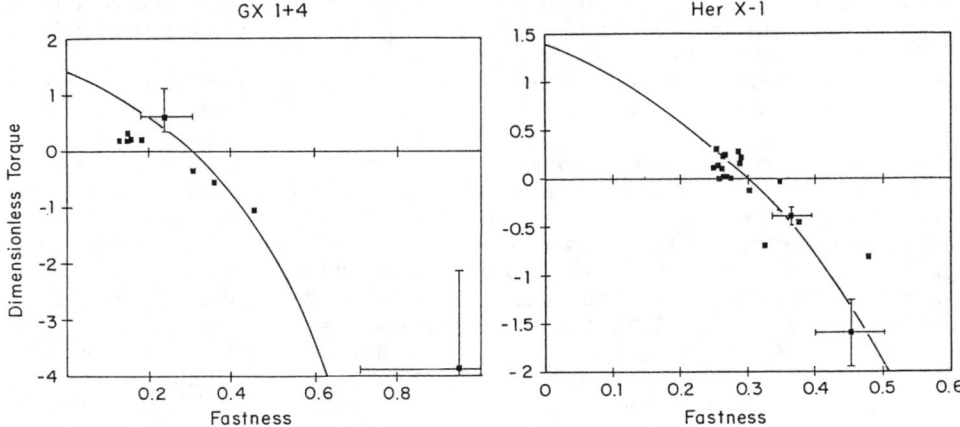

Fig. 2.–Torque curves for (a) GX 1+4, and, (b) Her X-1. Shown are the dimensionless torque (n) vs. fastness (ω_s) data, superposed on the curve given by eqn.[5] with $\omega_c = 0.3$. Also shown are the typical uncertainties in the observed values, dominated by luminosity variations (see text) during each observation. Such examples support the idea of universal scaling.

These examples are thus consistent with the idea of universal scaling, which implies the same ω_c for both sources. For the two sources above, this yields a relation between their stellar magnetic moments. Using the Her X-1 magnetic field inferred from cyclotron-feature observations (Nagase 1992), this indicates a surface magnetic field $\sim 10^{14}$ G for GX 1+4, in agreement with recent suggestions (Mony et al. 1991); proposals have been made for the observation of the corresponding cyclotron feature at ~ 1 MeV with OSSE on CGRO (Prince 1993). Finally, note that, although the normalization constant for the observed ω_s is unspecified in these examples, an *upper bound* on it is implied by the requirement that the maximum observed value of fastness must not exceed unity (see above). The use of this requirement for GX 1+4 leads to $\omega_c \lesssim 0.3$ (Fig.2a).

Similar arguments for Her X-1 lead to $\omega_c \lesssim 0.6$ (Fig.2b). Thus, spinup/spindown observations suggest a critical fastness of $\omega_c \sim 0.3 - 0.6$.

It is interesting to compare these values of ω_c with those inferred from frequencies of quasi-periodic oscillation (QPO) like features seen in the power spectra of some accreting pulsars, e.g.EXO 2030+375 (Angelini et al. 1989), Cen X-3, and 4U 1626-67 (Nagase 1992). Interpreted as the beat frequency between the stellar spin frequency and the Keplerian frequency at the disk's inner edge (in analogy with the situation for LMXBs; note that this interpretation is still controversial), these immediately yield $\omega_s \approx 0.11, 0.76$, and 0.85 for EXO 2030+375, Cen X-3, and 4U 1626-67, respectively. In the first case, the QPO feature was seen during spinup at high luminosities (Angelini et al. 1989), and the scaling from there using the equations given above implies $\omega_c \approx 0.3$, in excellent agreement with the above. For Cen X-3 and 4U 1626-67, it is not clear yet if there is a contradiction, as Cen X-3 is known to undergo strong spindown on short timescales (Finger et al. 1993), and 4U 1626-67 seems to have already started its spindown (Bildsten et al. 1993) when the QPO feature was observed.

3.3 Wind Accretion

In HMXBs, O/B supergiant companions drive strong stellar winds, gravitational capture from which drives the accretion to the neutron star. In addition, in Be star systems, the accretion in the quiescent state may be from a similar, weaker wind, and, even during outbursts due to episodic accretion from equatorial rings or disks shed by the Be star, this wind may continue to blow from the polar regions (Waters & van Kerkwijk 1989, hereafter WK). I outline here only the essential features of angular momentum transfer from stellar winds; for more details, see Blondin (1993).

3.3.1 Supergiant Systems. The angular momentum transferred to the neutron star per unit mass captured from the wind is usually parameterized in the form $l_w = \eta r_a^2 \Omega_{orb}$, where $r_a \equiv 2GM/v_0^2$ is the accretion radius, Ω_{orb} is the orbital angular velocity, and η is a dimensionless number. The value of η, which depends on the density- and velocity-gradient in the wind, has generally been calculated to be of order unity and both signs by various authors (Shapiro & Lightman 1976; Wang 1981; Anzer et al. 1987; Ho 1988). Alternatively, l_w can be expressed in terms of $r_a v_0$.

Numerical studies of mass capture from stellar winds carried out in recent years (Matsuda et al. 1987; Taam & Fryxell 1989; Blondin et al. 1990; Matsuda et al. 1991, Livio 1993) have shown that the flow patterns often do not approach a steady state; rather, the shock cone oscillates from side to side, producing circulation that reverses quasi-periodically. The origins of this "flip-flop" behavior, which occurs in 2D simulations, but sometimes appears in 3D calculations and sometimes does not, are not fully understood (Livio 1993), even apart from the question of whether numerical artifacts of the simulations, e.g., dimensionality and zoning, are responsible for it. Attempts to attribute it to asymmetries in the upstream flow (Blondin et al. 1990) have been in conflict with the result that uniform upstream flows can also develop this behavior when the size of the accreting body is relatively small (Matsuda et al. 1991).

The instantaneous magnitude of the specific angular momentum, l_w, given by these simulations can be $\sim 0.15 r_a v_0$ when a circulating "disk" develops, but its long-term average value is much smaller, as the flow has reversals, accompanied by outbursts of mass flux to the star (Taam & Fryxell 1989). This behavior may be related to the pulse period history of wind-fed HMXBs like Vela X-1, which do indeed show torque reversals on timescales of at least a few days

(Deeter et al. 1989). However, the typical timescales of reversal given by the current simulations are ~ 1 hr.

3.3.2 Be Star Systems. Accretion from the matter expelled episodically from the Be star in the form of equatorial rings, disks, or "disk-like winds" (WK) in these systems is thought to cause outbursts. This process of emission of a rotating, dense wind is just beginning to be studied quantitatively (Bjorkman & Cassinelli 1993), so that it is not surprising that little is known yet about the way in which the neutron star accretes from it. Indeed, it is not even certain yet if well-developed accretion disks always form in these systems, although the strong spinup torques often observed during outbursts (*e.g.*, in EXO 2030+375) are indirect indications that they do.

A useful probe into the spin evolution of these systems is the correlation, $P_{spin} \sim P_{orb}^2$, observed between their spin and orbital periods (Corbet 1986). This has been explained (WK) in terms of the concept of the critical spin period ($\Omega_s = \Omega_c \propto \Omega_K(r_m)$, where r_m is the outer radius of the magnetosphere), at which the net torque on the star vanishes, in analogy with what is done in disk accretion theory (see above). When one uses this concept, radial stellar winds give the (weaker) correlation, $P_{spin} \sim P_{orb}^{4/7}$, observed in wind-fed supergiant systems (WK). However, the observed relation for Be-star systems given above can only be reproduced by dense, equatorial, disk-like flows with a much more gradual velocity law (WK) than the standard radial stellar wind; such winds are now being studied in detail (Bjorkman & Cassinelli 1993).

4. FINAL SPIN UP

4.1 HMXB Evolution

The final evolution of HMXBs can go in two ways when the common-envelope (CE) phase begins (vdH92 and references therein). An initially wide binary (*e.g.*, a Be star system) can eject the entire envelope and produce a neutron star with a helium core companion. The system then evolves either (a) by a supernova explosion of the companion, or, (b) by evolution of the companion into a massive white dwarf. In the former case, the chances are high that the system remains bound, producing a double neutron star system like PSR 1913+ 16. In the latter, a system like PSR 0655+64 is thought to be produced. The spin evolution is qualitatively straightforward in both cases: the neutron star is spun up to short periods ($\sim 50-1000$ ms, say) determined by the strength of the full-scale Roche lobe overflow (that initiates the CE phase) and the somewhat smaller ($\sim 10^{10}$ G) field of the older neutron star. On the other hand, an initially narrow binary undergoes a complete spiral-in in the CE phase, producing a Thorne-Zytkow object (vdH92) with a disk-accreting neutron star in its core. The end product is a recycled, spun up, single radio pulsar.

4.2 LMXB Evolution

Neutron stars with low magnetic fields ($\sim 10^8 - 10^9$ G) in very bright (near-Eddington) LMXBs are thought to be spun up to $\sim$ millisecond periods by accretion torques, and these systems are then believed to evolve into binary pulsars of PSR 1953+29 class (vdH92, Webbink 92). This branch of the final spin evolution of accretion-powered pulsars is of crucial importance in current research on rotation-powered millisecond pulsars (Kulkarni 1993 and references therein), although millisecond pulsations from the putative LMXB progenitors have not been detected so far. Diagnostics of this spin evolution are thus indirectly obtained from the position of the so-called "spinup line" in the magnetic field vs. period diagram of rotation-powered pulsars (White & Stella 1988; Ghosh & Lamb 1992). These diagnostics can place severe constraints on models of the inner accretion disks in these systems, which are quite different from those in typical accretion-powered pulsars (Ghosh & Lamb 1992).

5. CONCLUSIONS

I conclude with a list of some of the most important problems on which research in this area is likely to focus in the near future. Foremost is the question of the nature of the accretion flow to the neutron star and the resultant accretion torque in Be star binaries: it is amazing that we have so little understanding of the class that forms the majority of known accretion-powered pulsars. A better understanding of the basic properties of the torque due to accretion from stellar winds is also necessary before we can account quantitatively for the spin behavior of wind-fed sources. In disk-fed sources, it would be important to identify the essential physical processes that determine the magnetic pitch and the spindown torque.

ACKNOWLEDGEMENTS

It is a pleasure to thank L. Angelini, P. Daumerie, R. L. Kelley, F. K. Lamb, L. Stella, J. H. Swank, and N. E. White for stimulating discussions, and to express gratitude to the BATSE team, particularly R. B. Wilson, M. H. Finger, and T. A. Prince, for communicating their results in advance of publication.

REFERENCES

Aly, J.-J. 1980, A&A, 86, 192
Aly, J.-J., & Kuijpers, J. 1990, A&A, 227, 473
Angelini, L., et al. 1989, ApJ, 346, 906
Anzer, U., et al. 1987, A&A, 176, 235
Arons, J. 1987, in The Origin & Evolution of Neutron Stars, ed. D. J. Helfand & J.-H. Huang (Dordrecht: Reidel), 207
Bhattacharya, D., & van den Heuvel, E. P. J. 1991, Phys Rep, 203, 1
Bildsten, L. 1993, Proc. Second Compton Symposium, in press
Bjorkman, J. E., & Cassinelli, J. P. 1993, ApJ, 409, 429
Blondin, J. M., et al. 1990, ApJ, 356, 591
Blondin, J. M. 1993, this volume
Campbell, C. G. 1987, MNRAS, 229, 405
Chakrabarty, D., et al. 1993a, ApJ, 403, L33
———. 1993b, Proc. Second Compton Symposium, in press
Cominsky, L. 1993, this volume
Corbet, R. H. D. 1986, MNRAS, 220, 1047
Corbet, R. H. D., & Day, C. S. R. 1990, MNRAS, 243, 553
Daumerie, P., Ghosh, P., & Lamb, F. K. 1993, in preparation
Davies, R. E., & Pringle, J. E. 1981, MNRAS, 196, 209
Deeter, J. E., et al. 1989, ApJ, 336, 376

Elsner, R. F., Ghosh, P., & Lamb, F. K. 1980, ApJ, 241, L155 (EGL)
Finger, M. H., et al. 1993, Proc. Second Compton Symposium, in press
Ghosh, P., & Lamb, F. K. 1979a, ApJ, 232, 259 (GL)
———. 1979b, ApJ, 234, 296 (GL)
———. 1991, in Neutron Stars: Theory & Observation, ed. J. Ventura & D. Pines (Dordrecht: Reidel), 363 (GL)
———. 1992, in X-ray Binaries & Recycled Pulsars, ed. E. P. J. van den Heuvel & S. Rappaport (Dordrecht: Kluwer), 487
Henrichs, H. 1983, in Accretion-Driven Stellar X-ray Sources, ed. W. H. G. Lewin & E. P. J. van den Heuvel (Cambridge: Cambridge Univ. Press), 393
Ho, C. 1988, MNRAS, 232, 91
Illarionov, A., F. & Kompaneets, D. A. 1990, MNRAS, 247, 219
Illarionov, A., F. & Sunyaev, R. A. 1975, A&A, 39, 185 (IS)
Israel, G. L., et al. 1993, IAU Circ. 5889
Iwasawa, K., et al. 1992, PASJ, 44, 9
Johnston, S., et al. 1992, ApJ, 387, L37
Johnston, S., et al. 1993, IAU Circ. 5794
Kulkarni, S. R. 1993, this volume
Kundt, W., & Robnik, M. 1980, A&A, 91, 305
Livio, M. 1993, private communication
Makishima, K., et al. 1988, Nature, 333, 746
Matsuda, T., et al. 1987, MNRAS, 226, 785
Matsuda, T., et al. 1991, A&A, 248, 301
Mihara, T. 1993, this volume
Mineshige, S., Rees, M. J., & Fabian, A. C. 1991, MNRAS, 251, 555
Mony, B., et al. 1991, A&A, 247, 405
Nagase, F. 1989, PASJ, 41, 1
Nagase, F. 1992, in Proc. Ginga Memorial Symposium, ed. F. Makino & F. Nagase (Tokyo: ISAS), 1
Parmar, A. N., et al. 1989, ApJ, 338, 359
Prince, T. A. 1993, Proc. Second Compton Symposium, in press
Rao, A. R., et al. 1993, in preparation
Shakura, N. I., & Sunyaev, R. A. 1973, A&A, 24, 337
Shapiro, S. L., & Lightman, A. P. 1976, ApJ, 204, 555
Stollberg, M. T., et al. 1993, IAU Circ. 5836
Taam, R. E., & Fryxell, B. A. 1989, ApJ, 339, 297
van den Heuvel, E. P. J. 1992, in X-ray Binaries & Recycled Pulsars, ed. E. P. J. van den Heuvel & S. Rappaport (Dordrecht: Kluwer), 233 (vdH92)
van Paradijs, J. 1993, in X-ray Binaries, ed. W. H. G. Lewin et al., (Cambridge: Cambridge Univ. Press), in press
Wang, Y.-M. 1981, A&A, 102, 36
Wang, Y.-M. 1987, A&A, 183, 257
Waters, L. B. F. M., & van Kerkwijk, M. H. 1989, A&A, 223, 196 (WK)
Webbink, R. F. 1992, in X-ray Binaries & Recycled Pulsars, ed. E. P. J. van den Heuvel & S. Rappaport (Dordrecht: Kluwer), 269
White, N. E., & Stella, L. 1988, Nature, 332, 416
Wilson, C. A., et al. 1993, Proc. Second Compton Symposium, in press
Wilson, R. B., et al. 1993, this volume

DISCOVERY OF THE HARD X-RAY PULSAR GRO J1008-57 BY BATSE

Robert B. Wilson, B. Alan Harmon, & Gerald J. Fishman
NASA/Marshall Space Flight Center-ES66, Huntsville, AL 35812
wilson@batse.msfc.nasa.gov

Mark H. Finger
Computer Sciences Corporation (NASA/MSFC-ES66),Huntsville, AL 35812

Mark T. Stollberg, Geoffrey N. Pendleton and Michael Briggs
Dept. of Physics, University of Alabama in Huntsville, Huntsville, AL 35899

Thomas A. Prince, Lars Bildsten, and Deepto Chakrabarty
Div. of Physics, Mathematics, and Astronomy, Caltech, Pasadena, CA 91125

Bradley C. Rubin and Nan S. Zhang
Universities Space Research Association (NASA/MSFC-ES66),
Huntsville, AL 35812

ABSTRACT

BATSE detection of a new hard x-ray pulsar is reported. The source was first observed on 14 Jul 1993, reached a maximum intensity on 25 Jul 1993, then declined smoothly until it became undetectable on 16 Aug 1993. The BATSE location was adequate to permit OSSE, ASCA, & ROSAT observations, leading to an improved source location. The observed period at the solar system barycenter was 93.548±0.002s, 93.5665±0.0005s, and 93.541 ±0.004 on 15 Jul 1993, 23 Jul 1993, and 10 Aug 1993, respectively. The source is detected between ~20 keV and 160 keV, with a spectrum fit by an optically thin thermal bremsstrahlung form, having a kT of 25.4±2.1 keV during the rise to maximum intensity, decreasing monitonically during the remainder of the outburst to 17.1±2.5 keV on 5 Aug 1993. The pulse profile in this energy range has a single broad peak, with the maximum occuring later at higher energies. The pulse profile, intensity history, and spectral behavior observed by BATSE are reported.

1. INTRODUCTION

Daily monitoring of the DISCLA Chl 1 (20-50 keV) data from the BATSE Large Area Detectors (LADs) is performed by the BATSE Instrument Pulsar Team, using FFT-based software developed by the joint MSFC/Caltech Key Project, under the CGRO Guest Investigator Program. The results of the analysis are available one day after receipt at MSFC of the data. A persistent 93.5 s periodic signal was observed starting on 14 Jul 1993 near the frequently observed third harmonic from Vela X-1, which was viewed by some of the same detectors. A location which excluded Vela X-1 and any other known x-ray source was obtained using the Earth occultation method (Stollberg, et al.1993), with an error

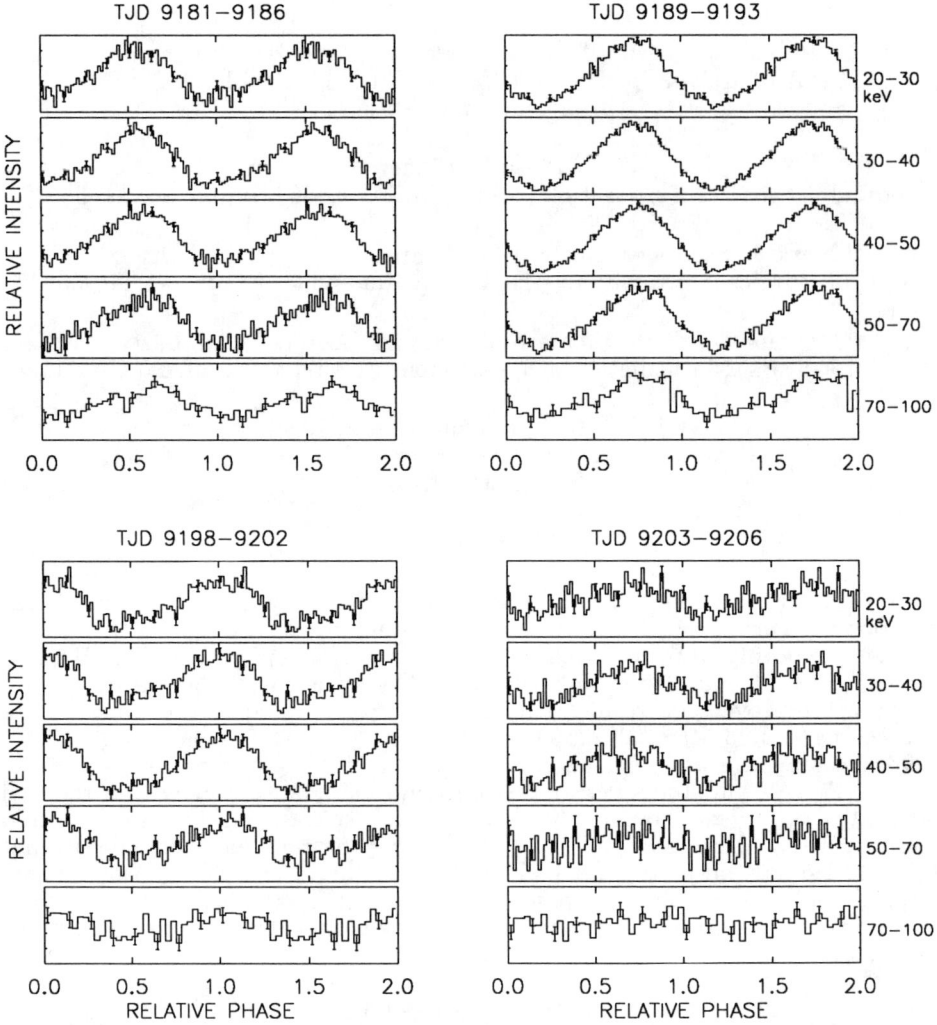

Figure 1. Pulse profile of GRO J1008-57 for four intervals during the outburst. The range of Truncated Julian Days are shown above each plot.

box of about 2.5 degrees. A CGRO Target of Opportunity was declared, with a reorientation of the Observatory performed to permit OSSE scans of the region, leading to the improved position reported in Grove, et al. (1993). Observations were then obtained by ASCA, which found the source within that error region (Tanaka 1993). ROSAT observations were also made (Petre and Gehrels, 1993) late in the outburst, yielding the 15 arcsec position error circle they report (RA = 10h 09m 46s, Decl. = -58° 17' 32", equinox 2000.0).

2. PULSE PROFILE

BATSE LAD CONT data (2.048s, 16 energy channels) have been epoch-folded at the approximate source period to obtain the profiles shown in Figure 1, for several time intervals during the outburst. Two cycles are shown. The peak of the emission occurs later at higher energies. It is reported by Tanaka (1993) that the pulse shape in 1-10 keV is much different, having a double-peaked structure.

3. PERIOD & INTENSITY HISTORY

DISCLA data (1.024s, 20-50 keV) have been epoch-folded with a constant-period model for 25000s intervals at the solar system barycenter, to obtain a set of pulse profiles. For each profile a pulse phase has been determined by fitting of a mean pulse template. Barycenter frequencies have been determined by first order linear fits to the observed pulse phases. The history of values obtained (converted to periods) is shown in Figure 2. The variations in period observed may be caused by Doppler shifts due to motion about a stellar companion, intrinsic changes due to torques applied by infalling matter to the neutron star responsible for the x-ray emission, or both. A unique orbital solution cannot be determined from this outburst alone. If the object producing this x-ray emission is in a binary system with a Be star, the orbital period is expected to be of the order of $\sim$ 100 days (Waters & van Kerkwijk 1989). No additional pulsed signals have been observed from this system as of 29 Nov 1993.

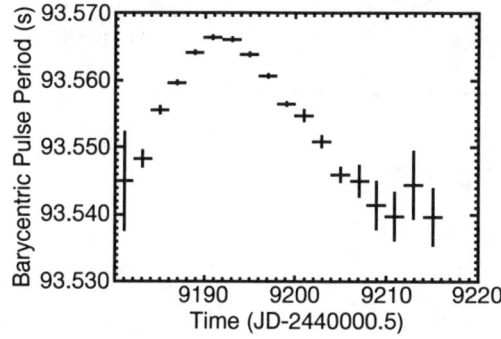

Figure 2. Period history of GRO J1008-57, corrected to the solar system barycenter. No binary orbital corrections are included, since the orbit cannot be determined from this single outburst.

For the phase of each pulse profile as determined above, the intensity of the source in counts per second is determined as the amplitude coefficient which gives the best fit to the mean pulse profile template. The intensity history obtained is shown in Figure 3.

4. SPECTRAL ANALYSES

Spectra have been fit for several time intervals during the outburst, corresponding approximately to the intervals for which pulse profiles are shown in Figure 1. An adequate fit is obtained using an optically thin thermal bremsstrahlung (OTTB) model of the form $A/E \cdot exp(-E/kT)$. The flux and temperature pa-

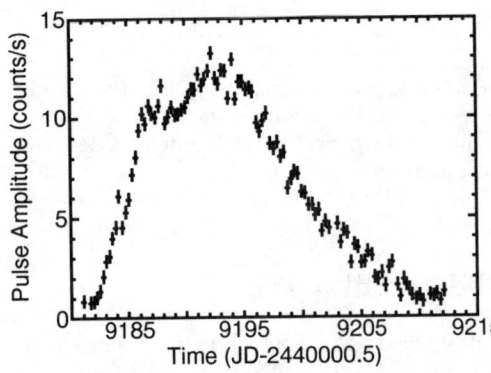

Figure 3. Intensity History – RMS pulsed amplitude in counts/s (20-50 keV). A sum of data from source facing detectors is used, weighted according to approximate effective area in 20-50 keV. A reorientation of CGRO occurred on TJD 9195, which may be partly responsible for the decrease shown on TJD 9196 (28 Jul 1993).

rameters obtained are shown in table I. The outburst is hardest in the rising portion of the lightcurve.

Table 1. Spectral Fit (OTTB) Parameters			
Start Day	End Day	Flux(45 keV) $(10^{-4} \times ph \cdot s^{-1} \cdot cm^{-2} \cdot keV^{-1})$	kT (keV)
9181.0	9186.1	1.33 ± 0.05	25.4 ± 2.1
9186.1	9189.1	2.59 ± 0.06	23.0 ± 1.1
9189.1	9194.0	3.03 ± 0.05	20.2 ± 0.7
9198.6	9202.5	2.32 ± 0.09	20.6 ± 1.6
9203.0	9206.4	1.20 ± 0.12	17.1 ± 2.5

REFERENCES

Crosa, L., & Boynton, P. E. 1980, ApJ, 235, 999
Grove, E. et al.1993, IAUC, 5838
Petre, R. , & Gehrels, N. 1993, IAUC, 5877
Stollberg, M. et al.1993, IAUC, 5836
Tanaka, Y. 1993, IAUC, 5851
Waters, L. B. F. M., & van Kerkwijk, M. H. 1989, A&A, 223, 196

BATSE OBSERVATIONS OF 4U1538-52: A 530 SECOND PULSAR

Brad C. Rubin (USRA/NASA/MSFC)

M. H. Finger (GRO Science Support Center/CSC)

R.B. Wilson, G. J. Fishman, C. A. Meegan
NASA/Marshall Space Flight Center

W. S. Paciesas
Dept. of Physics, Univ. of Alabama, Huntsville

T. Prince, J. Chiu, D. Chakrabarty
California Institute of Technology

ABSTRACT

The BATSE experiment on CGRO, a nearly continuous all sky monitor, is capable of detecting and monitoring pulsars in the hard X-ray/Gamma-ray energy range. Using data processed with a global background subtraction method, we are able to construct pulsed intensity and pulse period histories for the accreting binary X-ray pulsar 4U1538-52 in the 20-50 kev energy band. This pulsar, which has a pulse period of approximately 530 seconds, is believed to be a wind accretor with a high mass companion. It was observed by previous experiments to be in secular spin down during the 1976-1988 period. Here we report observing, for the first time, a secular spin up trend. We are able to use this data to improve the upper limit on orbital period changes.

1. INTRODUCTION

The pulse period and mideclipse epoch of the binary X-ray pulsar 4U1538-52 (pulse period $\sim$ 530 seconds, orbital period $\sim$ 3.7 days) has been observed sporadically over 12 years by Ariel V (Davison et.al. 1977), OSO-8 (Becker et.al. 1977), EXOSAT (Robba et.al. 1992), Tenma (Makashima et.al. 1987, hereafter MKHN87), and Ginga (Corbet, Woo, and Nagase 1993, hereafter CWN93). Cominsky and Moraes (1991, hereafter CM91) discussed archival results from Uhuru and HEAO A-1. Optical observations of the companion, QV Nor (Crampton et.al. 1978), have helped establish that it is a 14.5 mag B0 supergiant which is not overfilling its Roche lobe. The sporadic pulse period observations from these experiments indicated a long term (secular) spin down trend with rapid and probably random pulse period variations on shorter time scales. Thus, 4U1538-52 is probably similar to the wind fed source Vela X-1.

The BATSE experiment on CGRO is a continuous all sky monitor with 8 Large Area Detectors (effective area $\sim$ 1500cm^2 per detector in the 20-50 kev band) mounted at the faces of an octahedron (Fishman et.al. 1989) The gamma-ray background due to the Earth varies by about a factor of 2 over the course of an orbit (about 5600 seconds). It is therefore important when studying weak sources with long periods to subtract this background using a global model, which is continuous over periods as long as an orbit (Rubin et.al. 1992). We have performed an epoch folding analysis of the pulsed signal, obtaining the pulse profile, pulse period, and pulsed intensity for 4U1538-52 using background subtracted BATSE data in the 20-50 kev band at 1.024 second time resolution. We have further used this data to determine an orbital epoch for CGRO based on doppler timing analysis, and combining this with previous experiments, to make a determination of the orbital period.

2. PULSE DETERMINATION AND HISTORY

Background subtracted (count) data for each detector with normal less than 90 degrees to the source was weighted by the square of the cosine of the angle between the source and the normal and summed over detectors. This data was fit to a sixth order harmonic expansion in the pulse phase during each 40000 second interval. A solar system barycenter ephemeris and preliminary binary ephemeris from MKHN87 were used in constructing the phase model. Further details about the epoch folding method appear in Finger, et.al. (1992).

We performed a pulse frequency search by varying the pulse phases in each set of harmonic amplitudes according to the searched frequency and looking for a sharp peak in the significance of the coherently summed amplitudes. This is the Z_n^2 statistic where Z_n^2 is distributed as a χ^2 with $2 \times n = 12$ degrees of freedom. We found it necessary to sum over 20 days to obtain a sufficiently resolved peak. The frequency at maximum Z_n^2 is determined by quadratic interpolation on the bins surrounding the peak. The statistical error is determined from the width of the peak. The total error is corrected for systematic error due to changing pulse shape by multiplying the statistical error with the square root of the ratio of the median Z_n^2 to the number of degrees of freedom. The pulse frequency history was thus determined for one twenty day interval near the beginning of the CGRO mission, another interval about 500 days later, and 10 intervals during a recent stretch of 200 days of data. These results are converted to a pulse period history and displayed in Figure 1a and they are shown again in Figure 1b, together with the history from previous missions.

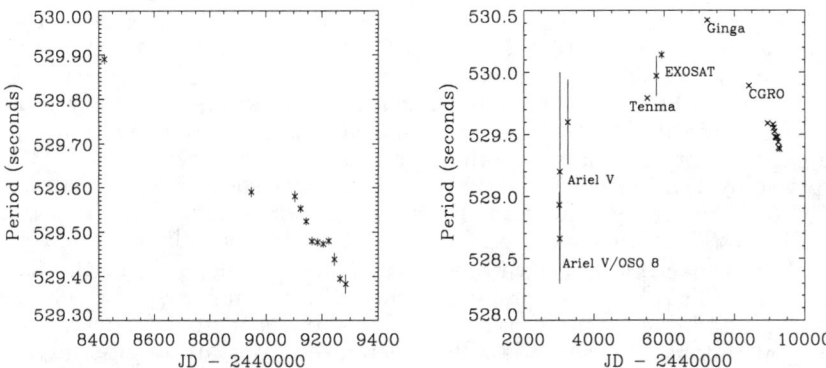

Fig. 1. 4U1538-52 Pulse Period History: a) BATSE results b) All observations

We have used the frequency history determined from the BATSE data during $JD2449094-2449293$ to construct an intrinsic phase model for the source by fitting those frequencies to a cubic polynomial in time. The intrinsic phase is the integral of this frequency in time. A light curve for the mean pulse profile is obtained by summing the harmonic amplitudes, rotated according to the phase model, over 200 days of data and transforming to the time domain. This mean pulse profile appears in Figure 2.

We have further converted this mean profile to a pulse template by normalizing it to a root mean square of 1.0. The source intensity, in counts per second, is then calculated by multiplying this template by the mean pulse light curve during shorter intervals and summing. We are interested either in contiguous intervals or binary orbit phases. The intensity history for a 200 day period is shown in Figure 3, where each data point is a sum over three complete binary orbits, and the average intensity as a function of orbital phase is shown in Figure 4. From Figure 3 we see that the intensity may vary by a factor of 2, but there are no large outbursts or long periods in which the source turns off. Figure 4 shows clear evidence for eclipse of the source at close to the expected binary phase, and little significant intensity variation throughout the rest of the orbit except for a possible dip in intensity after phase 0.4. The eclipse duration is consistent with 0.59 ± 0.01 days measured by CWN87.

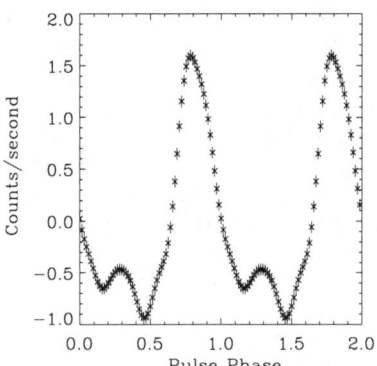 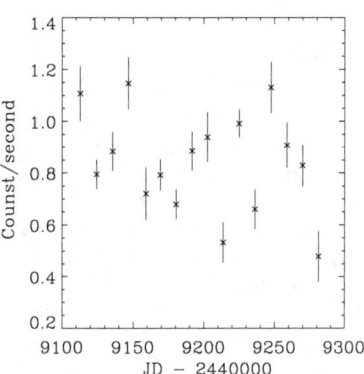

Fig. 2. Mean pulse profile. Fig 3. Intensity History

3. ORBIT DETERMINATION

The most recent orbit determination, by CWN93, indicated an orbit consistent with circular, but not well determined: $e = 0.082 \pm 0.047$, $a_x sini = 50.9 \pm 3.5$ lt-sec, and $P_{orb} = 3.72844 \pm 0.00002$ days. We have assumed a circular orbit, using the earlier determination by MKHN87: $a_x sini = 52.8 \pm 1.8$ lt-sec, and $P_{orb} = 3.72854 \pm 0.00015$ days. We doppler shifted the phases, simultaneously varying intrinsic frequency and orbital epoch by the Z_n^2 folding technique, to obtain a separate estimate of the epoch for each 20 day interval during $JD2449094-2449293$. Combining those ten intervals in a straight line fit gives the epoch $JD2449108.533 \pm 0.036$. Fitting this with epochs from previous experiments yields $P_{orb} = 3.72840 \pm 0.00003$ days, epoch $JD2445279.479 \pm 0.020$ and $\dot{P}_{orb} = -9.9 \times 10^{-7} \pm 2.3 \times 10^{-6} yr^{-1}$, with $\chi_\nu^2 = 1.01$. We are thus able to set a 3σ limit $\dot{P}_{orb} < 8.0 \times 10^{-6} yr^{-1}$, an improvement on the CWN93 limit by a factor of 2/3. The residuals of the epochs used are shown in Figure 5.

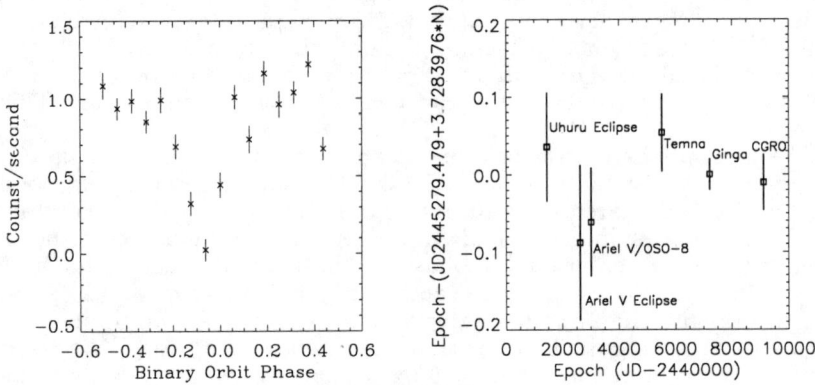

Fig 4. Intensity vs Orbital Phase   Fig. 5. Orbital Epochs

4. DISCUSSION

We observe that 4U1538-52 has been in secular spin up since the beginning of the CGRO mission, in contrast to its previous spin down behavior. The average $\dot{P}$ for the CGRO data as measured by a linear fit to the pulse periods is $\dot{P} = -6.3 \times 10^{-9} ss^{-1}$. The $\dot{P}$ with maximum magnitude over a shorter period is $-1.7 \times 10^{-8} ss^{-1}$ during $JD 2449224 - 2449264$. These are the same order of magnitude as period changes observed in previous missions, and these data appear consistent with a random walk in frequency, as observed in the similar system Vela X-1 (Boynton, et.al. 1984). Torques implied by data from previous missions (and this data) are consistent with the wind accretion model, but the long secular spin down has been somewhat of a mystery as is the cause of the random period changes. Here we have shown that the source was in secular spin down for not less than 12 years. We have obtained an upper limit on $\dot{P}_{orb}$ of magnitude a factor of 2.4 times larger than the tidal circularization rate of $-3.3 \times 10^{-6} yr^{-1}$ estimated by CM91. BATSE has been in operation for almost 1000 days. Processing these and future data over several more years will enable continuous pulse period monitoring and improved orbital determination.

REFERENCES

Davison, P. J., Watson, M. G., & Pye, J. P. 1977, M.N.R.A.S., 181, 73P
Becker, R. H., et. al. 1977, ApJ, 216, L11
Robba, R. H., et. al. 1992, ApJ, 401, 685
Makashima, K., et. al. 1987, ApJ, 314, 619 (MKHN87)
Corbet, R. H. D., Woo, J. W., & Nagase, F., 1993, A & A, 276, 52 (CWN93)
Cominsky, L. R., & Moraes, F. 1991, ApJ, 370, 670 (CM91)
Crampton, D., Hutchings, J. B., Cowley, A. P. 1978, ApJ, 225, L63
Fishman, G. J., et. al. 1989, GRO Science Workshop Proceedings, 2-39
Rubin, B. C., et. al. 1992 Compton Symposium Proceedings (AIP 280), 1127
Finger, M. H., et. al. 1992 Compton Symposium Proceedings (AIP 280), 386
Boynton, E. P., et. al. 1984, ApJ, 283, L53

HARD X-RAY OBSERVATIONS OF A 0535+262

Mark H. Finger
Compton Observatory Science Support Center, GSFC/CSC
finger@batse.msfc.nasa.gov

Lynn R. Cominsky
Sonoma State University

Robert B. Wilson, B. Alan Harmon, and Gerald J. Fishman
NASA/Marshall Space Flight Center

ABSTRACT

Three outbursts of the x-ray binary system A 0535+262 have been detected in the 20-120 keV energy range by the BATSE instrument onboard the Compton Gamma Ray Observatory. The system orbital parameters have been determined by a pulse timing analysis using data from these outbursts.

1. INTRODUCTION

A 0535+26 is a 103 s x-ray pulsar in a binary system with the Be star HDE 245770. Since its initial discovery in 1975 (Rosenberg et al. 1975), the source has been frequently observed to undergo transient outbursts. The outbursts show a variety of intensities, with the largest reaching 3 Crab at peak in the 2-10 keV band. For a review of past observations see Giovannelli and Graziati (1992).

Despite frequent observations, the system's binary orbit has remained poorly constrained. The spacing of weaker outbursts suggests a 111 orbital period (Nagase et al. 1982), which is confirmed by a Fourier analysis of the Vela 5B data (Priedhorsky & Terrell 1983). No published time-of-arrival analysis has yet confirmed this period or uniquely determined the other orbital parameters.

The BATSE instrument onboard the Compton Gamma-Ray Observatory (CGRO) has observed three outbursts from A 0535+26. The first outburst occurred from 1993 February 28 to March 26, and the second from 1993 June 20 to August 5 (Wilson et al. 1993). The third outburst, which is reported here for the first time, occurred from 1993 October 7 to November 16. After summarizing the observations, a timing analysis, including a determination of the binary orbital parameters, is presented.

2. OBSERVATIONS

The BATSE experiment is an all sky monitor having eight detector modules mounted on the corners of the CGRO spacecraft bus. The observations reported here use the Large Area Detectors (LAD's) which are 2025 cm^2 by 1.24 cm NaI(Tl) scintillators oriented in an octahedral geometry (Fishman et al. 1989).

Long period pulsar observations are made with either the 1.024 second, 4 energy channel DISCLA data, or the 2.048 second resolution, 16 energy channel

© 1994 American Institute of Physics

CONT data. In either case the data are combined from an appropriate set of detectors, and then epoch-folded using a trial ephemeris.

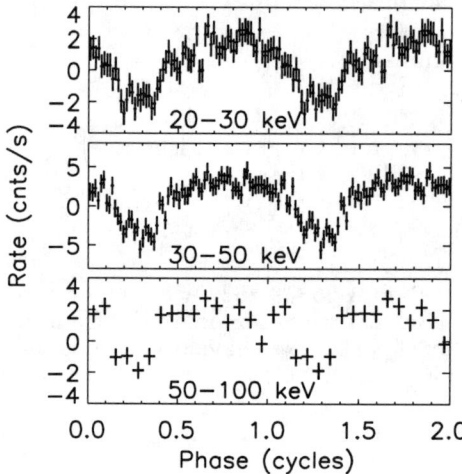

Figure 1. A 0535+26 epoch-folded pulse profiles for 1993 July 22.

Figure 1 shows the pulse profiles obtained by epoch-folding CONT data on July 22 at the peak of the outburst. To obtain good phase resolution, a binned representation of the pulse profile has been used. A quadratic spline fit to the rates was subtracted before folding. The spectrum of the pulsed flux was determined from the phase averaged count rates in each channel using the phase interval 0.125-0.375 as background. The count rates were fit with an optically thin thermal bremstrahlung model $F(E) = A \cdot exp(-E/kT)/E$ with values $kT = 26.3 \pm 2.7\ keV$ and $F(50\ keV) = 1.34 \pm 0.08 \cdot 10^{-4}$ photons/cm^{2}s·keV providing the best fit.

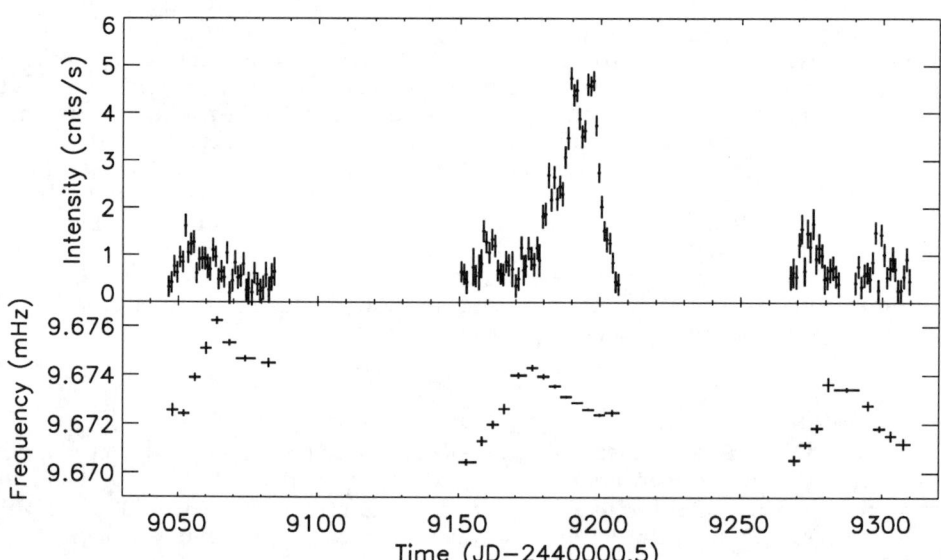

Figure 2. The top panel shows the A 0535+26 pulse amplitude obtained with epoch-folded 20-50 keV DISCLA data. The bottom panel shows the measured barycentric frequency obtained from the same data. The flux at the peak of the second outburst is about 60 mCrab.

Channel 1 (20-50 keV) of the DISCLA data were used for timing analysis. The data were summed over exposed detectors using a projected area weighting. The epoch-folding technique uses a harmonic representation of the pulse profile, with only the first six harmonics were used. A detailed description of the folding technique is given by Finger et al. (1992). Daily pulse profiles were obtained by folding with a trial phase model. Corrections to the phase model were then determined for each pulse profile by correlating the profile with a template which was obtained from the brightest portion of the second outburst. The pulse amplitudes determined from the template correlations are shown in the top panel of Figure 2. Frequencies obtained by fitting short intervals of the pulse phases are shown in the bottom panel of Figure 2.

3. PULSE TIMING ANALYSIS

The pulse phases, shown in the top panel of Figure 3, were fit with a model that included a phase and frequency at epoch, binary orbital motion, and accretion induced torque. Between outbursts pulse cycle count was assumed to be lost, and cycle slips were estimated. For simplicity these were estimated as real rather than an integer numbers.

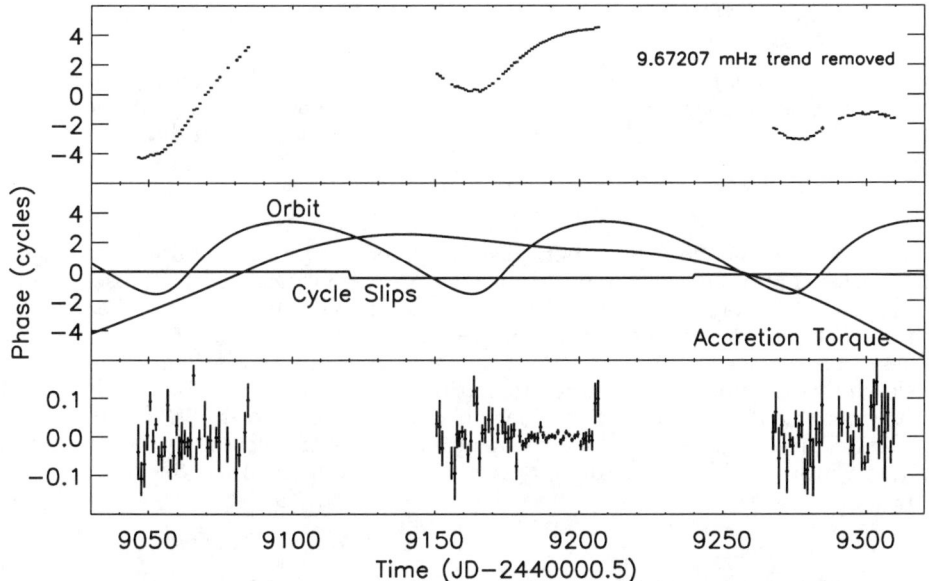

Figure 3. The top panel shows the A 0535+26 pulse phases obtained with epoch-folded 20-50 keV DISCLA data. These phases are relative to a constant barycentric frequency model. The middle panel shows the components of the model used to fit the measured phases. The bottom panel shows residuals (observed-model) of the fit.

To represent the accretion torque, independent frequency derivatives were measured in 6 intervals; TJD 9046-9085, 9085-9150, 9150-9180, 9180-9207, 9207-9267, and 9267-9310. The second and fifth intervals are between outbursts. Two intervals were used with the second outburst because of the large range of flux observed. The best fit model components are shown in the middle panel, and the residuals to the fit in the bottom panel of Figure 3. The estimated model parameters are given in Table 1.

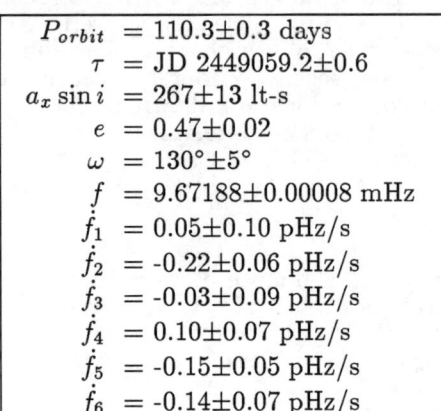

P_{orbit} = 110.3±0.3 days
τ = JD 2449059.2±0.6
$a_x \sin i$ = 267±13 lt-s
e = 0.47±0.02
ω = 130°±5°
f = 9.67188±0.00008 mHz
$\dot{f}_1$ = 0.05±0.10 pHz/s
$\dot{f}_2$ = -0.22±0.06 pHz/s
$\dot{f}_3$ = -0.03±0.09 pHz/s
$\dot{f}_4$ = 0.10±0.07 pHz/s
$\dot{f}_5$ = -0.15±0.05 pHz/s
$\dot{f}_6$ = -0.14±0.07 pHz/s

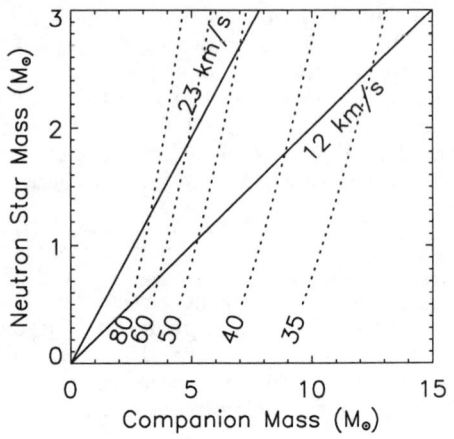

Table 1. Fit Parameters. The frequency epoch is TJD 9190. The $\dot{f}$ intervals are given in the text.

Figure 4. The optical constraints on system masses are given by solid lines. Dotted lines are for constant inclination labeled in degrees.

4. DISCUSSION

The A 0535+26 orbital period determined from pulse timing is in agreement with that deduced from outburst separations. Optical redshift observations folded at a 111 day period have been fit with orbital models (Janot-Pacheco et al. 1987), resulting in a constraint of the half amplitude of $12 < K_{opt} < 23$ km/s. These constraints are placed on the system's mass-mass diagram in Figure 4. Also shown are lines of constant inclination for the determined orbit. For a 1.4 solar mass neutron star, the inclination is constrained to $45° < i < 70°$. The optical orbital fit gives an eccentricity of 0.59±0.29 and an (x-ray) argument of periastron of 177°±29°, in rough agreement with our results.

REFERENCES

Finger, M.H. et al. 1992, Proc. Compton Observatory Science Workshop, 185
Fishman, G.J. et al. 1989, GRO Science Workshop Proc., GSFC, 2-39
Giovannelli, F. & Graziati, L.S. 1992, Space Sci. Rev., 59,1
Janot-Pacheco, E., Motch, C., & Mouchet, M. 1987, A & A, 177, 91
Nagase et al. 1982, ApJ, 263, 814
Priedhorsky, W.C. & Terrell, J. 1983, Nature, 303, 681
Rosenberg, F.D., Eyles, C.J., Skinner, G.K., & Willmore, A.P. 1975, Nature, 256, 628
Wilson, R.B., Finger, M.H., Gibby, L., & Fishman, G.J. 1993, IAUC 5833

CYCLOTRON ABSORPTION IN THE HIGH ENERGY SPECTRUM OF HER X-1

M. Kunz, R. Staubert
Astronomisches Institut der Universität, Tübingen, FRG

D. E. Gruber
Center for Astrophysics and Space Science, UCSD, San Diego, USA

W. Pietsch, J. Trümper
Max-Planck-Institut für extraterrestrische Physik, Garching, FRG

S. Kaniovsky, R. Sunyaev
Institute for Space Research, Moscow, Russia

ABSTRACT

We present the results of six Main-On observations of Her X-1, obtained by HEXE between 7/87 and 10/88 in hard X-rays. Pulse period measurements confirm the long term spin-up trend. A 16-60 keV pulse profile of excellent quality is given. The pulse-minus-off-pulse spectrum is best represented by a power-law with exponential cut-off and a cyclotron absorption line ($E_{cycl} = 34.1 \pm 2.5$ keV).

INTRODUCTION

The X-ray binary system HZ Her/Her X-1 was discovered in 1971 by Uhuru (Tananbaum et al. 1972). Among the established properties of the system are the 1.237 sec rotational period of the neutron star and the 1.7 day binary orbit. Various models have been proposed to explain the 35 day intensity cycle, such as precession of either the neutron star (Trümper et al. 1986) or the accretion disk (Petterson et al. 1991). The continuum X-ray spectrum of Her X-1 can be described by a thermal component at low energies, a power law continuum with photon index < 1 (e.g. Vrtilek et al. 1991), and an exponential cut-off above ~ 20 keV (Soong et al. 1990). A feature in the high energy spectrum is commonly interpreted as cyclotron absorption, implying a surface magnetic field strength of several 10^{12} Gauss (Trümper et al. 1978, Voges et al. 1982). The spectrum is strongly pulse phase dependent across the X-ray regime (0.1-100 keV). Altogether, this system has shown to be a model case for accretion onto and radiation emission by neutron stars. For a more detailed description of Her X-1, see e.g. Nagase (1989).

OBSERVATIONS AND RESULTS

The data reported here were obtained with the High Energy X-Ray Experiment (HEXE) on the MIR space station. HEXE is a system of four NaI(Tl) phoswich detectors sensitive in the energy range 16-200 keV, having a total effective area of 750 cm². The energy resolution is ≈ 30% (FWHM) at 60 keV, in normal mode the time resolution for individual photons is 25 msec. For a detailed instrument description see Reppin et al. (1985). The data have been presented in parts by Gilfanov et al. (1989).

We report on six observations of Her X-1 between 7/87 and 10/88 with a total on source time of 17 ksec. Every observation consists of several consecutive spacecraft orbits, called sessions.

Pulse period values have been established for six groups of consecutive sessions by means of epoch folding and, for comparison, the Rayleigh test (periodogram). Arrival times of photons in the energy range 16-60 keV were corrected for both binary and satellite motion, where the former is based on the Ginga ephemeris (Deeter et al. 1991):

$$T_{\pi/2} = \text{MJED } 43804.519980(14)$$
$$P_{orb} = 1.700167720(10) \text{ days}$$

The pulse period values obtained (Table 1) confirm the continued spin-up of the source ($\dot{P} = 2 \cdot 10^{-13}$ s/s). Deviations from the linear trend on short time scales are well known and explained in terms of a random walk process.

Pulse profiles in 16-60 keV for four out of the six observations have been added up after background subtraction, as they are similar in amplitude and shape. The composite pulse profile is shown in Fig. 1. The two remaining observations have been omitted here, as they were made towards the end of the Main-On phase, showing only marginal pulsation.

We have computed pulse-minus-off pulse spectra (P-OP) from pulse profiles in narrow energy bands to be able to compare the spectral results with those reported by other groups. The definition of pulse and off-pulse regions, resp., was made using the 16-60 keV pulse profile (*cf.* Fig. 1). For satellite borne scintillation detectors, this method avoids systematic errors in detector background reduction. As the pulse period of Her X-1 is much shorter than typical background fluctuation timescales, the latter are averaged out.

Continuum models without cyclotron resonance feature (CRF) are clearly rejected by the data (*e.g.* $\chi_r^2 = 9$ for a Thin Thermal Bremsstrahlung spectrum). Using a power-law with exponential cut-off (Soong et al. 1990) and multiplicative CRF (Mihara et al. 1990)

$$\frac{dN}{dE} = \begin{cases} I \cdot (E/E_0)^{-\alpha} \cdot CRF & \text{for } E \leq E_{cut} \\ I' \cdot (E/E_0)^{-1} \cdot CRF \cdot \exp\left(\frac{E_{cut}-E}{E_f}\right) & \text{for } E > E_{cut} \end{cases}$$

where $CRF = \exp\left(\frac{A \cdot W^2 \cdot (E/E_c)^2}{(E-E_c)^2 + W^2}\right)$

and I' such that dN/dE is continuous at $E = E_{cut}$

we obtain an absorption line centroid of $E_{cycl} = 34.1 \pm 2.5$ keV, $\chi_r^2 = 0.7$ for 10 degrees of freedom. Similar models have been widely used and yield values like $E_{cycl} = 36.2 \pm 0.2$ keV (Soong et al. 1990) and $E_{cycl} = 38.3 \pm 0.9$ keV (Voges et al. 1982). It should be noted, however, that the relatively small errors quoted by Soong et al. are not for a joint variation of the parameters in contrast to the ones given by us. Tueller et al. (1984) reported $E_{cycl} = 35.4^{+2.3}_{-1.9}$ keV, obtained with a high resolution Germanium spectrometer, a value in good agreement with our result. A second cyclotron harmonic is consistent with our data, but is not necessary for a satisfactory fit.

Table 1. HEXE Main-On Observations of Her X-1

Observation MJD	# of Sessions	Observing Time	Φ_{35}	$\Phi_{1.7}$	Pulse Period $-1237000\mu s$
6988.5	5	3030 s	0.23-0.24	0.63-0.79	768.93(0.15)
7021.7	20	6960 s	0.16-0.20	0.58-0.37	772.02(0.06)
7235.1	11	2290 s	0.30-0.34	0.37-0.23	767.00(0.30)
7302.9	6	2220 s	0.25-0.27	0.42-0.91	766.30(0.20)
7367.2	5	1130 s	0.10-0.11	0.40-0.55	769.45(0.20)
7441.4	6	1200 s	0.22-0.25	0.85-0.34	763.61(0.07)

REFERENCES

Deeter, J. E., et al. 1991, ApJ 383, 324
Gilfanov, M., et al. 1989, in: Proc. 23rd ESLAB Symp. on Two Topics in X-Ray Astronomy, p.71
Mihara, T., et al. 1990, Nature 346, 250
Nagase, F. 1989, PASJ 41, 1
Petterson, J. A., Rothschild, R. E., & Gruber, D. E. 1991, ApJ 378, 696
Reppin, C., et al. 1985, in: Non-Thermal and Very High Temp. Phenomena in X-Ray Astronomy, eds. Perola and Salvati, p.279
Soong, Y., et al. 1990, ApJ 348, 641
Tananbaum, H., et al. 1972, ApJ Lett. 174, L143
Trümper, J., et al. 1978, ApJ 219, L105
Trümper, J., et al. 1986, ApJ 300, L63
Tueller, J., et al. 1984, ApJ 279, 177
Voges, W., et al. 1982, ApJ 263, 803
Vrtilek, S. D., et al. 1991, ApJ Suppl. 76, 1127

Fig. 1. Main-On pulse profile of Her X-1 (in 16-60 keV for 3 out of 4 detectors). Phase regions denoted "Pulse" and "Off Pulse" indicate intervals used for the computation of the P-OP spectrum.

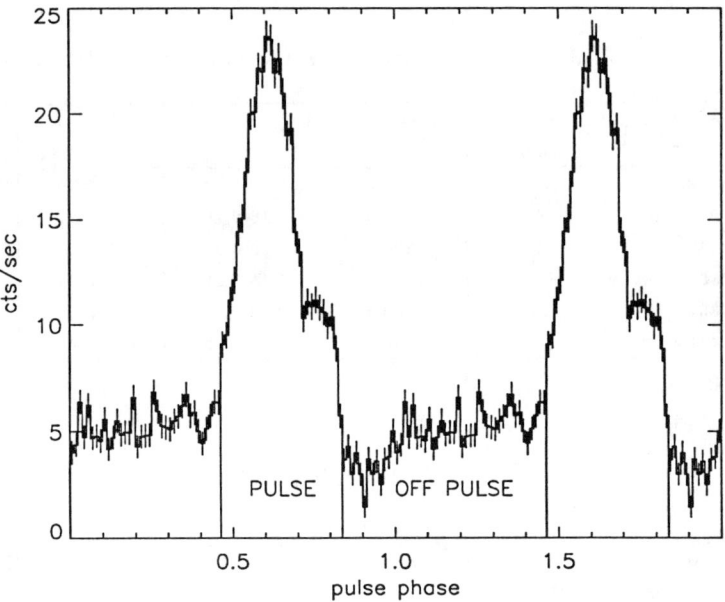

Fig. 2. P-OP spectrum for four HEXE Main-On observations in 16-200 keV (3 detectors). The theoretical photon spectrum used is a power-law with exponential cut-off and one multiplicative absorption line.

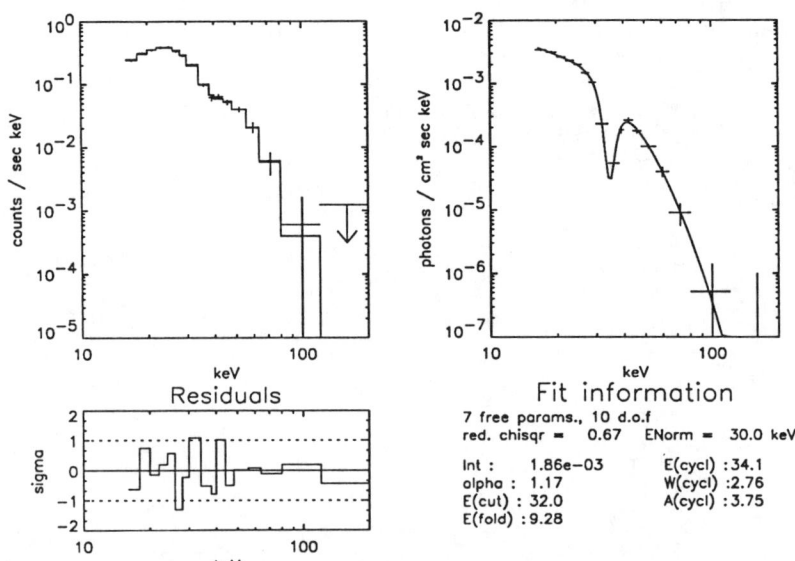

FLUX AND SPECTRUM VARIABILITY IN HER X-1

D.A. Leahy
Department of Physics and Astronomy,
University of Calgary, Calgary, Canada T2N 1N4

ABSTRACT

Hercules X-1 was observed by the Large Area Counters (LAC) on the Ginga satellite during several periods in 1988 and 1989. The spectrum of Her X-1 was measured on a timescale of several minutes, and the flux measured on a timescale of a few seconds. Large changes in flux and in spectral shape were observed, on short and on long timescales. The spectrum has been modelled consistently using a model with two continuum components plus iron emission line and absorption by cold matter. Spectral parameters are derived for pre-eclipse absorption dips and for eclipse ingress. The changes in spectral parameters are given a physical interpretation in terms of the emission region around the neutron star and in terms of the absorbing matter in the system.

1. INTRODUCTION

The X-ray pulsar Hercules X-1 has been studied in X-rays with several previous satellite instruments. Variations in absorption were seen in Tenma observations (Ohashi et al., 1984, Ushimaru et al., 1989). Soong et al. (1990) report on HEAO-1 observations of Her X-1. Mihara et al. (1990), give unambiguous evidence for the cyclotron absorption feature near 35 keV. Mihara et al. (1991) study the low-state spectrum of Her X-1 and find a two component emission model is needed. The high state spectrum was shown to also require a two component continuum model (Leahy et al. 1991). A 144 second periodicity in the flux decreases from Her X-1 was found (Leahy et al. 1992) during the pre-eclipse dip period in the 1988 GINGA observations. A detailed study of spectral changes during pre-eclipse dips for Her X-1 is presented in Leahy et al. 1993.

2. OBSERVATIONS AND DATA ANALYSIS

The data studied here is from observations of Her X-1 in 1988 and 1989 by the GINGA satellite. Fig. 1 shows the background subtracted LAC count-rate for the two observation periods in the lower panels. The upper panels show the hardness ratio (defined by 8-36 keV count rate divided by 1-8 keV count rate), useful for identifying spectral changes. The hardness ratio is different between the low state (Aug.17 and 19, 1988) the high state (e.g. Sept.1, 1988) and the pre-eclipse dips (Aug.18, 1988 and Apr.26, 1989).

Spectral model fitting shows that no single-continuum spectral model could fit either the individual spectra nor the set of spectra. Only one two-continuum spectral model was found to adequately fit the data. The model consists of: an unabsorbed continuum (power-law-plus-cyclotron-absorption-line); an absorbed continuum of the same form; and an iron emission line. The model has 10 parameters (P1 to P10):

$$P1 \times E^{-P2} \times \exp\left(-\frac{P3\ E^2}{(E-P4)^2+P5^2}\right) \times (1+P10 \times ABS(P6)) + GAUS(P7,P8,P9) \quad (1)$$

The exp function is used to represent the cyclotron absorption feature (e.g Mihara et al. 1990). ABS is absorption by neutral gas with log of column density given by P6. The iron emission line is represented by GAUS, a Gaussian with area P7, center energy P8 and width (1σ) P9.

Her X-1 was in the high state for the Aug. 28 1988 observations and a series of pre-eclipse absorption dips are clearly seen in the light curve and by the drop in the softness ratio, defined as (2- 4.7 keV rate)/(9.3- 14 keV rate). Seventeen spectra (labeled 1 to 17) (in groups of 6, 4 and 7 for the three dip periods) were selected manually based on the count rates and the softness ratio.

The 10 parameters for each spectrum were derived by fitting the above model. The spectra and best-fit models are shown for the second dip period (spectra 7 to 10) in Fig. 2. The crosses are the data points, the upper histogram is the sum of the three lower histograms, which show the two continuum components (absorbed and unabsorbed) and the iron line.

The eclipse ingress on Apr.30, 1989 can be seen in Fig. 1. Five spectra covering the eclipse ingress period were derived, corresponding to the pre-eclipse period, two periods during ingress, one period immediately after ingress, and one long period during mid-eclipse, respectively. Spectral parameters were derived for these five spectra.

3. DISCUSSION

The spectral changes during pre-eclipse dips are illustrated by Fig. 2. The gradual evolution of the observed spectrum in shape and in intensity can be seen. The absorbed component of these spectra is interpreted as the direct pulsar radiation absorbed by intervening cold matter (such as the accretion disk), and the unabsorbed component as radiation scattered to the observer by electrons in matter nearby the neutron star. The absorbed (direct) component remains almost constant for all of the spectra during the pre-eclipse dips, whereas the unabsorbed (scattered) component changes by a factor ~10. The ratio of scattered to direct normalization decreases steadily as a dip progresses, indicating a steadily increasing covering of the scattering region by optically thick matter.

The spectral fits are all consistent with absorption by cold matter, with no excess iron edge absorption. The increase of column density could be due to increasing path length through constant density matter or increasing density of the obscuring matter. Both effects can occur for either the case of obscuration by the accretion disk gradually moving into the line of sight as the dip progresses, or the case of a blob of matter moving into the line of sight. However a blob must be large enough to cause the significant fractional obscuration of the scattering region around the pulsar, as noted above.

The intensity decrease in the sharp dips occurs in about 12s. This puts a constraint of about 1000 km on the size, D, of the region over which column density increases significantly in the absorber. The scattered intensity also is seen to decrease in two stages: a fast stage and a slow stage, indicating a bright inner region of the scatterer which is small compared to the absorber scale height, D, and an extended component of the scatterer (larger than D).

The spectral parameters from the pre-eclipse spectra show real variations with time. These are studied in detail in Leahy et al. (1993). Some of the results are as follows. It is found that the power law spectral index is variable with time and that there is a strong positive correlation between power law index and absorbed (direct) normalization (i.e. pulsar luminosity). The fluorescence efficiency is anti-correlated with iron line energy and positively correlated with iron line intensity. The latter relation can be explained if the unobscured area of the fluorescing region is changing,

Ushimaru, N. et al., 1989, Publ. Astron. Soc. Japan, 41, 441.

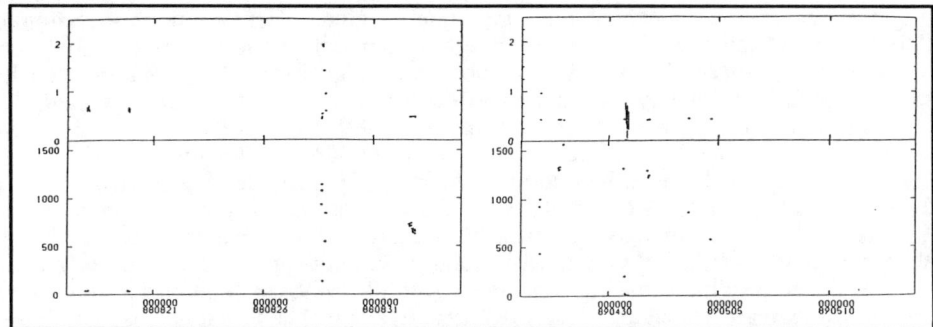

Figure 1 Light Curve and Hardness Ratio for the periods 1988.08.16 to 1988.09.04 and 1989.04.15 to 1989.05.20.

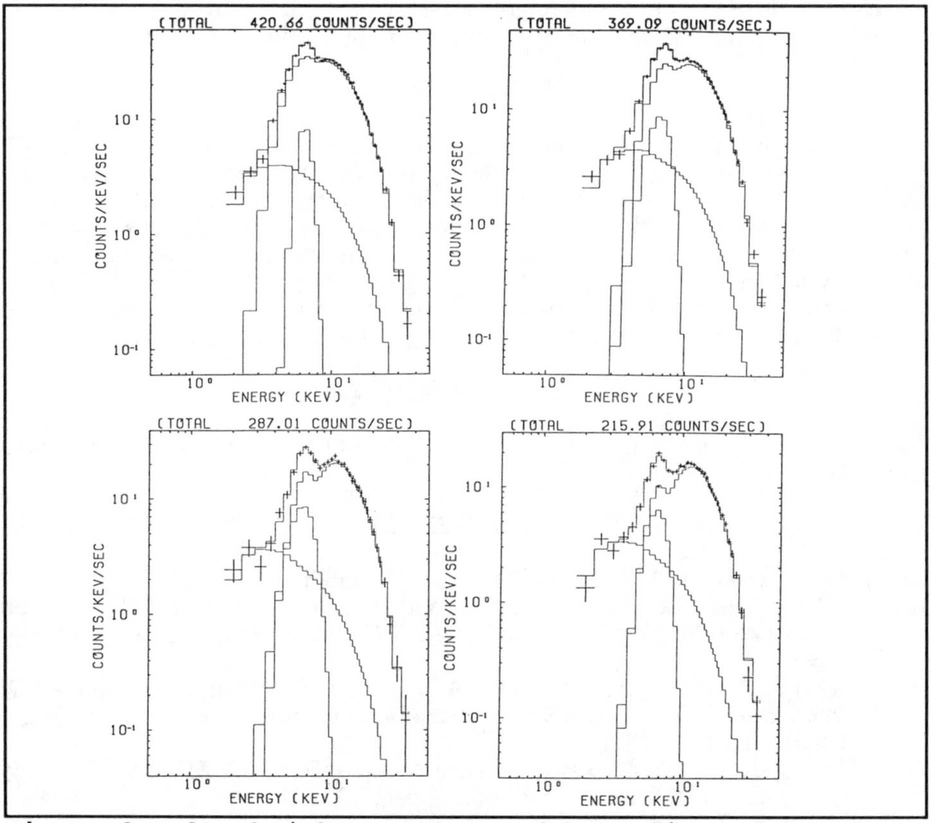

Figure 2 Pulse-height spectra and best fit model spectra for the second pre-eclipse dip period on 1988.08.28.

since both intensity and efficiency increase with greater area. The first relation is explained if the cooler gas is irradiated more than more highly ionized gas, i.e. the bright inner region is cooler than the outer region. There is no correlation of fluorescence efficiency with total, absorbed or unabsorbed continuum intensity. Iron line energy is not correlated with total, absorbed or unabsorbed continuum intensity, suggesting that luminosity variations and geometry variations are uncorrelated. The iron line energy is not correlated with column density which contrasts with the strong anticorrelation seen for the wind-fed X-ray binary GX301-2 (Leahy et al. 1989). The column density of the absorbed component is highly variable (log N_H from 22.0 to 24.4) and is strongly anti-correlated with the unabsorbed (scattered) normalization. This is likely due to increasing partial obscuration of the scattering region by the absorbing matter as the column density along the line of sight to the pulsar increases.

For the eclipse ingress which occurs on Apr.30, 1989, the light curve shows three regions of decrease which are clearly different in slope. The initial decrease is slower. the second decrease is rapid; the final decrease is slow again. The first decrease is interpreted as the small bright region around the pulsar which is undergoing increasing absorption. The hardness ratio rises during this period supporting this interpretation. The rapid decrease of count rate during the second interval is the eclipse of the bright region by the stellar surface. The bright region is eclipsed during the third interval of decrease, so there must be a less bright and more extended region around the neutron star which gives the ~7 counts/s still observed at mid-eclipse.

4. CONCLUSION

The data here cover many short periods over two 35-day cycles of Her X-1 (Fig. 1). Two continuum components are essential to fit the observed spectra (absorbed and unabsorbed) plus a fluorescent iron line. The interpretation of the two continuum components are direct and scattered radiation from the pulsar. Time resolved spectra have been studied for Her X-1 during a period with several pre-eclipse dips and for an eclipse ingress. Variations in the spectral parameters have been studied as a dip progresses. The eclipse ingress provides further evidence for two scattering regions around the neutron star: one compact and bright; one extended and dim.

ACKNOWLEDGEMENT

This work was supported by a grant from the Natural Sciences and Engineering Research Council of Canada.

REFERENCES

Leahy, D.A., Yoshida, A. & Matsuoka, M. 1993, submitted to ApJ
Leahy, D.A., Yoshida, A., Kawai, N. and Matsuoka, M. in "The Compton Observatory Science Workshop" (NASA Conference Publication 3137, 1992) p193.
Leahy, D.A., Matsuoka, M., Yoshida, A. and Kawai, N. in Proc. of the 22nd International Cosmic Ray Conference (Dublin Institute for Advanced Studies, Dublin Ireland, 1991) vol.1 p13
Leahy, D.A., Matsuoka, M., Kawai, N., and Makino, F., 1989, MNRAS, 237, 269
Mihara, T. et al., 1991, Publ. Astron. Soc. Japan, 43, 501
Mihara, T. et al., 1990, Nature, 346, 250.
Ohashi, T. et al., 1984, Publ. Astron. Soc. Japan, 36, 719.
Soong, Y. et al., 1990, Ap.J., 348, 641.

EXOSAT STUDIES OF DIPS IN HERCULES X-1

A.P. Reynolds & A.N. Parmar
Astrophysics Division, Space Science Department of ESA, ESTEC,
Postbus 299, Keplerlaan 1, 2200 AG Noordwijk, Netherlands.

ABSTRACT

EXOSAT observations of three absorption dips in Hercules X-1 are presented. A partial covering model is employed, and qualitative differences between the three dips are highlighted. The behavior of the pulse profile during the dips is discussed. A transition between different phase states of the absorbing medium is observed within one dip, consistent with the cooling and condensation of an ionized cloud into discrete absorbing clumps.

1. INTRODUCTION

Hercules X-1 is an X-ray binary system with a 1.7 d orbital period, exhibiting 1.24 s pulses; it has characteristics of both the massive and low-mass X-ray binaries. In addition to the well known 35 d cycle in the observed X-ray flux, usually ascribed to masking by a precessing accretion disk, the system shows recurrent intensity dips (period 1.6 d) which occur near orbital phases 0.8 and 0.4. These dips (pre-eclipse and anomalous, respectively) were first observed by Giacconi *et al.* (1973) and are assumed to arise from obscuration of the X-ray source by the material on the outer edge of the disk. Recently, Ushimaru *et al.* (1989) analyzed *Tenma* observations of Her X-1 which indicated the presence of different ionization states of absorbing material from one dip to the next. Here, we summarise EXOSAT observations of three dips made during 1984 April and June.

2. OBSERVATIONS

The three dips, which we shall refer to as Dips 1,2 and 3, occurred on 1984 April 5, June 11 and June 12 respectively, during the main-on state of the 35 d cycle (see Voges *et al.* 1987 for a fractal analysis of the same dips). The dips are shown in Fig. 1 together with the eclipse ingresses and egresses. Dip 2 was an anomalous dip, while the other two were of the pre-eclipse type. EXOSAT's on-board modes and detector configurations were similar during all these observations, with the Medium Energy experiment giving 32 channels of energy data between $1-20$ keV, with a time resolution of 0.03 seconds.

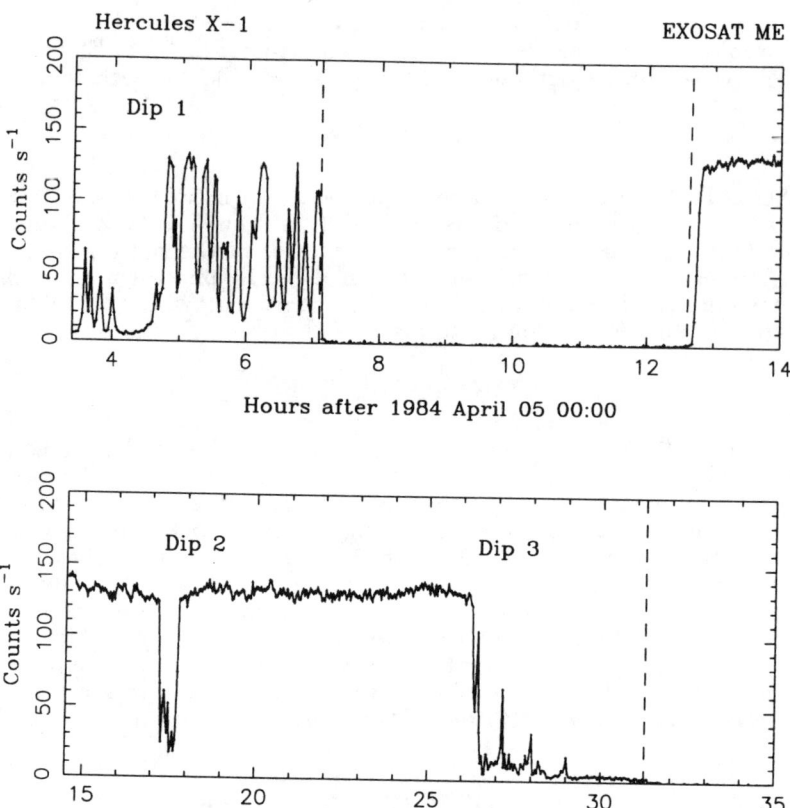

Fig. 1. The three dips studied.

3. ANALYSIS

Spectra have been accumulated at 5 minute intervals through each dip and then fitted with a partial covering fraction model. In this model the X-ray source is assumed to be partly obscured by cool material, leading to a two component (absorbed and unabsorbed) spectrum. For certain spectra, pulse profiles have been obtained over the same time interval. In Fig. 2, we display a pair of fitted spectra and models from before and during Dip 2 to illustrate the dramatic change in the shape caused by the low energy absorption. The first spectrum is fitted by a single unabsorbed power law, while the second is fitted by a power law with partial covering absorption.

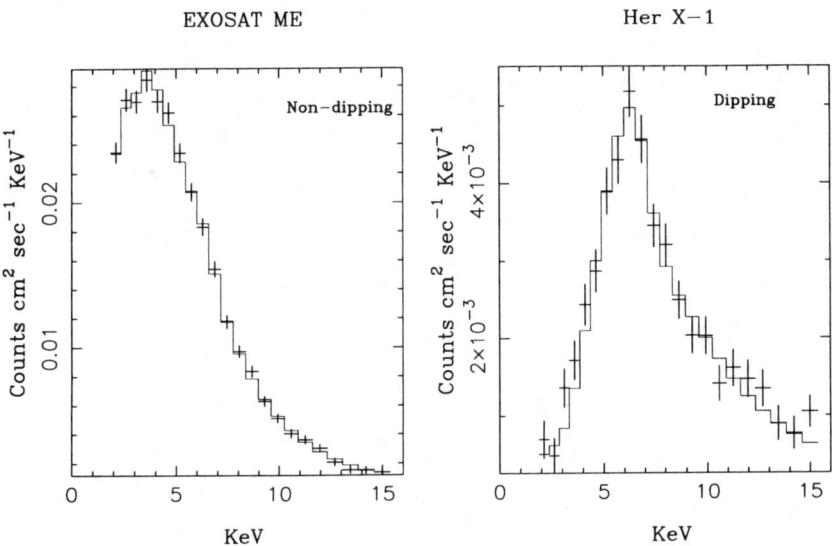

Fig. 2. Pre-dip and dip spectra.

4. RESULTS

Fig 3 displays Dip 1, the column density at three time intervals and the corresponding pulse profiles. Soon after the start of the observation, there is an interval of ~ 30 mins during which the spectrum shows very little low energy absorption despite a reduction in intensity by a factor of 20. We interpret this as due to scattering in an ionized medium, although we cannot exclude that the reduction in intensity is due to obscuration of the source by material which is totally opaque to X-rays. As the dip progresses, we see occasional returns to the normal intensity, punctuated by dipping episodes lasting a few minutes, which are believed to be due to the passage of cool absorbing clumps in front of the X-ray source. It is possible, therefore, that in this dip the scattering medium condensed into a cool, clumped cloud as the dip progressed. The pulse profile obtained during an interval of high absorption (profile 3) is narrower than that seen during the normal intensity interval (profile 2), which is due to the absorption of the smoother low energy pulse component.

The remaining 2 dips have been similarly analyzed. In general, when normal energy-dependent dipping is dominant, the properties of the absorber (clump size and density) do not differ significantly. The source metallicity lies between 0.2 and 0.8 of normal cosmic abundances.

For the material presumed responsible for the energy independent intensity reduction to be totally ionized, and assuming an absorber size of 0.1 of the distance to the source, it must be closer than 7×10^8 cm for a source luminosity of 2×10^{37} erg s^{-1}. This is much smaller than the radius of the accretion disk and comparable to the magnetospheric radius of 10^8 cm.

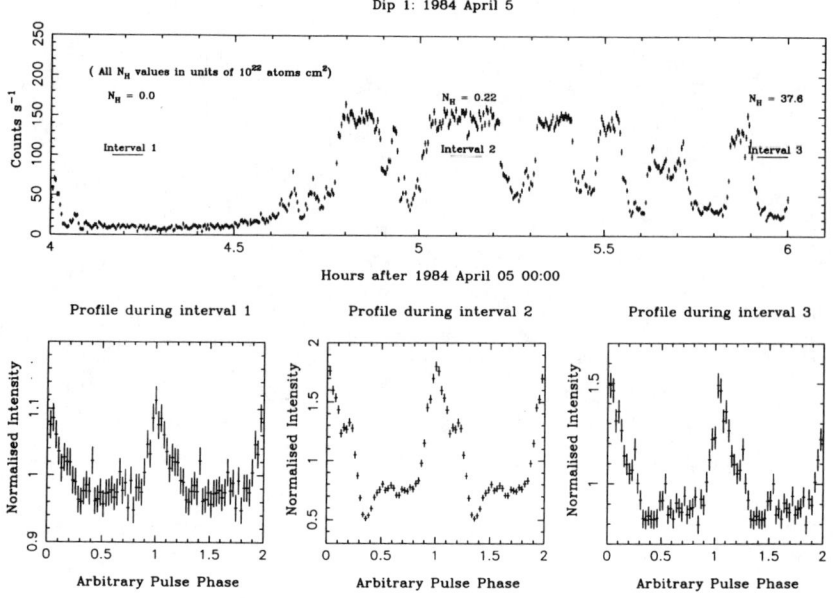

Fig. 3. Properties of Dip 1.

CONCLUSIONS

The "phase-change" in the nature of the absorber from dip to dip reported by Ushimaru et al. (1989) has been observed within a single dip. For the material to be significantly ionized, it must lie close to the inner edge of the disk.

ACKNOWLEDGEMENTS

A.P. Reynolds acknowledges funding via an ESA Research Fellowship. This work is derived from data and analysis packages maintained at ESTEC as part of the EXOSAT Database and Archive.

REFERENCES

Giacconi, R., Gursky, H., Kellogg E., Levinson, R., Schrier, E., Tananbaum, H. 1973, ApJ, 184, 227

Ushimaru, N., Tawara, Y., Koyama, K., Hayakawa, S., Kii, T., Nagase, F., Ohashi, T., Makishima, K. 1989, PASJ, 41, 441

Voges, W., Atmanspacher, H., Scheingraber, H. 1987, ApJ, 320, 794

OBSERVATION OF A CORRELATION BETWEEN MAIN-ON INTENSITY AND SPIN BEHAVIOR IN HER X-1.

Robert B. Wilson
NASA/Marshall Space Flight Center-ES66, Huntsville, AL 35812
wilson@batse.msfc.nasa.gov

Mark H. Finger
Computer Sciences Corporation (NASA/MSFC-ES66),Huntsville, AL 35812

Geoffrey N. Pendleton and Michael Briggs
Dept. of Physics, University of Alabama in Huntsville, Huntsville, AL 35899

Lars Bildsten
Div. of Physics, Mathematics, and Astronomy, Caltech, Pasadena, CA 91125

ABSTRACT

Using BATSE observations of Her X-1 for the entire CGRO mission to date, a correlation is observed between the hard x-ray pulsed flux at the peak of the main-on portion of the 35d cycle and the spin frequency derivative, a measure of the torque acting on the neutron star. An apparent correlation is also found between the 35 day cycle start times and the pulsed flux. Orbital parameters are also obtained, with evidence for orbital decay with a timescale slightly less than reported in Deeter et al.(1991).

1. INTRODUCTION

Long-term monitoring of the spin frequency of Her X-1 summarized in Nagase (1989) shows apparently stable values of spinup torque acting on the neutron star for as long as 10 years, changing to spindown behavior lasting for several years. On shorter timescales, as observed mainly with the *UHURU* spacecraft (Giacconi et al. 1973 and Deeter et al. 1981), "wandering" between spinup and spindown are observed. For much of the time since *UHURU*, observations of the source have not occurred often enough to ensure detection of short-term changes in main-on cycle peak intensity and spin frequency. We report here the observed spin behavior obtained throughout the CGRO mission by BATSE, and study its correlation with the hard x-ray luminosity.

2. DATA COLLECTION & ANALYSIS

Onboard-folded data has been collected from the BATSE Large Area Detectors (LADs) for Her X-1 during 780 days of the CGRO mission, covering all phases of the 35d cycle with a total livetime of $\sim 2 \cdot 10^6$s. The source is detected in a total of 165 days, only during the main-on portion of the cycle, with the observed intensity at the peak of a cycle varying by up to a factor of $\sim$4.

Onboard the CGRO, events in each of 16 energy channels are combined from the source-facing detectors and folded for $\sim$8s at the mean Doppler shifted period into 64 phase bins. In subsequent analysis on the ground, the solar system barycenter arrival of each on-board accumulation is computed, and the data are summed into phase bins fixed in the solar system barycenter. This analysis is more completely described in Wilson et al. (1992).

3. ORBIT DETERMINATION

Using orbital parameters from Deeter et al.(1991), hereafter DEE91, the pulse frequency, ν_s, at an epoch contained within the BATSE data set, and the historical long-term value of $\dot{\nu}_s$, data for ~150 days of exposure to Her X-1 during main-on portions of the 35d cycle have been epoch-folded to produce one pulse profile per CGRO orbit. A correlation analysis with a mean template formed from a portion of this data was used to determine the relative phasing of each of the profiles. A fit to the total phase (model + observed corrections) was made, allowing the epoch of Mean Orbital Longitude (MOL) = $90°$, $a_x \cdot \sin(i)$, $e \cdot \cos(\omega)$, and $e \cdot \sin(\omega)$, where ω is the longitude of periastron. Since torque variations are too large to permit cycle counting over the ~26 day gaps between detections, the fit permits separate ν_s and $\dot{\nu}_s$ and a break in phase for each 35d portion of data. The values obtained for the orbital parameters are shown in table I.

The BATSE value of the residual orbital epoch relative to a constant period is plotted in Figure 1 below, adapted from DEE91. The uncertainty shown includes statistical and our estimate of systematic errors, based on the scatter in values obtained for subsets of the data. The BATSE point indicates $\dot{P}_b$ may be slightly less in magnitude than the value of DEE91.

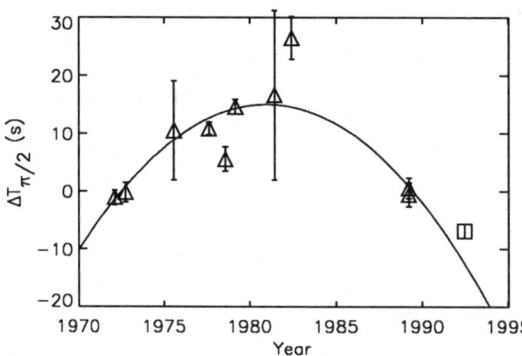

Figure 1. Residual orbital epoch of MOL = $90°$ relative to a constant orbital period (from DEE91, BATSE point at 1992.6 added). fixed at the DEE91 values).

4. PULSAR FREQUENCY DETERMINATION

We have obtained the mean ν_s for each main-on observation, by fitting the total phase with a 2nd order polynomial, where the 1st order term is the mean frequency, and the 2nd order term contains most of the effects of any pulse shape changes. The mean $\dot{\nu}_s$, has been computed as the frequency difference ν_s divided by the time between measurements. The values obtained are shown in Figure 2(a). Three episodes of spindown are evident, near TJD 8600, 8800 and 9175. The largest negative $\dot{\nu}_s$ corresponds to 1/6 of the maximum torque that would be applied to the pulsar for a mass accretion rate of $\dot{M} = 10^{-9} M_\odot$ per year from near the corotation radius of $2 \cdot 10^8$ cm.

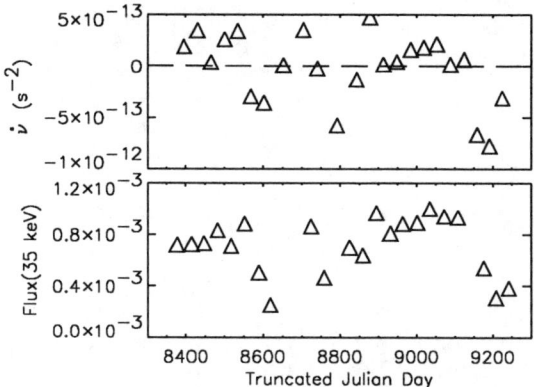

Figure 2. (a). $\dot{\nu}_s$ history (mean of each observed main-on cycle). (b). History of the flux at 35 keV at the peak of each observed main-on cycle in units of photons $\cdot cm^{-2} \cdot s^{-1} \cdot keV^{-1}$. The uncertainties are much smaller than the size of the plotted points in each plot.

Table I. Her X-1 Parameters	
Epoch(MOL = 90°)	JD 2448800.11235 (1) TDB
$a_x sin(i)$	13.1853 (2) light-s
Eccentricity	$< 1.3 \times 10^{-4}$ (3σ U. L.)
Epoch(Pulse Period)	JD 2448895.802276 TDB
Period	1.237746212 (3) s

5. SPECTRAL ANALYSES

The phase-averaged pulsed spectrum has been determined for a 3.4 day interval at the *peak* of each available main-on cycle, after removal of eclipse intervals. Backgrounds were obtained from an off-pulse interval. An optically thin thermal Bremsstrahlung (OTTB) model is an adequate fit, with kT typically 15 ±2 keV. Figure 2(b) shows the flux at 35 keV for each observed cycle as computed from the model parameters. It is evident that the outbursts are of low intensity during the 3 intervals of spin-down. This correspondence is not expected if the peak flux of each main-on cycle were determined solely by obscuration by a precessing accretion disk (Crosa & Boynton, 1980, Petterson et al. 1991).

Figure 3, a plot of the observed peak flux versus the frequency derivative, shows that times of low x-ray luminosity are correlated with spindown. These data are thus most consistent with flux being a good indicator of accretion rate ($\dot{M}$), so that the magnetospheric radius increases toward the co-rotation radius as $\dot{M}$ decreases, eventually leading to spindown (Ghosh & Lamb 1979).

6. 35D CYCLE OBSERVATIONS

Start times of main-on cycles were determined by inspection of residual phases of the pulsed signal (0.25d integrations) in comparison to the mean pulse template. For the *preliminary* values reported here, we have specified a range

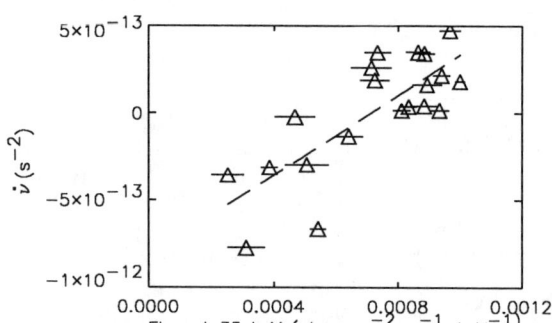

Figure 3. Correlation plot of the intensity of Her X-1 at the peak of the main-on portion of the 35d cycle, for 20 cycles observed by BATSE The value of the correlation coefficient is 0.783.

for the start times.

From previous observations (Tananbaum et al. 1972, Giacconi et al. 1973), it is known that the onset of the main-on of a 35d cycle occurs in two bands of orbital phases, around values of 0.2 and 0.7. The cycle interval appears to be 20.0, 20.5, or 21.0 times the binary orbital period, P_b (~1.70017d). We have folded our start times mod ($20.5 \times P_b$), as shown in Figure 4, for the entire CGRO mission to date. It appears that the repeat interval has been at the 20.5 multiple except during TJD ~8580 - 8820, when the cycles were shorter. This is also the time when we observe the intensity of the outbursts during main-on to be lower (times of low $\dot{M}$). This possible correlation of early turn-ons with decreased mass transfer rate has previously been suggested by Ögelman (1987).

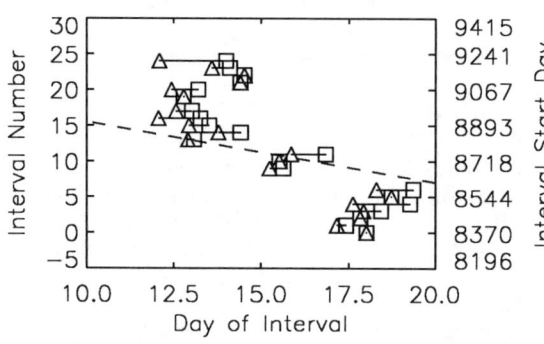

Figure 4. Start time of 35d cycle folded at $20.5 \times P_b$ (*preliminary* estimate range bounded by the triangles and squares. The dashed line is the slope for onsets spaced at $20.0 \times P_b$.

REFERENCES

Crosa, L., & Boynton, P. E. 1980, ApJ, 235, 999
Ghosh, P. , & Lamb, F. K. 1979, ApJ, 234, 296
Deeter, J. E. , Boynton, P. E., & Pravdo, S. H 1981, ApJ, 247, 1003
Deeter, J. E. et al.1991, ApJ, 383, 324 (DEE91)
Giacconi, R. et al.1973, ApJ, 184, 227
Nagase, F. 1989, PASP, 41
Ogelman, H. 1987, A&A, 172, 79
Petterson, J. A. , Rothschild, R. E. , & Gruber, D. E. 1991, ApJ, 696
Wilson, R. B. et al.1992, Isolated Pulsars, *Proceedings of the Los Alamos Workshop*, Cambridge Univ. Press, 380

HST/FOS OBSERVATIONS OF HOT GAS DURING THE TOTAL ECLIPSE OF THE NEUTRON STAR IN HZ HER/HER X-1

S. Wachter, S. F. Anderson, B. Margon
Dept. of Astronomy, FM-20, University of Washington, Seattle, WA 98195
wachter/anderson/margon@astro.washington.edu

R. A. Downes
STScI, 3700 San Martin Drive, Baltimore, MD 21218
downes@stsci.edu

1. INTRODUCTION

The pulsing binary X-ray source Hercules X-1 has been an intensively studied system since its discovery by *Uhuru* twenty years ago (Tananbaum et al. 1972), and the identification of the optical counterpart, HZ Herculis, shortly thereafter. The system exhibits periodic X-ray and optical light modulation at periods of 1.2^s, 1.7^d, and $\sim 35^d$, which are thought to represent the neutron star spin, binary orbital, and accretion disk precession periods, respectively. Prominent heating effects occur in the atmosphere of the cool (late A or early F) star; this effect is thought to explain the 1.7^d optical light curve, and the 1.2^s optical pulsations, for example. *In this paper, however, we report unambiguous observations of high excitation, hot gas at phases when the X-ray source is known to be eclipsed, and the X-ray illuminated side of HZ Her is not visible.*

We obtained ultraviolet spectrophotometry of HZ Her in 1992 April/May, using the *Hubble Space Telescope* Faint Object Spectrograph (FOS), as part of the FOS Investigation Definition Team's Guaranteed Time Observations program. The data were acquired at two different orbital phases, centered near maximum brightness and X-ray eclipse, and have spectral resolution $\lambda/\Delta\lambda \sim 1200$. Through the courtesy of Dr. Jules Halpern, low resolution CCD optical spectra of HZ Her were also obtained near in time, although not quite simultaneously, with the *HST* observations.

2. THE SPECTRA AT MAXIMUM LIGHT

HZ Her near maximum light has been observed on multiple occasions by IUE (*e.g.*, Gursky et al. 1980), and extensive discussion and modelling of these spectra may be found in Howarth & Wilson (1983) [hereafter HW]. Our FOS spectra (taken at binary phases 0.45–0.53) are, however, of substantially superior signal to noise and spectral resolution. Some important features include:

- *Emission line spectrum.* The large (4.3") FOS entrance aperture required to overcome the spherical aberration, unfortunately makes interpretation of the Lα spectral region almost impossible; most probably there is a complex mixture of geocoronal radiation, line emission, and underlying stellar absorption. The spectrum is dominated by strong, narrow emission lines (e.g. NV, SiIV, CIV) and indicates high excitation (note the very strong HeII λ1640, for example). The presence of these high excitation lines is not surprising, as we are observing the X-ray irradiated photosphere of the companion in addition to some contribution from the accretion disk. The CIII $\lambda 2297$ emission line appears to be previously unreported, and together with our strong detection of CIII $\lambda 1176$ and the total absence of CIII] $\lambda 1909$, yields interesting density diagnostics ($n_e \geq 10^{10}$ cm^{-3}), which will be discussed elsewhere.

• *Bowen fluorescence.* Many X-ray binaries, HZ Her included, show the blend of NIII/CIII $\lambda\lambda 4640$, 4650 emission lines at far greater strength than expected for reasonable abundances; McClintock et al. (1975) proposed that this feature is due to Bowen fluorescence, a pumping mechanism seen in gaseous nebula due to the near coincidence of HeII Lα $\lambda 304$ with an OIII $\lambda 304$ line. This hypothesis was supported by the detection of an additional predicted Bowen line, OIII $\lambda 3444$, in HZ Her (Margon & Cohen, 1978). However, the strongest predicted Bowen line ($4\times$ $\lambda 3444$) is OIII $\lambda 3133$, which is very awkward to observe from the ground. Note the prominence of this feature in our *HST* spectrum. (It has also been reported from IUE by HW).

3. THE SPECTRA AT MID-ECLIPSE

Although IUE attempted multiple observations of HZ Her near X-ray eclipse, the object is then really too faint for that instrument. Thus, the minimum light FOS UV spectra are particularly unique.

• *UV Emission line spectrum.* The spectra near mid-eclipse show a very surprising result: strong, high excitation emission is present, even though the neutron star is unquestionably totally eclipsed and the X-ray illuminated side of HZ Her is unobservable (the system inclination is known to be close to 90°). Current models of the geometry of the system render it extremely unlikely that we observe some unobscured portion of the accretion disk. Prominent NV, SiIV, NIV], and CIV emission is clearly evident even at binary phase 0.99! Again, Lα is obscured by geocoronal contamination in the large aperture. The physical conditions in the gas that dominates the emission at minumum light are obviously very different from those in the gas that dominates at maximum: note the different ratio of NV $\lambda 1240$ to NIV] $\lambda 1487$ between the two phases, as well as the absence of the OIII $\lambda 3133$ Bowen line at mid-eclipse.

• *Optical spectrum (not displayed).* As has been known for many years, when the X-ray source is eclipsed, the optical spectrum changes from B to late A. (Spectral type determination from the *HST* and optical spectra give a spectral type of A9 ± 1 spectral class). At minimum, with no source of $\lambda 304$ radiation to pump the Bowen process, the NIII/CIII $\lambda\lambda 4640$, 4650 complex is seen to rigorously disappear, confirming that this mechanism is indeed responsible for essentially all of the emission. Moreover, the Bowen emission evidently arises from either the eclipsed disk and/or the hidden X-ray illuminated side of the A star. However, very strong HeII $\lambda 4686$ emission (excitation potential 51 eV) is still present at mid-eclipse!

One incident of HeII $\lambda 4686$ emission during eclipse was reported previously in 1977 by Koo & Kron. It now seems likely that this is a common or perhaps even ubiquitous feature, possibly related to the strong UV emission we also observe at minimum with *HST*. (Unfortunately, we were unable to obtain a minimum light spectrum of the HeII $\lambda 1640$ region).

4. DISCUSSION

Indirect evidence has been available for many years indicating that hot gas is visible from HZ Her even during eclipse. The situation is complicated by the fact that multiple different hot components may be simultaneously visible. For example, Parmar et al. (1985) report soft (0.02–2 keV) X-ray emission observed by EXOSAT during Her X-1 eclipse. Mihara et al. (1991) present X-ray spectra taken during the 35^d minimum which they also attribute to this or a related component, the so-called "accretion disk corona". Although this hot component may well explain the observed HeII $\lambda 4686$ emission at minimum, the gas is probably too hot to be compatible with the UV-emission in our HST minimum-light spectra.

At least one additional warm component of gas seems to be associated with the Her X-1 system. Erratic "X-ray dips" have been attributed to absorbing (but still highly ionized, $T \sim 3 \times 10^4$ K, $n_e \sim 3 \times 10^{13}$ cm^{-3}) material in the line of sight (Crosa & Boynton 1980); these "blobs" are also indirectly inferred in some previous UV and optical studies (e.g., HW, Bochkarev & Karitskaya 1989). The blobs of material are thought to occupy an extended region above and around the accretion disk, and it has even been suggested (Bochkarev & Karitskaya 1989) that such blobs might be visible during eclipse.

The UV emission lines can be used to determine the physical conditions in the gas. The doublet ratios of both the NV and CIV lines are indicators of opacity. Although high resolution spectra of the NV doublet at maximum light have been previously obtained with IUE (HW, Boyle et al. 1986), the higher S/N of our FOS spectra allows a more accurate determination of the NV doublet ratio. In addition, our HST spectra provide (for the first time) the CIV doublet ratio. Under optically thin conditions, the doublet ratios should both be 2:1. Our data indicate ratios of about 1 at maximum light for both NV and CIV, thus implying that the emitting region which dominates at maximum is not optically thin. At minimum light, the S/N in the doublets is much lower and the results are less reliable. The doublets still appear resolved, and give a ratio of close to 2, which indicates that optically thin conditions may be approached at minimum light.

This work has been supported by NASA Grant NAG5-1630.

REFERENCES

Bochkarev, N. G., & Karitskaya, E. A. 1989, Ap & Sp Sci, 154, 189
Boyle, S., Howarth, I., Wilson, R., & Raymond, J., 1986, Proc. Joint NASA/ESA/SERC Conference, 471
Crosa, L. M., & Boynton, P. E. 1980, ApJ, 235, 999
Gursky, H., et al. 1980, ApJ, 237, 163
Howarth, I. D., & Wilson, B. 1983, MNRAS, 204, 1091
Koo, D. C., & Kron, R. G. 1977, PASP, 89, 285
Margon, B., & Cohen, J. G. 1978, ApJ, 222, L33
Mason, K. O., Murdin, P. G., Tuohy, I. R., Seitzer, P., & Branduardi-Raymont, G. 1982, MNRAS, 200, 793
Mihara, T., Ohashi, T., Makishima, K., Nagase, F., Kitamoto, S., & Koyama, K. 1991, PASJ, 43, 501
McClintock, J. E., Canizares, C. R., & Tarter, C. B. 1975, ApJ, 198, 641
Parmar, A. N., Pietsch, W., McKechine, S., White, N. E., Trümper, J., Voges, W., & Barr, P. 1985, Nature, 313, 119
Tananbaum, H., Gursky, H., Kellogg, E. M., Levinson, R., Schreier, E., & Giacconi, R. 1972, ApJ, 174, L143

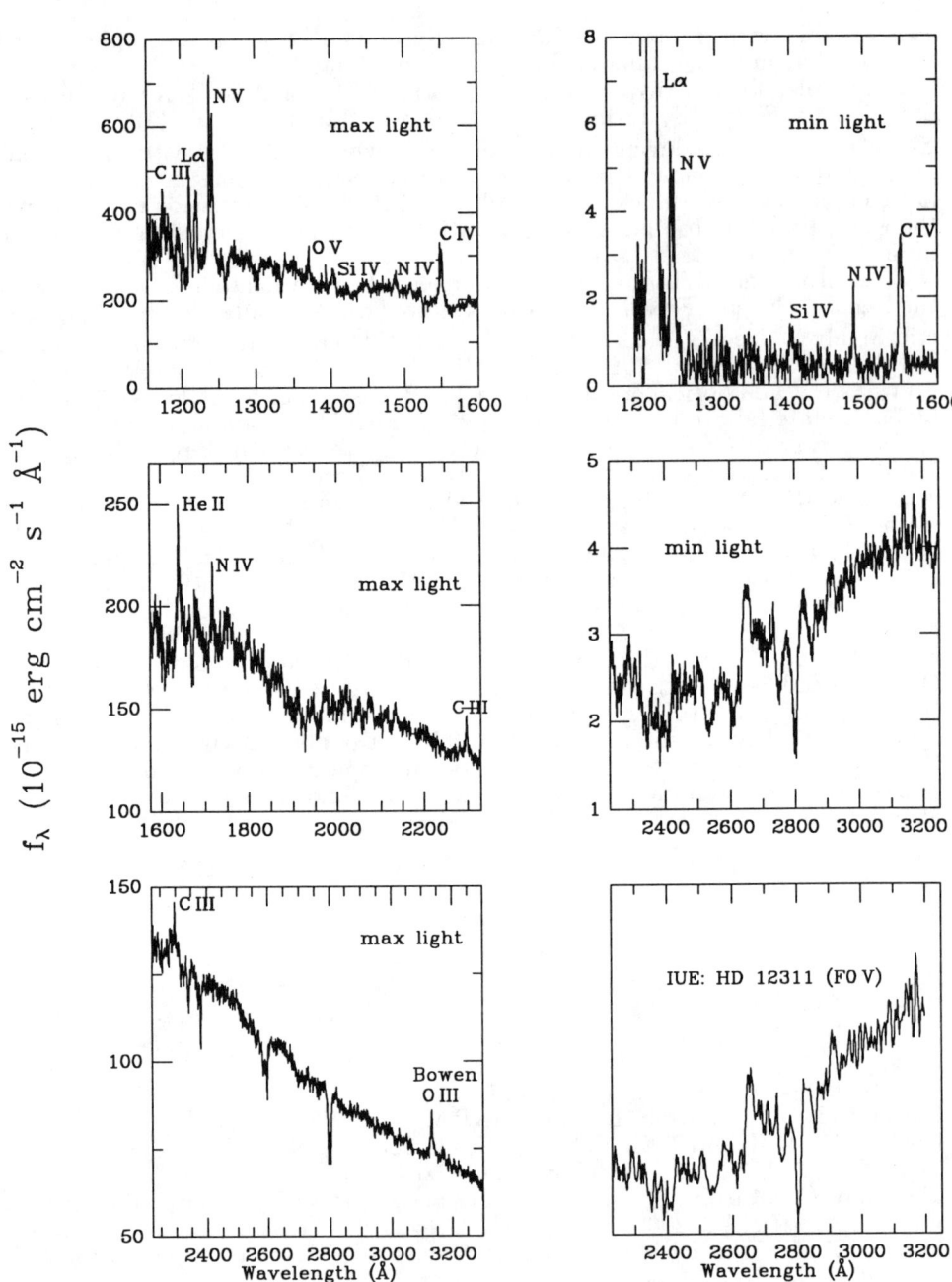

HST/FOS spectra of HZ Her. *Left column*, maximum light, *right column*, minimum light. An IUE spectrum of a F0 V star is shown for comparison.

SUB-SYNCHRONOUS ROTATION AND TIDAL LAG IN HD 77581/VELA X-1

R.E. Wilson and Dirk Terrell
Astronomy Department, University of Florida, Gainesville, FL 32611

1. INTRODUCTION

The ellipsoidal variable star HD 77581 is the optical counterpart of the pulsed, eclipsing X-ray source Vela X-1, with the system consisting of a B0.5 supergiant and a neutron star in a significantly eccentric orbit of 8.96 day period. As a result of self consistent parameter adjustment incorporating several kinds of observations, we demonstrate that the optical star rotates sub-synchronously and shows phase shifts which we interpret as tidal lags. The sub-synchronous rotation and phase shifts were evident in earlier work, but are established here on the basis of a new and comparatively rigorous kind of analysis. Both of these properties have major implications for understanding the system's evolution.

The X-ray eclipse duration provides information on the relative size (R/a) of the B star, while the X-ray pulses and the optical velocities provide information on the absolute dimensions, orbital eccentricity, and masses. The optical light curves give information on the figure of the B star and thus, together with the X-ray eclipse duration, on the star's size compared to its limiting lobe. Some information on star size and figure may be in the optical velocities, which are affected slightly by proximity effects due to tides and external heating. Potentially, this combination of kinds of observations promises to tie down the properties of this object very well, but in practice there are serious problems. For example, the optical observations have small amplitude and large scatter. Another problem is that neither the X-ray nor optical variations, taken separately, tell much about the inclination, except that the orbit is close enough to edge-on so as to give broad eclipses. In this situation it is helpful to treat the several types of variations together, so as to exploit inter-relationships and to ensure self-consistency of the overall solution. A formalism for doing this, even with eccentric orbits and non-synchronous rotation, was published earlier (Wilson 1979), but did not include treatment of the pulse arrival times. Pulse data fitting has now been incorporated, and we have applied the scheme in a simultaneous iterative solution of the B and V light curves, optical velocity curves, and pulse arrival time curves, subject to the constraint that the X-ray eclipse duration must be reproduced. Proper weighting is required in the least squares solutions. No assumptions about lobe-filling are involved, except that solutions with the lobe exceeded would not be allowed. The several types of curves complement one another nicely. For example, although the optical velocities measure the eccentricity only weakly, the pulse curves do it very well, and while only the light curves tell much about the figure of the optical star, the X-ray eclipse duration requires the figure to be consistent with the equipotential configuration. Distortions of the optical velocity curves can lead to incorrect orbital eccentricity, as discussed by Van Paradijs, et al. (1977). Here, however, almost the entire weight of the eccentricity determination comes from the pulse data, so that should not be a significant problem.

2. ANALYSIS AND RESULTS

Our full method of analysis, including parameter definitions and weighting scheme, cannot be given within the space of this note, nor has time permitted application to more than a small fraction of existing observations, but a more complete

© 1994 American Institute of Physics

paper will follow. The essence of the work is as follows. We fit X-ray pulse arrival times, optical velocities, and optical light curves *simultaneously* so as to ensure proper consistency, under the constraint that the observed X-ray eclipse duration must be reproduced. The binary star model is based on equipotentials and follows the supergiant's tidal deformation under assumptions that the star rotates at a constant angular rate, orbits in the proper Keplerian way, adjusts its figure instantaneously to the varying tidal field of the eccentric orbit, and maintains constant volume. The analysis by Van Paradijs, *et al.* (1977) is very thorough and statistically rigorous, but does not have the main advantages listed above. We include radial velocity proximity effects according to the relations in Wilson and Sofia (1976). The light curve consists of the B and V observations by Van Genderen (1981), the He I radial velocities by Van Paradijs, *et al.* (1977), and the pulse arrival times discussed by Rappaport, *et al.* (1980). Figures 1 shows fits to the data.

The rotation of the B0.5 star was estimated by Wickramasinghe, *et al.* (1974) and Zuiderwijk, *et al.* (1974) to lie in the range of 90 to 135 km/sec, based on spectral line broadening, which would be about 0.5 to 0.75 of the synchronous rate. Wilson (1979) showed that the supergiant's rotation must be considerably slower than synchronous to avoid overspilling the limiting lobe at periastron while satisfying the observed X-ray eclipse duration. However we can now make a much stronger statement about lobe filling and rotation, since we include the pulse arrival data and make a simultaneous solution of three kinds of observations. Estimates of the X-ray eclipse duration, θ_e range from 0.094 to 0.11 (Avni, 1976; Ögelman, *et al.*, 1977; Van der Klis and Bonnet-Bidaud, 1984). For the larger durations, the supergiant exceeds its gravitational-rotational limiting lobe at periastron even for zero rotation, so we set the rotation parameter to 0.00 and experimented to find the largest θ_e for which the star never exceeds its lobe. The result is 0.0986 or 35°.5. Then we went to the other extreme. We assumed the smallest published θ_e of 0.094, and solved (by our simultaneous least squares algorithm) for the ratio of angular spin rate to orbital angular rate, which is F_2 in the notation of Wilson, 1979. The result is F_2=0.55, with the star about 96% the size of its critical lobe, which is defined at periastron (Wilson, 1979). A tightened lower limit to θ_e would reduce the upper estimate of F_2. Thus, by following the phase-variable equipotential configuration and using the duration condition, we find sub-synchronous rotation for the B0.5 star, as did the earlier line broadening work, from sources of information which are independent of line broadening. We also find e=0.089 ± 0.007 (s.d.), ω=147°.8 ± 7°.5 (s.d.), $M_{Neutron}$=1.82 $M_\odot$, and $M_{B0.5}$=22.7 $M_\odot$.

3. PHASE SHIFTS AND TIDAL LAGS

The linear ephemeris on which the phases of all the observations depend, consists of the period 8.96443 days, adopted from Boynton, *et al.* (1986), and an initial epoch, HJD_0 = 2443768.05809, which was adjusted for best fit to the data. The ephemeris is likely to be refined for our later paper. In practice, almost the entire determination of the initial epoch is due to the pulse data. The figure shows an obvious phase shift of the observed light curve with respect to the computed one. The shift can be quantified by fitting the light curve separately, and turns out such that the observed curve lags the predicted one by about 0.043 in phase. The radial velocity shows no obvious shift, so there is no mismatch with regard to the orbit, but only with regard to the predicted light curve. The observed curve has about the right form of variation, so far as one can tell within the scatter, but has roughly twice the expected amplitude. We are able to make a meaningful comparison because we compute the phase-variable figure and surface brightness distribution of the B0.5 star in a way which is at least dynamically (although not necessarily structurally) correct. That is, we do not assume instantaneous co-rotation, which is the standard way in the literature and is certainly wrong, but rather constant angular rotation at constant volume (Wilson, 1979), which should be essentially correct. We interpret the phase and amplitude mismatches as evidence for tidal lags. These may be quite complicated, as the star tries to adjust to the varying gravitational field and cannot keep up, much like water sloshing gently in a tub. The phenomenon should be dissipative and thus help circularize the orbit. A detailed structural analysis is needed to decide whether this effect alone accounts for the increased amplitude, or whether there is some excitation of non-radial pulsations. Van Paradijs, et al. (1977) suspected pulsations because of erratic light and velocity variations. A phase lag similar to that in HD 77581 can be seen in the light curve of another supergiant-neutron star binary, HD 153919, although there the authors suggested a problem with their ephemeris (viz. Figure 3 of Strickland and Lloyd, 1993) rather than any real physical effect. It should be easy to settle that issue. In any case, HD 77581 appears to be a new type of laboratory for studying non-radial oscillations, and one can expect similar binaries also to serve as such laboratories.

4. HD 77581 AND COMMON ENVELOPE EVOLUTION

According to two independent estimates, HD 77581 is rotating at about half the synchronous rate, or perhaps slower. This is what one expects for a binary on the verge of common envelope evolution (Bodenheimer and Taam, 1986), as the orbit decays too rapidly for rotation to keep up with the increasing angular rate. There would seem to be no competing idea for explaining the lack of synchronism. The evolutionary expansion of the supergiant might help, but probably is too slow to be important with regard to non-synchronism. While the final remnant(s) may not be one of the standard bound objects which common envelope evolution ordinarily tries to explain (perhaps the result will be two runaway neutron stars), it seems inevitable that the neutron star will spiral into the envelope soon, because the observed slow rotation gives direct evidence for orbital decay. Therefore HD 77581 also should be a good laboratory in which to learn about the approach stage of common envelope evolution.

ACKNOWLEDGMENTS

We thank S. Rappaport for sending the pulse arrival time data on which his 1980 paper was based.

REFERENCES

Avni, Y. 1976, ApJ, 209, 574
Bodenheimer, P & Taam, R.E. 1986, In "The Evolution of Galactic X-ray Binaries", J Truemper, W.H.G. Lewin, & W. Brinkmann, eds., NATO ASI Series, Reidel, Dordrecht
Boynton, P.E., Deeter, J.E., Lamb, F.K., & Zylstra, G. 1986, ApJ, 307, 545
Ögelman, H, Beuermann, K.P., Kanbach, G., Mayer-Hasselwander, H.A., Capozzi, D., Fiordilino, E., & Molteni, D. 1977, A&A, 58, 385
Rappaport, S., Joss, P.C., & Stothers, R. 1980, ApJ, 235, 570
Strickland, D.J. & Lloyd, C. 1993, MNRAS, 264, 935
Van Genderen, A.M. 1981, A&A, 96, 82
Van der Klis, M. & Bonnet-Bidaud, J.M. 1984, A&A, 135, 155
Van Paradijs, J., Zuiderwijk, E.J., Takens, R.J., Hammerschlag-Hensberge, G., Van den Heuvel, E.P.J., & De Loore, C. 1977, A&A Suppl., 30, 195

Wickramasinghe, D.T., Vidal, N.V., Bessell, M.S., Peterson, B.A., & Perry, M.E. 1974, ApJ, 188, 167
Wilson, R.E. 1979, Ap J, 234, 1054
Wilson, R.E. and Sofia, S. 1976, Ap J, 203, 182
Zuiderwijk, E.J., Van den Heuvel, E.P.J., and Hensberge, G. 1974, A&A, 35, 353

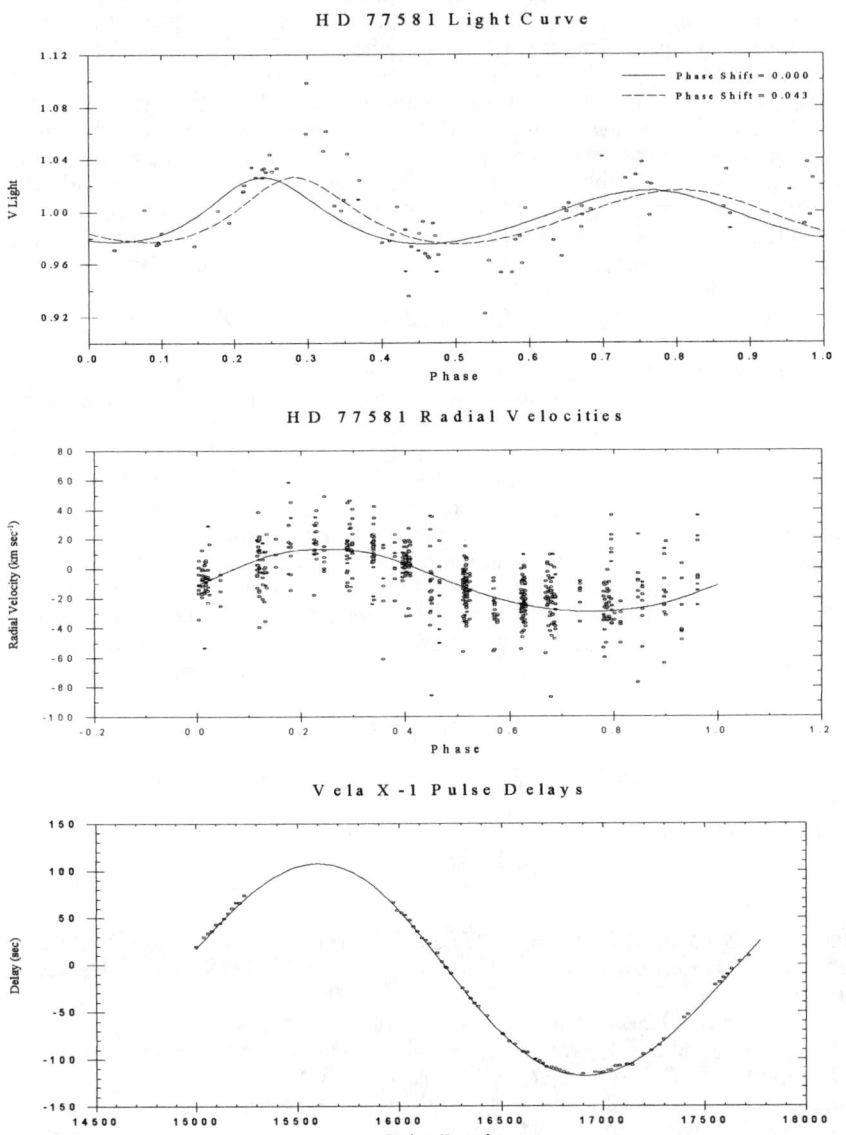

Figure 1. The observed and computed curves for V light (top), B0.5 star radial velocities (middle), and X-ray pulse delays (bottom). See the text for an explanation of the phase shifted light curve.

LONG TERM MULTIWAVELENGTH MONITORING OF HIGH MASS X-RAY BINARIES

Paul Roche, Malcolm Coe, Chris Everall
Astronomy Group, Physics Dept., Southampton University, England.
pdr/mjc/ce@uk.ac.soton.phastr

Juan Fabregat, Victor Reglero
Departamento de Astronomia, Universidad de Valencia, 46100 Burjassot, Spain.

Tom Prince, Deepto Chakrabarty, Lars Bildsten
Space Radiation Laboratory, California Institute of Technology, USA.

Andy Norton, Sarah Unger
Physics Dept., Open University, Milton Keynes, England

Dave Buckley
South African Astronomical Observatory, Cape Town, South Africa

ABSTRACT

We present a summary of the Southampton/Valencia/Caltech/SAAO campaign for long–term, multiwavelength monitoring of High Mass X-ray Binaries. A number of sources have been observed in unusual states, including detailed studies of the Be/X-ray transients X Persei (4U0352+309 : disk loss state, 1988-1990), 3A1118-616 (transient outburst, Jan. 1991), GX 1+4 (outburst Sept. 1993) and EXO2030+375 (periastron passage, June 1993).

1. INTRODUCTION

We present here details of the long–term monitoring programme of High Mass X-ray Binaries (HMXRBs) being carried out by Southampton, Valencia, Caltech and the South African Astronomical Observatory (SAAO). The sources are observed at ≈3 monthly intervals where possible, using optical spectroscopy (principally of the Hα line), infrared photometry (JHKL bands), optical photometry ($uvby\beta$ or UBVRI bands) and recently infrared spectroscopy (using the CGS4 detector on UKIRT; see Everall et al. 1993).

The recent (March 1993) involvement of the SAAO has greatly increased our access to southern hemisphere sources, making our programme "all–sky". Additional optical photometry is provided by Rene Hudec at Ondrejov Observatory (Czech Republic), with spectroscopy and IR photometry of some sources provided by Valeri Larionov (St. Petersburg) and Anatoli Tarasov (Crimean Astrophysical Observatory). Some of the brighter sources are monitored by amateur organisations (BAA and AAVSO) with optical photometry.

This combination of optical and IR data means that we are able to monitor the status of the circumstellar disk in the Be X–ray Binaries (BeXRBs), and observe long–term changes in the disk sizes. In addition, we include most of the Supergiant X–ray Binaries (SGXRBs), and the unusual symbiotic system

1728–247 (GX 1+4).

The combination of long–term optical/IR/X–ray data will allow the first multiwavelength studies of many of these objects. In addition, we can alert observers in all wavelengths to the prospect of a transient outburst during periastron passage of the compact companion through the Be star disk if, for example, our programme shows an unusually large circumstellar disk to be present.

To date, we have multi–wavelength observations of unusual behaviour in the following sources :

V0332+53 : outbursts 1983 and 1989 : optical/IR/X–ray (Coe et al. 1987)
3A1118–61 : outburst Jan. 1992 : UV/optical/IR/X–ray (Coe et al. 1994)
GX 1+4 : outburst Sept. 1993 : optical/IR/X–ray (Chakrabarty et al. 1994).
4U0352+309/X Persei : disk loss events : optical/IR/X–ray (Norton et al. 1991; Fabregat et al. 1992; Roche et al. 1993)
4U0728–25 : possible disk loss : optical/IR (Roche et al. 1994)

We have obtained multiwavelength observations of the *Rosat*–discovered HMXRB **J0146.9+6121** (LSI +61°235) (Coe et al. 1993), confirming its nature as a BeXRB. Observations of the unusual BeXRB **GT0236+61** (LSI +61°303) in collaboration with J. Paredes (Barcelona) have revealed the existence of the 26.5 day radio flaring period in our IR photometry (Paredes et al. 1994). We have also undertaken the first multiwavelength study of the (BATSE–predicted) periastron passage by the neutron star in the BeXRB **EXO2030+375** (Norton et al. 1994), involving observations of periastron–ingress from the Palomar 5m and WHT 4.2m (spectra), with coverage of the entire outburst from UKIRT (IR photometry) and BATSE.

Detailed papers on these observations are published or in preparation, (see References section) and here we only have space to summarise our source list and goals.

2. SOURCES OF DATA

Northern Hemisphere Observations from :

INT 2.5m & WHT 4.2m (Optical spectra), JKT 1m (Optical photometry), La Palma, Canary Islands. Lick Observatory 40″ (Optical spectra), California, USA. Palomar Observatory 1.5m and 5m (Optical spectra, photometry), California, USA. Calar Alto 1.5m (Optical photometry), Almeria, Spain. UKIRT 3.8m (IR spectra, photometry), Mauna Kea, Hawaii, USA. TCS 1.5m (IR photometry), Izana Observatory, Tenerife, Canary Islands.

Southern Hemisphere Observations from :

AAT 3.8m (Optical spectra, IR photometry), Siding Springs, Australia. SAAO 1.9m (Optical spectra, IR photometry), Sutherland, South Africa. CTIO 1.5m (Optical photometry), Cerro Tololo, Chile. Las Campanas 2.5m (Optical spectra), Las Campanas, Chile.

3. TARGET LIST

Source Name	Dataset Covers Period From :	Infrared Observations	Optical Spectra
Gamma Cas	Sep. 1987 – Oct. 1993	22 (JHKLM)	9
4U0114+65	Jul. 1990 – Oct. 1993	2 (JHK)	8
4U0115+63	Dec. 1990 – Oct. 1993	13 (JHK)	6
LSI +61°235	Aug. 1991 – Oct. 1993	10 (JHK)	4
LSI +61°303	Sep. 1987 – Oct. 1993	19 (JHKL)	6
V0332+53	Nov. 1983 – Oct. 1993	29 (JHKL)	11
X Persei	Sep. 1987 – Oct. 1993	29 (JHKL)	15
H0521+37	Sep. 1987 – Oct. 1993	24 (JHK)	10
A0535+26	Nov. 1987 – Oct. 1993	24 (JHK)	13
1H0556+286	Dec. 1991 – Aug. 1992	3 (JHK)	1
4U0728-25	Nov. 1987 – Mar. 1993	17 (JHK)	3
1H1907+09	Aug. 1991 – Jun. 1993	9 (JHK)	10
1H1936+541	Dec. 1991 – Oct. 1993	4 (JHK)	3
EXO2030+375	Jul. 1985 – June 1993	10 (JHK)	5
1H2202+501	Dec. 1991 – Oct. 1993	4 (JHK)	3
SMC X–3	May 1993 – Oct. 1993	2 (JHK)	1
SMC X–2	May 1993 – Oct. 1993	2 (JHK)	1
SMC X–1	Mar. 1993 – Oct. 1993	2 (JHKL)	2
LMC X–4	Mar. 1993 – Oct. 1993	2 (JHKL)	2
1H0739–53	Mar. 1993 – Oct. 1993	2 (JHKL)	2
1H0749–60	Mar. 1993 – Oct. 1993	2 (JHKL)	2
1E1024–573	May 1993 – Oct. 1993	1 (JHKL)	2
1H1034–56	Mar. 1993 – Oct. 1993	2 (JHKL)	2
3A1118–615	Jan. 1992 – Oct. 1993	3 (JHKL)	4
1145–614	Mar. 1993 – Oct. 1993	1 (JHKL)	2
1145–619	Mar. 1993 – Oct. 1993	2 (JHKL)	2
GX 301–2	Mar. 1993 – Oct. 1993	1 (JHKL)	2
1H1249–64	Mar. 1993 – Oct. 1993	1 (JHKL)	2
1H1253–76	Mar. 1993 – Oct. 1993	1 (JHKL)	2
1H1255–56	Mar. 1993 – Oct. 1993	1 (JHKL)	2
GX 304–1	Mar. 1993 – Oct. 1993	2 (JHKL)	2
2S1417–624	Mar. 1993 – Oct. 1993	2 (JHK)	–
4U1538–522	May 1993 – Oct. 1993	2 (JHKL)	2
1H1555–55	Mar. 1993 – Oct. 1993	2 (JHKL)	2
4U1700–377	Mar. 1993 – Oct. 1993	3 (JHKL)	3
GX 1+4	Jun. 1993 – Oct. 1993	6 (JHKL)	3

4. GOALS

It is hoped that by studying these sources over an extended period of time (several years), we may gain a better understanding of their transient behaviour, and the interaction of the compact object with the primary wind or disk. The transient existence of an accretion disk may also be observed during outbursts in some systems where Roche Lobe Overflow may occur at periastron.

In addition to studies of the long–term variability of these sources, we have been engaged in optical and IR studies of short timescale variability, which is often transient in nature (see Roche et al. 1993), i.e. search for periodic modulation of the Hα and/or HeI (6678Å) lines in X Per, searches for periodic IR variability (due to reprocessing of X-rays in the disk ?), searches for orbital modulation in the emission lines and/or IR photometry (as detected in GT0236+61/LSI +61° 303 by Paredes et al. 1994).

Data from the monitoring programme is also being used as input into a variety of Be star circumstellar disk models, such as modified versions of the CLOUDY program of Ferland (1991).

An additional feature of our programme has been the search for optical counterparts to several HMXRBs (i.e. OAO1657–415, GS0843–34, 1H1833–077) using optical and IR imaging. We also have a programme running at the 1.2m UK Schmidt telescope in Australia to obtain plates of the fields of new flaring sources discovered by BATSE.

5. CONCLUSIONS

The observations referred to in this short poster highlight the importance of a long–term, multiwavelength approach to the study of HMXRB systems, particularly the transient sources where pre–outburst data can reveal important details of the changes in the circumstellar disk. The all–sky, near–constant coverage provided by BATSE means that we now have simultaneous X–ray data to back up our optical and IR observations. We anticipate many more interesting observations will appear over the next few years as our southern hemisphere dataset builds up.

ACKNOWLEDGEMENTS

We wish to thank the various telescope time allocation committees who have supported our programme over the years - PATT, CATT, Palomar Observatory and the SAAO. Thanks also to various co-authors on several papers who have supplemented our data with additional observations of their own, in particular Robin Corbet and Brian Thomas (Penn State), Chris Shrader (GSFC), Francesco Polcaro and Franco Giovannelli (Rome), Raylee Stathakis and David Allen (AAO).

REFERENCES

Chakrabarty, D. et al., 1994, in prep
Coe, M.J., Longmore, B.J., Payne, B.J. & Hanson, C.G., 1987, MNRAS 226, 455
Coe, M.J., Roche, P., Everall, C. et al., 1994, in prep
Everall, C., Coe, M.J., Norton, A.J., Roche, P. & Unger, S.J., 1993, MNRAS 262, 57
Fabregat, J., Reglero, V., Coe, M.J. et al., 1992, A&A 259, 522
Ferland, G., 1991, Ohio State University Internal Report 91, 1
Norton, A.J., Coe, M.J., Estela, A. et al., 1991, MNRAS 253, 579
Norton, A.J., Bildsten, L., Chakrabarty, D. et al., 1994, in prep
Paredes, J.M., Marziani, P., Marti, J. et al., 1994, A&A submitted
Roche, P., Coe, M.J., Fabregat, J. et al., 1993, A&A 270, 122
Roche, P., Negueruela, I.N., Buckley, D. et al., 1994, A&A submitted

DYNAMICS OF ACCRETION SHOCKS IN AM HERCULIS SYSTEMS: MODELS FOR HIGH MASS WHITE DWARFS

Michael T. Wolff, Kent S. Wood
Code 7621, Naval Research Laboratory, Washington, D.C., USA

James N. Imamura
Institute of Theoretical Science
University of Oregon, Eugene, Oregon, USA

ABSTRACT

Several AM Her objects show low amplitude, 1 Hz quasi-periodic oscillations in power spectra of their optical emission. Wolff, Wood, and Imamura, in a recent series of papers, show that accretion noise driven radiative shocks are a plausible model for this phenomenon. Here, we discuss how two-temperature effects can play an important role in regulating the frequency of the shock oscillations within the formalism outlined by Wolff et al. We argue that two-temperature effects can be important in accretion flows onto white dwarfs with masses as low as 0.6 $M_\odot$ if cyclotron radiation is an important energy loss process in the shocked gas. The planned USA X-ray Timing Experiment on the ARGOS satellite should be able to detect the quasi-periodic X-ray oscillations predicted by our shock oscillation model.

1. INTRODUCTION

Several AM Her objects exhibit quasi-periodic oscillations (QPOs) in their optical emission. The QPOs appear as low amplitude, 1 - 3 %, broad excesses of power, $\nu/\Delta\nu_{FWHM} \sim 1 - 3$, centered on frequencies $0.3 - 1.4$ Hz in the power spectra of AN UMa (Middleditch 1982), V834 Cen (Mason et al. 1983), VV Pup (Larsson 1989), and EF Eri (Larsson 1987). Several authors have suggested that these QPOs are due to thermal instabilities in the radiative shocks which are thought to form in the accretion flows onto the white dwarfs (e.g. Middleditch 1982). If this conjecture is correct, then the observations of the QPOs can be used to probe the physical conditons of the radiating plasmas and accretion flows in the AM Her QPO sources (e.g. see Langer, Chanmugam, & Shaviv 1982). We examine the issue of the stability of shocks in accretion flows onto white dwarfs with masses $M_* \geq 1.0 M_\odot$. We point out important consequences of their two-temperature nature. We close with a suggested observational test of the shock oscillation model for AM Her QPOs.

2. SHOCK MODELS FOR HIGH M_*

We model the radiative shocks that form in the accretion flows onto the white dwarfs in AM Her systems. A number of authors (Imamura 1985; Wolff, Gardner & Wood 1989) showed that shocks formed in steady accretion rate flows onto nonmagnetic white dwarfs with $M_* \leq 0.4 M_\odot$ can oscillate in the first overtone mode. [See Chevalier & Imamura (1982) for a discussion of the mode notation that we use.] Chanmugam, Langer, & Shaviv (1985), and Wood, Imamura & Wolff (1992) showed that cyclotron cooling at the few percent level would stabilize the oscillations in steady accretion rate flows. These considerations led to the conclusion that if shock oscillations are responsible for the AM Her QPOs, then either $M_* \leq 0.4 M_\odot$ in the four AM Her QPO systems, or, some way must be found to excite oscillations in otherwise stable shocks. Wolff, Wood, & Imamura (1991), and Wood et al. (1992) have shown that accretion noise can excite oscillations in shocks in flows onto $0.6 M_\odot$ white dwarfs, even those which had $L_{cy}/L_{brem} \sim 1$, where L_{cy} and L_{brem} are the cyclotron and bremsstrahlung luminosities, respectively.

We have extended our calculations to consider accretion onto massive white dwarfs (i.e., $M_* \geq 1.0 M_\odot$). Table 1 gives the model parameters and resulting mode oscillation frequencies.

Table 1

High Mass White Dwarf Accretion Shock Models

M_* ($M_\odot$)	$\dot{M}$ ($\dot{M}_E f$)	B (MG)	ν_F (Hz)	ν_{1O} (Hz)	$\frac{L_{cy}}{L_{brem}}$	d_s (R_*)
1.0	0.10	10	4.5	12	0.011	0.012
1.0	0.01	10	0.40	1.2	0.12	0.12
1.0	0.004	10	0.20	0.65	0.25	0.25
1.0	0.10	30	6.0	17	0.21	8.8(-3)
1.0	0.01	30	1.3	3.8	1.1	0.029
1.0	0.004	30	1.0	2.8	1.7	0.039
1.2	0.10	10	5.6	15	0.015	0.018
1.2	0.10	30	7.1	22	0.26	0.013

Note:- The uncertainties in the frequencies are on the order of 10%

We find that even in these systems accretion noise can lead to quasi-periodic oscillations in the first overtone mode. Consider a model with $M_* = 1.0 M_\odot$, $B_* = 10$ MG, and $\dot{M}/f = 0.004 \dot{M}_E$, where f is the ratio of the cross sectional area of the accretion flow to the surface area of the dwarf, and $\dot{M}_E$ is the Eddington accretion rate, resulting in $L_{cy}/L_{brem} \sim 0.25$. The optical light curve power spectrum shows a substantial QPO bump at the first overtone frequency with only a small amount of power at the fundamental mode. On the other hand, the X-ray light curve shows a much broader QPO feature extending from just below the first overtone mode down to near zero frequency. This is consistent with the Wood et al. (1992) results for $0.6 M_\odot$ shocks. When the magnetic field strength is increased to $B_* = 30$ MG the optical cyclotron QPO strength is substantially weakened but the X-ray QPO tends to become narrower and centered on the first overtone frequency.

3. TWO-TEMPERATURE SHOCKS

In our models we allow for un-equal ion (T_i) and electron (T_e) temperatures in the postshock region. This is important because the dominant cooling processes are due to the electrons while the bulk of the thermal energy is stored in the ions. If the cooling processes are efficient the electrons can lose energy faster than they gain energy through collisions with the ions and a substantial separtation between T_e and T_i can occur. This effect is important for accretion onto nonmagnetic white dwarfs when Compton cooling is an important energy loss process and this occurs when $M_* > 1.2\ M_\odot$ (Imamura et al. 1987). In the case of accretion onto a magnetic white dwarf, substantial separation between T_e and T_i can occur even for white dwarf masses as low as $0.6\ M_\odot$ when cyclotron radiation is an important loss process (see Lamb and Masters 1979).

When two-temperature effects are important, T_e assumes a value where the radiative losses roughly balance the rate at which energy is transferred from the ions to the electrons. Our numerical calculations show that the post-shock T_e is substantially lower than the post-shock T_i even in models with white dwarfs of $M_* = 0.6 M_\odot$ for $B_* = 30$ MG (Imamura, Wolff & Wood 1993).

The separation between T_e and T_i leads to another effect, namely, the modification of the observed oscillation frequency. The frequency is determined by how fast the electrons can radiate thermal energy in the post-shock emission region. For high accretion rates, bremsstrahlung, by virtue of its strong density dependence, will dominate the cooling and the electron-ion separation will not be large except for high white dwarf masses. In this case the cooling time scale is longer than the exchange time scale and the frequency will be set by the cooling time. At low accretion rates, cyclotron radiation dominates due to its mild density dependence. Here, electron-ion energy exchange will be slower than cyclotron cooling and thus the frequency will be set by the exchange time. In low $\dot{M}$ models the oscillation frequency will be determined by the energy exchange time scale.

4. DISCUSSION

The fact that shock models of the type we present here yield time-varying optical and X-ray light curves suggests that an important observational test of the shock oscillation model for AM Her QPOs would be to simultaneously observe the optical and X-ray variations from QPO systems. Wood et al. (1992) showed that the optical and X-ray variations are strongly correlated. The USA X-ray experiment on the ARGOS satellite (see Wood et al. 1993, these proceedings) has sufficient collecting area to detect the QPO in AM Her systems. The expected count rate for EF Eri is 39 counts s^{-1} (source + background). If we search for a QPO with $\Delta\nu \sim 0.5$ Hz, an RMS amplitude of 5%, at the 4 sigma level, USA will detect the QPO in ~ 1 hour (see van der Klis 1989). The optical QPO can be observed in ~ 1 hour (e.g. Larsson 1987). Current plans call for USA to observe EF Eri for up to one month. Thus, simultaneous observations to determine if a correlation exists between the optical and X-ray variations becomes feasible for this AM Her system.

ACKNOWLEDGMENTS

This work was supported by the NASA Long-Term Space Astrophysics Research Program, the NASA Theory and Data Analysis Program, and a grant of computer time by the Naval Research Laboratory.

REFERENCES

Chanmugam, G., Langer, S.H., & Shaviv, G. 1985, ApJ, 299, L87
Chevalier, R.A. & Imamura, J.N. 1982, ApJ, 261, 543
Imamura, J.N. 1985, ApJ, 296, 128
Imamura, J.N., Durisen, R.H., Lamb, D.Q., & Weast, G.J. 1987, ApJ, 313, 298
Imamura, J.N., Wolff, M.T., & Wood, K.S. 1993, in preparation
Langer, S.H., Chanmugam, G., & Shaviv, G. 1982, ApJ, 258, 289
Lamb, D.Q. & Masters, A.R. 1979, ApJL, 234, L117
Larsson, S. 1987, A & A, 181, L15
Larsson, S. 1989, A & A, 217, 146
Mason, K.O., Middleditch, J., Cordova, F.A., Jensen, K.A., Reichert, G., Murdin, P.G., Clark, D., & Bowyer, S. 1983, ApJ, 264, 575
Middleditch, J. 1982, ApJL, 257, L71
van der Klis, M. 1989, Annu. Rev. Astron. Astrophys., 27, 517
Wolff, M.T., Gardner, J.H., & Wood, K.S. 1989, ApJ, 346, 833
Wolff, M.T., Wood, K.S., & Imamura, J.N. 1991, ApJL, 375, L31
Wood, K.S., Imamura, J.N. & Wolff, M.T. 1992, ApJ, 398, 593

EFFECTS OF SCATTERING ATMOSPHERES ON THE PULSE PROFILES OF ACCRETING X-RAY PULSARS

Steven J. Sturner[†] and Charles D Dermer
E. O. Hulburt Center for Space Research, Code 7653,
Naval Research Laboratory, Washington, DC 20375-5352

[†] National Research Council Research Associate

ABSTRACT

Resonant enhancements in the magnetic Compton cross section can produce regions of limited extent in an X-ray pulsar magnetosphere where the radiation force on electrons balances the gravitational force on protons. Infalling material outside the pulsar accretion column can collect in this region to form a radiation-supported scattering atmosphere. We propose that these scattering atmospheres can explain energy-dependent features in the observed pulse profiles of 4U 1626-67 and 4U 1538-52. Less complex pulse profiles will be observed at higher photon energies because the scattering atmospheres, which are supported by resonant Compton radiation pressure, become transparent to photons with energies greater than the cyclotron energy at the pulsar surface.

1. INTRODUCTION

X-ray pulsars are thought to be highly magnetized neutron stars accreting from a stellar companion either through Roche-lobe overflow or a stellar wind. The accreting material becomes entrained in the pulsar's magnetic field and is channeled onto the pulsar polar cap in an accretion column. Photons emitted from the base of the accretion column can resonantly interact with material outside the accretion column to form a radiation-supported scattering atmosphere (Dermer & Sturner 1991; Sturner & Dermer 1994). Here we present energy-dependent pulse profiles calculated with a Monte Carlo simulation.

2. MONTE CARLO SIMULATION

We have constructed a Monte Carlo simulation in which photons, emitted isotropically from the pulsar polar cap, interact with a scattering atmosphere via the resonant portion of the magnetic Compton cross section. In the strong magnetic fields near X-ray pulsars, resonant Compton scattering is normally much stronger than continuum scattering (Sturner & Dermer 1994). Hence for photons of a given energy, the scattering atmosphere can be accurately approximated by a thin spherical scattering layer at a height where the photon energy equals the local electron cyclotron energy. We assume that the angle-averaged resonant optical depth is unity. In the results presented here, the two polar caps are assumed to be identical.

The maximum size of the scattering atmosphere is determined by the spectrum and intensity of the radiation emitted from the polar cap and the pulsar's polar magnetic field strength. The actual size of the scattering atmosphere depends on the availability of material to form the atmosphere, which can vary from pulsar to pulsar. This is a free parameter in our model. For pulsars with field strengths that are derived from observations, the height of the scattering layer for a photon of a given energy is exactly determined for a dipole magnetic field geometry. The Monte Carlo simulation generates pulse profiles for different photon energies and sizes of the scattering atmosphere.

3. RESULTS

We compare our model results with the observed pulse profiles of 4U 1626-67 and 4U 1538-52 in Figures 1 and 2, respectively. Pulse profile modelling by Kii *et al.* (1986) imply that 4U 1626-67 has a surface magnetic field strength of 8×10^{12} G. For this magnetic field, the scattering layers for 9 and 22 keV photons would be at ~2.2 and 1.6 neutron-star radii, respectively. We take a value of 35° (as measured from the center of the neutron star) for the half-angular extent of the scattering layer for the 9 keV photons. For the 22 keV photons, we assume that no scattering atmosphere is formed at 1.6 neutron-star radii, which could occur if the radius of the unstable equilibrium point lies above this height (Dermer & Sturner 1991).

Clark *et al.* (1990) and Nagase (1990) have reported a 20 keV cyclotron line in the spectrum of 4U 1538-52, implying a surface magnetic field strength of 1.7×10^{12} G. The heights of the scattering layers for 2.25 and 9.15 keV photons, which are the mean energies of the observed X-ray bands, are at 2.1 and 1.3 neutron star radii, respectively. The angular extent of the scattering layer for the 2.25 keV photons was taken to be 60°. We again assume that no scattering atmosphere is formed at the height at which the 9.15 keV photons are in resonance.

4. CONCLUSIONS

We have shown that the existence of radiation-supported scattering atmospheres can explain the observed energy-dependent pulse profiles of 4U 1626-67 and 4U 1538-52. Electrons in the atmospheres resonantly scatter photons emitted near the base of the accretion column. Higher energy photons may not be scattered if the photon energy exceeds the cyclotron energy at the inner edge of the scattering atmosphere or the surface of the pulsar. Thus simpler pulse profiles are predicted at higher photon energies. Further details about this model are given in the paper by Sturner & Dermer (1994).

REFERENCES

Clark, G. W., Woo, J. W., Nagase, F., Makishima, K., & Sakao, T. 1990, ApJ, 353, 274
Dermer, C. D., & Sturner, S. J. 1991, ApJ, 382, L23
Kii, T., Hayakawa, S., Nagase, F., Ikegami, T., & Kawai, N. 1986, PASJ, 38, 751
Nagase, F. 1990, in Proc. 23rd ESLAB Symposium, ed. N. E. White and T. D. Guyenne, ESA SP-296 (Paris: European Space Agency), 45
Sturner, S. J., & Dermer, C. D. 1994, A&A, in press

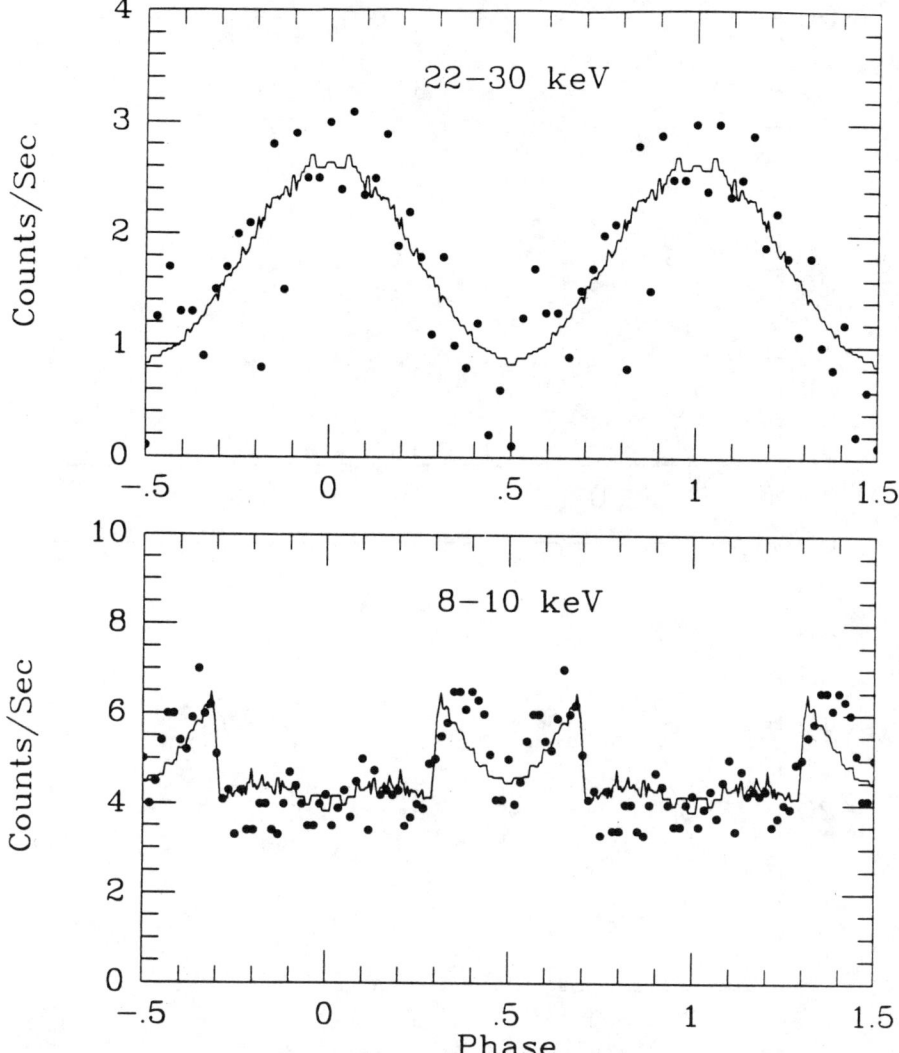

Fig. 1. Fit of the model (solid curve) to the observed pulse profiles of 4U 1626-67 in the energy ranges 8-10 keV and 22-30 keV (points). The scattering layer was at 2.175 neutron star radii for the 8-10 keV fit, consistent with a surface magnetic field strength of 8×10^{12} G. We assume that no scattering layer is present for 22-30 keV photons. The data were taken from Kii *et al.* (1986).

498 Effects of Scattering Atmospheres on the Pulse Profiles

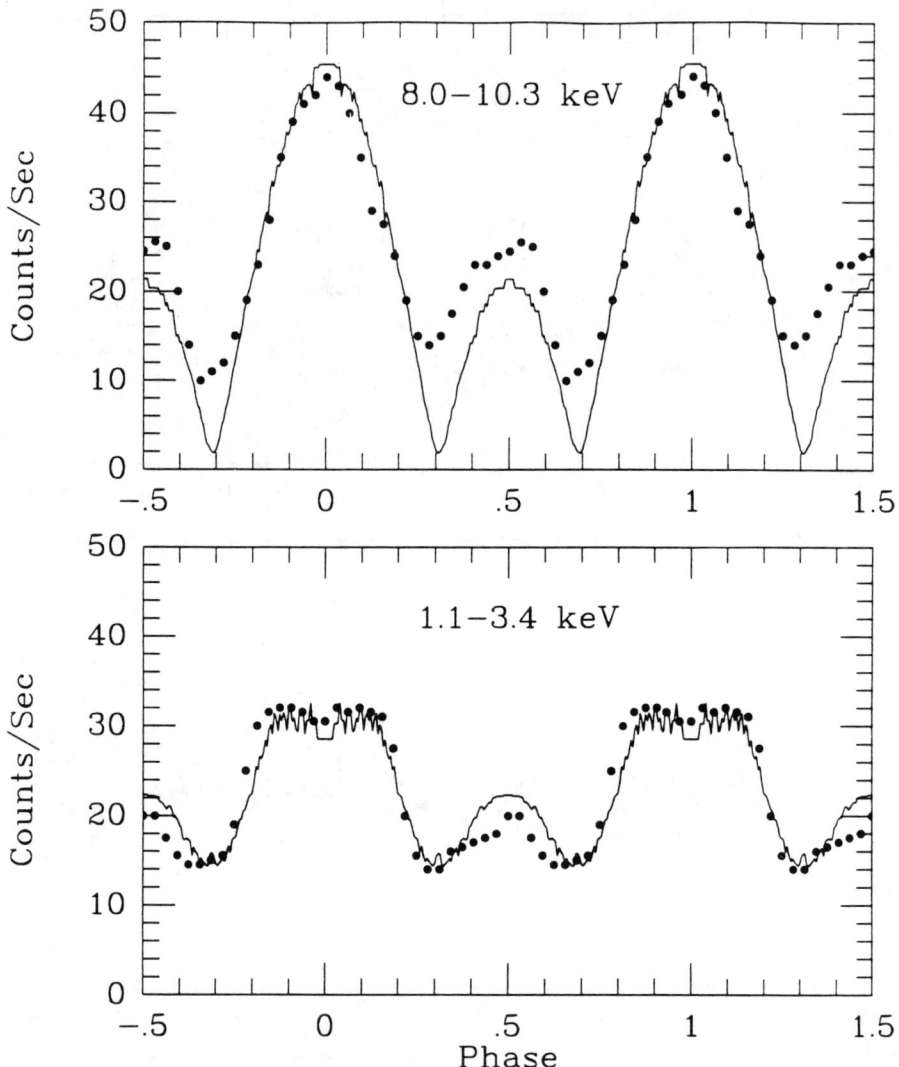

Fig. 2. Fit of the model (solid curve) to the observed pulse profiles of 4U 1538-52 in the energy ranges 1.1-3.4 keV and 8.0-10.3 keV (points). The scattering layer was at 2.1 neutron star radii for the 1.1-3.4 keV fit, consistent with the observed surface magnetic field strength of 1.7×10^{12} G. We assume that no scattering layer is present for 8.0-10.3 keV photons. A better fit can be achieved if the two polar caps have different temperatures. The data were taken from Clark *et al.* (1990).

X-RAY PULSAR HYDRODYNAMICS: COLLISIONLESS SHOCK WAVES AND STEADY STATE INFALL HYDRODYNAMICS

William K. Rose
Department of Astronomy, University of Maryland, College Park, MD 20742

ABSTRACT

X-ray pulsar emission is caused by mass infalling onto the magnetic polar regions of neutron stars. The collision stopping length caused by nuclear and coulomb collisions is too large for a standing shock to form above a neutron star surface. We show that collisionless excitation of plasma waves can probably produce a stopping length sufficiently small for such a shock to form. We also describe hydrodynamic models of mass infall under steady state conditions.

1. SHOCK WAVE FORMATION

For coulomb collisions the ratio, stopping length λ divided by neutron star radius R, is

$$\frac{\lambda}{R} = \frac{m_p}{R\rho\sigma_D} = \frac{m_p^3 v^4}{R\rho \, 4\pi e^4 \ln\Lambda} \qquad (1)$$

where the coulomb deflection cross section σ_D is uncorrected for particle symmetry. In equation 1 v is approximately .3c and $\ln\Lambda \sim 20$. To obtain the density ρ in equation 1 we must assume some mass accretion rate $\dot{M}$ and polar solid angle $\Delta\Omega = f\,4\pi$.

If we assume $\dot{M} = 2 \times 10^{-10} M_\odot/yr$ over each polar cap and $f = .1$ then the stopping distance becomes

$$\lambda \approx 2 \times 10^8 \text{cm (coulomb only)} \qquad (2)$$

The stopping distance λ must be much less than the neutron star radius $R = 10^6$cm for shock formation and therefore expression 2 shows that a shock does not form above the neutron star surface. We note that the density increase across a strong shock is only a factor of 4.

The corresponding nuclear and coulomb scattering cross section of 100 MeV protons is $\approx 3 \times 10^{-26}$ cm² which is appreciably greater than the coulomb cross section. However, the total collisional stopping length

$$\lambda \approx 5 \times 10^7 \text{cm (nuclear and coulomb)} \tag{3}$$

is larger than 10^6 cm.

The two stream instability between electrons causes excitation of plasma waves. Resonant plasma waves couple to nonresonnant plasma waves and ion acoustic waves. Heating and momentum transfer between counterstreaming electrons results. An electric field is generated by charge separation between protons and electrons. This induced electric field slows protons and thereby decelerates infalling gas.

Since kinetic energies are much greater than thermal energies we can use the well known cold electron gas two stream instability growth rate, which in our example, is approximately equal to the plasma frequency ω_p. Calculations by Rose Guillory, Beall and Kainer 1984 (Ap.J., 280, 550) indicate that nonlinear coupling between resonant plasma waves, nonresonant plasma waves and ion acoustic waves slow down electrons in $\sim 10^4/\omega_p$ seconds. Since protons are 1836 times more massive than electrons the protons slow down time becomes $\sim 2 \times 10^7/\omega_p$. Using the same physical input parameters as above the estimated stopping distance becomes

$$\lambda(\text{collisionless}) \sim 3 \times 10^3 \text{cm} \ll R = 10^6 \text{cm} \tag{4}$$

Although our estimated λ is clearly somewhat uncertain, it is plausible to conclude that a collisionless shock may form above a neutron star surface. The electron scattering optical depth above such a shock is $\gtrsim 1$ and therefore Comptonization of emitted thermal radiation is important.

II. STEADY STATE INFALL HYDRODYNAMICS

Under steady state conditions inflow hydrodynamics onto a neutron star polar cap are described by the equations of mass conservation, motion and energy conservation. Using the equation of mass conservation to eliminate the velocity v from the equations of motion and energy conservation we readily obtain the equivalent equations

$$\left(\frac{A^2}{r^4\rho^2} - \frac{5P}{3\rho}\right)\frac{dP}{dr} = \frac{2K\rho A}{r^2}\left(\frac{\mu m_p}{k}\right)^{1/2}\left(\frac{P}{\rho}\right)^{1/2} - \frac{10}{3}\frac{A^2}{r^5}\frac{P}{\rho^2} + \frac{5}{3}\frac{PGM}{r^2} - \frac{5P}{3\rho}F_R \tag{5}$$

$$\left(\frac{A^2}{r^4\rho^2} - \frac{5P}{3\rho}\right)\frac{d\rho}{dr} = \frac{2K\rho^3 r^2}{3A}\left(\frac{\mu m_p}{k}\right)^{1/2}\left(\frac{P}{\rho}\right)^{1/2} - \frac{2A^2}{r^5\rho} + \frac{\rho GM}{r^2} - F_R \qquad (6)$$

with $A = \dot{M}/f4\pi = -\rho v r^2$, $f = \Delta\Omega/4\pi$, $-\varepsilon_{ff} = K\rho T^{1/2}$ and F_R the radiative force. The radiative force F_R varies approximately as $1/r^2$ and free-free emission is the dominant energy loss mechanism above the shock.

By solving equations 5 and 6 numerically we determined how much energy was radiated from infalling mass (i.e. gas above the shock). For relevant physical conditions radiated energy is less than 1% of corresponding gravitation energy changes. This low efficiency of radiated energy is not surprising because similar low efficiencies have been obtained for spherical infall into a black hole.

The high x-ray luminosities of x-ray pulsars indicate that radiative fluxes from some of them are close to the Eddington limit. Since gravity and radiative force vary approximately as $1/r^2$ it might seem reasonable to assume that infall velocities in high luminosity pulsars might be substantially lowered. The amount of radiated energy above the shock would, therefore, be increased and the x-ray spectra somewhat softer. We solved equations 5 and 6 for mass accretion rates and corresponding radiative fluxes approaching the Eddington limit. We were successful in obtaining steady state solutions with infall velocities reduced by about 20%. However, we were unable to find steady state solutions closer to the Eddington limiting flux. Although our calculations do not prove that x-ray pulsars with accretion columns do not exist they do suggest that high luminosity x-ray pulsars might have characteristic x-ray intensity fluctuations that differ in some average sense from those whose x-ray radiative fluxes are significantly below the Eddington limit. The physical explanation for our result is that as the Eddington limit is approached radiation is scattered from the infalling mass and therefore only mass close to the neutron star is acted on by an appreciable radiative force.

Dynamics

RAPID VARIABLITY IN NEUTRON STARS AND BLACK HOLES- COMPARISON AND ATTEMPT AT UNIFICATION

M. van der Klis
Astronomical Institute "Anton Pannekoek", University of Amsterdam
Kruislaan 403, 1098 SJ Amsterdam, The Netherlands
michiel@astro.uva.nl

ABSTRACT

The rapid X-ray variability of black hole candidates and neutron stars with high, low and very low magnetic fields is compared. Similarities in the properties of the variability between source types suggest that similar physical mechanisms underly some power spectral components seen in common in neutron stars with various magnetic-field strengths and black hole candidates. Other components appear to be unique for magnetized neutron stars. Suggestions are made for the the sites of origin and generation mechnisms of these power spectral components.

1. INTRODUCTION

The X-ray spectrum and the rapid X-ray variability of accreting compact objects arise through processes in the same physical region (close to the compact object), so their properties may be expected to be coupled. The masses and dimensions of stellar mass black holes and neutron stars are similar, so their accretion phenomena may be expected to be similar as well. Consequently, much can be learned by studying X-ray spectra and rapid X-ray variability in correlation, and by comparing phenomena in neutron stars and black holes. In doing so, similarities emerge in the phenomenology that indicate that a unified description may be possible, although some of the perceived analogies between the various types of systems are, as yet, rather tenuous.

The power spectra of accreting compact objects can be described in terms of a small number of simple shapes. *Power law noise* has a power distribution $\propto \nu^{-\alpha}$. *Flat topped noise* and *peaked noise* are closely related, and together called *band limited noise*. Their power spectrum shows a fall-off (steepening) towards high ν and a roll-over (flattening) towards low ν. In flat-topped noise the maximum power density is reached at $\nu = 0$, in peaked noise at $\nu > 0$. The same power spectral component can be at one time flat-topped and at another time somewhat peaked; correlated shot models can produce such behaviour (Vikhlinin et al. 1993a). *Quasi-periodic oscillations* (QPO) are a type of peaked noise. Usually, the term is reserved for (often Lorentzian) peaks that are approximately symmetric and have a relative width that is *significantly* and *most of the time* less than 0.5. Table 1 lists the power spectral components that have been identified in X-ray binaries.

An important concept is that of *source state*, a generalization of X-ray spectral state. To identify source states one looks at X-ray spectra and power spectra *in correlation* and finds recurrent coincidences in their properties; these are indicative of recurrent source states. The idea is that the mass flux Ṁ towards the compact object varies with time and governs both the X-ray spectrum and

Table 1. Power spectral components in X-ray binaries

Compact object type (guess at magnetic field)	Power law noise (index)	Band limited noise (cut-off frequency[1])	QPO (frequency)
Black hole candidates (~ 0 G)	HS noise $\alpha \sim 1$	LS noise 0.03–0.3 Hz	Slow QPO 0.08–0.8 Hz
		VHS noise 1–10 Hz	Fast QPO 3–10 Hz
Atoll sources ($\lesssim 10^{9-10}$ G)	VLFN $\alpha \sim 1 - 1.5$	HFN 0.3–20 Hz	Slow QPO 0.4, 2 Hz
Z sources ($\sim 10^{9-10}$ G)	VLFN $\alpha \sim 1.5 - 2$	LFN 2–20 Hz	HBO 13–55 Hz
		HFN 50–100 Hz	N/FBO 6–20 Hz
Pulsars ($\sim 10^{12}$ G)	not well studied	Pulsar noise 0.001–3 Hz	Slow QPO 0.04–0.2 Hz

HS: high state; LS: low state; VHS: very high state; VLFN: very low frequency noise; HFN: high frequency noise; LFN: low frequency noise; HBO: horizontal branch oscillations; N/FBO; normal/flaring branch oscillations
[1] Various definitions used by original authors.

the power spectrum, so that these will show correlated variations. This hypothesis works well in explaining the data. A conclusion is that $\dot M$ is *not* proportional to the 1–20 keV X-ray count rate; sometimes the count rate even drops when inferred $\dot M$ increases.

In black hole candidates X-ray spectral fits often provide a good ingredient in determining source state. In Z and atoll sources, and also in black hole candidates in the very high state, the X-ray spectral changes are too subtle, and *colour-colour diagrams* (CDs) and *hardness-intensity diagrams* (HIDs), plots of X-ray hardness ratios *vs.* each other or *vs.* count rate are better at providing the necessary X-ray spectral information. Most sources produce a characteristic track in the CD/HIDs. Source position in the track is used as the X-ray spectral ingredient in the definition of source state.

2. Z AND ATOLL SOURCES

Z and atoll sources (Hasinger and van der Klis 1989, hererafter HK89) are low magnetic-field neutron stars. They have been extensively described previously (van der Klis 1989, 1993b and references therein), and only a brief summary of their properties is presented here, mainly for comparison to other source types.

Six *Z sources* are known. They produce Z-shaped tracks in X-ray CD/HIDs. $\dot M$ is inferred in increase following the Z track from upper left to lower right. Z source power spectra show three broad noise components, *very low frequency noise* (VLFN), *low frequency noise* (LFN) and *high frequency noise* (HFN) and two QPO components, *horizontal branch oscillations* (HBO) and *normal and flaring branch oscillations* (N/FBO).

VLFN is 1–6% amplitude power law noise that gets stronger with $\dot M$. HFN is band limited noise with cut off frequency $50 < \nu_{cut} < 100$ Hz. Its amplitude depends only weakly on $\dot M$. HBO and LFN are a QPO and a band limited noise component that appear and disappear together, and are likely physically related. They are strongest at low $\dot M$ and disappear at high $\dot M$. HBO frequency (13–55 Hz) and LFN cut-off frequency (2–20 Hz) increase with $\dot M$. LFN can be flat topped or peaked, depending on the source. N/FBO have a preferred

frequency near 6 Hz. In Sco X-1 and GX 17+2, their frequency increases from
$\sim$6 to $\sim$20 Hz when $\dot{M}$ increases.

The most succesful HBO model is the *magnetospheric beat frequency model*
(Alpar and Shaham 1985, Lamb et al. 1985), which requires Z sources to have
a magnetosphere. In most models for the N/FBO, *radiation pressure* plays the
key role (van der Klis et al. 1987, Hasinger 1987, Lamb 1989, Fortner et al.
1989, Miller and Lamb 1992, Alpar et al. 1992). Z sources have near-Eddington
luminosities, and the NBO frequency is roughly similar in each Z source, suggesting that the frequency is determined by the Eddington critical luminosity
L_{Edd}. Lamb (1991) proposed a comprehensive model for the QPO and X-ray
spectral properties of Z sources that uses the above ingredients.

Fourteen *atoll sources* and probable atoll sources are known (HK89,
van der Klis 1993b). They show one curved branch in the CD, often fragmented
due to observational effects. $\dot{M}$ is inferred to increase from left to right along
the branch. Their power spectra show two broad noise components called *very-low-frequency noise* (VLFN) and *high-frequency noise* (HFN). On one occasion,
slow (0.4 and 2 Hz) QPO were seen (Yoshida et al. 1993). Atoll source VLFN is
power law noise similar to that in Z sources. Atoll source HFN is *not* similar to
Z source HFN: it has a much lower cut-off frequency (0.3–20 Hz), can be much
stronger and depends strongly on $\dot{M}$. At low $\dot{M}$, the HFN is strong (up to 22%);
when $\dot{M}$ increases this decreases to <2% while the cut-off frequency increases
(Yoshida et al. 1993, Prins et al. 1993). Atoll HFN is sometimes flat-topped and
sometimes peaked.

HK89 proposed that the neutron stars in Z sources have *both* higher magnetic field strengths than atoll sources, *and* reach higher mass fluxes $\dot{M}$. The
higher field explains why the (magnetospheric) HBO are seen in Z sources (and
perhaps pulsars) only, and the higher maximum mass flux ($\sim \dot{M}_{Edd}$) explains
why the same is true for the (near-Eddington) N/FBO. Predictions are that an
atoll source that becomes bright will show Z source high-$\dot{M}$ properties (N/FBO
and appropriate spectral branches), but never HBO, and that a Z source that
becomes faint will show millisecond pulsations.

The properties of the low magnetic-field neutron star (Tennant 1986)
Cir X-1 fit the first prediction: Sometimes (at intermediate brightness levels)
its power spectrum and colour diagram behaviour are reminiscent of that of an
atoll source (Oosterbroek et al. 1993); when it becomes very bright, 6–20 Hz
QPO (Tennant 1987, Makino et al. 1992) and spectral branches reminiscent of
Z source N/FBO behaviour may appear. It seems likely that the source is a rare
example of an atoll source (a neutron star with a lower magnetic-field strength
than Z sources) that can reach the critical Eddington mass flux (van der Klis
1991). Cir X-1 also shares some characteristics with black hole candidates (see
below).

3. BLACK HOLE CANDIDATES

Three source states are distinguished in black hole candidates (see Tanaka
and Lewin 1993, van der Klis 1993b and references therein). In the *low state*
(LS) the X-ray spectrum is a flat power law with photon spectral index 1.5–2. In
the *high state* (HS) the 1–10 keV flux is much higher due to a soft component;
the power law is sometimes "sticking out" from under the soft component at
higher energies. In the *very high state* (VHS) the X-ray spectrum is similar to

in the high state (at higher 1–10 keV flux); it is distinguished from the HS by its rapid X-ray variability.

The transient black hole candidate GS 1124−68 (Nova Mus '91) in its decay went through all three states (Miyamoto et al. 1992b, Kitamoto et al. 1993), strongly suggesting that the states directly follow $\dot M$. I shall here use this assumption. The phenomenological correspondence between black hole candidates in each of the three states is good, and in the bright low magnetic-field neutron-star systems the same assumption worked very well. However, this classification of black hole candidate phenomenology is preliminary. It will be further scrutinized at the end of this section.

Fig. 1 summarizes the power spectra in the three states. The LS power spectrum shows strong (30–50% amplitude) band-limited noise with ν_{cut} between 0.03 and 0.3 Hz. This LS noise is usually flat-topped, but sometimes peaked (Vikhlinin et al. 1993b). A very characteristic property of LS noise is that the level of its flat top and, in anti-correlation with this, its cut-off frequency ν_{cut} vary, whereas the power spectrum above ν_{cut} remains approximately unchanged (Belloni and Hasinger 1990, Miyamoto et al. 1992a). In the HS power law noise with $\alpha \sim 1$ and an amplitude of a few % is present. Sometimes a weak remnant of the LS noise is present in the HS. Slow QPO with frequencies similar to the LS noise cut-off frequencies ($\sim$0.08–0.8 Hz; Motch et al. 1983, Ebisawa et al. 1989, Grebenev et al. 1991) sometimes occur in LS and HS. The VHS shows 3–10 Hz QPO and rapidly variable broad-band noise. This state has been reported in GS 1124−68 (Miyamoto et al. 1992a, 1993, Dotani 1992, Takizawa et al. 1993a, b, Grebenev et al. 1993), and in GX 339−4 (Miyamoto et al. 1991, 1993, Dotani 1992). In GX 339−4 the QPO frequencies were usually near 6 Hz, in GS 1124−68 between 3 and 10 Hz. In both sources the QPO show second harmonics and possible subharmonics. The noise alternates in shape between band-limited ($\nu_{cut}\sim$1–10 Hz), and power law shaped ($\alpha\sim$1). Transitions between the two noise shapes can occur within less than a second. Branches occur in the CD/HIDs, and the power spectral parameters seem to depend on position in the branches. The LS and VHS band limited noise cut off frequency and amplitude fit one relation (van der Klis 1993c), suggesting that the same phenomenon underlies them. The VHS power law noise is similar to that in the HS.

The situation with respect to the source states of black hole candidates, and their relation to $\dot M$ is not yet entirely clear. For example, in GX 339−4 the observable energy flux in the 1–200 keV band is higher in the low state than in the high state (see Fig. 3 in Grebenev et al. 1993) and GS 2023+337 remained in an apparent LS at very high X-ray brightness (17 Crab; Kitamoto et al. 1989, Inoue 1989, 1991, Terada et al. 1992). One way to explain this would be that a fourth state exists in black hole candidates, at higher $\dot M$ levels than the VHS, which mimicks the LS and which has erroneously been lumped together with the "real" low state. However, in the absence of any evidence for phenomenological differences between these two states, this seems unlikely.

Another possibility is that the emission from a black hole candidate shows a strongly $\dot M$-dependent anisotropy (see also van der Klis 1993a). If the reason for the disappearance of the hard LS X-ray spectral component in the HS is obscuration of a central, hot and rapidly variable region by matter in, e.g., a puffed-up accretion disk, then the $\dot M$ level at which the obscuration occurs would depend on the inclination. For a pole-on viewing geometry no obscuration would occur and the system would remain in a "LS" at all $\dot M$ levels; this might explain

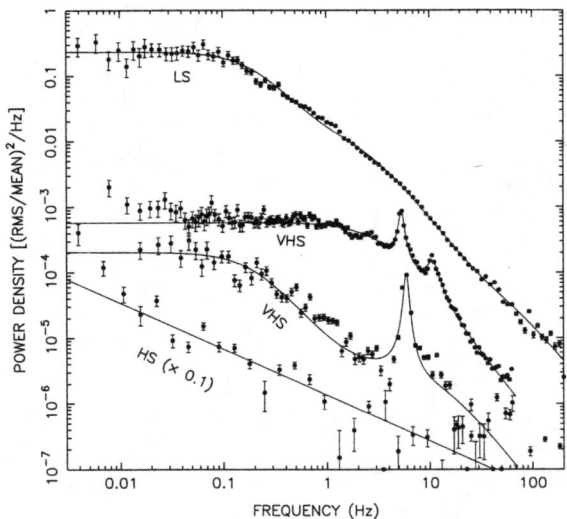

Fig. 1. Power spectra from Ginga data of black hole candidates in the low (LS; Cyg X-1), high and very high (HS and VHS; GS 1124−68) states.

the behaviour of GS 2023+337. The increasing concentration of the hard X-rays towards the (rotation) polar axes with increasing $\dot{M}$ would in this scenario explain why the apparent 1–200 keV luminosity of GX 339−4 in its LS seems to be (at least sometimes) higher than in its HS and VHS: most of the energy would be leaving the system in the HS and VHS along the polar axis and not be seen by us.

In the low magnetic-field neutron stars the X-ray flux is an unreliable indicator of $\dot{M}$; the same might turn out to be the case in an even stronger sense in the black hole candidates. Note, that the mass flux $\dot{M}$ that by hypothesis determines the state is the mass flux *towards* the compact object, just as is the case in the Z sources; at near- and super-Eddington rates, not all of this matter may actually be accreted; jets might for example be formed when $\dot{M}$ becomes high enough.

4. SIMILARITIES BETWEEN BLACK HOLE CANDIDATES AND LOW MAGNETIC FIELD NEUTRON STARS

There is a number of striking similarities between black hole candidate and neutron star phenomenology (see van der Klis 1993a, d).

The black hole candidate LS is very similar to the atoll source low $\dot{M}$ ("island") state. Both states occur at the lowest 1–10 keV count rates and inferred $\dot{M}$ levels. Both are dominated by strong (several 10%) band limited noise (LS noise and atoll HFN) which is sometimes flat-topped and sometimes slightly peaked. When an atoll source gets really faint, the power spectra are nearly indistinguishable from a black hole candidate in the low state (Fig. 2), and the 1–20 keV X-ray spectrum becomes hard (Langmeier *et al.* 1987, Yoshida *et al.*

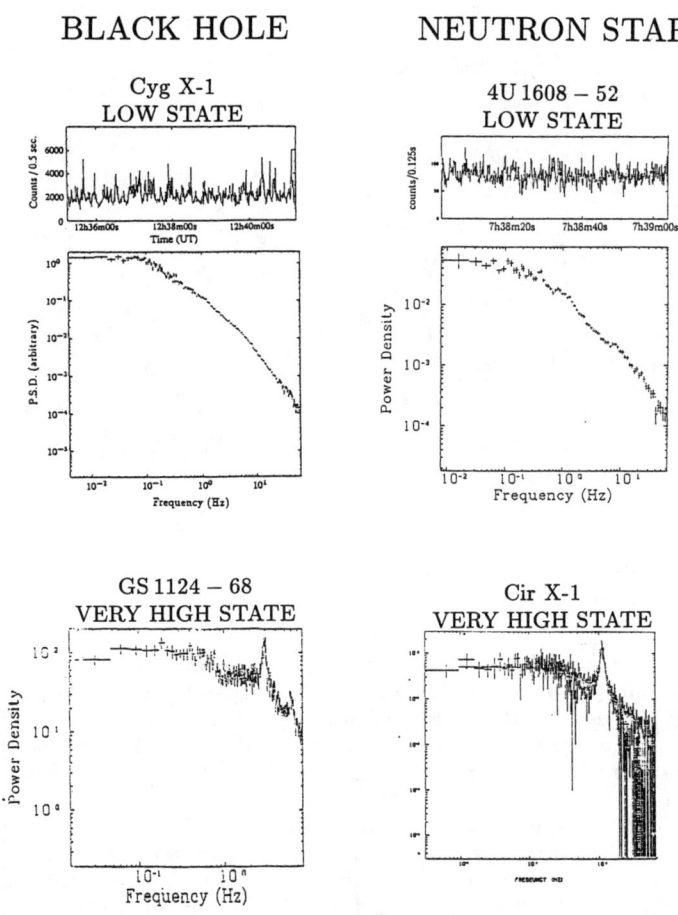

Fig. 2. Power spectra from the black hole candidates Cyg X-1 (*top left*) and GS 1124−68 (*bottom left*) in the low state and the very high state, respectively, and from the low magnetic-field neutron stars 4U 1608−52 (*top right*) and Cir X-1 (*bottom right*) in the atoll island state and a very high X-ray brightness state respectively, illustrating the similarity between neutron star and black hole candidate low and very high states. Compiled from Inoue (1992), Takizawa *et al.* (1993b) and Makino *et al.* (1991).

1993), like in black hole candidates in the LS. Even the inverse correlation between cut-off frequency and flat-top level, characteristic for black hole candidates in the LS, was seen in an atoll source, 4U 1608−52, at low $\dot M$ (Fig. 3 in Yoshida *et al.* 1993).

The black hole candidate VHS has strong similarities to the Z source high $\dot M$ ("normal/flaring branch") state. Both occur at the highest inferred $\dot M$ levels, and both show QPO, with similar frequencies (6–20 Hz in the neutron star systems, 3–10 Hz in the black hole candidates), that depend on the position of the source in branched tracks in the HID/CDs. Clearly different is the harmonic content of the QPO; black hole candidate VHS QPO show strong harmonics, Z source N/FBO do not. Another difference is that Z sources do not show the fast changes in broad band noise shape seen in black hole candidates.

The properties of Cir X-1 provide a further link between neutron stars and black holes. In some of its high states (Tennant 1987, Makino et al. 1992, Oosterbroek et al. 1993), the source shows a mix of characteristics of Z sources and black hole candidates in high $\dot{M}$ states. It shows 6–20 Hz QPO with a low harmonic content, branches in the CD/HID, *and* fast changes in the shape of the broad band noise. This suggests an interpretation where fast noise shape changes are related to high $\dot{M}$ accretion onto any very low magnetic field compact object (neutron star or black hole), but where a high harmonic content of the high $\dot{M}$ QPO is a black hole signature. The reason, then, that Cir X-1 sometimes resembles a black hole in its rapid variability characteristics, as was noted by Toor (1977) and Samimi et al. (1979), while its X-ray bursts show it to be a neutron star, is that it is the only neutron star that we know that has a magnetic field as low as in atoll sources that sometimes accretes a near- or super-Eddington rates.

Earlier this year, at the Integral meeting in Les Diablerets, Switzerland, I suggested that the phenomenology of the black hole candidates and low magnetic-field neutron stars could be described in terms of three $\dot{M}$-driven states that are common to accreting low magnetic-field neutron stars and accreting black holes (van der Klis 1993a). This was based on the similarities in the band limited noise of black hole candidates and atoll sources in low $\dot{M}$ states, and in the QPO of Z sources, black hole candidates and Cir X-1 in high $\dot{M}$ states. Since then, Yoshida et al. (1993) showed that the band limited noise of atoll sources is even more similar to black hole candidate LS noise than was known at that time, and an analysis of the band limited noise components across all X-ray binaries (van der Klis 1993c) showed that Z source LFN is similar to black hole candidate LS noise and atoll source HFN, and that Z source HFN is different (below). Fig. 3 presents a line-up of the three common states of black hole candidates and low magnetic-field neutron stars.

5. NOISE COMPONENTS

Band limited noise plays an important part in the above "unified" picture of the phenomenology of accretion onto black hole candidates and low magnetic-field neutron stars. There seems to be a similar band limited noise component in black hole candidates, Z sources and atoll sources, characterized by a large drop in strength and a simultaneous increase in cut off frequency when $\dot{M}$ increases. In black hole candidates, this is the LS noise and the VHS noise, in atoll sources the inappropriately named HFN and in Z sources the LFN.

Z sources exhibit a second type of band limited noise, HFN, with a 50–100 Hz cut off frequency, that varies much less with $\dot{M}$ than the LFN. This HFN may be similar to the band limited noise that is seen in accreting pulsars (van der Klis 1993c). This pulsar noise is also relatively independent of $\dot{M}$, surely over the $\dot{M}$ range covered by Z sources.

Takeshima (1992) reports a strong correlation between pulsar noise cut off frequency and pulse frequency among a sample of 10 pulsars covering three orders of magnitude in pulse frequency. The cut off frequency of Z source HFN is much higher than that of the pulsar noise. If the identification of Z source HFN with pulsar noise is correct, then the spin frequencies of the neutron stars in Z sources must be 50–100 Hz, in accordance with the beat frequency model (Section 2).

512 Rapid Variability in Neutron Stars and Black Holes

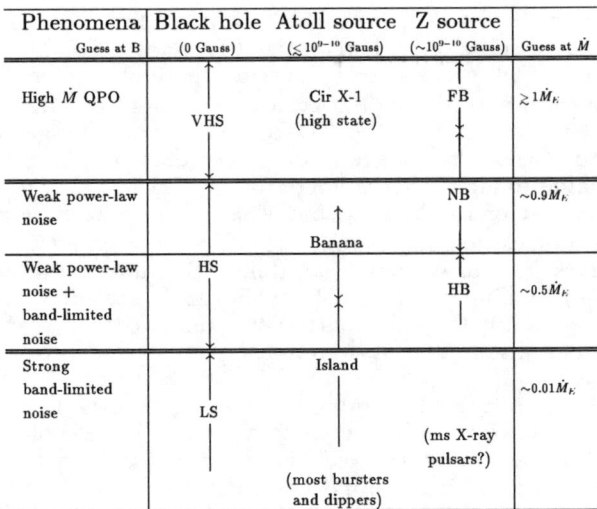

VHS: very high state; HS: high state; LS: low state; FB: flaring branch; NB: normal branch; HB: horizontal branch.

Fig. 3. Proposed classification scheme for X-ray binary source states. There a three states that are common to neutron stars and black holes; in a given source the mass transfer rate $\dot{M}$ towards the compact object determines the state. The power spectral shapes that are characteristic of each state are indicated in the leftmost column. The correspondence between the source states of each source type is indicated. Magnetic field strengths and mass fluxes are rough indications only. In particular, other source parameters might affect the $\dot{M}/\dot{M}_E$ levels at which state transitions occur.

The absence of strongly $\dot{M}$ dependent band limited noise in pulsars suggests that such noise arises in the inner, radiation pressure dominated part of the disk, which in pulsars is disrupted by magnetic stresses. The alternation of flat-topped and peaked power spectral shapes characteristic for this noise component is in qualitative accordance with a time series model involving slightly correlated shots (Vikhlinin et al. 1993a). The phenomenon might be related to the presence of, e.g., turbulent cells or magnetic loops in the inner disk.

The fact that the second, relatively $\dot{M}$ *insensitive* type of band limited noise is only seen in pulsars and Z sources, the only accreting compact objects where magnetospheric phenomena (pulsations and HBO) are seen, suggests that this noise requires a magnetosphere for its formation.

It was proposed recently that the *power law* noise seen in accreting neutron stars might be due to unsteady nuclear burning on the neutron star surface (Bildsten 1993). If correct, then the amplitude of this noise is constrained by the ratio of nuclear burning to accretion energy. For hydrogen, this is about ~ 0.04; for helium only ~ 0.01. Note that 4U 1820−30, which is believed to accrete hydrogen-poor matter, sometimes shows VLFN with a strength of 4.5% (HK89), in apparent violation of this. The power law noise of black hole candidates in the HS could not be caused by the same mechanism.

Table 2. Related variability components

Compact object type	Strongly $\dot{M}$ dependent band-limited noise	High $\dot{M}$ QPO	Magnetospheric QPO	Weakly $\dot{M}$ dependent band-limited noise
Black hole candidate	LS&VHS noise	Fast QPO in VHS	—	—
Atoll source	HFN	Fast QPO (Cir X-1)	—	—
Z source	LFN	N/FBO	HBO	HFN
Pulsar	—	—	Slow QPO	Pulsar noise
Model or site of origin	Inner (P_{rad}-dominated) disk	P_{rad} feed-back loop	Beat frequency model	Magnetosphere

Tentative classification of some neutron star and black hole rapid variability components and possible models. Four master types of variability are distinguished, two of which are common to neutron stars and black holes, and two of which occur only in magnetized neutron stars (Z sources and pulsars). See Table 1 for an overview of all known components and for abbreviations. P_{rad}: radiation pressure

6. CONCLUSION

Although some of the perceived analogies between the rapid X-ray variability characteristics of different types of accreting compact objects are as yet rather tenuous, a picture seems to be emerging where the millisecond fluctuations in black holes, and in neutron stars with high, low and very low magnetic fields can be understood in common terms (Table 2). Just two structures determine the basic physics of the accretion process, namely the magnetosphere and the inner (radiation pressure dominated) disk. Z sources have the most complex phenomenology, showing HBO and N/FBO, as well as LFN and HFN, because their magnetic field is weak enough to allow the presence of a mostly undisturbed inner accretion disk (like in black holes and atoll sources) and strong enough to allow the presence of a small magnetosphere (like in pulsars). X-ray pulsars have a magnetosphere and no inner disk and therefore show only an HBO-like and an HFN-like component, black holes and atoll sources have an inner disk and no appreciable magnetosphere and therefore only show an N/FBO-like and an LFN-like component.

The situation with respect to the power law components, and the slow QPO in atoll sources and black hole candidates is as yet unclear, but might be resolved with existing data. Further studies of bright transients, both black hole candidates and low magnetic-field neutron stars are very important to resolve outstanding issues.

I gratefully acknowledge stimulating discussions with participants in the 4th Maryland meeting. This work was supported in part by the Netherlands Organization for Scientific Research (NWO) under grant PGS 78-277.

REFERENCES

Alpar, M.A., Shaham, J., 1985, Nat 316, 239.
Alpar, M.A., et al., 1992, A&A 257, 627.
Bildsten, L., 1993, ApJ , in prep.

Belloni, T., Hasinger, G., 1990, A&A 227, L33.
Dotani, T., 1992, in: Frontiers of X-ray astronomy, Tanaka and Koyama (eds.), Universal Academy Press, Tokyo, p. 151.
Ebisawa, K., Mitsuda, K., Inoue, H., 1989, PASJ 41, 519.
Fortner, B., Lamb, F.K., Miller, G.S., 1989, Nat 342, 775.
Grebenev, S.A., et al., 1991, Sov. Astron. Lett. 17(6), 413.
Grebenev, S., et al., 1993, A&AS 97, 281.
Hasinger, G., 1987, A&A 186, 153.
Hasinger, G., van der Klis, M., 1989, A&A 225, 79. [HK89]
Inoue, H., 1989, Proc. 23d ESLAB Symp., ESA SP-296, p. 783.
Inoue, H., 1991, ISAS RN 469, paper presented at the Texas/ESO-CERN Symp., Brighton, UK, December 1990.
Kitamoto, S., et al., 1989, Nat 342, 518.
Kitamoto, S., et al., 1993, in prep.
Lamb, F.K., 1989, Proc. 23rd ESLAB Symposium, ESA SP-296, 215.
Lamb, F.K., 1991, NATO Advanced Study Institute 344, 445.
Lamb, F.K., et al., 1985, Nat 317, 681.
Langmeier, A., Sztajno, M., Hasinger, G., Trümper, J., 1987, ApJ 323, 288.
Makino, Y., Kitamoto, S., Miyamoto, S., 1991, poster presented at the 28th Yamada Conference, Nagoya, Japan, April 8-12 1991.
Makino, Y., Kitamoto, S., Miyamoto, S., 1992, see Dotani (1992), p. 167.
Miller, G.S., Lamb, F.K., 1992, ApJ 388, 541.
Miyamoto, S., et al., 1991, ApJ , 383,784.
Miyamoto, S., et al., 1992a, ApJ 391, L21.
Miyamoto, S., et al., 1992b, Proc. Ginga Memorial Symp., Makino and Nagase (eds.), ISAS, Tokyo, p. 37.
Miyamoto, S., et al., 1993, ApJ 403, L39.
Motch, C., et al., 1983, A&A 119, 171.
Oosterbroek, T., et al., 1993, A&A , in prep.
Prins, S., et al., 1993, A&A , in prep.
Samimi, J., et al., 1979, Nat 278, 434.
Takeshima, T., 1992, Ph.D. Thesis, Univ. Tokyo.
Takizawa, M., et al., 1993a, in prep.
Takizawa, M., et al., 1993b, in prep.
Tanaka, Y., Lewin, W.H.G., 1993, in: X-ray binaries, Lewin, van Paradijs and van den Heuvel (eds.), Cambridge University Press, in press.
Tennant, A.F., Fabian, A.C., Shafer, R.A., 1986, MNRAS 219, 871.
Tennant, A.F., 1987, MNRAS 226, 971.
Terada, K., et al., 1992, see Dotani (1992), p. 323.
Toor, A., 1977, ApJ 215, L57.
Van der Klis, M., 1989, ARA&A 27, 517.
Van der Klis, M., 1991, NATO Advanced Study Institute 344, 319.
Van der Klis, M., 1993a, in: Proc. INTEGRAL Meeting, Les Diablerets, Switzerland, Feb. 2-5 1993, AIAP preprint 1993-001; ApJS , in press.
Van der Klis, M., 1993b, AIAP preprint 1993-009; in: X-Ray Binaries, Lewin, van Paradijs and van den Heuvel (eds.), Cambridge University Press, in press.
Van der Klis, M., 1993c, AIAP preprint 1993-010; A&A , in press.
Van der Klis, M., 1993d, in: Proceedings of NATO ASI, Kemer, Turkey, 1993, in press.
Van der Klis, M., et al., 1987, ApJ 316, 411.
Vikhlinin, A., Churazov, E., Gilfanov, M., 1993a, preprint.
Vikhlinin, A., et al., 1993b, preprint.
Yoshida, K., et al., 1993, PASJ 45, 605.

TIDAL INSTABILITIES IN ACCRETION DISCS

A.R. King
Astronomy Group, University of Leicester, Leicester LE1 7RH, U.K.
ltvad::ark, ark@uk.ac.leicester.starlink

ABSTRACT

I briefly review the theory of tidal instabilities in the accretion discs of close binaries. Superhump–type disturbances result from the presence of the 3:1 resonance inside the disc, requiring mass ratios $q = M_2/M_1 < q_{\rm crit} \simeq 0.25-0.33$. Tidal instabilities seem to occur in all cataclysmic variables which satisfy this condition. A number of low–mass X–ray binaries also exhibit similar phenomena; their presence offers a new way of obtaining information about the component masses of these systems. This may be particularly useful for suspected black–hole binaries. The period excess associated with superhump phenomena may be rather small in LMXBs with very small mass ratios, and other signatures of tidal instabilities, such as possible vertical disc structure effects, should be looked for.

1. INTRODUCTION

The presence of accretion discs in several classes of accreting binary is now widely accepted; in some cases, notably cataclysmic variables (CVs), the discs appear to conform to simple theoretical ideas. For many purposes it is safe to neglect the effects of the companion star on the structure of these discs. However, it is now clear that in some circumstances the companion star can have a dramatic influence on the discs, chiefly through resonant tidal effects. The first observational manifestation of these was provided by optical photometry of a certain class of dwarf novae (the SU UMa systems). From time to time these systems undergo particularly long outbursts (superoutbursts): during these events photometric disturbances (superhumps) appear in the optical light curves with periods a few percent longer than the orbital period (cf Fig. 34 of Frank et al., 1992).

Several phenomenological models of this effect were proposed, but the first real insight into its cause was provided by Whitehurst's (1988) numerical particle simulations: these showed that under certain conditions the accretion discs in close binaries could become eccentric and undergo a slow prograde precession relative to inertial space. The dissipation pattern of the disc is affected by the relative position of the secondary star, so its luminosity has a period slightly longer than the orbit: one orbital revolution after maximum light, the secondary star finds that the major axis of the disc has precessed forward somewhat, and the star has to move slightly further on to recreate the configuration producing maximum light. Significantly, Whitehurst found that precessing eccentric discs appeared only in binaries with secondary–to–primary mass ratios $q = M_2/M_1$ less than a critical value $q_{\rm crit}$ which was of order $0.25 - 0.33$. Since such mass ratios among CVs are associated with short orbital periods $P_{\rm orb}$ this explained why superhumps are only observed in systems with $P_{\rm orb} \lesssim 2$ hr (see Section 3).

2. RESONANCE THEORY

Since Whitehurst's results were purely numerical it was important to understand their physical basis. Hirose & Osaki (1990) and Whitehurst & King (1991) independently suggested that the underlying cause was the presence in the disc, at small mass ratios, of the 3:1 resonance with the binary frequency. Both groups argued in terms of test–particle orbits, which is incorrect near a resonance in a fluid disc, since fluid orbits cannot cross themselves. However their identification of the 3:1 resonance as the basic cause of superhumps was correct, as shown by the detailed treatment by Lubow (1991a, b). Here I give a brief summary of the theory, and begin by explaining the idea of a resonance.

A resonance occurs in a disc when a particle orbit within it has a frequency of radial motion (about the accreting star) which is commensurable with the angular frequency with which it sees the secondary star move. This condition ensures that the same relative configuration of the particle and the two stars recurs periodically, so that the cumulative effect of quite small perturbations can build up and become significant. Although particle trajectories about the accreting star are quite similar to Kepler orbits, the deviation of the potential from the simple r^{-1} form means that they do not quite close, but precess with an apsidal frequency ω. Thus if the particle's average angular frequency is Ω, so that it takes a time $2\pi/\Omega$ to return to the same angular position, it will take slightly longer, i.e. a time $2\pi/(\Omega - \omega)$, to reach the same radial distance. This frequency $\Omega - \omega$ of radial motion is known as the epicyclic frequency. The particle sees the secondary moving in angle at a frequency $\Omega - \Omega_{\rm orb}$, where $\Omega_{\rm orb} = 2\pi/P_{\rm orb}$; the resonance condition is that these frequencies should be commensurable, i.e.

$$k(\Omega - \omega) = j(\Omega - \Omega_{\rm orb}), \tag{1}$$

where j, k are positive integers. Since $\omega \ll \Omega$, we see that resonances occur close to, but not precisely at, commensurabilities of the particle and orbital frequencies. Thus $j = 2, k = 1$ gives $\Omega \simeq 2\Omega_{\rm orb}$, while $j = 3, k = 2$ implies $\Omega \simeq 3\Omega_{\rm orb}$. These are known as the 2:1 and 3:1 resonances respectively.

Since the particle orbits are close to Keplerian, and indeed almost circular in most cases, it is easy to find out which resonances are accessible to disc orbits. Thus if Ω_{jk} denotes the value of Ω satisfying (1) the associated radius is

$$R_{jk} = (GM_1/\Omega_{jk}^2)^{1/3}. \tag{2}$$

By Kepler's law the binary separation is

$$a = (GM/\Omega_{\rm orb}^2)^{1/3}, \tag{3}$$

where $M = M_1 + M_2$ is the total mass. From (1, 2, 3) and the definition of q we find

$$\frac{R_{jk}}{a} = \left(\frac{j-k}{j}\right)^{2/3} (1+q)^{-1/3}. \tag{4}$$

The allowed resonances are given by requiring R_{jk} to be smaller than the largest possible disc radius, which is limited to about 90% of the accreting star's Roche lobe radius $R_{\rm Roche}$. This too is proportional to a, so the result is a condition on the mass ratio. Using

$$\frac{R_{\text{Roche}}}{a} = \frac{0.49 q^{2/3}}{0.6 q^{2/3} + \ln(1 + q^{1/3})} \qquad (5)$$

(Eggleton, 1983) we find that the presence of low–order resonances places significant restrictions on q. Resonances with $k = 1$ can only occur for very small mass ratios $q \lesssim 0.025$ (for j=2) or still smaller (for $j > 2$). The 3:1 resonance ($j = 3, k = 2$) can be inside the disc for

$$q < q_{\text{crit}} \simeq 0.33. \qquad (6)$$

Given that a resonance may be present in the disc, we still have to show that it causes the growth of the eccentric disc structure found by Whitehurst (1988). Lubow (1991a, b) demonstrates this as follows. He considers flow perturbations of the form $e^{i(r\theta - s\Omega_{\text{orb}})t}$, where r, s are integers. Thus an eccentric perturbation stationary in inertial space is described by $r = 1, s = 0$. Since the eccentric disc in Whitehurst's simulations precesses only slowly (i.e. the superhump period is only a few percent longer than P_{orb}) this is an adequate first approximation to the perturbation giving the superhumps. We can now ask if the initial $(1, 0)$ perturbation grows when acted upon by the tides exerted by the companion. These are a sum of terms of the form (j, j) ($j = 1, 2, 3, ...$). The interaction of each of these with the original $(1, 0)$ perturbation produces disturbances of the form $(j - 1, j)$; these waves interact in their turn with the (j, j) tides to produce a $(1, 0)$ disturbance which amplifies the original eccentric perturbation. The dominant term in the growth of the perturbation is indeed the $j = 3$ tide, provided that the 3:1 resonant radius is inside the disc. If this condition is satisfied the perturbation grows on a timescale $\sim q^{-2} P_{\text{orb}}$.

Of course this theory does not account for the prime observational characteristic of superhumps, namely the relative period excess $\Delta P / P_{\text{orb}} \sim$ a few percent, which turned up naturally in Whitehurst's (1988) simulations. Lubow (1992) shows that these arise from the combination of two principal effects. The first is that the local apsidal precession rate rise steeply with disc radius (i.e. as perturbations by the companion becomes more important): the precise size of the disc is therefore important in determining the precession. This contribution is always prograde, but the second effect, the contribution of pressure waves in the disc, produces a retrograde precession that can be comparable, although smaller. The disc size and pressure effects are both greatly affected by assumptions about the disc microphysics and external effects such as the impact of the gas stream from the companion. Not surprisingly therefore, the theory of the precession rate is a good deal less clear than that for the growth of the perturbations, although there is general agreement that $\Delta P / P_{\text{orb}}$ should increase with q. The results of Whitehurst (1988) and Lubow (1991b) agree well with observations of CVs, while those of Hirose & Osaki (1990) systematically overestimate the relative period excess (see Figure 7 of Lubow, 1993, adapted from Fig. 2 of Molnar & Kobulnicky, 1992).

Finally we should mention the possibility of the growth of a disc inclination to the orbital plane. In fact this can be excited by the 3:1 resonance also (Lubow, 1992), but its growth time is at least an order of magnitude greater than that of the eccentric instability, suggesting that it is unlikely to have much effect in most practical cases (but see the discussion of permanent superhumps below).

3. OBSERVATIONS

For the remainder of this article I shall consider the implications of the condition (6) for the appearance of tidal instabilities. For definiteness I shall fix on a value of $q_{\text{crit}} = 0.3$, but the range of uncertainty in this figure should be borne in mind.

(a) *Cataclysmic Variables.* The companion stars in most CVs are close to the main sequence, and so obey the approximate mass–period relation (e.g. King, 1988)

$$M_2 \simeq 0.11 P_{\text{hr}} M_\odot, \qquad (7)$$

where P_{hr} is P_{orb} measured in hours. Thus the resonance condition (6) becomes

$$M_1 \gtrsim 0.37 P_{\text{hr}} M_\odot. \qquad (8)$$

We can now understand why all non–magnetic CVs below the well–known period gap (i.e. with $P_{\text{hr}} \lesssim 2$) are observed to show superhumps. From (8) this requires

$$M_1 \gtrsim 0.7 M_\odot, \qquad (9)$$

which is very probably satisfied by the white dwarfs in CVs. In fact we know that white dwarfs can have masses as large as the Chandrasekhar value $M_1 = 1.44 M_\odot$. Using (8) this implies that non–magnetic CVs may show superhump behaviour for orbital periods up to about 4 hr. This is in reasonable accord with observation: the record for the longest period for a system showing photometric disturbances with a period longer than the orbit is currently held by V603 Aql (Nova Aql 1918) (Patterson & Richman, 1993) with $P_{\text{hr}} = 3.3$. This requires $M_1 \gtrsim 1.2 M_\odot$, in agreement with deductions from the speed class of the nova.

There is an important difference between the superhumps for systems on each side of the period gap. Below the period gap all non–magnetic systems are dwarf novae, probably because the average accretion rate is predicted to be low. Superhumps are only apparent during superoutbursts of these systems, probably because the disc does not spread out to the resonant radius for long enough at other epochs. On the other hand systems above the gap (and some rare ones within it) can have higher accretion rates, and appear as steady systems. In such cases any superhumps will be permanent. Patterson & Richman give a list of all systems with photometric periods differing from the orbital period.

Since there is ample time available for perturbations to grow in permanent superhump systems, a slow growth rate may be no hindrance to their ultimate appearance. Thus it would be interesting to check such systems for any signs of the inclination instability referred to above.

The system H0709–36 lies squarely in the period gap ($P_{\text{hr}} = 2.44$: it could for example have begun life as a mass–transferring system there), and so ought to be a good candidate for a system with permanent superhumps. Encouragingly, optical photometry shows a stable period 2% longer than the spectroscopically determined orbital period. However, the system would appear to fail the mass-ratio condition (6): as an eclipsing system, the eclipse width at half depth provides a direct measurement of the mass ratio q (Bailey, 1990). In H0709–36 the unusually wide eclipse implies $q > 0.6$, in flat contradiction to (6). But Bailey's method (devised before any suggestion of permanent superhump behaviour) uses the brightness distribution of a standard accretion disc to estimate the eclipse width expected for a given q. By hypothesis, if H0709−36 has a permanent superhump, its disc does *not* conform to this assumption: one would expect the superhump disturbance to enhance the radiation from the disc edges, necessarily making the eclipses wider. With this interpretation the eclipses of this system give a direct picture of the structure of the perturbed disc, and may thus repay further study.

(b) *Low–Mass X–ray Binaries.* The mass ratios of low–mass X–ray are in general smaller than those of CVs: the secondary star appears frequently to be undermassive for its radius and hence for the binary period, while the primary is a neutron star or black hole, either of which is usually assumed to be significantly more massive than a white dwarf. Tidal instabilities ought therefore to be quite common among them. To date there are four suggested candidates.

XB1916 − 05. White (1989) first suggested a superhump interpretation to explain the discrepancy between the X-ray period $P_X = 50.00$ min and the optical period $P_{\rm opt} = 50.46$ min of this system. The mass ratio condition (6) is certainly satisfied in this case: an orbital period near 50 minutes implies a very low mass ($M_2 < 0.1 M_\odot$) degenerate secondary star, while any reasonable neutron star mass is an order of magnitude larger. An interesting question here is the precise nature of the optical modulation. The optical emission from LMXBs is probably dominated by reprocessed X–rays, so the usual intrinsic mechanism based on tidal distortion of the disc flow pattern is unpromising. It may be that vertical disc structure generated by the inclination instability discussed above plays a role here.

The three remaining systems are all black hole candidates. This is hardly surprising, since only systems with large estimated primary masses are regarded as serious candidates. All three systems are soft X–ray transients; as pointed out by Mukai at this meeting, a black hole binary has larger inertia than a lower–mass system, and mechanisms such as magnetic stellar wind braking or nuclear expansion of the secondary therefore drive lower mass transfer rates, which are associated with transient behaviour. In satisfying analogy with the SU UMa systems, the systems tend to show different photometric periods in these two states.

GS2000 + 25. Charles et al (1991) and Callanan & Charles (1991) found slightly discrepant periods (8.3 and 8.2 hr) in outburst and quiescent photometry, and suggested a superhump origin. Chevalier & Ilovaisky (1993) combined these quiescent data with their own to find $P = 8.258$ hr, but with a "distortion wave" at a period of 10.1 hr. Clearly spectroscopy is required to narrow down the possibilities here, but there seems little doubt that at least two independent periods exist.

GRO J0442 + 32. Kato, Mineshige and Hirata (1993a) found a period of 5.18 hr in optical photometry during the outburst and decline, remarking that the shape of the modulation was reminiscent of superhumps. Later in the decline, the same authors (Kato et al, 1993b) saw dips in the light curve with a period of 5.091 hr.

GS1124 − 683 (= Nova Muscae 1991). Bailyn (1992) found a 10.5 hr photometric period during the decline and predicted a slightly shorter quiescent period on the basis of a superhump interpretation. Remillard, McClintock & Bailyn (1992) then measured a 10.398 hr period in both photometry and spectroscopy in quiescence.

Because of the interest in measuring the primary mass in these systems, there have been attempts to use the measured relative period excess to determine the mass ratio. As we have seen, current theory does not yet give a robust prediction of $\Delta P(q)$, although this may be possible in future. Certainly the data for CVs (see Figure 2 of Molnar & Kobulnicky 1992) do suggest a reasonably tight relationship. A further problem in applying this approach to estimate the primary mass in black–hole candidate systems is that the evolutionary status of the secondary, and hence its mass, is far from clear. Systems with 10 hr $\lesssim$ $P_{\rm orb} \lesssim$ 20 hr frequently appear to have evolved secondaries which have been stripped down almost to the mass of the helium core, which can be as low as $0.1 M_\odot$, and this may also be true of systems with shorter periods (see e.g. King, 1993, for a discussion of the black–hole system V404 Cygni).

4. CONCLUSIONS

Observations of CVs seem to offer very strong support for the proposal that superhumps result from the presence of the 3:1 resonance within the accretion disc. The idea that the superhumps may be permanent in steady systems with suitable mass ratios now also seems to be established. The anomalous photometric periods observed in certain LMXBs suggest that tidal disturbances may occur here too, as predicted by theory. There is scope for considerable progress in a number of areas.

1. Photometry of eclipsing systems such as H0709 − 36 offers potential insight into the radiation pattern of a tidally distorted disc. Doppler tomograms may help us understand the kinematics of such flows even in non–eclipsing systems.

2. There are a number of CVs in which the photometric modulation is slightly *shorter* than $P_{\rm orb}$, but the obvious interpretation as the spin period of a magnetic white dwarf is implausible for various reasons (e.g. in TV Col the spin period is known to be 1911 s and $P_{\rm orb} = 5.5$ hr; but the dominant optical modulation has a period of 5.2 hr). A tidal interpretation is very tempting here, but further theoretical work is needed. One possibility is that in some cases the white dwarf magnetic field forces matter on to orbits which it would otherwise avoid, allowing other resonances to play a role.

3. Cases where systems are presumed to satisfy the mass ratio condition (6) but apparently show no superhumps should be checked further.

4. Systems with permanent superhumps should be checked for any sign of unusual vertical disc structure. Demonstrating the growth of such structure theoretically may require large amounts of computer time.

5. Future theory may be able to predict an accurate relation between the relative period excess $\Delta P/P_{\rm orb}$ and the mass ratio q. Since we can often estimate lower limits to the companion mass M_2, this holds out the enticing prospect of getting estimates of the primary mass. This is of course particularly interesting for black hole candidates. It is worth bearing in mind that the rather small values of q expected in such systems probably mean that there will be only small relative period discrepancies $\Delta P/P_{\rm orb}$. As the period discrepancy is by far the most obvious signature of tidal phenomena, this may be the reason why

relatively few LMXB have as yet been found to show them. One might indeed speculate that resonant tidal effects in some way alter the vertical structure of discs in LMXBs; although deciding whether such effects can be strong enough to account for the large structures sometimes claimed from study of X–ray light curves clearly requires much more work.

ACKNOWLEDGMENTS

I thank the conference organizers for their kind support and hospitality, and many participants for useful discussions. I thank Drs H. Ritter and P.A. Charles for considerable help with Section 3.

REFERENCES

Bailey, J.A. 1990, MNRAS, 243, 57
Bailyn, C.D. 1992, ApJ, 391, 298
Callanan, P.J. & Charles, P.A. 1991, MNRAS 249, 573
Charles, P.A., Kidger, M.R., Pavenko, E.P., Prokofieva, V.V. & Callanan, P.J. 1991, MNRAS, 249, 567
Chevalier, C. & Iloviasky, S. 1993, A&A, 269, 301
Eggleton, P.P. 1983, ApJ, 268, 368
Frank, J., King, A.R., & Raine, D.J. 1992, Accretion Power in Astrophysics, 2nd Edition (Cambridge: Cambridge Univ. Press)
Hirose, M. & Osaki, Y. 1990, PASJ, 42, 135
Kato, T., Mineshige, S. & Hirata, R. 1993a, IAUC 5676
Kato, T., Mineshige, S. & Hirata, R. 1993b, IAUC 5704
King, A.R. 1988, QJRAS 29, 1
King, A.R. 1993, MNRAS, 260 L5
Lubow, S.H. 1991a, ApJ, 381, 359
Lubow, S.H. 1991b, ApJ, 381, 368
Lubow, S.H. 1992, ApJ, 398, 525
Lubow, S.H. 1993, to appear in "Theory of Accretion Disks 2", eds F. Meyer, W.Duschl, J. Frank & E. Meyer–Hofmeister, NATO ASI, Garching, 1993
Molnar, L. & Kobulnicky, H.A. 1992, ApJ, 392, 678
Patterson, J. & Richman, H. 1993, preprint
Remillard, R.A., McClintock, J.E. & Bailyn, C.D. 1992, ApJ, 399 L145
White, N.E. 1989, A&A Rev., 1
Whitehurst, R. 1988, MNRAS, 232, 35
Whitehurst, R. & King, A.R. 1991, MNRAS, 249, 25

FRACTAL ANALYSIS OF X-RAY EMISSION FROM CENTAURUS X-3 OBSERVED WITH GINGA

Reiko Kanetake and Mine Takeuti
Astronomical Institute, Tohoku University, Sendai 980 Japan.
reiko@astroa.astr.tohoku.ac.jp

ABSTRACT

X-ray time series data from the 4.8-second binary pulsar Cen X-3 observed with Ginga are analyzed. We study the fractal dimension of X-ray intensity of the total 1.7 ~ 37 keV energy band. The fractal dimensions of the temporal behavior on time scales ≥ 4.8 seconds are comparatively large. It implies that the source varies with small size perturbations compared with the scale of the system. The effect of noise on the fractal dimension is estimated bliefly. It is mentiond that there exists a low dimensional motion with thw time scale of 500s.

1. INTRODUCTION

The Galactic X-ray source Cen X-3 has been studied in various ways, since it was discovered in 1971 by *Uhuru* (Schreier et al. 1972).

The irregularity of X-ray radiation from these objects has been a suitable for nonlinear study. It is well known that the complexity of time-dependent variation can be expressed by the fractal dimension based on the nonlinear consideration. The study of the fractal dimension is useful to investigate the physical process of the system, because the dimension of temporal behavior of a system is restricted by the degrees of freedom of the system. The obtained dimension must be lower than the necessary number of variables of the system.

We examine whether or not the variability of X-ray pulse amplitude which depends on flowing-in accreting matter can be reconstructed as an attractor in a system with a few degree of freedom. If not, the variability is likely to be caused by the superposition of local variations in and near the X-ray emission region.

2. OBSERVATION

Cen X-3 was observed with Ginga from 22nd to 24th of March, 1989, using the Large Area Counters (LAC; Turner et al. 1989) which has 4,000 cm^2 effective area. The time resolution of the data used here were 0.25 s and 0.5 s. Further details of the observation are described in Takeshima et al. (1991), and Nagase et al. (1992). Takeshima et al. (1991) show that Cen X-3 was in a high state during the present observation.

3. ANALYSES AND RESULTS

The Fourier analyses show bumps at periods of several hundreds seconds, which we discuss later. We sum up the data of both 0.25 s and 0.5 s time resolution into 1-s bins. Hereafter we analyze the summed counts in the total energy band using the 1-s binning.

To reduce the effects of long-time variation, we determine both a straight line and a parabolic curve for data using the least square method, and choose one of them comparing the probable errors. The data hereafter used are corrected by subtracting the parabolic curve from the raw data.

We estimated the apparent pulse period of this object by using the folding method. The derived apparent period is 4.8167 s. Any other definite period has not been found in the present analysis.

Since we have found the pulse period, we next study the cycle-to-cycle deviation of original data from the mean pulse profile. We folded the whole data to obtain an averaged pulse profile by using the obtained period, 4.8167 s. Using this profile as a pulse template, we then calculate the cycle-to-cycle deviation of the data from the template. The deviation thus fluctuates by 500 counts/s, which is much larger than the Poisson 1σ fluctuation of 60~100 counts/s in the raw data. To examine the intrinsic variation of X-ray, we calculate again the Poisson 1σ fluctuation the data after the arrangement, and find that the fluctuation decreases the data before the modification.

The behavior of X-ray light curves in the phase space determines the property of strange attractor. Generally the correlation dimension is used to estimate the fractal dimension. To calculate the fractal dimension numerically, an enormous amount of memory is needed in the computer. Grassberger and Proccacia (1983) suggested a more convenient method. Their method was applied by Voges et al. (1987) to the X-ray light curves of the X-ray pulsar Her X-1 (See Voges et al. (1987)).

First we examine the data, which are corrected for the long period variation and from which the pulse template has been subtracted. Since we obtain the result that the correlation dimension (= fractal dimension) D_2 satulate only high dimension, it is not the white noise in spite of its appearance. To eliminate the short time-range fluctuation, we make a running average of the deviation over 5 bins that covers nearly one pulse period. This procedure is equivalent to using a kind of low-pass-filter.

The correlation dimension D_2 is 6.9, from our 5-bin running averages. This means that the variation will be expressed by a system defined by 7 independent variables.

4. DISCUSSION AND CONCLUSION

We calculated the correlational integral in 1.7-37 KeV band, and obtained that the correlational dimension is 6.9 which is compartively large if we suggest that this variations behave the low dimensional chaos. We suppose that the X-ray variation of Cen X-3 has a high dimension attractor.

We increased the span of running average, and have found that the data show a slow wavy change. The 100-second average, which is far longer than the 4.8-second period, shows two peaks with a 500-second time interval. The correlation dimension D_2 is 3.6. If there exists a process in the system with a time scale of about 500s, and if such a process affects the heights of individual pulse, the process would be described as a low dimensional system. On the other hand, the phenomenon occurring on the time scales of several tens seconds may have approximately 7-dimension attractor.

We have discussed interesting matters on the X-ray emission from Cen X-3. Even though the meaning of above discussion looks clear, the effect of noise and the dimension of phenomena would make a complex problem. To make clear that the effect of the noise additional to the harmonic oscillator, we analyzed the correlation dimension D_2 of several data sequences consisted of sine curve and noise.

Even though it is obvious that we have to judge the fractal dimension carefully, we may say that such a high fractal dimension obtained in the present paper shows, at least, that the data sequence of Cen X-3 is not a low dimensional phenomenon.

We performed the fractal analysis of the X-ray variability of Cen X-3 in the 1.7 $\sim$ 37 KeV energy band, after subtracting a mean pulse template from the raw data. The analysis did not give low-dimensional feature such as would be seen in 3- or 4-dimensional systems though the number of data seems sufficient to confirm such a low dimensional feature. This does not fit the idea that the variability of X-rays is caused by low-dimensional chaotic oscillations of the accretion disc. Even though the number of data we used seems not sufficient, it is likely that the dimension of the system controlling the variability of Cen X-3 is higher than 6 or 7. The oscillatory motion produced by a global hydrodynamical process is usually described in 2 or 3 dimensional systems (for example, see Kovacs and Buchler 1988 and Aikawa 1988). Therefore, such a comparatively high dimension of the system will not be a highly modified global oscillatory process. Then we have an implication that the observed irregularity will be produced by the superposition of local variability.

ACKNOWLEDGEMENTS

We wish to express our thanks to anonymous referees for their kind comments, and to Professors Y. Sawada and H. Mori for their discussions. A part of computations were carried out at the Computer Center of Tôhoku University.

REFERENCES

Aikawa, T. 1988, APS 149, 149
Grassberger, P. and Procaccia, I. 1983, Phys. Rev. Lett., 50, 346
Kanetake, R. and Takeuti, M. 1993 PASJ to be submitted
Kovacs, G. and Buchler, J.-R. 1988, ApJ, **334**, 971
Kolláth, Z. 1990, MNRAS, 247 377
Kolláth, Z. 1991, IAU General Assembly poster session
Nagase, F., Corbet, R. H. D., Day, C. S. R., Inoue, H., Takeshima, T., Yoshida, K., Mihara, T. 1992, ApJ, **396**, 147
Norris, J. P., and Matilsky, T. A. 1989, ApJ, **346** 912
Takeshima, T., Dotani, T., Mitsuda, K., and Nagase, F. 1991, PASJ. 43 L43
Turner, M. J. L., Thomas, H. D., Patchett, B. E., Readings, D. H., Makishima, K., Ohashi, T., Dotani, T., Hayashida, K., Inoue, H., Kondo, H., Koyama, K., Mitsuda, K., Ogawara, Y., Takano, S., Awaki, H., Tawara, Y., Nakamura, N. 1989, PASJ 41 345
Voges, W., Atmanspacher, H., and Scheingraber, H. 1987, ApJ, 320 794

PRECESSION AND LONG-TERM CYCLIC BEHAVIOR

Alan P. Smale
USRA Research Scientist,
Laboratory for High Energy Astrophysics, Code 668,
NASA/Goddard Space Flight Center, Greenbelt, MD 20771.
alan@osiris.gsfc.nasa.gov

ABSTRACT

In this paper I review the observational evidence for long-term super-orbital periodicities and quasi-periodic phenomena in X-ray binaries with time scales of ~30 days to ~10 years, and discuss their physical interpretation. Super-orbital periods may be directly measured from long-term X-ray monitoring, inferred by spectroscopic or "circumstantial" evidence, or (perhaps) deduced from changes in the binary orbital period. Possible causes of these periods are: precession processes within the binary; irradiation of the companion by the compact object and/or stellar activity in the companion star; or the presence in the system of a third body, making the source a hierarchical triple. Each of these processes has its own implications for binary evolution.

1. LONG TERM CYCLES IN MASSIVE BINARIES

Long-term cycles in high-mass and low-mass X-ray binaries were first discovered through the extended data sets provided by the All Sky Monitors flown on *Vela 5B* and *Ariel V*. Firm evidence is available for the existence of long-term cyclic behavior in the five massive systems listed in Table 1.

Source	*Compact Object*	*Long-term Cycle*	*Orbital period*	*Reference*
LMC X–4	NS	30.4 dy	1.4 dy	Lang et al 1981
Her X–1	NS	35 dy	1.7 dy	Giacconi et al 1973
SS433	BHC	164 dy	13.1 dy	Watson et al 1986
LMC X–3	BHC	198 dy	1.7 dy	Cowley et al 1991
Cyg X–1	BHC	294 dy	5.6 dy	Priedhorsky et al 1983

Table 1: Known long-term periods in MXRBs containing neutron star and suspected black hole compact objects.

Models proposed to attempt to explain these cycles have included the precession of a tilted accretion disk, the slaving of this disk to precession of the companion star, neutron star precession, mass transfer feedback, and triple systems (Priedhorsky & Holt 1987). The generally-accepted empirical model which has stood the test of time consists of a tilted and twisted accretion disk in retrograde precession. The observed variations would then arise from the changing aspect of the disk, and its obscuration both of its own inner regions and the central source. This basic physical scheme is supported by SS433, the one system for which precession is *directly* observed from the motion of the radio jets, the optical brightness variations, and the X-ray occultation of the jets by a tilting disk.

There is as yet no detailed physical model describing how such a tilted, twisted disk could be created, but possible twisting mechanisms include tidal, magnetic, or radiation pressure torques in the system. Most observed phenomena can be explained directly by such structure at the outer edge of the accretion disk, although the pulse profile variations in Her X–1 also require azimuthal structure at the inner disk edge, rotating synchronously with the neutron star (Petterson *et al.* 1991).

The stability of the precession clock has been measured at ~2% for Her X–1, with maximum deviations in *either* direction of 6 dy. The frequency characteristics of these deviations suggest that a white-noise or random-walk process is responsible, supporting the accretion disk precession model for the long-term variations (Baykal *et al.* 1983a). The clock noise in SS433 can be described using an almost identical formalism (Baykal *et al.* 1983b), another indication that the same phenomenon is responsible in both sources.

Some insight into the mechanics of accretion disk precession is provided by the numerical modeling performed by Iping & Petterson (1990). Among their conclusions, they found: (i) that tidal interactions between the annuli that made up their accretion disk model and the companion star led to the creation of a precessing, inclined disk; (ii) that radiation reaction forces could make the annuli precess in either direction; (iii) that differential precession between annuli twisted the disk; (iv) that tilted disks developed warped edges due to the influence of the mass transfer stream. Their models also showed that 10% variations in disk precession period could be made to occur rather easily. This last conclusion suggests that conventional period-searching algorithms may not be well suited to the task of searching for the quasi-periodic disk precession periods.

2. LONG TERM CYCLES IN LOW-MASS BINARIES

The original papers by Priedhorsky and Terrell, in which several long-term periodicities on LMXBs were reported based on analysis of *Vela 5B* data, are summarized in Table 2. The analysis for these papers was performed using 10-dy and 30-dy averages. Since then, the *Vela 5B* database has been reorganized to make the full 1s timing resolution readily available (Whitlock 1989). Using this database, Smale & Lochner (1992) performed a complete and consistent study of all the LMXBs detected by *Vela 5B* for which light curves could be extracted without source confusion. Their work confirmed the existence of the 175-dy period in X1820−303, with an extremely low false alarm probability (FAP), but showed that in the case of the 199-dy period in X1916−053 the situation was not so clear cut. The period was not significantly detected in the complete data set, but appeared when the data were filtered to remove the most variable intervals. Smale & Lochner concluded that the false alarm probability of the long-term period in X1916−053 was ~10−20%.

No other strong periodicities were discovered, and thus it is clear that pronounced cyclic variability is a rare occurrence in LMXBs.

Source	Long-term Cycle	Orbital period	Reference
X1820−303	175 dy	685 sec	Priedhorsky & Terrell 1984a
X1916−053	199 dy	50.1 min	Priedhorsky & Terrell 1984b
X1705−44	222 dy	?	Priedhorsky 1986
Aql X−1	1.19 yr	19.0 hr	Priedhorsky & Terrell 1984b
X1630−47	1.66 yr	?	Priedhorsky 1986

Table 2: Known long-term periodicities in LMXBs. In the first group the periods are detected in the long term emission; in the second, they are derived from preferred outburst times. The latter systems are not considered further in this review.

For the low-mass systems, the ratio P_{long}/P_{orb} is in the range 10^3–10^4 (cf. P_{long}/P_{orb} ~10–50 for MXRBs), so precession of the type seen in the massive systems is unlikely. The empirical model adopted for many years identified intrinsic variations in mass transfer as the probable cause of the cycles (Priedhorsky & Holt 1987), with no clear idea about why such variations should be periodic. The exchange of angular momentum between the companion and the disk was ruled out as a possible mechanism by Priedhorsky & Verbunt (1988), and disk instabilities of the type observed in dwarf novae are not viable as an explanation, as the higher temperatures of disks in LMXBs precludes the existence of the required thermal instabilities.

One promising idea is that the periodicities may be caused by mass transfer events analogous to the superoutbursts observed in SU UMa-type cataclysmic variables (Priedhorsky 1986). The bright-spot luminosity in the SU UMas is generally assumed to be caused by enhanced mass transfer from the companion, at comparable rates to those observed in LMXBs, and the typical 130–400 dy recurrence rates for superoutbursts are again similar to the time scales for long-term cycles and outburst cycles in the binaries. We will return to the possible LMXB–SU UMa connection in the next section.

3. X1916−053: A HIERARCHICAL TRIPLE SYSTEM?

Instead of detecting the effects of mass transfer variations, we may be seeing evidence for a different kind of accretion-driven system. Both of the sources which show ~200 dy periodicities also have very short binary periods, implying that the mass-donating companion star is either degenerate itself or at least hydrogen deficient and heading for degeneracy (a fate which may be shared by many of us who work in the field). There are other reasons why these sources may differ from the LMXB norm; X1916−053 displays an interesting timing phenomenon, discussed in detail below, and X1820−303 is located in a globular cluster (see 5.3).

For X1916−053, there is a significant discrepancy between the values of the orbital period measured at X-ray and optical wavelengths. The most accurate X-ray determinations come from the longer duration observations by OSO-8 and Ginga, yielding typical values of 50.06±0.03 min (White & Swank 1982),

50.08±0.11 min (Smale et al. 1989) and 50.00±0.08 min (Smale et al. 1992). The light curves from this source show a range of X-ray dip behaviors ranging from total absence, through a single, well-defined dip lasting 10% of the orbital period, to two broad, deep and very irregularly-shaped dips per orbital cycle (Smale et al. 1988). This variability in the dips precludes a more accurate estimation of the measured period. Variability is also seen in the optical light curves from the system (of which more later), but the more extensive data sets allow an extremely accurate estimate of the optical period of 50.4590±0.0030 min (Grindlay et al. 1988, Grindlay 1991). Two different physical models have been proposed to attempt to resolve this discrepancy.

3.1. Model 1: X1916−053 as a triple system

In this model (Grindlay et al. 1988, Grindlay 1989, 1991), the optical period is the true orbital period of the system and is caused by the partial eclipse of the accretion disk rim by a low-mass companion star. The X-ray dip periodicity is caused by clumps of matter near the outer disk, which intersect our line of sight, scattering and absorbing X-rays from the central source. The difference in periods is due to the presence in the system of a tertiary star in a retrograde 2.5-dy orbit around the central binary source; the 2.5-dy period represents 2/3 of the beat period between the X-ray and optical periods.

The system is then a hierarchical triple. The retrograde orbit implies that the tertiary star must have been captured through tidal interactions rather than having been formed primordially. The mass of the third body is probably <1 $M_\odot$, but is not well constrained (Bailyn 1987). The final twist is that the tertiary induces an eccentricity and precession in the inner X-ray binary, causing a long-term modulation of the mass transfer at a period P_{long}, where $P_{long} = KP_3^2/P_2$. With P_{long}=199 dy, P_3=2.5 dy and P_2=50 min, K is approximately unity, as required theoretically.

A principal argument for Model 1 is that the optical period is very stable. However, it is not proven that the X-ray period is unstable, merely that it appears to be irregular due to the structure at the edge of the accretion disk. Also, the stability of a period is not in itself a good reason for assuming that it is orbital in nature. Some CVs have non-orbital periodicities that are stable for many years; TV Col, for example, has a photometric period of P_{phot}=5.2 hr which differs from the spectroscopic period, P_{spec}=5.5 hr.

Against the model is the argument that it requires X1916−053 to have a very different geometry from other LMXBs such as X1822−371, X1658−298, and X0748−676, and a different root cause for its X-ray dipping behavior. In the latter sources it is known that the material causing the X-ray dips is fixed in the rest frame of the binary, which is defined clearly by the deep eclipses by the companion star, and is almost certainly located at the outer edge of the disk at the impact point of the accretion stream.

3.2. Model 2: a precessing disk in X1916−053.

The second model (Smale et al. 1992) has the X-ray period as the orbital period; the primary dip in the orbital cycle defines the period clock and is fixed in the rest frame of the system, as it is in other X-ray binaries for which the absolute phasing within the system can be determined. The principal defining feature of this model is that the accretion disk in the system is elliptical.

This interpretation is analogous to that proposed to explain the superhump phenomenon in SU UMa-type cataclysmic variables (Whitehurst 1988); during superoutbursts in SU UMas the intensity of the source plateaus, and regular superhumps are observed with a periodicity a few percent longer than the orbital period. The elliptical disk rotates in the rest frame of the system with a period P_p (of order days), and the observed superhump period P_s is the beat period between P_p and the orbital period P_{orb}. This behavior is predicted whenever the mass ratio is extreme (> 4:1) and the accretion rate is high, regardless of the nature of the compact object.

Model 2 thus proposes that this same mechanism is responsible for the longer optical period observed in X1916−053, with some additional details. The height of the azimuthal structure at the disk edge is determined by the balance between the vertical component of gravity and the gas pressure (Ko & Kallman 1991). When the semi-major axis of the system is aligned with the line of separation of the binary components, the disk edge will be at its closest to the companion star and the vertical component of gravity will be at a maximum. Thus, at this phase the height of the accretion disk structure will be at a minimum. The height of the accretion disk structure will then vary on P_p (as seen from within the binary rest frame). Both the depth and breadth of the X-ray dips and the amount of optical reprocessing in the accretion disk as seen by the distant observer will therefore be modulated on the beat between P_p and P_{orb}, which is exactly the observed behavior. In this model there is no third body and no linkage to the proposed 199-dy period.

3.3 Discussion

Both models predict systematic variations in the dipping behavior on a time scale of days, and this is in fact observed; the X-ray dip morphology was seen to vary over four days of *Ginga* observations (Smale *et al.* 1989), while the optical folded light curve is seen to increase dramatically in depth in two light curves separated by an interval of two days (Grindlay 1992).

A tricky feature of Model 1 is that the X-ray dip phase clock does not run free, but resets twice per outer orbit. Thus, X-ray dips are constrained to occur within a ±0.25 optical phase band. Obviously, the key to the puzzle is to observe X-ray and optical variations simultaneously and measure the offset between the dips in the two wavebands. This has been attempted once; a small number of simultaneous *Ginga*/optical dips showed that at that time the optical minima occurred 0.0-0.2 in phase after the X-ray minima. This is the correct phasing for Model 2, but does not rule out Model 1, and so a more extensive data set is required to choose between the models with any degree of confidence.

4. CIRCUMSTANTIAL EVIDENCE FOR TRIPLE SYSTEMS

In addition to the possible dynamic evidence from X1916−053, there may be circumstantial evidence from optical spectroscopy to support the existence of triple systems. In its "off" state, the 5.24-hr binary system X2129+47 should show radial velocity variations of ~ 300 km s^{-1} from the motion of the secondary, and a late K-M star spectrum. Instead, it shows an F7-8 spectrum and no orbital variations, although long-term variations of 40 km s^{-1} are observed over an interval of 15 months. The possibility of a positional coincidence is estimated at 10^{-3}. It has been suggested that the F star is the tertiary in a ~ 30 day orbit around the inner binary system (Thorstensen *et al.* 1988, Chevalier *et al.* 1988, Garcia *et al.* 1989). One appeal of this model is that the third body would induce a 45-year period in the eccentricity of the inner binary, corresponding to the separation between the observed extended "off" states.

X2129+47 is currently the best (i.e. least controversial) candidate for a triple system, and the only one in which radial velocity variations from the third body may have been detected. Two other sources display positional coincidences with stars of unexpected spectral types, incompatible with the expected companion masses for the X-ray binary. The radio position of the bright persistent X-ray binary GX17+2 lies within 0.5" of a 17th magnitude G star, and in addition there are unconfirmed periodicities of 5000s, 19.4 hr, and 6.5 dy in the system (Bailyn & Grindlay 1987). 4U1543−61 is coincident (within 0.1") with a A2V star with the correct distance and reddening (Chevalier 1989). The suggestion that X2023+338 (= V404 Cyg) might be a triple system with a G-K star tertiary (Casares *et al.* 1992) is now considered unlikely due to the discovery of ellipsoidal variations in this star (Wagner *et al.* 1992). In the future, radial velocity studies will help to show whether the proposed third members are indeed gravitationally bound to the X-ray binaries in X2129+47, GX17+2 and X1543−61.

5. ORBITAL PERIOD CHANGES

5.1 Massive systems

In X-ray binaries, eclipses provide the most reliable fiducial marker for determining period changes. It is thus no accident that most of the sources for which period change determinations have been made are eclipsing systems. Cen X−3 was the first massive X-ray binary for which eclipse time delays were used to derive an orbital period change, yielding a period decrease of $-(1.78\pm0.08)\times10^{-6}$ yr^{-1} (Kelley *et al.* 1983), and this value has been refined by recent *Ginga* observations (Nagase *et al.* 1992; see Table 3). The orbital period of SMC X−1 is also decreasing, on a time scale of 3×10^5 yr, a rate of change approximately twice that of Cen X−3 (see Nagase 1992).

Source	Orbital period	$\dot{P}/P$	
Cen X−3	2.1 dy	$-(1.738\pm0.004)$	$\times10^{-6}$ yr^{-1}
SMC X−1	3.89 dy	$-(3.36\pm0.02)$	$\times10^{-6}$
Cyg X−3	4.8 hr	$+(1.6\pm0.1)$	$\times10^{-6}$
Her X−1	40.8 hr	$-(1.32\pm0.16)$	$\times10^{-6}$
X1820−303	0.18 hr	$-(0.88\pm0.16)$	$\times10^{-7}$
X0748−676	3.82 hr	$-(0.82\pm0.02)$	$\times10^{-7}$
X1822−371	5.6 hr	$+(3.2\pm0.7)$	$\times10^{-7}$

Table 3: Summary of significant orbital period changes in massive, intermediate, and low-mass X-ray binaries.

In addition to these firm results there are a couple of possible period detections: X0115+634 shows a possible period decrease of $-(1.64\pm0.55)\times10^{-6}$

yr^{-1} (see Nagase 1992), and LMC X–4 shows very marginal evidence for an increase of $+(1.1\pm0.8)\times10^{-6}$ yr^{-1} (Levine et al. 1991).

For these systems it seems clear that the mechanism for the period changes is the tidal torque between the distorted supergiant companion and the neutron star (Kelley et al. 1983). In the course of the tidal evolution, angular momentum may be exchanged between the rotation of the companion and the orbit. It is worth noting that in MXRBs the rotation of the massive companion provides ~0.5 of the total angular momentum of the system, and so is a good source for angular momentum redistribution.

5.2 Intermediate-mass systems

Orbital period changes have been measured for two sources that might be considered intermediate between the massive and low-mass systems.

Cyg X–3 consists of a neutron star and probable Wolf-Rayet companion (van Kerkwijk et al. 1992), and has a measured $\dot{P}/P$ of $+(1.6\pm0.1)\times10^{-6}$ yr^{-1}, along with a probable cubic term of $-(1.6\pm0.4)\times10^{-10}$ yr^{-1} (van der Klis & Bonnet-Bidaud 1989). This effect may be due to a complex change in the mean mass loss rate from the system, but an alternative and perhaps more believable scenario is that the ephemeris is in fact sinusoidal. The latter is predicted by the apsidal motion theory and, if true, suggests an apsidal motion period of $P_{aps}=27^{+4}_{-3}$ yr and an orbital eccentricity of $e=0.25\pm0.01$.

The orbital period of Her X–1 is decreasing at a rate substantially larger than would be predicted from the mass transfer rate as derived from the X-ray luminosity (Deeter et al. 1991). X-ray heating of the companion, driving a magnetically-channeled stellar wind, may provide the angular momentum loss necessary to explain the period decrease.

5.3 Low-mass systems

A clear picture of the cause(s) of the period changes observed in LMXBs is hampered by the intrinsic differences between the systems displaying such changes and the differences in the sign of the period changes measured (Table 3). There is some evidence that these period changes may be cyclic. However, the proposed cycles are ~10 yr, and we will need to extend the observational baselines further before such evidence can become convincing.

X1820−303 has an extremely short orbital period, and thus its companion is almost certainly degenerate, probably a ~0.07 $M_\odot$ white dwarf. In evolutionary terms, one would expect conservative mass transfer through Roche-lobe overflow in this system, with the companion expanding as it loses mass to produce a period *increase* of $>+0.9\times10^{-7}$ yr^{-1}. What is actually observed is a secular *decrease* in the period of almost equal size (see Table 3, and van der Klis et al. 1993). (The sinusoidal variation in the orbital period proposed by Tan et al. (1991) was found to be not statistically required by the analysis of van der Klis et al.)

X1820−303 is located in the globular cluster NGC 6624, and favored models for the system make use of this; it has been suggested that the period change is apparent rather than actual, and is caused by the acceleration of the binary by the cluster potential. Another model has the binary accelerated by a third star in a hierarchical triple, a perhaps more likely scenario in a region with a high space density of stars. Thirdly, van der Klis et al. (1993) proposed that perhaps the companion in the system is a non-degenerate helium-burning star resulting from a stellar merger following a close encounter between two lower-mass stars, creating a $2.4M_\odot$ progenitor to the current $\sim 0.24 M_\odot$ companion.

For X1822−371 and X0748−676 the companions are expected to be normal (non-degenerate), and eclipses again form the fiducial markers for period timing. In this case, simple evolutionary theory suggests that the companion should fill its Roche lobe. Mass transfer would be driven by the loss of orbital angular momentum due to gravitational radiation, perhaps with a contribution from magnetic braking if the companion has a sufficiently large magnetic field. Both mechanisms suggest that standard LMXB periods should *decrease* on time scales of $\sim 10^9$ yr.

In fact, the periods of both systems are *increasing* on a time scale two orders of magnitude shorter than this (X1822: Hellier *et al.* 1990, Hellier & Smale, these proceedings. X0748: Asai *et al.* 1992). In addition it is becoming increasingly likely that the orbital period of X0748−676 is changing sinusoidally, with a period of 6.8 ± 0.2 yr and an amplitude of 10.0 ± 0.6 lt-sec (Asai *et al.* 1992; Asai and the ASCA PV Team, private communication). Asai *et al.* suggest that X0748−676 may be part of a hierarchical triple system, and that the Doppler motion induced by the third body modulates the eclipse timings. In this case the third body would need to have a very low mass of $\sim 0.01 M_\odot$ and the outer orbit would be inclined by $20°$ with respect to the plane of the inner binary.

Although it can be seductive to add a third body to attempt to explain various kinematic aspects of certain X-ray binaries, it seems unlikely that this is the explanation in all cases. There still remains the difficulty of how such triples are produced. Most models for tidal capture do not predict a high success rate for the process, so it would perhaps be surprising if a large number of X-ray binaries were found to have tertiary components.

If, over the long term, period changes in several LMXBs are confirmed to be cyclical, this may be due to solar-type cycles in the companion. On a time scale of years the radius of the companion may change due to the intrinsic nature of the star itself, rather than its location in an X-ray binary, and these changes would affect both the mass transfer rate and the angular momentum balance of the system, driving corresponding oscillations in the measured orbital period of the binary.

Another promising explanation concerns the X-ray heating of the companion by the flux from the compact object. There is a growing realization that the irradiation of the companion star cannot be neglected in models of X-ray binary evolution. Models in which irradiation does not play a part assume that the companion always fills its Roche lobe, which may not be the case; a large radiative flux from the compact object can drive a strong evaporative wind from the outer atmosphere of the companion, and high mass transfer rates which may be self-sustaining (Podsiadlowski 1991, Tavani & London 1991). Mass in excess of the Eddington limit may be lost from the system.

The sign of the period change is then determined by the fraction of the mass transfer accreted onto the neutron star. Evaluating this parameter physically for individual systems is difficult due to uncertainties in the system geometry, hydrodynamics and energy exchange, but as a rule of thumb if this fraction is small (<0.1) the period will decrease, while if it is large (>0.5) the period will increase (Tavani 1991).

Workers are now starting to take the effects of X-ray heating into account when building models for the evolution of LMXBs, with encouraging results. For sources containing companion stars of mass ~ 0.2–$0.3 M_\odot$ and periods similar to those of Cyg X–3 and X1822−371, X-ray heating can reproduce the observed large positive values of $\dot{P}/P$ (Harpaz & Rappaport 1991; Rappaport, these proceedings). Likewise, the observed negative values can be sustained assuming a large and steady mass loss from the system (Tavani 1991).

6. CONCLUSIONS

• Long-term super-orbital periodicities have been unambiguously detected in 5 MXRBs and 3 LMXBs.
— In the massive systems these periodicities are probably due to the precession of a tilted, twisted accretion disk.
— In the low-mass systems, mass transfer instabilities provide the traditional explanation for such periods.

• In X1916−053, which has a probable 199-dy period, a 1% difference is observed between the "orbital" optical and X-ray periods. This may indicate that X1916−053 is a hierarchical triple system, or that the accretion disk in the system is elliptical.

• For 3 other LMXBs there is some spectroscopic and/or dynamical evidence for a third body in the systems. None is unambiguously confirmed.

• Orbital period changes have been measured for a total of 7 X-ray binaries.
— In the massive systems tidal torques are probably responsible.
— For LMXBs a variety of possible mechanisms have been proposed. Variations in the radius of the secondary and the presence of evaporative winds due to X-ray heating seem the most viable processes in most cases. Some other options include cyclic variations due to third bodies or magnetic, solar-type stellar activity in the companion.

7. THE POTENTIAL OF FUTURE MISSIONS

X-ray astronomy is a young field and the baselines for observing long-term cyclic behavior are still tantalizingly short. Further advances in our understanding can be expected in the near future: the *XTE* ASM will provide high-quality long-term light curves for X-ray binaries with a sensitivity 30 times better than that of *Vela 5B*, and should provide more and better information about long-term periodicities; *ASCA* and *XTE* will extend the baseline for studies of orbital period changes, maybe allowing us to choose between secular and cyclic models; and the *USA* experiment on the *ARGOS* satellite (Wood et al., these proceedings) will provide long (∼1 month) looks at selected binaries and give valuable information about super-orbital variability. Five years from now, things may look very different in the study of precession and long-term cyclic behavior in X-ray binaries.

REFERENCES

Asai, K. et al. 1992, PASJ, 44, 633
Bailyn, C. D. 1987, ApJ, 317, 737
Bailyn, C. D. & Grindlay, J. E. 1987, ApJ, 312, 748
Baykal, A., Boynton, P. E., Deeter, J. E., & Scott, M. 1993a, MNRAS, in press
Baykal, A., Anderson, S. F., & Margon, B. 1993b, ApJ, in press
Casares, J., Charles, P. A., & Naylor, T. 1992, Nature, 355, 614
Chevalier, C. et al. 1988, A&A, 217, 108

Chevalier, C. 1989, in Proceedings of the 23rd ESLAB Symposium, eds. J. Hunt & B. Battrick, ESA SP-296, p. 341
Cowley, A. P. et al. 1991, ApJ, 381, 526
Deeter, J. E. et al. 1991, ApJ, 383, 324
Garcia, M. R., Bailyn, C. D., Grindlay, J. E., & Molnar, L. A. 1989, ApJ, 341, L75
Giacconi R. et al. 1973, ApJ, 184, 227
Grindlay, J. E. et al. 1988, ApJ, 334, L25
Grindlay, J. E. 1989, in Proceedings of the 23rd ESLAB Symposium, eds. J. Hunt & B. Battrick, ESA SP-296, p. 121
Grindlay, J. E. 1991, in Frontiers of X-ray Astronomy, eds. Y. Tanaka and K. Koyama, Universal Academy Press, Tokyo, Japan, p. 69
Harpaz, A. & Rappaport, S. 1991, 383, 739
Hellier, C., Mason, K. O., Smale, A. P., & Kilkenny, D. 1990, MNRAS, 244, 39P
Hellier, C. & Smale, A. P., these proceedings
Iping, R. C. & Petterson, J. A. 1990, A&A, 239, 221
Kelley, R. L. et al. 1983, ApJ, 268, 790
Ko, Y.-K. & Kallman, T. 1991, ApJ, 374, 721
Lang, F. L. et al. 1981, ApJ, 246, L21
Levine, A. et al. 1991, ApJ, 381, 101
Nagase, F. et al. 1992, ApJ, 396, 147
Nagase, F., 1992, in "Frontiers of X-ray Astronomy", eds. Y. Tanaka & K. Koyama, Universal Academy Press, Tokyo, Japan, p. 79
Petterson, J. A., Rothschild, R. E., & Gruber, D. E. 1991, ApJ, 378, 696
Podsiadlowski, Ph. 1991, Nature, 350, 136
Priedhorsky, W. C., Terrell, J., & Holt, S.S. 1983, ApJ, 270, 333
Priedhorsky, W. C. & Terrell, J. 1984a, ApJ, 284, L17
Priedhorsky, W. C. & Terrell, J. 1984b, ApJ, 280, 661
Priedhorsky, W. C. 1986, Ap&SS, 126, 89
Priedhorsky, W. C. & Holt, S. S. 1987, Spa. Sci. Rev., 45, 291
Priedhorsky, W. C. & Verbunt, F. 1988, ApJ, 333, 895
Smale, A. P. et al. 1992, ApJ, 400, 330
Smale, A. P. & Lochner, J. C. 1992, ApJ, 395, 582
Smale, A. P., Mason, K. O., White, N. E., & Gottwald, M. 1988, MNRAS, 232, 647
Smale, A. P., Mason, K. O., Williams, O. R., & Watson, M. G. 1989, PASJ, 41, 607
Tan, J. et al. 1991, ApJ, 374, 291
Tavani, M. 1991, Nature, 351, 39
Tavani, M. & London, R. 1993, ApJ, 410, 281
Thorstensen, J. R. et al. 1988, ApJ, 334, 430
van der Klis, M. et al. 1993, MNRAS, 260, 686
van der Klis, M. & Bonnet-Bidaud, J. M. 1989, A&A, 214, 203
van Kerkwijk, M. H. et al. 1992, Nature, 335, 703
Wagner, R. M., Kreidl, T. J., Howell, S. B., & Starrfield, S. G. 1992, ApJ, 401, L97
Watson, M. G. et al. 1986, MNRAS, 222, 261
White, N. E. & Swank, J. H. 1982, ApJ, 253, L31
Whitehurst, R. 1988, MNRAS, 232, 35
Whitlock, L. 1989, Ph.D. thesis, Univ. Florida

AN EPHEMERIS UPDATE FOR X 1822-371

Coel Hellier
Department of Astronomy, University of Texas, Austin, TX 78712

Alan P. Smale
USRA Research Scientist, Laboratory for High Energy Astrophysics,
NASA/Goddard Space Flight Center, Greenbelt, MD 20771

ABSTRACT

The increase in the orbital period of X 1822-371 with a timescale of 3×10^6 yrs is confirmed by a new *ROSAT* observation. In contrast to some other LMXBs, a simple quadratic ephemeris remains a good fit to the eclipse timings.

1. INTRODUCTION

Changes in the orbital periods of LMXBs are currently being monitored in Cyg X-3 (van der Klis & Bonnet-Bidaud 1989), X 0748-676 (Asai et al. 1992), X 1822-371 (Hellier et al. 1990), X 1820-30 (van der Klis et al. 1993) and Her X-1 (Deeter et al. 1991). So far we have seen increasing periods, decreasing periods and quasi-sinusoidal oscillations. This is inconsistent with a steady evolution to shorter periods, as required in conventional evolutionary models to keep the secondary in contact with its Roche lobe. Theorists (e.g. Podsiadlowski 1991, Tavani 1991) have pointed to the effect of X-ray irradiation on the secondary star and begun devising models in which radiatively driven evolution explains not only the observed $\dot{P}$s but also the different orbital period distributions of LMXBs and CVs, and the apparent discrepancy between the space densities of LMXBs and their touted descendants, the low-mass binary pulsars. An alternative (e.g. Warner 1988) is that magnetic cycles on the secondary could cause cyclical changes in the orbital periods. It is thus important to track the period changes in order to determine their long term significance and cause.

2. A *ROSAT* OBSERVATION OF X 1822-371

X 1822-371 is the proto-typical 'accretion disc corona' LMXB in which the X-rays we observe are scattered by a corona (White & Holt 1982). The secondary obscures part of the corona every 5.57-hr orbit, producing a partial eclipse, while a quasi-sinusoidal flux modulation results from a similar obscuration by disk structure. An analysis of observations made with *HEAO*-1, *Einstein*, *EXOSAT* and *Ginga* established that the orbital period of X 1822-371 was increasing on a timescale of 3×10^6 yrs (Hellier et al. 1990, Paper 1). We now update the ephemeris using a more recent *ROSAT* observation.

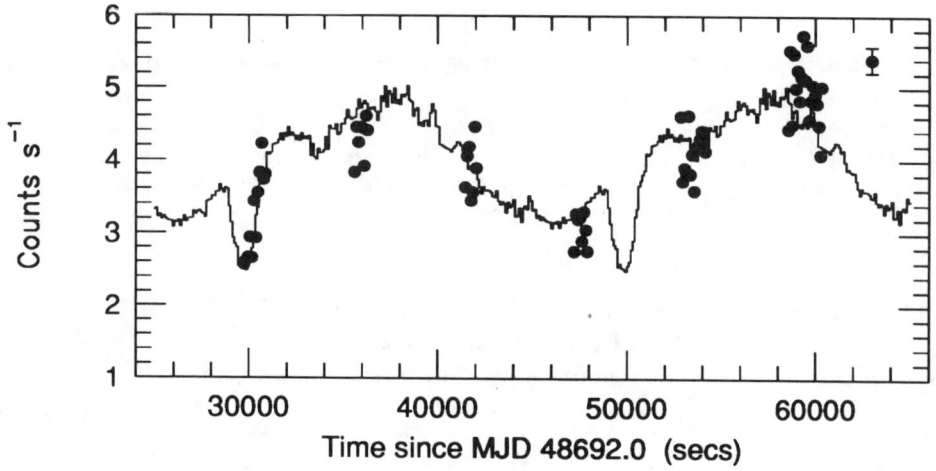

Fig. 1. The *ROSAT* data of X 1822–371 (dots) with the fitted *EXOSAT* template

X 1822–371 was observed with the *ROSAT* PSPC-B on 1992 March 11 08:10–16:45 UT, a total elapsed time of 30 Ksecs, or 1.5 binary orbits. The PSPC at the focus of the X-ray telescope provides a bandpass of 0.1–2.4 keV, over a 2 degree field of view. During the observation a slow spacecraft wobble was carried out, moving the pointing direction back and forth by several arcminutes to prevent shadowing of X-ray sources behind the wire grid in the PSPC detector window. X 1822–371 was placed off axis to increase the image size and so minimize the effect of the wobble. Appropriate exposure and vignetting corrections have been applied to the data where necessary. The data, binned at 100 secs, are shown as dots in Fig. 1.

3. THE TIMING AND EPHEMERIS

Timing analysis of the light curve is not straightforward because of the data gaps caused by the low-Earth orbit. In particular only part of one eclipse was seen. However, we can exploit our knowledge of the complete light curve by using the *EXOSAT* lightcurve template published in Paper 1. We found that the result of cross-correlating this template with the *ROSAT* data was too sensitive to the method employed for treating the data gaps. Instead, we performed a least-squares fit of the template to the data using 3 free parameters; a multiplicative amplitude, a mean offset, and the phase. The error in the phase was estimated by varying the weights assigned to the data points (for instance weighting the eclipse heavily relative to the other points) and recording the range of acceptable fits that resulted. The fitted template is shown in Fig. 1 (histogram).

Table 1. X-ray eclipse timings of X 1822–371.

JD$_\odot$	Uncertainty	Satellite
2443413.5272	0.0046	*HEAO*-1 Scan
2443591.5521	0.0046	*HEAO*-1 Scan
2443776.5459	0.0012	*HEAO*-1 Point
2443778.4065	0.0046	*HEAO*-1 Scan
2444133.5277	0.0030	*Einstein*
2445580.4932	0.0005	*EXOSAT*
2445615.30940	0.00038	*EXOSAT*
2445963.00914	0.00033	*EXOSAT*
2445963.24046	0.00030	*EXOSAT*
2445963.47254	0.00034	*EXOSAT*
2446191.63643	0.00031	*EXOSAT*
2446191.86768	0.00033	*EXOSAT*
2446192.10008	0.00029	*EXOSAT*
2447760.22900	0.00030	*Ginga*
2448692.84396	0.0007	*ROSAT*

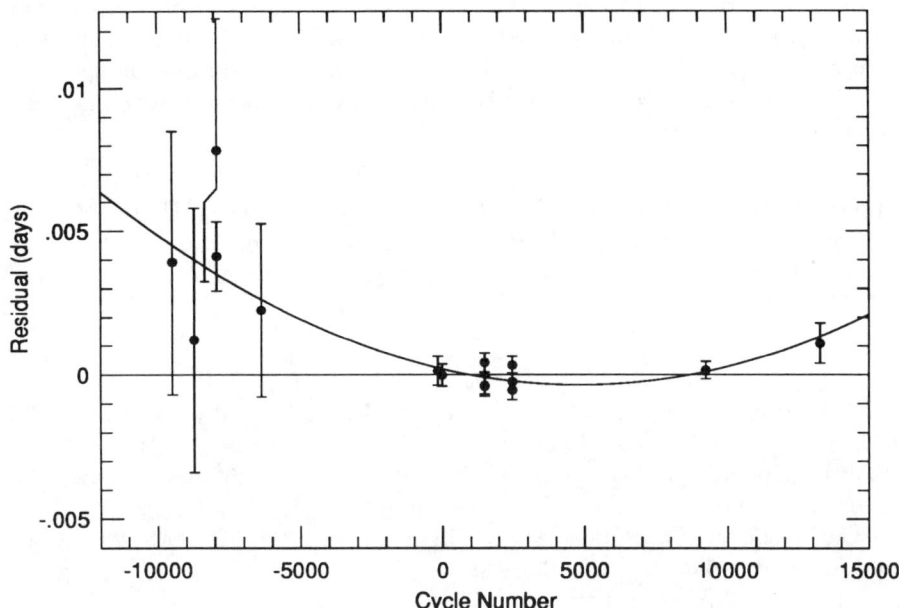

Fig. 2. Eclipse timings of X 1822–371 with the ephemerides of Table 2.

Table 2. Ephemerides of X 1822−371.

Linear ephemeris:		Quadratic ephemeris:	
$JD_\odot = T_0 + E\,P$.		$JD_\odot = T_0 + E\,P + E^2\,c$.	
T_0	2445615.30942(14)	T_0	2445615.30962(15)
P	0.232109017(33)	P	0.232108788(64)
Cov(T_0,P)	-3.1×10^{-12} d^2	c	$2.37 \pm 0.56 \times 10^{-11}$
$\chi_\nu^2 = 2.13$	$\nu = 13$.	Cov(T_0,P)	-5.7×10^{-12} d^2
		Cov(T_0,c)	2.7×10^{-16} d^2
		Cov(P,c)	-3.1×10^{-19} d^2
		$P/\dot{P}$	$3.1 \pm 0.7 \times 10^6$ yr
		$\chi_\nu^2 = 0.85$	$\nu = 12$.

We have recomputed the ephemeris using the eclipse timings of Paper 1 together with the *ROSAT* timing (Tables 1 & 2). The quadratic ephemeris of Paper 1 is still a good fit to the data, with only minor tweaking of the parameters required. The F-test implies that the quadratic term is now established with 99.9% confidence. So far, though, there is no need of a more complex representation such as that required for X 0748−676 (Asai et al. 1992). Note, however, that the eclipse timings of X 0748−676 are an order of magnitude more precise than those for X 1822−371, so we would not be sensitive to fluctuations of the ~ 10 sec amplitude seen in X 0748−676.

Support for this work was provided by NASA through grant HF-1034.01-92A awarded by the Space Telescope Science Institute which is operated by the Association of Universities for Research in Astronomy, Inc. for NASA under contract NAS5-26555.

REFERENCES

Asai, K., Dotani, T., Nagase, F., Corbet, R. H. D., Shaham, J., 1992, PASJ, 44, 633
Deeter, J. E., Boynton, P. E., Miyamoto, S., Kitamoto, S., Nagase, F., Kawai, N., 1991, ApJ, 383, 324
Hellier, C., Mason, K. O., Smale, A. P., Kilkenny, D., 1990, MNRAS, 244, 39p
Podsiadlowski, P., 1991, Nature, 350, 136
Tavani, M., 1991, Nature, 351, 39
van der Klis, M., Bonnet-Bidaud, J. M., 1989, A&A, 214, 203
van der Klis, M. et al. 1993, MNRAS, 260, 686
Warner, B., 1988, Nature, 336, 129
White, N. E., Holt, S. S., 1982, ApJ, 257, 318

GX 5–1 and GX 17+2 with EXOSAT: New insights in two Z-sources

E. Kuulkers[1], M. van der Klis[1], T. Oosterbroek[1], J. van Paradijs[1] & W. Lewin[2]

[1] Astronomical Institute "Anton Pannekoek", University of Amsterdam,
Kruislaan 403, 1098 SJ Amsterdam, The Netherlands,
[2] Massachusatts Institute of Technology, 37-627, Cambridge, MA 02139, U.S.A.

ABSTRACT

We found flaring branch behaviour in the Z-source GX 5–1. This source also shows secular variations in the position of the "Z" curve in its X-ray colour-colour and hardness-intensity diagram.

We report the detection of a new burst in GX 17+2. This burst and previous reported bursts with EXOSAT occurred when GX 17+2 was in the normal branch.

1. INTRODUCTION

GX 5–1 and GX 17+2 are two of the brightest low-mass X-ray binaries (LMXBs). Together with four other sources, Sco X-1, Cyg X-2, GX 340+0 and GX 349+2, they form a group of so-called "Z"-sources (Hasinger & Van der Klis 1989), after the approximate "Z" shape some of them describe in an X-ray colour-colour diagram (CD). The three branches of the "Z", corresponding to three distinct spectral states, are, from top to bottom, the *horizontal branch* (HB), *normal branch* (NB) and *flaring branch* (FB). The Z-sources do not jump from branch to branch, but move smoothly, however, irregularly along the "Z" on timescales of minutes to hours. The spectral states are closely connected to the temporal behaviour (Hasinger & Van der Klis 1989).

2. GX 5–1: HOW TO BECOME A Z-SOURCE

In the CD of GX 5–1 we see (for the first time in GX 5–1) the HB in three different positions (Fig. 1a). We also found secular variations in the position of the "Z" curve in the hardness-intensity diagram (HID). These secular variations are not due to instrumental systematic effects. We verified this by investigating the instrumental systematic effects with EXOSAT data of the Crab Nebula.

The secular variation of the "Z" curve in the CD and HID of GX 5–1 is likely within the uncertainties of the previous reported steady "Z" curve of GX 5–1 (e.g. van der Klis et al. 1991). Such secular variations are seen in a more dramatic form in the Z-source Cyg X-2 (e.g. Hasinger et al. 1990). The secular variations are thought to be due to the high inclination at which we view Cyg X-2 and GX 5–1 (Hasinger et al. 1990, Kuulkers et al. 1993).

One of the HBs shows an upward bend in the CD (Fig. 1a), a deviation from the "Z" shape. This has already been reported in Ginga data (Lewin et al. 1992). If we take a closer look at the lower NB in the CD we see a small limb up from the diagonal NB (Fig. 1b). We think that this limb is probably the missing "flaring branch". In this state the light curve varies with large amplitudes on short timescales, and therefore produces more very-low-frequency noise (VLFN) in power spectra, as compared to the NB (Fig. 1c), which is characteristic of the FB (Hasinger & Van der Klis 1989). This constitutes the first detection of flaring branch behaviour in GX 5-1. This behaviour is confirmed with Ginga-data of GX 5-1 (Kuulkers et al. 1993). We found no evidence for quasi-periodic oscillations (QPO) in the FB. For a full account on the EXOSAT data of GX 5-1 we refer to Kuulkers et al. (1993).

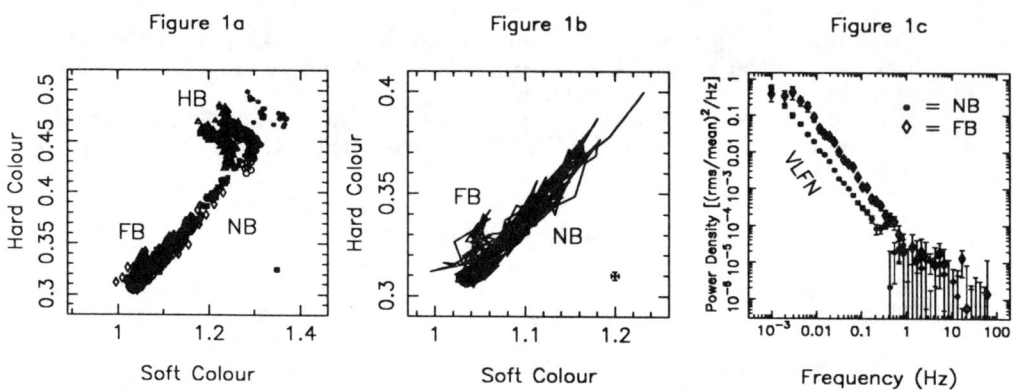

Fig. 1. (a.) X-ray colour-colour diagram of GX 5-1. Soft colour is defined as the ratio of the count rate in the 3.6-6.0 and 1.4-3.6 keV bands, respectively; hard colour as the ratio of the count rate in the 6.0-17.1 and 3.6-6.0 keV bands, respectively. Different symbols refer to different observation periods. Each point represents a 200 sec average. A typical error bar is given in the lower right of the frame. HB, NB, and the probable FB are indicated. (b.) as (a.), but only one observation period is shown. Points are connected for clarity. (c.) Typical power spectra of the normal and flaring branch. Note the flat-topped VLFN in the FB.

3. THE CRAB NEBULA, THE ONLY X-RAY "STANDARD STAR"

The Crab Nebula is a bright and steady X-ray source (see e.g. Toor & Seward 1974, and references therein). It is used for the calibration of X-ray satellites. Any changes in the observed spectrum of the Crab measured by an X-ray instrument are attributed to the instrument.

We did a detailed analysis of the systematic behaviour of the Crab-data observed with EXOSAT in the CD and the HID for the whole array (see Kuulkers et al. 1993 for a more detailed discussion). As the Crab Nebula is a steady X-ray source and it should not move secularly in the CD. During the EXOSAT-era the apparent soft colour of the Crab varied within ~3%, and the apparent hard colour within ~2% (Figs. 2a and b, respectively). Its intensity varied within ~4% (Fig. 2c). These variations are considerably below those of GX 5-1. They variations are due to changes in the gain of the ME-experiment.

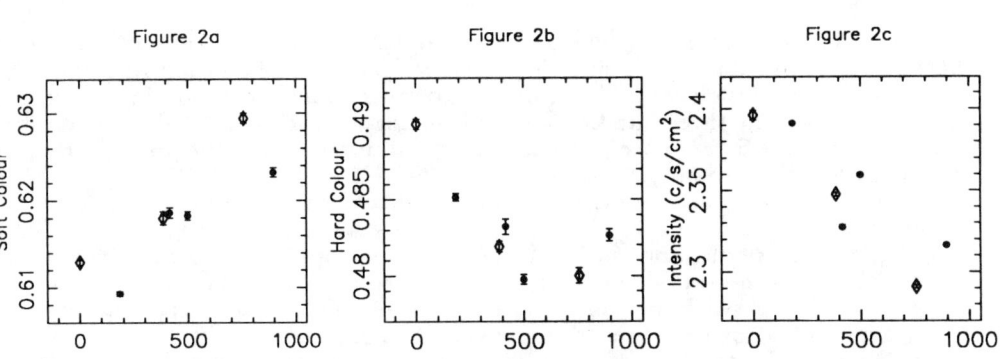

Fig. 2. Average soft (*a.*), hard (*b.*) and intensity (*c.*) points per observation period as a function of time. The variations are attributed to the ME-instrument. Diamonds represent observations which are obtained close to GX 5-1 in time.

4. GX 17+2: A NEW BURST IN OLD DATA

We report the detection of a burst in the EXOSAT data of GX 17+2, which was not reported before. This burst occurred at April 3 1986. Earlier reports of two other bursts observed with EXOSAT (September 6 1984 and August 20 1985) were done by Sztjano et al. (1986). The occurrence of bursts in GX 17+2 and characteristics similar to the other Z-sources signifies the presence of a neu-

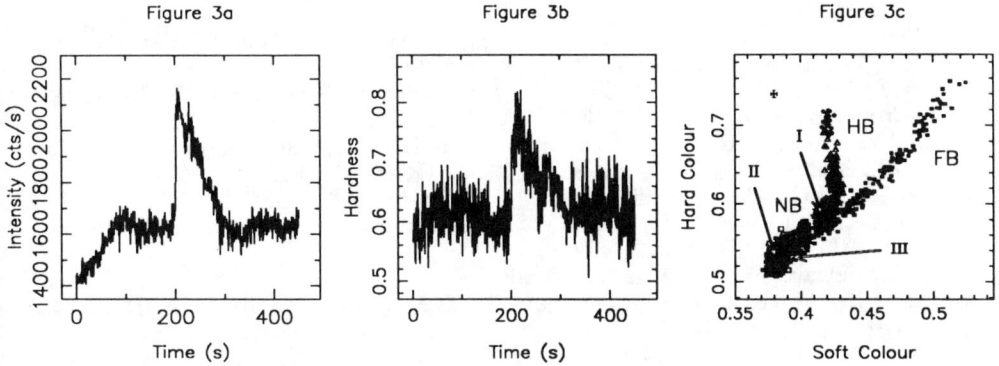

Fig. 3. Light curve (*a.*) and hardness curve (*b.*) of the "new" burst at April 3 1986. The intensity is the count rate in the 1.16–19.9 keV band, while hardness is defined as the ratio of the count rate in the 4.66–19.9 keV and 1.16–4.66 keV bands. (*c.*) X-ray colour-colour diagram of GX 17+2. Soft colour is defined as the ratio of the countrate in the 4.66–6.62 keV and the 1.16–4.66 keV bands, hard colour as the ratio of the count rate in the 6.62–19.9 keV and the 4.66–6.62 keV bands. Each point represents a 200 s average. A typical error bar is given in the upper left part of the frame. HB, NB and FB are indicated. The arrows point to the location of the bursts. The September 6 1984, the August 20 1985 and the April 3 1986 bursts are indicated by "I", "II" and "III", respectively.

tron star in these sources (see e.g. Lewin et al. 1988). The light curve and hardness curve of the "new" burst is presented in Figure 3a and b, respectively. A CD of GX 17+2 is given in Figure 3c. The corresponding position of the source in the CD when the three bursts occurred is indicated in Figure 3c. The smallest burst (September 6 1984) occurred at the highest intensities in the upper normal branch, while the two larger bursts (August 20 1985 and April 3 1986) occurred at the lowest intensities near the normal branch and flaring branch connection. Since bursts are only observed in lower luminosity LMXBs (see e.g. Lewin et al. 1988) the presence of bursts in GX 17+2 poses some questions to be solved:

- Why do we see bursts in such a luminous source as GX 17+2?
- Why do we see bursts in a branch in which GX 17+2 is thought to be accreting at or near the Eddington limit (e.g. Lamb 1989)?
- Why do the two large bursts occur near the normal branch and flaring branch connection, while the other smaller burst occurs in the upper normal branch?

ACKNOWLEDGEMENTS

This work was supported in part by the Netherlands Organization for Scientific Research (NWO) under grant PGS 78-277. WHGL acknowledges support from the National Aeronautics and Space Administration under Grant NAG8-674.

REFERENCES

Hasinger G., van der Klis M., 1989, A&A 225, 79
Hasinger G., van der Klis M., Ebisawa K., Dotani T. & Mitsuda K., 1990, A&A 235, 131
Kuulkers E., Van der Klis M., Oosterbroek T., Asai K., Dotani T., Van Paradijs J., Lewin W. H. G., 1993, A&A, submitted
Lamb F. K., 1989, in: 23rd ESLAB Symposium on Two-Topics in X-ray Astronomy, Bologna, Italy, 13-20 Sep. 1989 (ESA SP-296), p. 215
Lewin W. H. G., Lubin L. M., Tan J., van der Klis M., van Paradijs J., Penninx W., Dotani T. & Mitsuda K., 1992, MNRAS 256, 545
Lewin W. H. G., Van Paradijs J., van der Klis M., 1988, SSR 46, 273
Sztjano M., Van Paradijs J., Lewin W. H. G., Langmeier A., Trümper J., Pietsch W., 1986, MNRAS 222, 499
Toor A., Seward F. D., 1974, AJ 79, 995
van der Klis M., Kitamoto S., Tsunemi H. & Miyamoto S., 1991, MNRAS 248, 751

A SEARCH FOR CHAOS IN THE RAPID BURSTER

M. Bockrath
Department of Physics
University of California, Berkeley, CA 94720
marc@physics.berkeley.edu

R. Di Stefano
Harvard-Smithsonian Center for Astrophysics
MS 51, 60 Garden St., Cambridge, MA 02138
distefano@cfa.harvard.edu

S. Rappaport
Department of Physics and Center for Space Research
MIT, 37-551, Cambridge, MA 02139
sar@eagle.mit.edu

ABSTRACT

Progress is reported on an ongoing times series analysis of the Rapid Burster.

1. INTRODUCTION

This is a preliminary report on work in progress to use time series analysis, particularly phase space reconstruction techniques, to study the underlying dynamics of the type II bursts emitted by the Rapid Burster. We have studied data collected by a number of satellites, and have applied several different methods of analysis, including the Grassberger-Procaccia method (GP; Grassberger & Procaccia 1983), the computation of Lyapunov exponents, and forecasting techniques. Here we report on a GP analysis and on a Lyapunov exponent analysis of EXOSAT data taken during September of 1985. The details of the work summarized here may be found in Bockrath 1993, and in Di Stefano, Rappaport, & Bockrath 1993, where these results are extended.

Although our primary motivation is to learn more about the physics of the Rapid Burster, a second motivation is to use a comprehensive, multifaceted study of a single system to learn more about the strengths and weaknesses of these techniques as applied to astrophysical data sets. Although they have been successful in determining whether a wide range of systems are chaotic, and in helping to identify the number of active variables, it has proved to be difficult to use them to extract definitive results from astrophysical data sets, particularly for X-ray data (see, e.g., Norris & Matilsky 1989; Lochner, Swank, & Szymkowiak 1989). Yet, on general grounds, chaos is expected to be ubiquitous in astrophysical systems, and studies based on numerical simulations (see, e.g., Wisdom & Sussman 1993; Buchler 1990; Buchler, Perdang, & Spiegel 1985) have found convincing evidence of chaos. It therefore seems worthwhile to understand whether phase space reconstruction techniques—either those presently in use, or extensions of them—can be useful tools in the study of X-ray data.

2. THE RAPID BURSTER

The Rapid Burster (MXB 1730-335) is a transient X-ray source that reappears every six months and is active for a few weeks (Lewin et al. 1976; Grindlay and Gursky 1977). During its active phase it emits both classical type I X-ray bursts due to thermonuclear flashes of accreted matter on the surface of the neutron star, as well as type II bursts which are due to the release of gravitational energy as matter falls onto the neutron star (see Lewin, van Paradijs, & Taam 1992, and references therein). So far, no other sources have been found that emit most of their X-radiation in the form of type II bursts. In contrast to the type I bursts, which exhibit pronounced spectral cooling as they evolve, the type II bursts develop in a nearly "colorless" way. The type II bursts range in duration from a few seconds to more than 10 minutes, while their recurrence times range from ~ 10 seconds to ~ 1 hour. The temporal sequences of bursts vary from being nearly periodic to highly erratic. During the EXOSAT observations whose analysis is described here, the distribution of inter-burst arrival times is single-peaked, with an rms spread of $\sim 30\%$ (mode II; Hoffman, Marshall, & Lewin 1978). The one predictable property of the burst arrival times is that the integrated energy in a given burst is roughly proportional to the time interval to the next burst (the "$E - \Delta t$ relation"; see Lewin et al. 1992 and references therein). The $E - \Delta t$ relation leads to the inference that matter is stored in a "well" and accumulates at an approximately constant rate until the level reaches a critical value; this then triggers a release of the matter onto the neutron star surface. Even though the Rapid Burster has been the subject of intense study for 17 years, no clear understanding of the exact storage and release mechanism responsible for the bursts has emerged. However, it seems likely that the magnetic field and rotation of the neutron star play a central role. One of the main purposes of this study is to gain some insight into the enigmatic phenomenon of the rapid bursts via a very different line of investigation.

3. PHASE SPACE RECONSTRUCTION

From the time series of one observable, $f(t)$, it is possible to reconstruct the path of the system as it moves through its true m-dimensional phase space in a way that preserves its topological properties (Ruelle 1990; Packard et al. 1980). This can be done in an embedding dimension $N \leq 2m+1$ (Takens 1981). Since we don't know the dimensionality of the phase space which is associated with the observed burst behavior, we must search for an appropriate value of N. To construct a series of phase space points from the measured $f(t)$, we must also choose two time intervals, τ and Δt. The phase space points, $\vec{V}_i$, are given by $V_{i,j} = f(i\Delta t + j\tau)$, where i runs over the data set, and j runs from 1 to N. In what follows, we choose $\Delta t = \tau$. For the purposes of this study the important property of phase space vectors constructed in this way is that each vector represents a physical state of the system, and that vectors that are close together represent physical states that are similar.

In order to choose an optimal value for τ, we ran a series of tests in which, for each $N < 20$, τ was varied. These tests led us to choose a sample time of 1.8 s, or approximately 1/6 the duration of a typical burst during the EXOSAT observation. With this sampling, our time series consisted of 7600 values.

4. RESULTS

We carried out a GP analysis of the EXOSAT light curve; no dimension could be identified. Furthermore, simulated "data", designed to have the same gross features as the mode II bursts, gave similar results. To test the hypothesis that the bursts themselves might be sampling a relatively low-dimensional attractor, while the persistent emission between bursts might represent noise, we also performed the analysis for the light curves of the bursts only, treating the intervals between bursts as gaps in the data train. This also failed to yield a definitive dimensionality. Given the number of data points used, these failures indicate that the Rapid Burster is probably not sampling an attractor of less than 6 − 7 dimensions.

To test whether the time between bursts, T_i, might be governed by low-dimensional dynamics, we also carried out the GP analysis of the time series of burst intervals. Because the data set contained only 219 consecutive bursts, this analysis could have reliably identified a dimensionality ≤ 4; however no dimension was identified.

The GP technique is sensitive to the geometrical distribution of the phase space points. On the other hand, the Lyapunov exponents are sensitive to the way in which the distance between nearby points evolves in time. Consider an infinitesimal sphere of initial states; this sphere will deform and rotate as time progresses, becoming an ellipsoid. The evolution of the ellipsoid is described by the Lyapunov exponents, λ_i.

$$\lambda_i = \lim_{t \to \infty} \frac{1}{t} log_2 \frac{p_i(t)}{p_i(0)},$$

where $p_i(t)$ is the length of the i th principal axis. The Lyapunov exponents thus measure the long term evolution of a set of nearly identical physical states. A system is defined to be chaotic if it has at least one positive exponent.

Consider a set of phase space points, constructed from a time series. For any given point, a, find its nearest neighbor, and follow both points forward through a time δt. The log of the ratio of the distances between the time-evolved points and the initial points provides an estimate of the largest Lyapunov exponent. But since the Lyapunov exponent is meant to be a measure of the *local* rate of stretching, the initial distance must be as small as possible for *each* iteration of δt. Therefore, after each step, a point which is closer than the time-evolved nearest neighbor is sought. Finding the nearest neighbor to the time-evolved a, however is not the correct procedure, since, by definition, accurate results depend on the preservation of the relative orientation. In order to identify the best neighbor to choose, a search algorithm, which uses a set of input parameters (e.g., the radius of the search cone) is implemented after each time interval δt (Wolf et al. 1985). If this approach is to lead to a reliable estimate of the Lyapunov exponent, it must not be sensitive to the search parameters. Therefore, we calculated the Lyapunov exponent for a range of parameters, and found that, unlike the situation for a low-dimensional map (which we used for comparison), the results were strongly dependent on the choice of initial search distance. This is a symptom of having an inadequately covered attractor because, if the attractor is sufficiently well covered, increasing the search radius should not appreciably enhance the number of appropriate points found, since there will be appropriate points within the original radius. We therefore conclude that the EXOSAT data at our disposal cannot provide a reliable estimate of the largest Lyapunov exponent.

5. CONCLUSIONS

Application of the GP method and a calculation of Lyapunov exponents indicate that there is no low-dimensional attractor associated with the dynamics of the type II X-ray bursts, at least for the portions of the data that we have examined. This study also indicates that we do not have enough data at our disposal to make further investigation along these lines worthwhile.

However another set of techniques, based on "forecasting", may be more suitable for analysis of the Rapid Burster (see, e.g., Casdagli 1989; Farmer & Sidorowich 1987, 1988, Sugihara & May 1990). Like the computation of Lyapunov exponents, these methods are based on a study of the dynamics of the system; unlike Lyapunov exponents, one does not necessarily need to study long-term behaviour in order to derive an admittedly more narrowly focused set of results. Work along these lines is underway.

This work was supported in part by a NASA ADP grant. RD would like to thank J. Cannizzo, S. Eubank, Hao B-L, J. Lochner, J. Norris, and E.A. Spiegel for discussions. RD and SR would like to acknowledge the hospitality of the Aspen Center for Physics, where this work was begun.

REFERENCES

Arneodo, A., and Spiegel, E.A. 1986, private communication
Bockrath, M. 1993, MIT senior thesis
Buchler, J.R., Perdang, J.M., & Spiegel, E.A. 1985, Chaos in Astrophysics (Dordrecht:Reidel)
Buchler, J.R. 1990, The Numerical Modelling of Nonlinear Stellar Pulsations, Kluwer:Dordrecht
Casdagli, M. 1989, Physica D, 35, 335
Celnikier, L.M. 1977, Astr.Ap., 60, 421
Di Stefano, R., Rappaport, S., & Bockrath, M. 1993, in preparation
Farmer, J.D., & Sidorowich, J.J. 1987, Phys.Rev.Lett., 59, 845
Farmer, J.D., & Sidorowich, J.J. 1988, Evolution, Learning, and Cognition, ed. Y. Lee, (World Scientific Press) p.277
Grassberger, P., and Procaccia, I. 1983, Physica, 9D, 189
Grindlay, J.E., and Gursky, H. 1977, Ap.J. (Letters), 218, L117
Hoffman, J.A., Marshall, H.L., and Lewin, W.H.G. 1978, Nature, 271, 630
Lewin, W.H.G., van Paradijs, J., & Taam. R.E. 1992, Sp. Sci. Rev., 62, 223
Lewin, W.H.G., Doty, J., Clark, G.W., Rappaport, S.A., et al. 1976, Ap.J., 207, L95
Lochner, J.C., Swank, J.H., & Szymkowiak, A.E. 1989, Ap.J., 337, 823
Norris, J.P., and Matilsky, T.A. 1989, Ap.J., 346, 912
Packard, N.H., Crutchfield, J.P., Farmer, J.D., Shaw, R.S. 1980, Phys. Rev. Lett., 45, 712
Ruelle, D. 1990, Proc. R. Soc. Lond., A 427, 241
Sugihara, G., and May, R.M. 1990, Nature, 344, 734
Takens, F. 1981, Lecture Notes in Physics, 898, 336
Theiler, J. 1990, J.O.S.A., 7, 1055
Wisdom, J. & Sussman, G.J. 1993, Science, 257, 56
Wolf, A., Swift, J.B., Swinney, H.L., & Vastano, J.A. 1985, Physica, 16D, 285

MODELLING BLACK HOLE X-RAY POWER SPECTRA

Michael A. Nowak

CITA, 60 St. George St., Toronto, Ontario M5S 1A7

ABSTRACT

This paper presents a simple kinematic model of the so-called "very high state" of black hole candidates, during which several QPO observations have been made. The model is based upon the viscous and thermal instabilities that are believed to be present in accretion disks. In this model, the very high state is a transition phase between the high (thermal, quiescent) state and low (power law, highly variable) state that is characterized by a quasi-stable disk oscillating on the local instability timescales. The disk can be stabilized by a hot wind that could feed a Compton cloud, which in turn could produce the hard tail in the observed X-ray spectra. The model is able to reproduce both the overall shape and the amplitude of the power spectral density and the observed frequency dependent lags between the hard and soft X-rays. In addition, the required energetics of the hot wind is shown to be consistent with the energy required to feed the hypothesized Compton cloud. Specific comparisons between the model and data taken during the very high state of GX339-4 are made.

There are two extremes in the study of variability of black hole systems: detailed physical calculations, such as the MHD accretion disk stability studies of Balbus and Hawley (1991); or purely phenomenological models of the data, specifically shot noise models (cf. Lochner et al. 1991). The former is difficult to relate to observation, whereas the latter often lacks clear physical motivation. The philosophy of this work, explained in detail in Nowak (1994), is to add physical insight to the phenomenological models in an attempt to better understand the geometry and energetics that lead to the X-ray variability of black hole systems.

In this work we assume that for sufficiently high luminosities, the accretion disk will be *locally* unstable on either a viscous (τ_V) or thermal (τ_H) timescale, given by (*cf.* Shakura & Sunyaev 1976; Piran 1978):

$$\tau_V^{-1} \sim \alpha \mathcal{L}^2 \Omega \;, \quad \tau_H^{-1} \sim \alpha \Omega \;, \tag{1}$$

where Ω is the local keplerian frequency, $\mathcal{L}$ is the ratio of disk luminosity to Eddington luminosity, and α is the standard phenomenological viscosity parameter ($\lesssim 1$). The viscous and thermal instabilities can be supressed by a mass and energy loss, respectively, characterized by mass and energy fluxes of:

$$\dot{\mathcal{M}}_W/\Sigma \gtrsim \tau_V^{-1} \;, \quad \dot{E}_W/W \gtrsim \tau_H^{-1} \;, \tag{2}$$

where $\dot{\mathcal{M}}_W$ is the mass flux, Σ is the disk surface density, $\dot{E}_W$ is the energy flux, and W is the vertically integrated disk pressure. This mass and energy flux can be considered as arising from a wind emanating from the disk.

The instabilities will lead to fluctuations in the local temperature (T) and luminosity. Here, we assume that the luminosity, proportional to T^4, rises exponentially with the local instability timescale. After a rise time that is Poisson distributed about the same timescale, the disk instantaneously relaxes back to a rest value. (This could be explained, perhaps, by a sudden release of mass and energy via a flare.) The process then begins again. We use Monte Carlo calculations to simulate the X-ray light curve for such a disk. With this prescription, we can use standard Fourier methods to calculate the expected power spectral density of the total signal, and the (Fourier frequency dependent) time lag between the hard and soft X-ray photons. The results of such calculations are presented in Figures 1 and 2. Also presented in these figures are observational data taken during the "very high state" of GX339-4. The overall agreement between the theory (with α, $\mathcal{L} \sim 0.1$, and the central object mass $M \sim 4$ $M_\odot$) and observation is good.

Here we hypothesize that the very high state of GX339-4 is a transition between the lower luminosity high state and the higher luminosity low state (integrating the X-ray data presented in Figure 1 of Grebenev et al. [1991] shows the low state to have 5 times the total luminosity of the high state, though this is *not* a universal characteristic of all black hole candidates.) We assume that the disk is first dominated by the viscous timescale, then as the luminosity increases it becomes dominated by the thermal time scale in the inner regions. As the luminosity increases further, the entire disk becomes dominated by the thermal timescale. This transitional behavior will be characterized by $\dot{M}_W/\Sigma \lesssim \tau_V^{-1}$ and $\dot{E}_W/W \gtrsim \tau_H^{-1}$. This allows us to estimate the temperature and total energy of the hypothesized wind (*cf.* Nowak 1994). For a wind of electrons and positrons, the derived temperature is of order 5 – 80 keV. This is consistent with the temperature of the Compton cloud as found by Miyamoto et al. (1991). Furthermore, the total energy of the wind is consistent with the energy required to power the Compton cloud in this system (*cf.* Miyamoto and Kitamoto 1991).

It is interesting to note that at the transition radius between viscous and thermal instabilities the thermal time scale corresponds to 6 Hz, the observed QPO frequency in GX339-4. This leads us to speculate that the QPO is associated with either the wind that feeds the Compton cloud or the instability relaxation mechanism (*i.e.* flares). We will consider these mechanisms in a future work.

REFERENCES

Lochner, J. C., et al. 1991, ApJ, 376, 295.
Balbus, S. A., and Hawley, J. F. 1991, ApJ, 376, 214.
Grebenev, S. A., et al. 1991, Sov. Astron. Let, 17, 413.
Miyamoto, S., and Kitamoto, S. 1991, ApJ, 374, 741.
Miyamoto, S., et al. 1991, ApJ, 383, 784.
Nowak, M. A. 1994, ApJ, 422, *to be Published*.
Piran, T. 1978, ApJ, 221, 652.
Shakura, N. I. , and Sunyaev, R. A. 1976, MNRAS, 175, 613.

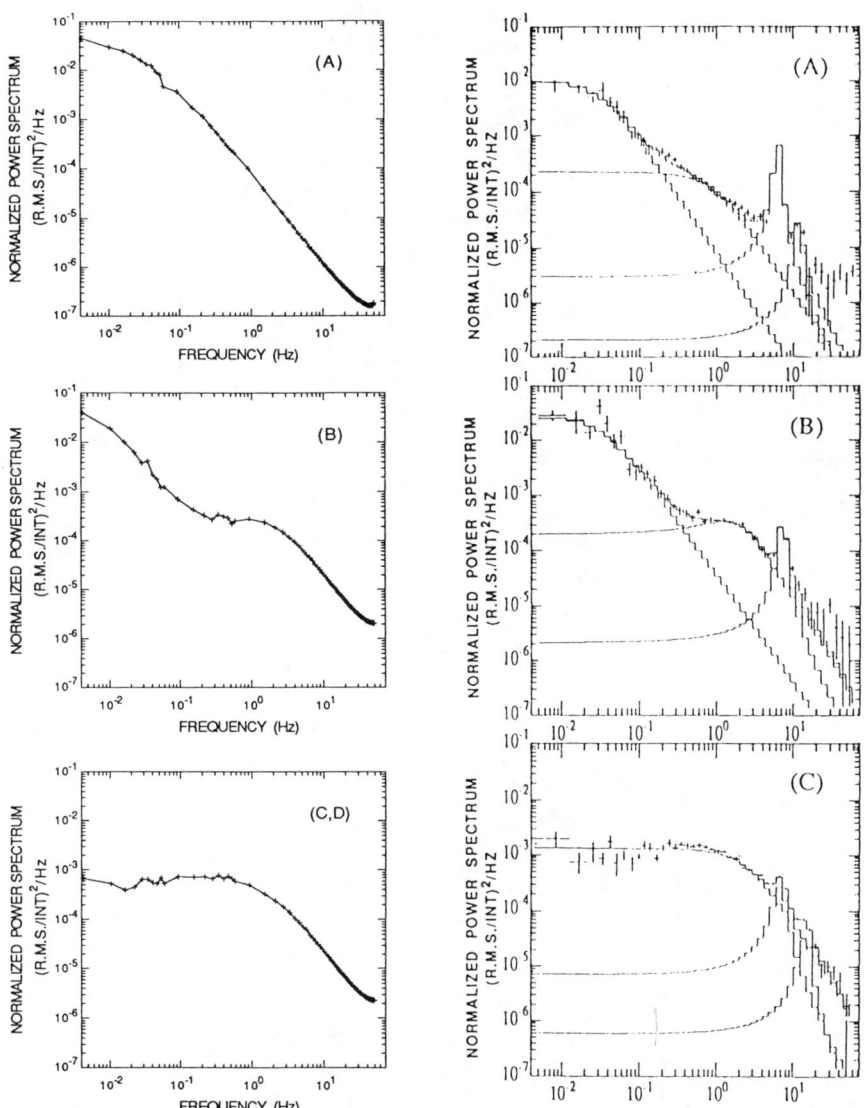

Figure 1: Power Spectral Densities for three substates of GX339-4. PSD's on the right are from Miyamoto et al. (1991), and the ones on the left are from Nowak (1994). (A) the disk is dominated by viscous timescales. (B) the disk is dominated by thermal timescales in the inner region, and viscous timescales in the outer region. (C,D) the disk is dominated by thermal timescales throughout.

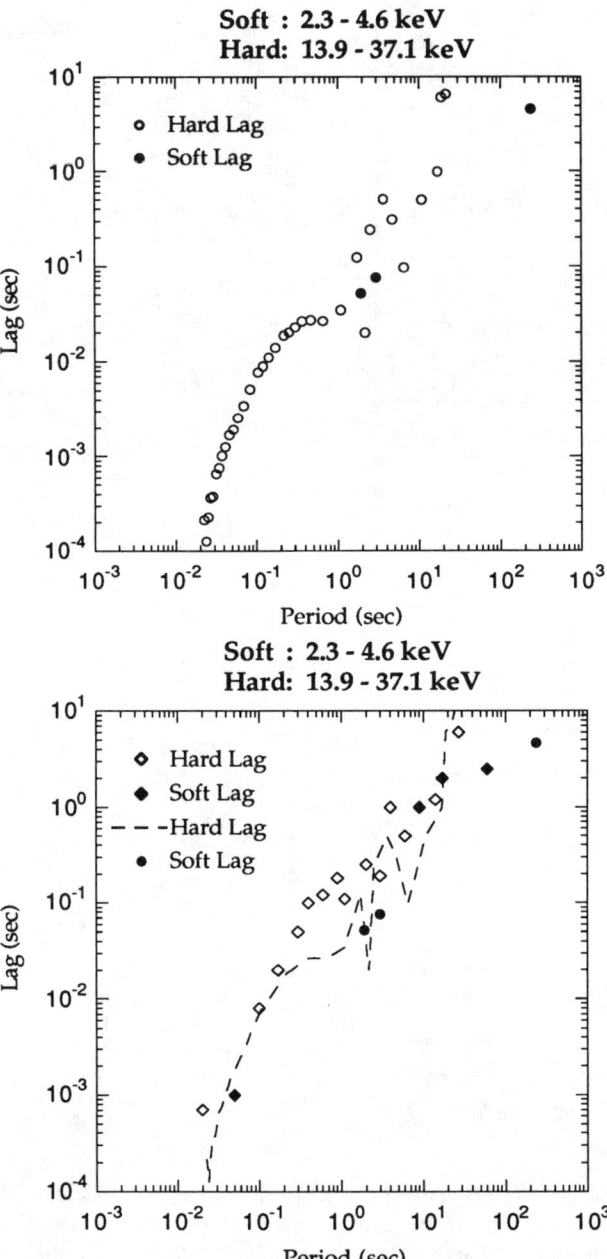

Figure 2: *Top:* Theoretical calculations of hard X-ray time lags as a function of Fourier period. Open circle indicates hard lag, closed circle indicates soft lag. *Bottom:* Theory (dashed line) vs. observations of GX339-4 taken from Miyamoto et al. (1991).

EFFECT OF NEW OPACITIES ON THE DISK INSTABILITY

Mine Takeuti and Reiko Kanetake
Astronomical Institute, Tohoku University, Sendai 980 Japan.
i4a0s4d@jpntohok.bitnet

ABSTRACT

The vertical oscillation of an accretion disk surrounding a neutron star is studied by using the OPAL opacities. We have found that the vertical structure of the model disk is vibrationally unstable when we use new opacities instead of the Los Alamos opacities.

1. INTRODUCTION

The use of new astrophysical opacities referred as the OPAL opacities (Iglesias and Rogers 1991) yields sometimes quantitative changes in results but sometimes qualitative changes. The vertical oscillation of an accretion disk was sometimes mentioned (Cox 1981), but the oscillation is usually ignored in the study of the accretion disk because it is stable. We examine the vibrational stability of such an oscillation by using the new opacity. Annuls of a disk surrounding a neutron star is chosen. When we use the Los Alamos astrophysical opacities, the model is stable against the vertical oscillation. On the other hand, the disk constructed with the new opacities is vibrationally unstable.

2. MODEL

We assumed the mass and the radius of the central neutron star as 2 $M_\odot$ and 10^4 m, respectively. The mass loss rate is $10^{-8} M_\odot$ yr^{-1}. The chemical composition is adapted to the population I. The new opacities are calculated by using a subroutine constructed by Ishida (1992) based on the data of Iglesias and Rogers (1991). We also calculate the annuls with the same parameters with the Stellingwerf formula of the Los Alamos opacities (Stellingwerf 1975a, b). The subroutine includes Alexander's opacities (Alexander 1975) for cool layers. The disk is geometrically thin, and we assume the standard α-disk model. The convective efficiency of the model is fixed as 0.5 to make numerical calculations easy. Taking the Smak's results on the structure of accretion disks (Smak 1982) into account, we assume that the viscous heat generation is effective only in the convection zone.

For calculating the linear vertical oscillation, we use the assumption that each annul is isolated from the neighboring annuls. This will be adequate because the disk is very thin. We only study the odd modes of vertical oscillation here for simplicity. The linear non-adiabatic (LNA) periods and the growth rates are calculated by Castor's method (Castor 1971). To avoid the numerical difficulty found in some modes, we use Worrell's procedure (Worrell 1986) which is useful to determine a few unresolved eigenvalues in the LNA calculations. The details of models are described in Kanetake & Takeuti (1993b).

© 1994 American Institute of Physics

3. RESULTS AND DISCUSSION

We calculate the annuls of which the effective temperatures are $6 \times 10^4 \sim 1.5 \times 10^3 \mathrm{K}$ where the radii are between 10^6 m and 10^8 m. The temperature of the inside of inner annuls reaches near $10^5 \mathrm{K}$, and it is found that the models are affected by the revision of opacities. The results are reported in Kanetake & Takeuti (1993a).

We have found the positive growth rate for some annuls by using the new opacities, although the growth rate of the annuls produced by the old opacities is near equal to zero. This means that the vertical oscillations can be enhanced at least at some annuls of the disk. The oscillations are excited more strongly at inner annuls. The periods would be several ten seconds, which are the same order as the quasi-periodic oscillation found in the Galactic X-ray sources (for example see Takeshima et al. 1991). Naturally it should be careful to judge the actual physical processes based on the apparent coincidence of periods. Especially, the radial oscillation without vertical oscillations also yields similar periods (Van Horn, Wesemael, and Winget 1980).

Anyhow we may expect the occurrence of very complex oscillatory motion, since there exist the exchanges of energy among a lot of oscillatory modes. It seems interesting to run hydrodynamic models of accretion disks, to study the fate of the enhanced vertical oscillations.

4. CONCLUSION

It is found that the effect of new astrophysical opacities on the vertical pulsation of the accretion disks. The results is interesting since the new opacities make the pulsations unstable. This seems a new source of instability possibly exciting the motion of accretion disks.

ACKNOWLEDGMENTS

Computations is performed at the Computer Center of Tôhoku University. The study was partly supported by funds from the Saito Ho-on Kai.

REFERENCES

Alexander, D. R. 1975, ApJS, 29 , 363
Castor, J, I. 1971, ApJ, 166, 109
Cox, J. P. 1981, ApJ, 247, 1070
Iglesias, C. A. & Rogers, F. J. 1991, ApJ, 371, L73
Ishida, T. 1992, private communication
Kanetake, R. & Takeuti, M. 1993a, PASJ, 45, in press
Kanetake, R. & Takeuti, M. 1993b, Saito Ho-on Kai Mus. Nat. Hist. Res. Bull. (Sendai), 61, in press
Smak, J. 1982 , Acta Astro. , 32, 199
Takeshima, T., Dotani, T., Mitsuda, K., & Nagase, F. 1991, PASJ, 43 , L43
Stellingwerf, R. F. 1975a, ApJ, 195, 441
Stellingwerf, R. F. 1975b, ApJ, 199, 705
Van Horn, H. M., Wesemael, F., & Winget, D. E. 1980, ApJ, 235, L143
Worrell, J. K. 1986, MNRAS, 223, 787

Searches for Millisecond Pulsations in Low-Mass X-ray Binaries with Ginga Data

Brian Vaughan[1], M. van der Klis[1], K. S. Wood[2], J. P. Norris[3], P. Hertz[2],

P. F. Michelson[4], J. van Paradijs[1], W. H. G. Lewin[5] & K. Mitsuda[6]

[1] Astronomical Institute "Anton Pannekoek", University of Amsterdam,
Kruislaan 403, 1098 SJ Amsterdam, The Netherlands.
[2] Code 4121, E. O. Hulbert Center for Space Research,
Naval Research Laboratory, Washington DC 20375.
[3] Code 690, Laboratory for High Energy Astrophysics,
NASA Goddard Space Flight Center, Greenbelt MD 20771.
[4] Physics Department, Stanford University, Stanford CA 94305.
[5] Massachusetts Institute of Technology, Center for Space Research,
Room 37-627, Cambridge MA 02139.
[6] Institute of Space and Astronautical Science, 3-1-1 Yoshinodai,
Sagimahara, Kanagawa 229, Japan.

ABSTRACT

Coherent millisecond X-ray pulsations are expected from low-mass X-ray binaries (LMXB), but remain undetected. Using the single-parameter Quadratic Coherence Recovery Technique (QCRT) to correct for unknown binary orbital motio we have performed Fourier transform searches for coherent oscillations in all long, continuous segments of data obtained at 1 millisecond time resolution during Ginga observations of LMXB. We have searched the six known Z-sources (GX 5-1, Cyg X-2, Sco X-1, GX 17+2, GX 340+0, and GX 349+2), seven of the ten known atoll sources (GX 3+1, GX 9+1, GX 9+9, 1728-33, 1820-30, 1636-53 and 1608-52), the "peculiar" source Cir X-1, and the high mass binary Cyg X-3. We find no evidence for coherent pulsations in any of these sources, with 99% confidence limits on the pulsed fraction between 0.3% and 5.0% at frequencies below the Nyquist frequency of 512 Hz.

1. INTRODUCTION

Because of their weak magnetic fields and high accretion rates the neutron star equilibrium spin periods in LMXB may be of order milliseconds (Ghosh & Lamb 1979a,b). Few LMXB have known spin periods, and those that do are all in some ways anomalous (Wood et al. 1991 and references therein). Detection of coherent millisecond X-ray pulsations from an LMXB would constitute strong evidence for the evolution of millisecond radio pulsars from LMXB (Helfand, Ruderman, & Shaham 1983; Joss & Rappaport 1983; Paczynski 1983; Savonije 1983), would support currently favored beat frequency models of horizontal-branch QPO formation (Alpar & Shaham 1985; Lamb et al.1985), and would provide a precise means to measure orbital periods and masses and to study accretion torques and neutron star dynamics on very short time scales.

The same weak magnetic fields and high accretion rates that can produce high neutron star spin periods in LMXB also make it likely that the pulsed fractions are low. With less channeling of accreted material along magnetic field lines, X-ray production is distributed over more of the neutron star surface, resulting in smaller modulations and smearing from gravitational lensing (Wood, Ftaclas & Kearney 1988; Meszaros, Riffert & Berthiaume 1988). Because the magnetic field is weak, the accretion disk may engulf the magnetosphere and perhaps reach the neutron star surface even at low ($\sim 10^{37}$erg $^{-1}$) luminosity, producing an environment with a large optical depth and resulting in further loss in observed mudulation depth from scattering that both dephases the pulsar signal and spreads the pulsar beam (Brainerd & Lamb 1987; Bussard et al.1988).

Most LMXB fall into two classes, Z-sources and atoll sources (Hasinger & van der Klis 1989). Z-sources are more luminous and exhibit three states, called the horizontal, normal and flaring branches. Z-sources are thought to have higher magnetic fields and accretion rates, and in general longer orbital periods than atoll sources (Hasinger & van der Klis 1989; van der Klis 1991). Atoll sources exhibit two spectral states, the island and the banana states, and sometimes show X-ray bursts. Because spectral varaiability in LMXB results from changes in the accretion environment, we wish to search for pulsations in all spectral branches.

2. Results

We searched for fast pulsations in all long, contiguous segments of 1 ms data obtained in pointed observations with Ginga during its observational life. Table 1 presents upper limits obtained in searching 51 independent segements of data. To correct for the usually unknown binary motion of the neutron star, we applied the Quadratic Coherence Recovery Technique (Wood et al.1991). The advance or retardation of the pulse phase due to orbital motion is approximated as a quadratic function of time. A grid of trial correction parameters is used. For each correction, an FFT of the data is calculated, and the power spectrum searched for peaks. See Wood et al.(1991) for details.

A key assumption used in determining upper limits in the past has been that the Poisson noise and the signal are independent. This assumption, although approximately valid if a large number of independent power spectral estimates are averaged together, is incorrect for searches performed using single FFTs, such as ours. See Groth (1975) or Vaughan (1994) for details. We incorporated the interaction between signal and Poisson noise in estimating all the upper limits in table 1. Upper limits in table 1 of this paper supercede those in table 2 of Wood et al.(1991).

ACKNOWLEDGEMENTS

WHGL gratefully acknowledges support from the National Aeronautics and Space Administration under grant NAG8-700. JvP acknowledges support from NATO through grant RG 331/88. This work was supported in part by the Netherlands Organization for Scientific Research (NWO) under grant PGS 78-277.

REFERENCES

Alpar, M. A., and Shaham, J. 1985, Nature, 316, 239.
Brainerd, J., and Lamb, F. K. 1987, ApJ, 317, L33.
Bussard, R. W., Weisskopf, M. C., Elsner, R. F., & Shibazaki, N. 1988 ApJ, 327, 284.
Ghosh, P., and Lamb, F. K. 1979a, ApJ, 232, 259.
Ghosh, P., and Lamb, F. K. 1979b, ApJ, 235, 296.
Groth, E.J. 1975, ApJS, 29, 285.
Hasinger, G., and van der Klis, M. 1989, AA, 225, 79.
Helfand, D. J., Ruderman, M. A., and Shaham, J. 1983, Nature, 304, 423.
Joss, P., and Rappaport, S., 1983, Nature, 304, 419.
Lamb, F. K., Shibazaki, N., Shaham, J., & Alpar, M. A., 1985, Nature, 317, 681.
Meszaros, P., Riffert, H., and Berthiaume, G. 1988, ApJ, 325, 204.
Paczynski, B. 1983, Nature, 304, 421.
Savonije, G. J. 1983, Nature, 304.
van der Klis, M. 1991, in Frontiers of X-ray Astronomy, ed. Y. Tanaka & K. Koyama (Tokyo: Universal Academic Press Inc.), 139.
Vaughan, B. A., van der Klis, M., Wood, K., Norris, J. P., Hertz, P., Michelson, P., van Paradijs, J., Lewin, W. H. G. & Mitsuda, K. 1994, submitted to ApJ.
Wood, K. S., Ftaclas, C., and Kearney, M. 1988, ApJ, 324, L63.
Wood, K. S., Norris, J. P., Hertz, P., Vaughan, B. A., Michelson, P. F., Mitsuda, K., Lewin, W. H. G., van Paradijs, J., Penninx, W., & van der Klis, M. 1991, ApJ, 379, 295.

Table 1 Caption: Pulsed fraction 99% confidence upper limits at 50 Hz (A 50) and 500 Hz (A 500). Abbreviations for source state are; H=Horizontal Branch, N=Normal Branch, F=Flaring Branch, FV=Flaring Branch Vertex, HV=Horizontal Branch Vertex, B=Banana State, I=Island State. RATE is the count rates in cnts/sec. T_{int} is the integration time in seconds. N_α is the number of correction parameters tried. Min. P_{orb} is the minimum orbital period for which the quadratic correction to the orbital motion is guaranteed to be valid.

[a] Known period (19.2 hr), [b] Known period ($9^d.8$), [c] Known period (4.2 hr.), [d] Low intensity soft state, orbital phase $\phi \sim 0.95$, [e] Known period ($16^d.6$), [f] Orbital phase $\phi \sim 0.65$, [g] Low state, $\phi \sim 0.6$, [h] Known period (4.8 hr.), [i] Low state, $\phi \sim 0.92$, [j] Low state, $\phi \sim 0.2$.

Table 1: Upper Limits on Pulsed Fractions

SOURCE		OBS. DATE		RATE	STATE	T_{int}	N_α	A 50	A 500	Min. P_{orb}
Sco X-1	(1617-15)	1989	Mar 9	14000	F	1024	100	0.0039	0.0072	a
		1989	Mar 10	10000	F	1024	100	0.0043	0.0079	
GX 340+0	(1642-45)	1988	Mar 30	2200	N	1024	215	0.0096	0.018	3
		1988	Apr 6	2200	N	1024	215	0.0096	0.018	3
GX 5-1	(1758-25)	1987	Apr 20	7000	H	1024	215	0.0054	0.0099	3
		1987	Apr 27	5000	N	2048	857	0.0045	0.0083	6
GX 9+1	(1758-20)	1988	Mar 29	3000	B	1024	215	0.0082	0.015	3
		1988	Mar 31	3250	B	1024	215	0.0079	0.015	3
GX 17+2	(1813-14)	1988	Mar 28	3500	F	1024	215	0.0076	0.014	3
		1988	Apr 1	2700	FV	1024	215	0.0087	0.016	3
Cyg X-2	(2142+38)	1987	Jun 7	3400	N	1024	3	0.0052	0.0095	b
		1987	Jun 8	3400	N	1024	3	0.0052	0.0095	
		1988	Jun 10	3000	H	2048	7	0.0050	0.0092	
		1988	Jun 10	2900	H	1024	3	0.0071	0.013	
		1988	Jun 11	3000	N	2048	7	0.0076	0.014	
		1988	Jun 11	3000	N	2048	7	0.0051	0.0094	
		1988	Jun 13	2900	N	1024	3	0.0073	0.013	
		1988	Oct 6	3600	F	1024	3	0.0063	0.012	
		1988	Oct 6	3600	FV	1024	3	0.0068	0.013	
		1988	Oct 7	4000	H	1024	3	0.0060	0.011	
		1989	Oct 22	4200	H	1024	3	0.0060	0.011	
		1989	Oct 22	6000	H	2048	7	0.0037	0.0068	
		1989	Oct 22	6100	HV	1024	3	0.0049	0.0090	
		1989	Oct 23	4600	H	1024	3	0.0055	0.010	
GX 9+9	(1728-16)	1988	Aug 8	2300	B	2048	961	0.0065	0.012	c
		1988	Aug 8	2300	B	1024	241	0.0089	0.016	
		1988	Aug 9	2200	B	2048	961	0.0071	0.013	
		1988	Aug 9	2100	B	1024	241	0.0093	0.017	
GX 349+2	(1702-36)	1989	Mar 5	5900	F	1024	301	0.0057	0.010	3
		1989	Mar 5	4300	FV	1024	301	0.0064	0.012	3
		1989	Mar 5	4000	FV	1024	301	0.0066	0.012	3
		1989	Mar 6	3900	F	1024	301	0.0071	0.013	3
Cir X-1	(1516-57)	1989	Mar 30	430	d	1024	301	0.013	0.024	e
		1987	Aug 25	2400	f	256	101	0.0099	0.018	
Cyg X-3	(2030+40)	1987	Aug 27	390	g	1024	201	0.023	0.043	h
		1987	Aug 27	280	i	1024	201	0.027	0.040	
		1987	Aug 27	330	j	1024	201	0.024	0.044	
GX 3+1	(1744-26)	1987	Apr 13	1250	B	1024	301	0.012	0.022	3
		1987	Apr 13	1400	B	1024	301	0.011	0.020	3
1608-52		1989	Aug 23	130	I	1024	301	0.042	0.078	3
		1989	Aug 23	130	I	1024	301	0.043	0.079	3
		1989	Aug 23	170	I	2048	961	0.024	0.044	6
		1990	Mar 7	310	I	512	161	0.034	0.063	1.5
		1990	Mar 7	270	I	512	161	0.037	0.068	1.5
		1991	Aug 13	300	I	512	161	0.034	0.063	1.5
		1991	Aug 13	400	I	512	161	0.029	0.054	1.5
1728-33		1987	Oct 12	650	I	512	161	0.023	0.043	1.5
		1987	Oct 12	770	I	1024	301	0.014	0.026	3
1636-53		1987	Aug 20	1150	B	1024	301	0.013	0.024	3
		1987	Aug 20	1300	B	1024	301	0.013	0.024	3
		1987	Aug 20	1200	B	1024	301	0.013	0.024	3

MICROSECOND TEMPORAL STRUCTURE FROM X–RAY BINARY PULSARS: OBSERVABILITY WITH XTE

Mauro Orlandini[1,2] and Elihu Boldt[1]

[1] Laboratory for High Energy Astrophysics, Code 666, NASA/GSFC, Greenbelt, MD 20771
[2] Institute TESRE/CNR, via de' Castagnoli 1, 40126 Bologna, Italy.

ABSTRACT

We discuss the possibility of detecting the microsecond granularity expected in the flux from wind-fed X–ray binary pulsars. This microsecond structure is predicted by the noisy accretion scenario; we derive the physical characteristics of such burst structure in the framework of this model. We investigate the temporal dispersion induced by the scattering of burst photons in their passage through the neutron star magnetosphere and conclude that microsecond structure could indeed be observable for energies less than the cyclotron energy. We show that the coincidence timing mode for the Proportional Counter Array on-board XTE can be used very effectively to statistically discriminate these microsecond bursts from accidental spurious detections.

1. INTRODUCTION

The physics of accretion in X–ray binary pulsars has been characterized by two fundamental assumptions on the accretion rate: homogeneity and stationarity. While these two constraints greatly simplify the mathematical treatment of the problem, they are far from being applicable to the class of objects we are interested in. Indeed, phenomena of flaring activity and inhomogeneity in the stellar wind, which can be associated with variability in the accretion rate, are very common in wind-fed pulsars (Nagase 1989).

In this framework it is easy to understand the need of some non-stationary, non-homogeneous model of wind accretion. And indeed, it is possible to find in the literature numerous models for non-homogeneous accretion onto neutron stars, essentially developed for explaining the super Eddington luminosity observed in some X–ray pulsars (see, *e.g.*, Nagase 1989). In the "noisy accretion" model by Orlandini & Morfill (1992) — which we will use in this paper — a "granularity" in the emission process is introduced due to the instability which occurs at the magnetospheric limit; such granularity manifests itself with time scales on the order of a microsecond. Time scales of this order of magnitude can be achieved only for processes which occur at the neutron star surface (Orlandini & Boldt 1993), therefore a microstructure in the emission can be associated only with processes which occur close to the magnetic polar caps of the neutron star.

2. THE PHYSICAL MODEL

In the noisy accretion model, the stellar wind coming from the optical companion of the neutron star is captured by its intense gravitational field and accreted. The strong magnetic field halts the incoming matter at a distance on the order of $r_m \approx 10^8$ cm from the neutron star surface. Here matter enters the magnetosphere by means of Rayleigh–Taylor instability, in the form of accreting blobs. The theory by Arons & Lea (1980) can describe the physical parameters of these blobs, as radius, density and temperature, in terms of quantities such as the X–ray luminosity, the neutron star mass and the magnetic field strength.

During their radial infall toward the neutron star surface, the blobs decrease their radius because of Compton cooling. At a distance called plasmapause (of the order of 10^7 cm) they are channeled by the magnetic field lines toward the neutron star polar caps; their physical properties remain "frozen" during their fall.

When these blobs arrive onto the magnetic polar caps, they release their kinetic energy into X–ray radiation which, after interacting with the strongly magnetized plasma, reaches the observer. We have therefore a "granularity" in the emission process due to the inhomogeneous accretion onto the neutron star.

In order to estimate the temporal structure of the X–ray emission due to the infalling blobs, we have estimated the duration of a single burst both from a dynamical point of view and a thermal point of view. In the former case, the numerical value of the dynamical time scale τ_{sq} which an unstable blob takes to be completely squashed onto the neutron star (in other terms, the duration of a shot in a shot noise process) is

$$\tau_{sq} = \begin{cases} 1.2 \times 10^{-5} \text{ s} & \text{for } L_{37} = 0.1 \\ 7.3 \times 10^{-7} \text{ s} & \text{for } L_{37} = 1 \end{cases}.$$

(Orlandini & Morfill 1992) where L_{37} is the X–ray luminosity in units of 10^{37} erg s^{-1}.

The duration of the "squashing" process of a blob depends on the way in which the kinetic energy of the infalling blob is converted into radiation. The time scale of conversion is determined by the transfer of momentum from the ions to the electrons, and by the photon production rate in the interaction of the electrons with the matter and the magnetic field.

Morfill et al. (1984) have shown that after a time of the order of 10^{-8} s, a deceleration front starts to move upward through the blob, *i.e.* the X–rays start to "diffuse" into the blob. This deceleration front will emerge from the blob after a time of the same order of magnitude as the dynamical time scale previously computed (Orlandini & Boldt 1993); both of these considerations predict microsecond structure in the emission from X–ray pulsars.

The other quantity we need in order to estimate the effect of this granularity on the observed flux is the average rate of shots per unit time, λ, which is

$$\lambda = \frac{1}{2}\frac{4\pi r_m^2}{\Lambda^2 \tau_{RT}} \approx \begin{cases} 10 \text{ s}^{-1} & \text{for } L_{37} = 0.1 \\ 100 \text{ s}^{-1} & \text{for } L_{37} = 1 \end{cases}$$

(Orlandini & Morfill 1992) where τ_{RT} is the characteristic time scale for the growth of the Rayleigh–Taylor instability, and Λ is the wavelength of the perturbation.

We now have all the information needed to quantitatively describe the bursting behavior of the X–ray emission. The luminosity during a single burst is

$$\mathcal{L}_b = \frac{\mathcal{L}_x}{\lambda \tau_{sq}} \approx \begin{cases} 8.3 \times 10^{39} \text{ erg s}^{-1} & \text{for } L_{37} = 0.1 \\ 1.4 \times 10^{41} \text{ erg s}^{-1} & \text{for } L_{37} = 1 \end{cases}$$

i.e. the luminosity produced by each instability blob is super Eddington (this is not surprising, because of the non stationarity of the emission process). Under the assumption of black-body emission from the blob, this would correspond to temperatures between 10 and 100 keV for the two luminosity states, respectively.

3. ON THE OBSERVABILITY OF MICROSECOND BURSTING

As we have shown in the previous section, the noisy accretion scenario predicts microsecond structure in the emission from wind fed X–ray binary pulsars. The passage of this radiation through the neutron star magnetosphere will tend to broaden the pulses. However, considering the photons as independent particles undergoing multiple scattering in a plasma of uncorrelated electrons, we have shown that the spread in transit time across a region could very well be comparable to what the transit time of that region would have been if there were no scattering (Orlandini & Boldt 1993).

The passage of the microsecond pulses through the neutron star magnetosphere, of scale height $r_m \approx 10^8$ cm, would cause a dispersion of the signal of the order of milliseconds, thereby smearing out the effect. However, the strong magnetic field of the neutron star has a pronounced effect on the Thomson scattering cross section. In particular, for photons moving in a direction parallel to that of the magnetic field, the quantum total cross section, σ_{tot}, is strongly reduced in comparison with the classical Thomson value, σ_{Th}, because of the transversality of the wave. Indeed, in the approximation $E \ll E_c$, we have $\sigma_{tot} \approx \sigma_{Th}(E/E_c)^2$ (Herold 1979), where E is the photon energy, and $E_c \equiv \hbar(eB/mc)$ is the cyclotron energy.

From the observed cyclotron resonance in the spectra of four X–ray binary pulsars (Makishima et al. 1990), we can say that for $E \lesssim 5$ keV, microsecond duration pulses from the surface of neutron stars such as those here represented should be detectable relatively undistorted.

4. FEASIBILITY FOR DETECTION OF MICROSECOND BURSTING WITH THE X–RAY TIMING EXPLORER

The PCA of five independent co-aligned X–ray detectors on the XTE is particularly well suited for detecting possible microsecond temporal structure in the X-ray emission from observed sources, even faint ones. Detecting such microsecond bunching of photons with a single proportional counter alone would be prohibitively difficult. In particular, a microsecond duration burst of multiple photon interactions all taking place in the *same* counter would tend to appear as a single "pileup" event or sometimes be vetoed. However, two or more such temporally correlated photon interactions, when distributed among *different* PCA detectors, could be readily identified as a genuine burst by means of microsecond coincidence timing among these independent PCA sensors. This has been experimentally verified for the PCA in our laboratory by noting that the recorded temporal jitter is less than 0.2 microseconds for the relative timing of detections in two separate XTE proportional counters for the simultaneous pair of 27 keV and 35 keV photons produced in the radioactive decay of ^{125}I (W. Zhang and K. Black, personal communication). At our suggestion, the temporal resolution to be available with XTE was adjusted to allow for the implementation of such a mode.

For estimating the limiting sensitivity of XTE for the detection of microbursts, we consider the hypothetical situation of an object such that the corresponding source count rate R_S in the composite PCA of 5 counters is

$R_S = 5 \times 10^2 L_{37}$ s^{-1} (Orlandini & Boldt 1993).

We assume that a fraction f of source emission appears in the form of discernible bursts of microsecond duration and that these bursts occur randomly at a rate λ s^{-1}. The expectation value for the number of counts (n) in the PCA from a characteristic burst is then $n = f R_S/\lambda$.

Table 1: XTE PCA Sensitivity to Microsecond Bursting							
Case #	L_{37}	λ (s^{-1})	f	R_S (s^{-1})	n	R_B (s^{-1})	R_A (s^{-1})
1	1	100	1	500	5	96	0.57
2	1	100	0.1	500	0.5	9.0	0.57
3	1	100	0.02	500	0.1	0.47	0.57
4	0.1	10	1	50	5	9.6	0.015
5	0.1	10	0.1	50	0.5	0.9	0.015

To exhibit the sensitivity of the XTE PCA for detecting microsecond bursting we have calculated the rate R_B for genuine burst detections and the rate R_A of accidental spurious detections for cases characterized by values of L_{37}, λ, and f of particular interest (see Table 1). From the rates listed in this table we conclude that in fact the accidentals considered here are sufficiently small for allowing the clear detection of bursts. Even for the worst case (#3) characterized by a small granular component of 2 percent, the number of detected *events* (bursts plus accidentals) would be greater than the expected number of accidentals by a statistically significant amount after an exposure of only 100 seconds.

For cases #4 and #5 we note that bursts of longer duration (*i.e.* 10 microseconds) could well be more appropriate. In such a situation the accidental rate would increase by a factor of ten. However, even for the worst-case scenario considered in this regime (Case #5) the average number of bursts detected in only a ten-second exposure would already be statistically significant. Hence, temporal granularity of the sort discussed here should be readily observable with a multi-detector instrument such as the XTE PCA. This unique capability of XTE will, for the first time ever, make it possible to directly probe the surface of a magnetized neutron star.

REFERENCES

Arons, J., & Lea, S.M. 1980, ApJ, 235, 1016
Herold, H. 1979, Phys. Rev. D, 19, 2868
Makishima, K., et al. 1990, ApJ, 365, L59
Morfill, G.E., Trümper, J., Bodenheimer, P., & Tenorio-Tangle, G. 1984, A&A, 139, 7
Nagase, F. 1989, PASJ, 41, 1
Orlandini, M., & Boldt, E. 1993, ApJ, in press
Orlandini, M., & Morfill, G.E. 1992, ApJ, 386, 703

THE USA EXPERIMENT ON THE ARGOS SATELLITE: A LOW COST INSTRUMENT FOR TIMING X-RAY BINARIES

K. S. Wood, G. Fritz, P. Hertz, W. N. Johnson, M. N. Lovelette, M. T. Wolff
Code 7621, Naval Research Laboratory, Washington, DC 20375

E. Bloom, G. Godfrey, J. Hanson, P. Michelson, R. Taylor, H. Wen
SLAC and Physics Dept., Stanford University, Stanford, CA 94305

ABSTRACT

The Unconventional Stellar Aspect (USA) experiment to be launched in September 1995 on the Advanced Research and Global Observations Satellite (ARGOS) is a low-cost, quick — yet scientifically ambitious — X-ray timing experiment. It is designed for the dual purpose of scientific research in X-ray timing and time resolved spectroscopy and also for exploration of applications of X-ray sensor technology. Bright galactic X-ray binaries are used simultaneously for both scientific and applied objectives.

1. SCIENTIFIC PROGRAM

The core observing program will involve viewing ∼ 30 bright X-ray sources, predominantly X-ray binaries, for a cumulative total of about a month apiece over a three year flight. Other selected targets will be observed for shorter intervals. USA will create a unique data base characterized by the highest exposure (area × time) and photon yields of any satellite to date. The precise (∼microsecond) time available on ARGOS will provide high accuracy for X-ray event times. Sub-millisecond time resolution will permit probing these sources at the timescale of processes near neutron star surfaces or the innermost stable orbit. A high priority goal is the discovery of millisecond binary X-ray pulsars that are likely sources of gravitational radiation.

States in low mass X-ray binaries. Coupled spectral and temporal states in LMXBs, with each state having a unique color–intensity relation, quasiperiodic oscillation (QPO), and low frequency noise (LFN), offer insight into the accretion mechanisms, physical properties, and geometrical configurations of these sources (Hasinger & van der Klis 1989; Lamb 1989). USA will enable detailed studies of Z-sources (e.g., Sco X-1, Cyg X-2) to be made on all three spectral branches. The large telemetry rates will mean that fast phenomena (e.g., QPOs) will always be accessible. The long exposures needed to overcome Poisson noise present in low signal-to-noise QPO and LFN signals will allow detailed tests (Norris et al. 1990) of prevailing QPO theories.

Coherent pulsations in low mass X-ray binaries. Long observations facilitate sensitive searches for the underlying spin periods of the neutron stars in LMXBs. The existence of QPOs, as well as favored models for the evolution of millisecond radio pulsars, imply that the neutron star in a LMXB has a weak magnetic field and a fast (1–10 millisecond) spin period. Accretion onto the magnetized neutron star should make the spin period detectable. A one month observation with USA of a bright LMXB, such as Sco X-1 or GX5-1, will yield sensitivity to pulse fluctuations well below current observation pulse fraction upper limits of 0.5% and well above current observational frequency limits of 500 Hz (Wood et al. 1991).

Orbital periods in low mass X-ray binaries. Long observing runs provide coverage to detect orbital periods in many LMXBs. Knowledge of binary orbit parameters also improves the sensitivity of searches for coherent millisecond pulsations. A one month observation with USA will contain 200–400 25 minute observations and will be sensitive to orbital modulations less than 0.5%.

Pulse period fluctuations in massive X-ray binaries. Binary pulsars such as Her X-1, Cen X-3, and Vel X-1 generate 500–2000 cts per pulse in the USA detectors. The resolution of individual pulses will reveal pulse-to-pulse variability in the sources. The power spectra of frequency fluctuations provide important constraints on the components of the spinning neutron star, e.g., the crust and core, and their coupling. Very small values of the period derivative will be within reach of USA during the initial three year observing program. Since essentially the entire sky is available every USA orbit, pulsars can be monitored at whatever frequency is necessary to preserve pulse count. Over three years, relative spin-up rates of 10^{-8} yr^{-1} are measurable.

Quasiperiodic variability in cataclysmic variables. Cataclysmic variables (CVs) exhibit a wide range of timing phenomena, including QPOs, X-ray transients, and complex light curves. While CVs are typically ~ 100 times fainter than LMXBs, their dynamical time scales are ~ 1000 times longer. USA's thin windows and soft X-ray sensitivity gives it the ability to observe more photons per dynamical time scale and may show that many of the phenomena observable in LMXBs are likewise observable in CVs. Highly correlated optical and X-ray luminosity variations are predicted (Wolff et al. 1991) and optical QPOs observed in several AM Her stars should also be accompanied by detectable X-ray QPOs.

Observational signatures of black hole candidates. Traditional observational characteristics of stellar mass black holes include spectral and luminosity states and high frequency flickering. The energy coverage of USA (1–15 keV) is good for studying both the soft and hard components of black hole candidate spectra. USA's fast time resolution capability (1 μs) and large area will allow flickering to be characterized at all relevant timescales, down to the innermost stable Keplerian orbital period, with highly significant amplitude measurements. From month-long observations of black hole candidates, USA will observe spectral state transitions and track the change in observable characteristics during these transitionss.

Simultaneous Observations. Simultaneous USA and Compton GRO observations will include targets such as the Crab Pulsar, Cyg X-1, and 3C279. In Cyg X-1 simultaneous USA and OSSE observations can be used for time lag studies to study the inverse Compton scattering region. Two other instruments alongside USA onboard ARGOS are capable of observing celestial targets. These are the far ultraviolet camera GIMI (P.I. G. Carruthers, NRL) and the extreme ultraviolet camera EUVIP (P.I. S. Bowyer, U. C. Berkeley). Coordinated observations of soft X-ray sources are possible and would provide simultaneous observations over a broad ultraviolet and X-ray bandwidth. USA can provide an enhancement of the simultaneous XTE mission in several ways. When USA and XTE simultaneously observe an X-ray source, they will constitute the largest collecting area ever pointed at an X-ray source, with USA substantially incrementing the collecting area below 3 keV. When the ARGOS and XTE orbits are out of phase, USA can provide monitoring observations of an X-ray source during the time it is in Earth occultation for XTE. The onboard GPS receiver will enable USA to make observations with absolute microsecond accuracy which can be used to check the time calibration of XTE.

2. USA INSTRUMENT AND ARGOS MISSION CHARACTERISTICS

The USA experiment will be built by the X-ray Astronomy Branch at Naval Research Laboratory and the Stanford Linear Accelerator Center, and the instrument will be integrated at NRL.

Key characteristics of USA include (i) long observing times on the brightest X-ray objects, (ii) large effective area and high time resolution capability (2000 cm^2 open area; 40 kbps telemetry, with 128 kbps available for short times; 1 μs time resolution), (iii) good low energy response (down to 1 keV), and (iv) high throughput computational capabilities onboard (Table 1).

USA is a reflight of Spartan-1 hardware, which was flown from the Space Shuttle in June 1985. The Spartan-1 detectors are being refurbished with the addition of a 1.2° field of view collimator, electronics capable of fast timing, and a two axis pointing system. The relatively large collecting area and thin front window will give USA the best low energy response ($E < 2$ keV) of any current or planned non-imaging X-ray mission. USA will have the telemetry rates necessary to support observations of bright sources with high count rates. The USA experiment will be mounted to a three-axis-stabilized nadir-pointed spacecraft. The X-ray detectors will be mounted on a 2-axis gimbaled platform to permit inertial pointing at celestial objects (Table 2).

Detailed design has begun, fabrication of USA subsystems will take place during early and mid 1993, and experiment integration will begin in the fall of 1993. Testing and calibration will occur in the summer of 1994 and USA will be delivered to Rockwell Industries in September 1994 for integration onto the ARGOS space vehicle. Launch will occur in September 1995 on a Delta II.

USA presents a rare opportunity to fly a scientific payload of considerable capability at a low cost to the astronomy community because it exploits the flight opportunity provided by the ARGOS mission now being prepared under the DoD Space Test Program (STP). The STP provides the space vehicle, including preflight tests and integration, launch, and data collection. NRL provides only the experimental payload. The ARGOS satellite will carry 8 experiments including three UV sensors and a GPS receiver. The orbit will be sun synchronous and circular at an altitude of 450 nautical miles and an inclination of 98.7°.

TABLE 1: Recent and Planned X-ray Timing Missions

	area	telemetry rate	typical observation	exposure	Crab count rate	
GINGA	4000 cm^2	8 kbps	< 1 day	10^8 cm^2 s	10^4 cts s^{-1}	‡
XTE	6250 cm^2	26 kbps	< 1 day	10^9 cm^2 s	10^4 cts s^{-1}	†
USA	2000 cm^2	40 kbps	1 month	10^{10} cm^2 s	10^4 cts s^{-1}	‡

† 2–30 keV count rate
‡ 1–30 keV count rate

3. TECHNOLOGICAL INNOVATION

ARGOS will be one of the first research satellites to fly a GPS (Global Positioning System) receiver, and USA will have access to the timing (1 μs) and positional (< 100 m) data from the receiver. This makes possible the determination of absolute UTC time with microsecond accuracy in orbit; similarly, barycentric correction is simplified as the spacecraft position is known onboard and this information becomes part of the downlinked data stream.

USA will provide the first flight test of the RH32 and RH3000 computers. These are radiation hardened, 32 bit processors with throughputs of ~ 20 MIPS. Two RH32 processors, manufactured by TRW and Honeywell, will be provided by Rome Laboratories, and a Harris RH3000 processor will be provided by the Naval Center for Space Technology at NRL. They will be mounted within the USA central electronics box and used for onboard data processing. This processing includes scientific applications, such as Fourier transforms and correlations, as well as applied navigational and fault tolerant algorithms.

TABLE 2: Operational Features of the USA Detector System

Gas:	P-10 at 1.1 atmosphere (baseline)
Flow rate:	low enough to allow for 3–6 year life
Window:	2.5 m Mylar
Energy range:	1–15 keV
Field of view:	collimation of 1.2° circular
Energy resolution:	0.17 (1 keV @ 5.9 keV)
Aperture (effective):	2000 cm^2 @ 3–6 keV
Background rate in orbit:	5-sided cosmic ray veto gives residual rate of 4×10^{-3} cts cm^{-2} s^{-1} @ 1–10 keV
Anode voltage:	2750 V
Calibration:	on command Fe55 source @ 5.9 keV
Temperature limits:	-10 to 50 C
Dimensions:	$75 \times 30 \times 75$ cm per detector
Mass:	40 kg per detector

4. COMMUNITY ACCESS

The USA consortium has established a small Science Working Group (SWG) with community representation. The SWG will optimize the scientific potential of USA and determine scientific priorities for observing targets, subject to certain constraints. Target sources must in general be compatible with the applied objectives of USA and its DoD funding agencies. USA will have the flexibility to respond quickly to some targets of opportunity with approximately 1–3 day turnaround after the decision to revise the observing plan; the SWG will establish guidelines for responding to potential targets of opportunity. The SWG will identify instances when coordinated observations with the Compton Observatory, XTE, or other ARGOS instruments are scientifically advantageous.

This work is supported by the Office of Naval Research, the Department of Energy, and the Ballistic Missle Defense Organization.

REFERENCES

Hasinger, G., & van der Klis, M. 1989, A&A 225:79
Lamb, F. K. 1989, in Proc 23rd ESLAB Symposium, ed. J. Hunt & B. Battrick, ESA SP-296 1:215
Norris, J. P., et al. 1990, ApJ 361:514
Wolff, M. T., Wood, K. S., & Imamura, J. N. 1991, ApJ 375:L31
Wood, K. S., et al. 1991, ApJ 379:295

Wind
Interactions

X-RAY SCATTERING IN X-RAY BINARY PULSARS

Fumiaki Nagase

The Institute of Space and Astronautical Science,
3-1-1 Yoshinodai, Sagamihara, Kanagawa 229, Japan

ABSTRACT

In X-ray pulsars, the emission from the bottom of the accretion column in the magnetic pole propagates to the observers through the surrounding environment: the magnetosphere, the accretion disk and the disk corona, the stellar wind, the companion's atmosphere and finally the interstellar medium. The photoionization zone of the stellar wind, which is formed surrounding the neutron star by X-ray irradiation, plays an important role as a reprocessing site. The spectrum originating from the polar caps is modified by the photoelectric absorption and subrequent fluorescent emission, Thomson/Compton scattering and synchrotron resonant scattering that occur in these environments. Modifications in the spectrum due to the reprocessing are briefly reviewed based on the recent results obtained from observations by *Tenma*, *EXOSAT* and *Ginga*. A preliminary result obtained from Vela X-1 with *ASCA* is presented, exhibiting a new aspect in studying X-ray pulsars.

1. INTRODUCTION

The accreting matter halted by the magnetic field at the magnetospheric boundary infalls onto either one or two poles of the neutron star along the dipole magnetic field. A shock front will be formed near the neutron star surface in the accretion column and the matter in the column between the shock front and the neutron star surface is thermalized, thus forming hot plasma caps at the magnetic poles. X-rays emitted from the region possess an energy spectrum that is characterized by a power-law at energies below ~ 15 keV and exponential break down above the cutoff energies (*e.g.*, White *et al.* 1983).

The X-rays emanating from the polar caps are first subjected to resonant scattering due to the strongly magnetized plasma in the magnetosphere near the bottom of magnetic poles. Thereafter the X-rays interact with the circumstellar matter, i.e., a stellar wind, an accretion disk and the companion atmosphere while propagating through the binary system. The interactions involve photoelectric absorption, subsequent fluorescent emission, and Thomson/Compton scattering. Thus, the pulsar spectra are modified from the power law shape and changed by orbital and pulse phases. Features of iron Kα emission and K-edge absorption, cyclotron resonant scattering and low energy absorption by cool circumstellar matter are required to explain the observed spectra.

The spectra of X-ray pulsars can be fitted by a function of the form:

$$f(E) = AE^{-\gamma}\exp\{-(H(E) + \sigma_a(E)N_H)\} + I_{Fe}(E), \tag{1}$$

where the emission line, $I_{Fe}(E)$, is usually simulated by a Gaussian distribution. The photoelectric absorption cross section σ_a is due to intervening cold matter—interstellar and circumstellar—of column density N_H, where cosmic abundances including an iron element are usually adopted (e.g., Morrison & McCammon 1983). The function $H(E)$ in equation (1) is a Lorentzian high-energy turn-over of the form (Makishima et al. 1990a):

$$H(E) = \frac{\tau_a W_a^2 (E/E_a)^2}{(E - E_a)^2 + W_a^2}, \tag{2}$$

where E_a, W_a and τ_a are, respectively, the energy, the width and the optical depth of the absorption line of the Lorentzian distribution. The $H(E)$ function resembles the cyclotron resonant scattering cross section (Herold 1979; Daugherty and Harding 1986). It is notable that this equation fits simultaneously both the features of high energy turnover and cyclotron resonant absorption when the cyclotron resonant absorption feature is visible, thus leading to $E_a = E_B$ (E_B the cyclotron energy; Makishima et al. 1990a).

The spectral features attributed to the cyclotron resonance have been observed at energies a few times larger than the cutoff energies in the spectra of several X-ray pulsars, such as Her X-1, X 0115+634, X 0331+53, and 4U 1538-52. A correlation between the cutoff energy in the continuum and the cyclotron energy is suggested from the *Ginga* observations.

Absorption of soft X-rays by circumstellar matter is observed from many X-ray pulsars, but it is most remarkably visible in the spectra of wind-fed X-ray pulsars, such as Vela X-1 and GX 301−2. It has been noted that the absorption column densities observed from individual pulsars have usually changed in correlation with the orbital phase but sporadic changes of N_H have been also detected occasionally. Even those pulsars which do not show significant absorption at the high-on non-eclipse phase, like Her X-1, do show soft X-ray absorption feature during specific orbital phases.

The fluorescent iron emission line at 6.4 keV is also prominent for many X-ray pulsars. The equivalent width usually ranges from100- 300 eV but occasionally increases upto 1 − 1.5 keV in some cases. Some pulsars such as V 0331+53 and 4U 1626-67, however, do not show significant iron line feature with an upper limit on the equivalent width of several tens of eV.

We can see occasionally the effect of scattered X-rays superposed on the heavily absorbed spectrum of the direct beam. During the eclipse phase of X-ray binaries, we are able to observe spectra due to purely reprocessed, i.e., scattering and fluoresced light. These are useful circumstellar matter diagnostics, which will be described in more detail in section 3.

Changes in the pulsar spectra are due to various combination of the effects of absorption, scattering and fluorescence. Some of these are intrinsic to the emission region, some due to the effect of radiation transfer in the strongly magnetized plasma, and the others are ascribed to the reprocessing by matter outside the magnetosphere: an accretion disk, its corona, stellar wind and companion atmosphere. Recent observations with ASCA have revealed previously unknown new features in the spectra observed during eclipse phases or low states of Vela X-1 Cen X-3 and Her X-1.

2. CYCLOTRON RESONANT SCATTERING

The propagation of photons in strongly magnetized plasmas in the pulsar magnetosphere is usually subject to anisotropic scattering, where the cross sections of scattering are different for the extraordinary and ordinary waves and show a resonant feature at the cyclotron energy:

$$E_B = \hbar\omega_B/(1+z_s) \sim 11.6(\frac{B}{10^{12}G})(1+z_s)^{-1} \text{ keV}, \tag{3}$$

where B is the magnetic field strength, $\omega_B = eB/m_e$ is the electron cyclotron frequency, and $z_s = (1 - 2GM_x/R_x c^2)^{-1/2} - 1$ is the gravitational redshift (M_x the mass and R_x the radius of the neutron star). Ignoring the continuum part of the cross section, the resonant part of Thomson scattering cross section of a photon with an angular frequency ω by an electron in the lowest Landau level is written in the nonrelativistic limit as (Herold 1979; Brainerd & Mészáros 1991):

$$\sigma_R = \frac{\sigma_T}{4}\frac{(1+\cos^2\theta)\omega^2}{(\omega-\omega_B)^2 + \Gamma^2/4}, \tag{4}$$

where σ_T is the Thomson scattering cross section, θ is the propagation angle relative to the magnetic field and Γ is the decay rate of the first Landau level, which is much less than ω_B. Noted that this function resembles the Lorentzian form of equation (2).

Thus, the detection of absorption line features due to cyclotron resonant scattering (CRSF) in the X-ray spectrum is a definitive method for estimating the strength of magnetic field at the neutron star surface. Before the Ginga observations, however, such measurements were obtained only for Her X-1 (Trümper et al. 1978) and 4U 0115+63 (Wheaton et al. 1979). From the Ginga observations, the CRSFs were discovered in the spectra of 4U 1538−52 (Clark et al. 1990), X0331+53 (Makishima et al. 1990b), Cep X-4 (Mihara et al. 1991a), etc. The results of the CRSF measurements with Ginga are summarized in Table 1 (see Mihara et al. in this volume for the details of the Ginga observations).

The following properties were obtained from a systematic search of the CRSF in X-ray pulsars with Ginga.

Table 1: *Ginga* Measurements of CRSFs in X-Ray Pulsar Spectrum [a]

Source	P_{pulse} (s)	P_{orb} (d)	E_c (keV) [b]	E_n (keV) [b]
Her X-1	1.24	1.7	17–21	35
4U 0115+63	3.6	24.3	7–9	12, 23
X 0331+53	4.38	34.2	14–17	28.5
1E2259+586	6.9	?	≤ 4	∼ 7?
Cep X-4	66.3	?	15–17	32
Vela X-1	283	8.96	15–20	27
4U 1907+09	438	8.4	14–16	21
4U 1538−52	530	3.74	14–16	21
GX 301−2	690	41.5	19–21	40

[a] Taken from Makishima and Mihara (1992).
[b] E_c and E_n represent the exponential cutoff energy (White *et al.* 1983) and the *n*-th cyclotron harmonic energy, respectively.

1. The cyclotron resonance energies measured so far from nine pulsars range between 7 keV and 40 keV. This corresponds to a magnetic field strength of $(0.6 - 3.5) \times 10^{12}$ G at the neutron surface, taking the gravitational redshift $(1 + z_s)$ to be approximately unity.

2. The luminosities of the nine pulsars in Table 1 were less than 4×10^{37} erg s^{-1} when the CRSFs were observed, while such a feature was not observed from very luminous pulsars such as Cen X-3 and SMC X-1 (Makishima & Mihara 1992).

3. There is a correlation of $E_B \simeq 2 \times E_c$ between the fundamental resonance energy (E_B) and the cutoff energy (E_c) of the power spectrum that is obtained through fitting the conventional exponential cutoff model (*e.g.*, White *et al.* 1983).

4. The absorption line features are, in general, broad (*i.e.*, much broader than the energy resolution of a proportional counter) and shallow. For some cases (*e.g.*, for the case of Vela X-1), the structure in the spectrum is visible only for a specific pulse phase.

5. The fundamental (and also the 2nd harmonic) resonance energies vary as much as 30 % along the pulse phase (Voges *et al.* 1982 and Soong *et al.* 1990 for Her X-1; Clark *et al.* 1990 for 4U 1538−52; Nagase *et al.* 1991 and Tamura *et al.* 1992 for 4U 0115+63).

The correlation in point 3 suggests that the cutoff of the spectrum at high energies is related to the cyclotron resonant scattering. The cutoff energy might then be a good measure of cyclotron energy and thus of the magnetic field strength for other X-ray pulsars from which the CRSF is not detectable. Thus the known X-ray pulsars in general seem to exhibit a narrow scatter in magnetic field strength, typically $B = (1-4) \times 10^{12}$ G. This fact provides an important clues for investigating magnetic field evolution of neutron stars.

These results also provides clues for investigating in detail the transport of radiation through strongly magnetized plasmas. As a result of these recent observations of CRSFs in the spectra of gamma-ray bursts (Murakami et al. 1988; Fenimore et al. 1988) and X-ray pulsars with *Ginga*, further theoretical calculations have been made (e.g., Pavlov et al. 1989; Lamb et al. 1990; Alexander & Mészáros 1991; Brainerd & Mészáros 1991).

Bulik et al. (1992) have successfully constructed phase-resolved model spectra, utilizing their magnetized radiation transfer code, and fit the model spectra with the 16-phase resolved spectra of 4U 1538−52 (Clark et al. 1990). This model predicts a distribution of magnetic field that is broader than that expected from a dipole field at constant radius. It also requires a significant difference between the two polar caps, both in terms of opening angle and temperature. This model can account for the phase dependent changes of the energy, the width and the depth of the cyclotron resonance.

3. SCATTERING BY A STELLAR WIND

Observations of X-ray pulsars covering large fractions of the orbital phase show a wide variety of changes in spectral shape, especially during the phase of eclipse transition, as well as the dip or the low intensity state of some pulsars such as Her X-1 and Cen X-3. The eclipse spectra of Cen X-3 and Vela X-1 show continuum shapes similar to those of non-eclipse spectra, with a reduction of intensity to a few % of the non-eclipse intensity. These X-rays observed during eclipses do not pulsate in all energy ranges, thus confirming the scattering origin by a stellar wind. Large iron line equivalent widths are observed in the eclipse spectra.

The spectrum of Her X-1 in the low state is much flatter than that of the high state non-eclipse spectrum. The X-ray intensity in the low state is about 2 % of the high state and it does not show pulsations. This spectrum can be interpreted by the sum of two kinds of scattering continua: scattering by a hot highly ionized plasma and forward scattering by cold dense matter (see Mihara et al. [1991b] for details). The pre-eclipse spectra in Her X-1 and Cen X-3 and the eclipse transition spectra in Vela X-1 can be interpreted as the sum of the highly absorbed direct beam and a soft component due to scattering by ambient matter (Mihara et al. 1991b; Nagase et al. 1992; Lewis et al. 1992). In these phases, the low energy X-rays below ~ 4 keV do not pulsate but X-rays in the higher energies do pulsate, thus suggesting a scattering origin of the soft X-rays. The role of the hot ionized region as an absorber and a scatterer has been also discussed using the *EXOSAT* observations of 4U1700−37 (Haberl et al. 1989) and those of Vela X-1 (Haberl & White 1990).

The properties of the stellar wind and the companion atmosphere in massive wind-fed pulsars were investigated from the light curves along the eclipse transitions (e.g., Clark et al. 1988; Day et al. 1988) and from the spectral analyses of data obtained during eclipse transition (e.g., Sato et al. 1986; Lewis 1992; Clark et al. 1993). From these studies, the atmospheres of massive early type companions such as the KRZ star V779 in Cen X-3, HD77581 in Vela X-1 and QV Nor in 4U 1538−52, are revealed to be extended over a scale height significantly larger than that expected from the hydrostatic equilibrium atmosphere. X-ray irradiation of the companion atmosphere may form a hot, gravitationally unbound corona region, giving rise to a thermally driven stellar wind from the X-ray heated face of the companion. The spectral changes along the eclipse transition in Vela X-1 and 4U1538−52 were well interpreted by the wind model consisting of the exponential corona region and the radiatively driven wind (Castor et al. 1975) using Monte Carlo simulations for the absorption and scattering (Lewis et al. 1992; Clark et al. 1993).

An "excess" of soft X-rays, compared to what is expected from photoelectric absorption by cold matter, is often visible at energies below $\sim$ 5 keV in the spectra obtained during high absorption phases, such as X-ray dips, eclipse transitions and low intensity states, of many X-ray binary pulsars. Haberl & White (1990) found from *EXOSAT* observations of Vela X-1 that X-ray pulsations in the low-energy excess during intervals of high absorption were greatly reduced, thus leading to the scattering origin of the excess component. Similar phenomena of soft excess spectra with no pulsation were also observed from GX 301−2 (Haberl 1991), Her X-1 (Mihara et al. 1991b) and Cen X-3 (Nagase et al. 1992) during their high absorption phases. Hence the scattering origin of the soft excess spectrum has been established.

Photoionization of circumstellar matter forms an ionized zone around the X-ray source (e.g., Hatchett & McCray 1977; Friend & Castor 1982; McCray et al. 1984) and reduces the X-ray opacity at low energies if the density of the matter is relatively low and X-ray irradiation is relatively strong (Krolik & Kallman 1984). This reduction of X-ray opacity will also contribute to an increase in the soft excess spectra. Haberl et al. (1989) found a correlation between the soft excess intensity and the source intensity in the spectra of 4U 1700−37 obtained from *EXOSAT* observations, suggesting the increase of partial ionization with the increasing intensity.

3. FLUORESCENT IRON LINE EMISSION

Evidence of iron fluorecsence emission line was first reported by Pravdo et al. (1977, 1978) from the spectrum of Her X-1. Thereafter emission line features at $\sim$ 6.5 keV were reoprted from several X-ray pulsars (Becker et al. 1978; Rose et al. 1979; White and Pravdo 1979; White et al. 1980). Spectroscopic studies using the gas scintillation proportional counters(GSPC) aboard *Tenma* and *EXOSAT* revealed that the iron line emission from X-ray pulsars is predominantly at 6.4 keV (Nagase 1985; Inoue 1985; Mkishima 1986; Nagase 1989), unlike the iron line emission from low-mass X-ray binaries which has been observed exclusively at 6.7 keV (White et al. 1985, 1986; Hirano et al. 1987). The 6.4-keV emission line in X-ray pulsars is attributed to fluorescent Kα emission due to reprocessing by cold (low ionization) circumstellar matter in the binary systems.

The first confident identification of the iron emission line as the 6.4 keV fluorescence line was made by Ohashi et al. (1984) using the GSPC data obtained with *Tenma* from Vela X-1. Detailed spectroscopic studies of iron lines were thereafter performed for Vela X-1 and GX 301−2 (Nagase et al. 1986; Sato et al. 1986; Leahy et al. 1989a, b). From the energy of the emission line, the ionization state of the line-emitting iron is estimated to be between Fe I and Fe XIX, whereas from the energy of the K-edge absorption, the ionization state of the X-ray absorbing iron is estimated to be between Fe V and Fe X for the two X-ray pusars (Nagase 1989). The line widths in X-ray the pulsars measured with *Tenma* are all smaller than the resolution of the GSPC (*i.e.*, 0.5 keV [FWHM] at 6.4 keV), in contrast to the broad line in the low mass X-ray binaries (White et al. 1985, 1986; Hirano et al. 1987).

The observed iron line intensity varied significantly from source to source, and from observation to observation. Yet a rough proportionality between the iron line intensity and the continuum intensity above 7.1 keV was observed for Vela X-1 (Ohashi et al. 1984) and for GX 301−2 (Leahy et al. 1989b). This relation and the observed energy of the iron line provide firm evidence that these iron lines are the result of the reprocessed fluorecsence emission of the X-rays at energies above 7.1 keV that are absorbed by relatively cold matter surrounding the source.

The observed relation between the equivalent width and the absorption column density toward the observer (N_H) is complicated and different from source to source. They are not always consistent with the estimates calculated from a simple spherically symmetric model (see Inoue 1985; Makishima 1986; Nagase 1989), thus suggesting complicated distribution of surrounding matter.

Pulsed modulation of the iron line intensity will be a good diagnostic for probing the distribution of reprocessing matter. No evidence, however, was measured from the previous *Tenma* observations (Ohashi 1984; Leahy and Matsuoka 1990). It is notable, however, that Day et al. (1992) recently found pulsations of the iron line intensity in Cen X-3 from the *Ginga* data. Since the amplitude of the iron line pulse is relatively small ($\sim$ 30 % at peak-to-bottom amplitude) compared with the amplitude of continuum pulse, this might imply the coexistence of a local reprocessing site that produces pulse modulation and a spherically (or axially symmetric) extended reprocessing site that smears pulsations of the iron line.

Apparent equivalent widths in spectra obtained during eclipse phases of a few pulsars such as Vela X-1 and Cen X-3 are relatively large: as large as $E.W. \simeq 1.0-2.0$ keV (Ohashi et al. 1984; Sato et al. 1986; Nagase et al. 1992). If the direct beam is completely occulted by the companion and only the scattering continuum is visible, the equivalent width is calculated to be $E.W. \simeq 1.5$ keV (Inoue 1985). Thus the spectra in eclipse can be explained by the scattering continuum and fluorescence iron line reprocessed by the extended region of the stellar wind.

Recent *Ginga* observations of Cen X-3 through the eclipse transition revealed a complex feature of iron emission lines. Co-existence of the 6.4 keV and 6.7 keV lines is strongly suggested from the dip and eclipse spectra (Nagase et al. 1992). From the orbital phase dependence of both the line intensities, they found that the 6.4 keV line originates from a local region smaller than the companion radius whereas the 6.7 keV line is produced in a wide region extended over the companion radius.

5. *ASCA* OBSERVATIONS OF VELA X-1

We have observed about 10 X-ray pulsars with *ASCA* during its performance verification phase in the first half year. Among these, Vela X-1 and Cen X-3 were observed throughout the eclipse transitions. Her X-1 was observed twice, once at the peak of main-on and again at the short-on of the 35-d cycle both during an extended low state.

Preliminary spectra obtained with the high resolution CCD camera on board *ASCA* are shown in Figure 1. The three spectra in the figure were observed from Vela X-1 in three different orbital phases; a post-eclipse phase, a pre-eclipse phase and an entire eclipse phase. In the spectra of pre-eclipse and post-eclipse, one can see clearly the two components; a heavily absorbed hard component and a relatively flat soft component. The former is attributed to the pulsating direct beam, whereas the latter is explained as the scattered light reprocessed by the stellar wind surrounding the neutron star.

A spectacular spectrum was obtained during the eclipse of Vela X-1, which is shown in the bottom of Figure 1 (note that this is an averaged spectrum of the whole eclipse interval). This spectrum should be purely due to the scattered light reprocessed by an extended stellar wind, as the direct beam is completely blocked by the companion star. The spectrum is characterized by the following properties: (1) iron emission lines are visible at 6.7 keV and 7.0 keV by resolving from the prominent 6.4 keV line, (2) the continuum spectrum is surprisingly ($\gamma \simeq 0$) flat between the energy range of 0.5 keV to 10 keV, (3) prominent lines are visible in the lower energies, at 0.92, 1.35, 1.85, and marginally at 2.4 keV, which are identified as the Kα lines from He-like ions of Ne, Mg, Si and S (note that Mg and Si lines are also visible in the pre-eclipse spectrum).

In addition to these features, a preliminary fit to the spectrum indicates possible intrinsic line broadening of about 100 eV, although we need further data reduction and analysis to confirm this. These features were never seen previously by *Tenma*, *EXOSAT* and *Ginga* (see the discussion in the previous section). We need to introduce new approaches to model the spectrum, since the

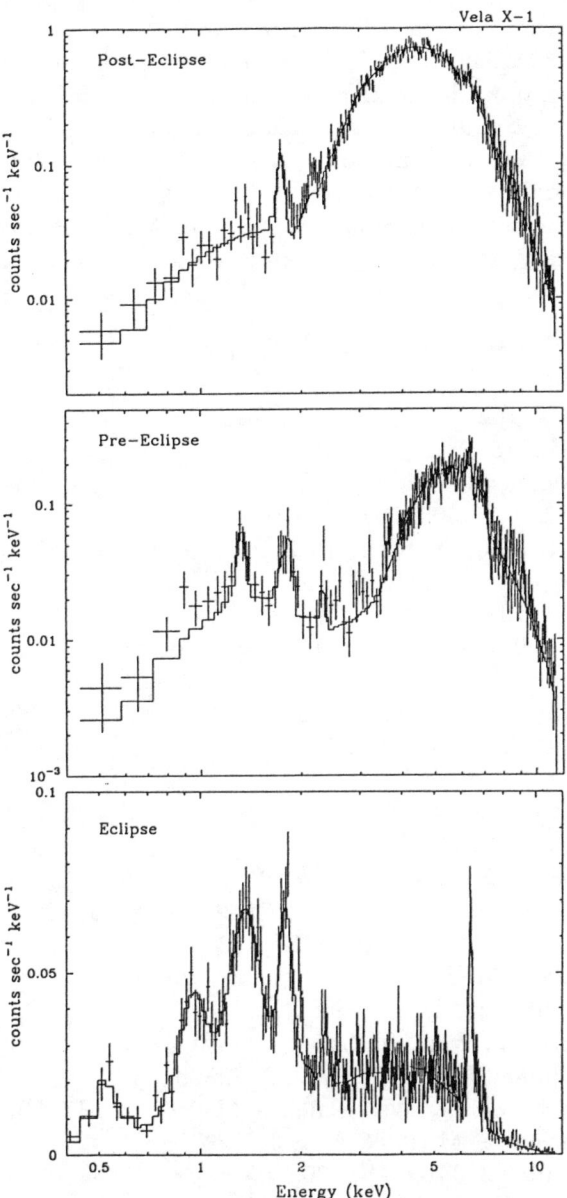

Figure 1: Pulse height spectra of Vela X-1 observed with the CCD camera on board *ASCA* on June 25 and July 6, 1993. The spectra obtained at the post-eclipse and pre-eclipse phases and during the eclipse interval are shown from top to bottom.

previous interpretations for the eclipse spectrum by Sato et al. (1986) and Lewis et al. (1992) do not explain these features. The photoionization structure of the stellar wind surrounding the neutron star (Hatchett & McCray 1977; Friend & Castor 1982; McCray et al. 1984) is crucially important. The He-like Kα lines may be caused by the recombination process. A delicate balance of scattering and self-absorption by the warm absorber (Krolick & Kallman 1984) will need to explain the flat continuum. Similar new features may be obtained for other massive X-ray binary pulsars using the high resolusion CCD cameras on board ASCA.

REFERENCES

Alexander, S. G. & Mészáros, P. 1991, ApJ, 372, 565
Becker, R. H., et al. 1978, ApJ, 221, 912
Brainerd, J. J., & Mészáros, P. 1991, ApJ, 369, 179
Bulik, T., et al. 1992, ApJ, 395, 564
Castor, J. I., Abbott, D. C., & Klein, R. I. 1975, ApJ, 195, 157
Clark, G. W., Minato, J. R., & Mi, G. 1988, ApJ, 324, 974
Clark, G. W., et al. 1990, ApJ, 353, 274
Clark, G. W., Woo, J. & Nagase, F. 1993, ApJ, in press
Daugherty, J. K. & Harding, A. K. 1986, ApJ, 309, 362
Day, C. S. R., Tennant, A. F., & Fabian, A. C. 1988, MNRAS, 231, 69
Day, C. S. R., et al. 1993, ApJ, 408, 656
Fenimore, E. E., et al. 1988, ApJ, 335, L71
Friend, D. B., & Castor, J. I. 1982, ApJ, 261, 293
Haberl, F. 1991, ApJ, 376, 245
Haberl, F., & White, N. E. 1990, ApJ, 361, 225
Haberl, F., White, N. E., & Kallman, T. R. 1989, ApJ, 343, 409
Hatchett, S., & McCray, R. 1977, ApJ, 211, 552
Hirano, T., et al. 1987, PASJ, 39, 619
Herold, H. 1979, Phys. Rev. D, 19, 2868
Inoue, H. 1985, *Space Sci. Rev.*, 40, 317
Krolik, J. H., & Kallman, T. R. 1984, ApJ, 286, 366
Lamb, D. Q., Wang, J. C. L., & Wasserman, I. M. 1990, ApJ, 363, 670
Leahy, D. A., & Matsuoka, M. 1990, ApJ, 355, 627
Leahy, D. A., et al. 1989a, MNRAS, 236, 603
Leahy, D. A., et al. 1989b, MNRAS, 237, 269
Lewis, W., et al. 1992, ApJ, 389, 665
Makishima, K. 1986, in *The Physics of Accretion onto Compact Objects*, ed. K.O. Mason, M.G. Watson & N. E. White (Springer-Verlag, Berlin), p. 249
Makishima, K., & Mihara, T. 1992, in *Frontiers of X-Ray Astronomy* (porc. of the 28th Yamada Conference, 1991 April 8-12 Nagoya, Japan), ed. Y. Tanaka & K. Koyama, (Uni. Acad. Press, Tokyo), p23
Makishima, K., et al. 1990a, PASJ, 42, 295

Makishima, K., et al. 1990b, ApJ, 365 , L59
McCray, R., et al. 1984, ApJ, 282, 245
Mihara, T., et al. 1991a, ApJ, 379, L65
Mihara, T., et al. 1991b, PASJ, 43, 501
Morrison, R., & McCammon, D. 1983, ApJ, 270, 119
Murakami, T., et al. 1988, Nature, 335, 234
Nagase, F. 1985, ASR, 5, 95
Nagase, F. 1989, PASJ, 41, 1
Nagase, F., et al. 1986, PASJ, 38, 547
Nagase, F., et al. 1991, ApJ, 375, L49
Nagase, F., et al. 1992, ApJ, 396, 147
Nagel, W. 1981a, ApJ, 251, 278
Ohashi, T., et al. 1984, PASJ, 36, 699
Pavlov, G. G., Shibanov, Yu. A. & Mészáros, P. 1989, Phys. Rep., 182, 187
Pravdo, S. H. 1979, in *X-ray Astronomy*, ed. W. A. Baity & L. E. Peterson (Pergamon Press, Oxford), p. 169
Pravdo, S. H., et al. 1978, ApJ, 225, 988
Rose, L. A., et al. 1979, ApJ, 231, 919
Sato, N., et al. 1986, PASJ, 38, 731
Soong, Y., et al. 1990, ApJ, 348, 641
Tamura, K., et al. 1992, ApJ, 389, 676
Trümper, J., et al. 1978, ApJ, 219, L105
Voges, W., et al. 1982, ApJ, 263, 803
Wheaton, Wm. A., et al. 1979, Nature, 282, 240
White, N. E., & Pravdo, S. H. 1979, ApJ, 233, L121
White, N. E., Swank, J. H., & Holt, S. S. 1983, ApJ, 270, 711
White, N. E., et al. 1980, ApJ, 239, 655
White, N. E., et al. 1985, ApJ, 296, 475
White, N. E., et al. 1986, MNRAS, 218, 129

HYDRODYNAMICS OF WINDS IN HIGH MASS X-RAY BINARIES

John M. Blondin
Department of Physics, North Carolina State University
Box 8202, Raleigh, NC 27695-8202
John_Blondin@ncsu.edu

ABSTRACT

The X-ray source in high mass X-ray binaries (HMXB's) provides a unique "active" probe of the winds of massive stars. As the binary system orbits around the center of mass, our line of sight changes continuously, allowing us to map the distribution of X-ray absorbing gas in the system. At the same time, the strong X-ray flux alters the local dynamics of the wind, providing us with an opportunity to study the physics of radiatively-driven winds. Some HMXB systems even provide a variable X-ray source, from which we can measure the effects of a local X-ray flux. New hydrodynamic models, coupled with recent observations, allow us to take advantage of this natural X-ray machine, providing new insight into the dynamics of radiatively driven winds. The various physical processes that affect the wind dynamics will be reviewed, as well as the observational consequences of these processes as gleaned from multidimensional hydrodynamic models.

1. INTRODUCTION

High mass X-ray binaries consist of a binary system with an OB giant or supergiant primary star in close orbit with a collapsed companion, either a neutron star or a black hole. These HMXB systems provide a convenient laboratory for studying the dynamics of winds from hot stars. The X-rays from the compact companion provide an active probe of the stellar wind, altering the ionization and thermal structure of the wind material and consequently changing the wind dynamics. In turn, the observed X-ray spectrum and the line of sight attenuation inferred from it, can provide information about the temporal and spatial structure of the winds in HMXB systems.

Multidimensional hydrodynamic simulations are beginning to provide us with the means for studying the dynamics of winds in HMXB's. Such models allow us to test a multitude of astrophysical theories, from the line-driving model for winds from hot stars to the ionization of X-ray irradiated gas. These simulations also provide valuable models of the structure of the gas in HMXB systems, a necessary ingredient in deciphering the observed X-ray spectra of HMXB's.

In this review we will focus on the dynamics of the stellar wind in HMXB systems. We will not consider the fate of the gas as it approaches the surface of the accreting neutron star, but rather concentrate on the gas dynamics on scales larger than $\sim 10^{10}$ cm. We will begin by detailing the individual physical processes that can effect the dynamics of the stellar wind in HMXB's, in effect presenting a "cookbook" for designing an HMXB model. In § 3 we will review some of the multidimensional hydrodynamic models of HMXB's that include some or all of these processes. Finally, we will conclude with a discussion of the observational implications of these models.

2. COMPONENTS OF AN HMXB MODEL

(a) Radiatively-driven winds. The theory of radiatively-driven winds from hot stars is a relatively mature theory in astrophysics. The force driving the mass outflow is the radiation pressure from the intense luminosity of OB-type stars. This radiation pressure is coupled to the wind material through numerous optical and UV lines (*e.g.*, Castor, Abbott, & Klein 1975). Advances on this standard model for hot-star winds predict wind speeds and mass-loss rates that are in quite good agreement with observationally inferred values for winds from single OB stars (*e.g.*, Friend & Abbott 1986; Pauldrach, Puls, & Kudritzki 1986). The radial velocity profile produced by this model can be fit by a simple expression, $v(r) = v_\infty (1 - R_*/r)^\beta$, where v_∞ is the terminal velocity of the wind, and the parameter β is typically ~ 0.8.

Of additional importance is the inherent instability of such a line-driven wind (*e.g.*, Owocki, Castor, & Rybicki 1988). In the HMXB models discussed in this review the radiatively-driven wind is modeled with the Sobolev approximation, which produces a smooth, steady wind. More accurate dynamical models of hot-star winds show that the line-driving force is unstable, leading to a very clumpy, time-variable wind. One must keep in mind the consequences of the Sobolev approximation when interpreting the results of these HMXB models.

(b) Wind accretion. Not long after the discovery of binary X-ray sources, Davidson & Ostriker (1973) pointed out that a compact companion in close orbit about an early-type star would capture some fraction of the stellar wind from such a star through its gravitational pull. If wind material passed within a distance $R_a \sim 2GM/v^2$ of the compact companion it would not have sufficient energy to escape the gravitational potential well of the companion and it would eventually be accreted. This simple theory leads to an estimated mass accretion rate onto the compact companion of

$$\dot{M}_a = \rho v \pi R_a^2 \sim \dot{M}_w \left(\frac{GM}{Dv^2}\right)^2,$$

where $\dot{M}_w$ is the wind mass-loss rate, D is the binary separation, and v is the relative velocity of the wind at the radius of the compact companion (including orbital motion). If one assumes the X-ray luminosity is driven by this mass accretion with some energy conversion efficiency η, such that $L_x = \eta \dot{M}_a c^2$, one finds that this simple relationship gives reasonable estimates of the observed X-ray luminosity for a few of the lower luminosity HMXB systems. However, the sensitivity of $\dot{M}_a$ to the relative wind velocity makes such conclusions very tentative. Nonetheless it suggests that at least some HMXB's may be powered by hydrodynamic accretion of the stellar wind from the primary star.

We have seen great strides in the understanding of hydrodynamic accretion in the past few years thanks to modern supercomputers. One of the principle highlights of this body of work is the discovery of the flip-flop instability. Virtually all 2D non-axisymmetric (*i.e.*, slab symmetry) hydrodynamic accretion simulations have shown evidence for this instability, in which the accretion wake swings from side to side in a quasi-periodic manner (Matsuda, Inoue, & Sawada 1987; Fryxell & Taam 1988; Taam & Fryxell 1989; Blondin et al. 1990, hereafter BKFT). However, we cannot, as yet, claim that this temporal behavior is present in unrestricted 3D accretion; Recent attempts to detect this instability in 3D simulations have so far been hampered by poor spatial resolution.

(c) Roche lobe overflow. While wind accretion provides a nice model for some HMXB's, for others it falls orders of magnitude short in producing the observed X-ray luminosity (Conti 1978). For some of these systems it appears that the primary star has filled its Roche lobe and is transferring mass to the compact companion directly via a tidal stream (Savonije 1983). Unfortunately we do not yet have a good quantitative understanding of the dynamics of the Roche lobe overflow process. In particular, we do not know the fraction of matter and angular momentum lost from the binary system. It is also important to point out that wind accretion and Roche lobe overflow are in fact just two extremes of a single process. There will always be some wind from the primary star, but as the binary separation decreases, such a wind becomes markedly peaked in the direction of the compact companion (Friend & Castor 1982; Blondin, Stevens, & Kallman 1991).

(d) X-ray photoionization. The X-ray luminosity of the accreting star can have a profound influence on the dynamics of the stellar wind by changing the line transitions available to absorb the momentum flux from the primary star (e.g., MacGregor & Vitello 1982). If the local flux of X-rays is sufficiently strong, the wind gas will be photoionized to such high levels of ionization that the most abundant ions will have most of their line transitions in the X-ray spectral range. However, the momentum flux from the primary star is radiated in the UV, so there will no longer be a significant overlap in frequency between the wind opacity and the photon flux from the primary.

The result of X-ray photoionization is a "Strömgren" region of superionization centered on the X-ray source, within which the stellar wind of the primary will no longer be continuously accelerated. The wind will simply coast through this Strömgren zone. This can have enormous effects on wind accretion because of the sensitivity of the mass-accretion rate to the relative wind velocity. If the Strömgren region extends a significant distance from the X-ray source to the primary, the wind will not be accelerated to high velocities as it passes the compact companion. The lower relative velocity of the wind at the companion then translates into a much higher mass accretion rate (Ho & Arons 1987).

In addition to slowing down the wind in the vicinity of the accreting star, X-ray photoionization can produce a wake of compressed wind material trailing the accreting star (Fransson & Fabian 1980). Gas that traverses the Strömgren zone will be moving with a slower radial velocity than gas that passes outside of this zone. Because the X-ray source is orbiting around the primary, the stalled wind will be "deflected" down-orbit by the Coriolis force, and the unaffected wind passing down-orbi of the Strömgren zone will run into and collide with this stalled wind. Hydrodynamic simulations including only this photoionization effect show that a significant column density can be built up in such a photoionization wake (BKFT).

The size of the Strömgren region is described by the critical value of the geometric parameter q, which defines surfaces of constant ionization parameter

(Hatchett & McCray 1977):

$$q_{cr} = \frac{\dot{M}_w \xi_{cr}}{4\pi \bar{m} v L_x} \approx \left(\frac{\dot{M}}{10^{-6} M_\odot/\text{yr}}\right)\left(\frac{L_x}{10^{37}\text{ erg/s}}\right)^{-1}\left(\frac{v_\infty}{10^3\text{ km/s}}\right)^{-1},$$

where v is the radial velocity of the wind at the radius of the companion, $\bar{m}$ is the average mass per ion, and ξ_{cr} is the critical ionization parameter at which the radiative driving force is removed. The value of ξ_{cr} is found to be of order 10^2 (Stevens 1993). The cutoff in the driving force is in fact relatively sharp, dropping by more than three orders of magnitude as ξ increases from 30 to 300. This rapid cut off is the result of a Helium ionization edge, beyond which the ionizing X-ray photons are strongly absorbed (Masai 1984; Stevens 1993). For small column densities in the wind ($N_H \ll 10^{22}\text{cm}^{-2}$), this radiative transfer effect is not as pronounced and the cut off in the driving force is much more gradual.

For $q_{cr} \gtrsim 10$ the Strömgren region is relatively small and X-ray photoionization has little effect on the wind. For $1 \lesssim q_{cr} \lesssim 10$ the Strömgren region covers a substantial fraction of the exposed stellar wind, producing a strong photoionization wake. For $q_{cr} \lesssim 1$, the entire wind exposed to the X-ray source is photoionized, and the radiatively-driven wind is completely quenched (save for the X-ray shadowed side of the primary).

(e) X-ray heating. In addition to changing the ionization state of the wind, a strong local X-ray flux will also change the thermal state of the wind. The temperature of the wind will range from the effective temperature of the OB star in regions far from the X-ray source, up to the inverse Compton temperature of the X-ray flux in regions where the local ionization parameter is greater than $\sim 10^5$. X-ray heating of the wind produces several effects: a lowering of the Mach number of the wind, a reduction in the accretion radius, and significant changes in the dynamics of the accretion flow. The lowering of the wind Mach number leads to both a decrease in the accretion radius and to a weaker bowshock. An important consequence of a weaker bowshock is a decrease in the compression of gas in the bowshock, resulting in less column density within the bowshock.

BKFT found that X-ray heating near the accreting star had a strong influence on the flow dynamics. The local ionization parameter very close to the X-ray source is very high ($\xi \gg 10^5$), for which the gas temperature is closely tied to the temperature of the X-ray source via Compton scattering. Simulations by BKFT suggest that this strong thermal coupling to the X-ray flux may contribute to the flip-flop phenomenon; An adiabatic simulation produced a steady, asymmetric accretion, but a similar simulation including X-ray heating and cooling produced a flip-flop accretion flow. For very low values of q_{cr} (high ratios of $L_x/\dot{M}_w$) the region of isothermal gas with $T = T_{IC}$ extends out beyond the accretion bowshock. In this case the wind accretion appears to be much more stable (BKFT).

(f) Thermally-driven winds. Another important consequence of strong X-ray heating is the possible presence of a thermally-driven wind. If the X-ray flux irradiating the surface of the OB star is sufficiently intense, the atmosphere can be heated up to the point where it is no longer gravitationally bound to the star, giving rise to a thermally-driven wind. The idea of an X-ray excited wind has been discussed in the context of Her X-1, low-mass X-ray binaries, and evaporating millisecond pulsars. More recently, Day & Stevens (1993) have suggested that an X-ray excited wind may also be relevant to HMXB's, and in particular to Cen X-3.

(g) Radiation Pressure. X-ray luminosities from HMXB's can approach the Eddington limit for solar-mass objects, at which point the momentum transfer from the X-ray flux to the surrounding gas is sufficient to counteract the gravitational pull of the accreting star. For the lower L_x systems the X-ray luminosity is 1 to 2 orders of magnitude smaller than the Eddington limit, and the radiation pressure of the X-ray flux is expected to do no more than slightly reduce the effective gravitational pull of the accreting star. This situation might be different, however, if the opacity of the gas is sufficiently larger than that of electron scattering, or if the emergent flux is not isotropic.

(h) Feedback loop. All of these processes by which the X-ray flux can affect the dynamics of the wind (photoionization, heating, momentum transfer) can give rise to a feedback effect by virtue of the fact that the X-ray luminosity is powered (at least in part) by accretion from the wind. If the X-rays can affect the dynamics of the wind they can, in effect, modulate the mass accretion rate from the wind, which in turn will modulate the X-ray luminosity itself.

Perhaps the strongest feedback is due to X-ray photoionization. Ho & Arons (1987) constructed a simple model to illustrate the feedback of a strong X-ray flux on the wind accretion fuelling the X-ray source. If the X-ray luminosity of a given system is increased, the region of super-ionization will be increased, and the radiative-driving force will be quenched closer to the surface of the primary star. As a result, the relative velocity of the wind in the vicinity of the accreting star will be lower, the effective accretion radius will be higher, and the total mass accretion rate will be higher. If the X-ray luminosity if fuelled by mass accretion, this implies that a higher X-ray luminosity will in turn lead to a higher accretion-driven X-ray luminosity: a feedback loop. Ho & Arons (1987) argued that this feedback can explain the relatively large X-ray luminosities of Cen X-3 and SMC X-1. However, the X-ray luminosity in these two systems is so high that the radiatively-driven wind is likely to be completely suppressed.

The feedback produced by X-ray heating of the wind and momentum deposition in the wind has recently been investigated using 2D axisymmetric models (Taam, Fu, & Fryxell 1991). At moderate accretion rates X-ray heating and radiative cooling become important and the accretion rate is suppressed below the value expected in adiabatic accretion. At high accretion rates, such that the accretion driven X-ray luminosity is of order the Eddington limit, the deposition of momentum by the X-ray photons greatly reduces the net mass accretion rate. In addition, the flow is found to be much more variable than accretion flows at lower mass accretion rates.

Another possible feedback effect in HMXB's can be traced back to theoretical work on Her X-1 (*e.g.*, McCray & Hatchett 1975), namely an X-ray excited wind. In this model the high X-ray luminosity state of an accreting X-ray binary is explained by the presence of a thermally-driven wind excited by the irradiation of the stellar surface by a strong X-ray flux. This excited wind in turn provides the high mass accretion rate needed to sustain the high X-ray luminosity. Such a feedback may play a role in the supergiant HMXB systems, providing a natural explanation for the observed high and low X-ray states of systems like Cen X-3 (Day & Stevens 1993). At low luminosity the X-ray source is powered by accretion from a normal radiatively-driven wind, while at high luminosity it is powered by accretion from an X-ray excited thermally-driven wind.

(i) Stellar rotation. Rapid rotation of the primary star in HMXB's can have a strong influence on the dynamics of the stellar wind. Bjorkman & Cassinelli (1993) have recently presented a simple but powerful explanation for Be stars, of how a radiatively-driven wind from a rapidly rotating star will be strongly focussed into the equatorial plane. If the OB primary is rotating at a significant fraction of the critical rotation rate ($\Omega_{cr} = \sqrt{GM/R^3}$), the radiatively-driven wind will be strongly focussed towards the equatorial plane forming a wind-compressed disk. Most of the supergiant HMXB systems have a primary star that is rotating at, or close to corotation. For these close binary systems corotation implies sufficiently high rotation rates that a wind-compressed disk will be formed. This is illustrated in a series of 2D simulations of the wind from the OB stars in several HMXB systems (see Owocki, Cranmer, & Blondin 1994). The OB star is assumed to be in corotation with the orbiting X-ray source, and the effects of the compact companion are completely ignored. The following table lists the ratio of the wind density in the equator to the density in the polar region for these 2D simulations.

HMXB	Ω/Ω_{cr}	v_∞/v_{esc}	ρ_{eq}/ρ_p
Vela X-1	.54	3.1	8.1
4U 1700-37	.58	2.8	8.4
Cen X-3	.58	2.6	5.7
SMC X-1	.45	3.0	9.0

3. HMXB MODELS

Of these various components of an HMXB model, each has been investigated individually and suggested as an explanation of certain observed properties of HMXB's. However, many of these physical processes affecting the wind dynamics can interact with each other, producing a situation which is more complicated than the sum of the individual components. Recent work (*e.g.*, BKFT, Blondin *et al.*1991) has begun to advance past the stage of simplified models, incorporating several of these components into a single multidimensional HMXB model.

An example of the current status of 2D hydrodynamic models of HMXB's is shown in Figure 1. This particular model is constructed with system parameters appropriate to Vela X-1, and as such represents the subclass of low X-ray luminosity systems. In this case the primary star is well within it's Roche lobe, but the wind is still peaked in the direction of the compact companion. The accretion flow is very unsteady and exhibits signs of the flip-flop phenomenon.

For the high-L_x, low-q_{cr} systems (*e.g.*, Cen X-3, SMC X-1) the entire wind exposed to X-rays is super-ionized, completely quenching the radiatively-driven wind from the exposed surface of the primary star. However, a stellar wind will still be driven from the shadowed side of the companion (Blondin 1994). Rotational effects can bring some of this wind material into the observers line of sight to the X-ray source, contributing to the attenuation of X-rays. Figure 2 shows a 2D hydrodynamic simulation of an HMXB system with $q_{cr} \approx 1$. Near

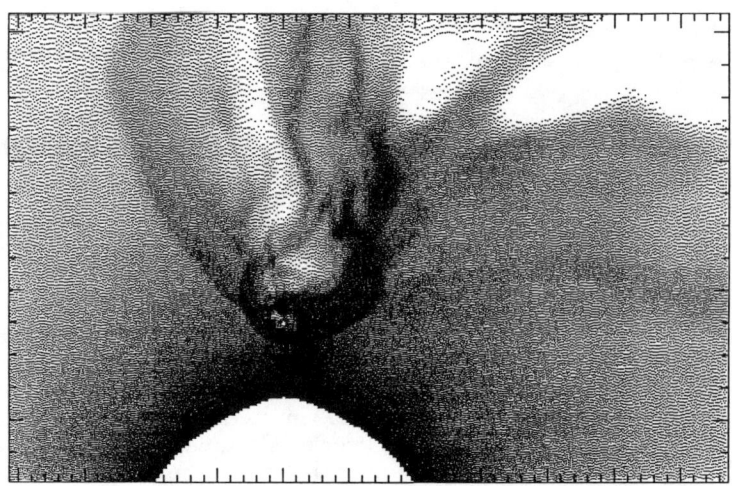

Figure 1. Two-dimensional hydrodynamical model of a low X-ray luminosity system such as Vela X-1. The gas density is plotted in halftone, with the highest density in black. The black dot marks the location of the X-ray source. Orbital rotation is counterclockwise.

eclipse egress the shadowed wind comes out of the X-ray shadow and is quickly ionized, stalling the wind at relatively low velocities. As the binary system continues to rotate, this stalled wind is deflected toward the X-ray source. Near eclipse ingress, gas that has been "boiled" off the heated surface of the primary makes its way into the X-ray shadow, where it is then slowly accelerated away from the system.

These 2D HMXB models are limited in several respects because they do not include flow into or out of the equatorial plane. This means that the density in the accretion and photoionization wakes may be an underestimate, that the mass accretion rate may be incorrect, and that the temporal behavior of the accretion flow may not represent the true behavior of HMXB's. Furthermore, some systems are seen at a relatively low inclination angle, for which the gas density in the equatorial plane is inadequate for predicting the line of sight column density.

Of all these anticipated differences between 2D and 3D hydrodynamic simulations of HMXB's, the most significant difference was not predicted: the wind-compressed disk due to stellar rotation. Here we present a new 3D HMXB simulation, which illustrates the importance of the Bjorkman-Cassinelli model of rotating winds for at least some HMXB's.

Figure 3 shows two views of a 3D model of SMC X-1. In this initial 3D model the gravitational pull of the compact companion has been removed in order to decrease the computational resources needed. In this particular system the accretion radius is expected to be relatively small. The view of the plane intersecting the rotation axis and the line of centers of the binary system shows the presence of a dense wind-compressed disk in the X-ray shadow, as predicted by the Bjorkman-Cassinelli model. This focussing of the wind into the equatorial plane can lead to a much larger line of sight column density when viewing the X-ray source through the equatorial plane.

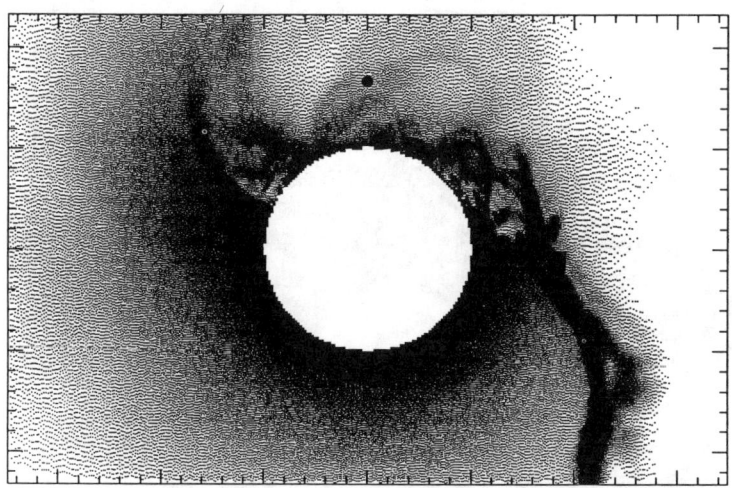

Figure 2. The shadow wind in a high X-ray luminosity HMXB. Rotation effects drag the shadow wind into the line of sight to the X-ray source. The radiatively-driven wind from the front side of the primary is quenched by photoionization, replaced by a thermally-driven wind excited by X-ray heating.

4. CONCLUSIONS DERIVED FROM HMXB MODELS

While there is still much to be done to improve the status of current HMXB models, we are now in a position where we can begin to draw meaningful conclusions by comparing these hydrodynamic models with X-ray, UV, and optical observations.

Perhaps the most straight-forward diagnostic of the wind dynamics is the line of sight column density, N_H, as revealed by X-ray attenuation. Hydrodynamic models have identified several sources of anomalous N_H including compression in an accretion bowshock and wake, a photoionization wake, a tidal stream, and the shadow wind. The bowshock/wake structure is almost always observed to vary in time, often displaying a quasi-periodic behavior associated with the flip-flop instability. This time scale is often only hours long, much shorter than an orbital period. Thus, the contribution to N_H made by the bowshock/wake can be expected to change from orbit to orbit.

In contrast, the tidal stream is expected to produce a large excess in N_H at phase $\Phi \sim 0.6$ that remains constant from orbit to orbit. The photoionization wake is also expected to remain relatively constant, but occurs at later orbital phases. The presence of a tidal stream implies a primary star close to filling it's Roche lobe, while the presence of a photoionization wake implies a value of $q_{cr} \sim 3$. However, the tidal stream and photoionization wake are mutually exclusive; if a tidal stream is present a photoionization wake will not form.

The presence of a shadow wind can profoundly change N_H near eclipse ingress and egress. This effect has not yet been studied quantitatively, but may help explain the previous poor fit of observationally inferred N_H to simple wind models (*e.g.*, Clark, Minato, & Mi 1988).

Another important diagnostic is the accretion rate of angular momentum, as reflected in changes of the rotational rate of the accreting neutron star.

586 Hydrodynamics of Winds in High Mass X-Ray Binaries

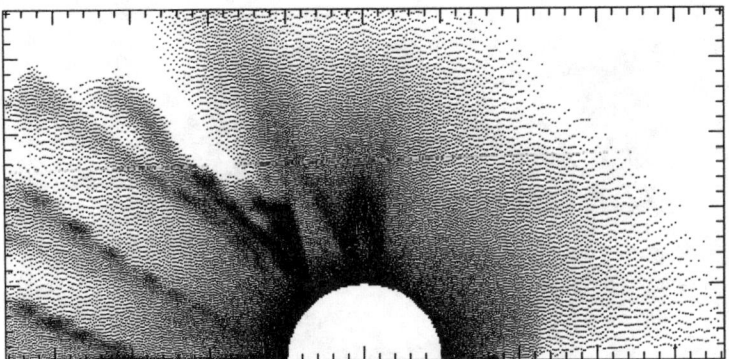

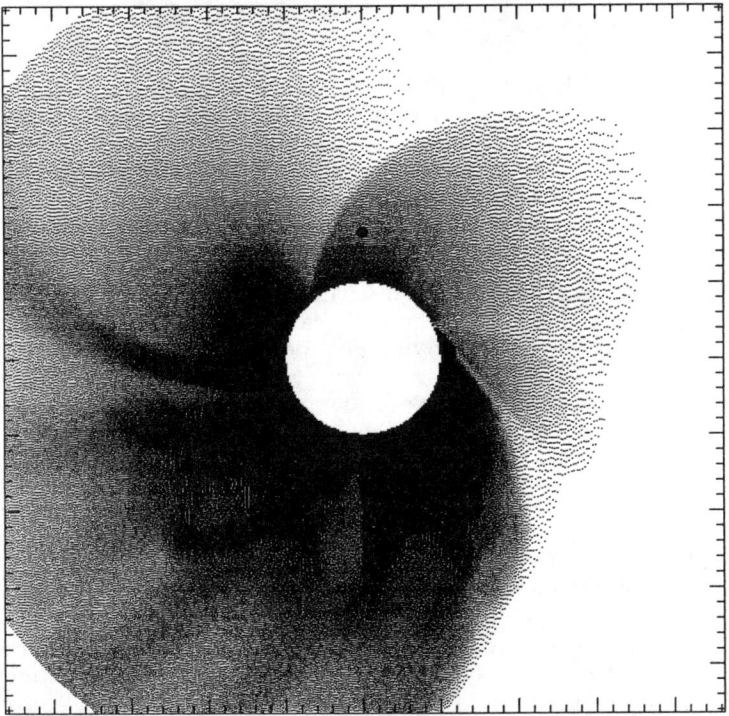

Figure 3. Three-dimensional hydrodynamic simulation of SMC X-1. The bottom panel shows the density of the wind in the equatorial plane. The top panel shows the wind density in the plane containing the rotation axis and the line of centers of the binary system. The X-ray source is marked by the black dot, and the orbital rotation is counterclockwise.

Contrary to conventional wisdom, wind accretion **can** produce large angular momentum accretion rates. The presence of the flip-flop instability implies that the sign of the accreted angular momentum will change on the time scale of hours. Such behavior appears consistent with observations of of Vela X-1 (Boyton *et al.* 1984), for which the time derivative of the X-ray pulsar period is seen to change sign on the shortest observed time scale.

The presence of a strong tidal stream appears to preclude the flip-flop instability, suggesting that observational evidence for a tidal stream would always accompany a relatively smooth change in the pulsar period. The reverse may not necessarily be true; a smooth change in the pulsar period need not imply the presence of a tidal stream, but could in fact be produced purely through wind accretion.

ACKNOWLEDGEMENTS

I would like to express my gratitude to Tim Kallman for introducing me to the fascinating world of HMXB's. This work was partially supported by NASA grant NAGW-3658. Many of the numerical simulations described herein were computed on a Cray Y-MP at the North Carolina Supercomputing Center.

REFERENCES

Bjorkman, J. E., & Cassinelli, J. P. 1993, ApJ, 409, 429
Blondin, J. M. 1994, in preparation
Blondin, J. M., Kallman, T. R., Fryxell, B. A., & Taam, R. E. 1990, ApJ, 356, 591
Blondin, J. M., Stevens, I. R., & Kallman, T. R. 1991, ApJ, 371, 684
Boyton, P. E., *et al.*1984, ApJ, 283, L53
Castor, J. I., Abbott, D. C., & Klein, R. I. 1975, ApJ, 195, 157
Clark, G. W., Minato, J. R., & Mi, G. 1988, ApJ, 324, 974
Conti, P. S. 1978, A&A, 63, 225
Davidson, K., & Ostriker, J. P. 1973, ApJ, 179, 588
Day, C. S. R., & Stevens, I. R. 1992, ApJ,
Fransson, C., & Fabian, A. C. 1980, AA, 87, 102
Friend, D. B., & Abbott, D. C. 1986, ApJ, 311, 701
Friend, D. B., & Castor, J. I. 1982, ApJ, 261, 293
Fryxell, B. A., & Taam, R. E. 1988, ApJ, 335, 862
Hatchett, S. P., & McCray, R. A. 1977, ApJ, 211, 552
Ho, C., & Arons, J. 1987, ApJ, 316, 283
MacGregor, K. B., & Vitello, P. A. J. 1982, ApJ, 259, 267
Masai, K. 1984, Ap& SS, 106, 391
Matsuda, T., Inoue, M., & Sawada, K. 1987, MNRAS, 226, 785
McCray, R., & Hatchett, S. 1975, ApJ, 199, 196
Stevens, I. R. 1991, ApJ, 379, 310
Owocki, S. P., Castor, J. I., & Rybicki, G. B. 1988, ApJ, 335, 914
Owocki, S. P., Cranmer, S. R., & Blondin, J. M. 1993, ApJ, in press
Pauldrach, A., Puls, J., & Kudritzki, R. P. 1986, A&A 164, 86
Savonije, G. J. 1983, in *Accretion Driven Stellar X-ray Sources*, eds. W. H. G. Lewin & E. P. J. van den Heuvel (CUP), p. 343
Taam, R. E., Fu, A., & Fryxell, B. A. 1991, ApJ, 371, 696
Taam, R. E., & Fryxell, B. A. 1989, ApJ, 338, 297

PROPERTIES OF INTERSTELLAR GRAINS DERIVED FROM X-RAY ECLIPSE OBSERVATION

Jonathan W. Woo and George W. Clark
Center for Space Research and Department of Physics
MIT, Cambridge, MA 02139

Fumiaki Nagase
ISAS, 3-1-1 Yoshinodai, Sagamihara, Kanagawa 229, Japan

ABSTRACT

By a Monte Carlo computation we have estimated the spectrum of X-rays scattered into the eclipse of 4U1538-52 by the atmosphere of its supergiant companion. The estimate fails to account for a soft component below 4.5 keV which amounts to approximately 1.4 % of the average uneclipsed flux in the same energy range. The intensity of the soft component exhibits an initial downward trend following eclipse ingress as expected of a component scattered by interstellar dust grains. Comparing the intensity with the optical extinction of QV Nor, we derive an upper limit on a quantity $R_{XV}(E)$ which we call the scattering/extinction ratio of interstellar dust grains and define as $(E/1 \text{ keV})^2$ times the ratio of the optical depth for scattering X-rays of energy E to the total optical extinction. In the Rayleigh-Gans approximation to the X-ray scattering efficiency, this quantity is independent of energy. Our upper limit on R_{XV} is 0.06 mag^{-1}, which implies that the X-ray scattering efficiency of interstellar dust is less than expected for solid grains with a size distribution of the form $n_g(a) \sim a^{-3.5}$ in the range from 0.005 μm to 0.25 μm and composed of silicate (R_{XV} =0.22 mag^{-1}) or a silicate-graphite mixture (R_{XV} =0.11 mag^{-1}) as derived from the calculations of Martin & Rouleau (1991). This lends support to the idea (Mathis & Whiffen 1989) that interstellar grains are "fluffy" aggregates with an average bulk density less than that of their constituent particles. Such aggregates would have a smaller ratio of X-ray scattering efficiency to optical extinction efficiency compared to solid grains of the same material.

1. INTRODUCTION

Xu, McCray, & Kelley (1986) pointed out that an eclipsing X-ray binary offers a unique advantage for the study of interstellar grains. Since the optical extinction and the intensity of grain-scattered X-rays depend on the size, composition, and density of the grains in quite different ways, a determination of the quantitative relation between these two observables can place a critical constraint on models of interstellar grains. The angular distribution of X-rays in the image of an X-ray source consists of a "core" component that comes directly from the star or its immediate vicinity, and a generally much fainter "halo" component that has been scattered by interstellar grains near the line of sight, as predicted by Overbeck (1965) and discussed by him and others (Hayakawa 1970; Martin 1970) before the availability in orbit of imaging X-ray telescopes. Measures of the intensities of Overbeck halos have since been derived from analyses of X-ray star images in which the halo is dissected from the much brighter core by subtracting the point-spread function of the telescope (e.g. Catura 1983; Mauche & Gorenstein 1986; Predehl et al. 1991). In addition, the halo intensity of a source near the Galactic center has been measured with a nonimaging detector during a brief time following a lunar occultation of the core component (Mitsuda et al. 1990). In the case of an eclipsing binary at a distance of several kiloparsecs, the halo will persist long after the neutron star is eclipsed by its primary because the average travel time of the grain scattered X-rays is many hours longer than the travel time of the undeviated X-rays in the core. Since the X-ray scattering efficiency of small particles varies as E^{-2}, the spectrum of the grain-scattered X-rays is much softer than the uneclipsed core spectrum. Consequently, when the core intensity is reduced by the eclipse, the grain-scattered X-rays may stand out in the spectrum as a "soft component" that decays with time after ingress. Analysis of high-resolution X-ray images of binary X-ray stars in eclipse will clearly be the best way to exploit such opportunities in the future. Nevertheless, we have obtained significant results from an analysis of nonimaging data obtained during the X-ray eclipse of 4U 1538-52/QV Nor, a system which is especially favorable for such a study because of its high optical extinction.

2. OBSERVATIONS

We observed 4U 1538-52 from 1988 February 29 to March 3 with the large-area gas proportional counter (LAC) of the *Ginga* satellite X-ray observatory. The observatory has been described by Makino et al. (1987), and details of the LAC detector have been given by Turner et al. (1989). The LAC had a $1°.1 \times 2°.0$ FWHM field of view and an effective area of 4000 cm^2. The energy resolution of the LAC detector was 20% FWHM at 5.9 keV.

3. DATA ANALYSIS

Estimates of the wind density as a function of the distance from the center of QV Nor are obtained from an analysis of the variation of X-ray attenuation during an eclipse egress (see Clark et al. 1994 for the details). The analysis takes account of the effects of X-ray ionization on the photoelectric absorption cross sections (Kallman & Krolik 1993) and yields a particle-number density described by the function

$$n(r) = \frac{\Psi}{4\pi\mu r^2}\left[1 + \left(\frac{r}{r_1}\right)^2 \exp\left(-\frac{(r-r_1)}{h}\right)\right]. \quad (1)$$

with $\Psi = 6.7 \times 10^{-10} M_\odot$ yr^{-1} km^{-1} s, $h = 4.3 \times 10^{10}$ cm, and $r_1 = 1.2 \times 10^{12}$ cm, where $\mu = 1.34 m_H$ is the average atomic mass per hydrogen atom.

To obtain a prediction of the spectrum of X-rays emerging from the system into the eclipse shadow, we developed a Monte Carlo code to compute the propagation of X-rays in three dimensions through an X-ray-ionized atmosphere. We have assumed that the density in the stellar atmosphere is given by the Equation (1) and the corresponding photoelectric absorption cross section at a given point is determined by the local ionization parameter $\xi = L_X/(nr^2)$. The local ionization parameter was calculated by propagating the source spectrum through the density model. The optically thin assumption ($E > 1$ keV) allowed us to determine the X-ray absorption cross section uniquely at a given position for a specific energy according to the local ionization parameter. At each interaction sight, we have determined the photoelectric absorption cross section by interpolating two dimensional look-up table for the various ionization parameters and energies; the table was constructed by running the XSTAR code (Kallman & Krolik 1993) for the measured X-ray source spectrum.

The average PHD for a deep eclipse phase is displayed in Figure 1. From the Monte Carlo calculation, we have obtained the predicted PHD of X-rays scattered into the umbra as shown as the dashed line in Figure 1. Based on the spherically symmetric hybrid density model determined from the fit using the egress PHDs, our initial attempt to predict the average deep eclipse PHD fell short by one third of the observed count rate. We attribute the extra third to X-rays scattered from an "accretion wake" like that suggested by the hydrodynamic calculations of Blondin et al. (1990). With a simple geometric model which mimics a typical accretion wake, we were able to predict a deep eclipse PHD that conforms in shape and magnitude with the observed eclipse PHD above 4.5 keV (solid line in Figure 1).

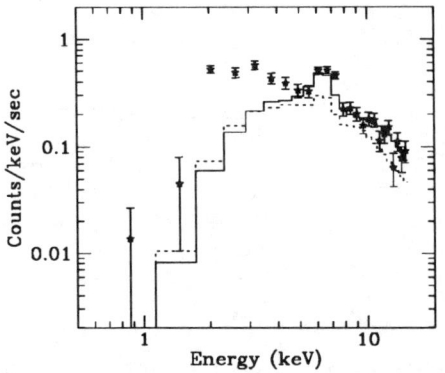
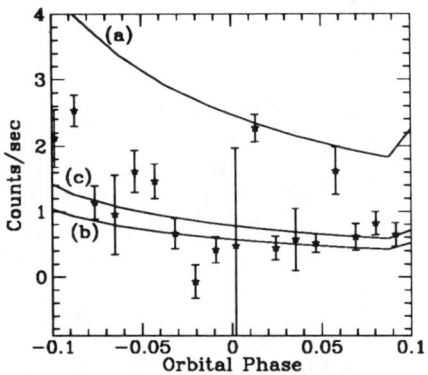

Fig. 1. Average PHD recorded during the eclipse. The dashed line is the PHD predicted by Monte Carlo computation for X-rays with the hybrid density function and the solid line is that plus an accretion wake.

Fig. 2. Plots of the soft component count rate against orbital phase during the eclipse. The curves are the predicted rates for grain-scattered X-rays of three grain models: (a) silicate, (b) graphite, and (c) silicate-graphite.

In channels below 4.5 keV, the Monte Carlo calculation with the density enhancements predicts rates that are substantially smaller than the observed rates. By subtracting the Monte Carlo rates, we obtain the average PHD of what we call the "soft component" whose average count rates are shown in Figure 2. There appears to be a downward trend during the first half of the eclipse, which may reflect the decaying intensity of grain-scattered X-rays. The solid lines are the predicted count rates of the grain-scattered component for "astronomical silicate", graphite, and a mixture of the two materials.

4. DISCUSSION

Taking the total intensity of the soft component as an upper limit on the intensity of grain-scattered X-rays during the eclipse, we derive a constraint on models of the composition and size distribution of interstellar grains. The constraint is in the form of an upper bound on a quantity we call $R_{XV}(E)$, which is a characteristic constant of a grain model defined as $(E/1 \text{ keV})^2$ times the quotient of the optical depth for scattering X-rays of energy E by the optical extinction at wavelength 5550 Å. In the Rayleigh-Gans approximation for X-ray scattering, R_{XV} is independent of the X-ray energy. Assuming that all of the soft component observed during the eclipse of 4U 1538-52 is grain-scattered, we find $R_{XV} < 0.06$ mag^{-1}. This limit is consistent with the results of previous grain-scattering study based on dissection of the halo and core components of images obtained with *ROSAT* (Predehl et al. 1991) and marginally consistent with an earlier image dissection study of images from the *Einstein Observatory* (Mauche & Gorenstein 1986). On the other hand, our upper limit is substantially less than the value implied by the calculations of Martin & Rouleau (1991) of the extinction and scattering cross sections solid grains with a power-law size distribution $n(a) \sim a^{-3.5}$, $0.005\mu m < a < 0.25\mu m$, and composed of "astronomical silicate" ($R_{XV} = 0.22$ mag^{-1}) or a mixture of silicate and graphite ($R_{XV} = 0.11$ mag^{-1}). This result lends support to the idea (Mathis and Whiffen 1989) that interstellar grains may be loose, "fluffy" aggregates of smaller solid particles. The R_{XV} value for such grains is less than for solid particles of the same material because the X-ray scattering efficiency decreases more rapidly with decreasing bulk density than does the optical extinction.

ACKNOWLEDGEMENTS

This research was supported in part by grant NAG8-701 from the National Aeronautics and Space Administration.

REFERENCES

Alcock, C., & Hatchett, S. 1978, ApJ, 222, 456
Blondin, J. M., Kallman, T. R., Fryxell, B. A., & Taam, R. E. 1990, ApJ, 356, 591
Catura, R. C. 1983, ApJ, 275,645
Clark, G. W., Woo, J. W., & Nagase, F. 1994, ApJ 422, 690
Hayakawa, S. 1970, Prog. Theor. Phys. 43, 1224
Kallman, T. R. & Krolik, J. H. 1993 XSTAR: A Spectral Analysis Tool, User'd Guide'
Martin, P. G. 1970, MNRAS, 149 221
Martin, P. G., & Rouleau, F. 1991, in Extreme Ultraviolet Astronomy, ed. R. F. Malina & S. Bowyer (New York: Pergamon), 341
Mathis, J. S., Rumpl, W., & Nordsieck, K. H. 1977, ApJ, 217, 425
Mathis, J. S., & Whiffen, G. 1989, ApJ, 341, 808
Mauche, C. W., & Gorenstein, P. 1986, ApJ, 302, 371
Mitsuda, K., Takeshima, T., Kii, T., & Kawai, N. 1990, ApJ, 353, 480
Overbeck, J. 1965, ApJ, 141, 864
Predehl, P., Bräuninger, H., Burkert, W., & Schmitt, J. H. M. M. 1991, A&A, 246, L40
Xu, Y., McCray, R., & Kelley, R. 1986, Nature, 319, 652

A POTENTIAL CYCLOTRON LINE SIGNATURE IN LOW LUMINOSITY X-RAY PULSARS

Robert W. Nelson[1], John C. L. Wang[2], E.E. Salpeter[3] and Ira Wasserman[3]

[1] CITA, 60 St. George St., Toronto, ON M5S 1A7, CANADA
nelson@cita.utoronto.ca
[2] JILA, University of Colorado, Campus Box 440, Boulder, CO 80309, USA
[3] CRSR, Cornell University, Ithaca, NY 14853, USA

ABSTRACT

Simple estimates indicate there should be $\gtrsim 10^3$ low luminosity X-ray pulsars ($L \lesssim 10^{34}$ erg s^{-1}) in the Galaxy undergoing "low-state" wind accretion in Be/X-ray binary systems, and $\sim 10^8 - 10^9$ isolated neutron stars which may be accreting directly from the interstellar medium. Despite their low effective temperatures ($kT_e \lesssim 300$ eV), we predict that low luminosity accreting neutron stars with magnetic fields $B \sim (0.7 - 7) \times 10^{12}$ G should emit a substantial fraction (0.5–5%) of their total luminosity in a narrow ($E/\Delta E \sim 2-4$) cyclotron emission line which peaks in the energy range $\sim 5 - 20$ keV. In sharp contrast to the underlying thermal emission, this *nonthermal* cyclotron component will not be strongly absorbed by the intervening H I gas, and consequently it may be the only observable signature for the bulk of these low luminosity sources. We propose a search for this cyclotron emission feature in long pointed observations by ASCA of the Be/X-ray transient pulsars V0331+53 and 4U0115+63 in their quiescent "low state".

1. INTRODUCTION

Most of the 30 known X-ray pulsars in the Galaxy are bright, with luminosities $L \gtrsim L_E = 1.4 \times 10^{35} M_{1.4} R_6^{-2} A_{cap}$ erg/sec, where L_E is the *effective* Eddington limit for magnetic polar cap accretion onto a neutron star of mass $M = 1.4 M_{1.4} M_\odot$, radius $R = 10^6 R_6$ cm and polar cap area $A = A_{cap}$ km^2. Nearly half of these pulsars undergo wind accretion from Be star companions, but are usually detected only during high luminosity transient outbursts. These bright sources may just be the "tip of the iceberg", however, of a much larger underlying population of lower luminosity magnetic accretors. Simple estimates suggest there are $\gtrsim 10^3$ similar Be/neutron star binaries in the Galaxy accreting at low luminosities $L << L_E$ (Rappaport & van den Heuvel 1982; Meurs & van den Heuvel 1989; King 1991). In addition, assuming a production rate from supernovae of $(10-100 \text{ yrs})^{-1}$ there should be $\sim 10^8$-10^9 isolated neutron stars in the Galaxy, of which some fraction may be accreting directly from the interstellar medium at detectable luminosities (Trevis & Colpi 1991; Nelson, Salpeter, & Wasserman 1991; Blaes & Madau 1993).

Here we present preliminary calculations of the emergent spectra of such low luminosity X-ray pulsars. In sharp contrast to their brighter counterparts, we find that these sources with low effective temperatures should display a rather unique nonthermal signature: a prominent cyclotron emission line, containing

0.5 − 5% of the total accretion luminosity, superposed on the Wien tail of the underlying soft thermal emission.

In section 2, we briefly describe our model for magnetic accretion and radiative transport. In section 3, we present spectra based on Monte Carlo cyclotron transport simulations. Finally, in section 4, as a test of our model, we propose a search for this spectral feature in some of the Be/X-ray binary transients with known magnetic field strengths. If found, this feature could be an important diagnostic for the physical processes taking place in low luminosity accreting magnetic neutron stars.

2 MAGNETIC ACCRETION AND RADIATIVE TRANSPORT

The physics of accreting magnetic neutron star atmospheres and the assumptions of our model are described in detail in Nelson, Salpeter, & Wasserman (1993). Accreting protons, channeled along magnetic field lines, free fall into the plasma atmosphere at the magnetic polar cap and decelerate by undergoing magnetic Coulomb collisions with atmospheric electrons. The presence of a strong magnetic field dramatically alters the microphysics of proton stopping and the subsequent conversion of accretion energy to radiation. For field strengths $B_{12} = (B/10^{12}$ G$) \ll 9M_{1.4}R_6^{-1}$, accreting protons have sufficient center of mass energy to excite large numbers of electron Landau transitions. These excited electrons then decay to their Landau ground state primarily through single step *radiative* transitions. Thus, a significant fraction of the accretion energy is converted directly to cyclotron photons distributed along the path of the decelerating proton. We emphasize that these cyclotron photons are produced in *nonthermal* collisions; their initial energy, E_B, reflects the Landau energies set by the magnetic field strength, and not the thermal energy, kT_e, of the atmosphere, that is,

$$E_B = 11.6 B_{12} \text{ keV} \gg k_B T_e = 190 \left(\frac{F_0}{10^{-4}}\right)^{1/4} M_{1.4}^{1/4} R_6^{-1/2} \text{ eV}, \qquad (1)$$

where $F_0 \equiv L_{accr}/L_E$.

Once produced, the cyclotron photons try to escape the atmosphere through multiple magnetic Compton scatters. We have used a Monte Carlo code to compute the polarized radiative transfer of these cyclotron photons through the magnetized plasma (Nelson et al. 1994, hereafter NWSW). The dominant transfer effects are (1) magnetic Compton scattering which results in both angle and frequency redistribution (Comptonization), (2) polarization mode switching, and (3) absorption via inverse magnetic Bremsstrahlung.

Of fundamental importance in our simulations is the low temperature of the atmosphere given by equation (1). This has several consequences: First, the atmosphere may be treated as a cold plasma and we use the cold plasma polarization modes in our treatment of polarized transfer (Gnedin & Pavlov 1974, Ventura 1979, Nagel & Ventura 1983). Secondly, unlike the case in hot magnetic atmospheres, the Doppler core of the resonant scattering line is narrow, so that cyclotron photons can quickly escape the line core and avoid resonant absorption (Wasserman & Salpeter 1980; Wang, Wasserman & Salpeter 1988; NWSW). This greatly increases the probability that they escape the atmosphere. Finally, despite significant energy loss to Compton recoil, the escaping cyclotron photons are much harder than the typical thermal photon, so that the emergent cyclotron emission line will be prominently displayed against the soft continuum.

3 EMERGENT LINE LUMINOSITY AND SPECTRUM

There are four parameters in our model: The neutron star's mass and radius, the field strength, B, and the dimensionless accretion luminosity, $F_0 = L_{accr}/L_E$. Without exception, we take $M_{1.4} = 1 = R_6$. We therefore construct a two-parameter family of models in B and F_0.

The shape of the emergent photon number spectra (summed over polarization) is shown in Figure 1 for $B = 10^{12}\,G$ (left) and $5 \times 10^{12}\,G$ (right). For both panels, $F_0 = 10^{-4}$; the spectral shape does not depend strongly on F_0. Note that for $B = 10^{12}\,G$, the entire line lies within the ASCA energy window (0.5 to 10 keV). We obtain the continuum by adopting a (unpolarized) black body spectrum with an effective temperature determined from equation (1). To this spectrum we add the emergent line spectrum, with the proper weighting, computed from the line transfer code. The spectrum thus consists of a soft thermal component plus a hard nonthermal cyclotron emission line superposed on the Wien tail of the underlying thermal continuum. This general form should persist even for harder (e.g., power law) continua as long as the bulk of atmospheric electrons have energies $\ll E_B$ (NWSW). Note, however, that the soft X-rays below 1 keV will generally be strongly absorbed by the intervening interstellar medium.

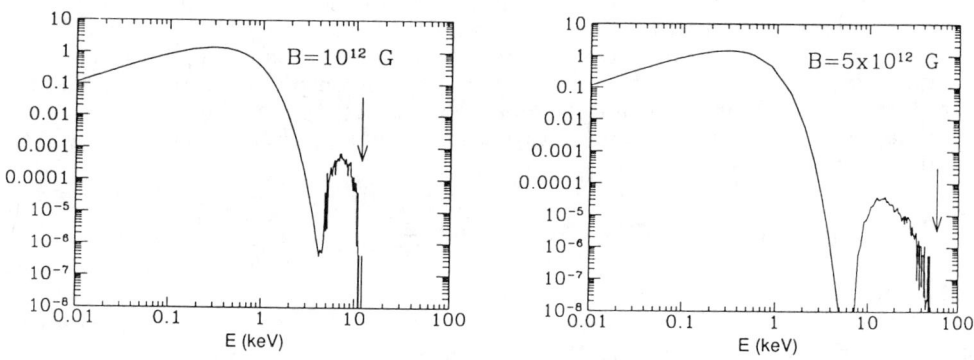

Figure 1. The emergent photon number spectra (summed over polarization and normalized to unit area) from a magnetized polar cap neutron star atmosphere with (left) $B = 10^{12}\,G$, $F_0 = 10^{-4}$, $L_{line} = 0.02 L_{accr}$, and (right) $B = 5 \times 10^{12}\,G$, $F_0 = 10^{-4}$, $L_{line} = 0.01 L_{accr}$. Arrows in figure give location of cyclotron energy, E_B, where line photons are born.

The vertical arrows in the figure denote the location of the cyclotron energy, E_B, where line photons are born. The photon energy degradation due to electron recoil (i.e., Comptonization) is clearly evident. In addition, owing to polarization mode switching and the strong energy dependence of the magnetic scattering cross sections, these line features are narrow ($E/\Delta E \sim 2\text{-}4$ for $0.7 \lesssim B_{12} \lesssim 7$; NWSW).

Figure 2 shows the fraction of accretion luminosity that escapes in the cyclotron line, L_{line}/L_{accr}, as a function of magnetic field strength. The dependence on F_0 enters weakly only through the absorption cross section, $\sigma_{abs} \propto 1/T_e \propto F_0^{-1/4}$. At high values of B, few cyclotron photons are produced so L_{line}/L_{accr} falls off rapidly. The rapid fall-off at low B arises from the transfer microphysics: The energy lost per scatter by photons to electron recoil along field lines is small compared to the thermal Doppler width, so that line photons become trapped inside the line core where the scattering cross section is large (cf. Wang, Wasserman & Salpeter 1988; Wasserman & Salpeter 1980) and are destroyed by absorption before they can escape.

The optimum conditions for line photon escape obtains when $0.7 \lesssim B_{12} \lesssim 7$ (where $0.005 \lesssim L_{line}/L_{accr} \lesssim 0.05$). In this regime, the field is sufficiently strong to avoid line trapping, but not strong enough to quench the initial cyclotron photon production.

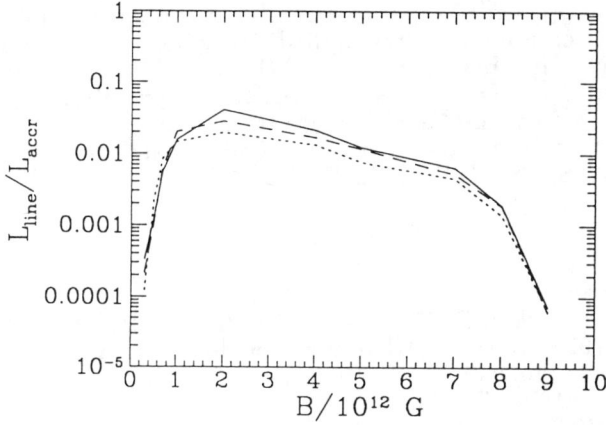

Figure 2. The fraction of the total accretion luminosity in the nonthermal cyclotron emission line as a function of magnetic field strength. The curves correspond to $F_0 = L_{accr}/L_E =$ (solid) 10^{-2}, (dashed) 10^{-4}, and (dotted) 10^{-6}.

4. AN OBSERVATIONAL TEST

Thus, if strongly magnetized neutron stars with $0.7 \lesssim B_{12} \lesssim 7$ accrete at luminosities $L \ll L_E$, they should give rise to a prominent cyclotron emission line which peaks between $\sim$ 5–20 keV. As an observational test, we propose a search for this nonthermal cyclotron emission feature in long pointed observations by ASCA of the two Be/X-ray transient pulsars systems, V0331+53 and 4U0115+63, in their quiescent "low state". These pulsars have known magnetic field strengths of $B_{12} \simeq 2.5$ and 1, respectively, as determined by cyclotron absorption features seen in their spectra during transient outbursts (Makishima et al. 1990; Nagase et al. 1991), and are located relatively nearby at $\sim$ 3.5 kpc (White, Swank and Holt 1983). Due to Compton recoil energy loss, the cyclotron emission lines predicted here should have energies lower than their associated absorption features. An 80 kilosecond integration with the Gas Imaging Spectrometer (GIS) onboard the X-ray satellite ASCA could detect these sources down to a flux level of about 10^{-13} erg/cm^2/sec (between 0.5 and 10 keV) (Tanaka, private communication). Thus, ASCA should be able to detect a cyclotron line feature containing 1% of the total flux down to a quiescent accretion luminosity of 1.4×10^{34} erg s^{-1}. According to conventional wisdom, at these low luminosities the centrifugal barrier at the spinning magnetospheric boundary should reduce the flow of material down the neutron star magnetic field lines (Illarionov & Sunyaev 1975). However, accretion will shut off completely only if this this so-called "propeller effect" is 100% effective. If detected, this cyclotron feature could be a clean diagnostic of the physics of magnetic accretion in a regime where radiation pressure is unimportant.

On a more speculative note, if so-called "classical" γ-ray bursts are associated with strongly magnetized Galactic neutron stars, then they may accrete from the interstellar medium after the burst. The cyclotron feature predicted here may then be present in their *quiescent* emission spectra (Wang & Nelson 1993). If long observations of unidentified X-ray sources in small γ-ray burst error boxes, such as has been proposed by Hurley et al.(1993) for GRB920501, were to reveal such an emission feature, it would strongly suggest that these sources are associated with Galactic magnetized neutron stars.

This work was supported in part by NASA grants NAGW-666, NAGW-766, NSF grants AST 91-19475, AST 91-20599 and the NSERC of Canada.

REFERENCES

Blaes O., & Madau, P. 1993, ApJ, 403, 690
Gnedin, Yu. N. & Pavlov, G. G. 1974, Soviet Phys. – JETP, 38, 903
Hurley, K. et al.1993, Proc. XXIII International Cosmic Ray Conference 1, 116
Illarionov, A. F., & Sunyaev, R. A. 1975 A&A, 39, 185
King, A. R. 1991, MNRAS, 195, 45P
Makishima et al.1990, ApJ, 365, L59
Meurs, E.J.A., & van den Heuvel, E.P.J. 1989, A&A, 226, 88
Nagase, F. et al.1991, ApJ, 375, L49
Nagel, W. and Ventura, J. 1983, A&A 118, 66
Nelson, R. W., Salpeter, E. E., & Wasserman, I. 1991, in Proc. of Taos Workshop on Physics of Isolated Pulsars, eds. K. A. Van Riper, R. I. Epstein, and C. Ho (Cambridge University Press) p. 145
Nelson, R. W., Salpeter, E. E., & Wasserman, I. 1993, ApJ, 418, 874
Nelson, R. W., Wang, J. C. L., Salpeter, E. E., & Wasserman, I. 1994 ApJ, in preparation (NWSW)
Rappaport, S. & van den Heuvel, E. P. J. 1982, in Be Stars, ed. M. Jaschek & H. G. Groth (Dordrecht:Reidel) 342
Trevis, A., & Colpi, M. 1991, A&A, 241, 107
Ventura, J. 1979, Phys. Rev. D 19, 1684
Wang, J. C. L. & Nelson, R. W. 1993, in Proc. of Huntsville Gamma Ray Burst Workshop, ed. G.J. Fishman, K. Hurley, & J.J. Brainerd (New York: AIP)
Wang, J. C. L., Wasserman, I. M. & Salpeter, E. E. 1988, ApJ Suppl., 68, 735
Wasserman, I. & Salpeter, E. E. 1980, ApJ, 241, 1107
White, N. H., Swank, J., & Holt, S. 1983, ApJ, 270, 711

NEAR–EDDINGTON WINDS FROM NEUTRON STARS

Luciano Nobili, Roberto Turolla
Department of Physics, University of Padova
Via Marzolo 8, 35131 Padova, Italy

Iosif Lapidus [‡]
Institute of Astronomy, University of Cambridge
Madingley Road, Cambridge CB3 0HA, UK

ABSTRACT

A model for the expansion phase of a neutron star atmosphere during strong X–ray bursts is presented. General relativistic hydrodynamical and radiative transfer equations are integrated from the star surface outwards, taking into account for helium burning in the dense, inner shells. The role of Compton heating in placing a lower limit on the mass loss rate for the existence of stationary, supersonic winds is analyzed. Comparison of our results with observational data of 4U/MXB 1820-30 provides an estimate of both the spectral hardening and the accretion rate in this source.

1. INTRODUCTION

The existence of winds from neutron stars during strong X–ray bursts is a well acknowledged fact (see e.g. Haberl *et al.* 1987; see also Lewin, Van Paradijs & Taam 1993 for a very recent review on X–ray bursters). In very strong bursts, in fact, the X–ray luminosity curve shows a rather flat maximum and the dip observed in the fitting temperature curve is regarded as the evidence of a photospheric expansion produced, most probably, by a super–Eddington energy release.

Steady–state winds from neutron stars were the object of several investigations in the past (see e.g. Kato 1983; Paczyński & Prószyński 1986; Joss & Melia 1987) and they are commonly thought to be propeled by the radiative force exerted on the atmospheric material by the large luminosity produced at the base of the envelope by thermonuclear helium burning. In the following we present a more complete model for neutron stars winds which overcomes some of the limitations introduced in previous investigations. In particular, radiative transfer through the outflowing gas is correctly handled using Thorne's (1980) PSTF moments, which give a satisfactory description of the radiation intensity also outside the diffusion regime. Although only the frequency–integrated transport is considered here, Comptonization is included along with free–free emission–absorption, both in the hydro and in the moments equations. The rate of energy injection at the base of the flow is consistently computed assuming that luminosity is generated by 3–α helium burning in the nearly–hydrostatic region close to the stellar surface.

[‡] The Royal Astronomical Society Sir Norman Lockyer Fellow

2. THE MODEL

We assume that the gas flow is steady and spherically symmetric and that the neutron star gravitational field can be described by the Schwarzschild vacuum solution; M_* and R_* denote the stellar mass and radius respectively. The wind radial evolution is governed by a system of six differential equations: energy, radial momentum and rest mass conservation, zeroth and first moments of the transfer equation plus a "phenomenological" equation for the radiation temperature T_γ (see e.g. Park & Ostriker 1989; see also Nobili, Turolla & Zampieri 1991), which is needed to evaluate the Compton heating term and can not be directly computed solving only the frequency-integrated transfer. Since the flow effective optical depth at the star surface is extremely large, LTE must hold there (radiation energy density = aT^4, $T_\gamma = T$) and also luminosity must vanish since all energy is released by nuclear burnings above R_*. The appropriate boundary condition at radial infinity is free streaming for the radiation field. Keeping in mind that one more condition must be specified at the sonic point, it follows that each solution is characterized by only one free parameter, which can be the mass loss rate $\dot M$ or the envelope mass $M_{env} = \int_{R_*}^{R_S} 4\pi \varrho R^2\, dR$, where R_S is the sonic radius.

3. RESULTS AND DISCUSSION

The coupled hydro and moments equations has been solved numerically for several values of $\dot M$, assuming $M_* = 1.5\, M_\odot$, $R_* = 13.5$ km and a nearly solar chemical composition for the outflowing gas; all the details about the technique used in calculations can be found in Nobili, Turolla & Lapidus 1993. Results are summarized in table 1, which gives M_{env}, the terminal velocity, the photospheric and last scattering radii, the photospheric temperature and the nuclear time (see below) for some selected models in the range $3 \lesssim \dot M/\dot M_{Edd} \lesssim 140$. In accordance with generally accepted scenarios, the energy released in excess of the Eddington luminosity is converted into the kinetic energy of the outflowing envelope, so that the radiative flux seen by a distant observer is always very close to the Eddington value.

Table 1

$\dot M$ ($\dot M_{Edd}$)	M_{env} (10^{22} g)	v_∞ ($10^{-3} c$)	R_{ph} (10^3 km)	R_{es} (10^3 km)	T_{ph} (keV)	t_{nuc} (s)
139.2	195.8	1.20	19.37	180.71	0.06	10
102.3	140.3	1.28	9.27	127.60	0.11	10
50.2	57.1	2.00	2.03	41.97	0.27	8
30.4	25.6	2.42	1.04	22.49	0.41	6
19.9	11.5	2.81	0.59	13.80	0.57	4
11.3	3.9	3.16	0.31	8.17	0.85	2
5.9	1.4	3.32	0.19	5.26	1.14	2
4.8	1.1	3.28	0.18	4.77	1.18	1
2.8	0.7	3.33	0.15	3.69	1.31	1

A quite unexpected result is that, owing to the presence of a Compton drag force in the momentum equation, there exists a lower limit for $\dot M$ below which no steady, supersonic wind can exist. For the present values of the model input parameters, the lower bound is $\dot{M}_{min} \sim 6 \times 10^{17}$ g/s $\sim \dot{M}_{Edd}$. Our series of quasi–stationary models with smoothly decreasing M_{env} can be thought to represent the time evolution of a X–ray burster during the envelope expansion phase. The decrease of M_{env} is due to the fact that the nuclear burning shell moves outwards leaving the products behind (and out of the wind region), the actual loss of mass in the wind itself being negligible. The characteristic duration of the quasi–stationary expansion phase can be assumed to coincide with the time scale needed to burn all the helium in the envelope $t_{nuc} \sim \epsilon Y M_{env} c^2 / \dot{E}$, where $\epsilon = 6.1 \times 10^{-4}$ is the efficiency of 3–α reactions and $\dot{E} \sim (1 + \dot{M}/\dot{M}_{Edd}) L_{Edd}$ is the total (radiative plus advected plus kinetic) luminosity. As can be seen from the table, the values of t_{nuc} for our models are always $\lesssim 10$ s, in agreement with observations.

Although our present model does not include frequency–dependent calculations, the existence of a lower limit for the envelope mass at which the steady wind phase terminates, allows us to estimate the spectral hardening factor $\gamma = T_{col}/T_{ph}$ in 4U/MXB 1820-30 by comparing our results with the observational data of Haberl et al.. The value of γ can be derived asking that

$$\gamma T_{ph}|_{\dot{M}_{min}} = T_{col}|_{max}$$

where $T_{col}|_{max}$ is the maximum observed color temperature, which is $\simeq 2.9$ keV for the bursts in 4U/MXB 1820-30 analyzed by Haberl et al. We emphasize that the choice of the last point as the fiducial one is based on the existence of a minimum value of $\dot{M}$ which is assumed to be reached at the end of the expanding envelope phase. The hardening factor obtained in such a way turns out to be $\gamma \sim 2.3$ and, since $\gamma > 1$, we expect a genuine hardening of the spectrum, due to Comptonization, as radiation propagates through the extended spherical shell $R_{ph} < R < R_{es}$. The knowledge of the envelope mass at the beginning of expansion, M_{env}^{in}, which can be derived from the fit of our T_{col}–R relation to the observed one, enables us to estimate also the accretion rate from the companion star between two successive strong type I bursts in this source: $\dot{M}_{acc} \sim (M_{env}^{in} - M_{env}^{min})/\Delta t \sim 10^{-6}$ $M_\odot$/yr. Although this value is rather high in comparison with accretion rates usually discussed in theoretical models of nuclear burnings on the surface of neutron stars (see e.g. Ayasli & Joss 1982; Fushiki & Lamb 1987), there seems to be no contradiction with the recent conclusion of Taam et al. (1993) that the unstable helium burning responsible for a strong X-ray burst can take place only if the accretion rate exceeds $\sim 10^{-9}$ $M_\odot$/yr. Moreover, even if the accretion rate is highly supercritical, no significant release of gravitational luminosity, $L_{acc} = GM_* \dot{M}_{acc}/c^2 R_*$, occurs in the interburst phase because the gas has no time to cool. The inner part of the flow, in fact, is optically thick and the appropriate cooling time is the adiabatic time (see e.g. Bildsten 1993)

$$t_{cool}^{ad} \sim \frac{c_p \kappa_{es} (\varrho_0 r)^2}{caT^3} \sim 3 \times 10^{11} \left(\frac{\varrho_0}{10^6 \text{ g}}\right)^2 \left(\frac{T}{10^8 \text{ K}}\right)^{-3} \text{ s}.$$

For 4U/MXB 1820-30 it is

$$t_{cool}^{ad} \sim 10^6 \text{ s} \gg t_{acc} \sim 10^4 \text{ s}.$$

It follows, then, that the heat produced by the conversion of gravitational potential energy can not be radiated away in a time t_{acc} and must go to increase the gas internal energy. This in turn implies that the accretion process can not be regarded as stationary. Only a tiny fraction of L_{acc} is expected to escape to infinity while the progressive heating of the inner gas layers produces, at the end, the helium flash. In our proposed scenario, the observed persistent X-ray luminosity, $\sim 0.1 L_E$, is most probably due only to stationary hydrogen burning on the surface of the neutron star.

4. CONCLUSIONS

Further work will be aimed to compute a whole grid of wind models, varying M_*, R_*, chemical composition, and to include a more complete treatment of nuclear reactions. Frequency-dependent calculations are in progress to get a more reliable determination of the hardening factor. It should be pointed out, however, that in order to follow the burst evolution outside the quasi-stationary phase a full, time dependent approach is needed.

REFERENCES

Ayasli, S., & Joss, P.C. 1982, ApJ, 256, 637
Bildsten, L. 1993, ApJ, in press
Fushiki, I., & Lamb, D. Q. 1987, ApJ, 323, L55
Haberl, F., Stella, L., White, N.E., Priedhorsky, W.C., & Gottwald, M. 1987, ApJ, 314, 266
Joss, P.C., & Melia, F. 1987, ApJ, 312, 700
Kato, M. 1983, PASJ, 35, 33
Lewin, W.H.G., Van Paradijs, J., & Taam, R.E. 1993, Space Sci. Rev., 62, 223
Nobili, L., Turolla, R., & Zampieri, L. 1991, ApJ, 383, 250
Nobili, L., Turolla, R., & Lapidus, I., 1993, ApJ, submitted
Paczyński, B., & Prószyński, M. 1986, ApJ, 302, 519
Taam, R.E., Woosley, S.E., Weaver, T.A., & Lamb, D.Q. 1993, ApJ, 413, 324
Thorne, K.S. 1981, MNRAS, 194, 439

THE OUTFLOWING REGIME OF QUASI-SPHERICAL ACCRETION ON TO X-RAY OBJECTS AND THE SPIN-DOWN MECHANISM FOR WIND-FED X-RAY PULSARS

A. F. Illarionov[1,2], I. V. Igumenshchev[3], D. A. Kompaneets [2]
[1] Johns Hopkins Univ. Dept. of Phys.& Astron. Baltimore, MD 21218
andrei@eta.pha.jhu.edu
[2] P.N.Lebedev Physical Inst, 84/32 Profsoyuznaya St, Moscow, Russia
[3] Institute of Astronomy, 48 Pyatnitskaya St, Moscow, Russia

ABSTRACT

We study numerically the quasi-spherical accretion of matter on to a compact object (neutron star or black hole). Anisotropic X-ray luminosity, powered by mass accretion, heats the accreting gas through Compton scattering. When the gas temperature increases above the local escape temperature, part of the accreting gas will flow outwards as a result of the action of buoyancy force. The direction of the outflow coincides with the maximum of the X-ray luminosity. The depth of outflow is correlated with the energy of X-ray quanta. In spite of its quantum nature, Compton heating markedly affects the gas, forcing the matter outflow at X-ray luminosities as small as three or four orders of magnitude less than the Eddington limit. The phenomenon of hot gas outflow takes place in the case of accretion on to a wind-fed X-ray source in a wide binary with massive OB or Be-star. We propose a new spin-down mechanism for accreting neutron stars that explains the existence of a number of long-period ($p \sim 100-1000$s) X-ray pulsars in these binaries. The spin-down is a result of efficient angular-momentum transfer from the rotating magnetosphere of the accreting star to an outflowing stream, when the outflow forms so deep as to capture the magnetic-field lines from the rotating magnetosphere. The balance between angular-momentum gain by accreting gas and loss by outflowing matter takes place at a particular value of the equilibrium spin period ($p_{eq} \sim 100$–1000s).

1. SCENARIO

In a wide massive X-ray binary the supergiant OB or Be-star loses mass in a wind at the rate $\dot{M}_w \sim 10^{-6}-10^{-9} M_\odot$ yr^{-1}. A wind-fed neutron star (or black hole) of mass $M \simeq M_\odot$ captures mass from the wind at the rate

$$\dot{M} \simeq \frac{\pi R_A^2}{4\pi A^2} \dot{M}_w \simeq 10^{-11}-10^{-9} M_\odot \text{ yr}^{-1}, \quad R_A = \frac{2MG}{w^2}$$

If we assume $\dot{M} \simeq 10^{-10} M_\odot$ yr^{-1} then the X-ray luminosity of this object is $L = e\dot{M}c^2 \simeq 10^{36}$erg s$^{-1} \ll L_{Edd}$. The radius of the wind capture cross-section, R_A, depends primarily upon the wind velocity, w. The characteristic wind velocities are $w \simeq 1000$ km s^{-1} for an OB star and $\simeq 300$ km s^{-1} for a Be star, implying $R_A \simeq 3 \cdot 10^{10}$cm and $R_A \simeq 3 \cdot 10^{11}$cm, respectively. The orbital separation $A \simeq 3 \cdot 10^{12}$cm for orbital period $P_{orb} \simeq 10^d$, and the Eddington luminosity $L_{Edd} \simeq 10^{38}$erg s^{-1}. The gravitational energy of the infalling matter is converted to X-rays with energy ~ 10 keV in the immediate vicinity of the neutron star ($r \simeq 10^6$ cm) with high efficiency $e = GM/rc^2 \simeq 0.15$ and with some anisotropy due to the flat geometry of polar caps.

The matter passing a bow shock is hot, additionally overheated by X-rays through the Compton effect and can not accrete spherically as in the Bondi's case (Ostriker et al.1976). Part of the overheated gas will outflow in the direction of the maximum of the X-ray luminosity as a result of the action of the buoyancy force. (Note that in our case $L \ll L_{Edd}$ and the effect of radiation pressure force is negligible; the angular momentum of matter captured from the wind is small and an accretion disk is not formed; the results are not applicable to white dwarf stars).

2. HYDRODYNAMICS OF ACCRETION

The numerical computation of the outflowing regime of quasi-spherical accretion on to an X-ray star (Igumenshchev et al.1993) was done by assuming azimuthal and mirror symmetry of the flows. We use an explicit second order Eulerian method and the Godunov sheme (200x200 computational cells per quadrant) to solve the time dependent hydrodynamical equations for a hydrogen 2-temperature quasi-neutral plasma which has an adiabatic index $\gamma = 5/3$. The viscosity is small, the Reynolds number $Re \sim 100$. The angular dependent Compton heating rate per electron is:

$$C = \frac{4k(T_C - T_e)\sigma_T}{m_e c^2 R^2} \mathcal{L}(\theta),$$

where T_e is the electron temperature, T_C is the Compton temperature of the X-rays, and we use $kT_C = 10$ keV in the calculations. $\mathcal{L}(\theta)$ is the anisotropic angular luminosity of the central source, say $\mathcal{L}(\theta) = L(3\cos^2\theta + 1)/8\pi$, where $L = 0.15\dot{M}c^2$ is the momentary luminosity, $\dot{M}$ is the momentary accretion rate through inner boundary. The other terms in the matter-energy equations are negligible.

At the inner boundary, $R_{in} = 6 \cdot 10^8$ cm, we set a condition of complete absorption. At the outer boundary, $R_{out} \simeq 3 \cdot 10^{10}$ cm $\simeq R_A$, we set a condition of matter infall (if $T \leq T_{esc}$) or outflow (if $T > T_{esc}$). Here $T_{esc} = MGm_p/5kR$ is the escaping temperature of matter at distance R. We use an inequality $T > T_{esc}$ (Illarionov & Kompaneets 1990) when the buoyancy force is greater than gravity as an approximate condition for the outflow formation. An estimate of the deepest stagnation point of the outflow is the Compton radius

$$R_{st} \simeq R_C = \frac{MGm_p}{5kT_C} \simeq 3 \cdot 10^9 \text{cm}.$$

Examples of maps of the plasma density, velocity, and Mach number distributions for a model with $\mathcal{L}(\theta) = L(3\cos^2\theta + 1)/8\pi$ at time $t \simeq 10\,R_A/w$ are shown in Fig.1. The stagnation point is at the proper place; the shock wave (round jam in Fig.1.b.) appearing near R_C propagates outwards. The accretion flows are not stable. A persistent outflow exists, but the mass-loss rate $\dot{M}_{out}$ in the outflow is highly variable, its time average is about 10% of the accretion rate, and the average solid angle of the outflow $\chi \sim 1$.

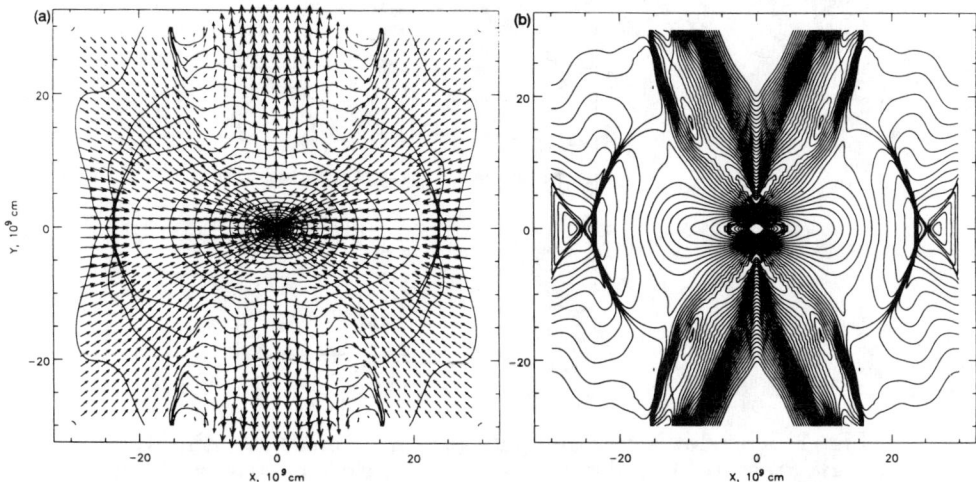

Fig. 1. (a) Arrows correspond to velocity field; curves of constant density are spaced with step $\triangle \lg \rho = 0.1$. (b) Mach numbers are spaced with $\triangle M = 0.05$.

3. SPIN-DOWN MECHANISM FOR WIND-FED X-RAY PULSARS

The deepest stagnation point of the outflow R_{st} is quite near the radius of the magnetosphere of the neutron star (Baan & Treves 1973, Lamb et al.1973) $r_H \simeq 10^9$ cm (for field $H_0 \simeq 3 \cdot 10^{12}$G). Imagine that the outer boundary of magnetosphere is dragged into the outflow (Illarionov & Kompaneets 1990). The spinning neutron star and its magnetosphere corotate, and the shear stress of the rotating magnetic field lines streching into the outflow causes angular momentum transfer from the star to the outflowing matter, just as in the Parker mechanism for the Sun. The rate of the angular momentum loss $\dot{J}_{out} \simeq \frac{\chi}{2\pi} \dot{M} r_H^2 \omega$, where ω is the pulsar rotation frequency.

The rate of the angular momentum supply by the accreting matter captured from the wind (Illarionov & Sunyaev 1975) $\dot{J}_{in} \simeq \dot{M} R_A^2 / P_{orb}$ (In wide binaries this rate is usually less then $\dot{M}\sqrt{MGr_H}$ required for a disk formation). Combining both rates we have for the dynamics of a spinning star $\dot{\omega} = (\dot{J}_{in} - \dot{J}_{out})/I = (\omega_{eq} - \omega)/\tau$. Here $I \simeq 10^{45}$gr cm^2 is a moment of inertia of a neutron star, the relaxation time $\tau = \frac{2\pi I}{\chi r_H^2 \dot{M}} \simeq 4 \cdot 10^4$yr for $\dot{M} \simeq 10^{-10} M_\odot$yr^{-1}. The equilibrium spin period $p_{eq} = 2\pi/\omega_{eq}$ of an X-ray pulsar

$$p_{eq} = \chi(\frac{r_H}{R_A})^2 P_{orb} \simeq 300 \text{s} \quad \text{for } R_A = 3 \cdot 10^{10} \text{cm and } P_{orb} = 10^d.$$

The distribution of massive X-ray binaries in the $p_{sp} - P_{orb}$ diagram is shown in Fig.2. Straight lines (1) and (2) correspond to the proportionality $p_{eq} \propto P_{orb}$ for the pulsars in wide binaries with OB (1) and Be stars (2). The considerable separation s between these lines is determined mainly by the large ratio $(w_{OB}/w_{Be})^4 \simeq 100$ of the characteristic wind velocities. The line (3) corresponds to the observational limit for detection of pulsars with $p > 15 - 20^{min}$.

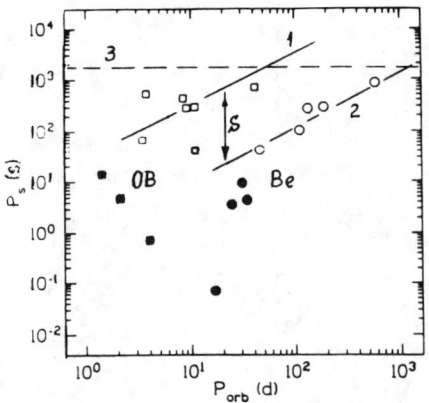

Fig. 2. The slowly rotating pulsars in wide binaries with OB (□) and Be (o) stars. Rapid disk-fed pulsars in close binaries with OB (■) and Be (●) stars (Waters & van Kerkwijk 1989, Chakrabarty et al.1993).

4. CONCLUSIONS

We propose a new spin-down mechanism for wind-fed X-ray pulsars in wide massive binaries and found the equilibrium spin-period $p_{eq} \sim 300$ sec for these cases. The old "propeller" spin-down mechanism (Illarionov & Sunyaev 1975) is applicable near equilibrium for fast $p_{eq} \sim 1$ sec pulsars in close binaries, where disk accretion takes place (Ghosh & Lamb 1979, Stella et al.1986). We hope that the both mechanisms can explain the entire range of observed spin-periods of the accreting X-ray pulsars.

ACKNOWLEDGEMENTS

This work was supported in part by the Summer Program of the Colorado University and by NASA Grant NAGW-3156.

REFERENCES

Baan, W., & Treves, A. 1973, A&A, 22, 421
Chakrabarty, D., Grunsfeld, J. M., Prince, T. A., Bildsten, L., Finger, M. H., Wilson, R. B., Fishman, G. J., Meegan, C. A., & Paciesas, W. S. 1993, ApJ, 403, L33
Ghosh, P., & Lamb, F. K. 1969, ApJ, 232, 259; 234, 296
Igumenshchev, I. V., Illarionov, A. F., & Kompaneets, D. A. 1993, MNRAS, 260, 727
Illarionov, A. F., & Kompaneets, D. A. 1990, MNRAS, 247, 219
Illarionov, A. F., & Sunyaev, R. A. 1975, A&A, 39, 205
Lamb, F. K., Pethick, C. J., & Pines, D. 1973, ApJ, 184, 271
Ostriker, J. P., McCray, R., Weaver, R., & Yahil, A. 1976, ApJ, 208, L61
Stella, L., White, N. E. & Rosner, R. 1986, ApJ, 308, 669
Waters, L. B. F. M., & van Kerkwijk, M. H. 1989, A&A, 223, 196

HOT AND COLD ATMOSPHERES AROUND NEUTRON STARS

R. Turolla [1], L. Zampieri [2], M. Colpi [3], A. Treves [2]

[1] Department of Physics, University of Padova
Via Marzolo 8, 35131 Padova, Italy

[2] International School for Advanced Studies, Trieste
Via Beirut 2–4, 34014 Miramare–Trieste, Italy

[3] Department of Physics, University of Milano
Via Celoria 16, 20133 Milano, Italy

ABSTRACT

Stationary, spherical accretion onto an unmagnetized neutron star is here reconsidered on the wake of the seminal paper by Zel'dovich & Shakura (1969). It is found that new "hot" solutions may exist for a wide range of luminosities. These solutions are characterized by a high temperature, $10^9 \div 10^{11}$ K, and arise from a stationary equilibrium model where the dominant radiative mechanisms are multiple Compton scattering and bremsstrahlung emission. For low luminosities, $\lesssim 10^{-2} \, L_E$, only the "cold" (à la Zel'dovich and Shakura) solution is present.

1. INTRODUCTION

Even before the observational evidence that Galactic X–ray sources are mostly binary systems containing an accreting neutron star, Zel'dovich & Shakura (1969, ZS in the following) studied in some detail the spectrum of radiation produced by stationary, spherical accretion onto an unmagnetized neutron star and compared their results with the (poor) data available at the time for Sco X–1. The pioneering paper of ZS shows that the resulting spectrum depends on two parameters, the accretion rate (luminosity) and the penetration length of the accreting ions in the outermost neutron star layers. The outcome can be described, in essence, as a black body with a high energy tail due to the Compton heating of thermal photons in the hot, external part of the atmosphere surrounding the neutron star (see also Alme and Wilson 1973).

In this investigation we reconsider the problem adopting a set of equations for the neutron star atmosphere which essentially coincide with those of ZS. We show that ZS solutions for the temperature profile are not unique and, moreover, for large enough accretion rates, there might be other solutions at considerably higher temperatures.

2. THE MODEL

We assume that the accreting flow impinges onto a spherically–symmetric, static atmosphere which surrounds the neutron star, and is decelerated as it penetrates into the atmosphere. The kinetic energy released by the incident protons goes mainly into electron thermal energy and is then re–emitted as free–free radiation. Following ZS, we treat the total column density of the atmosphere required to stop the incoming beam, y_0, as a free parameter. The heat injected by the infalling protons per unit time and per unit mass in the atmosphere is assumed to be constant and is related to the total luminosity observed at infinity by

$$W = \frac{L_\infty}{4\pi \int_0^{y_0} R^2 dy}. \tag{1}$$

In the inner region where $y > y_0$, $W = 0$; in all our models $5\,\mathrm{g\,cm^{-2}} < y_0 < 20\,\mathrm{g\,cm^{-2}}$.

The transfer of radiation in the atmosphere is governed by the two equations for the radiative luminosity L and the radiation energy density U which, in spherical symmetry and using the Eddington approximation, can be written as

$$\frac{dL}{dy} = -4\pi R^2 W \tag{2}$$

$$\frac{1}{3}\frac{dU}{dy} = \kappa_1 \frac{L}{4\pi R^2 c}. \tag{3}$$

The radiative processes taken into account are scattering and bremsstrahlung. The appropriate boundary condition for the radiation field at the outer edge of the non–illuminated medium is $U = L_\infty/2\pi R^2 c$. The inner boundary condition, at $y_{in} \gg y_0$, is fixed by the requirement that all the observed radiative flux must be generated within the atmosphere, that is, $L = 0$.

The runs of pressure, P, and temperature, T, are obtained from hydrostatic balance and radiative energy equilibrium

$$\frac{dP}{dy} = \frac{GM_*}{R^2} \tag{4}$$

$$\frac{W}{c} = \kappa_P (aT^4 - U) + 4\kappa_{es} U \frac{KT}{m_e c^2}\left(1 - \frac{T_\gamma}{T}\right), \tag{5}$$

where M_* is the mass of the neutron star. The boundary condition for equations (4) is $P = 0$ at $y = 0$. The matter density ϱ is calculated from the perfect gas equation of state assuming that the atmosphere is made by completely ionized hydrogen. In equation (5) κ_P is the Planck mean opacity and T_γ is the radiation temperature which is defined as the mean photon energy. In general T_γ can be computed only solving the full frequency–dependent transfer problem and will depend on y. In ZS, T_γ is set equal to $[U(y)/a]^{1/4}$, which is appropriate in LTE. Since we are interested in finding also solutions in which multiple Compton scattering becomes important, we derive the radiation temperature from the equation (see Wandel, Yahil & Milgrom 1984; Park & Ostriker 1989)

$$\frac{y}{T_\gamma}\frac{dT_\gamma}{dy} = 2Y_c\left(\frac{T_\gamma}{T} - 1\right) \tag{6}$$

where $Y_c = (4KT/m_ec^2)\max(\tau_{es}, \tau_{es}^2)$ is the Comptonization parameter and $\tau_{es} = \kappa_{es} y$. Use of equation (6) requires some care since it is meant to describe the variation of the radiation temperature when multiple Compton scattering is the dominant mechanism to exchange energy between photons and electrons. However, one can extend the validity of equation (6) to all optical depths, provided that no physical significance is attached to T_γ where the effective optical depth $\tau_{eff} > 1$. Models were obtained specifying a value of T_γ at $y = 0$. From the definition of the column density y, it follows immediately that $dR/dy = -1/\varrho$. The boundary condition is $R = R_*$ at $y = y_{in}$. The structre of the atmosphere was computed numerically and all models refer to a neutron star of $R_* = 10$ km and $M_* = 1$ $M_\odot$.

We find that two distinct kinds of solutions, "hot" and "cold", always exist for any y_0 provided that the luminosity exceeds a certain limit, which depends on y_0. The thermal properties of the "hot" atmospheres are illustrated in the figure, where the run of T versus column density is shown for $l_\infty = 7 \times 10^{-3}$ (continous line), $l_\infty = 2 \times 10^{-2}$ (dashed line) and $l_\infty = 7 \times 10^{-2}$ (dashed–dotted line); here $y_0 = 20 \,\mathrm{g\,cm^{-2}}$ and $l_\infty = L_\infty/L_{Edd}$. The "cold" solutions are just those already found by ZS and are obtained setting $T_\gamma(y = 0) = [L_\infty/(4\pi R_*^2 \sigma)]^{1/4}$. The "hot" solutions exist for values of T_γ satisfying the condition $T_\gamma > T_{crit}(L_\infty, y_0)$; here $T_\gamma(y = 0) = 2 \times 10^9$ K. Temperature is close to T_γ in the outer region ($y \lesssim 23 \,\mathrm{g\,cm^{-2}}$), while in the dense layers close to y_{in} LTE is attained.

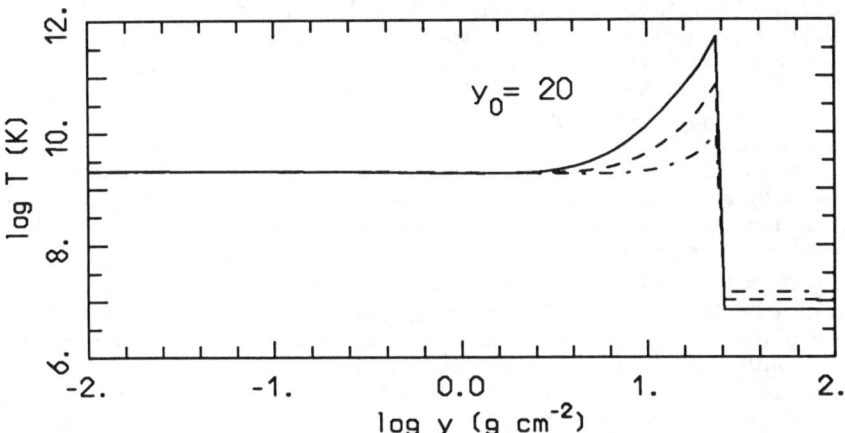

It is possible to get an insight on the existence of high–temperature solutions and to give an estimate of the limiting value T_{crit} by means of simple analytical considerations using a plane-parallel geometry for the atmosphere. For $y < y_0$ we can safely neglect free–free absorption in equation (3) and we get:

$$L = L_\infty \frac{y_0 - y}{y_0}; \quad U = \frac{W}{c}\left[2y_0 + 3k_{es}y\left(y_0 - \frac{y}{2}\right)\right]; \quad \varrho = D\frac{y}{T},$$

where $D = GM_* m_p/2kR_*^2 = 8.1 \times 10^5$, in c.g.s. units for $R_* = 10^6$ cm and $M_* = 1 M_\odot$. Neglecting the term $\kappa_P U$, the energy equation (5) becomes a cubic equation in $x = T^{1/2}$, that can be studied analytically for given values of y_0 and L_∞ and treating T_γ as a free parameter. For $T_\gamma \gg 10^8$ K and setting $y/U(y) \simeq y_0/U(y_0)$, equation (5) can be cast into the form

$$x^3 - T_\gamma x + B\frac{y_0}{U(y_0)} = 0, \tag{7}$$

where $B = 5.1 \times 10^{25}$, again in c.g.s. units. Once y_0 and l_∞ are fixed, equation (7) has one or three real roots, according to the sign of the discriminant, and it is easy to prove that all the roots are real only if

$$T_\gamma \geq \frac{2.5 \times 10^8}{\left(1 + \frac{3}{4} k_{es} y_0\right)^{2/3} l_\infty^{2/3}}. \qquad (8)$$

It can also be shown that if just one root is present it is $T \gtrsim T_\gamma$, while, when condition (8) is satisfied, the three roots have magnitudes $T \gtrsim T_\gamma$, $T \ll T_\gamma$ and $T \lesssim T_\gamma$, respectively. The solution $T \gtrsim T_\gamma$ is unacceptable since $T > T_\gamma$ will produce a negative radiation temperature gradient (see eq. [6]) and also $T \ll T_\gamma$ must be discarded because it is inconsistent with our starting assumption that absorption could be neglected because the plasma is very hot. Finally, the root $T \lesssim T_\gamma$ is the "hot" solution. It exists only when the radiation temperature exceeds the limit given by (8), which represents the analytical estimate of T_{crit}.

The numerical analysis indicates that high temperature solutions may exist only for high enough values of l_∞ and that the critical luminosity, l_{cr}, under which no "hot" solutions exist depends on y_0: for $y_0 = 5$, $l_{cr} = 2 \times 10^{-2}$, while for $y_0 = 20$, $l_{cr} = 6 \times 10^{-3}$. The existence of both "cold" and "hot" solutions for the same values of the flow parameters has been already found in black hole accretion (Park 1990; Nobili, Turolla & Zampieri 1991) when both free–free and Compton scattering are present.

In summary, we have found that, at very low accretion rates, there is a unique solution, the "cold" one discovered by ZS. This essentially indicates that the black body approximation at very low luminosities, as expected for instance in isolated neutron stars accreting the interstellar medium is reasonable (see e.g. Ostriker, Rees & Silk 1970; Treves & Colpi 1991; Blaes & Madau 1993). On the contrary, two different stationary states are possible for $l_\infty \gtrsim 10^{-2}$, a "hot" and a "cold" one. It can be shown (see Turolla et al. for details) that the mean photon energies of the two modes approach each other for increasing l_∞: the "cold" solution becomes hotter for increasing accretion rate while the "hot" one softens. This behaviour is opposite to the one exhibited by the hot, shocked solutions of Shapiro & Salpeter (1973). The transition between the "hot" and the "cold" regime, even at luminosities where the two solutions are rather different, may be expected in a time–dependent scenario. Non–stationary calculations are needed to explore the stability properties of the two solutions.

REFERENCES

Alme, M.L., & Wilson, J.R. 1973, ApJ, 186, 1015
Blaes, O., & Madau, P. 1993, ApJ, 403, 690
Nobili, L., Turolla, R., & Zampieri, L. 1991, ApJ, 383, 250
Ostriker, J.P., Rees, M.J., & Silk, J. 1970, Astrophys. Letters, 6, 179
Park, M-G. 1990, ApJ, 354, 64
Park, M-G., & Ostriker, J.P. 1989, ApJ, 347, 679
Shapiro, S.L., & Salpeter, E.E. 1973, ApJ, 198, 761
Treves, A., & Colpi, M. 1991, A&A, 241, 107
Turolla, R., Zampieri, L., Colpi, M., & Treves, A. 1993, ApJ, submitted
Zel'dovich, Ya., & Shakura, N. 1969, Soviet Astron.-AJ, 13, 175
Wandel, A., Yahil, A., & Milgrom, M. 1984, ApJ, 282, 53

X-Ray Sources in External Galaxies

SUPERSOFT X-RAY SOURCES

Günther Hasinger
Max-Planck-Institut für extraterrestrische Physik,
85748 Garching b. München, Germany.

ABSTRACT

Objects with temperatures on the order of 30 eV and luminosities around 10^{38} erg/s have been established as a separate class by the recent ROSAT observations. The prototype supersoft X-ray source (SSS) is the well-known X-ray binary CAL83 in the LMC, which has an orbital period of 1.04d. In the meantime about 10 SSS have been found in the Magellanic Clouds and at least 15 in the Andromeda Nebula. Two uncatalogued galactic supersoft sources have recently been discovered, but also the nova Muscae 1984 could be detected in a supersoft state 9 years after the explosion. In several cases dramatic time variability is found, with X-ray on- and off-states. The observations are summarized here.

Several models have been suggested to explain the high luminosity and low temperature of SSS, all of them require very high mass accretion rates, which are also indicated in some of the optical spectra. In one interpretation the compact object is a neutron star (or black hole) shrouded by super-Eddington accretion and the large radius and low temperature of the X-ray photosphere is due to a Compton scattering cloud. If this were true, a large number of undetected supersoft neutron star binaries could help to moderate the millisecond pulsar birthrate problem. In the more likely interpretation, the compact object is a white dwarf burning nuclear fuel on its surface at the Eddington accretion rate. In this case the long-sought predecessors for SN type-Ia explosions might be among the class of supersoft sources.

1. INTRODUCTION

The first systematic soft X-ray survey of the Magellanic Clouds has been performed using the IPC detector aboard the *Einstein* X-ray observatory (Long, Helfand & Grabelsky 1981; Wang & Wu 1992). Luminous X-ray binary sources with very soft spectra, such as the enigmatic objects CAL83 and CAL87 have been discovered in these observations. CAL83 has been identified with a 17^m variable blue stellar object with a low-mass X-ray binary (LMXB) type emission line spectrum and an orbital period of 1.04 days (Cowley et al., 1984; Pakull et al., 1985; Smale et al, 1988). CAL87, on the other hand, was found to be an eclipsing object of $\sim 19^m$ with an orbital period of $10.6h$ (Pakull et al., 1988; Callanan et al., 1989; Cowley et al., 1990) where the compact object may be a black hole. Originally the soft X-ray spectra, with peak photon energies more than an order of magnitude lower than those of conventional X-ray binaries, were interpreted in terms of scattering from an extended accretion disk corona, while the harder X-rays from the compact object were supposed to be almost completely shielded by the accretion disk (Fabian et al., 1987).

In the first-light observations with the ROSAT X-ray satellite (Trümper 1983) the central region of the LMC was surveyed again and in addition to CAL83 another bright object – RX J0527.8–6954 – with an X-ray spectrum similar to CAL83 was discovered (Trümper et al., 1991). With the improved energy resolution of the PSPC aboard ROSAT (Pfeffermann et al., 1986) it became obvious that the

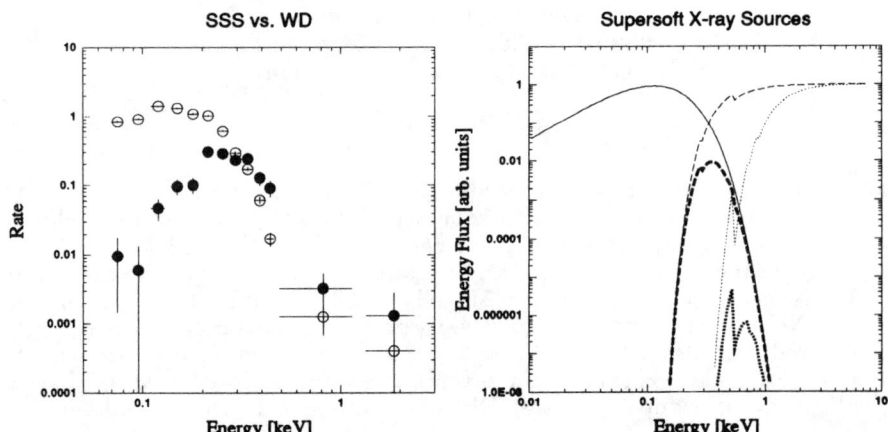

Figure 1: Comparison of the PSPC pulse height spectrum of RX J0527.8−6954 with that of the DA white dwarf RE J1629.2+7804 (Meaty), discovered during the first-light observations of the ROSAT extreme ultraviolet camera WFC (Cooke et al., 1992). Note the signature of absorption in the RX J0527.8−6954 spectrum and the fact that different temperatures can be clearly distinguished.

Figure 2: The effect of the interstellar medium on a 40 eV blackbody spectrum (solid line) is demonstrated assuming a column density of 10^{21} (dashed) and 10^{22} cm^{-2} (dotted), respectively.

two objects have an X-ray spectrum even softer than anticipated, with the bulk of the photons below 0.5 keV, but on the other hand show the signature of X-ray absorption below 0.2 keV which clearly distinguishes them from other objects with very soft X-ray spectra like e.g. white dwarfs (see fig 1). Formal spectral fits, assuming simple blackbody models folded through the PSPC response matrix, were able to constrain the temperatures to be smaller than ~ 40 eV, the line-of-sight absorption to be compatible with or larger than the galactic NH-value in the direction of the LMC and the bolometric luminosities to be larger than $\sim 10^{38}$ erg/s, i.e. the Eddington Luminosity of a 1 $M_\odot$ object (Greiner, Hasinger & Kahabka 1991). Only a very small fraction (~ 0.01) of the bolometric flux is actually observable in the soft X-ray range, so that the spectral parameters have huge uncertainties. Fig. 2 visualizes the dramatic attenuation of the SSS flux by the interstellar medium.

With the completion of the ROSAT all-sky survey (RASS) as well as further deep pointed observations of the LMC, SMC and M31, more and more similar objects were discovered, of which this review attempts to give an overview as complete as possible. This way the ROSAT observations were able to establish a new and distinct class of objects, now called the "Supersoft X-ray Sources" (SSS). Astronomical terminology has often been a bit heuristic and sometimes misleading. In particular terms like "soft" or "hard" are clearly related to an observer's preferred energy range. In the case of the supersoft X-ray sources, confusion could arise with the "Ultrasoft X-ray Sources", a class most likely connected to black holes and soft X-ray transients (cf. White & Marshall, 1984), which, however, stem from the "classical" 2–10 keV energy range and have X-ray temperatures at least a factor of 20 higher than SSS. Another sample of "ultrasoft sources" (USS) has been selected by Cordova et al. (1989) from the IPC archive, which in principle encompasses our SSS, but due to the relatively poor IPC energy resolution and problems in the data analysis seems to be partially contaminated (Cordova, 1992, priv. comm.). Judged from their (invisible) peak of the SSS energy distribution in the extreme ultraviolet we could also call them EUV–sources but this would again lead to confusion. Finally I would like to remark, that the PSPC X–ray spectrum of a SSS viewed through a substantially larger hydrogen column density (e.g. in our own galaxy) will be dramatically reduced in flux and peak at yet higher X–ray energies ($> 0.5\ keV$, see fig. 2). A very large number of SSS could therefore have escaped detection.

2. DEFINITION AND SELECTION OF SUPERSOFT SOURCES

Supersoft X–ray sources can simply be defined as luminous X–ray objects (i.e. with a bolometric luminosity close to the 1 $M_\odot$ Eddington limit) with blackbody temperatures between about 10 and 100 eV. However, observationally this definition is not very feasible because (a) given the almost monochromatic nature of the observed X–ray spectrum (see fig. 2) and the unknown absorption column density there are huge uncertainties in the determination of temperature and luminosity and (b) the distance is unknown to many potential candidates. Utilizing the improved spectral resolution of the PSPC and the almost perfect division of the two spectral bands above and below the carbon edge at 0.28 keV, another set of criteria turned out to be more useful to detect SSS candidates:

- More than 90% of the observed photons have to be below PSPC channel 50 (i.e. a pulseheight corresponding to 0.5 keV). The standard definition of PSPC hardness ratios is $HR1 = (H - S)/(H + S)$, where H (hard) is the sum of the counts in channel 52–201 (roughly 0.5–2 keV) and S (soft) is the sum of the counts in channel 11–41 (roughly 0.1–0.4 keV). (There is a "dead" energy band from 0.4–0.5 keV where only photons spilling over from the soft and hard band are detected.) In order to be able to significantly make the statement about the spectral softness usually a 1σ upper limit to the hardness ratio is calculated. The selection criterium then corresponds to $HR1 + \sigma HR1 \leq -0.8$.

- As far as possible it should be ruled out that the soft emission is due to a low–luminosity forground object (e.g. a white dwarf, a cool coronal star or a cataclysmic variable). One criterium selecting against nearby objects is a clear indication of X–ray absorption in the PSPC spectrum (see fig. 1) or at least the non–detection of the soft X–ray source in the EUV–range, e.g. the WFC all–sky survey (Pounds et al., 1993). However, some of the PSPC-detected white dwarfs did also not show up in the WFC data (Greiner, priv. comm.). Since the known SSS (like e.g. CAL83) have a relatively high f_X/f_{opt} flux ratio ($\simeq 1$), the existence of a bright ($< 15^m$) optical object in the X–ray error circle also argues against an SSS identification. In some cases AGN have shown very soft X–ray spectra; these would have to be sorted out by optical spectroscopy.

- Finally, in order to be able to put constraints on the source luminosity, there should be some distance information, either trough positional coincidence with a nearby galaxy or globular cluster or through the spectroscopic optical identification of the system.

Obviously this set of selection criteria is relatively fuzzy and it also biases severely against galactic SSS, in particular heavily absorbed SSS which have to be searched for using different methods (see section 4). Unless there is a solid optical identification there will always be doubts about the classification of an object as SSS.

3. THE LMC SUPERSOFT SOURCES

Table 1 gives a summary of all SSS discovered so far. A total of eight SSS has been announced in the LMC of which, however, two have been identified as foreground objects not belonging to the class of SSS. The following paragraph gives a short description of the individual objects.

CAL83: Discovered in the original *Einstein* survey of the LMC (Long, Helfand & Grabelsky 1981) with an IPC count rate of 0.137 c/s and shown to have a very soft IPC spectrum (Wang et al., 1991) this source is the prototype of the class of SSS. CAL83 has been identified with a 17^m variable blue stellar object with a low-mass X-ray binary (LMXB) type emission line spectrum and an orbital period of 1.04 days (Cowley et al., 1984; Pakull et al., 1985; Smale et al, 1988). The optical spectrum of CAL83 shows strong H_β and He II $\lambda 4686$ emission lines with a narrow peak component and a broad base, reflecting the high velocities ($\sim 1500\ km/s$) in the inner disk (Crampton et al., 1987). There is sometimes also an indication of the Bowen NIII/CIII line complex near $\lambda 4640$, often observed in galactic LMXB spectra (Cowley et al., 1993). The H_β line has an asymmetric shape with the blue wing being absorbed similar to a P Cygni profile, indicating a massive outflow of gas from the system (Cowley et al., 1993, Pakull et al., 1993). CAL83 is also surrounded by a very weak, but large photoionized nebula (Pakull & Angebault, 1986).

The ROSAT PSPC spectrum of CAL83 for the first time showed the absence of photons above 0.5 keV (Trümper et al., 1991). A blackbody fit to the spectrum indicates a temperature less than 40 eV, an absorbing column density compatible with the LMC value and a luminosity larger than 10^{38} erg/s (Greiner, Hasinger & Kahabka 1991). As a representative example for the large uncertainties in the determination of the SSS spectral fit parameters from PSPC data figure 3 shows the χ^2 contours of the CAL83 fit. The derived blackbody radii of the emission region (see Greiner et al., 1991) have correspondingly large errors. An additional point of concern is the uncertainty in the detailed shape of the assumed theoretical spectrum which causes large errors in the extrapolated bolometric luminosity. Detailed theoretical models, like e.g. NLTE–calculations for white dwarf atmospheres (see e.g. Jordan et al., 1994), have not yet been fit to many SSS spectra. First results indicate, however, that the bolometric luminosities derived from blackbody fits to the data in some cases may be overestimated by as much as a factor 10 (van Teeseling, Heise & Kahabka, 1994).

CAL87: Also discovered in the first *Einstein* survey (Long, Helfand & Grabelsky 1981) roughly a factor of 4 fainter than CAL83 and also having a very soft IPC spectrum, though slightly harder than CAL83 (Wang et al., 1991), CAL87 was found to be an eclipsing object of $\sim 19^m$ and $\Delta m \sim 1.3\ mag$ (Pakull et al., 1988) with an orbital period of $10.6h$ (Callanan et al., 1989; Cowley et al., 1990). The optical spectrum of CAL87 again shows a very prominent He II $\lambda 4686$ emission line, but in contrast to CAL83, H_β is practically absent (Pakull et al., 1988, Cowley

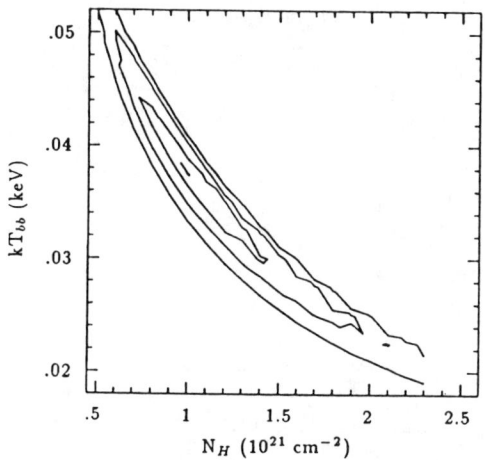

Figure 3: χ^2 contours of a blackbody fit to the PSPC data of CAL83 (reproduced from Greiner, Hasinger & Kahabka, 1991). The contours correspond to 68.3%, 95.4% and 99.7% confidence levels, respectively. The strong dependence between blackbody temperature and NH- value is apparent.

et al., 1990). From dynamical and lightcurve analysis Cowley et al. argued that the compact object may very well be a $> 6\ M_\odot$ black hole.

Detailed ROSAT observations showed this source to be considerably harder than CAL83 (Schmidtke et al., 1993; Cowley et al., 1993), with a hardness ratio $HR1 \approx 0.78$ (Kahabka, Pietsch & Hasinger, 1994) but a blackbody fit to the PSPC data indicates a very low temperature of $34 \pm 10\ eV$ too, however, viewed through a substantially higher column density of $\sim 1.5 \cdot 10^{22}\ cm^{-2}$ (Kahabka, Pietsch & Hasinger 1994; Schmidtke et al., 1993). This high value could be either due to a large LMC–intrinsic interstellar column (i.e. CAL87 lying deep inside the cloud) or due to circumstellar matter. Consequently CAL87 must be intrinsically much brighter than CAL83 to be detected at similar flux in the X-ray band (see e.g. figure 2). Assuming the above blackbody description, a lower limit to its bolometric luminosity is $5 \cdot 10^{38}\ erg/s$ (Kahabka, Pietsch & Hasinger 1994).

CAL87 displays a shallow X-ray eclipse coincident with the optical eclipse, the shape of which is not well constrained by the ROSAT data (Schmidtke et al., 1993; Kahabka, Pietsch & Hasinger 1994). The partial X-ray eclipse indicates an extended X-ray emission region, similar to the accretion–disk corona sources 4U 1822-37 (White et al., 1981) and 4U 2129+47 (McClintock et al., 1982), however its much softer X-ray spectrum and very high luminosity argue against this interpretation.

<u>RX J0527.8-6954</u>: This object was discovered during the ROSAT first–light observations of the LMC (Trümper et al., 1991) with a PSPC count rate of $\sim 0.3\ c/s$ and an extremely soft spectrum, similar to CAL83. The source was shown to be a supersoft transient, since it must have brightened at least by a factor of 10 compared to the previous *Einstein* observations where it was not detected. A blackbody fit to the PSPC pulseheight distribution yielded constraints on the temperature and the absorbing colum density very similar to CAL83 (Greiner, Hasinger & Kahabka, 1991).

In the meantime this source has been observed by ROSAT on several occasions,

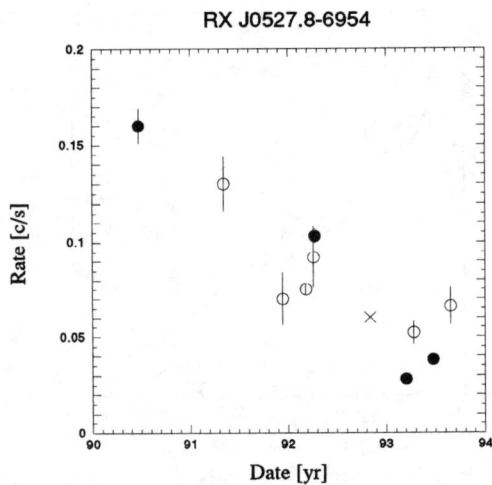

Figure 4: Complete record of the ROSAT count rate of RX J0527.8–6954 over 3.5 years. Filled symbols correspond to PSPC observations with off–axis angles less than 15', open circles to PSPC observations at larger off–axis angles (including the one of Orio & Ögelman 1993). The cross is from the HRI observation of Cowley et al. (1993), transformed to PSPC count rate.

first during the RASS, then in a series of dedicated pointed observations but also quite often serendipitously in the field of other pointing targets (it is e.g. $\sim 21'$ away from the bright LMC supernova remnant N132D which has been observed frequently for calibration purposes and also in the HRI field–of–view of a Cowley et al. pointing). As first noted by Orio and Ögelman (1993) its count rate has decreased substanially since its discovery (see also Cowley et al., 1993). Figure 4 gives a complete record of the count rates of RX J0527.8–6954 from its discovery in June 1990 to the end of 1993, showing an almost linear decrease over 3.5 years. Averaging the locations from all ROSAT pointings where the source was observed within 10 arcmin of the optical axis (i.e. 4 PSPC and 1 HRI observation) a refined X-ray position of $R.A. = 5^h27^m48.5^s$, $decl. = -69°54'10.6"$ ($J2000.0$) can be derived, consistent with the Cowley et al. (1993) location, however with an estimated 90%-error radius of only $\sim 5"$. According to Kürster and Hasinger 1992 (priv. comm.) on–axis PSPC and HRI positions have an irreproducible r.m.s. scatter of $\sim 5"$. Averaging several observations obtained with different roll angles therefore improves the positions (see also Hasinger et al., 1993). Trümper et al. (1991) have tentatively identified RX J0527.8–6954 with the Harvard variable star HV 2554 which, however, is difficult to localize from the historical publications (see discussion in Cowley et al., 1993). The relatively bright blue object that Cowley et al. tentatively identify as HV 2554 (in 1992 this star had $m_V = 15.8$), and which they suggested as a strong candidate for the optical counterpart of RX J0527.8–6954, is now outside the 90%–error circle. An optical spectrum of this object taken in 1992 has an A-star appearance without emission lines (Pakull 1994, priv. comm.). Both of these facts now make this star unlikely to be the counterpart. We therefore can conclude that the optical counterpart of RX J0527.8–6954 is fainter than $\sim 17^m$. There are, however, several fainter objects ($18^m - 19^m$) in the X-ray error circle visible on the finding charts. Further optical spectroscopy and a check of the historical plate records of the objects in the X-ray error box of RX J0527.8–6954 would be highly desired.

<u>RX J0513.9−6951</u>: This source was discovered by Schaeidt, Hasinger & Trümper (1993) through a time variability study of bright ROSAT all-sky survey sources near the ecliptic poles, where objects have been monitored several tens of days up to half a year. In 1991 October/November the source brightened by about a factor of 20 during a period of 10 days and showed a factor of 2 variability around a PSPC count rate of ~ 1.2 c/s during 10 days thereafter. The source was also in the PSPC field-of-view during the ROSAT first-light observation in June 1990 where it was not detected; the derived upper limit is a factor of ~ 1000 fainter than the observed peak count rate. RX J0513.9−6951 had a supersoft X-ray spectrum during all periods where it was detectable. A formal blackbody fit to the PSPC pulse height spectrum yields a temperature of 40 eV at a bolometric luminosity of $\sim 2 \cdot 10^{38}$ erg/s and a hydrogen column density consistent with the LMC (Schaeidt, Hasinger & Trümper, 1993).

Further monitoring observations roughly every three months between November 1992 and October 1993 showed the source three time in the off-state and once (July 1993) in an on-state at a brightness comparable to its peak count rate in November 1990 (Schaeidt 1993, priv. comm.). The X-ray source was not detected 14 years earlier during the *Einstein* LMC survey (Wang et al., 1991). This object is therefore the first supersoft source which shows repeating outbursts on a timescale of several years. Further X-ray monitoring would be highly desired.

RX J0513.9−6951 was readily optically identified with a $\sim 17^m$ variable blue object (Pakull et al., 1993), independently recognized as HV 5682 (Cowley et al., 1993). Unlike in the X-rays the optical star seems to be always at roughly the same magnitude, with long-term variations of about $\Delta m \sim 1$. The historical records give a magnitude range of $m_{pg} \sim 15.6$ to 16.6 between 1893 and 1906 (Leavitt 1908). The star was present on the 1958 LMC atlas plates and on the 1975 ESO J plate. The 1992–1993 photometry yielded a magnitude range of 16.6–16.8. By a lucky coincidence several ESO Schmidt plates of the LMC region containing RX J0513.9−6951 (taken for the Macho project) were exposed during the ROSAT all-sky survey encompassing the X-ray outburst of the object (although unfortunately none exactly simultaneous with the X-ray data). Surprisingly, the X-ray outburst was not accompanied by a large optical brightening as we are used to from other types of X-ray transients. The opposite seems to be true: while the X-ray source brightend, the optical star dimmed by about half a magnitude from $B \sim 16.5^m$ to 17^m (Pakull et al., 1993). Figure 5 shows the comparison of the optical and X-ray lightcurve of this object during the ROSAT all-sky survey.

Its optical spectrum is very similar to that of CAL83 with a very strong HeII $\lambda 4686$ line, the O VI $\lambda\lambda 3811, 34$ lines, an indication of the $\lambda 4640$ line complex and relatively strong H_β and H_γ Balmer emission lines with weak P Cygni profiles indicating substantial mass loss from the system (Pakull et al., 1993; Cowley et al., 1993). An IUE spectrum of RX J0513.9−6951/HV 5682 obtained in April 1992 shows strong HeII $\lambda 1640$ and N V $\lambda\lambda 1238, 40$ emission lines (Pakull et al., 1993). Surprising, however, is the almost complete absence of CIV which is otherwise a strong feature in LMXB and cataclysmic variables. Night-to-night variations between 190 and 300 km/s in the radial velocities derived from the emission lines (Pakull et al., 1993) as well as a range of 50 km/s during one night (Cowley et al., 1993) are most likely related to the binary nature of the system with an orbital period in the range of a few hours to a few days.

<u>RX J0534.6−7056</u> and <u>RX J0537−7034</u>: These two X-ray sources were detected serendipitously in a ROSAT PSPC pointing centered on an old nova in the di-

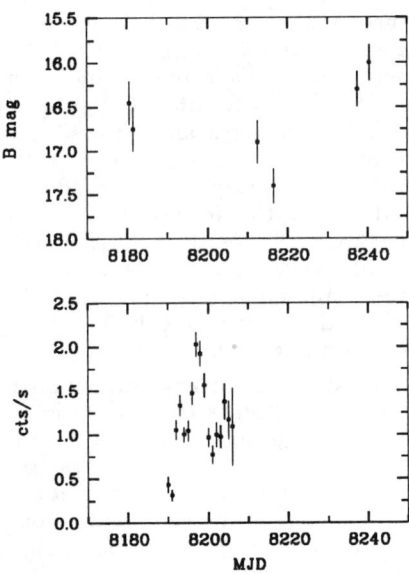

Figure 5: Comparison of the optical and X-ray lightcurve of RX J0513.9-6951 during the ROSAT all-sky survey (from Pakull et al., 1994). While a X-ray dramatic outburst is visible, the optical flux from the system becomes about 0.5^m fainter.

rection of the LMC (Orio and Ögelman 1993). Both of them are very faint, with PSPC count rates about a factor of 100 lower than the peak count rate of RX J0513.9-6951 and have been observed at large off-axis angles, so that the detailed nature of their soft X-ray spectra could not be investigated. Both sources have in the meantime been identified with with galactic foreground objects and should no longer be considered in the class of supersoft sources (see above): RXJ 0534.6-7056 is coincident with the 10.5m galactic star HD 37798 of spectral type K1 V (Cowley et al., 1993) and RX J0537.7-7034 has been identified as a galactic white dwarf (Pakull, Thomas and Bianchi, 1994).

RX J0550.0-7151: This object was discovered during the ROSAT all-sky survey (Beuermann 1992, priv. comm.) and serendipitously at 45' off-axis in a PSPC pointed observation centered on CAL87 (Cowley et al., 1993). It has a PSPC count rate of ~ 1 c/s and a blackbody temperature of ~ 32 eV at a column density of $2 \cdot 10^{21}$ cm^{-2}. The source was not detected during the *Einstein* LMC survey (Wang et al., 1991) although it is now substantially brighter than the Wang et al. detection threshold. The published PSPC position still has a relatively large uncertainty ($\sim 1'$) and therefore strong candidates for the otical counterpart are not yet available.

RX J0439.8-6809: This bright supersoft source in the direction of the LMC has been discovered most recently from the ROSAT all-sky survey data (Greiner, Hasinger & Thomas, 1994) in an attempt to look for galactic supersoft sources at high latitudes (see below). Its average PSPC count rate is ~ 1.3 c/s, a blackbody fit to the PSPC pulseheight distribution yields a temperature of 18 eV at an absorbing column density of $6 \cdot 10^{20}$ cm^{-2} (i.e. consistent with the LMC value). This temperature makes it one of the coolest SSS observed so far. The source was observed again in a pointed PSPC observation about one year after the survey detection with very similar count rate and spectral shape parameters. It was not detected during the ROSAT WFC all-sky survey.

The CCD finding charts for RX J0439.8−6809 show only objects fainter than $V \sim 19^m$ in the 15"-radius error circle (Greiner, Hasinger & Thomas, 1994). Spectra of the two brightest objects inside the error circle as well as of one $\sim 17^m$ object outside the error circle were taken, all three objects turned out to be K-stars. The optical counterpart of RX J0439.8−6809 should therefore be fainter than $V \sim 17^m$, so that the identification with a foreground coronal star or white dwarf can be excluded. The direction 1 degree off the rim of the LMC as well as the best-fit N_H-value are still consistent with an LMC membership which, however, can only be proven through radial velocity determination.

4. THE SMC SUPERSOFT SOURCES

The first soft X-ray survey of the SMC has been performed with the IPC aboard the *Einstein* observatory. Follow-up observations with the *Einstein* HRI improved the positions of most X-ray sources detected (Seward & Mitchell 1981). The SMC has recently been covered completely during the ROSAT all-sky survey (Kahabka & Pietsch, 1993) and in a series of deep PSPC pointed observations (Kahabka, Pietsch & Hasinger, 1994; see also Kahabka, 1994). Here I will describe the properties of the individual SSS detected in the SMC (see also table 1).

<u>1E0056.8−7154</u>: This bright, soft and point-like *Einstein* source (Seward & Mitchell 1981; Wang & Wu, 1992) correlates with the bright ($m_v = 16.6^m$) planetary nebula N67 in the SMC. The very soft IPC spectrum could be fit by a blackbody model with a temperature of $\sim 25 eV$, an X-ray luminosity (0.16−3.5 keV) of about 10^{37} erg/s and an absorbing column density of $7 \cdot 10^{20}$ cm^{-2} (i.e. consistent with the SMC). The soft spectrum and high luminosity was interpreted to be due to surface emission from the central star of the planetary nebula N67, a post asymptotic giant branch star evolving into a white dwarf (Wang 1991) or possibly a close cataclysmic binary (Iben & Tutukov, 1993).

The ROSAT observations of 1E0056.8−7154 show a PSPC count rate around 0.3 c/s, practically constant between the all-sky survey and three PSPC pointings spread over more than one year (Kahabka, Pietsch & Hasinger, 1994). A blackbody fit to the PSPC pulseheight distribution yields a temperature and N_H value consistent with the IPC parameters and a bolometric luminosity above 10^{38} erg/s.

<u>1E0035.4−7230</u>: The second bright and soft source in the SMC was also discovered during the first *Einstein* survey (Seward & Mitchell 1981; Wang & Wu, 1992). It appears point-like in the *Einstein* HRI and has a possible optical counterpart of $\sim 21^m$ in the HRI error box, however, no spectroscopic information is available for this candidate. The soft IPC spectrum indicates a temperature < 40 eV and an X-ray luminosity of $\sim 10^{37}$ erg/s in the *Einstein* band.

A blackbody temperature of $\sim 40 eV$ at an absorption column density of about $5 \cdot 10^{20}$ cm^{-2} was derived from ROSAT PSPC data (Kahabka, Pietsch & Hasinger, 1994). The source showed considerable variability on a timescale of 1−2 days, with count rates varying between 0.2 and 0.5 c/s, however, no general long-term trend. This could be a first indication of orbital variability or could be similar to the slow aperiodic variability observed for many LMXB (see eg. Hasinger & van der

Klis, 1989).

RX J0048.4-7332: This source has been discovered during the ROSAT all-sky survey with a PSPC count rate of 0.2 c/s (Kahabka & Pietsch 1993). It might have been missed during the *Einstein* SMC survey because there is unfortunately a small gap in the exposure map in this direction (Wang & Wu, 1992). The object has been re-observed twice with the PSPC in 1992 April and December, with a count rate compatible with the survey detection (Kahabka, Pietsch & Hasinger, 1994). The best-fit parameters for the ROSAT PSPC spectrum yields a very low blackbody temperature of $10-20\ eV$ and a relatively high absorbing column density of $\sim 5 \cdot 10^{20}\ cm^{-2}$ and, correspondingly, a very high bolometric luminosity. Since the spectrum is, however, relatively ill-constrained, parameters similar to those of 1E0035.4-7230 are also compatible with the data.

The X-ray position is consistent with the SMC cluster NGC 269, about 1' off the Nova SMC 1952 and is only 6" away from the third SMC symbiotic star (SMC 3) in the Morgan (1992) catalogue. This star showed a substantial brightening of about $\Delta V \sim 1-1.5$ and $\Delta U \sim 3$ some time between December 1980 and November 1981 and is a very strong candidate for the optical counterpart of RX J0048.4-7332. Further optical observations, in particular spectroscopic monitoring in order to determine the orbital period of this system are extremely important.

RX J0122.9-7521: A bright object ($\sim 0.8\ c/s$) with a supersoft X-ray spectrum was discovered during the ROSAT all-sky survey in the direction of the SMC, but originally associated with the background galaxy ESO 29-4 (Kahabka & Pietsch 1993). In the meantime an optical spectrum of a star in the X-ray error circle shows white dwarf signatures (Beuermann, 1993, priv. comm.) so that the status of this object in the class of supersoft X-ray sources remains unclear (Kahabka, Pietsch & Hasinger 1994).

RX J0058.6-7146: In a PSPC pointing centered on 1E0056.8-7154 a new transient SSS was discovered about 10' south of the *Einstein* source (Kahabka, Pietsch & Hasinger 1994). Over two days in March 1993 the X-ray count rate of this object showed a dramatic increase from practically zero to about 0.07 c/s, when unfortunately the PSPC coverage ended. It was not detectable during the all-sky survey nor during three other PSPC observations between October 1991 and December 1992. A very recent observation in October 1993 shows this object off again. Because of the relatively few photons detected from this object the spectral parameters are highly uncertain. Nothing is known yet about the optical counterpart of RX J0058.6-7146, so that its nature or even its SMC membership is still uncertain.

5. SUPERSOFT SOURCES IN M31

A soft X-ray survey of the Andromeda Nebula has been performed in July 1990 using the ROSAT PSPC (see Supper et al., 1994). Six contiguous pointings of $\sim 25000\ s$ observing time each covered the whole disk of the galaxy and its surroundings. A total of 396 X-ray sources has been detected in a solid angle of $\sim 5\ deg^2$, of which about 151 ± 22 are likely to belong to M31 and 22 lie in the confused bulge region (see Supper 1994). A hardness ratio analysis (see section 2) was performed on the whole set and a sample of 15 objects which fulfil the detection criteria for supersoft sources ($HR1 + \sigma HR1 \leq -0.8$) was selected. Figure

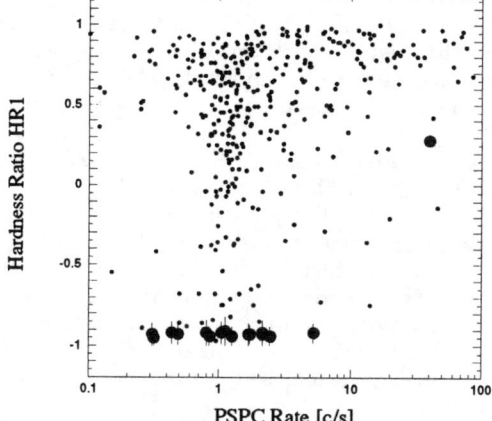

Figure 6: Hardness versus intensity diagram for all objects detected in the direction of M31 (Supper et al. 1994). Filled symbols indicate SSS candidates.

6 shows the hardness/intensity diagram for all objects detected in the direction of M31. For brighter objects, where the errors in the determination of hardness ratios are small, three different groups of sources can be clearly distinguished: (1) hard and relatively absorbed sources ($HR1 > 0$), mainly corresponding to the X-ray binaries in M31, many of them in the bulge region, (2) intermediate hardness, mainly corresponding to foreground stars or unabsorbed background AGN, and (3) very soft sources. Thick symbols indicate the objects selected as SSS. None of these coincides with a catalogued object or likely foreground star (see section 2). The potential number of soft sources is even higher (~ 30), but the statement about the supersoft nature of their spectra cannot be made with significance. Please note, that supersoft objects with high absorption column density would escape this selection. In particular, there is an additional X-ray source right in the direction of the M31 bulge (#208 in Supper's list), which has a much softer spectrum than the rest of the bulge sources and is very likely a confused SSS (also indicated by a thick symbol). One of the SSS candidates is bright enough to perform a detailed spectral fit. A blackbody model yielded a temperature of $kT \sim 30\ eV$ and a bolometric luminosity of $\sim 10^{38}\ erg/s$, i.e. right in the ballpark of the Magellanic SSS.

6. GALACTIC SUPERSOFT SOURCES

If galactic supersoft sources with a luminosity and interstellar extinction similar to those in the Magellanic Clouds would be visible from the solar system, their PSPC count rates would have to be $\sim 30 - 3000\ c/s$ for distances between 10 and 1 kpc. Practically no such object has been detected so far, the main reason is probably the substantially higher interstellar column density towards most distant galactic objects. Therefore several different methods described below have been proposed to search for galactic supersoft sources in the ROSAT all-sky survey data base, each of them selecting a different subset of the parameter space.

The bright and soft approach

A sample of bright and soft X-ray sources (i.e. PSPC count rates $> 0.5\ c/s$ and $HR1 < -0.5$ at galactic latitudes $> \pm 30°$ has been selected from the survey data base by Beuermann & Thomas (1993, priv. comm.) in an attempt to optically identify a complete, flux-limited X-ray sample and this way to detect ultrasoft AGN, Cataclysmic variables (see eg. Schwope, Thomas & Beuermann, 1993) and SSS. The sample is now spectroscopically relatively complete and contains mainly AGN, coronal stars, a few white dwarfs and CVs. There is one exciting discovery of a galactic supersoft source, RX J0019+21, found by this metod (Reinsch, Beuermann & Thomas, 1993). It has a PSPC count rate of $\sim 2\ c/s$ and a clearly supersoft spectrum with a temperature of $\sim 20 - 25\ eV$. Its luminosity is $(1-5) \cdot 10^{37}\ erg/s$ for an assumed distance of 3 kpc. The optical counterpart is a $\sim 12^m$ blue star with strong HeII $\lambda 4686$, H_α and H_β emission lines. Time-variable P Cygni profiles in the Balmer lines indicate fast mass outflow. An IUE spectrum reveals a bright ultraviolet source with a HeII $\lambda 1640$ emission line. An orbital period of 15.8h has been suggested. This object is the brightest known SSS example both in X-rays, UV and optical wavebands and therefore most promising for further detailed studies.

The fainter SSS approach

A systematic search specifically for supersoft sources with characteristics similar to CAL 83 has been performed by Greiner et al. (1994). In a first step objects which fulfil the criterion $HR1 + \sigma HR1 \leq -0.8$ have been selected. This list was then correlated with the WFC XUV all-sky survey data base (Pye et al., 1992, priv. comm.). Only those objects **not** detected by the WFC were maintained. The shortened list was then correlated against published catalogues (primarily SIMBAD, NED and the ST Guide Star Catalogue) in order to remove all known objects, and on the other hand to obtain magnitudes of possible optical counterparts. Objects with optical candidates fainter than 15^m were proposed for follow-up observations with the ROSAT PSPC in order to obtain high-quality X-ray spectra. A handful of objects selected this way later turned out to be white dwarfs. No new galactic SSS candidate could be found by this method, nevertheless the putative LMC SSS RX J0439.8-6809 (Greiner et al., 1994) was detected.

The absorbed SSS approach

As can be seen in figure 2, a supersoft source absorbed through a substantial column density is expected to have a very narrow, quasi monochromatic pulseheight distribution peaking in the range $0.5-1\ keV$. The soft hardness ratio HR1 should be relatively large for such an object, while a second hardness ratio, calculated at higher energies, should show a very soft source: $HR2 = (H2 - H1)/(H2 + H1)$, where H2 is the sum of the counts in channels 92-201 (roughly 0.9-2 keV) and H1 is the sum of the counts in channel 52-91 (roughly 0.5-0.9 keV)[1]. This was the motivation to search for heavily absorbed supersoft objects in the galactic plane. From the ~ 15000 sources detected in the ROSAT galactic plane survey (RGPS, Motch et al., 1991) at absolute galactic latitudes less than 20°, 98 point-like objects with $HR1 \geq -0.4$, $HR2 \leq -0.6$, and PSPC count rate $\geq 0.1\ c/s$ were selected (Motch, Hasinger & Pietsch, 1994). Some of these objects could

---

[1]This is the most recent definition of ROSAT PSPC energy bands while the preliminary output of the automatic Standard Analysis Software System (Voges, 1992) uses $S : 0.07 - 0.4\ keV$, $H : 0.4 - 2.4\ keV$, $H1 : 0.4 - 1.0\ keV$ and $H2 : 1.0 - 2.4\ keV$, so that slight inconsistencies in the currently available definitions of HR1 and HR2 occur.

be flagged out as spurious fragments of bright diffuse emission regions. Many of them turned out to be late–type coronal stars, either from a correlation with the SIMBAD database or from dedicated follow–up spectroscopy, and in 11 cases a bright star in the Guide Star Catalogue indicated a likely coronal counterpart. In addition one new AM Her binary was discovered this way.

Out of the remaining six candidates, Motch et al. selected the brightest X–ray source RX J0925.7–4758, having a count rate of ~ 1 c/s and hardness ratios $HR1 = 0.96$, $HR2 = -0.69$, for detailed X–ray and optical follow–up studies. (The second–brightest candidate has a count rate of 0.27 c/s). The peculiar pulseheight distribution of RX J0925.7–4758 could be confirmed in a pointed PSPC follow–up observation. A blackbody model with $kT \approx 30 - 55$ eV and a very high interstellar column density $N_H \approx 1.4-3.7\cdot 10^{22}$ cm^{-2} gives an acceptable fit to the data.

The object could be spectroscopically identified with a 17^m star with an H_α, He II $\lambda 4686$ and Bowen N III/C III $\lambda\lambda 4640 - 60$ emission line spectrum very similar to conventional LMXB (Motch, Hasinger & Pietsch, 1994). The emission line radial velocities vary by ± 80 km/s on a timescale of one day, indicating the binary nature of the object. Extended optical photometry in 1992 April and December finally revealed a likely orbital period of $\sim 3.5d$. Out of the four SSS with proposed optical periods (see table 1) this object has the longest orbit.

The reddening of RX J0925.7–4758 derived from equivalent widths of the interstellar absorption lines and the shape of the optical continuum gives an N_H-value consistent with the X–ray determination (an extinction of roughly 5 magnitudes). Through the known geometry and column density of the molecular clouds in this direction of the Milky Way a lower limit of $\sim 400pc$ can be placed on the distance of RX J0925.7–4758. Its bolometric luminosity can therefore be very likely comparable to the 1 $M_\odot$ Eddington limit, albeit with large systematic errors.

Optically selected supersoft sources

GQ Mus (Nova Muscae 1983) was the first classical nova from which X–rays were detected during outburst (Ögelman et al., 1984). Because of the limited energy resolution of those EXOSAT observations a spectral shape could not be determined, however, the data was consistent with thermonuclear burning on a white dwarf, producing a luminosity of $10^{37} - 10^{38}$ erg/s at a temperature of ~ 30 eV. As part of a program to monitor the soft X–ray emission of recent novae with the ROSAT PSPC, GQ Mus was detected again in 1992, i.e. 9 years after the optical outburst (Ögelman et al., 1993). It was the only detection out of a survey sample of 26 recent novae. The PSPC count rate was 0.11 c/s and the pulse height distribution could be fit with a blackbody temperature ≤ 30 eV, a column density $\geq 10^{21}$ cm^{-2} eV and a bolometric luminosity $\geq 10^{38}$ erg/s (assuming a distance of 4.7 kpc). These parameters are strikingly similar to those derived for the LMC and SMC supersoft sources (see above).

RR Tel is a symbiotic nova, i.e. an interacting binary system with an evolved red giant filling its Roche–lobe and losing mass to a hot compact component, most probably a white dwarf. In 1944 RR Tel had a slow nova outburst, brightening by 7 magnitudes for about 5 years and then slowly decaying. Out of a total of 7 such objects known, RR Tel was among the four symbiotic novae detected during the ROSAT all–sky survey and (together with AG Dra) displayed a very soft X–ray spectrum (Bickert et al., 1993). In the meantime Jordan et al. (1994) have obtained a high-quality PSPC spectrum from a pointed observation. They fit an elaborate NLTE model atmosphere to the combined ROSAT, IUE and Voyager data of this object, but qualitatively a simple blackbody fit to the PSPC data allows similar conclusions: the best-fit temperature is around 12 eV at an N_H-value of $3 \cdot 10^{20}$ cm^{-2} and a derived bolometric luminosity of $\sim 10^{37}$ erg/s. The relatively low N_H-value allows a substantially better determination of the temperature and bolometric luminosity than for most objects discussed above. In the PSPC spectrum there is also an indication of a separate hard component with a temperature of ~ 350 eV and comparable luminosity. The relatively low luminosity, the low temperature and the separate hard component in the spectrum are somewhat different to the findings for the objects discussed above.

The ROSAT all-sky survey of galactic globuar clusters yielded a very soft spectrum for the X-ray source in **M3** (Verbunt et al., 1993) which was independently indicated in a ROSAT HRI observation (Hertz et al., 1993). Actually this is the only one of the brighter globular cluster X-ray sources displaying a supersoft spectrum. Taking the known distance to this globular cluster (10.4 kpc) and assuming the relatively low absorption column density along the line of sight ($1.1 \cdot 10^{20}$ cm^{-2}) a luminosity of $\sim 10^{35}$ erg/s can be derived for this object, which is substantially lower than the 1 $M_\odot$ Eddington limit discussed for most systems above. Only if there would be a substantial source–intrinsic (or circumstellar) absorption column this object could be discussed in the context of SSS.

7. MODELS OF SUPERSOFT SOURCE EMISSION

A variety of models have been proposed to explain the nature of SSS. The original idea of X-rays scattered by an accretion disk corona which has an extraordinary height due to the low metallicity of the LMC (Fabian et al., 1987) is no longer tenable, given the extremely soft X-ray spectra and high luminosity of SSS, as determined by the ROSAT measurements. In the meantime the class of supersoft sources has been enlarged substantially to about 30 members and apparently contains a rather mixed bag of objects including the central star of a planetary nebula, symbiotic novae and a classical nova. It almost looks like the high sensitivity and moderate spectral resolution of the ROSAT PSPC in the $0.1 - 0.5$ keV band has opened a window in X-rays through which we now can discriminate new kinds of objects with a distinctly soft spectral nature and all radiating at a luminosity close to 10^{38} erg/s. A black–hole interpretation has been put forward by Cowley et al. (1990) for the compact object in the CAL87 system. Although the ROSAT data now indicate a very low temperature for this object too, its minimum luminosity is still substantially higher than that of the other SSS (Kahabka, Pietsch & Hasinger, 1994), so that a different nature might be actually considered.

Super–Eddington mass transfer rates

The bolometric luminosities of SSS, as determined from the spectral fits to the X-ray data under the assumption of simple blackbody radiation, are extremely uncertain and in most cases only yield lower limits. Even these limits might be overestimated, once realistic theoretical spectra are considered. Nevertheless there are independent indications that the energy consumption in SSS is extremely large indeed: the observed optical/UV flux distribution of the identified SSS, which is not heavily affected by the interstellar medium, yields optical/UV luminosities of several 10^{37} erg/s and is usually well above a simple blackbody extrapolation of the X-ray data (Greiner, Hasinger & Kahabka, 1991; Pakull et al., 1994; Reinsch et al., 1993). This indicates that we observe a highly luminous accretion disk in the optical, which is either heated through friction of a large mass transfer, or, more likely, reprocessing the luminosity produced by the central compact object (see e.g. Vrtilek et al., 1990). At any rate, the disk luminosity indicates an accretion rate which is substantially larger than that of even the highest luminosity LMXB (e.g. Sco X-1, Cyg X-2, LMC X-2). The substantial mass outflow signified by the P Cygni profiles observed in the spectra of practically all optically identified SSS (Cowley et al., 1993; Pakull et al., 1994; Reinsch, Beuermann & Thomas, 1993; Motch, Hasinger & Pietsch, 1994) is another proof for a very large, most likely super-Eddington, mass transfer rate.

These accretion rates are probably much larger than those which can be produced by nuclear evolution, gravitational radiation or magnetic breaking in a short-period binary system. They can, however, be provided by unstable mass transfer on a thermal timescale from a donor companion which is more massive than the accreting compact object, so that the orbital separation decreases as a function of time (Pacynski 1971; see also van den Heuvel et al., 1992). On the other hand, the companion stars in CAL83 and CAL87 can not have much more than $\sim 2\ M_\odot$, otherwise they would be visible in the optical spectra (see eg. Cowley et al., 1990). A grain of salt in this argument is also, that there are a number of LMC and SMC supersoft sources which are not yet optically identified and apparently have considerably fainter optical counterparts. Alternatively, the nuclear evolution of a low-mass evolved star in a long-period period system (like eg. Cyg X-2) could also provide a high enough mass transfer rate. The classical nova GQ Muscae, however, does not fit both of these scenarios, because it is a short-period system with a secondary much less massive than the accreting object. In this case a particular feedback mechanism, which is triggered by the nova explosion, might have to be invoked (Ögelman et al., 1993).

The time variability of SSS

Some of the known SSS are persistent X-ray sources. CAL83, the prototype SSS, seems to be very stable in X-rays on all observable timescales (however not in the optical, see Bianchi & Pakull, 1988). The same is true for 1E 0056.8-7154, the central star of the planetary nebula N67. CAL87 is constant apart from the eclipses observed in X-rays and optical. The majority of SSS, however, seem to be dramatically time variable, having X-ray outbursts or even on-off states. This can be judged from the comparison between *Einstein* and ROSAT data more than 10 years apart and from ROSAT monitoring observations over several years. The most dramatic time variability was seen for RX J0513.9-6951, where the X-ray outburst could be monitored during several days and which was observed in off- and on-states thereafter. An X-ray outburst on a similar time scale ($\sim 2\ days$) was observed for the putative supersoft source RX J0058.6-7146, while the decay of RX J0527.8-6954 lasted more than three years. Although not yet observed in any one object together, a rise time of a few days and a decay time of hundreds of days to years could be a characteristic of SSS outbursts. Also aperiodic variability on a timescale of days to months seems to be not uncommon: an example is 1E 0035.4-7230 in X-rays (Kahabka et al., 1994) and CAL83 in the optical and UV (Bianchi & Pakull, 1988).

It was a great surprise, that the X-ray outburst of RX J0513.9-6951 was not accompanied by a similar brightening of its optical counterpart, HV 5682 which, if anything, got optically fainter during the X-ray outburst. Schaeidt, Hasinger and Trümper (1993) argued, that the observed large variability might not be due to a large variation in mass–accretion rate, but might just be an "X–ray illusion", since a slight temperature change could result in a dramatic variation of the X–ray luminosity observed on the Wien tail of the blackbody distribution. Pakull et al. (1994) even assumed, that the observed X–ray outburst / optical depression of RX J0513.9–6951 could be entirely due to a decrease in radius and corresponding increase in the temperature of the X–ray photosphere while the mass transfer rate remains approximately stable. A similar situation, where the X–ray luminosity of the system does not follow the mass accretion rate, is observed in the brightest LMXB as they reach the Eddington limit (e.g. the Z–source Cyg X–2; see Hasinger et al. 1990), although at much higher temperatures.

LMXB models for SSS

A neutron star (or black hole) low–mass X–ray binary model for the LMC supersoft sources was originally supported by the orbital periods of CAL83 and CAL87, their observed aperiodic variability and the similarity of their optical spectra with those of galactic LMXB. The unusual softness of their spectra, compared to the $\sim 1\ keV$ temperatures of conventional LMXB, however, was a puzzle. The low temperature could only be reconciled with a higher central temperature, if a neutron star (or black hole) is surrounded by a large cocoon of matter, which is Compton–downscattering the original X–ray photons into the EUV (Greiner et al., 1991). The observed large blackbody radius ($> 10^4 km$) would then correspond to the photosphere at the last scattering surface. Ross (1979), in a different context, has calculated examples of shrouded neutron stars and predicted spectra which are quite similar to those observed for SSS. Naturally the neutron star should have a small magnetic field, like it is assumed for conventional LMXB too, so that the Alfven radius is well inside the cocoon. The scattering cloud could be supported by super–Eddington accretion and the associated mass outflow, which can also be observed in the P Cygni profiles of the Balmer lines. Recently, Kylafis and Xilouris (1993) have performed detailed calculations of spherical accretion onto unmagnetized neutron stars and found that the simulated spectra can be reconciled with supersoft sources if the accretion rate is near-Eddington and the radial flow extends to several thousand neutron star radii.

A criticism of this scenario comes from the fact that there is no known case of a conventional X-ray binary which becomes supersoft when the accretion rate exceeds the Eddington limit while there are at least two counter–examples of neutron stars which get harder and/or quench when they approach or exceed the Eddington limit (see van den Heuvel et al., 1992; Rappaport, Di Stefano & Smith, 1994). Also the stability of the assumed accretion flow is questionable. The systems used as counter–examples, LMC X–4 and Cen X–3, however, both contain highly magnetized neutron stars for which the shrouding scenario might not be applicable altogether, because the magnetic field influences the accretion flow already far away from the neutron star. On the other hand there are examples of conventional LMXB which get substantially softer and increase their photosphere when exceeding the Eddington limit, namely the X–ray bursters in their radius-expansion phase. There is a recent example of a very long burst of 4U 2129+11, measured with the Ginga LAC, where in the initial radius expansion phase the temperature dropped from $\sim 2 keV$ to less than $\sim 0.3\ keV$, below which the LAC proportional counter is no longer sensitive. The derived blackbody radius of the X–ray photosphere simultaneously grew to more than 1000 km, i.e. about 100 neutron star radii (van Paradijs et al., 1990). For a time interval of about one second, i.e. much longer than the dynamical timescale of the neutron star photosphere, this object therefore was on its way to be a supersoft X-ray source. The peculiar time–variability of RX J0513.9–6951, with an X–ray outburst while the optical light stays approximately constant or even drops, could be naturally explained in the context of a scattering cloud: when the accretion rate drops, the radius of the photosphere decreases with a corresponding increase in temperature which lets the X–ray source emerge above the interstellar hydrogen cutoff. In the meantime the optical radiation just reflects the bolometric energy output, i.e. drops corresponding to the accretion rate.

The possibility that some of the known SSS may actually be part of a large population of "unseen" neutron star LMXBs which could serve as additional progenitors for recycled radio pulsars opens the exciting possibility to moderate the millisecond pulsar birthrate problem (Kulkarni & Narayan 1988).

White–dwarf models for SSS

The mere fact, that among the known SSS there are at least four systems, where the compact object is believed to be a white dwarf (including one possible post asymptotic giant branch star), supports a white dwarf interpretaion for SSS in general (van den Heuvel et al., 1992): the nova GQ Muscae, the central star of the planetary nebula N67 and the symbiotic novae SMC3 and RR Tel fall into this category (see above). The SSS blackbody emission radii derived from the X–ray observations ($\sim 10^4 km$) actually fit the size of a white dwarf much better than that of a neutron star. In contrast to neutron stars, where the accretion luminosity per gram of accreted matter is about a factor of 100 higher than the nuclear burning luminosity, nuclear energy is about an order of magnitude more efficient in the case of a white dwarf. At relatively low accretion rates onto a white dwarf the thermonuclear reaction occurs in violent nova explosions. If, on the other hand, the accretion rate grows too high the white dwarf will develop a red–giant atmosphere. Only for a narrow range of accretion rates above $\sim 10^{-7}$ $M_\odot/yr$, stable nuclear burning can occur on the white dwarf surface (Pacynski & Zytkow 1978).

For relatively massive accreting white dwarfs ($M_{wd} = 0.7 - 1.2$ $M_\odot$) the nuclear fusion luminosity and temperature would be in the range allowed by the ROSAT observations of CAL83 and RX J0527.8-6954 (van den Heuvel et al., 1992). Because of the large white dwarf masses and the high mass accretion which is not expelled from the system by nova explosions, a fraction of these objects could actually grow beyond the Chandrasekhar limit and undergo an accretion–induced collapse. This opens the exciting evolutionary perspective that SSS may be the long–sought predecessors of supernova type Ia explosions.

Van den Heuvel et al. (1992) suggested a limit-cycle behaviour for the temporal variability of supersoft sources which would occur on a timescale of ~ 100 yr, during which the object would be in a supersoft state for ~ 10 yr. Detailed light curves for such "slow novae" have been calculated by Iben (1982) in the context of symbiotic novae like RR Tel as a function of mass accretion rate. For very hot white dwarfs and mass accretion rates on the order of 10^{-8} $M_\odot/yr$ outbursts with rise times on the order of tens of days and decay times of 10–50 yrs are predicted. The peculiar recurrent X–ray outbursts and optical persistence of of RX J0513.9-6951 can only be reconciled with this scenario, if the white dwarf photosphere is expanding and contracting considerably (roughly by a factor of 4) once the accretion rate approaches the Eddington limit (Pakull et al., 1994).

Detailed population studies for supersoft sources of this kind have been performed by Rappaport, Di Stefano & Smith (1994) using a Monte Carlo simulation technique for the formation and evolution of supersoft sources from a seed of main-sequence primordial binaries. A complicated network of evolutionary steps finally leads to close binary systems with white dwarfs and more massive donor stars with accretion rates in the required narrow range. Under a wide variety of input assumptions the authors come to the conclusion that at any moment there should be more than ~ 1000 supersoft systems in our galaxy or M31 with properties (i.e. luminosity, temperature and orbital periods) close to the observed distributions. About 100 and 15, respectively, should exist in the LMC and SMC. Taking into account the galactic scale height distribution, the interstellar absorption and detection efficiency these numbers are consistent with the actually observed systems (see e.g. Motch, Hasinger & Pietsch, 1994). Finally the authors try to constrain the fraction of supersoft accreting white dwarfs which produce type–Ia supernovae and come to interesting, however still quite uncertain production rates.

ACKNOWLEDGEMENTS

I would like to thank my collaborators in the various projects of studying supersoft sources: J. Greiner, P. Kahabka, Motch C., W. Pietsch, S. Schaeidt, H.–C. Thomas and J. Trümper. In particular I am indebted to R. Supper et al. for the permission to use the M31 catalogue data in advance of publication. I acknowledge helpful discussions with M. Pakull.

REFERENCES

Bianchi L. & Pakull, M.W., 1988, in: *'A decade of UV astronomy with IUE'*, ESA SP-281, p.145
Bickert K. et al., 1993, A&A submitted
Callanan P.J., Machin G., Naylor T. & Charles P.A., 1989, MNRAS 241, 37p
Cordova F.A., Rodriguez-Bell T., Harnden F.R., Kartje, J. & Mason K.O., 1989, in: *Proceedings of the Berkeley Colloquium on EUV Astronomy*, eds. R. Malina & S. Bowyer, Pergamon Press, 30
Cooke B.A., et al. 1992, Nature 355, 61
Cowley A.P., Crampton D., Hutchings J.B., Helfand D.J., Hamilton T.T., Thorstensen J.R. & Charles P.A., 1984, ApJ 286, 196
Cowley A.P., Schmidtke, P.C., Crampton D. & Hutchings J.B. 1990, ApJ 350, 288
Cowley A.P., Schmidte,P.C., Hutchings, J.B., Crampton D. & McGrath T.K, 1993, ApJ 418, L63
Crampton D., Cowley A.P., Hutchings, J.B., Schmidtke, P.C., Thompson, I.B. & Liebert J., 1987, ApJ 321, 745
Fabian A.C. et al. 1987, MNRAS 255, 29p
Greiner J., Hasinger G. & Kahabka P. 1991, A&A 246, L17
Greiner J., Hasinger G. & Thomas H.–C., 1994, A&A 281, L61
Hasinger G. & van der Klis M., 1989, A&A 225, 79
Hasinger G., van der Klis M., Ebisawa K., Dotani T. & Mitsuda K., 1990, A&A 235, 131
Hasinger et al., 1993, A&A 275, 1
Hertz P., Grindlay J.E. & Bailyn C.D., 1993, ApJ 410, L87
Iben I. Jr., 1982, ApJ 259, 244
Iben I. & Tutukov A., 1993, ApJ 418, 343
Jordan S., Mürset U. & Werner K., 1993, A&A submitted

Kahabka P. & Pietsch W., 1993, in: *Lecture Notes in Physics 416, New Aspects of Magellanic Cloud Research* eds. Baschek, Klare, Lequeux, 71
Kahabka P., Pietsch W. & Hasinger G., 1994, A&A in press
Kahabka P., 1994, MPE preprint 279
Kulkarni S.R. & Narayan R., 1988, ApJ 355, 755
Kylafis N.D. & Xilouris E., 1993, A&A 278, L43
Leavitt, H.S., 1908, Harvard Ann. 60, #4, 87
Long K.S., Helfand D.J. & Grabelsky D.A., 1981, ApJ 248, 925
McClintock J.E., London R.A. & Bond H.E., 1982, ApJ 258, 245
Morgan D.H., 1992, MNRAS 258, 639
Motch C., et al., 1991, A&A 246, L24
Motch C., Werner K. & Pakull M.W., 1993, A&A 268, 561
Motch C., Hasinger G., Pietsch W., 1994, A&A in press
Ögelman H., Beuermann, K. & Krautter, J., 1984, ApJ 287, L31
Ögelman H., Orio M., Krautter J. & Starrfield S., 1993, Nature 361, 331
Orio M. & Ögelman H., 1993, ApJ 273, L56
Pacynski B., 1971, ARA&A 9, 183
Pacynski B. & Zytkow A.N., 1978, ApJ 222,604
Pakull M.W., Ilovaisky S.A. & Chevalier C., 1985, Space Sci. Rev. 40, 229
Pakull M.W. & Angebault P., 1986, Nature 322, 511
Pakull M.W., Beuermann K., van der Klis M. & van Paradijs J., 1988, A&A 203, L27
Pakull M.W., Motch C., Bianchi L., Thomas H.-C., Guilbert J., Grison P. & Schaeidt S., 1994, A&A in press
Pakull M.W., Thomas H.-C. & Bianchi L., 1994, A&A in preparation
Pfeffermann E. et al., 1986 *Proc. SPIE* 733, 519
Pounds K.A. et al. 1993, MNRAS 260, 77-102
Reinsch K., Beuermann K. & Thomas H.-C., 1993, AG Abstract Series No. 9, 41
Rappaport S., Di Stefano R. & Smith J.D., 1994, ApJ submitted.
Ross, R.R., 1979, ApJ 233, 334
Schaeidt S., Hasinger G. & Trümper J., 1993, A&A 270, L9
Schmidtke P.C., McGarth T.K., Cowley A.P. & Frattare L.M., 1993, PASP 105, 863
Schwope A.D., Thomas H.-C. & Beuermann, K., 1993, A&A 271, L25
Seward F.D. & Mitchell M., 1981, ApJ 243, 736
Smale A.P. et al., 1988, MNRAS 233,51
Supper R., Hasinger G., Pietsch W., Trümper J., Jain. A., Lewin W.H.G., Magnier G. & van Paradijs J., 1994, A&A in prep.
Supper R., 1994, this volume
Trümper J. 1983, Adv. Space Res. 2, No. 4, 241
Trümper J., Hasinger G., Aschenbach B. et al., 1991, Nature 349, 579
van den Heuvel E.P.J., Bhattacharya D., Nomoto K. & Rappaport, S.A., 1992, A&A 262, 97
van Paradijs J., Dotani, T., Tanaka, Y. & Tsuru T., 1990, PASJ 42, 633
van Teeseling A., Heise J. & Kahabka, 1994, A&A (in preparation)
Verbunt F., Hasinger G., Johnston H., & Bunk W., 1993, Adv. Space. Res. 13, (12) 151
Voges, W., 1992, in: *Proc. of the ESA ISY Conference*, ESA ISY-3, p.9
Vrtilek S.D., Raymond J.C., Garcia M.R., Verbunt F., Hasinger G., & Kürster, M., 1990, A&A 235, 162
Wang Q., Hamilton, T.T., Helfand D.J. & Wu X., 1991, ApJ 374, 475
Wang Q., 1991, MNRAS 252, 47
Wang Q. & Wu X., 1992, ApJS 78, 391
White N.E. et al., 1981, ApJ 247, 994
White N.E. & Marshall F., 1984, ApJ 281, 354

Table 1: Supersoft X-ray Sources

Name	PSPC [c/s]	kT [eV]	opt. ID	opt. magnit.	Orb. Period	Remarks	Ref.
LMC							
RX J0439.8−6809	1.38	18	−	B>19			1
RX J0513.9−6951	<0.06−2	40	HV 5682	V∼17		repeating transient	2−4
RX J0527.8−6954	0.03−0.3	30	HV 2554??	V>17		transient, decline	4−8
RX J0534.6−7056	0.02		HD37798	V∼10		foreground K−star	4,7
RX J0537.7−7034	0.02		+			foreground WD	1,9
CAL 83	0.98	40	+	V∼17	1.04d	class prototype	6,10,11
CAL 87	0.12	31		V∼19	10.6h	opt.+X−ray eclipse	10,12,13
RX J0550.0−7151	0.9	32	−				4,14
SMC							
1E 0035.4−7230	0.38	41	−	V∼21		variable, eclipses?	15
RX J0048.4−7332	0.19	10	SMC3	dV∼1.5		symbiotic nova	13
RX J0058.6−7146	0.025	42	−			X−ray outburst	13
1E 0056.8−7154	0.33	28	N67	V∼16.6		planetary nebula	15
RX J0122.9−7521	0.81	15	?			foreground WD?	13
Milky Way							
GQ Mus	0.11	30	GQ Mus	+		nova after 9yr	16
RX J0925.7−4758	0.8	50		V∼17.2	3.5d	heavily absorbed	17
RX J0019+21	∼2	20		V∼12	15.8h		18
M3	1.25	45	−			glob. cluster source	19,20
RR Tel	0.18	12				symbiotic nova	21,22
AG Dra	∼8					symbiotic nova	21
RX J2117.1+3412						PG 1159−star	23
M31							
>15 SSS	∼0.001		−				24

References for table 1

(1) Greiner, Hasinger & Thomas, 1994; (2) Schaeidt, Hasinger & Trümper, 1993; (3) Pakull et al., 1994; (4) Cowley et al. 1993; (5) Trümper et al., 1991; (6) Greiner, Hasinger & Kahabka, 1991; (7) Orio & Ögelman, 1993; (8) this work; (9) Pakull, Thomas & Bianchi, 1994; (10) Long, Helfand & Grabelsky, 1981; (11) Smale et al., 1988; (12) Pakull et al. 1988; (13) Kahabka, Pietsch & Hasinger, 1994; (14) Beuermann, 1992, priv. comm.; (15) Wang & Wu, 1992; (16) Ögelman et al., 1993; (17) Motch, Hasinger, Pietsch, 1994; (18) Reinsch, Beuermann & Thomas, 1993; (19) Verbunt, Hasinger, Johnston 1993; (20) Hertz, Grindlay & Bailyn 1993; (21) Bickert et al., 1993; (22) Jordan, Mürset & Werner 1993; (23) Motch, Werner & Pakull 1993; (24) Supper et al., 1994;

ROSAT PSPC OBSERVATION OF M31

Rodrigo Supper
Max-Planck-Institut für extraterrestrische Physik,
85748 Garching b. München, Germany.
ros@mpeu27.rosat.mpe-garching.mpg.de

ABSTRACT

This article reports on the analysis of the first M31 survey with the ROSAT PSPC performed in July 1991. The spectral characteristics of the 396 individual X-ray sources detected will be discussed as well as their positional correlation with the 108 individual X-ray sources detected by the *Einstein* observatory. The optical identifications of the ROSAT X-ray sources are the result of a MIT - MPE - UvA collaboration. A detailed analysis of the integral flux distribution shows that just two third of the detected X-ray sources in the field of M31 must be background objects. The discovery of a significant absorption of background radiation is discussed, as well as an upper limit for the luminosity of a diffuse emission component. Finally a comparison of the luminosity function of 29 globular clusters sources with the one in our own galaxy reveals a good agreement.

1. INTRODUCTION

The proximity of M31, its size and spiral nature allows detailed study of X-ray sources with intensities comparable to those in our own galaxy. The *Einstein* observatory had discovered 108 individual X-ray sources (Trinchieri and Fabbiano 1991, hereafter TF) with $\sim$ 100 ksec IPC and $\sim$ 200 ksec HRI observations. The IPC observations reached a limiting sensitivity of $\sim 10^{37} erg\, s^{-1}$ (van Speybroeck et al. 1979; van Speybroeck and Bechtold 1981; Long and van Speybroeck 1983) and revealed isolated sources in the outer portion of the galaxy and a confused region surrounding the nucleus. The HRI observations resolved this region into individual sources too and improved the position accuracy of half of the IPC sources. Comparison with optical and HI-data allowed to assign a large group of X-ray sources to the bulge and a general association with the spiral arms of the galaxy. Additionally, Crampton et al. (1984) found a variety of possible optical counterparts within the *Einstein* error circles.

The first ROSAT PSPC deep survey of M31 covers the whole galaxy (1.3 times the D_{25}-diameter) in a total observation time of more than 200 ksec, leading to a sensitivity about a factor 10 fainter than that of the *Einstein* observations. This article gives an overview about the 396 X-ray sources detected, their spectral properties and results of optical identifications. The latter have been performed as part of a large collaboration involving a group at MIT (Lewin, Magnier), the University of Amsterdam (UvA - van Paradijs), and the Max-Planck-Institut für extraterrestrische Physik (MPE - Hasinger, Jain (now at the QAD Isro Satellite Center in Bangalore/India), Pietsch, Trümper, and the author); see Magnier (1994, these proceedings) same issue. Additionally, the luminosity distribution of the sources and the total luminosity of M31 is discussed, like a brief presentation of absorption and diffuse emission properties of this galaxy. A more complete discussion of the results presented here can be found in Supper et al. (1994).

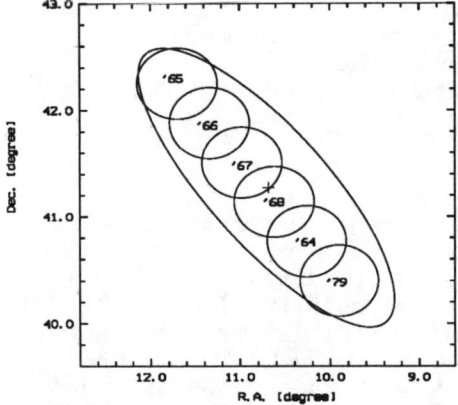
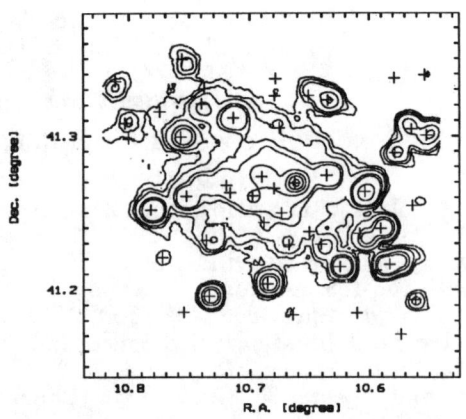

Figure 1: Locations of the central region of the 6 survey pointings plotted over the D_{25}-ellipse of M31 (Tully 1988). The cross marks the optical center of M31. The numbers identify the pointings (see Tab. 1).

Figure 2: Contour plot of the bulge of M31 in the H-band together with the X-ray sources found by ROSAT within this region and marked as crosses. The contour levels are: 8.4, 11.7, 16.7, 33.4, 83.6, 167.2, 334.5, 501.7, and 668.9 counts s^{-1} deg^{-2}. The few sources not sourronded by any contour line belong to soft or faint objects.

2. OBSERVATIONS

The analysis presented in this article is based on the pointed M31 survey with the ROSAT X-ray Telescope Position Sensitive Proportional Counter (ROSAT PSPC; Trümper (1991), Aschenbach (1988), Pfeffermann et al. (1986)) performed in July 1991. The results of a second PSPC survey of M31 carried out in August 1992 will be reported elsewhere. An isolated observation of the bulge region of M31 was performed in July 1990 with the ROSAT High Resolution Imager (HRI), and has been described by Primini et al. (1993).

The July 1991 M31 survey consists of 6 contiguous pointings, each with more than 25 000 seconds observation time, covering the whole disk at equidistant positions. Figure 1 shows an overlay of the central region of the 6 pointings over the D_{25}-ellipse of M31 (Tully 1988). Each circle marks the boundary of the inner area of the PSPC with a radius of 20' in which the instrument has it's highest angular resolution, whereas the total field of view of the PSPC is 57' in radius. The optical size of the disk of M31, its D_{25}-diameter with $\sim$ 192' (uncorrected), is somewhat larger than the merged central regions. The journal of observations is given in Tab. 1.

3. SOURCE DETECTION AND IDENTIFICATION

For the analysis, the broad energy band (B) of the ROSAT XRT (0.1 keV - 2.4 keV) has been divided into four energy bands: a soft band (S: 0.1 keV - 0.4 keV) and three hard bands (H: 0.5 keV - 2.0 keV; H1: 0.5 keV - 0.9 keV; H2: 0.9

Table 1: Log of the first ROSAT PSPC survey of M31.

Pointing	Date	RA (J2000)[1]			Dec (J2000)[1]			Exposure
		(h)	(m)	(s)	(°)	(')	('')	(s)
WG600065P	24.-25. July 1991	00	46	48.0	42	15	00	26 216
WG600066P	25.-26. July 1991	00	45	21.6	41	52	48	30 720
WG600067P	26.-27. July 1991	00	43	55.2	41	30	36	27 884
WG600068P	27.-28. July 1991	00	42	28.7	41	08	24	28 888
WG600064P	15.-16. July 1991	00	41	02.4	40	46	12	49 292
WG600079P	14.-15. July 1991	00	39	36.0	40	24	00	42 188

[1]The coordinates give the center of the field of view.

keV - 2.0 keV). For each pointing and energy band the data were binned into an image with a pixel size of $15'' \times 15''$. The source detection had then been carried out separately in each image using a maximum likelihood method (Cruddace et al. 1988). Only sources with an existence likelihood ≥ 10 were accepted. The resulting source lists were merged into one final list containing 396 individual X-ray sources. The whole procedure is described in detail in Supper et al. (1994), where the total source list with all characteristics of the individual sources can be found.

The detected X-ray sources in the field of M31 crowd around the bulge and additionally show a slight trend to be arranged with the spiral arms of the galaxy. Within the bulge there exists strong confussion which can be resolved into 22 individual X-ray sources by the ROSAT PSPC. The situation in the bulge region can be seen in Fig. 2, where the sources are plotted as crosses above a contour map of the H-band. Some sources are not surrounded by any contour line due to their soft energy spectrum or their faint flux lower than the lowest contour level.

For correlating the 396 detected X-ray sources with different published catalogs, the identification procedure described in Magnier (also in these proceedings) was used. There the most part of optical identifications is discussed, so that in this article only the correlation with the *Einstein* catalog and some special optical questions will be referred to.

From the 108 X-ray sources found by *Einstein* (TF) only 64 correlate in possition with sources found by ROSAT (within 2σ of their combined position error). Among them, 6 *Einstein* sources have more than one ROSAT identification. A number of 8.5 accidental correlation is expected. 44 *Einstein* sources don't correlate with any ROSAT source. An explanation for this is on one hand the fact, that the *Einstein* sources were determined in the most part with the HRI (see TF) which has a higher resolution than the ROSAT PSPC. Therefore, TF found ~ 50 sources within the confused bulge, whereas the ROSAT PSPC only resolved 22. On the other hand, the variability of X-ray sources is generally known. Besides this, 328 ROSAT sources couldn't be correlated to any *Einstein* source. For the most part, this is a result of the 10 times higher sensitivity of the ROSAT PSPC compared with the *Einstein* IPC/HRI.

Within the group of matching sources, 14 X-ray sources were determined as being variable with more than 3σ of their combined flux uncertainty when comparing *Einstein* with ROSAT fluxes. It should be mentioned, that the assumption of an uncorrect spectral model can result in a pretence of variability in flux when comparing IPC with PSPC observations. Among these variable sources, 7 are belonging to globular clusters in M31. Within the group of unmatched *Einstein* sources 2 bright objects with fluxes $\sim 10^{-12}$ erg cm^{-2} s^{-1} were found and have therefore be considered as bright transients. The flux calculations are based on a thermal bremsstrahlung spectrum with $kT = 5$ keV, $N_H = 7 \times 10^{20}$ cm^{-2}, and in the 0.2 - 4.0 keV energy band (used by TF). On the other hand, 2 additional ROSAT sources not correlating to any *Einstein* source and with fluxes $> 10^{-12}$ erg cm^{-2} s^{-1} were also determined as being bright transients.

From the correlations with optical databases, only the identifications with globular clusters shall be considered in this article. The others are discussed in Magnier (1994), also in these proceedings. 29 globular clusters could be identified, with 5 accidental correlations expected. Magnier et al. (1993) report their 490 globular clusters also completely lying within this most sensitive region. From the lists of Battistini et al. (1987 and 1993) 10 globular clusters not correlating with any of Magnier's one are also lying within this area. An estimation from the background data reveals, that any source must have at least (for the worst position in this area) 1.38×10^{-3} counts s^{-1} for being detected with a likelihood threshold of 10 elsewhere in this region (in the H band). This can be regarded as an upper limit for the 473 globular clusters not detected in this region and gives the capability to determine the luminosity function of the identified globular cluster sources using the Kaplan-Meier estimator to obtain a nonparametric estimation of the cumulative distribution function $\Phi(L)$ (Schmitt 1985). $\Phi(L)$ gives the fraction of detected sources among all detectable sources at a given flux. The result for GC's above 1.38×10^{-3} counts s^{-1} ($\simeq 2.4 \times 10^{36}$ erg s^{-1}) can be seen in Fig. 3 (thick line), where the luminosities have been calculated by using the spectral model and energy range for the listed globular cluster sources in the Milky Way reported by Hertz & Grindlay (1983) (with $N_H = 9 \times 10^{20}$ cm^{-2} and asuming 690 kpc for M31 distance). The distribution of the latter is also shown in Fig. 3 (thin line). As can be seen, both distributions are consistent with being drawn from the same parent distribution. Also the peak luminosities are comparable. Earlier investigations reported a significant difference between both distributions and raised the question whether the globular cluster sources in M31 are more luminous and/or more numerous compared with those in the Milky Way. Primini et al. (1993) report, that both distributions show the same peak luminosity, but the globular cluster sources in M31 are relatively more numerous. The reason for this earlier reports of a difference in the distributions is due to the fact of incomplete lists of globular clusters for M31. The list of Magnier et al. (1993) now enables us to get proper results of this matter. With this investigation, the difference in the distributions of globular clusters between M31 and the Milky Way vanishes and makes both galaxies more comparable.

4. SPECTRAL CHARACTERISTICS

The spectral characteristics of the 396 sources were investigated by calculating hardness ratios (HR = (H - S)/(H + S)) of the source counts. Most sources have hard spectra ($HR > 0$). On one hand, this is due to the fact of the hard spectrum of LMXBs and on the other hand it is the result of foreground absorption ($N_H > 6 \times 10^{20}$ cm^{-2}). There exists a small group of sources with clear negative hardness ratios consisting of foreground stars and possible candidates of Supersoft Sources (SSS), a new class of sources introduced by ROSAT. The criteria for the decision of being a candidate for a SSS are $HR + \sigma_{HR} \leq -0.8$; excluding any known (foreground-)object. Among all 396 detected sources 15 objects fulfill

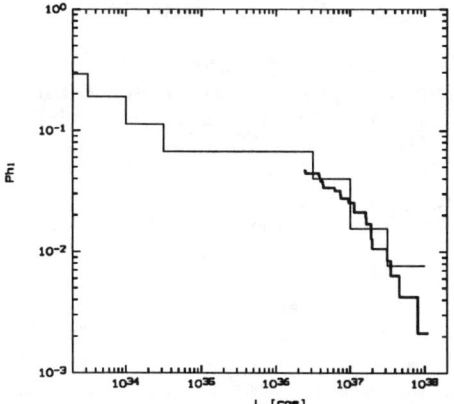

Figure 3: The integral luminosity distribution of the globular cluster sources in M31 (thick line) together with the distribution in our own galaxy (thin line) reported by Hertz and Grindlay (1983). Both presentation are normalized by the parameter-free estimator described by Schmitt (1985).

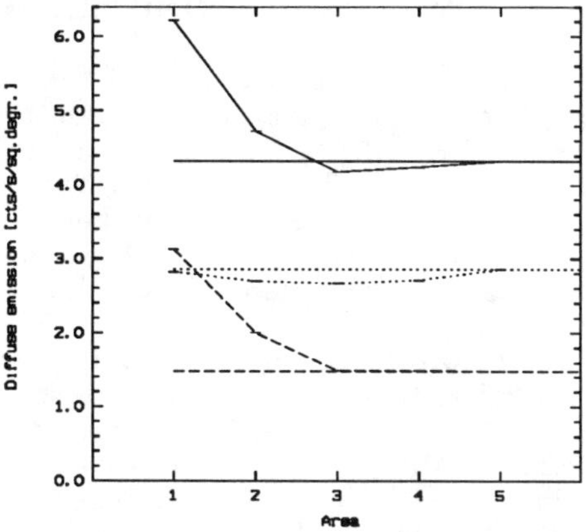

Figure 4: Background count rates for the different areas and energy bands. The solid line connects the data points for the B band, the dotted line for the S band, and the dashed line for the H band. The horizontal lines mark the extragalactical background count rate.

this requirements. About the properties of SSS in LMC and SMC and additional examples within our own galaxy, see Hasinger (1994; these proceedings).

The ROSAT PSPC energy resolution and sensitivity enables to derive spectral parameters for the 56 brightest sources with more than 200 counts each (omitting the also bright but highly confused bulge sources). Within this group the X-ray sources identified with foreground stars yield a value for the absorption column density which is less than the foreground absorption within our galaxy along the line of sight (6×10^{20} cm^{-2}) confirming the identification. As a mean spectral model for the X-ray sources belonging to M31 a power law with $\Gamma = -2.0$ and an absorption of $N_H = 9 \times 10^{20}$ cm^{-2} is asumed. These values are the mean of the best spectral fit results ($\chi^2/\nu < 2$) and the N_H-value is in good agreement with the mean HI-column density in M31 reported by Unwin (1980) including the absorption of our own galaxy. The individual fit parameters of all fitted sources are listed in Supper et al. (1994). One of the 15 SSS is bright enough to also allow a spectral fit. A black body model yielded a temperature of $kT \sim 30$ eV and a total luminosity of the order of $\sim 10^{38}$ erg s^{-1}, which is typical for the SSS. The spectral fit with a thermal bremsstrahlung model of a source correlating with M32 yielded $kT = 2.6 \pm 0.9$ keV and $N_H = (7.8 \pm 0.1) \times 10^{20}$ cm^{-2} which results together with the count rate of this source in a luminosity of $\sim 1.9 \times 10^{38}$ erg s^{-1}. Finally, a spectral fit of the bulge region in total (up to 5' around the optical center) with a thermal bremsstrahlung model yielded $kT = 0.74 \pm 0.04$ keV and $N_H = (6.3 \pm 0.2) \times 10^{20}$ cm^{-2}. With this model a luminosity of $\sim 1.3 \times 10^{39}$ erg s^{-1} could be derived for the bulge of M31.

5. FLUX DISTRIBUTION AND BACKGROUND SOURCES

It is likely, that a part of the 396 X-ray sources detected by the ROSAT PSPC does not belong to M31 but consists of background sources seen through the galaxy. In order to estimate the fraction of these sources an integral flux distribution of all sources lying within the D_{25}-ellipse of M31 (but outside the bulge) was compared with the known LogN-LogS relation from an analysis of a deep survey of the Lockman Hole reported by Hasinger et al. (1993). For the comparison, the different N_H-value within different regions of M31 had to be considered.

Unwin (1980) reports a large variation for N_H across the disk of M31. In particular, there exists a ring-like structure with a distance of ~ 10 kpc around the center within which the value for N_H rises up to 3.3×10^{21} cm^{-2} (in the plane of M31). In general, Unwin reports an increase of the N_H-value when going from the center to the edge of M31 up to a maximum within this ring and afterwards a decrease down to zero when leaving the galaxy. Because of this fact, the disk of M31 was divided into 3 elliptical regions: an inner ellipse with a low mean N_H-value, a middle ellipse with a relatively high value, and an outer ellipse with a low value again. The boarders of the middle ellipse were determined in such a way, that this ellipse is being nearly opaque (less than 0.1% transmission) for the soft X-ray componente. Using the spectral model of Hasinger et al. (1993) for the background sources (power law with $\Gamma = -2.0$) this condition yields $\sim 4 \times 10^{21}$ cm^{-2} for the limiting N_H-value (galactic absorption already included). Thus the inner ellipse is defined as going from the bulge (1 kpc around the optical center) up to a major half-axis of 8 kpc, the outer ellipse is going from the D_{25}-boarder of M31 down to ~ 15 kpc, and finally the middle ellipse is the region between the inner and the outer one. The minor half axes are always taken as 0.32 times the major half axes (the ratio of d_{25} to D_{25} reported by Tully (1988)).

In each of these three regions the determination of the fraction of background sources among all sources found within the considered region was carried out separately in the following way: First a raw integral flux distribution of the sources within the considered region was calculated. The division by the sensitive area as a function of flux and the application of an "efficiency function" to consider confusion effects (see Hasinger at al., 1993) yields the corrected integral distribution of sources per area as a function of flux. Before subtracting in a last step the known distribution function of background sources (DFBS), the DFBS had to be shifted in flux according to the mean N_H-value in the considered ellipse.

Figures 5a,b,c show the results for the inner, middle, and outer ellipse. For Fig. 5a,b the upper graph represents the corrected integral distribution function of the X-ray sources within the considered area. The middle graph gives the background subtracted distribution and is therefore considered as the distribution of the sources belonging to M31. The solid line shows the (shifted) analytical distribution of background sources reported by Hasinger et al. (1993). The statistical errors are plotted for three different flux values. For the outer ellipse (Fig. 5c) only the corrected integral distribution of the X-ray sources and the subtracted distribution is given for 6 different flux values due to the low statistics within this area.

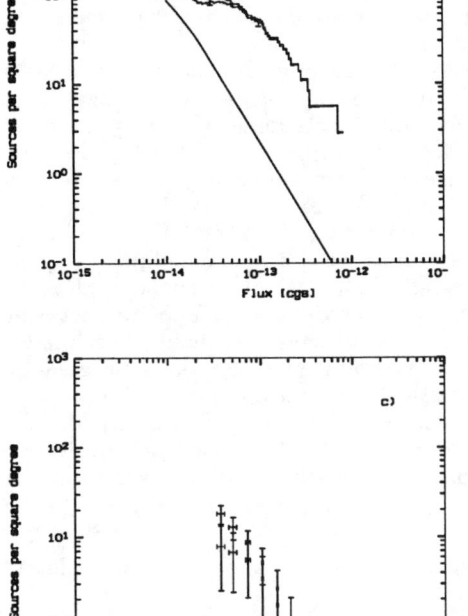
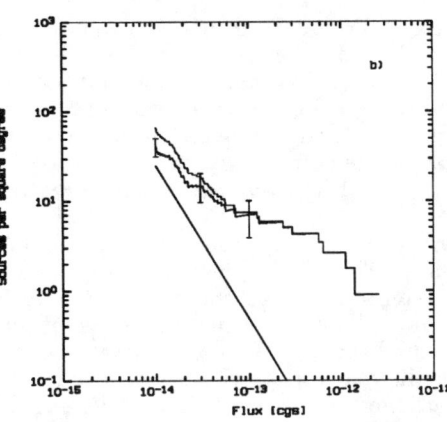

Figure 5: Integral flux distribution of the disk sources for the three different areas, divided by the sensitive area and corrected by the "efficiency function" (upper graph; a), b), c) = inner, middle, outer ellipse). The solid line shows the analytical distribution of background sources reported by Hasinger et al. (1993) and connected to the considered N_H value (not shown in c). The middle line gives the background reduced flux distribution of the sources belonging to M31.

With this analysis, one determines the fraction of X-ray sources belonging to M31 among all detected sources at a lower flux limit of 10^{-14} erg cm^{-2} s^{-1} (3×10^{-14} erg cm^{-2} s^{-1} in case c due to large statistical errors) as being 38.5 ± 12.9 out of 102 sources within the inner ellipse, 83.5 ± 13.6 out of 134 within the middle ellipse, and 7.4 ± 11.5 out of 70 within the outer ellipse (the errors are statistical). Counting on one hand all the 22 bulge sources as belonging to M31 and on the other hand all the sources outside the D_{25}-ellipse as being background sources reveals that only 151.4 ± 22.0 sources out of the 396 sources detected by the ROSAT PSPC really belong to M31, which is around one third.

6. DIFFUSE EMISSION AND ABSORPTION

The division of the disk of M31 into regions with different N_H-values also allows a brief estimate of the amount of diffuse emission and absorption by the galaxy. From the *Einstein* data it was pointed out (TF), that there exists a possible diffuse emission in the bulge region of M31 with a luminosity of $\sim 4 \times 10^{38}$ erg s^{-1}. An absorption of background flux by the galaxy, i.e. a possible shadowing effect was not reported. It is expected, that especially within the HI-ring with its high HI-column density such a shadowing effect should be seen – at least in the soft energy band.

For this, from the images created in each pointing the source photons of the detected sources have been subtracted, the resulting background images were divided by the exposure map and merged into one single image covering the total galaxy. The whole process is described in detail in Supper et al. (1994). Finally, an integration over these "count rate images" within the elliptic areas defined above yields the total count rates of any "residual luminosity" not belonging to detected sources. Figure 4 sketches the situation for the B, S, and H band within the different areas, where area 1 defines the bulge, areas 2 to 4 the elliptic regions mentioned above, and area 5 the surrounding of M31 which is considered to be representative for the pure extragalactic count rates. The solid line connects the data points for the B band, the dotted line for the S band, and the dashed line for the H band. The three horizontal lines show the count rates for the extragalactic flux, i.e. the values within area 5. The statistical errors are less than 1%.

As can be seen from Fig. 4, there exists an unresolved flux within the bulge and the inner ellipse in the hard energy band. This can be considered as the sum of faint unresolved sources and/or a possible gaseous emission. A quantitative analysis for the bulge yields $\sim 5 \times 10^{37}$ erg s^{-1} (applying a thermal bremsstrahlung spectrum with $kT = 5$ keV and $N_H = 6.9 \times 10^{20}$ cm^{-2}). This would indicate a gas mass of $\sim 6 \times 10^5 M_\odot$ when counting the flux as completely originating from gaseous emission.

On the other side an absorption effect within the disk area in the soft energy band can clearly be seen. This effect is strongest in the second ellipse which contains the highest mean N_H-value (the area with the HI-ring). A quantitative analysis of this effect yields $(446 \pm 93) \times 10^{-6}$ counts s^{-1} arcmin^{-2} for the absorbed extragalactic background flux in the S band. Such an absorption effect was nowhere reported up to now.

7. CONCLUSIONS

The first ROSAT PSPC survey of M31 yields 396 X-ray sources in the field of view, 22 of them belonging to the very confused bulge region. Most of the 396 sources have relatively hard spectra due to foreground absorption and a fair amount of LMXBs within all sources. 15 candidates for Supersoft Sources (SSS) with extremely soft but bright spectral characteristics have been found.

A comparison with the *Einstein* source list reported by TF confirms 64 *Einstein* sources, 14 of them seem to be variable on time scales of decades. Additionally, 4 bright transients were discovered, 2 of which appear in the ROSAT observations and 2 in the *Einstein* observations. 44 of the *Einstein* sources couldn't be found in the ROSAT observation (including the transients), whereas 328 ROSAT sources weren't seen during the *Einstein* observation.

A comparison of the integral luminosity distribution of the X-ray sources correlating with globular clusters with that one of our own galaxy leads to the conclusion, that both distributions are consistent with being drawn from the same parent distribution. This is against earlier reports, which described the globular cluster sources in M31 as being more numerous compared with the Milky Way.

The split up of the disk of M31 in three regions with different mean HI-column densities (according to radio measurements of Unwin (1980)) and the comparison with the distribution function of the background sources reported by Hasinger et al. (1993), enables to estimate the fraction of background sources within the total sample to be $\sim 62\%$. Therefore, only one third of the detected sources is expected to belong to M31.

Finally, a significant absorption of the background luminosity within the disk of M31 was found. Although there exists an interference of absorption and diffuse emission, it is possible to give a value of $(446 \pm 93) \times 10^{-6}$ counts s^{-1} arcmin^{-2} for the absorbed extragalactic background flux in the S band within the HI-ring) region. For the bulge region a first upper limit for a diffuse component in the range of $\sim 5 \times 10^{37}$ erg s^{-1} can be given. Counting this luminosity as completely originating from hot gas within the bulge region would indicate a gas mass of $\sim 6 \times 10^5 M_\odot$.

ACKNOWLEDGEMENTS

I would like to thank my collaborators at MIT, MPE, and UvA. Especially, I would like to thank Günther Hasinger for extensive discussions in general and belonging the question of background sources in special. I would also like to thank Eugene Magnier for the discussion of the identification method of sources, for putting his list of globular clusters, foreground stars and SNRs on my disposal and last but not least for his friendly support during the conference.

REFERENCES

Aschenbach, B. 1988, Appl. Opt., 27, No. 8, 1404
Battistini, P.L. et al. 1987, A&AS, 67, 447
Battistini, P.L. et al. 1993, A&A, 272, 77
Crampton, D. et al. 1984, ApJ, 284, 663
Cruddace, R.G., Hasinger, G.R. and Schmitt, J.H. 1988, Astronomy from Large Databases, ed. Murtagh, F. and Heck, A., 177
Hasinger, G. et al. 1993, A&A, 275, 1
Hertz, P. and Grindlay, J.E., ApJ, 275, 105
Long, K.S. and van Speybroeck, L.P. 1983, "Accretion Driven Stellar X-Ray Sources", ed. W.H.G. Lewin and E.P.J. van den Heuvel (Cambridge: Cambridge University Press), 117
Magnier, E.A. et al. "New Cluster Candidates in M31", 1993, A&A, submitted
Pfeffermann, E. et al. 1986, Proc. SPIE, 733, 519
Primini, F.A. et al. 1993, ApJ, 410, 615
Schmitt, J.H.M.M. et al. 1985, ApJ, 293, 178
Supper, R. et al. "ROSAT PSPC survey of M31", 1994, A&A, in prep.
Trinchieri, G. and Fabbiano, G. 1991, ApJ, 382, 82
Trümper, J. et al. 1991, Nat, 349, 579
Tully, R.B. 1988, "Nearby Galaxies Catalog", Cambridge University Press
Unwin, S.C. 1980, MNRAS, 192, 243
van Speybroeck, L.P. et al. 1979, ApJ, 234, L45
van Speybroeck, L.P. and Bechtold, J. 1981, "X-Ray Astronomy with the Einstein Satellite", ed. R. Giacconi (Reidel), 153

OPTICAL IDENTIFICATIONS OF M31 SOURCES

Eugene Magnier

Astronomical Institute "Anton Pannekoek" and Center for High Energy Astrophysics,
University of Amsterdam, Kruislaan 403, 1098 SJ Amsterdam, The Netherlands.
gene@space.mit.edu

ABSTRACT

The *Einstein* X-ray Observatory Satellite provided the first detection of isolated X-ray sources in the Andromeda Galaxy. Recent ROSAT observations have provided observations of the X-ray binaries in M31 complete to $L_x \sim 10^{34}$ erg sec^{-1}, and are sensitive enough to detect many supernova remnants and other X-ray sources. Optical identifications are crucial to identify or confirm the identifications of these sources. This article reviews the search for optical counterparts to the *Einstein* X-ray sources and discusses the ongoing efforts of the MIT - UvA - MPE group to identify optical counterparts to the ROSAT X-ray sources.

1. INTRODUCTION

In order to understand our Galaxy, it is important to study external galaxies for comparison. The Andromeda galaxy is fortunately quite similar to the Milky Way, and therefore allows for the study of various Galactic populations without the disadvantages inherent in studying such objects in the Milky Way. The distance to M31 is relatively well determined ($\sim$700 kpc - e.g. Welch et al. 1986) and the extinction for objects in M31 is small compared to that for objects in the plane of the Milky Way. Unlike the Magellanic Clouds, M31 is similar in size, metallicity and morphological type to the Milky Way.

The earliest X-ray observations of M31 were made with the Uhuru satellite and balloon-borne X-ray telescopes (see Long and Van Speybroek 1983), and were able to detect M31 as an X-ray source, but were unable to distinguish individual sources associated with the galaxy. It was not until the *Einstein* X-ray Observatory, with its imaging capabilities, was used to study M31 that the individual X-ray sources were detected. The first part of this review will briefly discuss the *Einstein* observations of M31 and related work on optical identifications, while the later half will describe in detail the search for optical counterparts to X-ray sources found by ROSAT. This later work has been performed as part of a collaboration involving a group at the Max-Planck-Institut für Extraterrestriche Physik (MPE – Günther Hasinger, Ashok Jain, Wolfgang Pietsch, Rodrigo Supper, and Joachim Trümper), the University of Amsterdam (UvA – Thomas Augusteijn, Jan van Paradijs, and Saskia Prins), and MIT (Zoltán Haiman, Walter Lewin, and the author – now at UvA).

2. X-RAY SOURCES AND OPTICAL COUNTERPARTS

It is important to make optical identifications of X-ray sources since this is one of the few ways we have of identifying their nature. In general, except for the brightest sources, too few X-ray photons are detected in these observations to make reliable identifications based on X-ray spectral or X-ray timing properties. It is useful to review the types of X-ray sources which may be detected in an

Table 1: Typical X-ray sources expected in observations of M31.

Source	L_x^a	N_{MW}^b	Appearance	N_t^d	N_a^d	N_c^d	σ^d
MXRB	$10^{36} - 10^{38}$	69^c	O/B star, $V \sim 18 - 20$	52	41	11	6.4
			OB assoc, group of O/B stars	(see note in text)			
LMXB	$10^{36} - 10^{38}$	124^c	K/M star, $V > 25$	0	0	0	0
			GC, $V \sim 14 - 20$, fuzzy	42	9	32	3.0
SNR	up to 10^{38}	>150	Emission-line Nebula	23	3	20	1.7
C.A. stars	$10^{28} - 10^{31}$		F/G/K star, $B - V > 0.4$	58	29	29	5.4
galaxies	up to 10^{47}		Galaxy	6	0	6	0

^{a}Luminosity in erg sec^{-1} ^{b}Number known in the Milky Way
^{c}From van Paradijs, 1993 ^{d}IDs in ROSAT observations, see text

observation of the Andromeda galaxy. Table 1 lists the most common classes of X-ray sources, and is divided into those which would detectable in M31 and those which would lie either in the foreground or background of M31. The next two columns list typical X-ray luminosities and, for those objects which would be detectable in M31, the number of these systems known in our Galaxy. These numbers can be used as an estimate of the number we should see in the Andromeda galaxy.

The *Einstein* observations were sensitive to about 10^{36} erg sec^{-1} while the ROSAT observations were sensitive to about 10^{35} erg sec^{-1}. It is clear then that many massive X-ray binaries (MXRBs or HMXBs) and low-mass X-ray binaries (LMXBs) should be detected by *Einstein*, and essentially all should be detected in the ROSAT observations. It should be noted that many of the MXRBs in our Galaxy are transient in nature (Be systems), so the number of 69 should be considered an upper limit. As for the supernova remnants (SNRs), the X-ray luminosity function continues to rise to rather faint levels, so deeper X-ray observations should reveal many more of these sources. It is very likely that observations of SNRs in our Galaxy are very incomplete, so the 150 known in the Milky Way can be taken as a lower limit. Cataclysmic Variables (CVs) are likely to be too faint to be detected in either the *Einstein* or ROSAT observations. Dwarf Novae in outburst can be detected, however, these are difficult to identify without simultaneous optical and X-ray observations, and will not be discussed further in this article.

Roughly 1000 Cornonally active stars (C.A. stars) have been identified as soft X-ray sources in the ROSAT All-Sky Survey (Haisch et al. 1991). Although they are much fainter than the X-ray binaries, those in our Galaxy – in the foreground of M31 – can be detected in the ROSAT and *Einstein* observations. Galaxies, AGN and galaxy clusters may be detected *behind* M31. However, since both *Einstein* and ROSAT are low-energy X-ray telescopes, the extinction through M31 may make many undetectable. In fact, those which are detected may be used to probe the ISM of M31.

Each of the preceding X-ray sources has a typical optical counterpart which is also listed in Table 1. For MXRBs, the massive secondary star should be detectable as a blue ($B-V \lesssim 0.4$), relatively bright ($V < 20$ if unreddened) star. Since O/B-type stars are often found in OB associations, these might also be associated with X-ray sources. Isolated LMXBs (ie. in the field) are unlikely to be detectable from ground-based observations. The low-mass stellar companions have apparent (unreddened) magnitudes of $V > 27$, and the accretion disks of these systems would generally be fainter than 23rd magnitude. The crowding at this depth is, in general, too severe to allow for identification with even the best ground-based telescopes. It is possible, however, that future HST observations in the UV or far-UV will allow for the identification of LMXB counterparts. Alternatively, many LMXBs are found in globular clusters (GCs) in our Galaxy. Globular clusters may be identified as slightly extended, approximately stellar objects with $V \sim 14 - 20$, though spectra are required to conclusively distinguish them from background galaxies. SNRs may be identified with optical narrow-band emission-line filter images or with radio observations. As with their X-ray luminosity function, the optical luminosity function continues to rise below current detection limits, so deeper optical observations in general reveal more SNRs. foreground C.A. stars may be identified by their colors ($B - V > 0.4$) and by the relative brightness ($V < 18$).

3. *EINSTEIN* OBSERVATIONS

The *Einstein Observatory* was used to study the X-ray sources in M31 (see Trinchieri and Fabbiano 1991, and references therein) and provided the first observations of individual M31 X-ray sources. Observations were performed 14 times between Jan 1979 and July 1980. The IPC was used to observe roughly 3 square degrees, and many HRI pointings were used, primarily to study the bulge regions. These observations were sensitive to the 5×10^{36} erg sec^{-1} level (in the 0.2-4.0 keV range, assuming the sources are at the distance of M31, 700 kpc), and had a resolution of $\sim 3''$ for the HRI observations and 45" for the IPC observations. A total of 108 sources were detected, with a large fraction located in the crowded bulge regions.

Optical identifications of the *Einstein* sources have been discussed by Battistini et al. (1982), Crampton et al. (1984), and by Trinchieri and Fabbiano (1991). Battistini et al. identified 21 of the sources with GCs. This work was performed as part of the (still ongoing) search for M31 GCs of the Bologna Group (Battistini et al. 1993, 1987, 1982, ETC). Crampton et al. 1984 used a set of photographic plates taken with a slitless low-dispersion grism at the Canada-France-Hawaii Telescope in good seeing for identifications. Using these plates, Crampton et al. identified 13 X-ray sources with blue stars in or near their X-ray error circles. However, the X-ray error boxes were too large to consider these strong identifications and not accidental coincidences. They also identified 11 X-ray sources which were coincident with bright, foreground stars in our Galaxy. However, they suggested that only one or two of the foreground stars were actually detections of coronally active stars. Crampton et al. 1984 suggested spectra of these optical counterparts, but there has been essentially no follow-up work along these lines.

4. ROSAT OBSERVATIONS

Recently, the ROSAT X-ray observatory has made several observations of M31 with enough sensitivity to ensure the detection of essentially all MXRBs and LMXBs. Supper (in these proceedings) discusses the various ROSAT observations (see also Supper et al. 1993). This review will only discuss the optical counterparts identified sources detected in the July 1991 PSPC observations. The analysis of these data has revealed 396 X-ray sources, with luminosities ranging from $\sim 9.0 \times 10^{34}$ to 2.8×10^{38} erg sec^{-1}, if the sources are at the distance of M31 (700 kpc).

In order to facilitate the identification of optical counterparts to ROSAT X-ray sources, we (the MPE – UvA – MIT collaboration) began a major program to

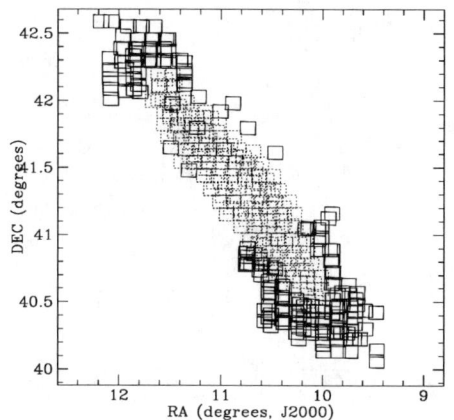
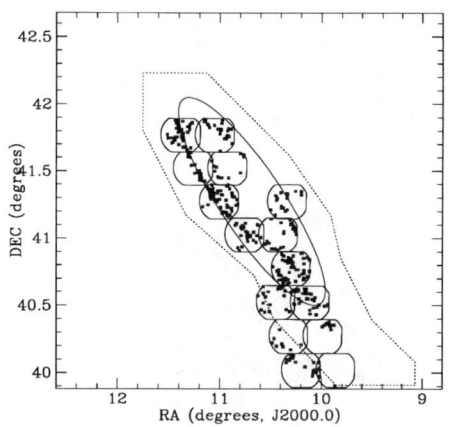

Figure 1. The locations of the *BVRI* CCD frames are shown. The dashed frames in the center are from 1990, while the bold frames are from 1991.

Figure 2. The locations of the Hα, [SII], V CCD images used for the SNR search. The dotted line shows the outline of the *UVBR* photometry study of Berkhuijsen et al. 1988, the solid ellipse shows the galactocentric 13kpc radius, and the dots show the locations of candidate SNRs.

observe M31 in detail with ground based telescopes and large-area CCDs. Using the Michigan – Dartmouth – MIT (MDM) 1.3m and 2.4m telescopes, we surveyed the disk portions of M31 with wide-band (*BVRI*) and narrow-band (Hα, [SII]) filter images. Figure 1 shows the locations of the CCD images used in the *BVRI* survey, with two line types referring to observations performed in 1990 and 1991. (Magnier et al. 1992; Haiman et al. 1993). The general aim of the wide-band observations was photometry for studies of the stellar components of M31 (the massive stars, globular clusters, etc). At that time, photometric studies of M31 consisted of either deep ($V \sim 22$), accurate ($\sim 1\%$) photometric observations with CCDs on very small regions (e.g. Massey et al. 1986; Hodge and Lee 1988), or photographic surveys with only fair accuracy (typically 15%) and of moderate depth ($V \sim 19$) covering a large area (Berkhuijsen et al. 1988). The narrow-band observations were performed primarily to search for SNRs. Figure 2 shows the locations of the narrow-band CCD images (roughly circular fields), with the 13kpc galactocentric radius shown as an ellipse and the locations of candidate SNRs marked as dots and the region surveyed by Berkhuijsen et al. 1988 marked as a dotted line.

To identify correlated sources, we use the "δRA - δDEC technique," which is demonstrated in Figure 3. In Figure 3A, every point represents the difference between the coordinates of an X-ray source and a possible optical counterpart (in this case, foreground stars, see below). Objects which are correlated between the two catalogs show up at near the center, with a distribution determined by the combined astrometric errors in the two catalogs. The chance of an accidental coincidence may be found from the distribution of points at a sufficiently large radius, which in general are not expected to be correlated. Figure 3B shows the cumulative number of objects within a circle of the given radius (thin solid line) as

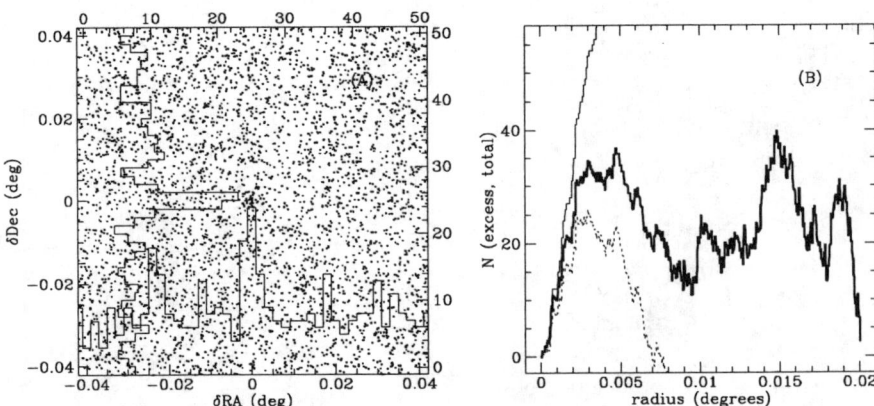

Figure 3. This figure illustrates the δRA - δDEC method. (A) shows the δRA - δDEC plot for the comparison between X-ray sources and foreground stars. (B) shows the cumulative number of objects as a function of radius (thin solid line), the calculated excess (thick solid line), and the 2σ lower-limit to this value.

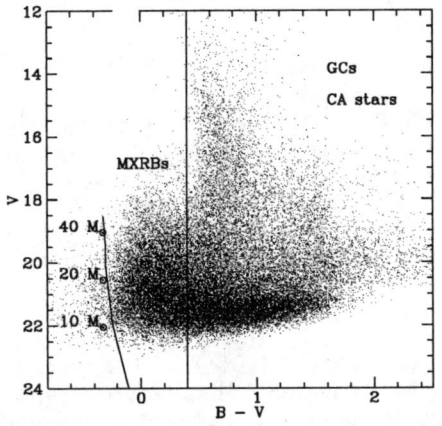

Figure 4. V, $B-V$ Color-Magnitude Diagram of $\sim$40 000 objects detected in B and V. The main sequence is shown (tilted solid line), along with the approximate locations of 40, 20, and $10 M_\odot$ main sequence stars. The vertical line divides the blue (primarily massive) stars from the red (primarily foreground) stars. The general locations of MXRBs, GCs, and CA stars are notes (see text).

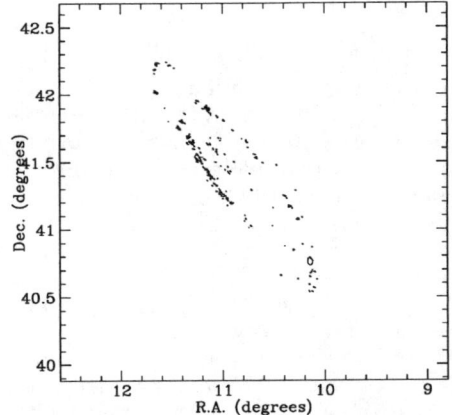

Figure 5. Map of the outlines of the 174 OB associations identified by Magnier et al. (1993a).

well as the excess objects above the expected number of accidentals (heavy solid line). The dashed line shows the 2σ lower bound to the number of excess objects, assuming Poisson statistics. From this plot, we determine the total number of counterparts within a given distance (14″) of the X-ray sources (N_t), the number of accidentals (N_a), the expected number of correlations (N_c), and the formal error on N_c. Notice that for determining the error, what matters is the Poisson distribution of the *accidentals*.

The wide-band survey resulted in accurate photometry ($\sim 2\%$ for $V \sim 19$) for stars complete to $V \sim 21.5$. Figure 4 shows a V, $B - V$ color-magnitude diagram of $\sim 40\,000$ objects detected in the B and V filter images. The main sequence, if projected to the distance of M31, is shown as a solid line, and the approximate location of 9, 20, and $40 M_\odot$ main sequence stars are noted. Also noted on this figure are the regions in which some of the optical counterparts are likely to be found.

We have identified C.A. stars by performing the δRA - δDEC technique with stars in the foreground of M31. These are found in Figure 4 as the "spur" of stars with $V < 18$ and $0.4 < B - V < 1.0$, and are supplemented with stars from several other catalogs (the HST Guide Star catalog, the catalog of Berkhuijsen et al. 1988, the SAO and AGK catalogs) which contained bright stars saturated in our CCD images. For this comparison, as noted in Table 1, $N_t = 58$, $N_a = 29$, and $N_c = 29 \pm 5.4$.

GCs are expected to lie within the region $B - V > 0.4$, $V \sim 14 - 20$. We have identified a sample of 417 GC candidates (of which ~ 150 have been previously identified) based on their color as well as by the fact that they have a finite extent (Magnier et al. 1993b). By combining our GC candidates with previous catalogs of GCs (e.g. Battistini et al. 1987, Crampton et al. 1985), we find $N_t = 42$, $N_a = 9$, and $N_c = 32 \pm 3.0$.

The large group of objects in Figure 4 with $B - V < 0.4$ are primarily massive (O/B-type) stars. These are the type of stars we expect to find as the counterparts to MXRBs. However, when these stars are compared to the X-ray sources, the number of accidentals is too high to make any individual identifications believable. We find $N_t = 52$, $N_a = 41$, and $N_c = 11 \pm 6.4$, from which it is clear that the number of blue stars correlated with X-ray sources is consistent with zero. However, the number is also consistent with the number expected in M31: Since the star formation rate in M31 is roughly 20 times smaller than that in our Galaxy, and since the number of MXRBs known in our galaxy is roughly 70, we expect there to be roughly 4 to 10 MXRBs in M31. In section 5, we address a way of improving the identification of specific MXRBs with another method.

We applied a correlation technique, the Path Linkage Criterion (PLC) of Battinelli 1991, to a subset of stars from Figure 4 with the colors of massive stars in order to objectively identify OB associations. Other identifications of M31 associations (van den Bergh 1964; Efremov et al. 1987) have been performed using subjective visual inspections of photographic plates to identify the clustering of blue stars. Using the PLC, we have identified 174 OB associations (Magnier et al. 1993a). Figure 5 shows the outline of the 174 associations listed by Magnier et al. (1993a).

When these OB associations are compared with the locations of the X-ray sources, an interesting effect is found. Figure 6 shows the δRA - δDEC plots for this comparison. For small radii, there is no large clump of excess objects. However, as the radius increases, the number of *excess* objects increases steadily with the increasing radius. This effect is interesting because one interpretation is that it provides supporting evidence to the hypothesis that MXRBs are runaway

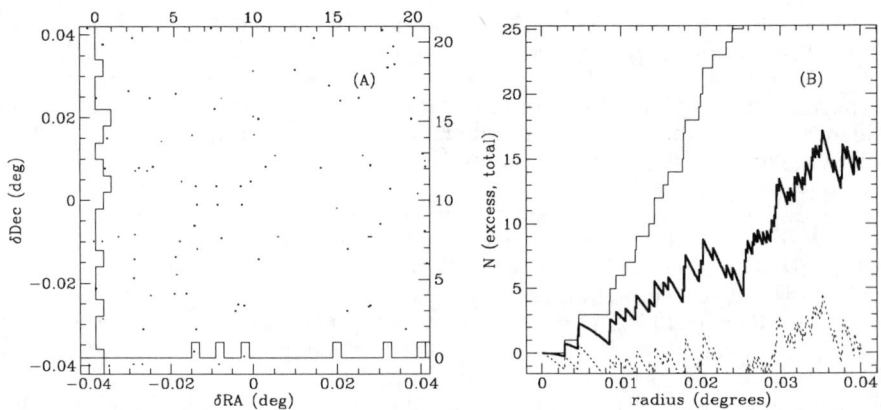

Figure 6. (A) The δRA - δDEC plot for the comparison between X-ray sources and OB associations. (B) The cumulative number of objects as a function of radius (thin solid line), the calculated excess (thick solid line), and the 2σ lower-limit to this value. Note that the excess increases steadily for increasing radius.

systems – i.e. the supernova explosion has given a "kick velocity", so they flee the OB associations in which they are assumed to form (e.g. van Oijen 1989; Wijers et al. 1992). However, there are several caveats to this interpretation. First, we have no *a priori* reason to believe these X-ray sources are MXRBs. Second, it is already well known that the OB associations cluster along the spiral arms (see Fig. 5). Therefore, such an effect may be seen if there were *any* population of X-ray sources which lies along the spiral arms as well. Since we know the SNRs lie along the spiral arms (see Fig. 2), it is quite possible that this is the cause of this effect. More investigation into this question is needed.

In addition to the *BVRI* survey, we also performed a Hα, [SII] survey to search for SNRs (see Fig. 2). While most of the SNRs known in the Galaxy have been identified using radio continuum observations, most of the extragalactic SNRs have been identified by their optical emission-line properties. SNRs appear as extended regions of bright Hα emission. The major difficulty is to distinguish SNRs from HII regions, which are also extended and bright in Hα. The usual strategy is to compare a strong permitted line (usually Hα) to a forbidden line ([OIII] and/or [SII] are the usual choices) (see e.g. Mathewson and Clarke 1973, Long et al. 1990). The ratio of a forbidden line to the permitted hydrogen line is used as a diagnostic to distinguish HII regions from SNRs: HII regions usually have suppressed forbidden line emission compared to SNRs since, in the HII regions, the forbidden line transitions are typically collisionally deëxcited by photons from the ionizing star before they may spontaneously deëxcite. In SNRs, the forbidden lines may be quite strong because the gas is heated quickly by the passing shock and there is no strong cooling mechanism other than the spontaneous atomic emission. We therefore made a survey of ~1.0 square degree of M31 with narrow-band Hα and [SII] images with V images to measure the off-line emission.

There have been several previous searches for SNRs in the field of M31 using Hα and [SII] images (e.g. Walterbos and Braun 1991, Blair et al. 1981, D'Odorico et al. 1980). The searches by Blair et al. and D'Odorico et al. used photographic plates to cover a large area, ~2.5 sq degrees. They found largely overlapping samples with 19 strong SNR candidates, of which 14 have been confirmed with spectra (Blair et al. 1981). Walterbos and Braun (1991) used CCDs to probe significantly deeper than the other two groups and found about 40 candidates, but they covered a much smaller total area (~0.3 sq degrees). Our CCD images are roughly as deep as those of Walterbos and Braun (1991), but they cover a larger area (~1.0 sq degrees). We have identified from these images roughly 150 candidates SNRs based on both the Hα to [SII] ratio as well as a test which depends on the presence or absence of O and B-type stars identified in our $BVRI$ survey (Magnier et al. 1993c). Applying the δRA - δDEC technique to these candidate SNRs and the X-ray sources, we find: $N_t = 23$, $N_a = 3$, and $N_c = 20 \pm 1.7$

There are three major areas where we expect to soon improve the identification of the ROSAT M31 X-ray sources. First, the sources in the central bulge are highly confused in the PSPC images. We will soon combine the PSPC observations with the HRI observations of Primini et al. (1993). This will allow the inner bulge regions to be more accurately studied. Second, deeper optical observations, radio observations, and optical spectral will be required of the Hα bright regions to improve the lists of SNR candidates. It is clear that this is a very rich field, and much progress can be made here.

Finally, there is the question of the MXRBs. It is clear that we are close to being able to identify the MXRBs based only on positional coincidence. However, the X-ray error boxes are somewhat too large. This is not expected to improve with the HRI observations which have been performed. The majority of the HRI observations have concentrated on the bulge regions. In order to identify the MXRBs, we have embarked on a project to detect variability in blue stars in the vicinity of the X-ray sources. MXRBs often exhibit ellipsoidal variability due to the Roche Lobe configuration. We hope to narrow the field of counterparts by identifying more likely candidates which exhibit variability. We have now (September 1993) performed observations at the required level of sensitivity with a sufficient baseline and we hope to shed light on this issue in the near future.

5. CONCLUSIONS

Similar to our own Galaxy, M31 contains a rich X-ray source population. 107 sources were found in the *Einstein* observations and 396 have been found with ROSAT. The optical identification of these sources is crucial to understanding the nature of these sources. However, in these crowded fields, with the spatial resolution of existing X-ray data, making optical identifications is quite difficult.

We have compared the ROSAT X-ray source list with our various catalogs of possible optical counterpart objects. We find 32 ± 3.0 globular clusters, 29 ± 5.4 foreground stars, 11 ± 6.4 O/B-type stars, and 20 ± 1.7 supernova remnants are associated with X-ray sources, *above* the number expected due to accidentals. We find a curious *anti-correlation* between the X-ray sources and the OB associations, which we may be due to the runaway nature of the MXRB systems, but is may also be due to the geometry of the distribution of the two populations. We have only begun the process of identifying the nature of the ROSAT X-ray sources, and it is clear there is much we can learn by studying these X-ray sources. Because of the crowding in the bulge regions, improvements in the X-ray data in this area will be unlikely until AXAF-I is launched, with its expect resolution capabilities. In the disk regions, however, the crowding is much less of a problem and much work can be done with the excellent spectral capabilities of ASCA.

ACKNOWLEDGMENTS

Research reported herein has been performed as part of a collaboration to study the Andromeda Galaxy. I would like to thank my collaborators at MIT, UvA, and MPE. I would also like to thank Frank Primini for discussions of the ROSAT HRI observations and Paolo Battinelli for collaborations on OB associations. Finally, I am grateful to Ed Cheng, Casey Inman, and Matt Kowitt for support during the conference. EAM acknowledges support by the Netherlands Foundation for Research in Astronomy (ASTRON) with financial aid from the Netherlands Organization for Scientific Research (NWO) under contract number 782-376-011.

REFERENCES

Battinelli P. 1991, A&A 244, 69
Battistini P., Bònoli F., Buonanno R., Corsi C.E. & Fusi Pecci F. 1982, A&A 113, 39
Battistini P., Bònoli F., Braccesi A., et al. 1987, A&AS 67, 447
Battistini P., Bònoli F., Casavecchia M., et al. 1993, A&A 272, 77
Berkhuijsen E.M., Humphreys R.M., Ghigo F.D. & Zumach W. 1988, A&AS 76, 65
Blair W.P., Kirshner R.P. & Chevalier R.A. 1981, ApJ 247, 879
Crampton D., Cowley A.P., Hutchings J.B., Schade D.J., & van Speybroek L.P. 1984, ApJ 284, 663
Crampton D., Cowley A.P., Schade D. & Chayer, P. 1985, ApJ 288, 494
D'Odorico S., Dopita M.A. & Benvenuti P. 1980, A&AS 40, 67
Efremov Yu.N., Ivanov G.R. & Nikolov N.S. 1987, Ap&SS 135, 119
Haiman A., Magnier E.A., Lewin W.H.G., Lester R.R., van Paradijs J., Hasinger G., Pietsch W., Supper R. & Trümper J. 1993, A&A in press
Hodge P.W. & Lee M.G. 1988, ApJ 329, 651
Long K.S., Blair W.P., Kirshner R.P., Winkler P.F. 1990, ApJS 72, 61
Long K.S. & Van Speybroek L.P. 1983, in "Stellar X-ray Sources", Eds. W.H.G. Lewin & E.P.J. van den Heuvel, Cambridge University Press, Cambridge
Magnier E.A., Lewin W.H.G., van Paradijs J., Hasinger G., Jain A., Pietsch W. & Trümper J. 1992, A&AS 96, 379
Magnier E.A., Battinelli P., Haiman Z., Lewin W.H.G., van Paradijs J., Hasinger G., Pietsch W., Supper R. & Trümper J. 1993a, A&A 278, 36
Magnier E.A., Lewin W.H.G., van Paradijs J., Hasinger G., Pietsch W., Supper R. & Trümper J. 1993b, A&A submitted
Magnier E.A., Lewin W.H.G., Prins S., van Paradijs J., Hasinger G., Pietsch W., Supper R. & Trümper J. 1993c, in prep
Massey P., Armandroff T.E. & Conti P.S. 1986, AJ 92, 1303
Mathewson D.S., Clarke J.N. 1973, ApJ 180, 725
Van Oijen J.G.J. 1989, A&A 217, 115
Primini F.A., Forman W. & Jones C. 1993, ApJ 410, 615
Supper R., Hasinger G., Pietsch W., Trümper J., Jain A., Lewin W.H.G., Magnier E.A., & van Paradijs J. 1993, in prep
Trinchieri G. & Fabbiano G. 1991, ApJ 382, 82
Van den Bergh S. 1964, ApJS 86, 65
Van Paradijs, J. 1993, in "X-ray Binaries," Eds. W.H.G. Lewin, J. van Paradijs, & E.P.J. van den Heuvel, Cambridge University Press, Cambridge
Walterbos R.A.M. & Braun, R. 1991 A&AS 92, 625

Welch D.L., McAlary C.W., LcLaren R.A. & Madore B.F., 1986, ApJ 305, 583
Wijers R.A.M.J., van Paradijs J., van den Heuvel E.P.J. 1992, å261145

LOW-MASS X-RAY BINARY MODELS FOR THE SUPERSOFT X-RAY SOURCES IN THE LARGE MAGELLANIC CLOUD

N. D. Kylafis and E. M. Xilouris
University of Crete, Physics Department, 714 09 Heraklion, Crete and
Foundation for Research and Technology-Hellas, 711 10 Heraklion, Crete
Greece
kylafis@iesl.forth.gr and xilouris@iesl.forth.gr

ABSTRACT

We propose that the supersoft X-ray spectra observed in CAL 83, CAL 87 and RX J0527.8-6954 can be explained as the result of near-Eddington accretion onto neutron stars. Our model is consistent with a recently proposed unified model for the Low-Mass X-ray Binaries (LMXRB). If the luminosity of the source is within 10% of the Eddington value, what determines whether the source appears as a supersoft X-ray source or as a "canonical" LMXRB is the nature of the accretion flow and its extend. If the radial flow is subsonic and extends to *at least* a few thousand neutron star radii the source behaves like a supersoft one. If on the other hand the flow is supersonic and extends to *at most* a few hundred neutron star radii, the source exhibits the characteristics of a "canonical" LMXRB.

1. INTRODUCTION

The supersoft X-ray sources seen in the LMC are the first examples of a new class of X-ray sources. They have high luminosities and spectra that are fitted with blackbody temperatures in the range 30 to 50 eV.

A promising model for their explanation was recently proposed by van den Heuvel *et al.* (1992). They explain the supersoft X-ray spectrum as arising from steady nuclear burning of hydrogen accreted onto massive white dwarfs.

Statements in the literature of the form,
  CAL 83 is the most luminous Low-Mass X-ray Binary (LMXRB) (Smale
    et al. 1988),
  CAL 87 has been identified as a LMXRB (Pakull *et al.* 1988),
  RX J0527.8-6954 has remarkably similar properties with CAL 83 (Greiner,
    Hasinger, & Kahabka 1991),
  the supersoft X-ray sources are neutron stars shrouded by massive accre-
    tion, probably highly super Eddington (Greiner, Hasinger, &
    Kahabka 1991),
led us to investigate the possibility of a neutron-star model for the supersoft X-ray sources. The requirement is that the photosphere of the accreting neutron star must be at a few thousand neutron star radii so that the blackbody temperature is in the tens of eV range.

Here we demonstrate that near-Eddington spherical accretion onto neutron stars can explain the supersoft X-ray sources (see also Kylafis & Xilouris 1993). Our model is consistent with a recently proposed unified model of LMXRB (Lamb 1989a, b; 1993; see also Hasinger *et al.* 1990 for a suggestion of a different model).

2. QUALITATIVE DISCUSSION

Consider a neutron star on which matter accretes radially with a rate close to the Eddington limit. There are two types of flow for the same accretion rate. One that starts out supersonic and continues supersonic (except possibly near the neutron star surface) and the other which starts out subsonic and continues subsonic throughout the flow. The second one is sometimes called a settling flow, because the velocity monotonically decreases from the outer boundary to the neutron star surface.

The supersonic flow is optically thin to absorption and moderately optically thick to electron scattering as the accretion rate approaches the Eddington limit. The X-ray spectrum is produced near the neutron star surface and it is hard.

The subsonic flow on the other hand piles up a lot of matter and it is optically thick to both scattering and absorption. It is therefore expected that at high accretion rates the photosphere will move out to large radii. If it happens that these radii are of order a few thousand neutron star radii, then the spectrum escaping from the photosphere is a blackbody with temperature in the tens of eV range. We propose that this is the case for the supersoft X-ray sources.

If our proposal for the supersoft X-ray sources is correct, then a natural prediction of our model is the following: Since a supersoft X-ray source has a luminosity close to the Eddington value and since both the accretion rate and the conditions at the outer boundary of the flow may vary with time, it is predicted that if the supersoft sources are monitored, in some of them the character of the flow may change and they will appear as hard X-ray sources.

3. CALCULATIONS

We have solved the hydrodynamic equations (mass, momentum and energy conservation) assuming LTE from a few thousand neutron star radii to the neutron star radius. Our calculations are similar to those of Vitello (1978) and as a check we reproduced his results.

Given the extend of the flow, the only remaining parameter is $\epsilon \equiv 1 - \dot{M}/\dot{M}_E$, where $\dot{M}$ is the radial accretion rate and $\dot{M}_E$ is the Eddington rate. Here ϵ is a measure of the relative importance of the radiation force on the radial flow because the luminosity at infinity is $L_\infty = (1 - \epsilon)L_E$, where L_E is the Eddington limit. For $\epsilon \lesssim 0.1$ we can always find radial subsonic flows which are optically thick to free-free absorption throughout the flow. In such a case, the emitted spectrum is a blackbody in the tens of eV range. Figures 1 - 3 are a typical example of such flows.

Figure 1 shows the subsonic (solid line) and supersonic (dashed line) velocity profiles of the radial flow for $\epsilon = 0.05$. Both velocities are divided by the local sound speed. The character of the flow is completely insensitive to the initial velocity. If the initial velocity is subsonic, the flow is subsonic throughout. Similarly for a supersonic initial velocity. In the subsonic flow the velocity decreases as the radius decreases and this is a typical example of a settling flow.

Figure 2 shows the LTE temperature profiles of the flows. The solid line corresponds to the subsonic flow and the dashed one to the supersonic. By construction the temperature of the photosphere is 30 eV, which is the blackbody temperature for $L \approx L_E$ and $r = 3 \times 10^3 R$, where R is the radius of the neutron star. The temperature reaches high values as the matter approaches the neutron star and nuclear reactions will take place there. Nevertheless, the increase in the luminosity is small and no qualitative changes in the flow occur due to this increase.

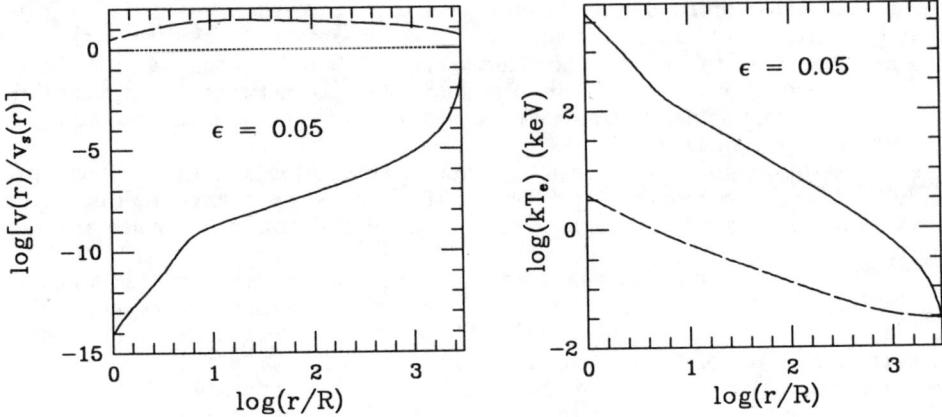

Fig. 1. *Subsonic (solid) and supersonic (dashed) radial flows for $\epsilon = 0.05$. Both velocities are divided by the local sound speed.*

Fig. 2. *Temperature profiles in the subsonic flow (solid) and the supersonic one (dashed) for $\epsilon = 0.05$.*

Figure 3 shows the profiles of the effective optical depth to free-free absorption from r to infinity $\tau_*(r) \equiv (3\tau_s \tau_a)^{1/2}$, where τ_s is the corresponding electron scattering optical depth and τ_a is the corresponding free-free absorption one. The solid lines indicate the subsonic flow and the dashed ones the supersonic.

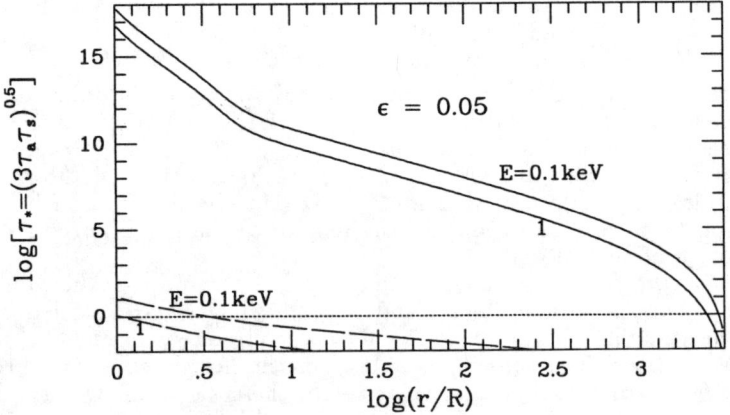

Fig. 3. *Effective absorption optical depth profiles in the subsonic flow (solid) and the supersonic one (dashed) for $\epsilon = 0.05$. Each curve is labeled with the energy used for its calculation.*

In each case, one curve is for $E = 0.1$ keV and the other for 1 keV. It is straightforward to verify in the subsonic flow that $\tau_*(r) > 1$ for all r and $E \lesssim 2kT_e(r)$. Therefore, our assumption of LTE is self-consistent.

4. CONCLUSIONS

We have demonstrated that if the luminosity of the LMXRB is near the Eddington limit, what determines whether the source appears as a "canonical" LMXRB or as a supersoft X-ray source is the extent of the radial flow and its character (supersonic or subsonic). If the luminosity of the source is within 10% of L_E and the extent of the radial flow is at least $3 \times 10^3 R$, one can always find flows for which the emergent spectrum is a blackbody of temperature in the tens of eV range.

If the extent of the radial flow is at most a few hundred neutron star radii, the source behaves as a "canonical" LMXRB.

One natural prediction of our model is the following: If *the nature and/or the extent* of the radial flow varies with time, a supersoft X-ray source may convert into a "canonical" LMXRB and visa versa.

ACKNOWLEDGEMENTS

We thank Brand Fortner, Peter Goldreich, Jean Marie Hameury, Günther Hasinger, Fred Lamb, Guy Miller and Kanaris Tsinganos for useful discussions and comments.

REFERENCES

Greiner, J., Hasinger, G., & Kahabka, P. 1991, A&A, 246, L17
Hasinger, G., van der Klis, M., Ebisawa, K., Dotani, T., & Mitsuda, K. 1990, A&A, 235, 131
Kylafis, N. D. & Xilouris, E. M. 1993, A&A, 278, L43
Lamb, F. K. 1989a, in Ann. NY Acad. Sci. 571, Proc. 14th Texas Symposium on Relativistic Astrophysics, ed. E. J. Fenyves (New York: NY Acad. Sci.), 347
———. 1989b, in Proc. 23rd ESLAB Symposium on X-ray Astronomy, ed. N. E. White (ESA SP-296), 215
———. 1993, ApJ, submitted
Pakull, M. W. et al. 1988, A&A, 203, L27
Smale, A. P. et al. 1988, MNRAS, 233, 51
van den Heuvel, E. P. J., Bhattacharya, D., Nomoto, K., & Rappaport, S. A. 1992, A&A, 262, 97
Vitello, P. A. J. 1978, ApJ, 225, 694

ROSAT OBSERVATIONS OF GLOBULAR CLUSTERS

Helen M. Johnston and Frank Verbunt
Astronomical Institute, P.O.Box 80000, 3508 TA Utrecht, The Netherlands

Günther Hasinger
Max Planck Institute for Extraterrestrial Physics, D-8046 Garching, Germany

ABSTRACT

We present deep pointed observations of ten globular clusters obtained with the ROSAT PSPC, and one with the ROSAT HRI. X-ray sources are detected in the cores of seven of them, at luminosities of $\sim 1-6 \times 10^{32}$ erg s^{-1}. At least two of these are multiple. Four have colours indicating soft spectra, with blackbody temperatures $kT \lesssim 0.3$ keV. Soft spectra can be excluded for the other three sources. The sources outside the cores are probably not associated with the cluster. The most likely counterparts for these sources are soft X-ray transients (SXTs).

1. INTRODUCTION

Globular cluster X-ray sources fall into two distinct classes: the "bright" sources, with X-ray luminosities $L_x \gtrsim 10^{36}$ erg s^{-1}, and "dim" sources, with $L_x \lesssim 10^{35}$ erg s^{-1}. The high luminosity of the bright sources implies that they must be accreting neutron stars or black holes; all well-studied sources have now also shown X-ray bursts, which confirm their identity as neutron stars. The nature of the dim sources is less clear. They have been suggested to be cataclysmic variables, quiescent X-ray transients, RS CVn stars, millisecond radio pulsars, and fore- or background objects.

2. RESULTS OF OBSERVATIONS

Eight of the ten clusters observed were found to contain X-ray sources in or near the core with luminosities in the region of 10^{32} erg s^{-1}. The properties of the sources are summarized in Table 1. The clusters were chosen so as to *(i)* have maximum sensitivity to detection of X-ray sources, which translates to (a) being nearby, and (b) having low Galactic reddening (because of ROSAT's soft X-ray response); and *(ii)* have high collision number. The clusters span a wide range of parameters, particularly central density; four of the ten clusters have undergone core-collapse.

In addition to these sources, we detected a serendipitous bright source in the NGC 6642 observation, which is almost certainly the source discovered by Ginga in 1988, GS 1826−24 (Makino et al. 1988).

Observations of some of the individual clusters are discussed below. Complete details of the observations will be published in Johnston et al. (1993) and Hasinger et al. (1993).

NGC 5139 = ω Cen — Seventeen sources were discovered within the tidal radius, one of which was within the core. All five *Einstein* low-luminosity sources in ω Cen were seen by ROSAT. Of the other sources, three can be identified with foreground stars: one K giant and two A dwarfs. X-ray emission from the latter presumably indicates the presence of an unseen white dwarf companion.

Table 1: Detected sources in all cluster cores

Cluster	Δ (″)	S_X	L_X
47 Tuc[b]	–	20.0	5.1
ω Cen	75	0.56 ± 0.12	1.6 ± 0.3
NGC 6341	11	0.47 ± 0.13	3.2 ± 0.8
NGC 6397	12	2.4 ± 0.6	1.4 ± 0.3
NGC 6544	–	< 1.0	< 0.77
NGC 6626	8	1.5 ± 0.3	6.2 ± 1.3
NGC 6642	–	< 0.58	< 4.4
NGC 6656	21	0.70 ± 0.20	0.76 ± 0.22
NGC 6752	6	1.5 ± 0.22	3.2 ± 0.5
NGC 7099	9	0.80 ± 0.13	5.3 ± 0.9

[a]Columns show cluster name, offset from core, 0.5–2.5 keV flux, ($\times 10^{-13}$ erg s^{-1} cm^{-2}) for a 0.3 keV blackbody spectrum, and 0.5–2.5 keV luminosity ($\times 10^{33}$ erg s^{-1}) for the same spectrum.
[b]The values for 47 Tuc are taken from the All-Sky Survey observations (Verbunt et al. 1993b).

The core source is almost certainly double, consisting of two separate sources of approximately equal brightness, separated by 44″ (see Figure 1). We cannot say much about the spectra of the sources as the number of photons is very small.

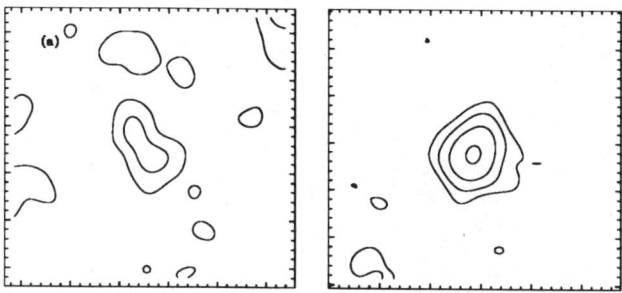

Figure 1: Contour maps of the core sources in (a) ω Cen and (b) NGC 6397. The data were binned into 3″ pixels and smoothed with a 4 pixel gaussian. Each box is 276″ on a side

NGC 6397 — Ten sources were detected; one is only 12″ from the cluster core. The core source appears to be multiple (see Figure 1). In HRI observations of the same cluster, Cool et al. (1993) resolved this source into between three and five separate objects.

NGC 6642 — No sources were detected in this cluster. However, the field contains a very bright source, with a count rate of 1.3 cts/s, in the field, whose position coincides with the error box of an X-ray source seen by Ginga in 1988, GS 1826−24 (Makino et al. 1988). The same source also appears in the ROSAT All-Sky Survey map of the field, in data taken in September 1990, 1.5 y before the pointed observation. Thus the source has probably been consistently bright for the past four years.

NGC 6752 — Eleven sources were detected, one coincident with the cluster core. There is some slight evidence of asymmetry in the contours of the core source, suggestive of two sources with fluxes in the ratio $\sim 2:1$.

NGC 104 = 47 Tuc — Two HRI pointings towards 47 Tuc were obtained. The sources appear to be variable between the two observations, with four sources visible in the first observation, but only three in the second.

3. DISCUSSION

Of the ten clusters observed, eight were found to contain X-ray sources with luminosities in the region of 10^{32} erg s^{-1} in or near the core. The properties of the sources are summarized in Table 1.

The spectra for the sources in NGC 6341, NGC 6397, NGC 6626 and NGC 7099 are soft, with $kT \lesssim 0.3$ keV for a blackbody spectrum, $\lesssim 1.5$ keV for a bremsstrahlung or optically thin spectrum. For three sources — ω Cen, NGC 6752 and NGC 6656 — we can exclude very soft spectra. One of these sources, the central source in NGC 6656, has observed colours incompatible with all three spectral models, unless a high intrinsic reddening is invoked.

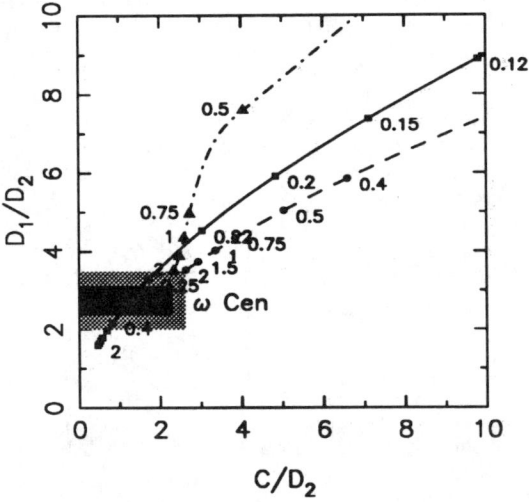

Figure 2: X-ray colour-colour diagram for the core source in ω Cen. The ratio of the counts in band D_1 to D_2 (channels 91–150 and 151–201) is plotted against the ratio of the counts in band C (channels 52–90) to D_2. The 1σ error box is shown with dark shading, the 2σ box with light. Also plotted are three theoretical spectral models for temperatures ranging between 0.1 and 5 keV: blackbody (solid line), thermal bremsstrahlung (dashed line) and optically thin (dot-dashed line); the temperature in keV is indicated next to the points. Clearly the observed colours are incompatible with an AM Her-type spectrum; the indicated temperature is 0.6 ± 0.2 keV for a blackbody spectrum, > 5 keV for the other two spectral models.

The observed temperature of the first four sources are similar to the spectra of the soft X-ray transient Aquila X-1 in quiescence (Verbunt et al. 1993a). The spectrum, observed on three separate occasions while the source was in quiescence, had a temperature of 0.3 keV at a luminosity of $\sim 4 - 6 \times 10^{32}$ erg s^{-1}.

Figure 3 shows the X-ray luminosities of the various classes of objects that have been proposed as counterparts for the dim cluster sources. The data for the cataclysmic variables are from the ROSAT All-Sky Survey (Bunk, pers. comm.); no AM Her systems have been included, since the distances to these systems are not well known. The RS CVn data are also from the Survey (Dempsey et al. 1993); the millisecond pulsars are from ROSAT pointed observations (Kulkarni et al. 1992; Becker & Trümper 1993); and the soft X-ray transients data are from Verbunt et al. (1993a).

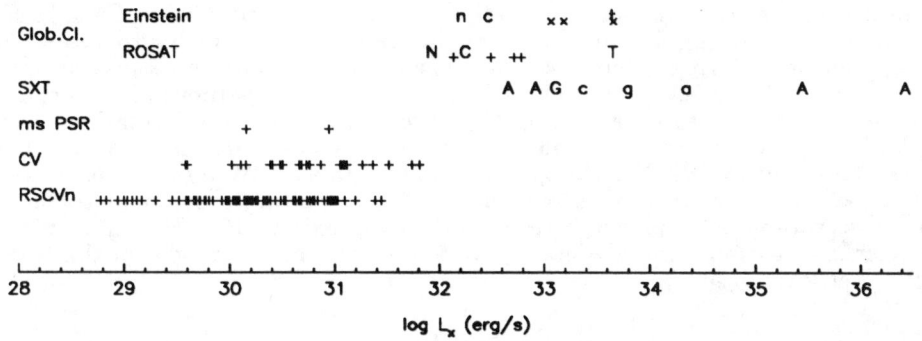

Figure 3: X-ray luminosities of disk and cluster X-ray sources. Uppercase letters refer to ROSAT observations, lowercase to observations by other satellites. Plotted is the 0.5–2.5 keV X-ray luminosity of globular cluster sources, soft X-ray transients, millisecond pulsars, cataclysmic variables, and RS CVn systems. Individual objects are indicated as follows: (a) Globular clusters: N,n = NGC 6656; C, c = ω Cen; T, t = 47 Tuc. (b) Soft X-ray transients: A, a = Aql X-1; G, g = GS 2023+38; c = Cen X-4. (From Verbunt et al., in prep.).

Of the sources in the galactic disk, then, only the soft X-ray transients in quiescence have luminosities comparable to those of the dim sources in the cores of globular clusters.

ACKNOWLEDGEMENTS

FV and HMJ are supported by the Netherlands Organization for Scientific Research NWO under grant PGS 78-277.

REFERENCES

Becker, W., Trümper, J. 1993, Nat, 365, 528
Cool, A., Grindlay, J., Krockenberger, M., Bailyn, C. 1993, ApJ, 410, L103
Dempsey, R., Linsky, J., Fleming, T., Schmitt, J. 1993, ApJS, 86, 599
Hasinger, G., Verbunt, F., Johnston, H. 1993, ROSAT HRI observations of 47 Tuc, in preparation
Johnston, H., Verbunt, F., Hasinger, G. 1993, ROSAT observations of globular clusters, A&A, submitted
Kulkarni, S., Phinney, E., Evans, C., Hasinger, G. 1992, Nat, 359, 300
Makino, F. et al. 1988, IAU circular 4653
Verbunt, F., Belloni, T., Johnston, H., van der Klis, M., Lewin, W. 1993a, ROSAT observations of soft X-ray transients in quiescence, A&A, in press
Verbunt, F., Hasinger, G., Johnston, H., Bunk, W. 1993b, in preparation

The Galactic High Mass X-ray Binary Population

William W. Dalton and Craig L. Sarazin
Department of Astronomy, University of Virginia
PO Box 3818, Charlottesville, VA 22903-0818

ABSTRACT

Modern stellar evolutionary tracks are used to calculate the evolution of a very large number of massive binary star systems ($M_{tot} \gtrsim 15 M_\odot$) which cover a wide range of total masses, mass ratios, and starting separations. Each binary is evolved accounting for mass and angular momentum loss (due to wind mass loss, mass loss during Roche lobe overflow of the primary, mass loss during a common envelope phase should it occur), through the supernova of the primary to the X-ray binary phase. Using the observed rate of star formation in our galaxy and the properties of massive binaries, we calculate the expected high mass X-ray binary (HMXRB) population in the Galaxy. We test various massive binary evolutionary scenarios by comparing the resulting predictions to the X-ray observations. Our principle result is that approximately 70% of the overflow matter is lost from a massive binary system during mass transfer in the Roche lobe overflow phase.

1. INTRODUCTION

The evolution of high mass binary systems ($M_{tot} \gtrsim 15 M_\odot$) from the main sequence to the HMXRB stage has been understood, at least conceptually, for some time (van den Heuvel 1978). Beyond these basic principles, detailed modeling is difficult because of the large uncertainties that remain in our understanding of the evolution of such systems. Chief among these uncertainties are the amount of mass and angular momentum lost from a close binary system when the primary (the more evolved star) overflows its Roche lobe (*e.g.* how "nonconservative" the system is; de Loore & De Greve 1992). By constructing several models for the Galactic HMXRB population using different overflow mass loss assumptions, and comparing the results with the observed Galactic HMXRB statistics, we hope to reduce the remaining evolutionary uncertainties.

Conventionally the observed HMXRB systems are divided into two categories: the persistent sources, and the transient Be sources. Generally speaking, the persistent systems contain a compact object (neutron star or black hole) accreting matter from an early type (O or B) giant or super giant star which fills or nearly fills its Roche lobe. The high X-ray luminosities observed may be powered either by accretion from the strong stellar winds produced by such stars, or by disk accretion in a Roche lobe overflow geometry. In the Be systems, the donor star is generally a main sequence B (or late O) emission star that is well within its Roche lobe. These stars show evidence for episodes of enhanced equatorial mass loss (Lamers & Waters 1987). Be X-ray binaries are characterized by periods of quiescence, with no detectable X-ray flux, followed by brief flaring events with very high peak luminosities, approaching or even exceeding the Eddington limit for spherical accretion onto a neutron star. This X-ray flaring is thought to arise from intermittent accretion as the compact object passes through the relatively dense, slow moving equatorial material that surrounds the emission star during outburst (Stella et al. 1986; Waters et al. 1989). The X-ray emission from a given Be system depends on the details of the system under scrutiny (Stella et al. 1986), and is therefore beyond the scope of this statistical treatment. Since the persistent sources are amenable to a more general analytic treatment, we calculate the expected emission from these systems in detail, and base our conclusions on comparisons with the observed persistent sources.

2. THEORY

Our binary evolutions are based on the most accurate theoretical stellar tracks currently available, those of Schaller et al. (1992). These single star evolutionary tracks contain several improvements over earlier efforts, including the use of the Rogers & Iglesias (1992) radiative opacities.

At the start of each evolution, the more massive star is defined to be the primary. For consistency we keep this definition through stages of mass loss, mass transfer, and even beyond the supernova event. The initially less massive star will likewise remain the secondary in our terminology.

Main Sequence Evolution: The prodigious stellar winds produced by massive stars carry both mass and angular momentum from the system. Huang & Taam (1990) studied non-conservative evolution in massive binary systems and we have followed their formalism in accounting for the effects of wind mass and angular momentum losses on the orbital elements of massive binaries. We assume the orbits are circular, and that both components are in synchronous rotation. In the close binaries that are the progenitors of the HMXRBs, tidal forces are expected to rapidly circularize any initially eccentric orbits (Tassoul 1990).

Mass Transfer: When the primary component ends its main sequence lifetime it overflows its Roche lobe, losing a substantial amount of matter. Some of this material is accreted by the secondary star, but a significant amount may be lost from the system. Let f be defined as the fraction of the overflow mass which is lost from the binary system. It is extremely difficult to explicitly calculate the mass and angular momentum losses during overflow, and f is very poorly known. By computing grids of massive binary evolutions for a range of values for f ($f = 0, 0.3, 0.5,$ and 0.7) and comparing our predictions with observations, we hope to reduce the uncertainty in this quantity.

Post Overflow Evolution: After overflow, the secondary is generally the more massive star, and therefore the superova explosion of the primary will eject less than half of the mass from the system. If the supernova explosion is spherically symmetric in the rest frame of the primary star, a bound main sequence star — compact object binary system results (Blaauw 1961). By assuming that the neutron star is given a $100\,\mathrm{km\,s^{-1}}$ kick in a random direction at birth (Dewey & Cordes 1987), we find that many of the long period HMXRB progenitor systems are disrupted. However, the short period systems that give rise to the powerful X-ray sources of interest here remain bound. In our simulations, we assume spherically symmetric explosions.

X-Ray Phase: At this point the compact object will begin to accrete matter from the stellar wind of the secondary. The resultant X-ray luminosity can be parameterized as $L_X = \eta \dot{M}_{acc} c^2$, where $\dot{M}_{acc}$ is the mass accretion rate onto the compact object and η is the efficiency for converting the accreted rest

Table I

f	Number of HMXRBs $\log L_X \geq 37$	$\log L_X \geq 36$	Percent with $P < 10$ days
0.0	15.1	45.8	16
0.3	15.5	56.9	40
0.5	14.5	51.9	54
0.7	10.6	49.9	71

mass energy into X-ray radiation. We assume that $\eta \approx 0.1$.

Friend & Castor (1982) modeled five HMXRB systems using the radiation-driven wind model of Castor et al. (1975). While the level of detail in this analysis is beyond the scope of our statistical work, their more general results are of use. Friend & Castor found that as the donor star expands and its surface approaches its Roche surface, the wind becomes stronger and begins to focus sharply toward the compact object. This process provides a smooth transition from a system with a compact object accreting from a spherically symmetric wind to accreting from a disk, as in the traditional Roche lobe overflow picture. Both observational data and the Friend & Castor models indicate that the wind's structure begins to depart significantly from spherical symmetry when the donor star's radius reaches about $0.9 R_{lobe}$. Rather than attempt to calculate the effect of the complex Friend & Castor potential on the wind structure, we employ a simpler model developed by Savonije (1983) to describe incipient Roche lobe overflow in HMXRB. We use a spherically symmetric wind fed accretion formula until the donor star's radius reaches $0.9 R_{lobe}$, when we switch to the Savonije formalism, using the wind-fed accretion rate as an initial condition.

Population Synthesis: Our library of massive binary evolutions has been used to predict the average statistical properties of the Galactic HMXRB source population, assuming a variety of models for the binary fraction in the Galaxy, the distribution of initial binary separations, the distribution of initial stellar mass ratios, the initial mass function (IMF) of star formation, and the Galactic star formation history. Specifically, we assume that the massive star binary fraction is 36% (Garmany et al. 1980; Abt & Levy 1978). We assume that the probability that a binary is formed with a separation in the range a_{init} to $(a_{init} + da_{init})$ is proportional to $(1/a_{init})da_{init}$ (Kraicheva et al. 1979). Similarly, the initial value of the binary mass ratio $q \equiv M_2/M_1$ is distributed so that the probability of a value in the range q to $(q + dq)$ is proportional to $q\,dq$ (Kraicheva et al. 1979; Garmany et al. 1980). We have considered several values for the slope of the IMF at the high mass end, but here we only present results for an intermediate value $\Gamma = 1.5$ (Scalo 1986). (Note that the Salpeter IMF corresponds to $\Gamma = 1.35$.) We determined the Galactic star formation rate from the observed Lyman continuum photon production rate, which is approximately 2×10^{53} photons s^{-1} (Mezger 1987). We assumed this rate has remained constant over the (rather short) time scale of HMXRBs (Basu & Rana 1992).

4. CONCLUSIONS

Meurs & van den Heuvel (1989) estimated the total number of strong $(L_X > 10^{36}$ ergs s$^{-1})$ HMXRB sources in the Galaxy by extrapolating from the 4 sources observed within 3.5 kpc (Cyg X-1, Vela X-1, 4U1700-37, and 4U1223-62). Assuming a uniform distribution of sources throughout the disk, and a disk radius of 13 kpc, they determined that the number of strong HMXRB in the Galaxy is 55 ± 27. Similarly, by extrapolating from the 2 sources within 3 kpc with $L_X > 10^{37}$ ergs s^{-1}, they estimated that the Galaxy contains 27 ± 14 of these very potent HMXRB sources. The results of our simulations are shown in Table 1, where we list the assumed mass loss parameter f in col. 1, the number of sources with $L_X > 10^{37}$ or 10^{36} ergs s^{-1} in cols. 2 and 3, respectively, and the percentage of the strong sources with orbital period < 10 days in col. 4. Since all of the known persistent HMXRB have orbital periods $\lesssim 10$ days (Verbunt 1993), we can strongly exclude the conservative scenarios ($f \sim 0$); in fact, we require $f \approx 0.7$.

ACKNOWLEDGEMENTS

Both W.W.D. and C.L.S. were supported in part by NASA Astrophysical Theory Program grant NAGW-2376. W.W.D. was also supported in part by NASA ROSAT grant NAG 5-1891. W.W.D. also gratefully acknowledges support from a Grant-in-Aid of Research from the National Academy of Sciences through Sigma Xi, the Scientific Research Society.

REFERENCES

Abt, H. A., & Levy, S. G. 1978, ApJS, 36, 241
Basu, S., & Rana, N. C. 1992, ApJ, 393, 373
Blaauw, A. 1961, Bull. Astr. Inst. Netherl., 15, 291
Castor, J. I., Abbott, D. C., & Klein, R. I. 1975, ApJ, 195, 157
de Loore, C., & De Greve, J. P. 1992, A&AS, 94, 453
Dewey, R. J., & Cordes, J. M. 1987, ApJ, 321, 780
Friend, D. B., & Castor, J. I. 1982, ApJ, 261, 293
Garmany, C. D., Conti, P. S., & Massey, P. 1980, ApJ, 242, 1063
Huang, R. Q., & Taam, R. E. 1990, A&A, 236, 107
Kraicheva, Z. T., Popova, E. I., Tutukov, A. V., & Yungelson, L. R. 1979, Sov. Ast., 23, 290
Lamers, H.J.G.L.M., & Waters, L.B.F.M. 1987, A&A, 182, 80
Meurs, E. J. A., & van den Heuvel, E. P. J. 1989, A&A, 226, 88
Mezger, P. 1987, in Starbursts and Galaxy Evolution, eds. T. X. Thuan, T. Montmerle & J. Tran Thanh Van (Gif Sur Yvette, France: Editions Frontières), 5
Rogers, F. J., & Iglesias, C. A. 1992, ApJS, 79, 507
Savonije, G. J. 1983, in Accretion-driven Stellar X-ray Sources, eds. W. H. G. Lewin & E. P. J. van den Heuvel (Cambridge: CUP), 393
Scalo, J. M. 1986, Fund. Cosmic Phys., 11, 1
Schaller, G., Schaerer, D., Meynet, G., & Maeder, A. 1992, A&AS, 96, 269
Stella, L., White, N. E., & Rosner, R. 1986, ApJ, 308, 669
Tassoul, J. L. 1990, ApJ, 358, 196
van den Heuvel, E. P. J. 1978, in Physics and Astrophysics of Neutron Stars and Black Holes, eds. R. Giacconi & R. Ruffini (Amsterdam: North Holland), 828
Verbunt, F. 1993, ARA&A, 31, in press
Waters, L.B.F.M., de Martino, D., Habets, G.H.M.J., & Taylor, A. R. 1989, A&A, 223, 207

Summaries

CREATOR'S RECORDS: THE PRE-HISTORY OF SINGLE AND BINARY NEUTRON STARS

Virginia Trimble
Astronomy Department, University of Maryland, College Park MD 20742
and
Physics Department, University of California, Irvine CA 92717

ABSTRACT

Single neutron stars were thought of more than 30 years before they were observed (and even then were not instantly recognized). Binary neutron stars, on the other hand, were observed before they had been though of as a separate, interesting class, and were also not immediately recognized for what they were. The traditional apportionments of credit for the various ideas are not fully supported by contemporaneous publications.

1. INTRODUCTION: DYSON'S LAW

Soon after the death of Richard Feynman, the American Physical Society and American Association for the Teaching of Physics honored his memory with a joint meeting session. Freeman Dyson spoke of their years together at Cornell, quoting letters he had written to his parents at the time and other documents. He began by saying that, in re reading these pieces of paper, he had been astonished both by how many things had happened that he had forgotten and by how many things he remembered that had never actually happened. Neutron star astronomy, undoubtedly in common with all other forms of human endeavor, shows examples of both phenomena. The sections following address single neutron stars, binary ones, and some special topics including accretion disks and globular cluster sources.

2. SINGLE NEUTRON STARS

Though Eddington was already sure in 1926 that stars must run on subatomic energy, the bandwagon was slow in gathering steam, and the literature of the 30's includes a mix of gravitational and nuclear processes. Milne (1931) attempted an early synthesis, toward the end of which he says, "It is possible that the passage of a configuration through the critical value $L = L_0$ may be discontinuous, as it certainly is for perfect-gas configurations of unlimited compressibility. If subsequent analysis confirms this, the passage of a configuration through $L = L_0$ would exhibit the phenomena of a nova, and we should have the suggestion that every star passes through a nova stage as it crosses from a configuration of ordinary density to one of great density, as its evolution of energy decays." This suggestion that novae are the transition from perfect-gas stars to degenerate white dwarf stars lingered in text books into the 60's.

Milne's idea was presumably still in the air when the words "neutron star" first appeared in print in the sentence, "With all reserve, we advance the view that a super-nova represents the transition of an

ordinary star into a neutron star, consisting mainly of neutrons...
[which] would represent the most stable configuration of matter as
such." (Baade & Zwicky 1934). They go on to suggest a connection with
cosmic rays and the Crab Nebula. The idea of the stability of neutron rich matter, in the context of nuclear transformations, is also
present in Sterne (1933).

Meanwhile, Chadwick in Cambridge had discovered the neutron in
1932. The news clearly diffused rapidly. Four decades later, in an
after dinner talk and conference proceedings, Leon Rosenfeld (1973)
recalled being in Copenhagen at the time, where Bohr received a letter from Chadwick and discussed it with him and Lev Landau, who invented the concept of neutron stars on the spot. This tale has become part of the oral folklore we relate to our students (being precisely the sort of thing one can imagine Landau doing). Unfortunately, "At the time in question (February or March 1932) none of the
three protagonists was in Copenhagen (Landau had moved to Kharkov).
The conversation may have taken place during a meeting in the Soviet
Union two years later." (Israel, 1987, making use of information unearthed by Gordon Baym). Baade and Zwicky apparently invented the
concept of neutron stars as well as the name and the connection with
supernovae.

They were to get precious little immediate glory out of it.
Revising a monograph in 1937, Gamow writes, "For still higher densities
electrons will probably be absorbed by the nuclei (an inverse β-decay process) and the mixture will tend to a state which can be described very roughly as a gas of neutrons." In the following calculations he cites only two papers (Landau 1932, which is shelved with JETP
in many libraries, and Chandrasekhar 1931). Both are white dwarf and
only white dwarf calculations (though Frenkel 1928 had considered the
consequences of degenerate protons). Baade and Zwicky are not mentioned, and Gamow does not have catastrophic collapse in mind. Rather
he says, "The question whether most stars at present actually possess such nuclei cannot, however, be answered definitely...but there
seems to be no reason why they should not...As to the liberation of
energy, one can easily see that pure gravitational energy liberated
in the contraction to such immense densities will already be quite
enough to secure the life of the star for a very long period of time."

Landau's (1938) much-cited neutron star paper is actually called
"Origin of Stellar Energy," and says "...we see that the conception of
a 'neutronic' state of matter gives an immediate answer to the question
of the sources of stellar energy...Even for such a bright star as β
Orionis, we find for the mass of the neutronic core only about $0.1\odot$."
Strangely, when Gamow and Teller (1938) come to say him nay, only the
Landau proposal, not the Gamow one, is mentioned. They correctly conclude, "... that in the neighborhood of the core temperature will rise
with r^{-1} and density with $r^{-3/2-n/2}$...[and] will reach the extremely
high values more than 10^9 degrees and 10^9 g cm^{-3}. Under such conditions all kinds of nuclear reactions will proceed at a great rate and
and will make the total energy production of the star many orders of
magnitude greater than the observed radiation. Therefore the core
model as well as any other model leading to such high temperatures

seems to be ruled out. This had to be rediscovered 35 years later when the solar neutrino problem led to a suggestion (attributed to Hawking) that the sun might have a small central black hole, accretion on to which was the primary energy source. The exceedingly high temperature and density actually result in more neutrinos than the standard nuclear sun should radiate!

Serious neutron star models begin with Oppenheimer and Volkoff (1939). They say quite correctly "that in sufficiently massive stars after all the thermonuclear sources of energy, at least for the central material of the star, have been exhausted a condensed neutron core would be formed." But who gets the credit? "G. Gamow... L. Landau,...and others," the first two of whom, at least, said no such thing. Not Baade and Zwicky, who did, more or less.

Stellar astrophysicists in the immediate post-war years largely focussed on main sequence and red giant structure, making use of the detailed nuclear reaction chains identified by Bethe and others in 1939. Neutron stars (and black holes) reappeared, however, as part of the revival of relativistic astrophysics in the groups centered around John A. Wheeler (in Princeton from about 1957 onward) and Yakov B. Zeldovich (in Moscow a few years later). Seminal speculations were rife, and there can hardly be an astrophysicist old enough to remember the period who is not still kicking himself for not having published more of them and more agressively (Compare elderly Californians trading stories about how they could have bought the site of "One Wilshire Blvd" for $350 back in 1931.) Two of these speculations are particularly germane to our story. "If a star contracts in a spherically symmetrical way and if flux is conserved [and] if neutron star densities are reached the field intensity would increase by a factor of 10^{10}, and thus stellar fields of up to $10^{14} - 10^{16}$ G could be reached...one may well speculate that such a theory could have a direct bearing on the problem of the origin and acceleration of the relativistic electrons in the Crab Nebula" (Woltjer 1964). And Pacini (1967) mused, "The problem therefore arises of finding out whether the energy stored in the neutron star plays an important part in connextion with the activity observed in some supernova remnants such as the Crab Nebula. The vibrations of the neutron star, however, do not last long enough for our purpose...It seems more rewarding therefore to look for some mechanisms by which the neutron star can release either its magnetic or its rotational energy or both." There follows the standard expression for emission by a rotating magnetic dipole (for which most of us cite Gunn and Ostriker, though they in turn credited Landau and Lifshitz).

Even as he wrote, Hewish and his colleagues (1968) were stringing the wires and reeling out the punched paper tape and strip charts whose contents would soon be announced as, "Observations of a Rapidly Pulsating Radio Source [whose] radiation may be associated with oscillations of white dwarfs or neutron stars." White dwarfs were definitely ruled out by the Crab pulsar's 0.033 sec period and oscillations by the positive sign of the first-measured period change (also the Crab). The last fundamentally wrong paper on the subject may well have been Trimble (1969) on the "Frequency of Events Producing Pulsars."

3. BINARY NEUTRON STARS

Herbert Gursky has described on several occasions the events surrounding the discovery of what we now call Sco X-1 (Giacconi et Italia 1962), including the first attempt at optical identification: "So we got a star atlas and opened it up...and discovered there were a lot of stars in the sky; so we closed it again." Not surprising for a 10 X $10°$ error box. Meanwhile, Hawakawa and Matsuoka (1964) considered the possibility that known classes of early type binaries might be detectable X-ray sources, "If the plasma stream ejected from one star hits the atmosphere of the other, a shock wave can be produced." Even their most optimistic stream velocity of 1000 km/sec achieved a luminosity of only 2 X 10^{34} erg/sec. The proposed mechanism is roughly that we now associate with X-rays from RS CVn stars and cataclysmic variables, though very few reach even this modest luminosity.

Successful optical identification of Sco X-1 (Sandage et al. 1966) followed closely upon an improvement of the X-ray position, the photometry and spectroscopy having been shared between Japanese and American observers in June-July 1966. The optical counterpart was summarized as having "certain of the properties of an old nova, even though its spectral characteristics cannot be identified with any one class of old nova." Bruno Rossi carried the news to Europe, to IAU Symposium 31 ("Radio Astronomy and the Galactic System") in Nordwijk, almost simultaneously with the 27 August submission of the identification paper.

At this point, relationships become both strained and numerous. The contemporary account of the discussion at Nordwijk (Burbidge 1967) mentions Rossi, Ginzburg, Shklovskii, Woltjer, himself and other (unnamed) participants. Strong connections with old novae were assumed. Discussion focussed on energy sources, especially gravitational energy, either "the gravitational energy of the binary system [or] we must derive the energy from the internal energy of one or the other of the stars...one might suppose that the highly evolved component of the binary system is a neutron star, and that the internal energy of this star is being slowly released in the form of high-velocity gas." (It is worth recalling that informed opinion by then recognized, largely on the basis of work by R.P. Kraft, that old novae were binaries with white dwarf components). Ginzburg is quoted as saying, "...the potential and kinetic energy of the double system is of the order of 10^{49} erg. This is an enormous amount, and we must find ways to use it for maintaining the X-ray emission. Tidal effects may produce streams of matter or stellar winds, and they may hit the surface of a star or produce a transient atmosphere." Notice first that the proposed energy reservoir will keep a bright XRB going only 3000 yr, and second that the description of the hypothetical systems seems to have all the right pieces, but put together in slightly the wrong order. There are no comments from Shklovskii in the published proceedings, and no one else comes closer than Ginzburg to hitting the mark.

Nevertheless, the invention of X-ray binaries is nearly always attributed to Shklovskii (1967). The Russian and English version of the paper are not identical (the latter was the first-ever ApJ Letter published separately from the main Journal). The Russian one was submitted slightly later, and the quotes are from it (in translation).

Shklovskii envisions "a neutron star forming a comparatively massive component of a close binary system. A stream of gas flowing out of the second component is permanently incident on the neutron star... It is suggested that the optical object accompanying the x-ray source might be a cool dwarf star, with half of its surface heated by a strong flux of hard x rays from the source." Footnote 3 reports "The hypothesis that the source Sco X-1 is a close binary system, one of whose components is a neutron star, with the hot plasma in the neighborhood of the neutron star being formed through accretion of a gas stream flowing out of the normal component, was formulated by us during discussions at the Nordwijk syposium on August 28, 1966."

It is impossible to be sure whether "we" is a euphemism for "me" (implying that he had sat quietly thinking while the others were all talking around the correct concept) or is meant to be a sharing of credit with other participants. Burbidge (1972) retelling the story 5 years later, recalls additional participants (Savedoff, M. Burbidge Prendergast, and Herbig), but sheds no further light on the evolution of the model, again saying that "Dissipation of rotational energy of the system" was the source contemplated. Ginzburg's (1990) still later recollections add no new information, and he does Shklovskii something of an injustice by ignoring the footnote about Nordwijk.

Most remarkable of all is a paper (Novikov & Zeldovich 1966) submitted in late 1965 and accepted on 16 April 1966. One of three footnotes added in proof declares, "The baryon or hardened star can be a component of a double star. In this case it can be discovered by the perturbation of the other, normal star...In a double system, the accretion on the collapsed star is felt by the stellar wind from the normal star. It shall led (sic) to strong X-ray and γ-ray emission. In the case of accretion the falling gas is heated by a stationary shock wave and radiates like an optically thin hot layer (brehmsstralung) instead of the black-body radiation. The last investigations of Scorpius XR1 seem to be in accordance with this picture." Clearly they corrected their proofs some time after 16 April (and the journal issue did not reach library shelves until the end of the year), but it is impossible to be sure whether they had heard of either the optical identification or of the Nordwijk discussion when those words were written.

In either case, Zeldovich and his associates were in the best possible position to appreciate the significance of close binaries with compact components, for they had already begun a search (Guseynov & Zeldovich 1966; Zeldovich & Guseynov 1966) for "Collapsed Stars in Binaries" (meaning primarily black holes). The strategy was to examine known single-line spectroscopic binaries for ones where the orbit paramaeters suggested an unseen star more massive than the visible one. They found seven candidates for which "the hypothesis is put forward that the second unobserved star is a collapsed star, or, as in case 7, an old neutron star." X-ray emission is not mentioned, indicating that Sco X-1 had not yet come to their attention in November 1965. None of their candidates is now thought to harbor a compact component.

A few years later, Trimble and Thorne (1969), examining a much larger catalog of SBs, had a clearer idea of what to look for and summarized the situation as, "The absence of a secondary spectrum in these systems could, in principle, result from the secondary star's being either a collapsed star or a massive neutron star. For all these systems, however, other explanations are possible in the light of present observation. Statistical considerations suggest that few, if any, of the systems in these lists contain collapsed or neutron-star secondaries. None of these binary systems coincide with any published X-ray position." Given the size of pre-Uhuru error circles, this in itself borders on the surprising. Every couple of years, a serious binary star astronomer turns attention on one of these systems and finds that, indeed, other explanations are preferable, e.g. Wonnacutt et al. 1993 on IK Peg.

The last wrong paper on X-ray binaries seems to have been Cameron and Mock (1967) who preferred white dwarf accretors, on the grounds that gas hitting a neutron star would produce photons of excessive hardness (at one photon per protron they do, but thermalized over the surface at Eddington luminosity or lower, they do not)

4. BLACK HOLES, ACCRETION DISKS, AND GLOBULAR CLUSTER SOURCES

The first black hole "candidate" (see A.P. Cowley elsewhere in this volume) was Cyg X-1 (Gursky et al. 1971) and its optical identification HDE 226868 (Bolton 1972). The X-ray observers initially put it at about a kpc from us, on the basis of low-energy absorption and average ISM density, and Bolton at 2 kpc on the assumption that the visible star was a reasonably normal OB supergiant. He was, of course, right, but, as Cowley notes, the mass function for the system is relatively small, and the resulting compact mass therefore exceedingly dependent upon the assumed M_1. Thus arose the last fundamentally wrong paper on this subject (Trimble, Rose & Weber 1972), "A Low-Mass Primary for Cygnus X-1?", propounding the curious notion that "If the primary of HDE 226868 (Cyg X-1) is a low mass (0.3 = 0.5 $M_\odot$), low surface gravity B star of the type of which the primary of HZ 22 is the prototype, then the secondary falls well within the mass range of stable neutron stars (and white dwarfs) and need not be a black hole." We were last because observers flocked instantly to their telescopes to trace out the reddening of HDE 226868 and the velocities of interstellar absorption lines in its spectrum, thereby firmly placing it at the larger distance and ruling out a low mass (and low luminosity) primary. This is the only paper with which the author has ever been involved that was in print before the postcard acknowledging receipt came. This accounts for the misspelling of Mal Ruderman's name in the text, for which we belatedly apologize to him.

The significance of accretion disks and accretion luminosity was first appreciated in the context of cataclysmic variables. Crawford and Kraft (1956) wrote of AE Aquarii, "If, in fact, as the foregoing considerations suggest, mass is accreted by the blue star at a rate of the order of 10^{25} g/year, we would expect that the kinetic energy of the tenuous infalling material would be dissipated into heat by viscosity in the turbulent motion and by impact with the denser matter of the star's atmosphere. The heat thus liberated should appear as a contribution to the luminosity of the blue star of about the amount $L = GM\sigma R^{-1}$," where σ is the accretion rate. With M and R appropriate to a white dwarf, they thereby account for the blue, nonstellar luminosity of the system. Accretion is explicity from a ring or disk where, "The velocity gradient will cause a very rapid turbulent exchange of angular momentum, so that the inner part of the ring will lose angular momentum and move in closer, while the outer part gains angular momentum and spreads out...and the net effect is that matter accretes onto the surface of the blue star, which acts as a sink for the angular momentum as wel." The rate of mass transfer is found independently from a model of the expansion of the K donor and from recombination of hydrogen. That all three agree the authors (righly) regard as strong support for the model. We can only regret that neither Crawford nor Kraft was at Nordwijk!

Application of spherical accretion to the powering of active galaxies came later (Salpeter 1964; Zeldovich & Novikov 1964) and disk accretion in that context (Lynden-Bell 1969) and for X-ray binaries (Prendergast and Burbidge 1968) still later.

One globular cluster X-ray source predates the Uhuru catalog. A rocket flown by NRL on 25 April 1965 recorded a source which they called SGR XR-4. Called L16 by the Lockheed group, whose rocket flew on 30 September 1965, it is now catalogued as 4U 1820-30 and sits near the center of NGC 6624, though no one knew this for several years thereafter (Seward 1970 and pr. comm.). The 3U catalog, submitted in August 1973, included two firm and one possible globular cluster IDs. Who should be credited with first noticing that this is far more than the clusters' fair share is nearly lost in the mists of folklore. E. van den Heuvel (elsewhere in this volume) remembers hearing it from H. Gursky at a July 1973 meeting in Cambridge UK (Physics and Astrophysics of Compact Objects). According to my lengthy and still extant notes from that meeting, Gursky's public talks discussed properties of a number of individual sources, but not the globular cluster ones, while J. Ostriker remarked that there seemed to be X-ray sources in globulars, but did not mention an excess (He also asked rhetorically whether the clusters contain U Gem stars, which have similar orbit periods to the XRBs then known, an issue not fully resolved to the present time, J. Grindlay elsewhere in this volume). In his remarks here, Gursky expressed doubts about ever having gotten around to publishing the point. It appears, however, in the printed version of a February 1974 talk (Gursky & Schreier 1975). At least three other sets of notes from the 1973 meeting probably exist, and there is, therefore, some hope still of determining whether this bit of history now incorporates a Dyson error of the first kind or of the second!

5. AFTERTHOUGHTS

The oral version of this presentation decribed the historical period under consideration as Paleolithic, or pre-Dutch. But in fact the scenario for X-ray binary formation described by van den Heuvel in   1973 used as one input statistics of main sequence binaries compiled by Kuiper (1935). There was, in other words, no pre-Dutch period. My notes from that talk include the gloss "what is this for Batten's Catalogue" next to Ed's choice of mass-ratio distribution. The answer (Trimble 1974) disagreed with what Kuiper had found, and was the precursor of a binary-system-mass-ratio-statistics cottage industry that now produces about a dozen papers per year. And I was probably wrong again (Duquennoy & Mayor 1991)

REFERENCES

Baade, W. & F. Zwicky 1934. Proc. NAS 20, 263, and PR 45, 138
Bolton, C.T. 1972. Nat. 235, 271, and Nat. Phys. Sci. 240, 124
Burbidge, G.R. 1967. IAU Symp. 31, P. 465ff
Burbidge, G.R. 1972. Comm. Astrophys. Space Sci. 4, 105
Cameron, A.G.W. & M. Mock 1967. Nat. 215, 464
Chandrasekhar, S. 1931. ApJ 74, 81
Crawford, J.A. & R.P. Kraft 1956. ApJ 123, 44
Duquennoy, A. & M. Mayor 1991. A&A 248, 485
Frenkel, Y. 1928. Zs. f. Ap. 50, 234
Gamow, G. 1937. The Structure of Atomic Nuclei and Nuclear Transformations. Oxford Univ. Press. p. 234-8
Gamow, G. & E. Teller 1938. PR 53, 929
Giacconi, R. et al. 1962. PRL 9, 440
Ginzburg. V.I. 1990. ARA&A 28, 31
Gursky, H. & E. Schreier 1975. in H. Gursky & R. Ruffini (Eds.), Neutron Stars, Black Holes, and Binary X-Ray Sources (Reidel), p. 163
Gursky, H. et al. 1971. ApJ 167, L15
Guseynov, O. Kh. & Ya. B. Zeldovich 1966. AZh 43, 313
Hayakawa, S. & M. Matsuoka 1964. Prog. Theor. Phys. Sup. 20, 218
Israel, W. 1987. in S.W. Hawking & W. Israel (Eds.) 300 Years of Gravitation. Cambridge University Press, p. 276
Kuiper, G. 1935. PASP 47, 38
Landau, L. 1932. Zs. f. Sov. Phys. 5, 285
Landau, L. 1938. Nat. 141, 333
Lynden-Bell, D. 1969. Nat. 223, 610
Milne, E. 1931. MNRAS 91, 54
Novikov, I.D. & Ya. B. Zeldovich 1966. Nuovo Cim. Sup. 4, 827
Oppenheimer, J.R. & G.M. Volkovv 1939. PR 55, 375
Pacini, F. 1967. Nat. 216, 567
Prendergast, K. & G.R. Burbidge 1968. ApJ 111, L53
Rosenfeld, L. 1973. Seizieme Conseil de Physique Solvay (Stoops, Grussels), p. 174
Salpeter, E.E. 1964. ApJ 140, 796
Sandage, A.R. et al. 1966. ApJ 146, 321
Seward, F.D. 1970. Illustrated Catalogue of Cosmic X-ray Sources
Shklovskii, I.S. 1967. AZh 44, 930 (Sov. AJ 11, 749) and ApJ 148, L1
Sterne, T.E. 1933. MNRAS 93, 750
Trimble, V. 1969. Nat. 221, 1038
Trimble, V. 1974. AJ 79. 967
Trimble, V.L., W.K. Rose, & J. Weber 1973. MNRAS 162, Lp
Trimble, V.L. & K.S. Thorne 1969. ApJ 156, 1013
Woltjer, L. 1964. ApJ 140, 1312
Zeldovich, Ya. B. & O. Kh. Guseynov 1966. ApJ 144, 840
Zeldovich, Ya. B. & I.D. Novikov 1964. DAN 155, 67 & 158, 811

X-RAY BINARIES: WHAT PROGRESS HAVE WE MADE?

Julian H. Krolik
Department of Physics and Astronomy, Johns Hopkins University
Baltimore MD 21218
jhk@gauss.pha.jhu.edu

ABSTRACT

A summary is given of progress reported at this meeting on six basic questions having to do with X-ray binaries: which stellar systems create them; what drives mass transfer; the nature of the accretion disk, if one exists; how the X-rays are made; neutron star life-cycles; and analogous systems in other galaxies.

After a comprehensive conference such as this one, it's appropriate to step back for a moment and ask just how far we have come. In the rest of this review, I will present an evaluation (inevitably idiosyncratic and superficial) of how much we actually know about X-ray binaries, as represented by the reports given at this meeting. It will be organized according to six basic questions which together span much of the essential information we should have in hand before we can legitimately claim to understand this subject.

1. WHICH BINARIES BECOME X-RAY SOURCES?

Perhaps the single most fundamental question we may ask is why X-ray binaries are made at all. This amounts to investigating why it is that certain stellar binaries evolve into X-ray sources, and others do not. After more than twenty years of work, we are hardly without ideas, and indeed there is a well-established conventional story to answer this question.

According to this story, X-ray binaries in which the companion is a high mass star arise in Population I, while those with low mass companions are created predominantly in Population II. In Population I, several cuts on the complete binary population separate those destined to become X-ray sources from those which will not. First, of course, at least one of the members must be massive enough to leave either a neutron star or black hole remnant. Second, the binary must be close enough for the neutron star or black hole to capture a significant amount of mass lost from the secondary (see §2). Third, because sudden loss of more than half a binary's mass would result in disruption of the binary, it is essential in the high mass systems for there to be significant mass transfer from the initially more massive star to the initially less massive before the first star explodes as a supernova (van den Heuvel and Heise 1972).

The conventional picture is less clear about the low mass systems. It has been known for twenty years (Clark 1975, Katz 1975; but see also the historical comments in Trimble's article in these proceedings) that Population II is more efficient than Population I at producing X-ray binaries (measured per unit stellar mass) by one to two orders of magnitude, while globular clusters are yet another order of magnitude more efficient than the rest of Population II. While the conventional story can explain the fertility of globular clusters by tidal capture and binary exchange interactions (Fabian, Pringle, and Rees 1975), it has no ready answer to the question of why the rest of the Galactic bulge produces X-ray sources with such fecundity.

Clearly, this conventional story raises at least as many questions as it answers. In Population I, can we ascertain quantitatively what initial masses, orbital periods, and eccentricities result in eventual X-ray emission? What is the birthrate of such systems? When an accretion binary is made, under what circumstances is the accreting member of the pair a neutron star, a black hole, or a white dwarf? In the case of Population II, what explains the formation of many X-ray binaries outside globular clusters? Can accretion induced collapse create neutron stars long after the massive stars are gone? For both populations, what is the effect of supernova dynamics on the remnant binary? That is, if supernovæare not perfectly symmetric, which binaries survive, and in what form?

Several interesting new results bearing on these questions were presented in the three days of this meeting. Perhaps the most interesting of these is the first Galactic map of accreting black hole candidates (White, these proceedings). Choosing black hole candidates by their X-ray spectra (looking for a combination of a thermal component with $T \simeq 1 - 2$ keV and a hard power-law component), White found several dozen candidates. Their galactic distribution is distinctly that of Population I: a definite concentration toward the Galactic plane, and no concentration toward the Galactic Center. If these are indeed black holes, the question immediately arises: why is it that Population II is much more efficient at creating neutron star X-ray binaries, while Population I has at least an equal edge at producing black hole X-ray binaries? Virtually all the mechanisms discussed for the creation of accretion binaries provide at most only weak distinctions between neutron stars and black holes. Many of these (in particular, many of those thought to be relevant to globular clusters, such as tidal capture or binary exchanges) distinguish between the two sorts of collapsed stars only by mass, and, if anything, favor the black holes, which are more massive.

Ron Webbink reported some new steps toward making X-ray binary evolution a predictive science, rather than one in which creation myths are invented *post hoc*. If we are to understand which binary systems evolve into X-ray binaries, it will be necessary for us to be able to calculate forward in time from a given initial stellar binary, and predict which ones evolve into a binary containing a neutron star or a black hole able to accrete from its companion. Webbink showed how to do a part of this exercise, computing the consequences for close binaries of supernova explosions whose mass loss and asymmetries are parameterized. Unfortunately, in the application of these ideas to low mass X-ray binary progenitors outside globular clusters, it appears necessary to go through a common envelope stage; their dynamics are so difficult to understand that this crucial step is likely to remain a gap in our understanding for quite a while.

We are also beginning to learn about the population of accreting white dwarfs. They are, of course, much harder to see, for their luminosities are ~ 100 times smaller than those of accreting neutron stars or black holes. In addition, the characteristic energy of the escaping photons is generally much lower for they also have much larger radiating areas. Nonetheless, recent surveys with greater effective area, lower background, and sensitivity to lower energy photons have succeeded in discovering sizable numbers of systems which are most likely explained as accreting white dwarfs. A few of these "Super Soft Sources" had been found in the Magellanic Clouds in observations using the *Einstein Observatory*; *Rosat* now sees many dozens in M31 (Grindlay, these proceedings; Hasinger, these proceedings). In the long run, we may hope to define surveys with sufficient statistical power to calibrate stellar evolution pictures predicting the relative numbers of accretion binaries with white dwarfs and neutron stars.

Where does this leave us? While we do seem to have a firm grasp on several qualitative points (*e.g.* massive binaries with sufficient mass transfer can create high-mass X-ray binaries), we are still missing several crucial ingredients that are necessary to give us true *predictive* power. In the context of the high-mass systems, there are two big gaps: a quantitative theory of intrinsic stellar mass loss (*e.g.* in radiation pressure-driven winds), and a similarly quantitative command of the degree of asymmetry in supernova explosions. If these two gaps were filled, it would be possible to follow the evolution of a given binary forward from its creation, and successfully predict whether it would ultimately become an X-ray source.

In the case of the low-mass systems, we are in a similar condition, but for different reasons. In globular clusters, tidal captures and exchange interactions provide a plausible way of creating sufficient numbers of neutron star–main sequence star close binaries. Outside of globular clusters, common envelope evolution provides a potential mechanism for bring a neutron star close to its future donor, but transforming this qualitative thought into quantitative predictions remains elusive.

2. WHAT DRIVES MASS TRANSFER?

After learning which binaries will ultimately spend time transferring mass from one star to a collapsed companion, we would next like to know what controls the rate at which mass is transferred. Once again there is a classical picture, and just as before, it has two branches. In Option A the companion loses mass by Roche lobe overflow, while in Option B the collapsed star accretes from a wind generated by the companion (Davidson and Ostriker 1973). Generally speaking (though not without exceptions), Option A is thought to operate in low-mass systems, and Option B in the high-mass case. In both cases, however, the dynamics which cause the mass transfer are either unknown or controversial. In the former case, is the Roche lobe overflow due to orbital shrinkage or stellar expansion? If orbital shrinkage, what is the mechanism causing the orbit to lose angular momentum—gravitational radiation (first proposed for white dwarf binaries by Kraft, Mathews, & Greenstein 1962; for neutron star binaries by Paczyński and Sienkiewicz 1981; and Rappaport, Joss, & Webbink 1981), magnetic braking (again, first proposed for white dwarf binaries: Huang 1966 and Mestel 1968; initially applied to neutron star binaries by Verbunt & Zwaan 1981), or something else entirely? In the latter case (a stellar wind), is the mass-loss driven by intrinsic mechanisms (*e.g.* line radiation pressure, or the thermal expansion of the red giant phase) or by extrinsic influences (*e.g.* X-ray heating)?

While first suggested many years ago (Alme and Wilson 1974, Arons 1973), renewed attention (Ruderman *et al.* 1989) has been focussed on the last possibility (but in the context of *low-mass* binaries) by the discovery of a number of radio pulsar systems in which the companion is losing mass at a significant rate (*e.g.* Fruchter *et al.* 1990; Nice *et al.* 1990). Ironically, while most of the theories invented to explain this mass-loss invoke either X-rays or γ-rays as the heating agents, in all the examples currently known, the only radiation definitely detected from the neutron star is in the radio band. It remains possible that self-excited winds exist in neutron star binaries, but not in those making X-rays (though Day and Stevens 1993 suggest Cen X-3 may actually be an example of this process in an X-ray binary).

On the other hand, there has been some genuine progress in the area of more conventional mass transfer schemes in high-mass systems. High mass X-ray binaries appear to divide fairly cleanly into two classes: those in which the companion is an OB supergiant driving a strong wind entirely by its own radiation presure, and those in which the companion is a Be star expelling mass by a combination of rotation and radiation. Particularly striking results relevant to the former class were presented by John Blondin.

It has been known for quite a long time that there may be strong feedback between wind-fed X-ray sources and their accretion flows. Both the mass loss rate in the wind and its terminal velocity are controlled by the force exerted on the wind gas by a long list of ionic resonance lines scattering the stellar UV continuum (Castor, Abbot, & Klein 1976; Kudritzki et al. 1989). Whereas the mass loss rate is determined in a rather complicated way quite close to the stellar photosphere, the terminal velocity is set by continuing acceleration at considerably greater distances from the star. X-ray photoionization can influence the terminal speed in a very direct way by controlling the ionization balance in the wind (Hatchett and McCray 1977). Generally speaking, increasing photoionization decreases the wind opacity, and therefore the radiative acceleration. Because the ability of the neutron star to capture material from the wind decreases rapidly with increasing relative velocity between the neutron star and the wind, there is a strong positive feedback between the strength of the X-ray emission at any particular time and the X-ray luminosity an accretion time later (*e.g.* Ho and Arons 1987 found linearly growing waves driven by this effect).

Blondin et al. (1990) presented the first 2-d time-dependent hydrodynamical simulations studying these effects; at this meeting, Blondin reported new simulations in full 3-d. These simulations clearly showed a very strong interaction of just the sort predicted. While the results shown are still quite preliminary, the general program will clearly have a major impact on our understanding of how accretion works in these systems. For example, it is likely that the X-ray luminosity–accretion rate feedback loop will have significant consequences for the spin history of the neutron star (see §3).

The situation with regard to low-mass systems has been almost static for ten years now. While magnetic braking is a plausible angular momentum loss mechanism, we have no physical theory for it. All we know how to do at the moment is derive empirical scaling laws from isolated stars with magnetized winds. Thus, in this respect, progress in understanding accretion binaries depends on progress in understanding the mechanics of ordinary stars.

3. WHAT HAPPENS TO THE TRANSFERRED MASS? OR, WHAT IS AN ACCRETION DISK?

If the transferred mass has been delivered by Roche lobe overflow, then its specific angular momentum with respect to the neutron star is, to order of magnitude, the binary orbital angular momentum, which is far too great to allow quick accretion onto the neutron star. Because the matter can easily lose energy by radiation, it is forced to settle down into a flattened configuration orbiting the neutron star at a fraction of the orbital separation. After many orbital periods, the material spreads in radius, as angular momentum is transported outward and matter drifts inward. The result is, of course, a classical accretion disk as originally described by Prendergast & Burbidge (1968), Shakura & Sunyaev (1973), and Lynden-Bell & Pringle (1974).

Depending on circumstance, one of several mechanisms defines the inner boundary of the disk: collision with the physical surface of the neutron star (or white dwarf); magnetic stress if the collapsed star has a sufficiently strong magnetic field; or relativistic orbital mechanics when the center of the disk is occupied by a black hole.

When the mass transfer arises from a stellar wind, the situation looks rather different. If the wind density and velocity did not vary across the neutron star's accretion radius r_{acc}, the accreted matter's net specific angular momentum j (with respect to the neutron star) is only $\sim j_{orb}(r_{acc}/a)^2 \sim 10^{-3} j_{orb}$, where j_{orb} is the specific angular momentum of the neutron star's orbit, r_{acc} is the accretion radius, and a is the orbital separation (Shapiro & Lightman 1976). To zeroth order, accretion would proceed via a Bondi-Hoyle accretion column trailing downstream from the neutron star. If one of the barriers listed in the preceding paragraph were to interrupt the flow at a distance at least $10^{-6}(M_c/M_{ns})a$ from the neutron star, there would be no disk at all.

Real life is, of course, more complicated. Real stellar winds are generally strongly inhomogeneous. Even a modest asymmetry relative to the neutron star can give the accretion flow a net angular momentum which is at least comparable to that required for a Keplerian orbit at a radius defined by any of the possible inner disk interruption mechanisms. In addition, it is possible that even in homogeneous Bondi-Hoyle accretion the column is unstable to flapping back and forth (Matsuda, Inoue, & Sawada 1987; Taam & Fryxell 1988), so that the instantaneous angular momentum of the accreted material is significant. Consequently, we can expect that even when the magnetic field is strong enough to interrupt the disk at fairly large radius, a disk will usually form in this case as well.

The internal mechanics of disks are still, to use the scientific diplomatic code-phrase, "a subject of active research". We can easily point out many effects which may be important but whose influence we cannot reliably evaluate. If the gas supply has a net angular momentum whose direction changes on a timescale short compared to the time required for material to make its way through the disk (as is quite likely when accreting from a wind), the disk will contain warps and twists (Petterson 1975). The temperature of the disk is determined by its ability to radiate away the heat created by dissipative angular momentum transport. This, in turn, is controlled by radiation transfer effects in the disk atmosphere which can sometimes be quite complicated. If the disk flares sufficiently for X-rays to irradiate its surface, a corona may form, and substantial mass may be driven off (White and Holt 1982; Begelman, McKee, & Shields 1983). If the inner edge is defined by magnetic stresses, just how well does the disk couple to the stellar magnetosphere (Ghosh & Lamb 1977, Ghosh in these proceedings)? On the other hand, if the central object is a black hole, how fast does material move radially inward in the vicinity of the marginally stable orbit?

The most critical lacuna in our knowledge of accretion disk physics is the absence of a theory for angular momentum transport. Twenty years ago Shakura & Sunyaev (1973) estimated by dimensional analysis the magnitude of the torque required to drive the accretion rates we see, and also showed that ordinary viscosity is far too weak to explain it. Since then, many ideas have been proposed to explain the angular momentum transport, but only in the past two years has a suggestion been put forward which appears to be grounded in comprehensible physics—the MHD instability whose significance to accretion disks was first pointed out in the pair of papers Balbus & Hawley (1991), Hawley & Balbus (1991). If Balbus and Hawley are correct, any weak poloidal field initially present in the disk grows radial bulges (exponentially amplified on roughly the orbital timescale) which immediately begin to create a torque in the right direction.

While basic physics issues such as the nature of angular momentum transport in the disk were not discussed at the meeting, other disk phenomena did receive much attention at the meeting. In these areas there is also much promise for rapid progress in the near future. Our power to map the physical state of disks is growing rapidly, due to such techniques as "eclipse-mapping" and allied arts (Rutten, these proceedings). By making use of precise orbital ephemerides, we can infer the progress of the companion's shadow line across the disk. It is already possible to obtain color and emission line maps of the disk surface, and in some circumstances, the illuminated face of the companion star; it may be possible in the future (if we but knew how to make the bolometric corrections) to use this method to return to fundamental issues, such as mapping the accretion rate as a function of radius within the disk.

Similarly rapid progress may be expected from coupling the data now streaming out of ASCA (for samples, see Tanaka in these proceedings) with photoionization models (Kallman, these proceedings). On the one hand, X-ray spectroscopy has now arrived at such a state that high signal/noise measurements may be made of numerous X-ray lines; on the other, the computing power sitting on the average astronomer's desk allows extremely detailed and physically accurate modelling to be done with comparatively easily-used codes (*e.g.* the recently distributed XSTAR: Kallman and Krolik 1993).

However, there is one important area of disk studies in which the passage of time seems only to have made the problems more acute: the nature of X-ray pulsar spin evolution. Once again, qualitative suggestions made very early on have provided a framework for nearly everyone's thinking, but advances beyond that level have been difficult.

When X-ray pulsars accrete from a disk whose inner edge is determined by the edge of the neutron star's magnetosphere, it is easy to estimate the characteristic spin-up rate because it is simply given by supposing that when the accreting matter reaches the edge of the magnetosphere, all its angular momentum is delivered to the neutron star (Pringle & Rees 1972; Lamb, Pethick, & Pines 1973). If one guesses that all neutron stars have magnetic dipole moments $\sim 10^{30}$ G cm^3, the relation between mean luminosity, spin period P, and mean $\dot{P}$ predicted by this estimate is confirmed by the data for many X-ray pulsars, particularly that subset for which $\dot{P}$ is nearly constant (Joss and Rappaport 1984).

Problems appear when one takes a closer look. First, there are a few objects with essentially constant $\dot{P}$ (*e.g.* Her X-1 and LMC X-4) whose values of $\dot{P}$ are much smaller than would be predicted by this simple scaling. Second, there are some others (*e.g.* Vela X-1 and GX 1+4) for which a good fit is obtained only for magnetic moments as large as $\sim 10^{31} - 10^{32}$ G cm^3 (Ghosh, these proceedings). It may or may not be significant that $\dot{P}$ is far from constant in both these sources. Third, there are a number of pulsars for which $P/\dot{P}$ is as small as a few decades, and yet have rather long spin periods, > 100s (Fabian 1975; Bildsten, these proceedings). Fourth, $\dot{P}$ fluctuates, and in some cases goes negative (Fabbiano and Schreier 1977; Ögelman et al. 1977; Prince, these proceedings; Ghosh, these proceedings). This point is particularly troubling because the fluctuations in $\dot{P}$ can be correlated over timescales of many years, far longer than the dynamical timescale for the companion's wind, or the likely residence time of material in the accretion disk. Fifth, while this theory predicts a definite relation between $\dot{P}$ and luminosity for individual objects, it is unclear how well they adhere to this relation. GX1+4, for example, had roughly the same luminosity in 1980 and 1992, but $\dot{P}(1992) = -\dot{P}(1980)$ (Prince, these proceedings).

Some of these difficulties may be solved by the theory of Ghosh and Lamb (1978, 1979; Ghosh, these proceedings), in which the disk is thought to penetrate far enough into the magnetosphere to possess regions exerting both signs of torque. In this picture, the objects with small $\dot{P}$ are cases in which the two contributions nearly balance. Their answer to the problem of small spin-up times and long spin periods is to point out that high accretion rates produce both high luminosity and rapid spin-up, while low accretion rates produce dim sources and either slower spin-up or, in extreme cases, spin-down. Consequently, we can only see the sources when $\dot{P}/P$ is disturbingly large; a sufficiently small duty cycle for high luminosity would then allow P to stay in its observed range.

Unfortunately, this theory does not solve all the problems, and its solutions are not without criticism. The balance between positive and negative torque depends on a number of plasma transport processes which are no better understood today than when the theory was proposed fifteen years ago. In those objects for which a very large magnetic moment is inferred, there are no other signs indicating such a strong field. Our only other indicators of field strength are cyclotron line features and the exponential break in the continuum thought to be due to a cyclotron thermostat (see §4). However, these indicators are spread over only a factor of several, not over a range of two orders of magnitude. The hardest known spectrum in an X-ray pulsar is found in GX1+4, whose cut-off energy is $\simeq 50 - 100$ keV (Mony et al. 1991; Prince, these proceedings), not ~ 1 MeV, as would be required to explain its spin-up rate by this theory. While many X-ray pulsars are indeed transient, there is as yet too little information on either the duty cycle of high luminosity, or the actual value of $\dot{P}$ during periods of low luminosity, to verify the duty cycle suggestion.

Whether or not we understand the dynamics by which the current accretion rate influences the current spin-up rate, we have as yet no explanation for what causes $\dot{P}$ to fluctuate on timescales of years. Older data (Deeter et al. 1987) indicated that the power spectrum of $\dot{P}$ fluctuations was well fit by white noise over more than six orders of magnitude in frequency. This was, in a sense, the "expected" result because it corresponded to uncorrelated fluctuations with characteristic timescale a few hours, i.e. the dynamical time for the companion's wind. However, new BATSE data shows at least two examples (GX1+4 and 4U1626-67) in which the noise spectrum is clearly much "redder", i.e. the fluctuations' autocorrelation has a width of at least several years. This is very surprising because the only dynamical timescales in these systems of which we are aware are many orders of magnitude shorter. It is possible that there are dynamical processes in the mass-donor star which fluctuate on multi-year timescales, or that there are feedback loops in the accretion process which stabilize the sign of the accreted angular momentum over times much longer than the basic timescale. In the case of GX1+4, because the companion is a red giant, some believe its mass loss is driven by Roche lobe overflow. However, we know of no dynamical mechanism which would modulate the orbital angular momentum loss rate on a timescale of years. Moreover, in this system we have only a lower bound on the orbital period. It is therefore possible that the orbital separation is large enough that the companion does not fill its Roche lobe, and wind dynamics may be of comparable importance to gravitational dynamics.

Still another thorny issue involving disk dynamics and disk–magnetosphere coupling (Alpar and Shaham 1985; Lamb et al. 1985) is raised by the existence of QPO's (van der Klis, these proceedings). Interestingly, van der Klis has been able to find a single phenomenological framework linking the aperiodic variability properties of X-ray pulsars, low mass X-ray binaries, and black hole candidates. This link is made plausible by the fact that the accretion disk penetrates very deep into the central object's gravitational well in both Population II neutron star binaries (because the magnetic field is relatively weak), and in black hole systems (because there is no stationary magnetic field at all).

Thus, while the diagnostic power we possess for accretion disks is growing rapidly, our understanding of their dynamics remains quite weak. We now have some hope that the problem of angular momentum transport inside accretion disks may be one its way to a solution; however, there are still many open questions having to do with how much of the accreted matter's angular momentum is delivered to the neutron star.

5. HOW ARE THE X-RAYS MADE?

That the energy for X-ray emission in binary systems is derived from accretion onto neutron stars or black holes has been conventional wisdom since the late 1960's (Shklovsky 1967; but see Trimble, these proceedings, for a more detailed account of the origin of this idea). If a neutron star radiates as a black body over its entire surface, the resulting temperature is

$$T_{bb} \simeq 2(M/1M_\odot)^{-1/4}(L/L_E)^{1/4} \text{keV},$$

where L/L_E is, as usual, the luminosity in Eddington units. Because the surface area of the innermost rings of an accretion disk around a few $M_\odot$ black hole is not too different from the surface area of a neutron star, the same estimate applies to black hole accretion; the only difference is that, by definition, we only consider a system to be a good candidate for "black hole-hood" if its mass is rather greater than that of a neutron star.

In fact, the simple thermal picture is not a bad first approximation both for low mass X-ray binaries and for the soft component in the spectra of black hole systems. However, the spectral shape it predicts is grossly wrong both for high mass X-ray binaries and for the hard component in black hole systems. High mass systems generally seem to have spectra better described by the product of a power law and an exponential in which the characteristic energy of the exponential is $\simeq 20 - 30$ keV, while the fluctuating hard component radiated by black hole systems is closer to a pure power law all the way from energies of a few keV to several hundred keV. In both of these cases, clearly the matter and radiation are not so closely coupled thermodynamically, and detailed questions about the radiation mechanisms become relevant.

By the evidence of this meeting, this is a subject where the quality of observations continues to outstrip the quality of theoretical understanding. Parmar (these proceedings), for example, reported striking changes in the pulse profiles of individual objects as their luminosity changed. There are also more and more examples of cyclotron absorption features (Mihara, these proceedings). In some black hole systems, there are transient 511 keV $e^{\pm}$ annihilation features (White, these proceedings).

Unfortunately, before we can get a grip on this spectroscopy, we have some genuinely hard physics problems to solve. Understanding the dynamics of the innermost rings of accretion disks around black holes is clearly a prerequisite to a better understanding of their spectra, for that is where most of the energy is released. Ebisawa (these proceedings) has demonstrated that it is possible to model the variable spectra of black hole candidates under the assumption that the innermost radius producing significant radiation does not change, while its temperature does. This is an intriguing speculation, but we surely need a much better account of the actual surface brightness distribution near the marginally stable orbit before making strong predictions about this matter.

The existence of annihilation lines makes it clear that perfect local thermal equilibrium cannot always apply in the accretion disks around black holes—at a temperature of 1 – 2 keV, the rate of production of $e^{\pm}$ pairs is minuscule. Possibly there is another region where the temperature is higher by a factor ~ 100; possibly there is another region with truly nonthermal plasma. In either case, we will need to find a way to transport a significant amount of energy away from the dense, optically thick, and therefore near-equilibrium immediate environs of the black hole to a more distant location where the density and optical depth can be small enough to allow substantial departures from thermal equilibrium.

Just as most of the energy in a black hole system is liberated in the last few orbits outside the radius of marginal stability, most of the energy released in accretion onto X-ray pulsars comes from the region close to the neutron star where the flow is decelerated. As Harding (these proceedings) discussed, when the luminosity is greater than $\sim 10^{36}$ erg s^{-1}, the deceleration is either completely mediated by, or at least strongly influenced by, the pressure of the outgoing X-rays. Thus there is automatically a strong coupling between the transfer of escaping radiation and the gas dynamics. To make matters worse, because the dominant photon-matter interaction is through the cyclotron resonance, which is strongly angle-, energy-, and polarization-dependent, this coupling is a very messy business (*cf.* Arons, Klein, & Lea 1987). Hints about the only computations ever done on this problem are given in Burnard, Arons, & Klein (1991).

Finally, it seems quite plausible that generating cyclotron emission lines provides a strong thermostat on the electron temperature, limiting it to $\sim 10(B/10^{12}\text{G})$ keV. However, demonstrating that in detail, and defining the shape of the electron distribution function for motion both parallel and perpendicular to the field, are jobs that remain to be attempted, much less accomplished. There are already observational hints (Parmar, these proceedings) that something of the sort is going on, for the characteristic energy of the exponential cut-off is generally about half the cyclotron energy, when a cyclotron feature exists. Unfortunately, there are complications to this story: the cut-off energy in a particular source can change by a factor of two; this can only be made consistent with the thermostat picture if the continuum photosphere can move far enough along the polar field lines. Only after these and related problems have been solved will we have the background information to begin the sort of atmosphere calculations which will be necessary for pinning down the formation mechanisms for the cyclotron lines and other X-ray pulsar spectral phenomena. Given the fact that these systems are interesting primarily because they emit X-rays, it is surprising that relatively little effort has been devoted to improving our understanding of just how these X-rays are made.

5. IS THERE LIFE AFTER X-RAYS? OR, THE HISTORY OF A NEUTRON STAR

In the discussion so far we have focussed exclusively on the portion of a neutron star's life during which it is an X-ray source. However, particularly for the high mass sources, this is a very small fraction of its lifetime. We would surely like to know what happens to these neutron stars both before and after their brief epoch of X-ray glory. In addition, the character of the neutron star's magnetic field has a strong influence on the nature of its X-ray emission, and to understand the field we must surely look to events prior to the beginning of accretion.

Several ways exist by which we may (approximately) infer neutron star field strengths. Radio pulsar spindown may be used to estimate the magnetic dipole moment (or, in the case of millisecond pulars, possibly higher multipole moments: Krolik 1991); if they are present, X-ray cyclotron lines directly measure the field in the region where the lines are made; and, as discussed in §3, the spin-up rate of X-ray pulsars indicates the magnetic dipole moment (and may also limit higher multipole moments: Arons 1993). In addition, in the case of low mass X-ray binaries, an upper bound may be put on the surface field by requiring that the accreted matter spread over a large enough portion of the neutron star surface to suppress pulses, and to permit explosive nuclear burning leading to X-ray bursts of the observed sort (Joss & Rappaport 1984). Crudely speaking, the results may be summarized as: all ordinary radio pulsars and at least some X-ray pulsars have surface fields $\sim 10^{12}$ G; millisecond radio pulsars have surface fields (counting only the dipole component) $\sim 10^8 - 10^9$ G; and low mass X-ray binaries have surface fields sufficiently weak and/or axisymmetric that the X-ray luminosity shows no sign of rotational modulation (*e.g.* the papers by Verbunt and by Kulkarni in these proceedings).

This last point may have repercussions beyond the issue of neutron star magnetic field evolution: if the magnetic field in low mass X-ray binaries cannot produce pulsations in the X-ray luminosity, one wonders how it is able to do so after accretion stops, and, as is widely believed, the neutron star becomes a millisecond pulsar. Fields so weak that they cannot channel the accretion flow will be weaker than the observed spindown rates indicate; fields so axisymmetric they produce negligible rotational modulation of the X-ray luminosity can hardly produce strong rotational modulation of the radio signal after accretion ceases.

Because the populations with weaker fields in this list tend to be associated with older stellar populations, there has long been a prejudice in the field that magnetic fields in neutron stars slowly decay. Likewise, the peaks in the distribution function near 10^{12} G and 10^8 G suggest there is something special about these field strengths. Unfortunately, any quantitative description of either the field distribution at neutron star birth, or its time evolution, is extremely controversial: as Verbunt's paper demonstrated, population evolution studies organized in very similar fashions can nonetheless reach strikingly different conclusions due to subtly different detailed choices (compare Narayan & Ostriker

1990 with Bhattacharya *et al.* 1992).

The long-term fate of neutron stars is a subject about which there is much speculation, but little in the way of genuine calculation. As Blaes (these proceedings) reported, there are a number of scenarios by which high mass X-ray binaries will eventually evolve into double neutron star, or neutron star-black hole systems. Some of these systems may eventually merge, emitting bursts of gravitational radiation, or possibly γ-ray bursts. Extrapolating from the three systems in our Galaxy whose merger time is less than a Hubble time, Narayan, Piran, & Shemi (1991) and Phinney (1991) arrived at similar estimates of the rate of such events: $\sim 10^{-6}$ per galaxy per year. This estimate, however, is extremely uncertain because it almost entirely depends on the one system with a particularly short predicted lifetime. In the case of low-mass systems, many outcomes are possible: the entire companion may be used up (or evaporated, as suggested by Ruderman *et al.* 1989); if the companion becomes degenerate, the orbit may expand to make the mass transfer rate fall asymptotically to zero (Rappaport, Joss & Webbink 1982); enough mass may be transferred to the neutron star to make it collapse to a black hole; or, if the binary is in a globular cluster, it may suffer an interaction in which the companion is smeared into a disk around the neutron star (Krolik, Meiksin, & Joss 1984; Krolik 1984).

6. ARE OTHER GALAXIES LIKE OURS?

Finally, as with almost any question in stellar astronomy, we would like to know whether our Galaxy is a fair sample of the rest of the galaxies in the Universe. Of course, at our present stage of technology, the only external galaxy we are able to study in detail is M31. Depending on whether one finds it more important to test the question of homogeneity of Hubble type, or to explore what may happen in different sorts of galaxies, it is either a happy or a sad coincidence that M31 has a stellar population rather similar to our own.

Despite this similarity, there are still difficulties in making the comparison. First, of course, for a fixed flux limit, we sample luminosities in M31 $\sim$ 2500 times greater than we do in our own Galaxy. Second, because our line of sight to M31 does not lie in its equatorial plane, we do not have to contend with strong soft X-ray absorption in its disk the way we must in our own Galaxy. On the other hand, the identification problem is far harder. As Supper (these proceedings) points out, we can expect 2/3 of those sources projected on M31 to actually be in the background!

Now that a deep Rosat image complements the Einstein image, progress is beginning to be made in this subject. It was heartening to learn from Supper that the X-ray luminosity function of M31 globular clusters is very similar to that of globular clusters in the Milky Way.

For the future, one possible avenue which has not yet been much explored is to search for X-ray pulsars by studying the Fourier power spectra of the X-ray arrival times in individual sources. Simulations suggest that if the amplitude of the Fourier component at the pulse period fundamental is as large as half the mean amplitude, as few as $\sim$ 300 photons suffice to discover it reliably in the power spectrum (Krolik & Griffiths 1994). However, the program with the greatest likely pay-off is the continuing effort to find optical identifications for M31 sources (Magnier, these proceedings). When that work is farther along, it should be possible to make detailed comparisons of the X-ray binary luminosity functions in Populations I and II of M31, and compare them to their local analogs.

ACKNOWLEDGEMENTS

This work was partially supported by NASA Grant NAGW-3129. I also wish to thank Steve Holt and the other organizers of this workshop for much hard work and careful thought.

REFERENCES

Alme, M.L. and Wilson, J.R. 1974, Ap.J. 194, 147
Alpar, M.A. and Shaham, J. 1985, Nature 316, 239
Arons, J. 1973, Ap.J. 184, 539
Arons, J. 1993, Ap.J. 408, 160
Arons, J., Klein, R.I., and Lea, S.M. 1987, Ap.J. 312, 666
Balbus, S.A. and Hawley, J.F. 1991, Ap.J. 376, 214
Begelman, M.C., McKee, C.F., and Shields, G.A. 1983, Ap.J. 271, 70
Bhattacharya, D., Wijers, R.A.M.J., Hartman, J.W., and Verbunt, F. 1992, A&A 254, 198
Blondin, J.M., Kallman, T.R., Fryxell, B. and Taam, R. 1990, Ap.J. 356, 591
Burnard, D., Arons, J., and Klein, R.I. 1991, Ap.J. 367, 575
Castor, J., Abbott, D., and Klein, R.I. 1975, Ap.J. 195, 157
Clark, G.W. 1975, Ap.J.Lett. 199, L143
Davidson, K. and Ostriker, J.P. 1973, Ap.J. 179, 588
Day, C.S.R. and Stevens, I.R. 1993, Ap.J. 403, 322
Deeter, J.E., Boynton, P.E., Lamb, F.K., and Zylstra, G. 1989, Ap.J. 336, 376
Fabbiano, G. and Schreier, E.J. 1977, Ap.J. 214, 235
Fabian, A.C. 1975, M.N.R.A.S. 173, 161
Fabian, A.C., Pringle, J.E., and Rees, M.J. 1975, M.N.R.A.S. 172, 15p
Fruchter, A.S., Berman, G., Bower, G., Convery, M., Goss, W.M., Hankins, T.H., Klein, J.R., Nice, D.J., Ryba, M.F., Stinebring, D.R., Taylor, J.H., Thorsett, S.E., and Weisberg, J.M. 1988, Ap.J. 351, 642
Ghosh, P. and Lamb, F.K. 1978, Ap.J.Lett. 223, L83
Ghosh, P. and Lamb, F.K. 1979, Ap.J. 234, 296
Hatchett, S.P. and McCray, R.M. 1977, Ap.J. 211, 552
Hawley, J.F. and Balbus, S.A. 1991, Ap.J. 376, 223
Ho, C. and Arons, J., 1987, Ap.J. 321, 404
Joss, P.C. and Rappaport, S.A. 1984, A.R.A.A. 22, 537
Kallman, T.R. and Krolik, J.H. 1993, NASA Internal Report
Katz, J.I. 1975, Nature 253, 698
Krolik, J.H. 1984, Ap.J. 282, 452
Krolik, J.H. 1991, Ap.J.Lett. 373, L69
Krolik, J.H. and Griffiths, R.E. 1994, in preparation
Krolik, J.H., Meiksin, A., and Joss, P.C. 1984, Ap.J. 282, 466
Kudritzki, R.P., Pauldrach, A., Puls, J., and Abbott, D.C. 1989, A&A 219, 205
Lamb, F.K., Pethick, C.J., and Pines, D. 1973, Ap.J. 184, 271
Lamb, F.K., Shibazaki, N., Alpar, M.A., and Shaham, J. 1985, Nature 317, 681
Lynden-Bell, D.O. and Pringle, J.E. 1974, M.N.R.A.S. 168, 603
Matsuda, T., Inoue, M., and Sawada, K. 1987, M.N.R.A.S. 226, 785
Mony, B., Kendziorra, E., Maisack, M., Staubert, R., Englhauser, J., Döbereiner, S., Pietsch, W., Reppin, C., Trümper, J., Churazov, E.M., Gil'fanov, M.R.,

Sunyaev, R.A. 1991, A&A 247, 405
Narayan, R. and Ostriker, J.P. 1990, Ap.J. 352, 222
Narayan, R., Piran, T., and Shemi, A. 1991, Ap.J.Lett. 379, L17
Nice, D.J., Thorsett, S.E., Taylor, J.H., and Fruchter, A.S. 1990, Ap.J.Lett. 361, L61
Ögelman, J., Beuermann, K.P., Kanbach, G., Mayer-Hasselwander, H.A., Capozzi, D., Fiordilino, E., and Molteni, D. 1977, A.&A. 58, 385
Petterson, J. 1975 Ap.J.Lett. 201, L61
Phinney, E.S. 1991, Ap.J.Lett. 380, L17
Prendergast, K. and Burbidge, G.R. 1968, Ap.J.Lett. 151, L83
Pringle, J.E. and Rees, M.J. 1972, A & A 21, 1
Rappaport, S.A., Joss, P.C., and Webbink, R.F. 1982, Ap.J. 254, 616
Ruderman, M., Shaham, J., and Tavani, M., 1989, ApJ 336, 507
Shakura, N. and Sunyaev, R.A. 1973, A&A 254, 22
Shapiro, S.L. and Lightman, A.P. 1976, Ap.J. 204, 555
Shklovsky, I.S. 1967, Ap.J.Lett. 148, L1
Taam, R.E. and Fryxell, B.A. 1988, ApJ Lett. 327, L73
van den Heuvel, E.P.J. and Heise, J. 1972, Nature Phys. Sci. 239, 67
White, N.E. and Holt, S.S. 1982, ApJ 257, 318

Appendix A:

Conference Programme

Programme

Monday 11 October 1993

1. Introduction — S. Holt, chair

Historical Perspecitve and Current Issues

0900	W. Lewin	.....from the Point of View of an Observer
0945	E. van den Heuvel	...from the Point of View of a Theoretician
1030	Coffee	

2. Black Holes — L. Stella, chair

1100	A. Cowley	Transients and Dynamical Evidence
1130	N. White	X-Ray Signatures
1200	K. Mukai	Application of Magnetic Braking Theories to Black Hole Binaries
1210	Short Contributions and Discussion	
1230	Lunch	

3. Accretion Discs — J. Swank, chair

1400	K. Ebisawa	Disc Spectra
1430	T. Kallman	Spectral Diagnostics/ADC Reprocessing
1500	R. Rutten	Mapping the Discs of CVs
1530	L Titarchuk	X-Ray Burst Spectra
1540	Short Contributions and Discussion	
1600	Tea	

4. Binary Interactions — R. Taam, chair

1630	S. Rappaport	Orbital Evolution
1700	T. Prince	Be Systems
1730	O. Blaes	Coalescence
1800	M Tavani	Irradiation Driven Evolution of X-Ray Binaries
1840	Short Contributions and Discussion	
1830	Wine and Cheese with poster perusal (to 2200)	
1900	Y. Tanaka	Special Presentation: New Results from ASCA

Tuesday   12 October 1993

5. Evolution I — Mario Livio, chair
0830	R. Webbink	Origin of LMXBs
0900	A. Lyne	Recycled Radio Pulsars
0930	P. Podsiadlowski	The Fate of Thorne-Zytkow Objects
0940	Short Contributions and discussion	
1000	Coffee	

6. Evolution II — J. Dolan, chair
1030	J. Grindlay	Globular Cluster Sources
1100	F. Verbunt	Field Decay (?)
1130	S. Kulkarni	Origin of Millisecond Pulsars
1200	Short Contributions and Discussion	
1230	Lunch	

7. X-Ray Pulsars — F. Lamb, chair
1400	A. Parmar	Properties of X-Ray Pulsars
1430	A. Harding	Emission Processes
1500	P. Ghosh	Spin Evolution
1530	Short Contributions and Discussion	
1600	Tea	

8. Dynamics — M. Abramowicz, chair
1630	M. van der Klis	QPO and Noise in Black Holes and Neutron Stars
1700	A. King	Accretion Disc Instabilities
1730	A. Smale	Precession and Long-Term Cyclic Behavior
1800	Short Contributions and Discussion	
1830	Banquet	
	H. Gursky	Some Events in X-ray Astronomy that Aren't in the Journals

Wednesday 13 October 1993

    9. Wind Interactions — **G. Clark,** chair

0830	F. **Nagase**	Cyclotron/Scattering Features
0900	J. **Blondin**	Hydrodynamics of Winds in High Mass X-ray Binaries
0930	L. **Nobili**	Near-Eddington Winds from Neutron Stars
0940	J. **Woo**	Atmospheric Structure of the B0 I Companion of 4U1538-52
0950	K. **Wolinski**	UV Polarimetry of 4U0900-40
1000	Coffee	

    10. X-Ray Sources in External Galaxies — **M. Watson,** chair

1030	G. **Hasinger**	Supersoft Sources
1100	R. **Supper**	Rosat Observations of M31
1130	E. **Magnier**	Optical Counterparts to M31 X-ray Sources
1200	Short Contributions and Discussion	
1230	Lunch	

    11. Summary — **S. Holt,** chair

1330	V. **Trimble**	Historical Retrospective on Pulsars and X-Ray Binaries
1400	J. **Krolik**	Conference Rapporteur

Appendix B: List of Attendees

Abramowicz, M.	University of Göteborg (Sweden)
Angelini, Lorella	USRA/Goddard Space Flight Center
Antunes, Alex	ISAS (Japan)
Araya, Rafael	Johns Hopkins University
Arzoumanian, Zaven	Princeton University
Audley, Damian	University of Maryland
Bailyn, Charles	Yale University
Balbus, Steven A.	University of Virginia
Ballet, Jean	DAPNIA/Sap Saclay (France)
Baptista, Raymundo	STScI
Barrett, Paul	USRA/Goddard Space Flight Center
Bartlett, Lyle	University of Maryland/GSFC
Bazzano, Angela	C. N. R. (Italy)
Begoly, Zsolt	Eötvos University (Hungary)
Bennett, Charles	NASA/Goddard Space Flight Center
Bildsten, Lars	Cal Tech
Blaes, Omer	Univ. of California - Santa Barbara
Blondin, John	North Carolina State University
Boggs, Steve	University of California - Berkeley
Boldt, Elihu	NASA/Goddard Space Flight Center
Boroson, Bram	University of Colorado/JILA
Bradt, Hale	M. I. T.
Bunn, Jenny	Oxford University (UK)
Callanan, Paul	Harvard-Smithsonian C f A
Cannizzo, John K.	NASA/Goddard Space Flight Center
Casares, Jorges	Oxford University (UK)
Castro-Tirado, Alberto J.	Danish Space Research Institute
Cavallo, Rob	University of Maryland
Chakrabarty, Deepto	Cal Tech
Charles, Phil	Oxford University (UK)
Chen, Wan	USRA/Goddard Space Flight Center
Chen, Xingming	Northwestern University
Cheng, Fuhua	STScI
Cheung, Cynthia	NASA/Goddard Space Flight Center
Chipman, Eric	Compton GRO Science Support Center

Appendix B—List of Attendees

Chiu, Hong-Yee	NASA/Goddard Space Flight Center
Christian, Damian	University of Maryland
Clark, George	M. I. T.
Cline, Thomas	NASA/Goddard Space Flight Center
Cominsky, Lynn R.	Sonoma State University
Cowley, Anne	Arizona State University
Dalton, William	University of Virginia
Daly, Ruth	Princeton University
Danner, Rudolf	Cal Tech
Davies, Melvyn	Cal Tech
Davis, Stanley P.	CUA/Goddard Space Flight Center
Day, Charles	USRA/Goddard Space Flight Center
Dermer, Chuck	Naval Research Laboratory
Dingus, Brenda	USRA/Goddard Space Flight Center
Dolan, Joseph	NASA/Goddard Space Flight Center
Ebisawa, Ken	USRA/Goddard Space Flight Center
Eckhardt, Karinn A.	NSSDC/Goddard Space Flight Center
Edberg, Timothy	University of California - Berkeley
Eracleous, Mike	STScI
Esteban, Ernesto P.	University of Puerto Rico
Evans, Charles R.	University of North Carolina
Fahey, Richard	NASA/Goddard Space Flight Center
Finger, Mark H.	Compton GRO Science Support Center
Focke, Warren	University of Maryland/GFSC
Frank, Juhan	Louisiana State University
Frost, Kenneth J.	NASA/Goddard Space Flight Center
Gall, Clarence A.	Universidad Del Zulia (Venezuela)
Gehrels, Neil	NASA/Goddard Space Flight Center
Ghosh, Pranab	Tata Inst. of Fund. Research (India).
Giles, Barry	USRA/Goddard Space Flight Center
Gilfanov, Marat	Space Research Institute IKI (Russia)
Glendenning, Norman K.	Lawrence Berkeley Laboratory
Grebenev, Sergei	Space Research Institute IKI (Russia)
Grindlay, Josh	Harvard Observatory
Grove, J. Eric	Naval Research Laboratory

Appendix B—List of Attendees

Guarnieri, Adriano	Bologna University (Italy)
Gull, Theodore	NASA/Goddard Space Flight Center
Gursky, Herbert	Naval Research Laboratory
Halpern, Jules	Columbia University
Hardee, Philip	University of Alabama
Harding, Alice	NASA/Goddard Space Flight Center
Harlaftis, Emilios	Royal Greenwich Observatory (UK)
Hartman, Robert	NASA/Goddard Space Flight Center
Hasinger, Guenther	Max Planck-Institut MPE (Germany)
Haswell, Carole	STScI
Hauser, Michael G.	NASA/Goddard Space Flight Center
Hawley, John	University of Virginia
Heap, Sally	NASA/Goddard Space Flight Center
Heise, John	Space Research Org. (Netherlands)
Hellier, Coel	University of Texas
Hertz, Paul	Naval Research Laboratory
Holt, Stephen S.	NASA/Goddard Space Flight Center
Horne, Keith	University of Utrecht (Netherlands)
Hoshino, Masahiro	RIKEN (Japan)
Iping, Rosina	University of Guam
Irit, Idan	Technion (Israel)
Jahoda, Keith	NASA/Goddard Space Flight Center
Johnston, Helen	University of Utrecht (Netherlands)
Joss, Paul C.	M. I. T.
Kallman, Tim	NASA/Goddard Space Flight Center
Kalogera, Vicky	University of Illinois - Urbana
Kanetake, Reiko	Tohoku University (Japan)
Kaspi, Victoria	Princeton University
Kawai, Nobu	RIKEN (Japan)
Kazanas, Demosthenes	NASA/Goddard Space Flight Center
Kelley, Richard L.	NASA/Goddard Space Flight Center
Kerr, Frank	University of Maryland/USRA
Kim, Soon-Wook	University of Texas - Austin
King, Andrew	University of Leicester (UK)
Kluth, Edward	

Ko, Yuan-Kuen — Goddard Space Flight Center/CEA
Kolman, Michiel — Kluwer Academic Publishers
Kondo, Yoji — NASA/Goddard Space Flight Center
Kouveliotou, Chryssa — USRA/Marshall Space Flight Center
Kretschmar, Peter — Universitat Tübingen (Germany)
Kroeger, Richard — Naval Research Laboratory
Krolik, Julian — Johns Hopkins University
Kuin, Paul — ADC/NSSDC & Hughes STX
Kulkarni, Sri — Cal Tech
Kume, Kaori — Tohoku University (Japan)
Kunz, Mathias — Universitat Tübingen (Germany)
Kuulkers, Erik — Univ. of Amsterdam (Netherlands)
Kylafis, Nick — University of Crete (Greece)
Lai, Dong — Cornell University
Lamb, Fred — University of Illinois - Urbana
Lanning, Howard H. — Computer Sciences Corp./STScI
Larsson, Stefan — Stockholm Observatory (Sweden)
Leahy, Denis — University of Calgary (Canada)
Lee, Geunho — Johns Hopkins University
Leighly, Karen — NASA/Goddard Space Flight Center
Leiter, Darryl — NASA/Goddard Space Flight Center
Leonard, Peter — Los Alamos National Laboratory
Leventhal, Marv — University of Maryland
Levine, Alan — M. I. T.
Lewin, Walter — M. I. T.
Lewis, Wayne — St. John Fisher College
Liang, Edison P. — Rice University
Liedahl, Duane — LLNL
Livio, Mario — STScI
Lochner, James — USRA/Goddard Space Flight Center
Luo, Chuan — Rice University
Lyne, Andrew — Nuffield Radio Astronomy Labs (UK)
Macomb, Daryl — Compton GRO Science Support Center
Magnier, Eugene — M. I. T.
Mangus, John D. — NASA/Goddard Space Flight Center

Appendix B—List of Attendees

Maran, Stephen	NASA/Goddard Space Flight Center
Marshall, Frank	NASA/Goddard Space Flight Center
Martin, Andrew	Oxford University (UK)
Mattox, John R.	Compton GRO Science Support Center
Matz, Steven M.	Northwestern University
Mazeh, Tsevi	Smithsonian Institute/NASM
McCormick, Patrick	Louisiana State University
Mendenhall, Jeffrey A.	Penn State University
Michalitsianos, Andy	NASA/Goddard Space Flight Center
Mihara, Tatehiro	RIKEN (Japan)
Milsom, John A.	Northwestern University
Minamitani, Takhisa	Hughes STX
Miyaji, Takamitsu	University of Maryland/GSFC
Morgan, Ed	M. I. T.
Moscoso, Michael	University of Texas - Austin
Mukai, Koji	USRA/Goddard Space Flight Center
Murray, Stephen D.	University of California - Berkeley
Muslimov, Alexander	University of Rochester
Nagase, Fumiaki	ISAS (Japan)
Nelson, Robert W.	CITA (Canada)
Niedner, Mal	NASA/Goddard Space Flight Center
Nobili, Luciano	University of Padova (Italy)
Norman, Colin A.	JHU/STScI
Norris, Jay P.	NASA/Goddard Space Flight Center
Nowak, Michael A.	CITA (Canada)
Okada, Rika	Tokyo Metropolitan Univ. (Japan)
Oliversen, Nancy	Hughes STX/NSSDC/GSFC
Orlandini, Mauro	GSFC & TESRE CNR/Bologna (Italy)
Ormes, Jonathan	NASA/Goddard Space Flight Center
Orosz, Jerry	Yale University
Paciesas, Bill S.	University of Alabama - Huntsville
Paerels, Frits	Universtiy of California - Berkeley
Parmar, Arvind	ESTEC/ESA (The Netherlands)
Parsons, Ann M.	NASA/Goddard Space Flight Center
Patterson, Joe	Columbia University

Pavlenko, Elena	Crimean Astrophys Observatory (Uk.)
Pesce, Joe	SISSA/ISAS
Petre, Robert	NASA/Goddard Space Flight Center
Petro, Larry	STScI
Philp, Colin	University of North Carolina
Pisarski, Rysard L.	NASA/Goddard Space Flight Center
Podsiadlowski, Philipp	Inst. of Astronomy-Cambridge (UK)
Polidan, Ronald	NASA/Goddard Space Flight Center
Popham, Bob	University of Illinois
Primini, Frank A.	SAO
Prince, Thomas	Cal Tech
Rajasekhar, Aruna M.	Louisiana State University
Rappaport, Saul	M. I. T.
Rathnasree, N.	University of Vermont
Raymond, John	Harvard-Smithsonian C f A
Reisenegger, Andreas	Institute for Advanced Study
Remillard, Ron	M. I. T.
Reynolds, Alastair	ESTEC/ESA (The Netherlands)
Rho, Jeonghee	University of Maryland/GSFC
Rhoads, James	Princeton University
Richman, Hayley	Columbia University
Roche, Paul	Southampton University (UK)
Rose, William	University of Maryland
Rosenbaum, Doris	SMU
Rosenberg, Duane	University of North Carolina
Rothschild, Richard	University of California - San Diego
Rubin, Brad	USRA/Marshall Space Flight Center
Rutten, Rene	Royal Greenwich Observatory (UK)
Sambruna, Rita	STScI
Schaefer, Brad	USRA/Goddard Space Flight Center
Schlegel, Eric M.	USRA/Goddard Space Flight Center
Schmidt, Ed	National Science Foundation
Schmidtke, Paul	Arizona State University
Selvelli, Pierluigi	CNR-Astron. Observatory (Italy)
Serlemitsos, Peter	NASA/Goddard Space Flight Center

Appendix B—List of Attendees

Shahbaz, Tariq	Keele University (UK)
Shapiro, Maurice	University of Maryland
Shaw, Lisa Susan	Internat Scien. Data Analysis, Inc.
Shrader, Chris	Compton GRO Science Support Center
Silverberg, Robert	NASA/Goddard Space Flight Center
Sion, Edward	Villanova University
Skillman, Dave	NASA/Goddard Space Flight Center
Smale, Alan	USRA/Goddard Space Flight Center
Sonneborn, George	NASA/Goddard Space Flight Center
Soong, Yang	USRA/Goddard Space Flight Center
Stahle, Caroline K.	NASA/Goddard Space Flight Center
Starr, Christopher H.	Compton GRO Science Support Center
Starrfield, Sumner	Arizona State University
Stecher, Ted	NASA/Goddard Space Flight Center
Stella, Luigi	Brera Observatory (Italy)
Stiller, Bertram	
Stollberg, Mark T.	University of Alabama/MSFC
Stone, James	University of Maryland
Strickman, Mark	Naval Research Laboratory
Stringfellow, Guy S.	Penn State University
Sturner, Steven	Rice University
Supper, Rodrigo	Max Planck-Institut MPE (Germany)
Swank, Jean	NASA/Goddard Space Flight Center
Taam, Ronald	Northwestern University
Takeshima, Toshiaki	RIKEN (Japan)
Takeuti, Mine	Tohoku University (Japan_
Tanaka, Yasuo	ISAS (Japan)
Tavani, Marco	Princeton University
Teplitz, Doris	
Titarchuk, Lev	NASA/Goddard Space Flight Center
Trasco, John D.	University of Maryland
Trimble, Virginia	University of Maryland/U C - Irvine
Truran, Jim	University of Chicago
Turolla, R.	University of Padova (Italy)
Ubertini, Pietro	C.N.R. (Italy)

Appendix B—List of Attendees

Umemura, Rika	
van den Heuvel, E.	Univ. of Amsterdam (Netherlands)
van der Klis, Michiel	Univ. of Amsterdam (Netherlands)
van Steenberg, Michael E.	NASA/Goddard Space Flight Center
van Teeseling, Andre	Sterrekwundig Inst. Utrecht (Neth)
Vaughan, Brian	Univ. of Amsterdam (Nethelands)
Vaughan, Eva	
Verbunt, Frank	University of Utrecht (Netherlands)
Volz, Stephen	NASA/Goddard Space Flight Center
Vrtilek, Saeqa D.	University of Maryland
Wachter, Stefanie	University of Washington
Wade, Richard	Penn State University
Wagner, R. Mark	Ohio State University
Wang, John C. L.	JILA
Watson, Michael	University of Leicester (UK)
Webbink, Ron	University of Illinois - Urbana
Wen, Han	Stanford Linear Accelerator Center
White, Nick	NASA/Goddard Space Flight Center
Wijers, Ralph	Princeton University Observatory
Wilkerson, Jeff	University of California - Berkeley
Wilson, R. E.	University of Florida
Wilson, Robert B.	NASA/Marshall Space Flight Center
Wilson-Hodge, Colleen A.	NASA/Marshall Space Flight Center
Wolff, Michael T.	Naval Research Laboratory
Wolinski, Karen G.	Purdue University & NASA/GSFC
Woo, Jonathan	M. I. T.
Wood, Kent	Naval Research Laboratory
Yilmaz, Aysegul	Middle East Technical Univ. (Turkey)
Yoshida, Kenji	Kanagawa University (Japan)
Yuan, Weimin	RIKEN (Japan)
Zampieri, L.	SISSA (Italy)
Zhang, Weiping	USRA/Goddard Space Flight Center
Zhao, Ping	Harvard-Smithsonian C f A

Appendix C:

Physical Constants

TABLE OF PHYSICAL CONSTANTS

CONSTANT	SYMBOL	MKS		CGS		OTHER
speed of light	c	$3.00 \cdot 10^8$	m/s	$3.00 \cdot 10^{10}$	cm/s	(2.997925)
electron charge	e	$1.60 \cdot 10^{-19}$	coul	$4.80 \cdot 10^{-10}$	esu	
Planck constant	h	$6.63 \cdot 10^{-34}$	J·s	$6.63 \cdot 10^{-27}$	erg·s	
	$\hbar$	$1.05 \cdot 10^{-34}$	J·s	$1.05 \cdot 10^{-27}$	erg·s	
	hc	$1.99 \cdot 10^{-25}$	J·m	$1.99 \cdot 10^{-16}$	erg·cm	
	$\hbar c$	$3.15 \cdot 10^{-26}$	J·m	$3.15 \cdot 10^{-17}$	erg·cm	200 MeV·fm
Boltzmann constant	k	$1.38 \cdot 10^{-23}$	J/K	$1.38 \cdot 10^{-16}$	erg/K	$8.6 \cdot 10^{-5}$ eV/K
	k/h	$2.08 \cdot 10^{10}$	s^{-1}/K	$2.08 \cdot 10^{10}$	s^{-1}/K	
	k/hc	67.5	m^{-1}/K	0.675	cm^{-1}/K	
Gravitational constant	G	$6.67 \cdot 10^{-11}$	N·m^2/kg^2	$6.67 \cdot 10^{-8}$	dy·cm^2/gm^2	
Gas constant	R	8.314	J/K·mole	8.31·10^7	erg/K·mole	
Avogadro's number (= R/k)	N	$6.02 \cdot 10^{26}$	amu/kg	$6.02 \cdot 10^{23}$	amu/kg	$6 \cdot 10^{23}$ molecules/mole
electron mass	m_e	$9.11 \cdot 10^{-31}$	kg	$9.11 \cdot 10^{-28}$	gm	0.51 MeV
proton mass	M_p	$1.67 \cdot 10^{-27}$	kg	$1.67 \cdot 10^{-24}$	gm	938 MeV
neutron mass	M_n	$1.67 \cdot 10^{-27}$	kg	$1.67 \cdot 10^{-24}$	gm	939 MeV
pion mass (=270·m_e)	m_π	$2.46 \cdot 10^{-28}$	kg	$2.46 \cdot 10^{-25}$	gm	140 MeV
muon mass (=207·m_e)	m_μ	$1.89 \cdot 10^{-28}$	kg	$1.89 \cdot 10^{-25}$	gm	106 MeV
classical elect radius (=e^2/mc^2)	r_c	$2.82 \cdot 10^{-15}$	m	$2.82 \cdot 10^{-13}$	cm	
Compton wavelength (=h/mc)	λ_c	$2.43 \cdot 10^{-12}$	m	$2.43 \cdot 10^{-10}$	cm	0.02 Å

Appendix C—Physical Constants

Quantity	Symbol	Value	Unit	Value	Unit
Thomson cross-section	σ_T	$6.65 \cdot 10^{-29}$	m^2	$6.65 \cdot 10^{-25}$	cm^2
Planck length ($=\sqrt{\hbar G/c^3}$)	l_{Pl}	$1.61 \cdot 10^{-35}$	m	$1.61 \cdot 10^{-33}$	cm
Planck time ($=\sqrt{\hbar G/c^5}$)	t_{Pl}	$5.39 \cdot 10^{-44}$	s	$5.39 \cdot 10^{-44}$	s
Planck density ($=c^5/\hbar G^2$)	ρ_{Pl}	$5.16 \cdot 10^{96}$	kg/m^3	$5.16 \cdot 10^{93}$	gm/cm^3
Bohr radius ($=\hbar^2/me^2$)	r_B	$0.53 \cdot 10^{-10}$	m	$0.53 \cdot 10^{-8}$	cm	0.5 Å
Fine structure constant ($=e^2/\hbar c$)	α	$7.30 \cdot 10^{-3}$		$7.30 \cdot 10^{-3}$		1/137
Bohr magneton ($=e\hbar/2m_e c$)	μ_B	$9.27 \cdot 10^{-24}$	J/T	$9.27 \cdot 10^{-21}$	erg/gauss
Nuclear magneton ($=e\hbar/2M_p c$)	μ_N	$5.05 \cdot 10^{-27}$	J/T	$5.05 \cdot 10^{-24}$	erg/gauss
Permittivity of vacuum	ε_o	$8.85 \cdot 10^{-12}$	fd/m			$1/4\pi\varepsilon_o = 9.0 \cdot 10^9$
Permeability in vacuum	μ_o	$4\pi \cdot 10^{-7}$	Hen/m		
Stefan-Boltzmann constant	σ	$5.67 \cdot 10^{-8}$	W/m$^2 \cdot$K^4	$5.67 \cdot 10^{-5}$	erg/s$\cdot$cm$^2 \cdot$K^4
Rydberg ($=m_e e^4/2\hbar^2$)	R_∞	$2.18 \cdot 10^{-18}$	J	$2.18 \cdot 10^{-11}$	erg	13.6 eV
1 amu		$1.66 \cdot 10^{-27}$	kg	$1.66 \cdot 10^{-24}$	gm	931.5 MeV
1 calorie		4.19	J	$4.19 \cdot 10^7$	erg
1 year		$3.16 \cdot 10^7$	s	$3.16 \cdot 10^7$	s
1 atmosphere		$1.01 \cdot 10^5$	N/m^2	$1.01 \cdot 10^6$	dyne/cm^2	14.2 lbs/in^2
		$1.01 \cdot 10^5$	Pascal			760 Torr
1 eV		$1.6 \cdot 10^{-19}$	J	$1.6 \cdot 10^{-12}$	erg	11,605 K
		$1.24 \cdot 10^{-6}$	m	$1.24 \cdot 10^{-4}$	cm
		1	Tesla	10^4	gauss

Appendix C—Physical Constants

ASTROPHYSICAL CONSTANTS

CONSTANT	SYMBOL	MKS		CGS		OTHER
astronomical unit	AU	$1.50 \cdot 10^{11}$	m	$1.50 \cdot 10^{13}$	cm	
	AU/year					4.74 km/s
parsec	pc	$3.09 \cdot 10^{16}$	m	$3.09 \cdot 10^{18}$	cm	3.26 LY
solar mass	$M_\odot$	$1.99 \cdot 10^{24}$	kg	$1.99 \cdot 10^{33}$	gm	
solar luminosity	$L_\odot$	$3.90 \cdot 10^{26}$	J/s	$3.90 \cdot 10^{33}$	erg/s	
solar effective temperature	$T_{eff\odot}$	5780	K	5780	K	
solar radius	$R_\odot$	$6.96 \cdot 10^{8}$	m	$6.96 \cdot 10^{10}$	cm	
Earth radius	$R_\oplus$	$6.38 \cdot 10^{6}$	m	$6.38 \cdot 10^{8}$	cm	
Earth mass	$M_\oplus$	$5.98 \cdot 10^{24}$	kg	$5.98 \cdot 10^{27}$	gm	
Earth density	$\rho_\oplus$	5520	kg/m^3	5.52	gm/cm^3	
Jansky	Jy	$1.0 \cdot 10^{-26}$	W/m$^2 \cdot$Hz	$1 \cdot 10^{-23}$	erg/s$\cdot$cm$^2 \cdot$Hz	
Hubble constant	H_0	$3.24h \cdot 10^{-18}$	s^{-1}	$3.24h \cdot 10^{-18}$	s^{-1}	$100h$ km/s$\cdot$Mpc
critical density (=$3H_0^2/8\pi G$)	ρ_0	$1.88h^2 \cdot 10^{-26}$	kg/cm^3	$1.88h^2 \cdot 10^{-29}$	gm/cm^3	
plasma frequency						$8.98\sqrt{n_e(\text{cm}^{-3})}$ kHz/gauss
radian						$57.29578° = 206,265''$
CMB photon density	n_γ	$4.15 \cdot 10^{5}$	m^{-3}	415	cm^{-3}	

Author Index

A

Abbott, T. M. C.	201
Anderson, S. F.	479
Arzoumanian, Z.	287

B

Bailes, M.	295
Ballet, J.	131
Bartolini, C.	103, 271, 275, 279
Bell, J. F.	295
Beskin, G.	103
Bessell, M.	295
Bildsten, L.	235, 451, 475, 487
Blaes, O.	245
Blondin, J. M.	578
Bloom, E.	561
Bockrath, M.	543
Boldt, E.	557
Boyd, P. T.	367
Briggs, M. S.	115, 451, 475
Buckley, D.	487

C

Callanan, P. J.	99, 127, 375
Campana, S.	387
Cannon, R. C.	403
Casares, J.	107, 371
Castor, J. I.	209
Castro-Tirado, A. J.	135
Chakrabarty, D.	235, 451, 455, 487
Charles, P. A.	91, 107, 123, 127, 371
Chen, W.	67
Chen, X.	217, 225
Cheng, F. H.	197
Chiu, J.	455
Christian, D. J.	193
Churazov, E.	131
Clark, G. W.	588
Coe, M.	487
Colpi, M.	387, 605
Cominsky, L. R.	291, 363, 459
Cowley, A. P.	45

D

Dalton, W. W.	658
D'Amico, N.	295
Davies, M. B.	315
Dermer, C. D.	495
DiStefano, R.	543
Dolan, J. F.	367
Done, C.	181
Downes, R. A.	479
Dyachkov, A.	131

E

Ebisawa, K.	143, 181
Elliot, J. L.	367
Evans, C. R.	299
Everall, C.	487

F

Fabregat, J.	487
Finger, M. H.	235, 255, 259, 451, 455, 459, 475
Finoguenov, A.	131
Fishman, G. J.	115, 255, 259, 451, 455, 459
Frail, D. A.	299
Frank, J.	379, 383, 395
Fritz, G.	561
Fruchter, A. S.	287

G

Gall, C. A.	87
Gallego, J.	135
Garcia, M. R.	99
Ghosh, P.	439
Gilfanov, M.	131
Godfrey, G.	561
Goldreich, P.	311
Grabelsky, D. A.	263
Grebenev, S. A.	61
Grindlay, J. E.	339
Gruber, D. E.	463
Guarnieri, A.	103, 271, 275, 279
Gursky, H.	39

H

Han, X. H.	95
Hanson, J.	561
Harding, A. K.	429
Harlaftis, E.	91, 127
Harmon, B. A.	115, 255, 259, 451, 459
Hasinger, G.	611, 654
Haswell, C. A.	201
Hellier, C.	535
Hertz, P.	363, 553, 561
Hess, C. J.	166
Hjellming, R. M.	95, 111
Holt, S. S.	v
Horne, K.	197
Howell, S. B.	111
Hubeny, I.	197

I

Igumenshchev, I. V.	601
Illarionov, A. F.	601
Imamura, J. N.	491

J

Johnson, W. N.	263, 561
Johnston, H. M.	654
Johnston, S.	295
Jones, D.	91

K

Kahn, S. M.	166
Kallman, T. R.	155, 205
Kalogera, V.	321
Kanetake, R.	522, 551
Kaniovsky, S.	463
Kaspi, V. M.	295
Kawai, N.	267
Khavenson, N.	131
Kim, S.-W.	213
King, A. R.	379, 395, 515
Kinzer, R. L.	263
Klein, R. I.	209
Ko, Y.-K.	205
Kompaneets, D. A.	601
Kreidl, T. J.	111
Krolik, J. H.	673
Kunz, M.	463
Kurfess, J. D.	263
Kuulkers, E.	539
Kylafis, N. D.	650

L

Lai, D.	303
Lapidus, I.	597
Laurent, P.	131
Leahy, D. A.	467
Lebrun, F.	131
Lee, G.	221
Lee, U.	307
Leonard, P. J. T.	299, 315
Lewin, W. H. G.	3, 539, 553
Liang, E.	75
Liedahl, D. A.	166
Livio, M.	67
Lovelette, M. N.	561
Luo, C.	75
Ly, Y.	363
Lyne, A. G.	295, 331

M

Magnier, E.	640
Malet, I.	131
Manchester, R. N.	295
Mandrou, P.	131
Margon, B.	479
Marsh, T. R.	197
Martin. A. C.	91, 127
Martín, E. L.	371
Matsuoka, M.	267
Matz, S. M.	263
Mazeh, T.	283
McClintock, J. E.	99
McCormick, P.	379, 395
McKee, C. F.	209
Meegan, C. A.	455
Meirelles, C.	75
Mendelson, H.	283
Mereghetti, S.	387
Michelson, P. F.	553, 561
Milsom, J. A.	225
Minarini, R.	275
Mineshige, S.	213
Mitsuda, K.	553
Moscoso, M. D.	83
Mukai, K.	79
Murray, S. D.	209
Muslimov, A. G.	307, 391

N

Nagase, F.	567, 588
Naylor, T.	123
Neizvestny, S.	103
Nelson, M. J.	367
Nelson, R. W.	592
Nobili, L.	597
Norris, J. P.	553
Norton, A.	487
Nowak, M. A.	547

O

Oosterbroek, T.	539
Orlandini, M.	557
Ortiz, J. L.	135

P

Paciesas, W. S.	115, 255, 455
Paerels, F.	166
Panferova, I.	103
Parmar, A. N.	415, 471
Paul, J.	131
Pavlinsky, M. N.	61
Pendleton, G. N.	115, 451, 475
Percival, J. W.	367
Philp, C.	299
Piccioni, A.	103, 271, 275, 279
Pietsch, W.	463
Plokhotnichenko, V.	103
Podsiadlowski, P.	71, 403
Prince, T. A.	235, 451, 455, 487
Purcell, W. R.	263

R

Rajasekhar, A.	379, 383
Rappaport, S.	543
Rasio, F. A.	303
Rebolo, R.	371
Rees, M. J.	71, 403
Reglero, V.	487
Reisenegger, A.	311
Remillard, R. A.	99
Rengelink, R.	375
Reynolds, A. P.	471
Roche, P.	487
Rogers, R. D.	166
Roques, J. P.	131
Rose, W. K.	499
Rubin, B. C.	451, 455
Rutten, R. G. M.	171

S

Salpeter, E. E.	592
Sarazin, C. L.	658
Sarna, M. J.	307, 391
Schlegel, E. M.	119
Schmidtke, P. C.	229
Schmitz-Fraysse, M. C.	131
Shahbaz, T.	123
Shapiro, S. L.	303
Sheikhet, A.	131
Shrader, C. R.	67, 95
Silber, A.	99
Silingardi, R.	279
Smale, A. P.	525, 535
Starrfield, S. G.	95, 111
Staubert, R.	463
Stella, L.	387
Stollberg, M. T.	255, 451
Strickman, M. S.	263
Sturner, S. J.	495
Sunyaev, R. A.	61, 131, 463
Supper, R.	631

T

Taam, R. E.	217, 225
Takeuti, M.	522, 551
Tavani, M.	387, 407
Taylor, J. H.	287
Taylor, R.	561
Teodorani, M.	271, 275, 279
Terrell, D.	483
Townsley, L. C.	367
Treves, A.	605
Trimble, V.	665
Trümper, J.	463
Turolla, R.	597, 605

U

Ueda, Y.	181
Ulmer, M. P.	263
Unger, S.	487

V

van Citters, G. W.	367
van den Heuvel, E..P. J.	18
van der Klis, M.	505, 539, 553
van Paradijs, J.	375, 539, 553
van Teeseling, A.	189
Vaughan, B.	553
Verbunt, F.	189, 351, 654
Vikhlinin, A.	131

W

Wachter, S.	479
Wagner, R. M.	95, 111
Wang, J. C. L.	592
Wasserman, I.	592
Webbink, R. F.	321
Wen, H.	561
Wheeler, J. C.	83, 213
White, N. E.	53
Whitehurst, R.	383
Wijers, R. A. M. J.	399
Wilson, C.A.	115, 255, 259
Wilson, R. B.	235, 255, 259, 451, 455, 459, 475
Wilson, R. E.	483
Wolff, M. T.	491, 561
Wolinski, K. G.	367
Woo, J.	588
Wood, K. S.	363, 491, 553, 561

X

Xilouris, E. M.	650

Y

Yoshida, K.	185
Yuan, W.	267

Z

Zampieri, G.	103
Zampieri, L.	605
Zhang, N. S.	451
Zhang, S. N.	115
Zhao, P.	99
Zhuravkov, A.	103

Subject Index

Page numbers refer to the *first* page of the contribution in which the subject appears

A

Accretion and accretion disks 3, 18, 61, 67, 75, 79, 83, 91, 103, 135, 143, 155, 166, 171, 185, 189, 193, 197, 205, 209, 217, 221, 225, 229, 235, 321, 379, 387, 395, 403, 415, 429, 439, 467, 491, 505, 515, 525, 547, 551, 553, 557, 561, 567, 578, 601, 605, 611, 650, 673
   accretion disk coronae 155, 189, 193, 205, 209, 673
   boundary layers 189, 193, 439, 673
   dipping behavior 185, 467, 471, 525
   disk evaporation 379, 395
   disk mapping 171
   disk precession 525
   instabilities and turbulence 67, 75, 79, 103, 135, 213, 217, 221, 225, 321, 505, 515, 547, 551, 557, 673
   oscillations 18, 217, 505, 515, 547, 551, 553, 561
   shocks 429, 491, 499, 505
   spectra 61, 111, 119, 155, 166, 181, 185, 189, 197, 205, 229, 471, 495
   super Eddington 75, 611
   viscosity 75, 217, 221, 225, 547, 673
   winds 83, 439, 499, 557, 567, 578, 601, 673
ASCA observations 267, 567
Asteroseismology 307

B

BATSE observations 115, 235, 255, 259, 451, 455, 459, 475
BBXRT observations 119
Be-star X-ray binaries 18, 235, 255, 259, 271, 275, 279, 283, 291, 387, 415, 451, 459, 487, 592, 658
Black holes 3, 18, 45, 53, 61, 67, 75, 79, 95, 103, 107, 111, 115, 119, 123, 127, 131, 135, 143, 181, 213, 217, 225, 371, 403, 505, 515, 547, 561, 665
AGN black holes 18, 71
   dynamics and temporal behavior 45, 79, 111, 127, 213, 217, 505, 515, 547, 553, 561
   galactic distribution 53, 673
   mass determinations 3, 45, 107, 123, 665
   spectra 53, 61, 111, 119, 131, 143, 181, 547
   X-ray novae and transients 3, 45, 53, 61, 67, 75, 79, 91, 95, 99, 103, 111, 115, 127, 135, 181, 213, 371, 515
Bursts, Gamma-ray 18, 245, 307
Bursts, X-ray 3, 18, 307, 539, 543, 597

C

Cataclysmic variables 18, 171, 189, 197, 339, 383, 399, 491, 515, 640
Chaos 543
Coalescence and collisions 245, 303, 311, 315, 321
Companion stars 3, 71, 95, 123, 127, 171, 271, 275, 283, 291, 299, 307, 321, 399, 415, 479, 483, 578, 640
Cyclotron features 415, 429, 463, 495, 567, 592, 673

D

Dust haloes 567, 588

E

Einstein Observatory observations 193, 640
Emission mechanisms 3, 18, 83, 155, 166, 171, 181, 193, 197, 209, 229, 267, 371, 415, 429, 495, 499, 547, 567, 597, 605, 611, 673
Comptonization 155, 181, 193, 209, 429, 495, 547, 567, 597, 605, 611, 673

Subject Index 711

Lines 3, 83, 91, 119, 155, 166, 171, 181, 193, 197, 205, 229, 267, 271, 279, 371, 415, 495, 567, 673
Pair plasmas 83, 673
EXOSAT observations 471, 539, 543
External Galaxies, X-ray sources 18, 611, 640, 650, 673
Large Magellanic Cloud 18, 611, 650, 673
M31 18, 611, 631, 640, 673
Optical counterparts 640

F
Fractal analysis 522

G
Ginga observations 181, 185, 267, 467, 522, 553, 588
Globular cluster X-ray sources 18, 39, 315, 331, 339, 631, 640, 654, 665, 673
Granat observations 61, 115, 131

H
HEXE observations 463
High-mass X-ray binaries (HMXB) 3, 18, 45, 53, 131, 235, 299, 367, 415, 429, 439, 455, 463, 471, 475, 479, 483, 487, 495, 522, 525, 561, 567, 578, 588, 601, 640, 658, 673
eclipses 415, 479
relationship to black hole candidates 45, 53, 131, 505
spectra 131, 367, 415, 429, 471, 479, 495, 567, 588, 673
Hubble Space Telescope observations 197, 339, 479

I
Infra-red observations 123, 487
Instabilities (see Accretion, instablities and turbulence)
Irradiation and photoionization 3, 18, 67, 71, 91, 155, 205, 213, 375, 391, 395, 407, 525, 578, 673
IUE observations 229, 271, 307

J
Jets 3, 18, 267

L
Long-term cyclic behavior 3, 267, 475, 525
Low-mass X-ray binaries (LMXB) 3, 18, 45, 127, 135, 155, 166, 185, 193, 201, 205, 229, 307, 321, 331, 339, 363, 371, 383, 387, 391, 395, 403, 407, 415, 439, 515, 525, 535, 539, 553, 561, 640, 665, 673
dippers (see Accretion and accretion disks)
origin 18, 321, 391, 395, 403, 407, 673
relationship to black holes 45, 53, 127, 135, 371
relationship to radio pulsars 3, 18, 307, 331, 383, 387, 391, 399, 407, 439, 553
spectra 18, 155, 166, 185, 193, 205, 229, 673

M
Magnetic braking 79, 395, 439, 525, 673

N
Neutron stars 3, 291, 311, 351, 505, 561, 650, 665
accreting isolated neutron stars 3, 291, 650, 665
dynamics and temporal behavior 311, 505, 561
magnetic fields 351, 505
New missions 491, 561
Noise 505, 522, 557, 561
Novae, X-ray (see Black holes)

O
Optical observations 67, 91, 95, 99, 103, 107, 127, 135, 201, 275, 279, 283, 295, 339, 371, 375, 479, 483, 487, 640
Orbits, binary 18, 71, 91, 99, 107, 127, 201, 235, 245, 255, 259, 267, 287, 363, 375, 383, 395, 399, 407, 415, 459, 475, 479, 483, 525, 535, 561
period changes 18, 71, 235, 245, 287, 383, 395, 399, 407, 415, 525, 535

period determinations 18, 91, 99, 107, 127, 201, 235, 255, 259, 267, 287, 363, 375, 415, 459, 475, 479, 483, 525, 535, 561
period precession 267, 525
Oscillations 395, 491, 505, 547, 551, 561
  in accretion discs 505, 547, 561
  in stars 307
OSSE observations 263

P

Photoionization (see Irradiation)
Polarization 367
Population synthesis 351, 658, 673
  radio pulsars 351, 673
  X-ray binaries 658
Pulsars, radio 3, 18, 287, 291, 295, 299, 307, 331, 351, 375, 379, 383, 387, 391, 399, 403, 439, 665, 673
  magnetic fields 18, 351, 665, 673
  millisecond 18, 287, 331, 351, 375, 391, 673
  periods 3, 18, 287, 291, 295, 351, 379, 673
  relationship with high-mass X-ray binaries 18, 291, 299
  relationship with low-mass X-ray binaries 3, 18, 295, 307, 331, 351, 387, 391, 399, 439
Pulsars, X-ray 3, 18, 39, 235, 245, 255, 259, 367, 387, 415, 429, 439, 451, 455, 459, 463, 467, 471, 475, 483, 491, 495, 499, 522, 525, 553, 557, 567, 588, 592, 601, 673
  magnetic fields 18, 235, 387, 415, 429, 439, 463, 495, 499, 567
  pulse periods 3, 18, 235, 255, 259, 387, 415, 439, 451, 455, 459, 463, 475, 483, 601, 673
  pulse profiles 415, 429, 459, 471, 495
  spectra 415, 451, 463, 467, 471, 495, 567, 592, 673

Q

Quasi periodic oscillations (QPO) 3, 18, 119, 217, 491, 505, 539, 547, 553, 561, 673

R

Radio observations 95, 295, 299, 331
Relativity 87
Roche lobes 18, 171, 235, 287, 303, 321, 375, 379, 395, 525, 578, 658, 673
ROSAT observations 111, 189, 291, 339, 363, 535, 611, 631, 640, 654
Runaway OB stars 299, 640

S

Supernovae 18, 321, 439, 611, 640
Supersoft X-ray sources 18, 611, 631, 650, 673

T

Thorne-Zytkow objects 403
Transients, X-ray (see Black holes)
Triple systems 525
Turbulence (see Accretion, instablities and turbulence)

U

UV observations 95, 229, 271, 307, 367, 479

W

White dwarfs 18, 189, 197, 339, 611, 673
Winds 83, 287, 291, 383, 395, 439, 547, 557, 578, 597, 601, 605, 673
  dynamics 291, 383, 557, 578, 601, 673
  evaporative 287, 395, 673
  in accretion disks 83, 547
  in HMXB companions 439, 557, 567, 578
  in neutron stars 383, 597, 601, 605
  spectral features 567, 578

X

XTE observations 557

AIP Conference Proceedings

		L.C. Number	ISBN
No. 170	Nuclear Spectroscopy of Astrophysical Sources (Washington, DC, 1987)	88-71625	0-88318-370-6
No. 171	Vacuum Design of Advanced and Compact Synchrotron Light Sources (Upton, NY, 1988)	88-71824	0-88318-371-4
No. 172	Advances in Laser Science—III: Proceedings of the International Laser Science Conference (Atlantic City, NJ, 1987)	88-71879	0-88318-372-2
No. 173	Cooperative Networks in Physics Education (Oaxtepec, Mexico, 1987)	88-72091	0-88318-373-0
No. 174	Radio Wave Scattering in the Interstellar Medium (San Diego, CA, 1988)	88-72092	0-88318-374-9
No. 175	Non-neutral Plasma Physics (Washington, DC, 1988)	88-72275	0-88318-375-7
No. 176	Intersections Between Particle and Nuclear Physics (Third International Conference) (Rockport, ME, 1988)	88-62535	0-88318-376-5
No. 177	Linear Accelerator and Beam Optics Codes (La Jolla, CA, 1988)	88-46074	0-88318-377-3
No. 178	Nuclear Arms Technologies in the 1990s (Washington, DC, 1988)	88-83262	0-88318-378-1
No. 179	The Michelson Era in American Science: 1870–1930 (Cleveland, OH, 1987)	88-83369	0-88318-379-X
No. 180	Frontiers in Science: International Symposium (Urbana, IL, 1987)	88-83526	0-88318-380-3
No. 181	Muon-Catalyzed Fusion (Sanibel Island, FL, 1988)	88-83636	0-88318-381-1
No. 182	High T_c Superconducting Thin Films, Devices, and Applications (Atlanta, GA, 1988)	88-03947	0-88318-382-X
No. 183	Cosmic Abundances of Matter (Minneapolis, MN, 1988)	89-80147	0-88318-383-8
No. 184	Physics of Particle Accelerators (Ithaca, NY, 1988)	89-83575	0-88318-384-6
No. 185	Glueballs, Hybrids, and Exotic Hadrons (Upton, NY, 1988)	89-83513	0-88318-385-4
No. 186	High-Energy Radiation Background in Space (Sanibel Island, FL, 1987)	89-83833	0-88318-386-2
No. 187	High-Energy Spin Physics (Minneapolis, MN, 1988)	89-83948	0-88318-387-0

No. 188	International Symposium on Electron Beam Ion Sources and their Applications (Upton, NY, 1988)	89-84343	0-88318-388-9
No. 189	Relativistic, Quantum Electrodynamic, and Weak Interaction Effects in Atoms (Santa Barbara, CA, 1988)	89-84431	0-88318-389-7
No. 190	Radio-frequency Power in Plasmas (Irvine, CA, 1989)	89-45805	0-88318-397-8
No. 191	Advances in Laser Science—IV (Atlanta, GA, 1988)	89-85595	0-88318-391-9
No. 192	Vacuum Mechatronics (First International Workshop) (Santa Barbara, CA, 1989)	89-45905	0-88318-394-3
No. 193	Advanced Accelerator Concepts (Lake Arrowhead, CA, 1989)	89-45914	0-88318-393-5
No. 194	Quantum Fluids and Solids—1989 (Gainesville, FL, 1989)	89-81079	0-88318-395-1
No. 195	Dense Z-Pinches (Laguna Beach, CA, 1989)	89-46212	0-88318-396-X
No. 196	Heavy Quark Physics (Ithaca, NY, 1989)	89-81583	0-88318-644-6
No. 197	Drops and Bubbles (Monterey, CA, 1988)	89-46360	0-88318-392-7
No. 198	Astrophysics in Antarctica (Newark, DE, 1989)	89-46421	0-88318-398-6
No. 199	Surface Conditioning of Vacuum Systems (Los Angeles, CA, 1989)	89-82542	0-88318-756-6
No. 200	High T_c Superconducting Thin Films: Processing, Characterization, and Applications (Boston, MA, 1989)	90-80006	0-88318-759-0
No. 201	QED Structure Functions (Ann Arbor, MI, 1989)	90-80229	0-88318-671-3
No. 202	NASA Workshop on Physics From a Lunar Base (Stanford, CA, 1989)	90-55073	0-88318-646-2
No. 203	Particle Astrophysics: The NASA Cosmic Ray Program for the 1990s and Beyond (Greenbelt, MD, 1989)	90-55077	0-88318-763-9
No. 204	Aspects of Electron-Molecule Scattering and Photoionization (New Haven, CT, 1989)	90-55175	0-88318-764-7
No. 205	The Physics of Electronic and Atomic Collisions (XVI International Conference) (New York, NY, 1989)	90-53183	0-88318-390-0
No. 206	Atomic Processes in Plasmas (Gaithersburg, MD, 1989)	90-55265	0-88318-769-8
No. 207	Astrophysics from the Moon (Annapolis, MD, 1990)	90-55582	0-88318-770-1

No. 208	Current Topics in Shock Waves (Bethlehem, PA, 1989)	90-55617	0-88318-776-0
No. 209	Computing for High Luminosity and High Intensity Facilities (Santa Fe, NM, 1990)	90-55634	0-88318-786-8
No. 210	Production and Neutralization of Negative Ions and Beams (Brookhaven, NY, 1990)	90-55316	0-88318-786-8
No. 211	High-Energy Astrophysics in the 21st Century (Taos, NM, 1989)	90-55644	0-88318-803-1
No. 212	Accelerator Instrumentation (Brookhaven, NY, 1989)	90-55838	0-88318-645-4
No. 213	Frontiers in Condensed Matter Theory (New York, NY, 1989)	90-6421	0-88318-771-X 0-88318-772-8 (pbk.)
No. 214	Beam Dynamics Issues of High-Luminosity Asymmetric Collider Rings (Berkeley, CA, 1990)	90-55857	0-88318-767-1
No. 215	X-Ray and Inner-Shell Processes (Knoxville, TN, 1990)	90-84700	0-88318-790-6
No. 216	Spectral Line Shapes, Vol. 6 (Austin, TX, 1990)	90-06278	0-88318-791-4
No. 217	Space Nuclear Power Systems (Albuquerque, NM, 1991)	90-56220	0-88318-838-4
No. 218	Positron Beams for Solids and Surfaces (London, Canada, 1990)	90-56407	0-88318-842-2
No. 219	Superconductivity and Its Applications (Buffalo, NY, 1990)	91-55020	0-88318-835-X
No. 220	High Energy Gamma-Ray Astronomy (Ann Arbor, MI, 1990)	91-70876	0-88318-812-0
No. 221	Particle Production Near Threshold (Nashville, IN, 1990)	91-55134	0-88318-829-5
No. 222	After the First Three Minutes (College Park, MD, 1990)	91-55214	0-88318-828-7
No. 223	Polarized Collider Workshop (University Park, PA, 1990)	91-71303	0-88318-826-0
No. 224	LAMPF Workshop on (π, K) Physics (Los Alamos, NM, 1990)	91-71304	0-88318-825-2
No. 225	Half Collision Resonance Phenomena in Molecules (Caracas, Venezuela, 1990)	91-55210	0-88318-840-6
No. 226	The Living Cell in Four Dimensions (Gif sur Yvette, France, 1990)	91-55209	0-88318-794-9
No. 227	Advanced Processing and Characterization Technologies (Clearwater, FL, 1991)	91-55194	0-88318-910-0

No. 228	Anomalous Nuclear Effects in Deuterium/Solid Systems (Provo, UT, 1990)	91-55245	0-88318-833-3
No. 229	Accelerator Instrumentation (Batavia, IL, 1990)	91-55347	0-88318-832-1
No. 230	Nonlinear Dynamics and Particle Acceleration (Tsukuba, Japan, 1990)	91-55348	0-88318-824-4
No. 231	Boron-Rich Solids (Albuquerque, NM, 1990)	91-53024	0-88318-793-4
No. 232	Gamma-Ray Line Astrophysics (Paris-Saclay, France, 1990)	91-55492	0-88318-875-9
No. 233	Atomic Physics 12 (Ann Arbor, MI, 1990)	91-55595	088318-811-2
No. 234	Amorphous Silicon Materials and Solar Cells (Denver, CO, 1991)	91-55575	088318-831-7
No. 235	Physics and Chemistry of MCT and Novel IR Detector Materials (San Francisco, CA, 1990)	91-55493	0-88318-931-3
No. 236	Vacuum Design of Synchrotron Light Sources (Argonne, IL, 1990)	91-55527	0-88318-873-2
No. 237	Kent M. Terwilliger Memorial Symposium (Ann Arbor, MI, 1989)	91-55576	0-88318-788-4
No. 238	Capture Gamma-Ray Spectroscopy (Pacific Grove, CA, 1990)	91-57923	0-88318-830-9
No. 239	Advances in Biomolecular Simulations (Obernai, France, 1991)	91-58106	0-88318-940-2
No. 240	Joint Soviet-American Workshop on the Physics of Semiconductor Lasers (Leningrad, USSR, 1991)	91-58537	0-88318-936-4
No. 241	Scanned Probe Microscopy (Santa Barbara, CA, 1991)	91-76758	0-88318-816-3
No. 242	Strong, Weak, and Electromagnetic Interactions in Nuclei, Atoms, and Astrophysics: A Workshop in Honor of Stewart D. Bloom's Retirement (Livermore, CA, 1991)	91-76876	0-88318-943-7
No. 243	Intersections Between Particle and Nuclear Physics (Tucson, AZ, 1991)	91-77580	0-88318-950-X
No. 244	Radio Frequency Power in Plasmas (Charleston, SC, 1991)	91-77853	0-88318-937-2
No. 245	Basic Space Science (Bangalore, India, 1991)	91-78379	0-88318-951-8

No. 246	Space Nuclear Power Systems (Albuquerque, NM, 1992)	91-58793	1-56396-027-3 1-56396-026-5 (pbk.)
No. 247	Global Warming: Physics and Facts (Washington, DC, 1991)	91-78423	0-88318-932-1
No. 248	Computer-Aided Statistical Physics (Taipei, Taiwan, 1991)	91-78378	0-88318-942-9
No. 249	The Physics of Particle Accelerators (Upton, NY, 1989, 1990)	92-52843	0-88318-789-2
No. 250	Towards a Unified Picture of Nuclear Dynamics (Nikko, Japan, 1991)	92-70143	0-88318-951-8
No. 251	Superconductivity and its Applications (Buffalo, NY, 1991)	92-52726	1-56396-016-8
No. 252	Accelerator Instrumentation (Newport News, VA, 1991)	92-70356	0-88318-934-8
No. 253	High-Brightness Beams for Advanced Accelerator Applications (College Park, MD, 1991)	92-52705	0-88318-947-X
No. 254	Testing the AGN Paradigm (College Park, MD, 1991)	92-52780	1-56396-009-5
No. 255	Advanced Beam Dynamics Workshop on Effects of Errors in Accelerators, Their Diagnosis and Corrections (Corpus Christi, TX, 1991)	92-52842	1-56396-006-0
No. 256	Slow Dynamics in Condensed Matter (Fukuoka, Japan, 1991)	92-53120	0-88318-938-0
No. 257	Atomic Processes in Plasmas (Portland, ME, 1991)	91-08105	0-88318-939-9
No. 258	Synchrotron Radiation and Dynamic Phenomena (Grenoble, France, 1991)	92-53790	1-56396-008-7
No. 259	Future Directions in Nuclear Physics with 4π Gamma Detection Systems of the New Generation (Strasbourg, France, 1991)	92-53222	0-88318-952-6
No. 260	Computational Quantum Physics (Nashville, TN, 1991)	92-71777	0-88318-933-X
No. 261	Rare and Exclusive B&K Decays and Novel Flavor Factories (Santa Monica, CA, 1991)	92-71873	1-56396-055-9
No. 262	Molecular Electronics—Science and Technology (St. Thomas, Virgin Islands, 1991)	92-72210	1-56396-041-9
No. 263	Stress-Induced Phenomena in Metallization: First International Workshop (Ithaca, NY, 1991)	92-72292	1-56396-082-6

No. 264	Particle Acceleration in Cosmic Plasmas (Newark, DE, 1991)	92-73316	0-88318-948-8
No. 265	Gamma-Ray Bursts (Huntsville, AL, 1991)	92-73456	1-56396-018-4
No. 266	Group Theory in Physics (Cocoyoc, Morelos, Mexico, 1991)	92-73457	1-56396-101-6
No. 267	Electromechanical Coupling of the Solar Atmosphere (Capri, Italy, 1991)	92-82717	1-56396-110-5
No. 268	Photovoltaic Advanced Research & Development Project (Denver, CO, 1992)	92-74159	1-56396-056-7
No. 269	CEBAF 1992 Summer Workshop (Newport News, VA, 1992)	92-75403	1-56396-067-2
No. 270	Time Reversal—The Arthur Rich Memorial Symposium (Ann Arbor, MI, 1991)	92-83852	1-56396-105-9
No. 271	Tenth Symposium Space Nuclear Power and Propulsion (Vols. I–III) (Albuquerque, NM, 1993)	92-75162	1-56396-137-7 (set)
No. 272	Proceedings of the XXVI International Conference on High Energy Physics (Vols. I and II) (Dallas, TX, 1992)	93-70412	1-56396-127-X (set)
No. 273	Superconductivity and Its Applications (Buffalo, NY, 1992)	93-70502	1-56396-189-X
No. 274	VIth International Conference on the Physics of Highly Charged Ions (Manhattan, KS, 1992)	93-70577	1-56396-102-4
No. 275	Atomic Physics 13 (Munich, Germany, 1992)	93-70826	1-56396-057-5
No. 276	Very High Energy Cosmic-Ray Interactions: VIIth International Symposium (Ann Arbor, MI, 1992)	93-71342	1-56396-038-9
No. 277	The World at Risk: Natural Hazards and Climate Change (Cambridge, MA, 1992)	93-71333	1-56396-066-4
No. 278	Back to the Galaxy (College Park, MD, 1992)	93-71543	1-56396-227-6
No. 279	Advanced Accelerator Concepts (Port Jefferson, NY, 1992)	93-71773	1-56396-191-1

No. 280	Compton Gamma-Ray Observatory (St. Louis, MO, 1992)	93-71830	1-56396-104-0
No. 281	Accelerator Instrumentation Fourth Annual Workshop (Berkeley, CA, 1992)	93-072110	1-56396-190-3
No. 282	Quantum 1/f Noise & Other Low Frequency Fluctuations in Electronic Devices (St. Louis, MO, 1992)	93-072366	1-56396-252-7
No. 283	Earth and Space Science Information Systems (Pasadena, CA, 1992)	93-072360	1-56396-094-X
No. 284	US-Japan Workshop on Ion Temperature Gradient-Driven Turbulent Transport (Austin, TX, 1993)	93-72460	1-56396-221-7
No. 285	Noise in Physical Systems and 1/f Fluctuations (St. Louis, MO, 1993)	93-72575	1-56396-270-5
No. 286	Ordering Disorder: Prospect and Retrospect in Condensed Matter Physics: Proceedings of the Indo-U.S. Workshop (Hyderabad, India, 1993)	93-072549	1-56396-255-1
No. 287	Production and Neutralization of Negative Ions and Beams: Sixth International Symposium (Upton, NY, 1992)	93-72821	1-56396-103-2
No. 288	Laser Ablation: Mechanismas and Applications-II: Second International Conference (Knoxville, TN, 1993)	93-73040	1-56396-226-8
No. 289	Radio Frequency Power in Plasmas: Tenth Topical Conference (Boston, MA, 1993)	93-72964	1-56396-264-0
No. 290	Laser Spectroscopy: XIth International Conference (Hot Springs, VA, 1993)	93-73050	1-56396-262-4
No. 291	Prairie View Summer Science Academy (Prairie View, TX, 1992)	93-73081	1-56396-133-4
No. 292	Stability of Particle Motion in Storage Rings (Upton, NY, 1992)	93-73534	1-56396-225-X
No. 293	Polarized Ion Sources and Polarized Gas Targets (Madison, WI, 1993)	93-74102	1-56396-220-9
No. 294	High-Energy Solar Phenomena — A New Era of Spacecraft Measurements (Waterville Valley, NH, 1993)	93-74147	1-56396-291-8
No. 295	The Physics of Electronic and Atomic Collisions: XVIII International Conference (Aarhus, Denmark, 1993)	93-74103	1-56396-290-X

No. 296	The Chaos Paradigm: Developments an Applications in Engineering and Science (Mystic, CT, 1993)	93-74146	1-56396-254-3
No. 297	Computational Accelerator Physics (Los Alamos, NM, 1993)	93-74205	1-56396-222-5
No. 298	Ultrafast Reaction Dynamics and Solvent Effects (Royaumont, France, 1993)	93-074354	1-56396-280-2
No. 299	Dense Z-Pinches: Third International Conference (London, 1993)	93-074569	1-56396-297-7
No. 300	Discovery of Weak Neutral Currents: The Weak Interaction Before and After (Santa Monica, CA, 1993)	94-70515	1-56396-306-X
No. 301	Eleventh Symposium Space Nuclear Power and Propulsion (3 Vols.) (Albuquerque, NM, 1994)	92-75162	1-56396-305-1 (Set) 156396-301-9 (pbk. set)
No. 302	Lepton and Photon Interactions/ XVI International Symposium (Ithaca, NY, 1993)	94-70079	1-56396-106-7
No. 303	Slow Positron Beam Techniques for Solids and Surfaces Fifth International Workshop (Jackson Hole, WY 1992)	94-71036	1-56396-267-5
No. 304	The Second Compton Symposium (College Park, MD, 1993)	94-70742	1-56396-261-6
No. 305	Stress-Induced Phenomena in Metallization Second International Workshop (Austin, TX, 1993)	94-70650	1-56396-251-9
No. 306	12th NREL Photovoltaic Program Review (Denver, CO, 1993)	94-70748	1-56396-315-9
No. 307	Gamma-Ray Bursts Second Workshop (Huntsville, AL 1993)	94-71317	1-56396-336-1
No. 308	The Evolution of X-Ray Binaries (College Park, MD 1993)	94-76853	1-56396-329-9
No. 309	High-Pressure Science and Technology—1993 (Colorado Springs, CO 1993)	93-72821	1-56396-219-5 (Set)
No. 310	Analysis of Interplanetary Dust (Houston, TX 1993)	94-71292	1-56396-341-